ELECTROMECHANICAL CONTROL TECHNOLOGY AND TRANSPORTATION

PROCEEDINGS OF THE 2ND INTERNATIONAL CONFERENCE ON ELECTROMECHANICAL CONTROL TECHNOLOGY AND TRANSPORTATION (ICECTT 2017), 14–15 JANUARY 2017, ZHUHAI, CHINA

Electromechanical Control Technology and Transportation

Editors

Xiaoling Jia
Tongji University, Shanghai, China

Feng Wu
Wuhan University, Wuhan, China

CRC Press
Taylor & Francis Group
Boca Raton London New York

CRC Press is an imprint of the
Taylor & Francis Group, an **informa** business

A BALKEMA BOOK

Published by:
CRC Press/Balkema
P.O. Box 447, 2300 AK Leiden, The Netherlands
e-mail: Pub.NL@taylorandfrancis.com
www.crcpress.com – www.taylorandfrancis.com

First issued in paperback 2020

Typeset by V Publishing Solutions Pvt Ltd., Chennai, India

ISBN 13: 978-0-367-73619-4 (pbk)
ISBN 13: 978-1-138-06752-3 (hbk)

Visit the Taylor & Francis Web site at
http://www.taylorandfrancis.com

and the CRC Press Web site at
http://www.crcpress.com

Table of contents

Signal processing and computer science

Traffic and transportation

Preface

The First International Conference on Electromechanical Control Technology and Transportation (ICECTT 2015) took place from October 31–November 1, 2015 in Zhuhai, China. All accepted papers were published in Advances in Engineering Research (ISSN 2352-5401) and have been indexed by CPCI and submitted to Ei Compendex and Scopus.

The 2017 2nd International Conference on Electromechanical Control Technology and Transportation (ICECTT 2017) was held on January 14–15, 2017 in Zhuhai, China. ICECTT 2017 brought together academics and industrial experts in the field of electromechanical control technology and transportation to a common forum. The primary goal of the conference was to promote research and developmental activities in electromechanical control technology and transportation. Another goal was to promote exchange of scientific information between researchers, developers, engineers, students, and practitioners working all around the world. The conference will be held every year thus making it an ideal platform for people to share views and experiences in electromechanical control technology and transportation and related areas.

In order to organize ICECTT 2017, we have sent our invitation to scholars and researchers from all around the world. Eventually, over 93 papers were submitted for publication. These papers have gone through a strict reviewing process performed by our international reviewers. All the submissions were double-blind reviewed, both the reviewers and the authors remaining anonymous. First, all the submissions were divided into several sections according to the topics, and the information of the authors, including name, affiliation, email and so on, removed. Then the editors assigned the submissions to reviewers according to their research expertise. Each submission was reviewed by two reviewers. The review results were requested to be sent to chairs on time. If two reviewers had conflicting opinions, the paper would be transmitted to a third reviewer assigned by the chairs. Only papers that were approved by all reviewers were accepted for publication.

With the hard work of the reviewers, only 104 papers were finally accepted for publication. These papers were divided into three sections:

– Mechanic Manufacturing System and Automation
– Signal Processing and Computer Science
– Traffic and Transportation

To prepare ICECTT 2017, we have received a lot of help from a lot of people.

We thank all the contributors for their interest and support to ICECTT 2017. We also feel honored to have the support from our international reviewers and committee members. Moreover, the support from CRC Press / Balkema (Taylor & Francis Group) is also deeply appreciated. Without their effort, this book would not have been able to come into being.

The Organizing Committee of ICECTT 2017

Electromechanical Control Technology and Transportation – Jia & Wu (Eds)
© 2017 Taylor & Francis Group, London, ISBN 978-1-138-06752-3

Organizing committees

GENERAL CHAIRS

Prof. Z.H. Wang, *Taiyuan University of Technology, China*
A. Prof. X.L. Jia, *Tongji University, China*

TECHNICAL PROGRAM COMMITTEES

Dr. N. Pombo, *Department of Computer Science, University of Beira Interior, Portugal*
Dr. G. Sunitha, *Sree Vidyanikethan Engineering College, India*
Prof. A. Rostami, *University of Tabriz, Iran*
Prof. M.A. JABBAR, *Department of Computer Science and Engineering, Vardhaman College of Engineering, India (Senior Member, IEEE)*
Dr. X.W. Zhang, *Software Engineering Institute, Chongqing University of Posts and Telecommunications, China*
A. Prof. J.W. Liu, *China University of Petroleum, Beijing Campus (CUP), China*
Dr. Y.L. Gao, *Chongqing University, China*
Prof. L. Li, *College of Computer Science, Hengyang Normal University, China*
Dr. F.M. Chen, *The Fourth Millitary Medical University, China*
Prof. F. Wu, *Wuhan University, China*
Dr. C.B. Li, *Inner Mongolia University, China*
Dr. G.Q. Ma, *Changchun University of Science and Technology, China*
PhD. M. Landowski, *Maritime University of Szczecin, Poland*
Dr. T. Pumpoung, *Rajamangala University of Technology Isan, Thailand*
Dr. Bhagwanjee Jha, *Gujarat Technological University, India*
Dr. C.H. Chang, *Nanyang Technological University, Singapore*

Mechanical manufacturing system and automation

Electromechanical Control Technology and Transportation – Jia & Wu (Eds)
© *2017 Taylor & Francis Group, London, ISBN 978-1-138-06752-3*

Experimental research on the catalytic synthesis of hydrocarbon fuel from *Jatropha curcas* oil

Y.B. Chen, Y.J. Hao, Y.Y. Zhao, S.P. Yang, Y.N. Gao & D. Souliyathai
China-Laos Joint Laboratory for Renewable Energy Utilization and Cooperative Development, Yunnan Normal University, Kunming, Yunnan, China

J.C. Du & A.M. Zhang
Kunming Institute of Precious Metals, Kunming, Yunnan, China

ABSTRACT: In this paper, we introduce the research that takes *Jatropha curcas* oil as raw material, puts it on Pd-based catalyst through catalytic cracking process for hydrodeoxygenation and catalytic cracking, and eventually converts it into a type of hydrocarbon fuel. We focused on the effect of temperature, hydrogen partial pressure, and catalyst dose during the catalytic cracking process under optimal catalytic cracking conditions, including a reaction temperature of 300 °C, a hydrogen partial pressure of 2.5 MPa, 30 m(oil)/m(catalyst), a reaction time of 5 h, and a stirring speed at 100 rpm. The percentages of effective components of converted hydrocarbon fuel and C_8–C_{16} can reach 99.52% and 73.91%, respectively. Chemical components of the converted oils are similar to petroleum-based fuels. Furthermore, the product can contain a large amount of hydrocarbons from C_8 to C_{16} and thus can be regarded as the source of abundant raw material for further development of aviation biofuel.

1 INTRODUCTION

Because of the foreseeable exhaustion of fossil resources and the increasing requirements of global environmental protection, it has become one of the major focuses in strategic study by many countries to turn biomass feedstock into fuel and industrious chemicals (Zhang et al. 2008, Geng et al. 2014). Biomass energy is abundant and renewable and can be recognized as a zero-emission source. Therefore, it is an important supplement of fossil energy and will replace fossil resources when the crucial technology achieves maturity in the future (Wang et al. 2012).

In response to the government's policy of biomass energy development, which dictates "not taking away grains from people" and "not taking away land from farming", there is great significance to conduct studies on the conversion of *Jatropha curcas* oil into biofuel (Li et al. 2012, Li et al. 2014). Plant oil can be converted into biofuel by transesterification, F-T synthesis, cracking, and so on (Qin et al. 2010, Liu et al. 2013, Du et al. 2014, Chen 2015). Vegetable oils can be turned into biodiesel through transesterification process, can substitute diesel oil, and have the characteristics of cleanness and renewability (Yao et al. 2010, Huang et al. 2014). The biodiesel production is expensive, has high cloud and condensation points, lacks fluidity

under low temperature, and the engine may be frequently damaged (Liu et al. 2013, Berenblyum et al. 2010). It is also prone to oxidation and deterioration because of its low energy density and poor stability for storage (Carioca et al. 2009, Demirbas 2011). Therefore, scholars worldwide have conducted a series of experiments on these defects, and relevant studies to convert nonedible oil into renewable hydrocarbon fuel through catalytic cracking technology have earned attention widely. Recent studies also indicate that the converted oil by catalytic pyrolysis of vegetable oil possesses the characteristics of high energy density, low oxygen content, and can be recycled and used as a substitute for fossil fuels (Santillan-Jimenez et al. 2013, Morgan et al. 2012). Researchers from the National Renewable Laboratory in the United States have found that, using nickel as a stable carrier or noble metal catalysts can reduce hydrogen consumption and improve deoxygenation rate in the process of catalytic cracking. Transitional metal catalysts have already been widely applied in the catalytic cracking of oils, which has already been reported by numerous studies. Deepak Verma et al. (Verma et al. 2011) used NiMo as an active metal supported on a layered mesoporous molecular sieve under the conditions of 410°C, 5 MPa, and 1 h^{-1} airspeed, directly added hydrogen catalytic transformation of *Jatropha curcas* oil (54% of the liquid-phase product was kerosene component (C_9–C_{15}); seaweed was taken as raw material under the same condition, 78.5% of the product was kerosene component), and has a high heterogeneous selectivity (isoparaffin/alkanes = 2.5). Tian Weiqian et al. (Tian et al. 2013) found that the main product was C_{15}–C_{18} hydrocarbon, which accounts for 75.6% in the liquid-phase product under the optimal conditions, with *Jatropha curcas* oil as raw material and CoMoS/Al_2O_3 as catalyst.

In this paper, we develop the Pd/SiO_2–Al_2O_3 catalyst and use Jatropha oil as raw material to produce hydrocarbon fuels through one-step hydrogenation; discuss the effects of temperature, pressure, oil/catalyst ratio, reaction time, and stirring speed on the total hydrocarbon hydrogenation reaction product of Jatropha oil in one-step catalytic content and the content of C_8–C_{16} hydrocarbons; and open up more broad application prospects. Furthermore, it furthers the research on biology fuel for aviation.

2 MATERIALS AND METHODS

2.1 *Equipment and material*

High-temperature and high-pressure reactor system (GS-1 L): Zhengwei Mechanical Equipment Corporation in Weihai; electric-heated thermostatic water bath (DZKW-D-6): Yongguangming Medical Instrument Factory in Beijing; electric-heated air dry oven (DHG-9203 A): electric-heated air dry oven; GC-MS (TRACE ISQ): American Thermo Fisher Scientific Corporation; double-layer vapor-bathing constant temperature vibrator (HZQ-C): Dadi Automation Instrument Factory in Jintan.

The major fatty acid in *Jatropha curcas* oil produced in Shuangbai, Yunnan Province measured by GC-MS is shown in Table 1.

2.2 *Properties of the catalyst*

The active component of 2% Pd was loaded on the SiO_2–Al_2O_3-based molecular sieve by the method of equal-volume impregnation and through high-temperature calcinations and cooling grinding under specific conditions of catalyst preparation (Du et al. 2015, Rabaev et al. 2015). It has

Table 1. Relative content of major fatty acid in *Jatropha curcas* oil.

Composition	$C_{16:0}$	$C_{16:1}$	$C_{18:0}$	$C_{18:1}$	$C_{18:2}$
Relative content (wt.%)	17.01	0.85	5.83	39.01	37.23

an excellent cracking selectivity when the factors, including temperature, hydrogen partial pressure, and time, are under control.

Calculated by BET method, the Pd/SiO_2–Al_2O_3 catalyst surface area was 225 m^2/g, Kong Rong 0.43 cm^3/g, and the average pore diameter was 3.75 nm. To activate the Pd/SiO_2–Al_2O_3 catalyst, it needed to be activated in the hydrogen atmosphere programmed from room temperature to 300°C and kept for 6 h.

2.3 *Catalytic cracking of rubber seed oil*

The whole process of catalytic cracking was conducted in a closed-tank reactor (Figure 1). A certain proportion of the catalyst and *Jatropha curcas* oil was mixed at a specific ratio in the reactor. Then, hydrogen was added into the reactor and heated to the reaction temperature. Catalytic cracking took place correspondingly; through heat preservation, the temperature was held and the reaction rate was decreased. The reaction ended after cooling. The tank reactor was opened and the materials were extracted. The produced liquid consisted of liquid fuel components and water (the water analyzed accounts for about 3–5% of the rubber seed oil when raw materials were put into the reactor), and the reaction products were analyzed with GC–MS after a series of treatments.

2.4 *Analysis of liquid fuel*

The liquid product taken out from the tank reactor needed to undergo methyl esterification first. Then, it was sealed and sent for analysis. The composition of the product was measured by THRACE ISQ Gas Chromatograph–Mass Spectrometer. The procedure and conditions of methyl

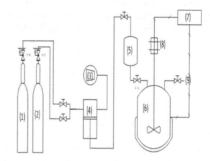

(1) hydrogen; (2) argon; (3) booster pump; (4) air compressor; (5) buffer tank; (6) high-temperature and high-pressure reactor; (7) controller (adjusting reaction temperature and rotation speed); (8) magnetic rotating stirrer; (9) gas collection and exhaust port

Figure 1. Schematic diagram of experimental device.

esterification include: measure and pick up three to five raw initial processing product; add it into 20 ml of 1% H_2SO_4-methyl alcohol solution and shake well; place the uniformly mixed solution into a water bath kettle to conduct backflow for 60 min under the temperature of 70°C; then, perform extraction by adding 15 ml of dichloromethane and add 30 ml of distilled water for stratification; remove the distilled water in the upper layer of the solution after backflow comes to an end; then, add 30 ml of distilled water into the solution and remove the distilled water on the upper layer after stratification was completed; then, add 5–10 ml of dichloromethane once more for secondary extraction and wash it with distilled water two to three times; finally, pour out the sample on the upper layer, add enough anhydrous Na_2SO_4 into it, shake well for desiccation, and take out the sample for analysis by filtration after 30 min standing. Gas chromatographic analysis conditions include: chromatographic column CETM-5 (30 m × 0.25 mm × 0.25 μm); injection temperature, 200°C; for temperature raising procedure: initial temperature, 80°C; heat preservation time, 3 min and heating to 280°C at the rate of 10°C/min, then hold the temperature for 3 min; and flow rate of carrier gas, 1.5 mL/min. Mass spectrum analysis conditions include: Electron Impact (EI) ion source (electron energy 70 eV); temperature of transmission line, 280°C; detection voltage, 0.9 kV; mass scan range, m/z 32–500; and the data acquisition time range, 1–25 min.

2.5 *Data processing*

Components of the product could be identified by standard mass spectrum on the basis of data analysis through searching the spectrogram database by GC-MS automatic; detection and verification of the components of the product were conducted by data processing system automatic retrieval spectrogram database combined with the standard mass spectrum; after quantitative analysis, the percentages of the components were calculated by the analysis of "area normalization method" through data processing system. Analysis was performed by data processing system on the basis of "area normalization method", and then the proportion of each component was determined respectively.

3 RESULTS AND DISCUSSION

3.1 *Effect of temperature on the reaction*

Experiments were conducted to test the effect of the reaction temperature on the composition of hydrocarbon fuels in the liquid product generated by catalytic cracking, with reaction temperature between 290 and 340°C. The results under different temperatures are shown in Fig. 2.

The contents of hydrocarbons and C_8–C_{16} in the liquid product were 99.31% and 70.67%, respectively, when the temperature was maintained at 300°C. However, with the increase of temperature from 300 to 340°C, the proportion of hydrocarbons component decreased accordingly. As the temperature rose, it would promote the occurrences of both cracking reaction and other side effects, and the proportion of each component changed with the increasing temperature. It can also be seen from the morphology of products that they exhibited solid or semisolid state, not fully converted and cracking extent below 300°C, and not completely converted and the cracking extent was not enough; when the temperature was maintained above 300°C, the product mainly consisted of short-chain alkanes in liquid state. However, if the temperature gets too high, the catalyst would end up with an obvious coking phenomenon. Furthermore, as the coking normally takes place at the bottom of the reactor, more and more catalyst would adhere to the inner wall of the reactor as the temperature increases, which is shown in Fig. 2. When the temperature was in the range of 290–340°C, the content of each hydrocarbon compound increased first, then decreased, and finally became stable with increasing reaction temperature. Consequently, the optimal reaction temperature was 300°C.

3.2 *Effect of hydrogen partial pressure on the reaction*

Experiments were conducted to test the effect of hydrogen partial pressure on the content of

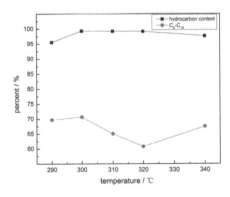

Figure 2. Effect of reaction temperature on the composition of each hydrocarbon fuel in liquid product (P: 3 MPa; ratio of oil and catalyst: 10; reaction time: 5 h; stirring speed: 100 rpm).

hydrocarbon fuels in the liquid product, with hydrogen partial pressure from 1.5 to 4.0 MPa. The results of materials under different hydrogen partial pressures are shown in Fig. 3.

When the hydrogen partial pressure was 2.5 MPa, the contents of liquid hydrocarbon and C_8–C_{16} were 100% and 72.89%, respectively. During the experiment, it was obvious that the higher the hydrogen partial pressure, the higher the severity caused by coking of the catalyst in the reactor tank. Iva Kubičková et al. (Kubičková et al. 2010) studied this phenomenon and concluded that the reaction can proceed in an inert atmosphere without H_2. However, during the experiment, the hydrogen partial pressure had certain effects on coking of the catalyst during the reaction process of selection, recombination, and polymerization of unsaturated hydrocarbon. The saturation and deoxygenation reaction of double bond generally proceeds in acid sites on the surface of the catalyst, where the cracking reaction of hydrocarbons proceeds in the channels of the catalyst. Consequently, if the particle size of the formed molecules is smaller than the pore size of catalyst, then the molecules can flow through the channels when the reaction completes. However, if the particle size of molecules is too large, generated molecules would tend to block the channel of the catalyst, resulting in inactivation of the catalyst, and the inactivated catalyst would congeal to form a clot and will be adhered at the bottom of the reactor.

The key point of this study was to get maximum fuel components, namely the C_8–C_{16} hydrocarbon components. In this experiment, C_8–C_{16} hydrocarbon components obtained included alkanes as the majority and a little of aromatics, tiny amount of cycloparaffin hydrocarbon, and methyl isomerization alkane, as shown in Table 2.

Table 2. Product components under different hydrogen pressure distributions.

Hydrogen pressure/ MPa	*Jatropha curcas* oil		
	C_8–C_{16} content of hydrocarbons/ %	C_8–C_{16} alkane/ %	Aromatic hydrocarbon/ %
1.5	72.59	49.97	22.62
2	71.97	51.69	20.28
2.5	72.89	56.47	16.42
3	70.67	54.84	15.83
3.5	68.8	52.51	16.29
4	66.09	48.77	17.32

Table 2 shows that the change of hydrogen partial pressure has no obvious effect on C_8–C_{16}. Under the condition of low hydrogen partial pressure, the proportion of C_8–C_{16} components is high; however, the proportion of hydrocarbons in C_8–C_{16} hydrocarbon is also high. Therefore, when the hydrogen partial pressure is high or low, experiments cannot make ideal amount of product. At 2.3 and 3 MPa, C_8–C_{16} total hydrocarbon and product of alkanes are relatively high in the peak, and the content of aromatic hydrocarbons is low. Therefore, 2.5 MPa is the optimal hydrogen partial pressure to produce the most content and the least aromatic content.

3.3 Effect of catalyst dose on the reaction

Experiments were conducted to test the effect of catalyst dose on the content of the hydrocarbon fuels in the liquid product, from 10 to 90 of m(oil)/m(catalyst). The effects of different catalyst doses on the reaction products are shown in Fig. 4.

When m(oil)/m(catalyst) was 30, the contents of liquid hydrocarbons in product and C_8–C_{16} were 99.16% and 78.06%, respectively. As shown in Fig. 4, if the ratio was lower than 30, that is to say, even if the dose of catalyst is increased, rather than increasing continuously, the content of each component of hydrocarbons would not match the increasing but tend to decrease slightly. That is why the reaction would not be promoted obviously when the dose of the catalyst reaches certain extent. Consequently, the optimal ratio of oil and catalyst is 30.

3.4 Effect of reaction time on the reaction

Experiments were conducted to test the effect of reaction time on the component content of the hydrocarbon fuels in the liquid product, from 4 to 8 h of reaction time. The effects of different reaction times on the reaction products are shown in Fig. 5.

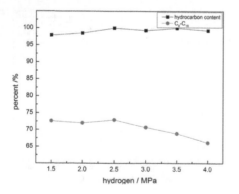

Figure 3. Effect of hydrogen partial pressure on the content of hydrocarbon fuels in liquid product (temperature: 300 °C; ratio of oil and catalyst: 10; reaction time: 5 h; stirring speed: 100 rpm).

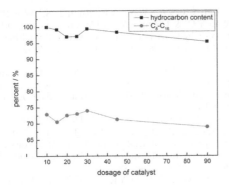

Figure 4. Effect of the dose of catalyst on the content of hydrocarbon fuels in liquid product (temperature: 300°C; P: 2.5 MPa; reaction time: 5 h; stirring speed: 100 rpm).

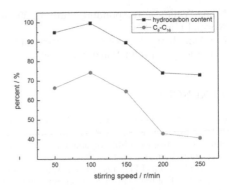

Figure 6. Effect of stirring speed on the content of hydrocarbon fuels in liquid product (temperature: 300°C; P: 2.5 MPa; ratio of oil and catalyst: 30; reaction time: 5 h).

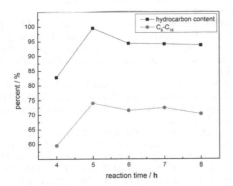

Figure 5. Effect of reaction time on the content of the hydrocarbon fuels in liquid product (temperature: 300°C; P: 2.5 MPa; ratio of oil and catalyst: 30; stirring speed: 100 rpm).

As can be seen from Fig. 5, when the reaction time was 5 h, the total hydrocarbon content and C_8-C_{16} content reached a peak volume. With the increase of reaction time, the content of each hydrocarbon compound decreased slowly, and when finally the fluctuation was relatively stable, the content of the product was not changed. Therefore, the optimal reaction time was 5 h.

3.5 Effect of stirring speed on the reaction

Experiments were conducted to test the effect of stirring speed on the content of hydrocarbon fuels in liquid product, with the stirring speed from 50 to 200 rpm. The effects of different stirring speeds on the reaction products are shown in Fig. 6.

Stirring speed affects the effective contact time of the catalyst and the grease. The higher the speed, the higher the centrifugal force; the shorter the effective contact time, the lower the conversion rate. Fig. 6 shows that, when the speed was 100 rpm, the contents of hydrocarbon in the liquid product and C_8-C_{16} were 99.45% and 73.96%, respectively. On the basis of the reaction results of Jatropha oil in Fig. 6, the higher speed led to lower conversion rate and target component content, which complies with the results of laboratory experiment. Therefore, the optimum speed is 100 rpm.

To conclude, the idealist conditions include: reaction temperature, 300°C; hydrogen pressure, 2.5 MPa; 30 m(oil)/m(catalyst); reaction time, 5 h; reaction speed, 100 rpm; and by catalytic hydrogenation reaction under the conditions, the content of hydrocarbon oil in the total hydrocarbons was up to 99.52% and that of C_8-C_{16} hydrocarbons was up to 73.91%.

4 CONCLUSION

With the new Pd/SiO$_2$–Al$_2$O$_3$ catalyst, we studied the effects of temperature, hydrogen pressure, catalyst dose, reaction time, and stirring speed on the distribution of the product and deoxidation and identified the optimal parameters of every factor: reaction temperature, 300°C; hydrogen partial pressure, 2.5 MPa; 30 m(oil)/m(catalyst); reaction time, 5 h; and stirring speed, 100 rpm. Under these parameters, catalytic cracking yields 99.52% of the total hydrocarbon and 73.91% of C_8-C_{16} in the liquid product.

ACKNOWLEDGMENTS

1. This study was supported by the National Natural Science Foundation of China (No. 21266032).

2. *Hao Yajie; corresponding author; Yunnan Normal University.
3. Liu Tiancheng; School of Chemistry and Biotechnology, Yunnan Minzu University.

REFERENCES

Berenblyum A.S., Danyushevsky V.Y., Katsman E.A., Podoplelova T.A. & Flid V.R. (2010). Production of engine fuels from inedible vegetable oils and fats. *J. Petroleum Chemistry, 50 (4)*, 305–311.

Carioca J.O.B., Filho J.J.H., Leal M.R.L.V. & Macambira F.S. (2009). The hard choice for alternative biofuels to diesel in Brazil. *J. Biotechnology Advances, 27 (6)*, 1043–1050.

Chen H.M. (2015). Research progress on modification of bio oil by catalytic esterification. *J. Chemical Intermediate, (12)*, 125–126.

Demirbas A. (2011). Competitive liquid biofuels from biomass. *J. Applied Energy, 88 (1)*, 17–28.

Du J.C., Zhang A.M., Xia W.Z., Zhao Y.K., Chen Y.B. & Tao F. (2014). Progress in the catalysts for preparation of bio-fuels through hydrogenation of animal fats and vegetable oils. *J. Journal of Functional Materials, 45(9)*, 8–12.

Du J.C., Zhao Y.Y., Xia W.Z., Chen Y.B., Tao F. Zhang A.M. (2015). Efects of precious metals and supports on catalytic hydrodeoxygenation and hydrocracking performance over catalysts. *J. Rare Metal Materials and Engineering, 44(9)*, 2210–2215.

Geng Y. & Lai S.P. (2014). Research progress in technology for refining biomass oil into fuel oil. *J. Gas & Heat, 2014, 34 (9)*, 17–20.

Huang D.S., Lv P.M., Cheng Y.F., Pan H., Luo W., Yang L.M. & Yuan Z.H. (2014). Improving the cold flow properties of palm oil biodiesel. *J. Acta Energiae Solaris Sinica, 35 (3)*, 391–395.

Kubičková I. & Kubička D. (2010). Utilization of triglycerides and related feedstocks for production of clean hydrocarbon fuels and petrochemicals: A review. *J. Waste & Biomass Valorization, 1 (3)*, 293–308.

Li C.Z., Li P.W., Xiao Z.H., Chen J.Z. & Zhang L.B. (2012). Research status and industrialization prospect of woody bio diesel raw material in China. *J. Journal of China Agricultural University, 17 (6)*, 165–170.

Liu Y.H, Liu Y.Y., Wang Y.P., Ruan R.S., Wen P.W., Wan Y.Q. & Cheng F.Y. (2013). Progress of production of hydrocarbon fuel by cracking non-edible oil. *J. Chemical Industry and Engineering Progress, 32 (11)*, 1–6.

Li L., Quan K.J., Xu J.M. & Ge X P. (2014). Liquid hydrocarbon fuels from catalytic cracking of rubber seed oil using USY as catalyst. *J. Fuel, 123 (1)*, 189–193.

Morgan T., Santillan-Jimenez E, Harman-Ware A.E., Ji Y.Y., Grubb D. & Crocker M. (2012). Catalytic deoxygenation of triglycerides to hydrocarbons over supported nickel catalysts. *J. Chemical Engineering Journal, 189 (2)*, 346–355.

Qin X.X., Wang T.J., Li Y.P., Wu C.Z., Ma L.L. & Li H.B. (2010). Fischer-tropsch synthesis from simulated biomass-derived syngas on Co/SiO_2 catalyst. *J. Acta Energiae Solaris Sinica, 31 (6)*, 671–675.

Rabaev M., Landau M.V., Vidruk-Nehemya R., Koukouliev V., Zarchin R. & Herskowitz M. 2015, Conversion of vegetable oils on $Pt/Al_2O_3/SAPO$-11 to diesel and jet fuels containing aromatics. *J. Fuel, 161*, 287–294.

Santillan-Jimenez E., Morgan T., Lacny J., Mohapatra S. & Crocker M. (2013). Catalytic deoxygenation of triglycerides and fatty acids to hydrocarbons over carbon-supported nickel. *J. Fuel, 103 (1)*, 1010–1017.

Tian W.Q., Liu J., Liu C.L., Fan K. & Rong L. (2013). Hydrotreatment of jatropha oil over $CoMoS/γ-Al_2O_3$ catalys. *J. Journal of Fuel Chemistry and Technology, 41 (2)*, 207–213.

Verma D., Kumar R., Rana B.S. & Sinha A.K. (2011). Aviation fuel production from lipids by a single-step route using hierarchical mesoporous zeolites. *J. Energy & Environmental Science, 4 (5)*, 1667–1671.

Wang D.J., Liu H.Y., Liu Y.X., He C.H., Liu C.G. & Zhang B.J. (2012). Research progress in catalysts for biofuel hydrodeoxygenation. *J. Petrochemical Technology, 41 (10)*, 1214–1219.

Yao Z.L. & Min E.Z. (2010). Harm and resource utilization of waste edible oil. *J. Natural Gas Industry, 30 (5)*, 123–128.

Zhang, B.L., Song H.M. & Li S.X. (2008). Bioenergy development and opportunity of technological innovation. *J. Transactions of the Chinese Society of Agricultural Engineering, 24 (2)*, 285–289.

Measurement of the lithium battery internal resistance using the sampling integral method

J.A. Lou & D.J. Yang
Mechanical Engineering College, Shijiazhuang, China

ABSTRACT: Lithium battery is the most potential EV battery in the field of EV. Accurate detection of the state of battery charge is a problem to be solved. The detection of the battery's internal resistance is of the utmost importance for the detection of the battery state. In this paper, the weak signal detection technology is introduced. On the basis of AC impedance method, the measurement of the lithium battery's internal resistance is based on the sampling integral method. This method is modeled by using the DSP builder module in Simulink. Under the disturbance of 20 dB Gauss noise, the measured signal detected by the sampling integration method is close to the original signal. The method of hardware circuit is realized by software, which reduces the hardware noise and makes the measurement more precise.

1 INTRODUCTION

In the 1950s, American scientist Klein achieved sampling integral in California and named it as BOXCAR integrator. Sampling integrator is used to extract the signal submerged in strong noise. The principle is to use the signal accumulation method to amplify useful signals and suppress noise signals. With the development of science and technology, the use of sampling integration is becoming more and more mature. It is very effective for the identification of fast changing signals with strong noise interference. Moreover, the method is widely used in the fields of physics, biology, and medicine. According to this principle, focusing on the question of the research and application of lithium battery is more and more extensive, and its future prospect is very broad. However, the measurement of internal impedance is also a serious technical problem. The study found that there are great relationships between the size and the internal structure of internal resistance. In this paper, the internal structure of lithium battery is analyzed. The equivalent impedance of lithium battery is obtained by the principle of electrochemical impedance spectroscopy. Moreover, on the basis of the AC impedance method, the internal resistance is detected by sampling integration method.

2 RESISTANCE MODEL OF A LITHIUM BATTERY

At present, electrochemical impedance method is the best way to measure the internal resistance of the battery, which plays an important role in the study of lithium-ion batteries and its materials. Many experts worldwide have made great achievements in the research of battery using the AC impedance method. The principle of AC impedance method is summarized as follows. Under the normal working condition of battery, input a small sinusoidal voltage (current) signal to the battery system at different frequencies. Because the batteries are working in a DC state, the input of small amplitude AC signal is not enough to affect the normal operation of the battery system. Then, the four-line method is used to measure the output current (voltage) signal, and the frequency impedance spectrum of the lithium battery is obtained from the ratio of the different voltage signal to the current signal. Electrochemical Impedance Spectroscopy (EIS) can reflect the subtle changes of lithium batteries, especially the study of battery materials; therefore, it is very important to establish the equivalent model of the battery.

2.1 Impedance spectra of a lithium battery

Under different frequencies, the size of internal resistance is different, but the resistance change is very little. The small signal waveform of AC impedance method can avoid other nonlinear interference, so the result will be more accurate. Electrochemical impedance spectroscopy is shown in Figure 1. It can be seen from the diagram that the impedance spectrum of the lithium battery is divided into three parts. The value of each section changes with the frequency, and the magnitude of impedance changes nonlinearly. However, the small signal waveform of the AC impedance method can avoid other nonlinear interferences, which can further improve the accuracy of measurement results. Through research and analysis, at low-frequency bands, it mainly includes the diffusion impedance of the lithium-ion in electrode; at middle-frequency bands, it mainly includes the impedance of the electrolyte; at high-frequency bands, it mainly includes the impedance of solid electrolyte interface. Different impedances have different structures and exhibit different characteristics in different environments. In this paper, the equivalent model of Li-ion battery is constructed from the structural characteristics of each part, which can more accurately reflect the characteristics of lithium-ion battery.

2.2 *Equivalent impedance of a lithium battery*

Before the analysis of lithium battery, it is necessary to analyze the existing equivalent model of lithium battery. The existing equivalent models usually consist of the series–parallel system of capacitance and resistance. The internal resistance of lithium battery is a resistor in the actual operation, so it can be equivalent to a value in simulation. From the above analysis, the impedance of lithium batteries mainly includes Ohmic resistance, solid electrolyte interface impedance, and charge transfer impedance. According to the characteristic of the impedance spectrum, the composite elements of different series–parallel systems are

respectively used to replace the impedance of different frequency bands. As shown in Figure 2, R_Ω is the Ohmic resistance that represents the lithium battery electrolyte resistance, interface resistance, and so on. The parallel circuit composed of Q and R_f represents the capacitance effect and diaphragm resistance SEI. The series–parallel circuit composed of C_h, R_{sol}, and Z_w represents the transfer resistance and diffusion impedance of lithium ion.

3 SAMPLING INTEGRAL METHOD

3.1 *Principle of the sampling integral method*

Sampling integral method is one of the weak signal detection technologies. And it is also a branch of synchronous accumulation method. In general, it is mainly used to detect weak signals in noise. The principle is sampling the same phase of the different periods of the input signal. The sampling signals obtained by time-sharing are accumulated in turn. Because of the influence of noise, the result of the integration is different each time. However, accurate results can be obtained by repeated sampling. The principle is shown in Figure 3.

3.2 *Algorithm used in this paper*

The working of sampling integral can be divided into single point and multipoint. This paper uses single point, and the circuit is relatively simple because the method takes only one time for each signal cycle. Single point is divided into fixed and scanning. Fixed point operation is used in this paper, whose characteristic is that the peak value of the measured signal is repeated. In this paper, on the basis of AC impedance method, a small amplitude sinusoidal signal is input to the system. According to the principle of sampling integral method, the voltage value of the phase near the peak of weak sinusoidal signal is sampled. Therefore, the peak signal with noise is obtained. However, through integration and averaging, the noise signal is filtered. Finally, the peak value of sine signal is obtained. The advantages of this algorithm are relatively less sampled data, quickness, and

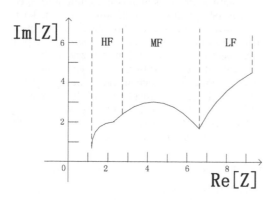

Figure 1.   Impedance spectra of a lithium battery.

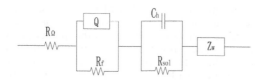

Figure 2.   Impedance equivalent model.

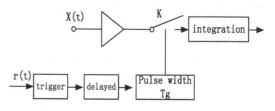

Figure 3.   Principle of the sampling integral method.

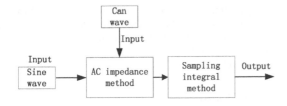

Figure 4.   Flowchart of the method used.

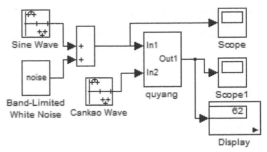

Figure 5.   Simulation of the sampling model.

high precision. The flowchart of the method used in this paper is shown in Figure 4. In the diagram, the input signal of the system is based on the AC impedance method, sampling method, and output.

4   SIMULATION OF THE BATTERY INTERNAL RESISTANCE MEASUREMENT

4.1   *Simulation model of the impedance of the lithium battery measurement*

The simulation model of the lithium battery resistance is shown in Figure 5. According to the actual measurement, the simulation model is composed of a signal module, sampling module, and display module. According to the principle of sampling integration, we add a reference signal to the system. The picture is shown in the Cankao waveform module. The Cankao signal is the same as the measured signal. The main function of the quyang module is sampling and integrating, which is equivalent to a filter. The output signal is displayed by Scope 1. Because the simulation is established by using the module of DSP builder, the model generates digital signals. In other words, it can be implemented in the actual FPGA system and lay a good foundation for follow-up study.

The system-mixed signals are collected from the two poles of the battery. The signal is composed of a sine wave module and a band-limited module. The size of white noise is 20 dB. The waveform of mixed waveform is displayed by Scope in Figure 5. The actual waveform is shown in Figure 6.

4.2   *Subsystem of quyang*

According to the principle of sampling integral method, the quyang simulation model is shown in Figure 6. The amplitude of the input sine wave is 63 and hence the trigger condition is 62. When the amplitude of the reference signal is larger than 62, the system begins to sample. That is to say, the peak signal is the only signal to sample. The quyang module structure is shown in Figure 7. Because the

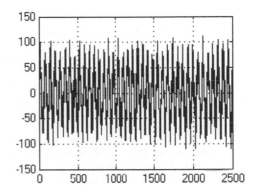

Figure 6.   Measured signals with noise.

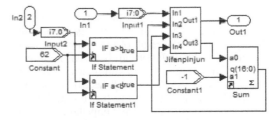

Figure 7.   Internal structure of the quyang model.

model system is established in DSP Builder, there is digit position in the graph. As we know, this is the basis of digital signal simulation.

4.3   *Output signal waveform*

The amplitude of the input sine wave signal is 63. It can be seen from the Display module that the output signal is 62. The waveform of the simulation process is shown in Figure 8. It can be seen that there are fluctuations in the presampling data. However, after a period of acquisition and integration, the fluctuation will be decreased in the process. Through the analysis, there are differences to the amplitude of the same phase in each

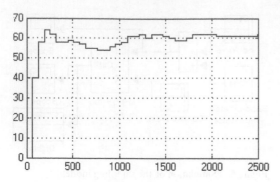

Figure 8.   Output signal.

cycle because of the interference of noise. After a period of sampling, the noise interference will be decreased and the true amplitude of the measured signal can be detected. Through the comparison of the output signal and the input signal, it can be proved that the method has high accuracy. From the above analysis, we know that the sampling integral method is the same as the accumulation method. The more the accumulation, the greater the Signal-to-Noise Ratio (SNR). However, the sampling data of sampling method is very little, so too many sampling integrations are not necessary.

5   CONCLUSION

In this paper, we studied the measurement of internal resistance. The feasibility of this method has been verified by the result of the simulation. First, according to the relationship between the impedance of the lithium battery and internal structure, the impedance model is established. It lays a foundation for further measuring the internal impedance. Then, on the basis of AC impedance method, the sampling integral method is used to detect the mixed signal. Finally, a simulation of the digital module using Simulink shows that the sampling integral method has a high filtering ability. It can accurately restore the amplitude information of the original signal. From the result of the simulation, it is possible for the algorithm to be implemented in FPGA.

REFERENCES

[1] Wang. F.L. 2014. The summary of the status and application development of the global lithium battery market [J]. Journal of power sources.
[2] Jin. K. 2012. The optical integrator design based on Virtual Instrument Technology [D]. Changchun: Changchun University of Science and Technology.
[3] Peng. F. 2014. Research and implementation of lithium ion battery state estimation [D]. Chengdu: University of Electronic Science and technology of China.
[4] Li. F.P. & Mao. J.G. 2009. Battery internal resistance measurement based on AC impedance method [J]. Journal of Chongqing Institute of Technology.
[5] [America] Mark. O. (Mark E. Orazem); Yong. X.X. Zhang. X. Y et al. 2014. electrochemical impedance spectroscopy [M]. Beijing: Chemical Industry Press.
[6] Li. Y.F. & Ge J.F. 2015. Micro ohm battery internal resistance measurement method [J]. Instrument Technique and Sensor.
[7] Jiang. J.C. & Wei. W. et al. 2014. Analysis of impedance parameters of LiFePO$_4$ battery [J]. Journal of Beijing Institute of Technology.
[8] Jiang. Z.J. & Yu. S.B. 2007. Data sampling integral technology acquisition system [J]. Journal of Jilin University: Based on Information Science Edition.
[9] Gao. J.Z. 2011. Weak signal detection [M]. Beijing: Tsinghua University press.
[10] Sun S.P. 2013. Weak signal detection and application [M]. Beijing: Publishing House of electronics industry.
[11] Yang. Q & Zhou. H. Q, et al. 2012. The design of resistance test system of lithium ion battery based on LabWindows/CVI [J]. Journal of power sources.
[12] Jiang. X.Y. 2013. The application of MATLAB/FPGA/DSP Builder in the teaching of "digital signal processing" [J]. Software Guide.

Electromechanical Control Technology and Transportation – Jia & Wu (Eds)
© 2017 Taylor & Francis Group, London, ISBN 978-1-138-06752-3

The influence of fiber angle on the mechanical performances of a composite cylindrical part

Hanjun Gao, Yidu Zhang, Qiong Wu & Wenbing Zhou
*State Key Laboratory of Virtual Reality Technology and Systems, School of Mechanical Engineering
and Automation, Beihang University, P.R. China*

ABSTRACT: Carbon fiber composites have been widely used in aerospace, automotive, shipbuilding and other fields. As an anisotropic material, Carbon fiber angle has a significant effect on the mechanical properties of the composite. In this paper, the Finite Element Model (FEM) of a typical cylindrical thin-walled composite part (cylindrical part) is established in ANSYS. By using the FE model, the vibration modes and buckling modes of the cylindrical part are calculated with the fiber angle 0°, 30°, 45°, 60° and 90° respectively. The results showed that when the fiber angle is 90°, the cylindrical part has optimal dynamic performance among the five fiber angles. And the buckling resistance is the best when the fiber angle is 0°.

1 INTRODUCTION

Carbon fiber composite materials have been widely applied to aerospace, automobile, shipbuilding and other fields (Bao L R, 2002; Sastri S B, 1996 & Marbán G, 2003) because of the light weight, high strength, and anti-fatigue capability. There have been many studies on the effects of different carbon fibers on the properties of the composites (Kumar S, 2002; Shen G, 2006 & Patton R D, 1999), influence of composition of materials on the properties of composites (Sheehan J E. 1989 & Xiao Y, 2000), and effects of carbon fiber on the kinetics of composites (Kenny J M, 1991). Rahmani H, Najafi S H M, Saffarzadeh-Matin S, et al. (Rahmani H, 2014) studied the effect of fiber content, the number of layers, and the orientation on the mechanical properties of Multi-axial Multiply Fabric (MMF) composites. In this paper, the effect of fiber angle on the natural frequency and buckling behavior of carbon fiber composites are investigated.

2 MODELLING AND SIMULATIONS

2.1 Cylindrical part material properties and geometry

The cylindrical part is made of the carbon fiber composite material, the mechanical properties of the carbon fiber composite material are presented in Table 1.

Table 1. Mechanical properties of the carbon fiber composite material.

Young's modulus (MPa)	E1	120000	Shear modulus (MPa)	G12	8200
	E2	8800		G13	8200
	E3	8800		G23	8200
Poisson's ratio	$\mu12$	0.31	Density (ton/mm³)	DENS	1.62E-9
	$\mu13$	0.31			
	$\mu23$	0.31			

The workpiece is a hollow cylindrical part and manufactured by the layer technology of carbon fiber composite material. Its main geometric parameters include cylinder radius, cylinder height and wall thickness. The specific geometric dimensions were 900 mm, 1000 mm, 4 mm.

2.2 Development of the cylindrical part FEM

The FE model is established by using the shell element (SHELL181). In order to study the effect of fiber angles on the vibration modes and buckling modes of the cylindrical part, the FE models of the cylindrical part is built with 5 fiber angles, which are 0°, 30°, 45°, 60° and 90°. Fig. 1 shows the specific values of the 5 fiber angles. The angle between the carbon fiber and the axis of the cylindrical part is defined as the fiber angle.

The thickness value of the shell element is the wall thickness of the cylindrical part. The FEM is established with 0° fiber laying angle as shown in

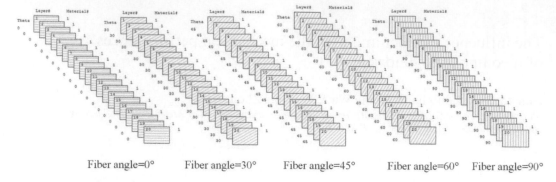

Fiber angle=0°   Fiber angle=30°   Fiber angle=45°   Fiber angle=60°   Fiber angle=90°

Figure 1.   The specific values of the 5 fiber angles.

Figure 2.   The FEM of 0° fiber angle.

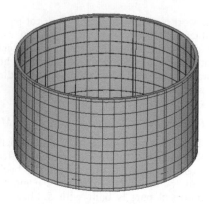

Figure 3.   Meshing results of the cylindrical part.

Fig. 2. The cell size is an important parameter for cell division. For the cylindrical part, several experiments show that it has both high computational efficiency and sufficient calculation accuracy when the grid size is 20 mm. The meshing result is shown in Fig. 3.

3 FEM SIMULATIONS RESULTS AND DISCUSSIONS

3.1 *Modal analysis*

The most commonly method for calculating the undamped mode is the block Lanczos method. The method will also be used to analyze the cylindrical part. The boundary condition is to constrain all degrees of all nodes on the cylindrical part bottom surface. Finally, the first 10 order natural frequencies and modes of the cylindrical part are extracted from the simulation results. The first 10 order natural frequencies values for 5 fiber angles of the cylindrical part are shown in Table 2.

In order to avoid the cylindrical part resonance phenomenon occurs, the cylindrical part 1st order natural frequency needs to be enhanced. As is seen from the simulation results in Table 2, the 1st order natural frequency of the workpiece is the largest when the fiber angle is 90 degrees. In addition to the 4th order natural frequency, the natural frequencies of the workpieces were determined to be the largest with the 90 degree fiber angle.

3.2 *Buckling analysis*

Buckling analysis is a technique that determines the critical load and buckling mode shape when the structure begins to become unstable. By setting boundary conditions, the buckling analysis of the cylinder is simulated and the simulation results are analyzed. The first 10 order buckling critical loads for 5 fiber angles of the cylindrical part are shown in Table 3.

The 1st order buckling critical load is the maximum load that the workpiece can bear without buckling, so it is an important index to measure the structural stability of a workpiece. The 1st order buckling of the cylindrical part occurs when the actual load is between the 1st order and 2nd

Table 2. The first 15 order natural frequencies values for 5 fiber angles.

Fiber angles (°)	0	30	45	60	90
1st order natural frequency (Hz)	64.25	61.78	62.66	65.25	68.75
2nd order natural frequency (Hz)	65.64	65.09	64.81	65.52	70.76
3rd order natural frequency (Hz)	67.04	66.41	65.29	69.16	71.84
4th order natural frequency (Hz)	71.29	66.62	70.51	73.17	72.81
5th order natural frequency (Hz)	73.51	70.96	76.50	74.26	82.75
6th order natural frequency (Hz)	73.56	71.39	77.34	80.67	98.81
7th order natural frequency (Hz)	73.95	72.28	80.18	91.46	98.91
8th order natural frequency (Hz)	74.02	81.45	80.22	92.25	99.19
9th order natural frequency (Hz)	79.17	90.04	85.06	105.00	114.62
10th order natural frequency (Hz)	88.34	90.78	99.37	106.49	115.94

Table 3. The first 10 order buckling critical load for 5 fiber angles.

Fiber angles (°)	0	30	45	60	90
1st order buckling critical load (N)	1518473.18	1012101.95	1022074.27	1075348.36	1428938.21
2nd order buckling critical load (N)	1521345.03	1012752.56	1022832.7	1076206.42	1431626.98
3rd order buckling critical load (N)	1527735.89	1036425.07	1054288.94	1098584.35	1432047.17
4th order buckling critical load (N)	1529721.30	1037050.83	1054845.11	1099185.21	1434316.92
5th order buckling critical load (N)	1530101.24	1042105.54	1066944.72	1132236.68	1439492.37
6th order buckling critical load (N)	1536100.71	1042486.19	1067779.17	1133342.06	1440399.64
7th order buckling critical load (N)	1537753.49	1114907.09	1167055.9	1225170.17	1440625.97
8th order buckling critical load (N)	1538285.78	1115747.47	1167964.54	1226756.44	1442317.93
9th order buckling critical load (N)	1540500.99	1125894.71	1184953.28	1234965.61	1450896.38
10th order buckling critical load (N)	1543702.14	1126533.11	1185475.24	1236428.59	1452274.52

order critical loads. From the simulation results in the table, the 1st order buckling critical load of the cylindrical part is the highest when the fiber angle is 0 degrees.

4 CONCLUSIONS

In this paper, the FE model of a thin-walled carbon fiber cylindrical part are established with five fiber angles, which are 0°, 30°, 45°, 60°, and 90°. And the vibration modes and buckling of the cylindrical parts are calculated. The results show that when the fiber angle is 90°, the cylindrical part has optimal dynamic performance among the five fiber angles, and the buckling resistance is the best when the fiber angle is 0°.

ACKNOWLEDGEMENTS

This work is supported by, National Science and Technology Major Project (2014ZX04001011), State Key Laboratory of Virtual Reality Technology Independent Subject (BUAA-VR-16ZZ-07), Beijing Municipal Natural Science Foundation (3172021) and Defense Industrial Technology Development Program (A0520110009). The authors thank the referees of this paper for their valuable and very helpful comments.

REFERENCES

Bao L R, Yee A F. 2002. Effect of temperature on moisture absorption in a bismaleimide resin and its carbon fiber composites [J]. *Polymer* 43(14):3987–3997.

Kenny J M, Maffezzoli A. 1991. Crystallization kinetics of poly(phenylene sulfide) (PPS) and PPS/carbon fiber composites [J]. *Polymer Engineering & Science* 31(8):607–614.

Kumar S, Doshi H, Srinivasarao M, et al. 2002. Fibers from polypropylene/nano carbon fiber composites [J]. *Polymer* 43(5):1701–1703.

Marbán G, Antuña R, Fuertes A B. 2003. Low-temperature SCR of NOx, with NH$_3$, over activated carbon fiber composite-supported metal oxides [J]. *Applied Catalysis B Environmental* 41(3):323–338.

Patton R D, Jr C U P, Wang L, et al. 1999. Vapor grown carbon fiber composites with epoxy and poly (phenylene sulfide) matrices [J]. *Composites Part A Applied Science & Manufacturing* 30(9):1081–1091.

Rahmani H, Najafi S H M, Saffarzadeh-Matin S, et al. 2014. Mechanical properties of carbon fiber/epoxy composites: Effects of number of plies, fiber contents, and angle-ply layers [J]. *Polymer Engineering & Science* 54(11):2676–2682.

Sastri S B, Armistead J P, Keller T M. 1996. Phthalonitrile-carbon fiber composites [J]. *Polymer Composites* 17(6):816–822.

Sheehan J E. 1989. Oxidation protection for carbon fiber composites [J]. *Carbon* 27(5):709–715.

Shen G, Xu Z, Li Y. 2006. Absorbing properties and structural design of microwave absorbers based on W-type La-doped ferrite and carbon fiber composites [J]. *Journal of Magnetism & Magnetic Materials* 301(2):325–330.

Xiao Y, Wu H. 2000. Compressive Behavior of Concrete Confined by Carbon Fiber Composite Jackets [J]. *Journal of Materials in Civil Engineering* 12(2):139–146.

Electromechanical Control Technology and Transportation – Jia & Wu (Eds)
© 2017 Taylor & Francis Group, London, ISBN 978-1-138-06752-3

SmartFES: Reliable retreat route selection and navigation for indoor firefighters using exit signs

Jian Tang, Zhen Zhang, Disheng Yang, Yang Huang, Liang Hu,
Zhengang Zhao & Hengchang Liu
School of Software Engineering, University of Science and Technology of China, Hefei Shi, China

ABSTRACT: In this paper, we present a novel indoor communication system that leverages the Fire Exit Sign (FES) for firefighting. We named it SmartFES. This work is motivated by the increasing firefighter deaths that happen because of their inability to find proper routes within the building on fire. This unfortunately remains a challenging task. SmartFES uses exit signs as anchor nodes to provide landmarks for indoor localization and navigation. Despite most pre-deployed sensors failing to work in buildings on fire, we argue that a fire exit sign is a good choice, as they have independent power sources and can be made available in almost all buildings. SmartFES dynamically seeks an optimal retreat route for each firefighter, which consists of a sequence of FESs. The technical novelty of SmartFES mainly lies in a reliable heading estimation approach, with the help of body-shadowing effect and an optimal route selection algorithm in real time. We fully implemented SmartFES and conducted extensive experiments in an office building environment. Evaluation results show that SmartFES provides efficient and reliable route selection and navigation.

1 INTRODUCTION

Our society relies on a multitude of public safety personnel, for example, firefighters. Unfortunately, firefighter safety remains a challenging task worldwide. The 9/11 disaster in New York killed 343 American firefighters in 2001, and the 8/12 explosion in Tianjin took 95 Chinese firefighters' lives away in 2015. A fair amount of tragedies are because firefighters are unable to find the route out of the building in time (R. F. Fahy, 2002). For firefighters inside on-fire large commercial buildings, searching retreat routes generally is not accommodated with standard GPS technology (H. Liu, 2010). Current standard approach for firefighter retreat is mainly based on memorizing the routes they entered the building, while it is not easy to distinguish landmarks along the route, especially when the building is filled with smoke. Moreover, those landmarks may be burned up by fire, which is often the case. A slightly better way is to take a rope for marking the trace. However, it is still questionable, as the rope may easily tangle when walking deep into the building, especially when going upstairs/downstairs. Because heat and fumes are common in burning environments, firefighters have to wear breathing apparatus, which provide support only for a short period (L. Ramirez, 2012). Finding a way out of the danger zone before complete exhaustion of the air supply can be a matter of life

and death in some situations, while getting lost is a common cause of accidents in firefighting.

If we could build a reliable route map, update it in real time, and localize firefighters on the map, incident commander outside the building could then give more precise orders according to the map information. Although previous work has taken initial steps toward this goal, it remains limited to vocal reports from firefighters inside the building, which has poor reliability due to time-varying harsh environments.

Academic Research utilizes Wi-Fi (S. Kumar, 2014) and lights (Z. Yang, 2015) to help localization and navigation; infrastructure-free solutions like Pedestrian Dead Reckoning (PDR) (J.-O. Nilsson, 2012; A. R. Jimenez, 2009) have also been explored, however, neither of them is applicable in firefighting scene.

In this paper, we argue that Fire Exit Sign (FES) (S. Pu, 2005) is a good candidate to help achieve this goal as Fig. 1. FESs are mandatory devices in almost all buildings in modern societies. They have separate power sources and thus work well even when power in the building is down due to fire. Typically, they provide LED lights to show the direction to get out of the building (M. Kobes, I, 2010). We envision that FESs have huge potentials to play a more important role in firefighting with additional sensors equipped inside. Especially, a large amount of FESs in the building form a sensor network that senses the dynamic indoor environment, localizes firefighters when they are close, and selects an optimal and reliable route for each firefighter.

With this insight, we present in this paper Smart-FES, a novel indoor communication system that leverages the Fire Exit Signs (FESs) inside buildings for firefighting. By accounting for exit signs as landmarks and the interactions between firefighters and FESs, the system significantly increases the efficiency and robustness of indoor navigation. Furthermore, it utilizes a novel body shadow-effect-based approach that accurately estimates the heading, providing an invisible rope for directing firefighters along the route. In addition, SmartFES employs an optimal route selection algorithm in real time that addresses the dynamic changes in the indoor environment. We fully implemented a prototype of the SmartFES system, including the wireless FES system and the heading estimation components. We evaluate the system by comparing our solution to the state-of-the-art pedestrian dead reckoning approach. The results demonstrate that our system provides efficient and reliable route selection and navigation.

The main contributions of this paper include:

- A practical and reliable solution for retreat route selection and navigation to help save lives of firefighters.
- A novel accurate heading estimation strategy that utilizes the body-shadowing effects and an optimal retreat route map updating algorithm in real time.
- A prototype of hardware and software system, which further can be embedded in FES for a quicker and smarter firefighting blueprint.
- Empirical results show that our solution outperforms alternative solutions by reducing accumulative error and is capable of adapting

to dynamic environment by updating the retreat map in real time.

The remainder of this paper is organized as follows. A detailed description of SmartFES is provided in Section 2. The implementation and evaluation for our system are discussed in Section 3. Finally, we conclude this paper in Section 4.

2 SYSTEM DESIGN

2.1 *Overview*

Figure 2 demonstrates a use case of SmartFES, where exit signs function as landmarks, combined with intersections of route segments, giving a full route map of the on-fire building. When the firefighter inside the on-fire building receives the retreat instruction from the incidence commander, SmartFES locates the firefighter to the nearest exit sign and provides a retreat route with minimum effort. The retreat route is represented as a sequence of exit signs for firefighter to follow. As soon as he/she comes across an exit sign, his/her location can hence be calibrated accordingly. When he/she leaves for the next exit sign, SmartFES tells the heading and decides if he/she is on the route toward it.

2.2 *FES network*

Exit signs are distributed at known positions of the building, which can be used as landmarks just as mentioned before. SmartFES extends those as wireless sensor nodes functioning as reference signs, which are further employed to represent the route on the retreat map. The system also contains other two types of devices: the mobile sniffer and the base station. The base station or coordinator is a terminal device, which gathers information collected

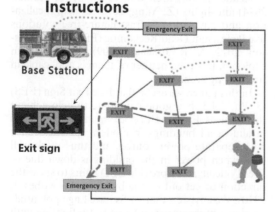

Figure 1. SmartFES generates the optimal retreat route map for incidence commander use to provide an optimal retreat route for firefighter inside the on-fire building.

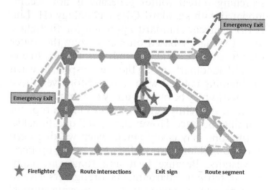

Figure 2. When a firefighter moves close to an available exit sign, his/her location is captured and the system quickly finds an optimal retreat route as a sequence of signs.

by the sensor network and provides visualization results for commander use. The mobile sniffer is a wearable device carried by firefighters, which sniffs wireless signals from the reference sign. The sniffing process can be described as follows: First, the sniffer broadcasts querying messages. After the reference sign receives the message, it sends back query contents together with the signal strength to the sniffer. In this way, the reference sign functions in a passive mode, avoiding multibroadcasting. Besides, it makes the communication easy and smart.

2.3 Route selection

Route map Because of volatile decorations such as furniture and shelves, floor plan cannot describe real-time distribution accordingly, making most floor-plan-based localization system unavailable (S. Beauregard, 2008). We do not put much emphasis on the details of the building, and just focus on the route map consisting of lines and intersections, shaping the indoor navigation problem into a binary choice of walking either forward or backward, which we would describe later.

As SmartFES provides firefighter a route as a sequence of exit signs, a natural thought is to represent the graph with exit signs as vertexes; however, this would make the problem rather difficult. First, this would lead to judge the existence of edges among signs; second, weights of the edges would be very hard to decide, as some edges contain several turning points while others not. Finally yet importantly, the derived graph cannot reflect the physical routes correspondingly. To address the problems above, we propose a graph model, which treats intersections as vertexes and exit signs along route segments as landmarks.

To model the map mathematically, we denote route segments by edges E, intersections of route segments by V, length and status of route segments by weights W, and get a weighted graph $G(V, E, W)$. Then, we search the cost-based optimal route from the graph we get. We represent the cost for starting from an exit sign t in the middle of edge (v_i, v_j) by the following formula:

$$cost(t) = \min\left\{cost(v_i) + w_{it}, cost(v_j) + w_{jt}\right\} \quad (1)$$

where w_{it}, w_{tj} and w_{ij} are defined as:

$$
\begin{cases}
w_{it} = \dfrac{w(d_{it}, m_{it})}{w(d_{it}, m_{it}) + w(d_{jt} + m_{jt})} w_{ij} \\[2ex]
w_{tj} = \dfrac{w(d_{jt}, m_{jt})}{w(d_{it}, m_{it}) + w(d_{jt}, m_{jt})} w_{ij} \\[2ex]
w_{ij} = w_{it} + w_{tj}
\end{cases}
\quad (2)
$$

where $cost(v_i)$ and $cost(v_j)$ are the cost to retreat from intersection vi and v_j, w_{ij} is the weight of the edge (v_i, v_j), and wit and w_{jt} are the cost for exit sign t to achieve v_i and v_j, respectively. It is obvious that if an exit sign coincides with an intersection, then it should have the same cost as the intersection for retreating.

Next, we present an algorithm for building a retreating map embedded in the physical route map and represent the retreat route as a sequence of exit signs and intersections.

Route Search Dijkstra algorithm is a classical and powerful algorithm for searching the shortest path between one single source and many destination vertexes; however, it cannot be applied here directly. As there are usually more than one emergency exits, it is tedious to search routes toward each emergency exit for every exit sign. Besides, some path segment of the on-fire building may be destroyed while the firefighter is walking, which would lead to update the graph and start over the entire route searching process before. As firefighting is a critical real-time issue, the above problem would call for much effort, leading to time delay. To address the problems, an alternative solution is to search the optimal route with the exit sign as the start point, which works better, but iterating emergency exit is not smart enough, too. To make it smarter, we propose a concept as destination set as Fig. 3.

Algorithm 1 first merges all the exits as a big vertex, which is denoted by *start*, and treats all the edges adjacent to emergency exits as the adjacent ones to *start*; hence, *start* is a destination set *DS*. Then, it searches the shortest path toward all the other vertexes from the *start* as Dijkstra algorithm processed. The reverse order of vertexes along the result path is hence the shortest path toward the emergency exits accordingly. Now we get a retreat map with the minimum cost attached to each intersection vertex. Getting a retreat map with exit sign denotes the retreat

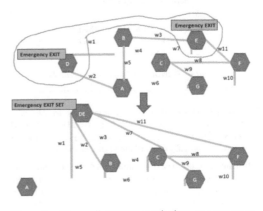

Figure 3. Merge all emergency exits into one vertex as destination set.

routes, we should do more effort. Since the exit signs are distributed along the retreat route, the cost can hence be determined by the two neighbor intersections at the end of the route segment; formula 1 is hence applied here. The algorithm avoiding iterating emergency exits one by one, making it smart and efficient as Figure 5. If any route segment shall fail, algorithm 1 removes the route segment and starts over the path searching process for one time, then all the optimum retreat routes for all the firefighter are acquired. Hence, the algorithm is capable of adapting the violate environment and provide optimal retreat map in real time.

2.4 Heading estimation

After we derived the retreat route map, the retreat process for each firefighter shall be as follows.

Algorithm 1: Cost based optimal retreating route map

Data: Weighted graph $tt(V, E, W)$, Landmarks M,
   $EmergencyExit \subseteq V$.

Result: A cost based retreating map with exit signs as landmarks

begin

   Merge EmergencyExit into one vertex as start. $T \leftarrow V \setminus EmergencyExit \cup start\ adjacent(T, start) \leftarrow \emptyset$

   foreach vertice $es \in EmergencyExit$ **do** adjacent$(T, start) \leftarrow adjacent(T, start) \cup adjacent(V, es)$
   foreach vertex $v \in T$ **do**
   $cost(start, v) \leftarrow \infty$

      $S \leftarrow \emptyset$
      $Q \leftarrow T$

         $cost(start, start) \leftarrow 0$

         $parent(start, start) \leftarrow start$
   while $Q \neq \emptyset$ **do**
      $u \leftarrow min\text{-}cost\text{-}vertex(Q, start)\ S \leftarrow S \cup u$
         $Q \leftarrow Q \setminus u$
         for $v \in adjacent(Q, u)$ **do**
            $newD \leftarrow cost(u) + w(u, v)$

            if $newD < cost(start, v)$ **then**
               $cost(start, v) \leftarrow newD$
               $parent(start, v) \leftarrow u$

   $result \leftarrow RetreatRouteMap(cost, parent, M, tt)$
   rn $result$

Definition 1 (Destination Set). *A destination set DS is a set of destination vertexes* $\{dv_i | i = 1, 2, ..K. DS\}$ *is treated as a big vertex, the adjacent edges of the big vertex are the union of adjacent edges for all the vertex in the set.*

Definition 2 (Cost based Optimal retreating Graph). *For a given graph $G(V, E)$ with cost $w(d_{ij}, m_{ij})$ attached to the edge, where d_{ij} is edge length, m_{ij} is collected information by exit signs on the route (v_i, v_j). DestinationVertexes $= \{dv_1, dv_1, ..., dv_k\}, |V| = n, |E| = m$, search a route as a sequence of exit signs on the route towards an emergency exit $dv \in DestinationVertexes$ with minimum cost.*

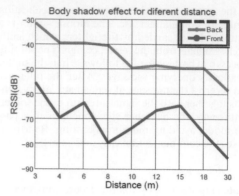

(a) **With back side towards the intended sensor exit sign**

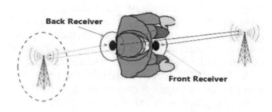

(b) **Schematic diagram of heading estimation from top view**

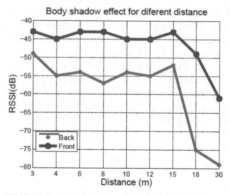

(c) **With front side towards the intended sensor exit sign**

Figure 4.   Body shadow effect for heading estimation.

First, the firefighter would select an accessible exit sign with minimum cost when received the retreating command. After reaching the exit sign, he could follow a sequence of exit signs just as the optimal retreat map provided, and the retreat process is hence be described as hopping from one exit sign to another and eventually can be divided

20

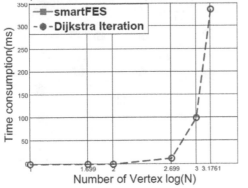

Figure 5. Comparison of SmartFES and Dijkstra iteration.

into several repeating hopping steps. To address hopping problem, we proposed a novel strategy, which utilizes body shadow effect of wireless signal propagation.

Received signal strength exit sign (RSSI) is an important parameter in describing the quality of wireless link, which can also be used as the measurement of distance. Several studies in indoor application use this technology to fulfill the goal. Previous works mainly focus on the position to locate, while paying little attention on heading, but heading is very essential to accurate and robust navigation, knowing whether the intended exit sign ahead or behind is very useful for fire fighters to take the first step. To the best of our knowledge, currently, the most available solution counts on compass or gyroscope; however, none of them is stable or reliable for indoor application. Others can only conclude heading from the localization information, like a trace, but it would bring a time delay for navigation, especially when the initial heading is wrong.

Considering the body shadow-effect on wireless signal, we present a novel strategy for determining heading. Figure 4(b) gives a schematic diagram of the strategy as two receivers are set separately at the front and the back of the body, sniffing wireless signals from anchor nodes. Because of the body shadow effect, two receivers would result in different signal strengths, which can easily tell if the anchor nodes is ahead or behind. We denote RSSI of front receiver by frontRSSI and that as the back by *backRSSI*.

When the firefighter is on the route, only two directions remain for determining as front and back. We examine the effect by carrying two receivers in the front and back. The results are shown in Figure 4. It is obviously from the figure that the side facing to the anchor node results in a greater

RSSI value; hence, it is easy to distinguish the direction toward the intended exit sign.

3 EVALUATION

We have built a prototype of our proposed system to navigate firefighters for retreat routes in harsh indoor environments. As shown in Figure 6(a), the hardware set includes a mobile sniffer, reference exit signs, and a base station. The communication is based on 2.4 GHz chips and the Zigbee protocol. The mobile sniffer and heading estimator are wireless sensor nodes with the same hardware design but different embedded software. Next, we present the performance evaluation details. We first describe the experimental setup and then examine our system in several aspects. Finally, we discuss several observations and explain the evaluation results in detail.

3.1 *Experimental setup*

We conducted extensive experiment in an office building. Figure 6(b) displays the distribution of the floor, with yellow lines for route segments, capital letters for intersections, and green diamond for exit signs. The whole floor is 120 m long and 80 m wide, with 13 intersections (A, M) and 3 of them are exit stairs (L, H, M).

3.2 *Methodology*

We evaluate our system in an office building from several aspects as efficiency, accuracy, and robustness. SmartFES updates route segment states according to the collected environment information. The result is then delivered to the base station to present the retreat route map in real time. Algorithm 1 calculates the cost-based optimal routes and assigns each exit sign a direction to next route segment and a value to represent the minimum cost to reach the emergency exit. The result

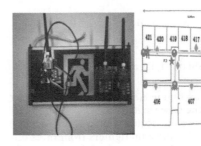

(a) Hardware prototype   (b) Floor plan and distribution

Figure 6. Experimental setup.

21

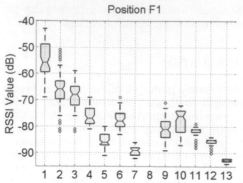

(a) RSSI between exit signs and position F1

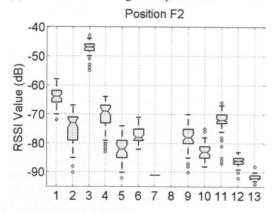

(b) RSSI between exit signs and position F2

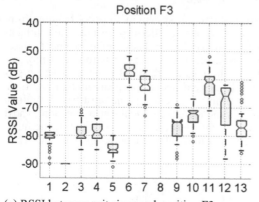

(c) RSSI between exit signs and position F3

Figure 7.   RSSI of F1, F2, and F3.

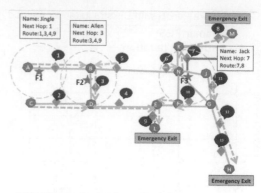

(a) Searching the retreat route according to the map

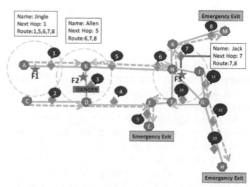

(b) Exit sign 3 sensed the danger, SmartFES updated retreat map

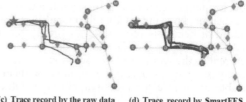

(c) Trace record by the raw data of IMU   (d) Trace record by SmartFES

Figure 8.   Benefit of retreating exit sign for calibration.

is presented as Figure 8(a), which is also the user interface of the base station.

Then, we use the acquired map to provide the retreat routes for three firefighters from three different positions, denoted as F1, F2, and F3. We record the retreat trace of F1 and compare it with PDR method. We also set exit sign 3 broken so as to test if our system is capable of adapting to the situation in real time. F3 is used to verify the selection strategy when several exit signs are available.

3.3 Results

The mobile sniffer on firefighter's body keeps sniffing the wireless network to detect the signal strength so as to determine the accessible exit sign. Figure 8 shows the retreat trace with different methods. SmartFES is far better than the PDR method mainly because exit signs can calibrate the navigation errors and give firefighters continues notification toward the emergency exits, while the PDR method can only locate the firefighter by recording the walked trace, which is easily affected by drift error.

Figure 7 shows the RSSI value from different sensor exit signs for F1, F2, and F3. We observe that, for position F1, exit sign 1 gives the maximum RSSI mainly because the exit sign 1 is the nearest one, and so is exit sign 3 for position F2. The result implies that we can use the RSSI value to determine which exit sign is the nearest one to reach. Situation for F3 is a little different: the system chooses exit sign 7 as the next one; this is mainly because it takes minimum cost to reach the emergency exit from sign 7, and it is also not difficult to reach.

According to the collected sensed danger information, the system can update the retreat map in real time, as the map search problem is as efficient as the Dijkstra algorithm. We remove indicator 3 to simulate sensing the danger on the route segment. The system updates the map accordingly and presents the result in Figure 8(b). The result shows that the updated map also provides an optimal retreating route selection.

4 CONCLUSIONS

In this paper, we presented the design, implementation, and evaluation of a robust communication system motivated by the importance, while challenging firefighter safety issues. All previous work failed to provide an efficient, accurate, and reliable solution to navigate firefighters to a safe retreat route in real-time. We also proposed a novel wireless sensor network system to achieve this goal, by leveraging exit signs as landmarks combined with body shadow effect for heading use. We described the details of an efficient algorithm for building and updating the retreat map. We fully implemented this system and evaluated it in an office building environment.

We compared our solution to the state-of-the-art pedestrian dead reckoning approach. Evaluation results show that our approach reduces the maximum firefighter location errors and provides a reliable navigation guidance.

ACKNOWLEDGMENT

This work was financially supported by Qingdao Innovation and Entrepreneurship Leading Talent project (13-cx-2), Qingdao Strategic Industry Development project (13-4-1-15-HY), and Shandong Province Science and Technology project (2013GHY11519).

REFERENCES

Beauregard S., M. Klepal, et al., "Indoor pdr performance enhancement using minimal map information and particle filters," in 2008 IEEE/ION Position, Location and Navigation Symposium, pp. 141–147, IEEE, 2008.

Fahy R.F., US Fire Service fatalities in structure fires, 1977–2000. NFPA, 2002.

https://en.wikipedia.org/wiki/2015 Tianjin explosions.

https://en.wikipedia.org/wiki/September 11 attacks.

Jimenez A.R., F. Seco, C. Prieto, and J. Guevara, "A comparison of pedestrian dead-reckoning algorithms using a low-cost mems imu," in Intelligent Signal Processing, 2009. WISP 2009. IEEE International Symposium on, pp. 37–42, IEEE, 2009.

Kobes M., I. Helsloot, B. de Vries, J.G. Post, N. Oberijé, and K. Groenewegen, "Way finding during fire evacuation; an analysis of unannounced fire drills in a hotel at night," Building and Environment, vol. 45, no. 3, pp. 537–548, 2010.

Kumar S., S. Gil, D. Katabi, and D. Rus, "Accurate indoor localization with zero start-up cost," in Proceedings of the 20th annual international conference on Mobile computing and networking, pp. 483–494, ACM, 2014.

Liu H., J. Li, Z. Xie, S. Lin, K. Whitehouse, J.A. Stankovic, and D. Siu, "Automatic and robust breadcrumb system deployment for indoor firefighter applications," in Proceedings of the 8th international conference on Mobile systems, applications, and services, pp. 21–34, ACM, 2010.

Nilsson J.-O., I. Skog, P. Händel, and K. Hari, "Foot-mounted ins for everybody-an open-source embedded implementation," in Position Location and Navigation Symposium (PLANS), 2012 IEEE/ION, pp. 140–145, IEEE, 2012.

Pu S. and S. Zlatanova, "Evacuation route calculation of inner buildings," in Geo-information for disaster management, pp. 1143–1161, Springer, 2005.

Ramirez L., T. Dyrks, J. Gerwinski, M. Betz, M. Scholz, and V. Wulf, "Landmarke: an ad hoc deployable ubicomp infrastructure to support indoor navigation of firefighters," Personal and Ubiquitous Computing, vol. 16, no. 8, pp. 1025–1038, 2012.

Yang Z., Z. Wang, J. Zhang, C. Huang, and Q. Zhang, "Wearables can afford: Light-weight indoor positioning with visible light," in Proceedings of the 13th Annual International Conference on Mobile Systems, Applications, and Services, pp. 317–330, ACM, 2015.

A new detection method for short-circuit current based on a combination of criteria

Qiong Wu
Shenyang University of Technology, Shenyang, Liaoning, China
Shenyang University, Shenyang, Liaoning, China

Shaohua Ma, Zhiyuan Cai & Weichen Wu
Shenyang University of Technology, Shenyang, Liaoning, China

Zhijian Ni
State Grid Shenyang Electric Power Supply Company, Shenyang University of Technology, Shenyang, Liaoning, China

ABSTRACT: The main research content is the fault detection circuit for high-voltage current-limiting fuse. This paper analyzes the current fault current detection methods including its advantages and disadvantages, and proposes a new detection method for short-circuit current by using the comprehensive criterion of absolute value of current, rate of change, and curvature. According to the experimental data, it has been verified that this method can accurately detect the short-circuit current in a very short time.

1 INTRODUCTION

The method of setting current threshold has been often used for the detection of short-circuit in traditional current limiter, considering the on-off difficulty of current caused by the rising of short-circuit current, and the problem of the increase of electric power. Previous work has put up a detection method based on the raising rate of current. Once the current changes, the short-circuit will be cut off immediately, which can avoid huge thermal shock and electric power effectively for the protected equipment and current limiter itself. However, the practice has proved that the current limiter is unable to avoid the high order harmonic component of power grid, and may generate false action. In order to effectively avoid the system flow and the interference of higher harmonic, while avoid adding system testing time and reduce the burden of circuit, the paper introduces the curvature which is the derivative of the current, using the absolute value of current, the rate of change of current and curvature as a criterion to detect the coming of the short-circuit current. For the different periods of short-circuit, we propose to use different combinations in different angles of the short-circuit current by analyzing the rate of change and the size of the impact for detection the rate of change of current.

2 GETTING STARTED: THE BASIC WORKING PRINCIPLES OF CLF

CLF consists of the following four parts, as shown in Figure 1.

- Rogowski Coil.
  In the proposed method, Rogowski coil and the basic principle of electromagnetic is used to sample the current in a wide range of changes, and to complete the change of huge voltage and high current from the primary side of coil to the secondary side of low voltage, low current signal.
- The Detection Circuit of Current.
  The main research of this topic is about completing the recognition function for the fault identification of sampling current signal, sending control signal to capacitor discharge circuit and driving the control pole (K) of thyristor when recognizing the short-circuit current.

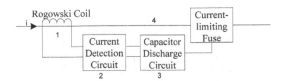

Figure 1. The structure principle of current-limiting fuse.

- The Capacitor Discharge Circuit.

  The capacitor discharge circuit consists of thyristor, capacitor and sub-closing coil. After the control pole receives the drive signal, the thyristor is conducted, and capacitor discharges to the coil through the thyristor, to generate, electric repulsion and drive, that drive the vacuum switch.
- Current-limiting Fuse.

  The current-limiting fuse is made up of vacuum switch and arc-suppressing fuse in parallel. Under normal circumstances, a pair of contacts of vacuum switch is often in the closed state by the external force and vacuum self-closing force. When a fault occurs, the vacuum switch will get electromagnetic repulsion under the function of discharge coil; separates actuator quickly; compete the transfer process of current; transfers the current to the arc-suppressing fuse in parallel; compete the recovery process of dielectric strength; and breaks the circuit successfully.

  The key of the reliable action of the CLF is the response time of the detection current. The accurate detection for the occurrence of the short-circuit current in the shortest time has an important practical significance for the effective circuit protection and avoiding the damage of circuit by huge current.

3 THE CALCULATION METHOD OF SHORT-CIRCUIT CURRENT

When the short-circuit fault occurred, the equivalent circuit of the system is shown in Fig. 2. In this figure, U_S is the equivalent voltage source of system. R_1, L_1 are the equivalent resistance and the equivalent inductance in the power side respectively. R_2, L_2 are the equivalent resistance and inductance of load respectively.

Before short-circuit fault occurs, the voltage is set at $u = U \angle 0°$, and the initial phase is zero. The current in the system can be expressed as

$$\dot{I} = \frac{\dot{U} \angle 0°}{Z} = \frac{U \angle 0°}{j\omega(L_1 + L_2) + (R_1 + R_2)}$$

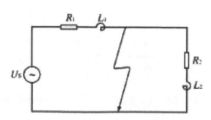

Figure 2. Equivalent circuit diagram of the short-circuit.

If $\phi_1 = \arctan \dfrac{\omega(L_1 + L_2)}{R_1 + R_2}$, then

$$\dot{I} = \frac{\dot{U} \angle - \phi_1}{[\omega(L_1 + L_2)]^2 + (R_1 + R_2)^2}$$

$$i = \frac{\sqrt{2}U \sin(\omega t - \phi_1)}{\sqrt{[\omega(L_1 + L_2)]^2 + (R_1 + R_2)^2}} = \sqrt{2}\,I \sin(\omega t - \phi_1)$$

In the equation, i is rated current Effective value

$$I = \frac{U}{\sqrt{[\omega(L_1 + L_2)]^2 + (R_1 + R_2)^2}}$$

The derivative and second derivative of I are

$$\frac{di}{dt} = \sqrt{2}\,I\,\omega\cos(\omega t - \phi_1)$$

$$\frac{d^2 i}{dt^2} = -\sqrt{2}\,I\,\omega^2 \sin(\omega t - \phi_1)$$

When the fault of the short-circuit occurred at the time $\omega_t = \alpha$, then

$$i_{0^-} = \sqrt{2}\,I \sin(\alpha - \phi_1)$$

If $t' = t - \dfrac{\alpha}{\omega}$, then the short-circuiting time is $t' = 0$, according to single-phase short-circuit current calculation formula

$$i = \frac{\sqrt{2}U \sin(\omega t - \phi_2)}{\sqrt{(\omega L_1)^2 + R_1^2}} + A e^{-\frac{R_1}{L_1}t'}$$

$$= \sqrt{2}\,I' \sin(\omega t - \phi_2) + A e^{-\frac{R_1}{L_1}t'}$$

In the equation, $\phi_2 = \arctan\frac{\omega L_1}{R_1}$, the short-circuit current effective value is

$$I' = \frac{U}{\sqrt{(\omega L_1)^2 + R_1^2}}$$

The short-circuit occurred time is $i_{0^+} = \sqrt{2}\,I' \sin(\alpha - \phi_2) + A$, because of $i_{0^-} = i_{0^+}$ then

$$A = \sqrt{2}\,I \sin(\alpha - \phi_1) - \sqrt{2}\,I' \sin(\alpha - \phi_2)$$

Thus

$$i = \sqrt{2}\,I' \sin(\omega t' + \alpha - \phi_2)$$
$$+ \left[\sqrt{2}\,I \sin(\alpha - \phi_1) - \sqrt{2}\,I' \sin(\alpha - \phi_2)\right] e^{-\frac{R_1}{L_1}t'}$$

If the ratio of the steady-state value of the short-circuit current to the steady-state value of the short-circuit current is N, then

$$N = \frac{I'}{I} = \frac{\sqrt{[\omega(L_1 + L_2)]^2 + (R_1 + R_2)^2}}{\sqrt{(\omega L_1)^2 + R_1^2}}$$
$$= \frac{R_1 + R_2}{R_1} \frac{\cos\phi_2}{\cos\phi_1}$$

The expression of short-circuit current, rate of change, change the curvature are

$$i = \sqrt{2}\,\mathrm{NI}\sin(\omega t' + \alpha - \phi_2)$$
$$+ \left[\sqrt{2}\,\mathrm{I}\sin(\alpha - \phi_1) - \sqrt{2}\,\mathrm{NI}\sin(\alpha - \phi_2)\right]e^{-\omega\cot\phi_2 \times t'}$$

$$\frac{di}{dt} = \sqrt{2}\,\mathrm{NI}\,\omega\cos(\omega t' + \alpha - \phi_2)$$
$$- \left[\sqrt{2}\,\mathrm{I}\sin(\alpha - \phi_1) - \sqrt{2}\,\mathrm{NI}\sin(\alpha - \phi_2)\right]$$
$$\times e^{-\omega\cot\phi_2 \times t'}\,\omega\cot\phi_2$$

$$\frac{d^2 i}{dt^2} = -\sqrt{2}\,\mathrm{NI}\,\omega^2 \sin(\omega t' + \alpha - \phi_2)$$
$$+ \left[\sqrt{2}\,\mathrm{I}\sin(\alpha - \phi_1) - \sqrt{2}\,\mathrm{NI}\sin(\alpha - \phi_2)\right]$$
$$\times e^{-\omega\cot\phi_2 \times t'}\,\omega^2 \cot^2\phi_2$$

We can normalize the above equation and obtain the following expression

$$i = \mathrm{N}\sin(\omega t' + \alpha - \phi_2)$$
$$+ [\sin(\alpha - \phi_1) - \mathrm{N}\sin(\alpha - \phi_2)]e^{-\omega\cot\phi_2 \times t'}$$
$$\frac{di}{dt} = \mathrm{N}\,\omega\cos(\omega t' + \alpha - \phi_2)$$
$$- [\sin(\alpha - \phi_1) - \mathrm{N}\sin(\alpha - \phi_2)]e^{-\omega\cot\phi_2 \times t'}\,\omega\cot\phi_2$$

4 THE ANALYSIS OF SHORT-CIRCUIT CURRENT

4.1 *Impact of different short-circuit angle on short-circuit current*

Short-circuit current, the raising rate of short-circuit current and the second derivative of the current have different forms at different short-circuit time. For the convenience of analysis, temporarily take I = 1250 A, N = 5, $\phi_1 = 60°$, $\phi_2 = 80°$, short-circuit angle α for the case 0°, 60°, 120°, 180°, 240°, 300° calculating i, di/dt, d^2i/dt^2 with the time change, calculation time of short-circuit current is taken as 20 μs, and it can be shown in Figure 3.

In Figure 3(a), the reactions of current, current change rate, and current change curvature on different short-circuit current is different, when the short-circuit angles are 60, 120, 180, 240, 300, and 360 respectively. When the short-

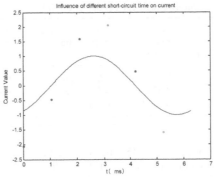

(a) Influence of different short-circuit time on current

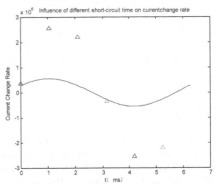

(b) Influence of different short-circuit time on current-change rate

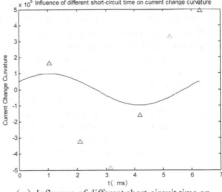

(c) Influence of different short-circuit time on current change curvature

Figure 3. Comparison for rate of change of current and path of current.

circuit angle ranges from 60 to 240, the change is relative small. Otherwise, the change is more intense which can faster reflect the arrival of short-circuit current.

As shown in Figure 3(b), the change rate of current change rate is small when the short-circuit

27

Table 1. Short-circuit current detection time.

Parameter	Normal max value	1.5 times of the normal max value	Short-circuit angle is different, more than normal time						
			0	30	60	90	120	150	180
i	0.1	$0.015 * e + 2$	0.2	0.2	–	0.6	0.2	0.2	0.2
di/dt	3.1397	$4.5 * e + 2$	0.4	0.01	0.01	0.01	0.01	0.01	0.4
d^2i/dt^2	0.9864	$0.15e + 6$	0.013	0.01	0.011	0.68	0.01	0.01	0.01

angle ranges from 180 to 360. Otherwise, the current change rate varies dramatically, which can accelerate the reflection of short-circuit current coming.

In Figure 3(c), the second derivative of current change has no obvious change when the short-circuit angles are from 60 to 240. Otherwise, the current change rate varies dramatically, which can accelerate the reflection of short-circuit current coming.

4.2 Criterion analysis of short-circuit current

When $\phi_1 = 60°$, $\phi_2 = 80°$, $N = 5$, in order to avoid the influence and operational errors to the circuit of the inrush current and high harmonics, under normal circumstances, we can use the same order of magnitude to calculate the maximum value of the current, derivative and second derivative in the normal range. At the same time, we increase the value for the discriminant threshold, taking more than 1.5 times the normal operation as the judge criteria, and improving the anti-interference performance to ensure the any short-circuit angle can go through the i, di/dt, d^2i/dt^2. The results are shown in Table 1.

As shown in Table 1, for most of the short circuit current, the circuit can react in 0.2 ms. For extremely unfavorable conditions such as when the short circuit angle is 90 degrees, the system can make correct reactions within 0.6 ms.

4.3 The impact of short-circuit current multiple N on detecting short-circuit current

Under different short-circuit current where $\phi_1 = 60°$, $\phi_2 = 80°$, $\alpha = 90°$, we explore the impact of different N including 5, 10, 15. The dynamics of i, di/dt, d^2i/dt^2 are shown in Figure 4.

We can see from this figure that a higher short-circuit current will increase the absolute value, rate of change, and rate of curvature correspondingly. The determination speed will also be accelerated. Such observation indicates that the proposed method is suitable for judging short-circuit current with N > 5.

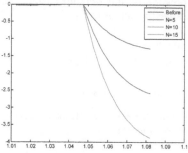

(a) The impact of N on the short-circuit current

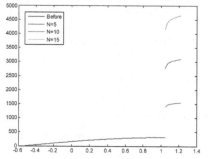

(b) The impact of N on the change rate of short-circuit current

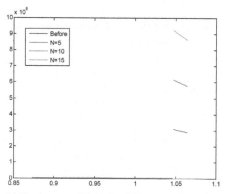

(c) The impact of N on the change curvature of short-circuit current

Figure 4. The impact of N on the current, current change rate, and current change.

5 CONCLUSION

In this paper, we introduce the current curvature, which is the derivative of current change rate to detect the short-circuit current. Our method take advantage of current absolute value, current change rate, and current change curvature to judge the arrival time of short-circuit current. We analyze the change rate and the impact of various factors under different short-circuit time. Meanwhile, we adopt different combined methods regarding to different short-circuit angles and realize the fast detection of current change rate. This method is very suitable and can judge the short-circuit current within extremely short time based on both theoretical analysis and actual simulation testing.

Detecting the short-circuit current through utilizing the absolute value of a current, current change rate, and current change curvature, not only can it effectively avoid the system inrush current and higher harmonic components, but also detect the short-circuit current within an extremely short time. Our method possesses important practical significance in effectively protecting the circuit and avoiding the harm of large current to a circuit.

REFERENCES

Jin, J.X. et al. (2000). Magnetic saturable reactor type HTS fault current limiter for electrical application. J. Physical Superconductivity. 341–348, Part 4, November 2000, 2629–2630.

Kulicke, B., H.H. Schranm. (1980) Clearance of short-circuits with delayed current zeros in the Itaipu 550 kV-substation. J. IEEE Transactions on Power Apparatus and Systems. 99, 1406–1414.

Wang Jimei. (1988). Vacuum switch theory and its application. J. Xi'an: Xi'an Jiaotong University Press, 1988.

Zhang Xiaoqing, Ming Li. (2008). Using the Fault Current Limiter With Spark Gap to Reduce Short-Circuit Currents Power Delivery. J. IEEE Transactions on, Volume 23 Issue 1, Jan. 2008, 506–507.

Zhao Cuixia, et al. (2007). Transient simulation and analysis for saturated core high temperature superconducting fault current limiter. J. IEEE Transactions on Magnetics. 2007, 43(4), 1813–1816.

Zyborski J, et al. (2000). A DC Hybrid Arcless Low voltage AC/DC Current Limiting Interrupting Device. J. IEEE Trans on Power Delivery. 2000. 15(4), 1182–1187.

Study on high-temperature characteristics of a 1200 V 40A SiC JBS diode

Lin Wang & Runhua Huang
Nanjing Electronic Devices Institute, Nanjing, China

Ao Liu
State Key Laboratory of Wide-Bandgap Semiconductor Power Electronic Devices, Nanjing, China

Gang Chen & Song Bai
State Key Laboratory of Wide-Bandgap Semiconductor Power Electronic Devices, Nanjing, China
Nanjing Electronic Devices Institute, Nanjing, China

ABSTRACT: A 1200 V/40A SiC JBS diode is fabricated on the basis of 4H-SiC epitaxial wafer and a newly developed processing sequence. The 4H-SiC epitaxial wafer is grown by Chemical Vapor Deposition (CVD) on commercial 4H-SiC substrates. The epitaxial structure and manufacturing process were optimized. By using Ti Schottky metal and Ti/Ni/Au multilayer metal bond pad structure, the double-side Au process 4H-SiC JBS diode is formed. A good balance between blocking voltage and conduction current was successfully obtained, especially at a high temperature. The leakage current of the 1200 V/40A SiC JBS diode was less than 1 µA at the reverse voltage of 1200 V at room temperature, which increased to 250 µA at 1200 V at 250°C.

1   INTRODUCTION

SiC has a wide band gap, high breakdown electric field, high saturation drift velocity, and high thermal conductivity, which make it an ideal material for making high-power, high-frequency, high-temperature, and radiation-resistant devices. The SiC devices in the blocking performance compared to the traditional Si power electronic devices have great advantages. The breakdown voltage and switching frequency of SiC Schottky diode have surpassed those of the Si PiN diode. The JBS structure in the active region is formed by p^+ ion implantation, which greatly reduces the electric field intensity on the surface of the device, resulting in a reduced leakage current at a high reverse voltage. The JBS structure is formed along with the Floating Field Limiting Ring (FFLR) in the same process of high-energy ion implantations (K. Vassilevski, 2006 & Eugene A, 2009).

SiC devices often work under extreme conditions such as high temperature. It is very valuable to study the device characteristics in different temperature environments. The lower the conduction loss, the smaller the switching loss and higher the thermal conductivity, making SiC devices more suitable to work at high temperatures. According to theory, SiC devices can operate at a very high temperature. However, in fact, there are many process and technical difficulties that limit their abil-

ity of high-temperature applications. In this paper, the design and fabrication of 1200 V/40A SiC Schottky diode, which can work at a high temperature of 250°C are introduced, and the blocking and conducting performance at different temperatures are analyzed.

2   STRUCTURE DESIGN

The basic structure of a SiC JBS diode is shown in Figure 1. In order to reduce the blocking-state leakage current, active region is designed with

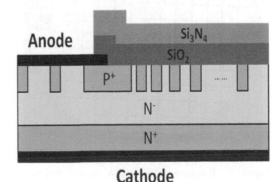

Figure 1.   Schematic of JBS diode.

a JBS structure, in which the pn junctions in the active area can reduce the electric field intensity of surface area and thus reduce the leakage current under high reverse voltage.

The edge protection uses an FFLR structure, by adding a plurality of p+ doped rings in the periphery of an active area to reduce the electric field strength. The FFLR termination structure is currently widely used in SiC power electronic devices with breakdown voltage under 3000 V. Although the theoretical efficiency of the FFLR termination is lower than that of some other structures, such as JTE, the protection efficiency is not affected by the interface charge. For the 1200 V SiC JBS diode, the total width of the FFLR termination area is about 50 μm, the single ring width is about 2.5 μm, and the ring spacing increases gradually from inside to outside.

3 EXPERIMENT

SiC epitaxial wafers (4 inch) were grown in-house using a hot-wall CVD by high-temperature and low-pressure homoepitaxial growth technology. The SiC epitaxial layers were grown on n-type conducting SiC substrates. The thickness of the epitaxial layer is 12 μm, and the doping concentration is 6E15 cm^{-3}. The highest Al$^+$ ion implantation energy is 320 keV for the formation of both the p$^+$ region in the active region and the floating field limiting rings. The activation of ion implantation was completed by high-temperature annealing at 1850°C. The Ti and Ni metal was used as the front-side Schottky contact metal and the back-side Ohmic contact metal, respectively. Both the front-side and back-side electrodes were finished with thick gold. The passivation layers were formed by using the Silicon Nitride (Si$_3$N$_4$) and the Silicon Dioxide (SiO$_2$) multilayer composite structure. The process of SiC JBS diode mainly includes lithography, ion implantation, etching, oxidation, passivation, and metal lift off. The key processes include Al$^+$ ion implantation, Schottky contact, Ohmic contact, and passivation (H. Bartolf, 2014). A finished 4-inch wafer of 1200 V/40A SiC JBS diodes is shown in Figure 2. The size of the single die is 5 mm * 5 mm.

4 RESULTS AND DISCUSSIONS

The static characteristics of the packaged devices were tested using a Tektronix 370B transistor tester. To obtain the high-temperature characteristics of the SiC JBS diodes, special designed heating equipment was used to heat the devices. The reverse and forward characteristics were obtained at room temperature and elevated temperature. As shown in Figure 3, the leakage current of the 1200 V/40A SiC JBS diode was less than 1 μA at a reverse voltage of 1200 V at room temperature. With the gradual increase of the temperature, the leakage current increased accordingly. Even at a high temperature of 200°C, the device showed a fairly low leakage current of less than 50 μA at a reverse voltage of 1200 V. The leakage current increased rapidly at both high and low reverse voltages, when the temperature was over 200°C. At 250°C, the leakage current increased to 250 μA at the reverse voltage of 1200 V, which is still too small to cause reliability concerns. The results of the reverse characteristics showed that the

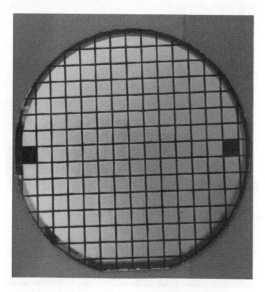

Figure 2. Fabricated 1200 V/40A 4H-SiC JBS diode wafer.

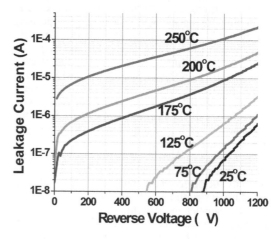

Figure 3. Temperature-dependent reverse I–V characteristics of 1200 V/40A SiC JBS diode.

Schottky barrier of SiC diode based on Ti metal can work at 250°C (Runhua Huang, 2014).

Figure 4 shows the forward conduction performance of the 1200 V/40A SiC JBS diode at temperatures covering the range of 25–250°C. The decrease of forward voltage with increasing temperature showed a clear positive temperature-dependent coefficient. At 25°C, the forward voltage at 40A was 1.6 V, and the turn-on voltage was about 0.8 V. When the temperature was increased to 250 °C, the forward voltage increased to 2.7 V at 40A, and the turn-on voltage reduced to 0.5 V.

Figure 5 shows a dynamic test curve of the 1200 V/40A SiC JBS diode at room temperature. The test result showed a reverse recovery time of 44ns and a reverse recovery current of 5.5A. Because

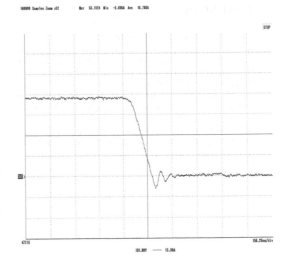

Figure 6. Reverse recovery waveform of the 1200 V/40A SiC JBS diode measured at 150°C.

of the limitation of the test equipment, high-temperature dynamic test was conducted only at 150°C. The test result is shown in Figure 6, which showed a reverse recovery time of 45ns and a reverse recovery current of 6.0A. As the SiC JBS diode is a majority carrier device, it does not have any stored minority carriers. Therefore, there is no reverse recovery current associated with the turn-off transient of the diode, but only a small amount of displacement current required to charge the Schottky junction capacitance, which is independent of the temperature.

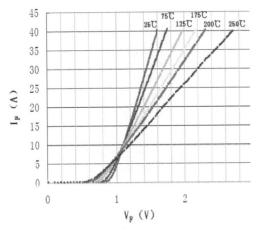

Figure 4. Temperature-dependent forward I–V characteristics of 1200 V/40A SiC JBS diode.

5 CONCLUSION

In summary, we have fabricated SiC JBS devices on n+ conductive 4H-SiC substrates. Design and fabrication of 1200 V/40A SiC JBS diodes were reported, and both static and dynamic characteristics were obtained at different ambient temperatures. At temperatures as high as 250°C, a good combination of blocking and conducting characteristics is achieved. At 250°C, the forward voltage drop was controlled to be below 2.7 V, and the blocking voltage is more than 1200 V. The experimental results show that the SiC material is a good candidate for making devices working in high-temperature environments.

ACKNOWLEDGMENT

This work was supported by the National High Technology Research and Development Program of China (No. 2014AA052401).

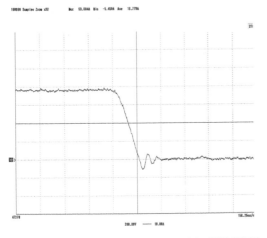

Figure 5. Reverse recovery waveform of the 1200 V/40A SiC JBS diode measured at 25°C.

REFERENCES

ATLAS User's Manual, www.silvaco.com

Bartolf, H., V. Sundaramoorthy, A. Mihaila, Study of 4H-SiC Schottky Diode Designs for 3.3 kV Applications[c], Materials Science Forum Vols. 778–780 (2014) pp 795–799.

Eugene A. Imhoff and Karl D. Hobart, High-Current 10 kV SiC JBS Rectifier Performance[c], Materials Science Forum Vols. 600–603 (2009) pp 943–946.

Huang Runhua, Tao Yonghong, Cao Pengfei et al, Development of 10 kV 4H-SiC JBS diode with FGR termination[c], Journal of Semiconductors Vol. 35, No.7.

Runhua Huang, Gang Chen, Song Bai et al, Simulation, Fabrication and Characterization of 4500 V 4H-SiC JBS diode[c], Materials Science Forum Vols. 778–780 (2014) pp 800–803.

Runhua Huang, Yonghong Tao, Gang Chen et al, Simulation, Fabrication and Characterization of 6500 V 4H-SiC JBS diode[c], Advanced Materials Research Vols. 846–847 (2014) pp 737–740.

Vassilevski, K., I. Nikitina, P. Bhatnagar et al, High temperature operation of silicon carbide Schottky diodes with recoverable avalanche breakdown[c], Materials Science Forum Vols. 527–529 (2006) pp 931–934.

Electromechanical Control Technology and Transportation – Jia & Wu (Eds)
© 2017 Taylor & Francis Group, London, ISBN 978-1-138-06752-3

Thermal design and simulation analysis of rectifier cabinet for nuclear power generating stations

Qiang Yin, Ze-cheng Xiong, Chuan-tao Zhu & Yong-feng Wang
XJ Power Co. Ltd., State Grid Corporation of China, Xuchang, Henan, China

Rong-huan Li
XuChang Third Senior High School, Xuchang, Henan, China

ABSTRACT: In order to obtain directly and accurately the distribution characteristics of temperature field and the temperature of main elements of rectifier cabinet for nuclear power generating stations, the components composition of the topological structure are analyzed, the loss of main components are calculated, the thermal resistance and heat dissipation effect of the radiator is discussed, the overall layout of the device and fan selection are considered, and the 450 kVA physical model is proposed. The operating process of ANSYS software in handling such thermal problems is summarized. According to the simulation results, the thermal design is analyzed and optimized, and the reference scheme for thermal design is given.

1 INTRODUCTION

With the development of power electronic technology, especially the development and successful application of high power thyristor devices, it brings a new prospect to the power system. However, in the case of high power, the power dissipation is very large and the heat energy generated is very high. If the way of the heat dissipation is blocked, it will seriously affect the safe operation of rectifier cabinet for nuclear power generating stations.

In this paper, in order to obtain directly and accurately the distribution characteristics of temperature field and the temperature of main elements of rectifier cabinet for nuclear power generating stations, based on the analysis of the components composition of the 12 pulse rectifier topology, the loss of main components are calculated and the thermal resistance and heat dissipation effect of the radiator is discussed. The 450 kVA physical model is proposed. The operating process of ANSYS software in handling such thermal problems is summarized. According to the simulation results, the stability and reliability of the system are ensured.

2 THE TOPOLOGICAL STRUCTURE

The principle diagram of 12 pulse rectifier circuit is shown in Figure 1. The 12 pulse rectifier circuit is consisted of three phase phase-controlled rectifier bridge in parallel, with the same structure

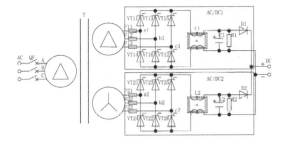

Figure 1. The principle diagram of 12 pulse rectifier.

and parameters, which has the advantages of low cost, a transformer and a switch. where T is the fundamental frequency phase shifting transformer, AC/DC1 and AC/DC2 are the converter circuit of the two groups rectifier bridge, VT11~VT16 and VT21~VT26 are thyristors, L1 and L2 are the filter inductor, C1 and C2 are the filter capacitor, R1 and R2 are the false load, and D1 and D2 are the blocking diode.

3 THE THERMAL DESIGN

3.1 *The loss calculation*

3.1.1 *The transformer loss*
The transformer loss is composed of the iron loss and copper loss. The loss generated in the core excitation resistor is called iron loss, related materials and quality of transformer. When the current enters into the transformer winding, the loss gener-

ated in the one or two winding resistance is called copper loss.

It is difficult to calculate the transformer iron loss, so the iron loss can be approximately calculated through the table based on per kg core in the magnetic induction intensity and frequency. The transformer iron loss is approximately proportional to its working frequency. The transformer copper loss is proportional to the square of the current and the winding resistance.

The formula of the transformer copper loss is given.

$$P_{T\_Cu} = \left(I_{1\_rms}\right)^2 * R_{1Cu} + 2*\left(I_{2\_rms}\right)^2 * R_{2Cu} \tag{1}$$

where I_{1_rms} is the effective value of the transformer primary current; R_{1_Cu} is the copper resistance from the transformer primary winding; I_{2_rms} is the effective value of the transformer secondary current; and R_{2_Cu} is the copper resistance from the transformer secondary winding.

The formula of the transformer iron loss is given.

$$P_{T\_Fe} = P_V * V_{Fe} \tag{2}$$

where PV is the loss from the unit volume of the transformer magnetic core; and VFe is the volume of the transformer magnetic core.

3.1.2 The thyristor loss
The formula of the thyristor loss is given.

$$P_{T(AV)} = V_{T0} \times I_{T(AV)} + F^2 \times I_{T(AV)}^2 \times r_{T0} \tag{3}$$

where V_{T0} is the threshold voltage; r_{T0} is the slope resistance; F is the wave factor which is the ratio between the RMS current and average current on state; and $I_{T(AV)}$ is the average current on state.

3.1.3 The filter inductor loss
The filter inductor loss is divided into the copper loss and iron loss. The filter inductor copper loss is proportional to the square of the inductor effective current and the inductor winding resistance. The effective value of the inductor current is given.

$$I_{L\_rms} = \sqrt{I_o^2 + \left(\Delta I_o\right)^2} \tag{4}$$

where I_O is the output current; and ΔI_o is the ripper current.

And then, The formula of the inductor copper loss is given.

$$P_{L\_Cu} = \left[I_o^2 + \left(\Delta I_o\right)^2\right] * R_{Cu} \tag{5}$$

The formula of the inductor iron loss is given.

$$P_{L\_Fe} = P_V * V_{Fe} \tag{6}$$

where P_V is the loss from the unit volume of the inductor magnetic core; V_{Fe} is the volume of the inductor magnetic core; and R_{Cu} is the inductor copper resistance.

3.1.4 The filter capacitor loss
The filter capacitor loss is mainly caused by the ripple current and the capacitor equivalent series resistance, and the effective value of the ripple current can be obtained by the integral method.

$$I_{C\_rms} = \sqrt{\left(\Delta I_o\right)^2} \tag{7}$$

And then, the formula of the capacitance loss can be given.

$$P_C = \left(I_{C\_rms}\right)^2 * R_C \tag{8}$$

where I_{C_rms} is the ripple current; and R_C is the equivalent series resistance of the capacitor.

3.1.5 The diode loss
The diode has the forward voltage and forward current which is exist on state, at the same time it has the on state resistance. so the formula of the diode loss is given.

$$P_D = V_{F0} \times I_F + R_D \times I_F^2 \tag{9}$$

where V_{F0} is the voltage of the diode pressure drop; R_D is the diode resistance on state; and I_F is the current flowing the diode.

3.2 The selection of radiator and fan

3.2.1 The selection of radiator
The component junction temperature is can't be measured in fact, so it can be determined with the operating temperature T_C and thermal resistance R_{ja}. The formula of the thermal resistance can be given.

$$T_j = PR_{ja} + T_C \tag{10}$$

where T_j is the component junction temperature; and P is the component heat dissipation.

$$R_{ja} = R_{jc} + R_{cs} + R_{sa} \tag{11}$$

where R_{jc} is the thermal resistance from the component core to the component surface; R_{cs} is the thermal resistance from the component surface to the radiator; and R_{sa} is the thermal resistance from the radiator to the air.

The value of Rcs is relative to the component surface, the fastening force, the radiator surface and the radiator installation, which is more complicated. So, The value of Rcs is ratio to the R_{jc} about $1/3 \sim 1/5$ in formerly.

3.2.2 *The selection of fan*

The wind pressure has a great influence on the heat dissipation effect. Firstly, in view of the existence of the pressure head loss, when the wind pressure is lower than the loss of the radiator pressure head loss, the cooling wind is not blowing into the passageway. Secondly, there is a big difference between the gap of the radiator fin space and the gap between the wind passageway and the radiator. When the wind pressure is too low, the cooling wind does not flow through the radiator fin space. It can be flow through the big gap, resulting in the "short circuit" phenomenon. Finally, the heat exchange rate can be increased between the radiator and the air, with heightening the air pressure which the wind speed of the radiator micro surface can be increased.

The wind passageway has a great influence on the heat dissipation effect. According to the principle of heat transfer, the heat transfer efficiency is much higher than the laminar flow through the air in the form of turbulent flow. At the same time, it can be careful to reduce the debris attached the wind passageway in order to reduce the wind resistance of the wind passageway. In view of the enlargement of the wind resistance, the velocity flow can be decreased.

The wind resistance of the single wind passageway can be given.

$$\Delta P_1 = (0.314 * \mathrm{Re}^{-0.25} * H * \omega^2 * \rho)/d^2 \qquad (12)$$

where Re is the Reynolds number; H is the forced air cooling coefficient; ω is the flow speed in the cabinet; ρ is the air density; and d is the equivalent diameter of single wind passageway.

$$\Delta P_1 = \omega^2 * \rho/2 \qquad (13)$$

The wind passageway distributing layout is used to enlarge the heat dissipation in the cabinet, with increasing series fan to improve wind pressure in the cabinet, at the same time, the external suction fan can be fixed on the top of the cabinet.

4   SIMULATION AND ANALYSIS

The 450kVA physical model of rectifier cabinet for nuclear power generating stations is proposed. The operating process of ANSYS software in

handling such thermal problems is summarized. The system parameters can be given. The fundamental frequency transformer primary voltage is 380VAC ± 15%, the fundamental frequency transformer secondary voltage is 250VAC ± 15%, the output voltage is from 198VDC to 254 VDC, and then, the output current is 1200 A. At the same time, the filter inductance is 1 mH, and then, the filter capacitor is 10 mF.

The loss specification table of the main components of the rectifier cabinet is shown in Table 1.

The fan specification and quantity in the rectifier cabinet are shown in Table 2, and then, the P-Q diagram of fan specifications is given from the corresponding relation in the P-Q panel of Table 2.

The external temperature of the rectifier cabinet is set 40°C, and then, the output power is the

Table 1.   The loss of the main components.

Name	Loss	Number	Temperature
Transformer	$4.6(P_{T\_Cu})+1.4(P_{T\_Fe})$	1	140
Thyristor	$0.356(P_{T(AV)})$	2	105
Filter inductor	$1.1(P_{L\_Cu})+0.3(P_{L\_Fe})$	2	120
Filter capacitor	$0.03(P_C)$	2	70
Diode	$0.6(P_D)$	2	130

Table 2.   The specification and quantity of fan.

Positation	Type	Number	P-Q diagram
Thyristor	SKF 16 A-230–11	2	Figure 2.
Cabinet top	W2E200-HK38–01	2	Figure 3.
Diode	AA1752HB-AW	3	Figure 4.

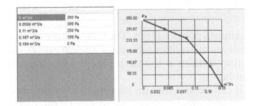

Figure 2.   The P-Q diagram of the thyristor fan.

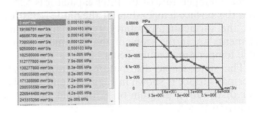

Figure 3.   The P-Q diagram of the cabinet top fan.

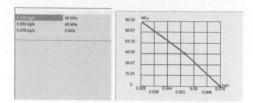

Figure 4.   The P-Q diagram of the diode fan.

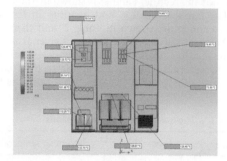

Figure 5.   The temperature of main elements.

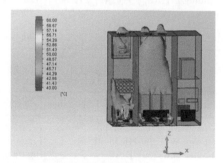

Figure 6.   The stereoscopic distribution diagram of iso-thermal surface.

maximum power output. The temperature of main elements of rectifier cabinet for nuclear power generating stations is shown in Figure 5 by computer simulation.

It can be seen that the maximum temperature of the transformer core is close to 125°C, the surface temperature of the thyristor is 75°C, the surface temperature of the filter inductor is 110°C, the surface temperature of the filter capacitor is 52°C, and the surface temperature of the diode is 120°C. So, the surface temperature of the main heating elements are in the scope of the maximum allow-able temperature requirements.

The stereoscopic distribution diagram of iso-thermal surface is shown in Figure 6, and the pla-nar distribution diagram of isothermal surface is shown in Figure 7.

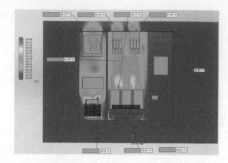

Figure 7.   The planar distribution diagram of isother-mal surface.

5   CONCLUSIONS

In this paper, the components composition of the 12 pulse rectifier topological structure are ana-lyzed, the loss calculation of the main components are given, the thermal resistance and heat dissipa-tion effect of the radiator is discussed, and the 450 kVA physical model of rectifier cabinet for nuclear power generating stations is proposed. The oper-ating process of ANSYS software in handling such thermal problems is summarized. It can be seen from the simulation results that the surface temperature of the main heating elements are in the scope of the maximum allowable temperature requirements. It provides some references for engi-neering application.

This work was financially supported by Science and Technology Project of State Grid Corporation of China and Science and Technology Project of XU JI GROUP CO.,LTD..

REFERENCES

Chang Bing-quan, Wu Guang-jun. Heat emission analyses of heavy duty rectifying cabinet[J]. Heilongjiang Electric Power, 2005.27(3):204–209.

Fan Feng-xin, Cao Biao, Zhang Shu-hao, etal. Analysis and solution: Loss of high power spot welding machine[J]. Electirc Welding Machine, 2009.39(11):37–41.

He Wei-chao. Analysis on the selection method for heat dissipation fans operating in non-safety level DCS cabinets of nuclear power station[J]. Process Automation Instrumentation, 2013.34(10):32–36.

Liu Lu, Zhu Guo-rong, Chen Hao, et al. Power loss analysis and thermal design of three phase four wire VIENNA rectifier[J]. Transactions of china Electrotechnical Society, 2014.29(Sup.1):282–290.

Qing Xiao-dong, Yang Wei-hong. Design of single transformer 12 pulse rectification system with large range of regulation[J]. Transformer, 2015.52(10):1–5.

Ren Xiao-yong, Yao Kai, Kuang Jian-jun, etal. Design of output inductor for minimal power loss[J].Proceedings of CSEE, 2008.28(27):84–88.

Wang Zhao-an, Huang Jun. Power electronics technology[M]. Beijin: China Machine Press, 2000:67–70(in Chinese).

Electromechanical Control Technology and Transportation – Jia & Wu (Eds)
© *2017 Taylor & Francis Group, London, ISBN 978-1-138-06752-3*

Design of supplementary excitation damping controller for damping SSR based on projective theorem

Dapeng Xu, Lei Li & Lanming Zhao
Shandong Electric Power Engineering Consulting Institute Corp., Ltd., Jinan, China

Shi Chen
School of Electrical Engineering and Information, Chengdu, China

Kuan Li
State Grid Shandong Electric Power Research Institute, Jinan, China

ABSTRACT: The series capacitive compensation is added to AC transmission line which can improve the power transmission capacity effectively. However, the series capacitive compensation may cause subsynchronous resonance and damage of the generator shaft system. Supplementary excitation damping controller is widely used in practical engineering which could restrain SSR as an economic and effective way. This paper proposed a SEDC based on projective theorem which is from the perspective of system transfer functions. Firstly, the system transfer functions of every natural torsional vibration frequency need to be identified. Then, the corresponding output feedback damping controller will be designed based on projective theorem. IEEE second benchmark model is used and simulated by PSCAD/EMTDC. The simulation results show that the designed SEDC based on TLS-ESPRIT and projective theorem can restrain subsynchronous resonance effectively, and the controller order is lower which could be applied to the practical projects easily.

1 INTRODUCTION

Series capacitive compensation is widely used in long-distance AC transmission as a kind of effective method which could improve transmission capacity of power system. However, the risk of Subsynchronous Resonance (SSR) would be increased when the series capacitive compensation is added to the power system (Kundur P, 1994). When the system was disturbed, the energy exchange may be caused between electrical system and generator unit in a frequency which is less than the normal operation frequency. And then the generator shafting fatigue accumulation would be caused. Finally, the generator shaft fracture may be caused, in which that phenomenon would affect the safe and stable operation of power system greatly (Cheng Shijie, 2009)

SSR could be restrained through the FACTS devices effectively (Xiang Changming, 2013 & Zhao Xin, 2012). However, the FACTS devices may cause adverse effect to power system and increase the project investment. So the FACTS devices have a lower economic benefit (Jiang Zhenhua, 200 & Lu Jiancheng, 2015). Supplementary Excitation Damping Controller (SEDC) is used for the excitation system as an economic and effective way. The synchronization component was added to the exciting voltage through SEDC, which could produce damping of subsynchronous frequency, and the SSR would be restrained (Liu Shiyu, 2008 & Tang Fan, 2010). Reference (Tang Fan, 2010) designed a SEDC based ant colony algorithm, the simulation results on the IEEE second benchmark model showed that the SEDC was effective and robust. However, this SEDC is difficult to applied in practical projects which is based on intelligence algorithm. Reference (Guo Xijiu, 2008) adjusted parameters of SEDC based on the field testings, the designed SEDC had good damping effect to the SSR through little interference test. However, this reference did not demonstrate the effectiveness of SEDC with large disturbance. Reference (Wu Xi, 2013) designed SEDC based on the genetic algorithm, simulation results showed that this SEDC could restrain SSR effectively. However, the genetic algorithm needs multi-iterations to obtain an optimized solution, so this method had little benefit to engineering practicality. Reference (Dong Baifeng, 2011) designed SEDC and Supplementary Subsynchronous Damping Controller (SSDC) based on phase compensation method, the Subsynchronous

Oscillation (SSO) could be restrained through these controllers which caused by HVDC. This reference got conclusions that SSDC could produce greater electrical damping than SEDC. Most literatures on the design of SEDC are based on the mathematical algorithm or mechanism of SSR. However, there is little research about the system transfer function based on system state equation, and the SEDC which designed based on transfer function are undertaken.

The generator shafting torsional vibration frequency between the cylinder blocks can be calculated by stiffness coefficient, inertia time constant and torsion coefficient among the cylinder blocks (Cheng Shijie, 2009). SEDC is designed in this paper based on projective theorem, and IEEE second benchmark model is built for simulation by PSCAD/EMTDC. Firstly, the system transfer functions of every natural torsional vibration frequency need to be identified which are based on TLS-ESPRIT. Then, the SEDC transfer functions will be calculated by projective theorem. Finally, in order to verify the effectiveness of the SEDC, the damping characteristics of main vibration modes with and without SEDC are identified. The simulation results show that the designed SEDC can restrain SSR effectively.

2 FUNDAMENTALS

2.1 Projective theorem

Because of the generator shafting torsional vibration frequencies between the cylinder blocks can be calculated, so the vibration frequency could be used as the retained characteristics root of SEDC.

The state equation of the controlled system is denoted by

$$\begin{cases} \dot{x} = Ax + Bu \\ y = Cx \end{cases} \tag{1}$$

where x, y and u represent state vector, output vector and control vector. A, B and C are state matrix, control matrix and output matrix.

The state feedback is added to closed loop system which can be denoted by

$$\begin{cases} \dot{x} = (A - BK)x \\ y = Cx \end{cases} \tag{2}$$

Characteristic root decomposition of system (2) is denoted by

$$(A - BK)X = X\Lambda \tag{3}$$

where Λ represents the characteristic root triangle array, and X represents eigenvector matrix.

The state equation of SEDC which is based on projective theorem can be denoted by

$$\begin{cases} \dot{w} = A_c w + B_c y \\ u = C_c w \end{cases} \tag{4}$$

where w represents the state vector of SEDC, A_c, B_c and C_c represent state matrix, control matrix and output matrix of SEDC

Simultaneous equation (1) and equation (4)

$$\begin{cases} \dot{x}' = A'x' \\ y' = C'x' \end{cases} \tag{5}$$

where

$$x' = \begin{bmatrix} x \\ w \end{bmatrix} y' = \begin{bmatrix} y \\ u \end{bmatrix} \tag{6}$$

$$A' = \begin{bmatrix} A & BC_c \\ B_c C & A_c \end{bmatrix} C' = \begin{bmatrix} C & 0 \\ 0 & C_c \end{bmatrix} \tag{7}$$

The equation (5) represents closed loop feedback system of controlled system which is based on the projective theorem. And the characteristic root decomposition of A' is denoted by

$$\begin{bmatrix} A & BC_c \\ B_c C & A_c \end{bmatrix} \begin{bmatrix} X' \\ W' \end{bmatrix} = \begin{bmatrix} X' \\ W' \end{bmatrix} \Lambda' \tag{8}$$

where Λ' represents the keep characteristic root of system, which is the generator shafting natural torsional vibration mode. X' is eigenvector of the shafting inherent characteristic root.

$$(A - BK)X' = X'\Lambda' \tag{9}$$

The system order is increased because of the state feedback is introduced and form closed loop system. Where W' represents the introduced eigenvector, and define $P = W'^{-1}B_c$.

Simultaneous equation (8) and equation (9)

$$\begin{cases} A_c = W'(\Lambda' - PCX')W'^{-1} \\ B_c = W'P \\ C_c = -KX'W'^{-1} \end{cases} \tag{10}$$

It can be seen from equation (10) that the state matrix, control matrix and output matrix of SEDC are obtained through state feedback gain matrix K, and then the transfer function of SEDC is obtained.

2.2 Solving the state feedback matrix

The concept of SEDC is to design the corresponding controller respectively of every shafting natural torsional vibration frequency. Every independent controller could restrain one or two shafting natural frequencies. The state feedback matrix K can be solved through the transformational matrix T.

In the controlled system (1), the characteristic polynomial of matrix A is denoted by

$$|sI - A| = s^n + a_1 s^{n-1} + \cdots + a_{n-1} s + a_n \quad (11)$$

The value of $a_1, a_2, ..., a_n$ are obtained through equation (11).

The system state equations can be changed to the transformational matrix which is controllable and standard.

$$T = MW \quad (12)$$

where

$$M = \begin{bmatrix} B & AB & ... & A^{n-1}B \end{bmatrix} \quad (13)$$

$$W = \begin{bmatrix} a_{n-1} & a_{n-2} & ... & a_1 & 1 \\ a_{n-2} & a_{n-3} & ... & 1 & 0 \\ \vdots & \vdots & \ddots & \vdots & \vdots \\ a_1 & 1 & ... & 0 & 0 \\ 1 & 0 & ... & 0 & 0 \end{bmatrix} \quad (14)$$

The characteristic root polynomial is denoted by

$$(s - \mu_1)(s - \mu_2) \cdots (s - \mu_n) \\ = s^n + \alpha_1 s^{n-1} + \cdots + \alpha_{n-1} s + \alpha_n \quad (15)$$

The value of $\alpha_1, \alpha_2, ..., \alpha_n$ can be obtained through equation (15).

The state feedback gain matrix is denoted by

$$K = \begin{bmatrix} \alpha_n - a_n & \alpha_{n-1} - a_{n-1} & ... & \alpha_2 - a_2 & \alpha_1 - a_1 \end{bmatrix} T^{-1} \quad (16)$$

2.3 Controller design

Firstly, the generator speed deviation is got. Then the natural torsional vibration frequencies can be obtained when the speed deviation pass the fourth order Butterworth filter. The transfer functions of SEDC could be got which are based on projective theorem. Finally, the output of the controller is put into the generator exciter. The structure of SEDC is shown in Figure 1.

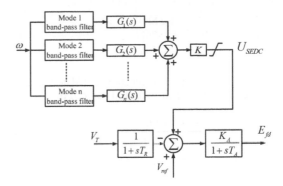

Figure 1. The structure of SEDC.

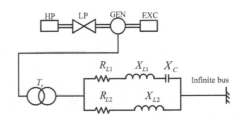

Figure 2. IEEE second benchmark model.

3 SIMULATION EXAMPLE

The IEEE second benchmark model for simulation is built based on PSCAD/EMTDC, and the topological structure diagram is shown in Figure 2. It can be seen from Figure 2 that the model has four cylinder blocks. So the natural torsional vibration frequencies can be calculated, and the frequencies are 24.4 Hz, 32.3 Hz and 51.7 Hz.

A step is added to the V_{ref} of the generator excited system when the power system running to 3 s stably. Then the generator cylinder speed deviation is treated as the observed signal, and the transfer functions of shafting natural torsional vibration frequencies are identified based on TLS-ESPRIT. The system transfer functions of IEEE second benchmark model are denoted by

$$\begin{cases} G_{mod1} = \dfrac{-1.793e^{-5}s^2 - 0.01203s}{1s^2 - 1.273s + 2.411e^4} \\ G_{mod2} = \dfrac{1.889e^{-5}s^2 - 0.0005261s}{1s^2 - 0.08802s + 4.14e^4} \\ G_{mod3} = \dfrac{6.375e^{-8}s^2 + 5.777e^{-8}s}{1s^2 - 1.638s + 2.41e^4} \end{cases} \quad (17)$$

The state feedback gain matrix K could be calculated through the system transfer functions which correspond to natural torsional vibration frequencies based on transformational matrix T.

$$\begin{cases} \boldsymbol{K}_{\text{mod1}} = \begin{bmatrix} 2.6131 & 1.2228 \end{bmatrix} \\ \boldsymbol{K}_{\text{mod2}} = \begin{bmatrix} 4.088 & 3.4886 \end{bmatrix} \\ \boldsymbol{K}_{\text{mod3}} = \begin{bmatrix} 3.0378 & -0.7369 \end{bmatrix} \end{cases} \quad (18)$$

The controller transfer functions could be deduced based on projective theorem and denoted by

$$\begin{cases} G_1(s) = \dfrac{8.608s - 8.376}{s^2 + 2.2s + 2.339e^4} \\ G_2(s) = \dfrac{-358.7s + 36.66}{s^2 + 1.409s + 2.414e^4} \\ G_3(s) = \dfrac{-302.2s + 216.1}{s^2 + 3.085s + 2.411e^4} \end{cases} \quad (19)$$

3.1 *Without SEDC*

A three-phase grounding fault happened on the AC line at $t = 2.5$ s, and the fault duration is 0.075 s. The generator cylinder speed deviations are shown in Figure 3 without SEDC.

It can be seen from Figure 3 that the generator shaft speed deviations show a trend of divergence when the system is disturbed.

3.2 *With SEDC*

The designed SEDC which shown as equation (19) is added to the excitation system. The generator cylinder speed deviations are shown in Figure 4 with SEDC.

It can be seen from Figure 3 and Figure 4 that SEDC could restrain SSR effectively, and the generator shaft speed deviation decrease rapidly.

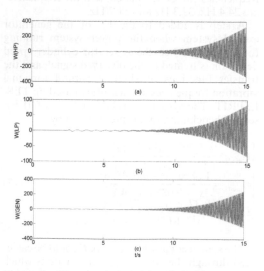

Figure 3. The speed deviation of shafting without SEDC.

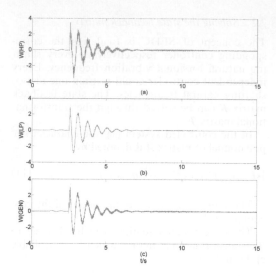

Figure 4. The speed deviation of shafting with SEDC.

Table 1. Identified by TLS-ESPRIT method.

	Mode	f(Hz)	attenuation factor	damping ratio(%)
Without SEDC	Mode1	24.6239	−0.0064	0.0041
	Mode2	32.8375	−0.0084	0.0041
	Mode3	51.1735	0.0071	−0.0022
With SEDC	Mode1	24.3139	−1.1849	0.7756
	Mode2	32.9479	−0.7294	0.3523
	Mode3	51.7498	−0.0023	0.0007

The SEDC controller order is lower and could be applied to actual engineering easily.

3.3 *Characteristic root analysis*

The system main vibration mode parameters are identified based on TLS-ESPRIT with and without SEDC. The identification results are shown in Table 1.

It can be seen from Table 1 that the damping ratio of main vibration modes are greatly improved with SEDC, and the identification results consistent with the results of Figure 3 and Figure 4.

4 CONCLUSION

The transfer functions of generator shafting natural torsional vibration frequencies are obtained based on TLS-ESPRIT. In order to restrain the SSR, this paper designed the damping controller aim at every transfer function based on projective theorem. IEEE second benchmark model is used for the simulation and results show that designed SEDC could restrain SSR effectively. The characteristic root analysis also indicates that, the system damping could be improved significantly if SEDC was added into the system

The added SEDC could restrain SSR, which avoids increasing primary equipment and reduces the engineering investment. This controller based on projective theorem has a lower order, facilitates the actual engineering and has practical engineering significance.

REFERENCES

Cheng Shijie, Cao Yijia, Jiang Quanyuan. Subsynchronous oscillation theory and method of the power system [M]. Science press, 2009.

Dong Baifeng, Li Zhou, Wan Qiulan. Suppress SSO of HVDC system using SEDC and SSDC [J]. Power System Protection and Control, 2011, 39 (9): 77–82.

Gu Yilei, Wang Xitian, Zhao Dawei, et al. Optimization design of supplementary excitation damping controller based on improved particle swarm optimization algorithm [J]. Automation of Electric Power Systems, 2009, 33 (7): 11–16.

Guo Xijiu, Xie Xiaorong, Liu Shiyu, et al. Field test of SEDC for subsynchronous torsional damping enchancement at shangdu power plant [J]. Automation of Electric Power Systems, 2008, 32(10): 97–100.

Jiang Zhenhua, Cheng Shijie, Fu Yuli, et al. Analysis of subsynchronous resonance of power system with TCSC [J]. Proceedings of the CSEE, 2000, 20 (6): 47–52.

Kundur P. Power system stability and control [M]. New York: McGraw-hill, 1994.

Liu Shiyu, Xie Xiaorong, Zhang Donghui, et al. Mechanism andmitigation of multi-mode subsynchronous resonance [J]. Journal of Tsinghua University: Science & Technology Edition, 2008, 48 (4): 457–460.

Lu Jiancheng, Li Xiaocong, Haung Wei, et al. Linear optimal controller of static series synchronous compensator and excitation to suppress subsynchronous oscillation [J]. Power System Protection and Control, 2015, 43 (1): 21–27.

Tang Fan, Liu Tianqi, Li Xingyuan. Optimal design of supplementary excitation damping controller for damping SSRbased on ant colony algorithm [J]. Power System Protection and Control, 2010, 38 (19): 170–174.

Wu Xi, Jiang Ping. Subsynchronous Damping Controller Design Based on Practical Stability Region [J]. Proceedings of the CSEE, 2013, 33 (25): 123–129.

Xiang Changming, Wu Xi, Jiang Ping. Suitability study of sample-data model of TCSC for SSO analysis [J]. Power System Protection and Control, 2013, 41 (23): 55–60.

Zhao Xin, Gao Shan, Zhang Yuyu. A practical damping formula of SSR used in series compensated system including SVC [J]. Power System Protection and Control, 2012, 40 (20): 94–100.

Electromechanical Control Technology and Transportation – Jia & Wu (Eds)
© *2017 Taylor & Francis Group, London, ISBN 978-1-138-06752-3*

Analysis of the loss in the region of damper winding in a large tubular hydro-generator by 3D electromagnetic field calculation

De-wei Zhang, Jing Zhang & Wu-jin Li
Skill Training Center of Sichuan Electric Power Corporation, Chengdu, Sichuan, China

Zhen-nan Fan
School of Electrical Engineering and Electronic Information, Xihua University, Chengdu, China

ABSTRACT: In order to study the eddy current losses in the end region of damper winding thoroughly, a 3D moving electromagnetic field FE model of tubular hydro-generator is built. Furthermore, the flux density and the eddy current loss in the damper winding end region of a 36 MW tubular hydro-generator are calculated and analyzed. The study also provides a more rational calculation model to assess the eddy current loss in the damper winding end region of large hydro-generators, which helps improve the design standard of tubular hydro-generators.

1 INTRODUCTION

Damper winding is one of the key components of hydro-generator and plays an important role in the safety and stability of the generator and power system. Tubular hydro-generators have then been applied widely in hydro-power stations with water head is lower than 20 m (Chao-Yang Li, 2006). In order to avoid the end region overheat failures of damper winding and improve the generator design of tubular hydro-generator, it is necessary to carry out an in-depth study on the amount and distribution of loss within the damper winding end region of tubular hydro-generators.

2 CALCULATION MODELS

To solve the eddy current loss in the end region of damper winding, the 3D moving electromagnetic field model of the hydro-generator end region is built.

According to the periodicity of the magnetic field, a pair of poles at the end region is chosen as the 3D electromagnetic field calculation region, as shown in Fig. 1.

Considering the core saturation, by Coulomb norm $\nabla \cdot A = 0$ and the boundary condition of the problem region, the 3D boundary value problem of nonlinear time-varying moving electromagnetic field is then obtained.

$$\begin{cases} \nabla \times (\nu \nabla \times \mathbf{A}) + \dfrac{1}{\rho}\left[\dfrac{\partial \mathbf{A}}{\partial t} - \mathbf{V} \times (\nabla \times \mathbf{A})\right] = \mathbf{J}_s \\[2mm] \left.\begin{array}{l} \mathbf{n} \times \mathbf{A} = 0 \\[1mm] \dfrac{\partial A_n}{\partial n} = 0 \end{array}\right\} S_1, S_2, S_3 \\[4mm] \mathbf{A}\big|_{S_4} = \mathbf{A}\big|_{S_5} \end{cases} \tag{1}$$

where A_n is the normal component of A, S_1 is the symmetric boundary surface, which is parallel to the flux line, S_2 is the end shade, S_3 is the end cover idealized boundary with $\rho = 0$, which contains only the tangential component of magnetic field, and S_4 and S_5 are periodic boundary surfaces, which meet the cyclic boundary condition.

For solving the 3D moving electromagnetic field model of the hydro-generator end region, the reasonable consideration of the stator current phasors is the key point of the stator winding excitement. Only if the phasors are accurate, can the operating state of the generator be simulated correctly.

Therefore, the stator currents and their phasors at the rated load can be obtained as follows:

In Fig. 2, when the load resistance R_L and inductance L_L have rated values, the rated load operating state can be simulated by the multislice moving electromagnetic field–circuit coupling model built in Zhen-nan Fan (2013) and then the rated currents and their phasors are obtained, as shown in Fig. 3.

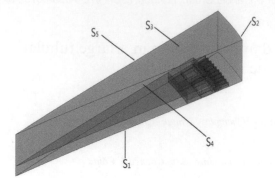

(a) Problem region of 3D moving electromagnetic field computation

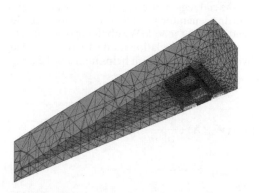

(b) FE meshes of 3D moving electromagnetic field computation

(c) Coils of generator

Figure 1. Problem region, FEM meshes and coil distribution of the end region of hydro-generator.

The waveform and the stator current phasor in Fig. 3 are used as the excitement on the stator winding for the 3D FE analysis. And with the application of DC exciting current in the rotor

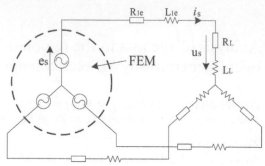

Figure 2. Coupling circuit of the stator.

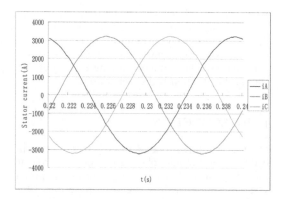

Figure 3. Waveforms of stator currents at the rated load.

field winding, the 3D moving electromagnetic field model of the hydro-generator end region can be solved.

According to the computation of 3D moving electromagnetic field model of end region, the instantaneous value of eddy current density and eddy loss density per element are obtained:

$$J = -\frac{1}{\rho}\frac{\partial A}{\partial t} \tag{2}$$

$$P_e = \rho(J.J) \tag{3}$$

And the average value of eddy loss density within a period is as follows:

$$P_{eav} = \frac{1}{T}\int_0^T p_e dt \tag{4}$$

The average value of eddy loss in the volume is as follows:

$$P_{eddy} = \iiint_V dP_{eav} dv \tag{5}$$

3 RESULTS AND DISCUSSION

Some calculation results are shown in Figs.4–6 and Table 1.

Figures 3 and 4 show that, in the calculation results of this paper, the closer to the end, the air gap magnetic field density is smaller and the magnetic density of the slot is less than that of the teeth. These indicate that the calculation result S of magnetic density is in accordance with the physical reality of generator.

Figures 5 and 6 show that whether it is the end of the damper bars, or the damper winding end ring, the eddy current distribution is not uniform, so we can see that the calculation result of this paper is more reasonable than that of the magnetic circuit method.

Table 1 shows that, in the rated operating conditions, the eddy current losses of the damper winding end region components in pole of the 36 MW

Figure 6.   Eddy current density distributions in the end of damper bar.

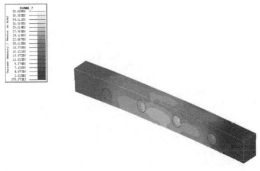

Figure 7.   Eddy current density distributions in the end ring.

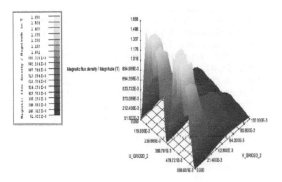

Figure 4.   Magnitude of air gap flux density of the end region.

Table 1.   Calculation results of the eddy current loss in damper winding end region.

	End of damper bar	End ring
Loss (W)	22	9

tubular turbine generator are not more than 30 W, which for a 36 MW large tubular turbine generator, is very small.

4 CONCLUSIONS

In this paper, a 3D moving electromagnetic field FE model is established for the calculation of the eddy current losses in the end region of a 36 MW tubular turbine generator. Then, the eddy losses in the damper winding end region are obtained. The research work is helpful to improve the level of electromagnetic analysis and design of hydro-generator.

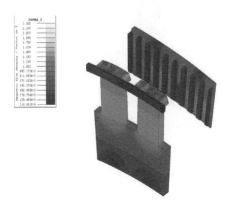

Figure 5.   Electromagnetic field distributions in the end region of generator.

REFERENCES

Boglietti, A., A. Cavagnino, D. Staton, M. Martin Shanel, M. Mueller, C. Mejuto, "Evolution and Modern Approaches for Thermal Analysis of Electrical Machines," Industrial Electronics, IEEE Transactions on, Vol. 56, no. 3, 2009, pp. 871–882.

Chao-Yang Li, "The Application of Bulb-type Hydro-Generator Set at Low Head Hydropower Station," *Developing*, no. 9, pp. 145–146, 2006.

Dong Deng, Yu-Tian Sun, Guo-Wei Tan, Jin-Xiang Li, and Zhi-Hua An, "Analysis on Operating Faults of Generator in Feilaixia Hydropower Station," *Large Electric Machine and Hydraulic Turbine*, no. 6, pp. 13–17, 2003.

Guang-hou Zhou, Li Han, Zhen-nan Fan, Yong Liao, Song Huang. 3D loss and heat analysis at the end region of 4-poles 1150 MW nuclear power turbine generator. Archives of Electrical Engineering. 2014, 63(1), 47–61.

Hai-xia Xia, Ying-ying Yao, and Guang-zheng Ni, "Analysis of ventilation fluid field and rotor temperature field of a generator," *Electric Machines and Control*, vol. 11, no. 5, pp. 472–476, October 2007.

Jing-Bin Guo, "Analysis of Damaged Damping Winding and Magnetic Pole in Bulb Type Generator," *Chinese Power*, vol. 34, no. 7, pp. 63–67, May 2001.

Kaehler, C., G. Henneberger, "Transient 3-D FEM computation of eddy-current losses in the rotor of a claw-pole alternator," Magnetics, IEEE Transactions on, Vol. 40, no. 2, Part: 2, 2004, pp. 1362–1365.

Kay Hameyer, Johan Driesen, Herbert De Gersem, and Ronnie Belmans. The Classification of Coupled Field Problems[J]. IEEE Transactions on Magnetics, 1999, 35(3):1618–1621.

Li Weili, Zhou Feng, Hou Yunpeng, Cheng Shukang. Calculation of Rotor Temperature Field for Hydro-generator as Well as the Analysis on Relevant Factors[J]. Proceedings of the CSEE, 2002, 22(10): 85–90.

Li-xin Fu, *GB/T 1029-2005: The test measures of three phases synchronous machine*, Bei Jing: Standards Press of China, 2006.

Vong, P. K., D. Rodger, "Coupled electromagnetic-thermal modeling of electrical machines Magnetics," IEEE Transactions on, Vol. 39, no: 3, Part: 1, 2003, pp. 1614–1617.

Weili Li, Chunwei Guan, and Yuhong Chen. Influence of Rotation on Rotor Fluid and Temperature Distribution in a Large Air-Cooled Hydrogenerator, *IEEE Trans. On Energy Conversion*, vol. 28, no. 1, pp.117–124, March, 2013.

Weili Li, Yu Zhang, Yonghong Chen, "Calculation and Analysis of Heat Transfer Coefficients and Temperature Fields of Air-Cooled LargeHydro-Generator Rotor Excitation Windings," *IEEE Trans. On EnergyConversion*, vol. 26, no. 3, pp. 946–952, September, 2011.

Wen Jiabin, Meng Dawei, Lu Changbin. Tynthetic calculation for the ventilation and heating of large water-wheel generator[J]. Proceedings of the CSEE, 2000, 20(11): 115–119.

Yong Liao, Zhen-nan Fan, Li Han, Li-dan Xie. Analysis of the Loss and Heat on Damper Bars in Large Tubular Hydro-Generator base on the 3D Electromagnetic-Temperature field Calculation. *PRZEGLAD ELEKTROTECHNICZNY.* 2012, 88(5B), 97–100.

Zhen-nan Fan, Yong Liao, Li Han, Li-dan Xie. No-load Voltage Waveform Optimization and Damper Bars Heat Reduction of Tubular Hydro-Generator by Different Degree of Adjusting Damper Bar Pitch and Skewing Stator Slot. *IEEE Transactions on Energy Conversion.* 2013, 28(3), 461–469.

Electromechanical Control Technology and Transportation – Jia & Wu (Eds)
© *2017 Taylor & Francis Group, London, ISBN 978-1-138-06752-3*

Loss and heat calculation of damper winding in a damper winding shifted 600 MW large hydro-generator

De-wei Zhang, Jing Zhang & Wu-jin Li
Technology and Skill Training Center of Sichuan Electric Power Corporation, Chengdu, Sichuan, China

Zhen-nan Fan
School of Electrical Engineering and Electronic Information, Xihua University, Chengdu, China

ABSTRACT: In order to solve the problem of loss and heat of shifted damper winding of a large hydro-generator for a 600 MW large hydro-generator, a 2D moving electromagnetic field–circuit coupling model of a generator and a 3D temperature field model of the rotor are built, and the influences of the structure on the damper winding loss and heat are computed and analyzed. This study helps enhance the design standard of the hydro-generator.

1 INTRODUCTION

As a kind of structure design scheme for optimizing the no-load voltage waveform of the hydro-generator, the damper winding shifting structure is used in the design and manufacture of the integer slot hydro-generator (Xi-fang Chen, 2011 & Zhesheng Li, 1983). However, the influences of the structure on the damper winding loss and heat, so far, has not been fully studied (Lingyun Shao, 2014; Wang Ting-ting, 2013; Qudsia, J., 2010; Bastawade, P., 2012; Ranlöf, M., 2010; Whei-Min Lin & Bruzzese, C., 2011; De-Wei Zhang, 2012; Stefan Keller, 2006; Zhen-nan Fan, 2013; Yong-gang Luo, 2012; Hong-lian Wang, 2013 & Guang-hou Zhou, 2009).

In order to solve the above problems, in this paper, for a 600 MW large hydro-generator, a 2D moving electromagnetic field-circuit coupling model of generator and a 3D temperature field model of rotor are built, and the influences of the structure on the damper winding loss and heat are computed and analyzed. This study helps enhance the design standard of hydro-generator.

2 CALCULATION MODELS

2.1 *Basic data of the tubular hydro-generator*

The basic data of the generator are shown in Table 1.

In this paper, t_1 is the stator tooth pitch, which is invariant, and t_2 is the pitch of damper bars.

Table 1. Basic data of the generator.

Parameter	Value
Rated power (MW)	600
Rated voltage (kV)	18
Rated current (A)	21383
Power factor	0.9
Number of magnetic poles	64
Number of slots per pole per phase	4
t_2/t_1	0.83,0.87,1.0

2.2 *Boundary value problem of moving electromagnetic field*

To solve the eddy current loss of damper winding, a 2D moving electromagnetic field-circuit coupling model of the hydro-generator is built.

According to the periodicity of magnetic field, the area of a pair of poles is chosen as the electromagnetic field calculation region, as shown in Fig. 1.

In the 2D moving electromagnetic field model, the current density and vector magnetic potential have only the axial z components, and the speed has only the axial x component. By Coulomb norm $\nabla \cdot \mathbf{A} = 0$ and the boundary condition of the problem region, the 2D boundary value problem of nonlinear time-varying moving electromagnetic field for the generator is then obtained:

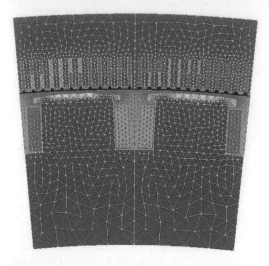

Figure 1. Problem region and meshes of electromagnetic field.

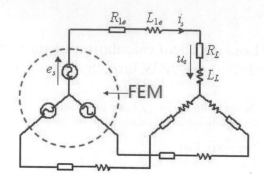

Figure 2. Coupling circuit of stator winding.

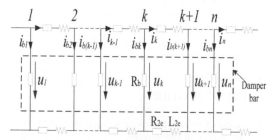

Figure 3. Coupling circuit of damper winding.

$$\begin{cases} \dfrac{\partial}{\partial x}(\nu\dfrac{\partial A_{slz}}{\partial x}) + \dfrac{\partial}{\partial y}(\nu\dfrac{\partial A_{slz}}{\partial y}) = -J_{slz} + \sigma\dfrac{\partial A_{slz}}{\partial t} \\ \quad + V_x\sigma\dfrac{\partial A_{slz}}{\partial x} \\ A_{slz}\big|_{arc\_in} = A_{slz}\big|_{arc\_out} = 0 \\ A_{slz}\big|_{cyclic\_boundary\_start} = A_{slz}\big|_{cyclic\_boundary\_end} . \end{cases} \quad (1)$$

2.3 Coupling circuits

To considering the influence of the end winding of the stator and damper end rings of rotor, the coupling circuit models are established. The external circuit equation and electromagnetic equation should be combined in the calculation (Sarikhani, A., 2013 & Youpeng Huangfu, 2014).

And the coupling circuits of the stator and damper windings are shown in Figs. 2 and 3, respectively.

The stator and rotor coupling circuit equation of generator and the electromagnetic equation are combined, the magnetic vector (A_z) is calculated in time-step finite-element method and then the flux density, voltage, and losses can be obtained.

2.4 Boundary value problem of rotor 3D temperature field

Considering the anisotropic heat conduction of the rotor core, the boundary value problem of 3D steady-temperature field can be expressed as follows:

Figure 4. Problem region and meshes of 3D temperature field.

$$\begin{cases} \dfrac{\partial}{\partial x}(\lambda_x\dfrac{\partial T}{\partial x}) + \dfrac{\partial}{\partial y}(\lambda_y\dfrac{\partial T}{\partial y}) + \dfrac{\partial}{\partial z}(\lambda_z\dfrac{\partial T}{\partial z}) = -q_V \\ \lambda\dfrac{\partial T}{\partial n}\bigg|_{S_2} = 0 \\ \lambda\dfrac{\partial T}{\partial n}\bigg|_{S_3} = -\alpha(T - T_f) \end{cases} \quad (2)$$

where T is the temperature, λ_x, λ_y, and λ_z are heat conductivities on each direction, q_v is the heat source density, which is obtained by the loss calculation mentioned above, S_2 the rotor middle profile and the interface between rotor core and rim related with the thermal insulation boundary condition, S_3 is the outside surface of the rotor related with the heat dissipation boundary condition, α is the heat dissipation coefficient of S_3, and T_f is the environmental air temperature. The problem regions are illustrated in Fig. 4.

3 RESULTS AND DISCUSSION

Some calculation results are shown in Figs. 5–7 and Table 2.

In Table 2, P_{DB} is the loss of damper winding of the two poles in the problem region of electromag-

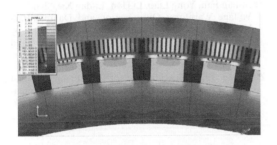

Figure 5. Distribution of flux density.

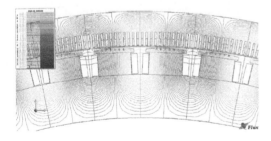

Figure 6. Distribution of flux line.

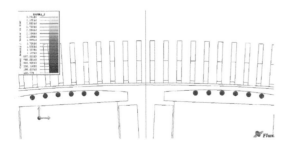

Figure 7. Distribution of eddy current density.

Table 2. Losses and heat of damper winding.

t_2/t_1	Shift degree	P_{DB} (W)	T_{max} (°C)
0.83	0	864	66.2
	0.25 t_1	846	66.07
0.87	0	790	66.06
	0.25 t_1	776	65.99
1.00	0	576	65.58
	0.25 t_1	567	65.44

netic field, and the T_{max} is the highest temperature of the two poles.

It is evident from the table that, when the center line of damper winding is shifted, the eddy current loss and the highest temperature of damper winding have almost no change.

The above results indicate that, as a kind of structure design scheme, the damper winding shifting scheme not only improves the no-load voltage waveform quality of the integer slot hydro-generator, but also not increase the loss and heat of the damper winding.

4 CONCLUSIONS

In this paper, a 2D moving electromagnetic field–circuit coupling model and a 3D temperature field model of the rotor of a 600 MW hydro-generator are built, and the influences of the damper winding shifted structure on the damper winding loss and heat are computed and analyzed. The study helps enhance the design standard of a hydro-generator.

REFERENCES

Bastawade, P., Reza, M.M., Pramanik, A., Chaudhari, B.N. "No-load magnetic field analysis of double stator double Rotor radial flux permanent magnet generator for low power wind turbines" *in Proc.* Power Electronics, Drives and Energy Systems (PEDES), 2012 IEEE International Conference on, 2012, pp. 1–6.

Bruzzese, C., Joksimovic, G. "Harmonic Signatures of Static Eccentricities in the Stator Voltages and in the Rotor Current of No-LoadSalient-Pole Synchronous Generators." *IEEE Trans. on Industrial Electronics*, vol. 59, no. 5, pp. 1606–1624, May, 2011.

De-Wei Zhang, Yuan-juan Peng, Zhen-nan Fan. "No-Load Voltage Waveform Optimization and Rotor Heat Reduction of Tubular Hydro-Generator" *in Proc.* Electromagnetic Field Problems and Applications (ICEF), 2012 Sixth International Conference on,2012, pp. 1–4.

Guang-hou Zhou, Li Han, Zhen-nan Fan, Xiao-quan Hou, Yi-gang Liao. "No-load Voltage Waveform Optimization of Hydro-generator With Asymmetric

Poles," Proceedings of the CSEE, vol. 29, no. 15, pp. 66–73, May, 2009.

Hong-lian Wang, Zhen-nan Fan. No-load Voltage Waveform Optimization of Integral Number Slots Large Hydro-Generator by shift Damper winding.in *Proc.* AMSMT Conference, 2013, pp. 8–12.

Lingyun Shao, Wei Hua, Ming Cheng. "Design of a twelve-phase flux-switching permanent magnet machine for wind power generation,"*in Proc.* Electrical Machines and Systems (ICEMS), 2014 17th International Conference on, 2014, pp. 435–441.

Qudsia, J., Junaid, I., Byung-Il Kwon. "Analytical analysis of the magnetic field and no-load voltage for the double sided axial flux permanent magnet synchronous generator," *in Proc.* Electromagnetic Field Computation (CEFC), 2010 14th Biennial IEEE Conference on, 2010, pp. 1.

Ranlöf, M., Perers, R., Lundin, U. "On Permeance Modeling of Large Hydrogenerators With Application to Voltage Harmonics Prediction," *IEEE Trans. on Energy Conversion*, vol. 25, no. 4, pp. 1179–1186, Dec, 2010.

Sarikhani, A., Nejadpak, A., Mohammed, O.A. "Coupled Field-Circuit Estimation of Operational Inductance in PM Synchronous Machines by a Real-Time Physics-Based Inductance Observer," *IEEE Trans. on Magnetics*, vol. 49, no. 5, pp. 2283–2286, May, 2013.

Stefan Keller, Mai Tu Xuan, Jean-Jacques Simond, "Computation of the No-Load Voltage Waveform of Laminated Salient-Pole Synchronous Generators," *IEEE Trans. on Industry Applications*, vol. 42, no. 3, pp. 681–687, May/June 2006.

Wang Ting-ting, Lu Mei-ling, Zhao Xiao-Zhong, Wang Hui-zhen, Meng Xiao-li. "Magnetic field analy-sis and structure optimization of high speed EEFS machine," *in Proc.* Industrial Electronics Society, IECON 2013–39th Annual Conference of the IEEE, 2014, pp. 978–983.

Whei-Min Lin; Tzu-Jung Su; Rong-Ching Wu. "Parameter Identification of Induction Machine With a Starting No-Load Low-Voltage Test," *IEEE Trans. on Industrial Electronics*, vol. 59, no. 1, pp. 352–360, Jan, 2011.

Xi-fang Chen, *Electromagnetic Calculation of Hydro-Generator*, Beijing: China Water Power Press, 2011.

Yong-gang Luo, Zhen-nan Fan. "Optimization of no-load voltage waveform in a large hydro-generator by the schemes which shift the damper winding and skew stator slots." in *Proc.* EESD Conference, 2012, pp. 397–403.

Youpeng Huangfu, Shuhong Wang, Jie Qiu, Haijun Zhang, Guolin Wang, Jianguo Zhu. "Transient Performance Analysis of Induction Motor Using Field-Circuit Coupled Finite-Element Method," *IEEE Trans. on Magnetics*, vol. 50, no. 2, pp. 2283–2286, Feb, 2014.

Zhen-nan Fan, Yong Liao, Li Han, Li-dan Xie, "No-load Voltage Waveform Optimization and Damper Bars Heat Reduction of Tubular Hydro—Generator by Different Degree of Adjusting Damper Bar Pitch and Skewing Stator Slot," *IEEE Trans. on Energy Conversion*, vol. 28, no. 3, pp. 461–469, September, 2013.

Zhesheng Li, "Measurement of Improving No-load Voltage Waveforms of Salient-pole Synchronous Generator", *Journal of Harbin Institute of Electrical Technology*, vol. 6, no. 3, pp. 1–16, Sept 1983.

Design and implementation of feeder automation

Penghou Liu
Qingdao University of Technology, Shandong Sheng, China

ABSTRACT: With the development of the Chinese economy, the demand for electricity from all walks of life has gradually increased. People also put forward higher requests regarding the reliability of power supply. Feeder automation can quickly cut off the faulty line to ensure the safe and stable operation of the line. Under the sustainable development of modern society, people attach great importance to all aspects of smart power grid construction, which is also significant for the smart grid distribution work. To ensure the reliable and high-quality power supply environment, the scientific and reasonable design and planning of the power distribution automation system shall pay attention to arranging the power distribution, and the appropriate feeder automation operation mode shall be selected. Only in this way can we ensure that the power outage is short once a fault occurs in the power grid system. At the same time, the line loss rate can be reduced, the efficiency of the power supply can be increased, and the issues in power equipment investment can be effectively treated. This paper mainly discusses the design and implementation of feeder automation, puts forward some practical application measures, and provides a reference for the stable operation of feeder automation.

1 INTRODUCTION

The developing Chinese economy has promoted the further development of power system modernization, and the continuous improvement and development of distribution network automation system has been the development trend of modern power enterprises. The monitoring management function equipped with the distribution network automation system of control backstage has an excellent integration and perfection. The feeder automation and the combination of intelligent terminal can effectively measure the operation results of distribution network automation system. The continuous development of power system makes the problems of distribution network become more and more serious, and the development of power demand and power facilities is not coordinated. For instance, we can easily find the problem in recovery, processing, and load of distribution network fault, which cannot adapt to the needs of the development of modern society. Therefore, it is of great practical significance for the safe operation of feeder automation to discuss the design and implementation of feeder automation.

2 PRINCIPLES OF DISTRIBUTION NETWORK AUTOMATION SYSTEM

2.1 *Principles of fault diagnosis*

We can determine whether any fault current passes through the switches on feeder line on the basis of the judgment of other fault signals. If the fault in the feeder line is single, the fault section is the power supply in the last switch side direction that flows from the fault current to the end, and the first switch section that did not flow any fault current is important.

2.2 *Failure estimation algorithm*

Distribution main station system constitutes its elements by using distribution network defined by the user, such as the relationship between power point, the subsection switch, feeder section, sectional switch, and power point connection. Afterward, the matrix for describing the relation of distribution network topology structure shall be automatically generated. Then, use various fault signals collected by SCADA system that flow through the segmentation switch to ensure the timely formation of the appropriate fault matrix. Results obtained by multiplying the two matrices shall be normalized and then the fault section shall be accurately located.

2.3 *Starting conditions of the failure estimation algorithm*

The distribution main station system uses the SCADA system to collect the state of the outlet switch of each power point and the total signals of the fault. As a result, if the power point is faulty and the tripping problem occurs, the fault location algorithm module can be started immediately.

2.4 Power restoration by the distribution main station system

When the distribution network system fails, the power switch will trip, which results in nonfault lines outage. However, in order to ensure the normal and stable operation of the circuit, and the improvement of the safety and efficiency of power supply, we need to resume power supply of the nonfault lines in time. The distribution main station system puts forward relevant ways to restore power automatically and manually by combining specific problems with graphs.

3 DESIGN OF FEEDING SYSTEM IN DISTRIBUTION NETWORK

3.1 Design scheme

The design scheme should be based on the following aspects in designing feeding system in the distribution network:

1. Layers of master station generally receive FTU terminal data, in which the integration work of SCADA system data and FTU terminal data are very critical. We can use the network hardware equipment to realize the integration of two kinds of data, but the software system shall be built based on Windows/UNIX hybrid platform so as to carry out deep analysis of all kinds of information in the system, monitor system fault conditions in real time, and carry out effective coordination of all aspects of the work of the power. In realizing the system optimization, layers of master station should focus on improving the application functions and functional modules, storing the data in the relevant functional modules, and generating reports with relevant data by graphics and model-based software, which shall make it possible for the user to access the information at any time.
2. Slave station of distribution network automation. In designing slave station of distribution network automation, we should fully integrate the WEB server and the company MIS network so as to provide the corresponding WEB browsing service. Afterward, we can use them in substation, the open and close operation, which can effectively control the WEB server in the information area, meanwhile collecting and filtering all kinds of information during this period. Furthermore, it will put forward the corresponding warning signal when the information being integrated, which shall provide CSOE, fault location, and other functions for the operation of the distribution network and achieve the purpose of fault isolation.
3. Distribution terminal layer. In designing this layer, we should strictly integrate data information, and strictly screen information,

and send the useful information. Traditional distribution network construction mode cannot meet the needs of modern social development, so we should build feeder automation system according to this and create a good system to monitor the environment.

3.2 Terminal design of fault detection

We should focus on creating a good fault detection terminal environment when designing the feeder automation system. Sheathing materials should be PC/ABS alloy materials, which shall be stored at −60 to −120°C, and electroplated with TUNGK-OUPC/ABC so as to improve fault detection. We should take single chip as the core in the terminal design of fault detection, whose operation speed is 5 million times per minute. Afterward, we carry out the sampling with 10 digits to judge the existence state of fault. We should use solar energy storage battery and supply power with lithium thionyl chloride battery to ensure the accuracy of the results, and power consumption shall not be more than 0.1 mA in dormancy. We use GSM network to collect and screen fault information and effectively transfer it to supervision center to solve problems in the system. We should strictly analyze the data to determine whether the line is in the case of short circuit, or if the ground-phase voltage drop is greater than 3 kV. Zero sequence current mutation is also greater than 10 A, then some faults exist in the operation of the system. Determine whether the alarm signal needs to be issued according to the results, and if there is a power failure. If the voltage is 0 and the current is 0, power failure occurs in the monitoring points.

3.3 Communication structure design

GPRS has been widely used in major industries with its own economic advantages. When building the GPRS data communication mode, we connect related terminal in communication structure, pack the data into IP packets, and send GPRS to power system and the data to SCADA. Finally, we make reasonable decisions and strictly implement terminal tasks after making a full analysis. Design of GPRS correspondence network should be connected with special-line and router, and transmission time delay should be strictly controlled within 1 s. We forward the corresponding data with the GPRS data center service system to form a perfect communication space.

3.4 Automation system design

We should focus on the following aspects in designing automation system:

1. Regard GIS as master station layers of feeder automation so as to ensure that it shall have

telemetry, remote control, remote communication, and other functions. Furthermore, regard intelligent permanent magnetic mechanism coincidence device as the basement so that when a fault occurs, GPS collects data about the voltage, current, power, and composite power flow simultaneously, strictly implements the corresponding data instructions, and carries out a comprehensive analysis of the differences between different states. As a result, we ensure that the fault information in GIS can have timely feedback.

2. Use SGSN to design mobile terminal, use SGSN to send APN to HLR to query mobile terminal, and make a chart with the feedback data so that users can query related data on the network. At this point, we can form a one-to-many or many-to-many data delivery model.

3. In designing automation system, we should combine GPRS with GGSN and set APN point in GGSN so as to achieve the purpose of passing private information data.

4 IMPLEMENTATION OF FEEDER AUTOMATION

4.1 Fault in main loop

If a fault occurs between systems, the outlet position of the substation will immediately trip, then re-close it. If the fault is transient, re-closing starts, while if the fault is a permanent fault, it will trip again, at the same time, it will be in a closed state. The information collected by the master station system is the basis of the network topology, by which we can quickly determine the fault section and send the remote command promptly to achieve the purpose of opening and then closing up. We can achieve the purpose of separating and transferring power supply from 45 to 60 s after determining the location of the fault. When a fault occurs in the relevant section of the main loop, the fault treatment process is similar to that of the above operation. Thus, it is desired to repeat the above operation.

4.2 Fault in branch line

If a fault occurs between systems, the circuit breaker in the front position of the outlet of the substation will take immediate actions. The substation outlet after tripping will produce returns when the circuit breaker will overlap again. If the fault is transient, re-closing starts, while if the fault is a permanent fault, it will trip again, at the same time, it will be in a closed state. After the failure is cleared, it can run stably; thus, there is no need to isolate and transfer the power supply.

Main loops of the distribution network transmission line of X city are equipped with load switches, and the branch circuit is equipped with the corresponding circuit breaker. The principle of dealing with the fault of the circuit structure is mainly that the branch line fault is eliminated by the tripping of branch line circuit breaker; while faults in the main loop are controlled by a centralized mode. Therefore, if there is a fault in the system, the substation will start tripping thus to determine the location of the fault according to the information collected by the main station and promptly issue a command to remove the fault as well as achieve the purpose of separation and transfer of power supply. The design scheme is simple, and because master station involves little work, the overhead line fault handling can be fast and accurate. Table 1 refers to the project evaluation index of distribution feeder automation in power enterprises in X city.

This is the project evaluation index of distribution feeder automation in power enterprises in X city, from which we can find that when it comes to the popularization and application of feeder automation, people lay stress on the output value per unit of electricity, cost of the unit substation capacity, cost of unit capacity, and load ratio of main transformer and regard them as important evaluation indexes so as to create a stable supply space; while feeding system in distribution network operates, the corresponding fault state variables are all installed in the intelligent terminal unit. If the fault current exceeds the setting value, the state variables shall be represented as 1, and if the fault current is less than the value of the whole set, the state variable shall be represented as 0. Furthermore, the system protection end will give timely variable feedback to the terminal, where fault

Table 1. Project evaluation index of distribution feeder automation in power enterprises in X city.

First-level indexes	Secondary indexes
	Output value per unit of electricity B11
Secondary indexes	Power supply load net asset B12
	Cost of the unit substation capacity B13
	Load ratio of main transformer B14
Technical efficiency index U2	Average outage time of users B15
	Voltage deviation B16
Social benefit index U3	One-family one-standard rate B17
	Power consumption rate B18

location is analyzed after changing the number of state parameters, and then immediately send trip command to the next protection system to be activated. When judging the state variables, we can transmit information through GPRS, calculate the related power and voltage value, and take corresponding protection measures. Furthermore, we should improve the operating efficiency of feeder automation to ensure stable operation of power system.

5 CONCLUSIONS

With the rapid development of modern society and economy, the level of information technology has also been promoted, and traditional feeder design of power enterprises cannot meet the needs of modern social development. Therefore, in order to create a good environment for a distribution network, we should pay much attention to the design of feeder automation and focus on the strict analysis of the automation system structure, system communication structure, fault detection terminal, and so on in order to ensure that the design is scientific and effective, scientifically controls problems in the distribution, and ensures the safe and stable operation of the distribution network.

In addition, we should rapidly remove all the faults occurring in the operation of the feeder system to improve its operating efficiency.

REFERENCES

Chen Lei, Liu Wei. Design and Implementation of Distribution Network Automation System in Tiedong Substation [J]. Coal Mine Modernization, 2011(02).

Du Haodong. Analysis and Research on Communication Technology of Distribution Network Automation [J]. China New Telecommunications, 2013(14).

Fu Xichen. Analysis and Application of Distribution Network Automation in Isolating Distribution Network Fault [J]. Technology Wind, 2015(24).

Li Jiaxu, Gao Mingbo. Design and Implementation of Distribution Network Automation System [J]. Bengang Technology, 2015(02).

Pan Qing. Exploration on realization of automatic line fault isolation function of distribution network automation system [J]. Electrical Applications, 2012(8).

Peng Hui, Ren Yuan, Song Xin, Chen Ning, Ge Yiyong. System Design of Distributed Integrated with County Technical Support Based on Dual Core Architecture [J]. Automation of Electric Power Systems, 2013(04).

Zheng Liji. Discussion on Common Problems of Planning and Construction of Distribution Network Automation [J]. Straits Science, 011(11).

Distance protection operation characteristics analysis of the AC/DC system based on MMC-HVDC interconnection

Maolin Tang, Rui Yu & Xi Zhang
State Grid Southwest Division, Chengdu, China

Chao Xiao, Jinxin Ouyang & Xiaofu Xiong
State Key Laboratory of Power Transmission Equipment and System Security and New Technology, Chongqing University, Chongqing, China

ABSTRACT: With the attractive advantages of high stability and controllability, the MMC-HVDC system has become a key development direction for the HVDC. Dynamic operation and controllability features of the MMC-HVDC system differ significantly from those of conventional AC and DC transmissions. During the grid fault, the MMC-HVDC system exports various short-circuit currents to the AC system, which can be a threat to the correct operation of distance protection. On the basis of the control principle of the MMC-HVDC converter station, the voltage and phase features of the MMC-HVDC short-circuit current are analyzed. The measured impedance variation of the MMC-HVDC distance protection in the AC side is also analyzed. The influence and interrelated factors of the MMC-HVDC system to distance protection are proposed. Digital simulation of PSCAD/EMTDC is done to verify the validity of the analysis.

1 INTRODUCTION

The MMC-HVDC system can supply power to passive network, adjust the active and reactive powers separately, and does not have the problem of commutation failure. All of these advantages have made it develop quite fast worldwide. It is widely applied to send renewable energy and interconnect large-scale AC systems (Xu 2012). Although the application of MMC-HVDC increases the stability and flexibility of the power system, its large amount of power electronic equipment makes the system more complicated and bring potential risk to the safe operation of power system. Especially under the condition of fault, current AC/DC system protection is equipped only on the basis of simple AC/DC system fault features. As the spread of fault impact in AC system and the influence of power electronic equipment control and blocked effect to protection features are not taken into consideration, the reliability of protection operation may be decreased, and cascading blackouts may even occur.

With the development of DC transmission, the relay protection of AC/DC hybrid interconnected power system gains more and more attention. However, related research is mainly focused on those based on conventional DC transmission (Shen 2015, Fei 2015). In conventional DC system,

when AC system faults on inverter side especially on near side occur, the inversion station will have a consequential commutation failure and is difficult to restore by itself. Thus, the low-voltage limit current instruction is usually adopted to reduce the current temporarily, and the AC short-circuit current may be smaller than the load current (Dong 2014, Zhang 2012). In the MMC-HVDC system, as the converter station and the control mode are different, during the AC system fault, the system exhibits different fault characteristics. Therefore, the AC/DC system protection characteristic analysis theory for conventional DC transmission is not suitable for the MMC-HVDC interconnection system.

At present, the research on MMC-HVDC system protection is focused on the protection principle of converter station and DC line. A new principle of MMC-HVDC transmission pilot protection based on modulus model identification was proposed by Jin (2014). Feng (2015) established a multilevel converter valve protection system from the submodule and the valve control system to the DC control protection. Xu (2012) presented a control and protection strategy for multiterminal transmission system DC-side fault. At present, the AC-side fault characteristic analysis of MMC converter station is mainly based on simulation analysis, and no analytical work has been done to analyze the AC side short-circuit fault characteristic of the MMC converter station. Thus, it is unable to guide relay protection and LVRT control research and implementation of flexible DC interconnected system. There are no reports on the operation characteristics of AC-side protection for MMC-HVDC systems. Distance protection is a commonly used principle, which is based on the distance between the fault point to the protection equipment to estimate the transmission line malfunction. Its correct action is crucial to the safe operation of the entire power system (Zhang 2010). However, the influence of the converter station on the AC short-circuit current in the MMC-HVDC system changes the measurement impedance characteristics of the distance protection, which may lead to a mistaken action of the distance protection.

Therefore, in view of the AC system distance protection operation characteristics under MMC-HVDC interconnection, the characteristics of short-circuit current in the converter station of the MMC-HVDC system under AC system fault are analyzed on the basis of MMC-HVDC system operation and control principle. In this paper, we analyze the change of distance protection characteristics of AC line and converter station and its influence on distance protection. The theoretical analysis is verified by PSCAD/EMTDC.

2 THE STRUCTURE AND CONTROL OF MMC-HVDC SYSTEM

The MMC-HVDC interconnection system is shown in Fig. 1. It includes the rectifier-side converter station, inverter-side converter station, DC lines, and both sides of the AC system. The converter station consists of a converter valve, DC capacitor, phase reactor, AC-side filter, connection transformer, AC circuit breaker, DC isolation switch, and control system (Lu 2015).

The MMC-HVDC converter station generally adopts vector control method, which includes inner current loop control and outer loop control. Common outer loop control methods are constant DC voltage control, constant active power control, constant reactive power control, constant AC voltage control, and so forth. The control block diagram of the converter station is shown in Fig. 2.

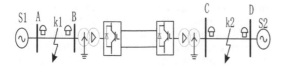

Figure 1. AC–DC interconnected system based on MMC-HVDC.

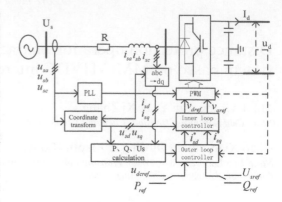

Figure 2. Control diagram of converter substation.

For a double-terminal MMC-HVDC transmission system, there must be one side of the converter station using the constant DC voltage control strategy (Chen 2007).

In the d,q coordinate system, the dynamic state of the inverse converter station can be described as:

$$\begin{cases} L\dfrac{di_{sd}}{dt} = -Ri_{sd} + \omega Li_{sq} + u_{sd} - u_{cd} \\ L\dfrac{di_{sq}}{dt} = -Ri_{sq} - \omega Li_{sd} + u_{sq} - u_{cq} \end{cases} \quad (1)$$

where u_{cd} and u_{cq} are the d and q components of the converter station AC side voltage, respectively; u_{sd} and u_{sq} are the d and q components of the receiving end AC system S2, respectively; i_{sd} and i_{sq} are the d and q components of the converter station AC-side current; ω is the synchronous angular velocity, R and L are the AC-side filtering inductance and equivalent resistance.

In order to eliminate the coupling influence between the d, q axis of the converter, the feedforward decoupling and PI control is adopted:

$$\begin{cases} u_{cd} = -\left(K_p + \dfrac{K_i}{s}\right)(i_{sd}^* - i_{sd}) + \omega Li_{sq} + u_{sd} \\ u_{cq} = -\left(K_p + \dfrac{K_i}{s}\right)(i_{sq}^* - i_{sq}) - \omega Li_{sd} + u_{sq} \end{cases} \quad (2)$$

where i_{sd}^* and i_{sq}^* are the d and q current instructive values, respectively; and K_p and K_i are the current inner loop proportional gain and the integral gain, respectively.

The inner loop current command value is provided by the outer loop controller, and the outer loop controller tracks and controls according to the reference values, such as active power, reactive

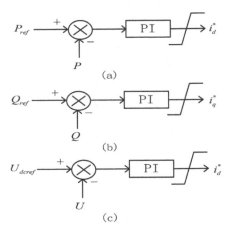

(a)

(b)

(c)

Figure 3. Outer loop active power, reactive power, and DC voltage controller.

power, and DC voltage. The d-axis is oriented to the receiving-end grid voltage (U_s), according to the instantaneous power theory, the inverter output active power and reactive power can be expressed as:

$$\begin{cases} P = U_s i_{sd} \\ Q = -U_s i_{sq} \end{cases} \tag{3}$$

Therefore, the active and reactive powers can be independently controlled by i_{sd} and i_{sq}, respectively. The PI regulator is instructed to eliminate the steady-state error, and the active power and reactive power controller structures are shown in Fig. 3 (a) and 3 (b), respectively.

The constant DC voltage controller is used to balance the DC system active power and keep the voltage stable. Ignoring R and converter losses, the active power on AC/DC sides of converter station is equal. Therefore:

$$P = U_s i_{sd} = P_{dc} = u_{dc} i_{dc} \tag{4}$$

In steady state:

$$i_{dc} = \frac{U_s i_{sd}}{u_{dc}} \tag{5}$$

When the active power on both sides of the converter station is unbalanced, the DC voltage will fluctuate. At this point, the active current will charge (discharge) the DC capacitor until the DC voltage stabilized at the reference value. Fig. 3 is the schematic diagram of the outer loop DC voltage controller.

3 FAULT CHARACTERISTICS OF CONVERTER STATION IN THE MMC-HVDC SYSTEM

Ignoring the signal sampling, delay time of filtering, converter switching, and other factors, decoupled converter station control system control block diagram shown in Figure 4, where $1/(R + sL)$ represents the transfer function of the AC filter. In this case, the closed-loop transfer function of the d-axis current is:

$$G(s) = \frac{i_{sd}}{i_{sd}^*} = \frac{K_p s + K_i}{Ls^2 + (R + K_p)s + K_i} \tag{6}$$

Set $K_p/K_i = L/R$, the closed-loop transfer function of the current control can be simplified as:

$$G(s) = \frac{K_p}{K_p + Ls} = \frac{1}{1 + \dfrac{s}{\omega_c}} \tag{7}$$

where $\omega_c = K_p/L$.

At this time, the current inner loop is a first-order inertia link. Let the pre-fault active command be i_{d0}^*; the active current command after the voltage dip is i_{d1}^*, and the d fault component of the converter output is:

$$\Delta i_{df}(s) = G(s)i_{d1}^*(s) - G(s)i_{d0}^*(s)$$
$$= \frac{1}{1 + \dfrac{s}{\omega_c}} \cdot \frac{i_{d1}^* - i_{d0}^*}{s} = (i_{d1}^* - i_{d0}^*)\left(\frac{1}{s} - \frac{1}{s + \omega_c}\right) \tag{8}$$

By the inverse Laplace transform, the time-domain expression of the fault current component can be obtained as:

$$\Delta i_{df}(t) = (i_{d1}^* - i_{d0}^*)(1 - e^{-\omega_c t}) \tag{9}$$

When the switching frequency is high enough, the current inner loop has a high response speed, so the transient process can be ignored

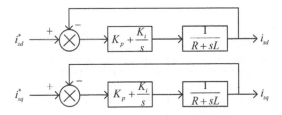

Figure 4. Simplified control diagram of inverter substation inner current loop.

(Kong 2013). In this case, the d axis component of fault current is:

$$i_{df}(t) \approx i_{d1}^*$$ (10)

As the d, q axis has the corresponding control structure, the q axis component fault current can also be expressed as:

$$i_{qf}(t) \approx i_{q1}^*$$ (11)

where i_{q1}^* is the reactive current command after the voltage dip.

Before the fault, the inverse converter station runs in unity power factor, and the output current of the converter station is:

$$\begin{cases} i_{ca0}(t) = i_{d0}^* \cos(\omega_1 t + \theta_1) \\ i_{cb0}(t) = i_{d0}^* \cos\left(\omega_1 t + \theta_1 - \dfrac{2\pi}{3}\right) \\ i_{cc0}(t) = i_{d0}^* \cos\left(\omega_1 t + \theta_1 + \dfrac{2\pi}{3}\right) \end{cases}$$ (12)

where θ_1 is the initial phase of a phase current.

After the three-phase voltage drops in the AC side, the estimated d and q components of the output short-circuit current of the inverter substation are:

$$\begin{cases} i_{d1}^* = k_p(P_{ref} - P) + k_i \int (P_{ref} - P)\,\mathrm{dt} \\ i_{q1}^* = k_p(Q_{ref} - Q) + k_i \int (Q_{ref} - Q)\,\mathrm{dt} \end{cases}$$ (13)

As the inertia time constant of the converter station is very small, the output transient component of the converter station can be neglected. Therefore, the output three-phase short-circuit current of the inverter station is:

$$\begin{cases} i_{ca1}(t) = I_m \cos(\omega_1 t + \theta_1 + \varphi) \\ i_{cb1}(t) = I_m \cos\left(\omega_1 t + \theta_1 - \dfrac{2\pi}{3} + \varphi\right) \\ i_{cc1}(t) = I_m \cos\left(\omega_1 t + \theta_1 + \dfrac{2\pi}{3} + \varphi\right) \end{cases}$$ (14)

where $\varphi = \arctan(i_{q1}^*/i_{d1}^*)$ and $I_m = \sqrt{(i_{d1}^*)^2 + (i_{q1}^*)^2}$.

Through a comparison of the short-circuit current before and after fault, it can be seen that the output fault current of the MMC-HVDC system inverse converter station is quite different from that of the traditional AC system and conventional DC converter station. In the traditional AC system, the amplitude of the short-circuit current provided by the servo power is large, and its phase keeps unchanged. The amplitude of the short-circuit

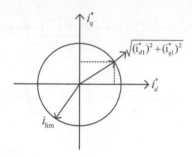

Figure 5. Short-circuit current limiting measure of substation.

current provided by the conventional DC converter station may be less than the steady-state operating current, and its phase varies with time and the severity of the fault. According to equation (14), when the converter current limit is not reached, the short-circuit current amplitude of the MMC converter station is determined by the converter station control command, and its phase is affected by the reactive power control mode during converter station failure. In addition, as the MMC-HVDC system has weak overload ability, excessive short-circuit current may lead to the damage of converter station IGBT components. Fig. 5 is the short-circuit current restriction formula diagram of the converter station, when the magnitude I_m of the fault current exceeds the short-circuit current limit value i_{lim}, the reference value of the active and reactive current components will be limited.

Therefore, when considering the current-limiting control of the converter station, the output short-circuit current expression of the converter station can be expressed as:

$$\begin{cases} i_{ca1}(t) = i_{lim} \cos(\omega_1 t + \theta_1 + \varphi) \\ i_{cb1}(t) = i_{lim} \cos\left(\omega_1 t + \theta_1 - \dfrac{2\pi}{3} + \varphi\right) \\ i_{cc1}(t) = i_{lim} \cos\left(\omega_1 t + \theta_1 + \dfrac{2\pi}{3} + \varphi\right) \end{cases}$$ (15)

After the current-limiting control, the converter station output short-circuit current is limited to a certain amplitude to ensure the safe operation of the converter station.

4 ANALYSIS OF AC SYSTEM DISTANCE PROTECTION OPERATION CHARACTERISTICS

After the AC system failure, based on the short-circuit current expression of the converter station in AC system failure, depending on whether the current-limiting protection start condition is reached or not, the output short-circuit current amplitude of the converter station can be expressed as:

$$I_m = \begin{cases} \sqrt{(i_{d1}^*)^2 + (i_{q1}^*)^2}, & I_m < i_{\lim} \\ i_{\lim}, & I_m \geq i_{\lim} \end{cases} \quad (16)$$

When the converter current-limiting condition is not reached, the output short-circuit current of the converter station is determined by the outer loop power reference value. The MMC converter station can be considered as the voltage source. When the fault current reaches the current-limiting condition, the converter station output short-circuit current is constant. In this case, the MMC-HVDC commutation station and the inverter station can be equivalent to a current source with respect to the AC system. The influence of MMC-HVDC on the distance protection of AC system is analyzed with the system as shown in Fig. 1, where k1 and k2 indicate different fault positions and A, B, C, and D bus outlets are all equipped with distance protection. They are called the sending-end system side, rectifier station side, inverter station side, and receiving-end system side distance protection, respectively.

Assuming that a ground short-circuit fault occurs on the receiver AC line, the equivalent circuit of the inverter and the receiver AC system is shown in Fig. 6. Because of the introduction of the transition resistance at the short-circuit point, the measured voltage and impedance of the distance protection measuring elements of the inverter station C are:

$$\dot{U}_C = \dot{I}_d Z_{LC} + (\dot{I}_d + \dot{I}_2)R_g = \dot{I}_d(Z_{LC} + R_g) + \dot{I}_2 R_g \quad (17)$$

$$Z_C = \frac{\dot{U}_C}{\dot{I}_d} = Z_{LC} + \left(1 + \frac{\dot{I}_2}{\dot{I}_d}\right)R_g$$
$$= Z_{LC} + R_g + \left|\frac{\dot{I}_2}{\dot{I}_d} R_g\right| \angle \theta \quad (18)$$

$$Z_f = \left|\frac{\dot{I}_2}{\dot{I}_d} R_g\right| \angle \theta \quad (19)$$

where $\dot{U}_C$ is the measured voltage of the relay location of C side bus; $\dot{I}_d$ and $\dot{I}_2$ are the short-circuit current flowing into the short-circuit point of the inverter station and receiving-end system, respectively; Z_{LC} is the line impedance of the C-side protection to the failure point; Z_c is the measured impedance of the distance protection elements; R_g is short-circuit transition resistance; θ is the phase difference between $\dot{I}_2$ and $\dot{I}_d$; and Z_f is the additional measured impedance.

After the fault, the amplitude of system-side short-circuit current $\dot{I}_2$ is much larger than the converter station current $\dot{I}_d$; meanwhile, the phase difference θ of $\dot{I}_2$ and $\dot{I}_d$ also changes a lot. This may cause the measured impedance Z_c to exceed the boundary of the quadrangle impedance relay setting, resulting in the refused operation of distance protection. Fig. 7 shows the influence of the inverter station distance protection operation characteristic. Without additional measured impedance, the measurement impedance is only affected by the size of the transition resistance. The additional measured impedance may cause the measured impedance not to be within the impedance setting range.

Factors affecting the additional measurement impedance of converter station distance protection mainly include the following two points:

1. Short-circuit ratio of AC/DC system

When a short-circuit fault occurs on the receiving-end AC transmission line, there will be great over-current. To keep the safety of the converter station components, the converter station control system generally restricts d, q current reference value to limits short-circuit current. During the fault, the converter station does not supply great short-circuit current to the short-circuit point (Li 2009), so the AC/DC system based on MMC-HVDC interconnection has the characteristics of weak infeed system. The weak-feeding degree of the system is related to the Short-Circuit Ratio (SCR) of the AC system to which the DC transmission is connected. It represents the ratio of the short-circuit capacity A to the rated DC power B of the converter station AC bus:

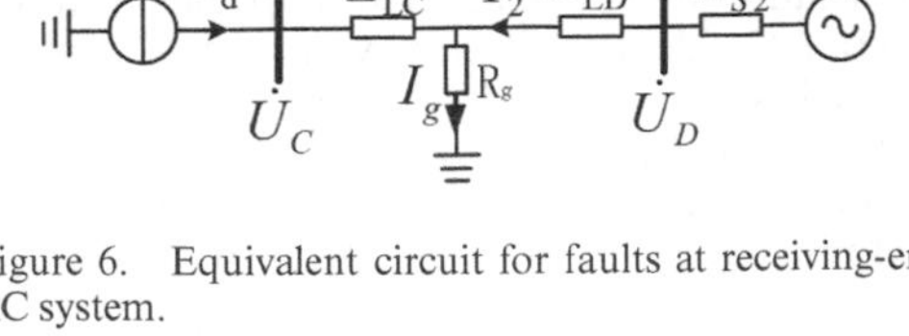

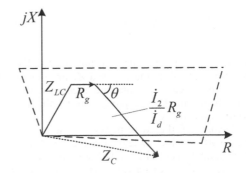

Figure 6. Equivalent circuit for faults at receiving-end AC system.

Figure 7. Operating characteristic chart of distance protection at inverter substation side.

$$SCR = S_{ac} / P_{dN} \qquad (20)$$

The larger the short circuit ratio, the greater the short-circuit capacity of the AC system, the more obvious the weak-feeding characteristics of the system, the greater the amplitude ratio of the short-circuit current provided by the system to the short-circuit current provided by the converter station, the greater the magnitude of the additional measured impedance, and the more serious is the influence on the distance protection of the inverter station. On the contrary, the smaller the short-circuit current ratio, the more similar the short-circuit capacity of both sides; the smaller the magnitude of the additional measured impedance, the less the influence on the distance protection of the inverter station.

2. Reactive power control of converter station during fault

Converter stations generally transmit power in unity power factor when under normal operating condition. As the active power and reactive power of the MMC-HVDC converter station can be adjusted independently, the station can provide some reactive power support during fault. That is to say, the phase angle of the fault current is related to the reactive power control mode of the converter station.

When the converter station only generates active power during the fault, as the converter station and other equipment all need to consume some reactive power, then the short circuit current of the converter station is phase-leading, and $\theta < 0$, the additional measurement impedance is capacitive. When the converter station generates reactive power during the fault, the short-circuit current phase of the inverse converter station lags behind the short-circuit current of the receiver AC system. When the reactive power of the converter station is high enough, it may result in $\theta > 0$.

For the distance protection components of the system receiver side D, the protection measurement impedance is:

$$Z_D = \frac{\dot{U}_D}{\dot{I}_2} = Z_{LD} + R_g + \left|\frac{\dot{I}_d}{\dot{I}_2}R_g\right| \angle -\theta \qquad (21)$$

where $\dot{U}_D$ is the measured voltage of the D-side bus bar protection and Z_{LD} and Z_D are the actual line impedance and protection measured impedance from the D-side protection to the failure point, respectively.

Because of the weak power characteristics of the MMC-HVDC system side, the amplitude of $\dot{I}_2$ is much larger than $\dot{I}_{dc}$. Although the phase difference of the short-circuit current on both sides will result in an additional inductive or capacitive

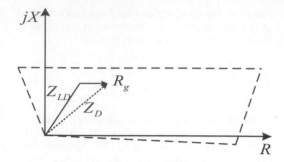

Figure 8. Operating characteristic chart of distance protection at inverter substation side.

measured impedance, the amplitude of the additional measured impedance is relatively small. Figure 8 shows the influence schematic diagram of the operating characteristics of the system-side distance protection components. The measured impedance error is mainly due to the transition resistance. It is only slightly affected by the MMC-HVDC system.

5 SIMULATION VERIFICATION

In order to verify the above theoretical analysis, a simulation model of the two-level converter MMC-HVDC is built using PSCAD/EMTDC simulation software. Among them, the AC system phase voltage is 110 kV, rated frequency is 50 Hz, rated DC voltage is ±10 kV, rated transmission active power is 10 MW, the AC overhead transmission lines between the AB bus and the CD bus are 20 km long, and the impedance parameter of AC transmission line is z1 = 0.105 + j0.383 Ω/km, z0 = 0.315 + j1.149 Ω/km. The distance protection adopts quadrilateral characteristic impedance relays. Before the fault, the converter station mainly aims at transmitting electric energy and runs with unit power factor, considering that the converter station has certain reactive power support capability.

At t = 3 s, the single-phase fault occurs at the midpoint of the receiving AC line, the reactive power reference of the converter station is set to zero during the fault, and the full-wave Fourier algorithm is used to extract the frequency (Zhang 2010). When the single-phase short-circuit occurs at the midpoint k2 of the receiving-end AC line, the change of the measured impedance at the inverter C side and the receiving-end AC system D side is shown in Fig. 9. The red, black, and blue lines represent the measured impedance values of the distance protection elements when the transition resistance is 5, 10, and 15 Ω, respectively.

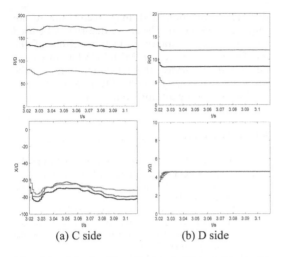

(a) C side   (b) D side

Figure 9. Measured impedance of different sides under single-phase fault at receiving-end AC system.

From Fig. 9, it can be seen that the measured impedance at the C-side distance protection of the inverter station varies greatly with respect to the actual impedance value. At the transition resistances of 5, 10, and 15 Ω, the measured resistances of the C-side converter station are about 71, 134, and 175 Ω, respectively. The measured resistance value increases significantly while the measured reactance is negative, decreasing with the increase of the transition resistance, and it shows a capacitive measured impedance characteristic. The D-side receiving-end system basically reflects the impedance value of the actual line and the transition resistance. Table 1 shows the distance protection operating condition of the inverse converter station and the receiving-system AC side when different types of short-circuit faults occur at the midpoint of the receiving AC line.

It is evident from Table 1 that because of the introduction of the MMC-HVDC system, the distance protection operating characteristics of the inverter station is affected relatively seriously, the measured impedance of the receiving system side distance protection operates correctly, while the inverter station side distance protections all refuse to operate for the impact of additional measured impedance.

At t = 3 s, single-phase fault occurs at the midpoint k2 of the receiving AC line, and the inverter station provides 3 Mvar reactive power during the fault. Figure 10 shows the change of the distance protection measured impedance at the C-side inverter station and the D-side receiving AC system. The red, black, and blue lines represent the measured impedance values of the distance protection elements when the transition resistance is 5, 10, and 15 Ω, respectively.

Table 1. Zone-I action situation of distance protection at AC receiving side.

Protection device location	Fault resistance	Protection situation		
		AG	BG	ABC
C side	5	Failure	Failure	Failure
	10	Failure	Failure	Failure
	15	Failure	Failure	Failure
D side	5	Action	Action	Action
	10	Action	Action	Action
	15	Action	Action	Action

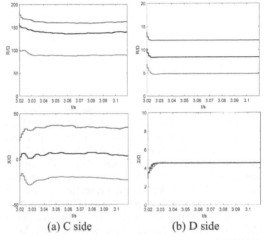

(a) C side   (b) D side

Figure 10. Measured impedance of different sides under single-phase fault at receiving AC system.

As Fig. 10 shows, when the converter station provides certain reactive power during a fault, under the condition of 5, 10, and 15 Ω transition resistors, the measured resistances of the C-side converter station are 90, 138, and 172 Ω, respectively, higher than those under the condition of rated power factor, while the measured reactance values, respectively, become 24, 4 Ω, and 31, which change from capacitive measured impedance to inductively measured impedance. For the impedance relay of quadrangle characteristic, after the converter station sends out certain reactive power, the resist-operation risk of the distance protection of the MMC AC side system is reduced. The measured impedance value of the D side receiving system basically reflects the impedance value of the actual lines and the transition resistance.

From the comparison of Figs. 9 and 10, it can be seen that the different reactive power control modes of the converter station during the fault mainly affect the measurement impedance of the converter station distance protection.

From Fig. 12, it can be concluded that, at different reactive power reference values, the measured impedance of the AC-side converter station exhibits different amplitudes. It has certain influence on the operation characteristics of the AC-side MMC distance protection, that is, when the converter station operates in unit power factor, the measured impedance is capacitive during an internal fault, and the distance protection of AC-side MMC system is likely to refuse operation. As the transition resistance value increases, the measured impedance is closer to the operating range of the impedance relay. When the converter station emits certain reactive power during the fault, the measured impedance appears as an inductive impedance, and the distance protection characteristic of the AC-side MMC system is improved.

6 CONCLUSION

Because of the nonlinearity and the control mode of the converter station power electronic devices, the AC/DC system based on MMC-HVDC exhibits different short-circuit current characteristics compared to the traditional AC system and the AC/DC system based on conventional DC transmission during the short-circuit fault, which may have an adverse effect on the protection system. In this paper, the fault current and the distance protection operating characteristic of the receiving AC system of the AC/DC system based on MMC-HVDC interconnection are analyzed, and the following conclusions are obtained:

1. Because of the introduction of the MMC-HVDC converter station, the distance protection of the converter station is likely to refuse operating because of the capacitive measured impedance. The operational characteristics of the AC system distance protection are affected less by the introduction of the MMC-HVDC converter station.
2. In the AC/DC system, the distance protection of the converter station is affected mainly because the polymorphism of the short-circuit current from the converter station produces an additional measured impedance during the AC system failure.
3. The degree to which the converter station distance protection is affected is related to the short-circuit current ratio of the converter station distance protection of the AC/DC system and the reactive power control mode of the converter station during a fault.

The above characteristics of the AC/DC system based on the MMC-HVDC interconnection will affect the operation characteristics of the existing distance protection and may even lead to an improper operation. The above theory can be used to improve the distance protection setting of the AC/DC system based on MMC-HVDC interconnection and provide a theoretical reference for the protection strategy of the AC/DC power system.

Corresponding author: Xiao Chao

REFERENCES

Chen, Hairong. 2007. Transient Model and Controller Design for VSC-HVDC Based on Synchronous Reference Frame. *Transactions of China Electrotechnical society*, 22(2): 121–126.

Dong, Yang. 2014. Coordinated optimization for controlling short circuit current and multi-infeed DC interaction. *Journal of Modern Power Systems and Clean Energy*, 2(4): 374–384.

Fei, Bing. 2015. Impact of AC-DC interconnected system on distance protection and countermeasure. *Electric Power Automation Equipment*, 35(8): 15–21.

Feng, Yadong. 2015. Valve Protection Design of Modular Multilevel Converter for VSC-HVDC. *Automation of Electric Power Systems*, 39(11): 64–68.

Jin, Xingfu. 2014. A Novel Pilot Proteciton for VSC-HVDC Transmission Lines Using Modulus Model Identification. *Automation of Electric Power Systems*, 38(10): 100–106.

Kong Xiangping. 2013. Study on Fault Current Characteristics and Fault Analysis Methods of Power Grid With Inverter Interfaced Distributed Generation. *Proceeding of the CSEE*, 33(34): 65–74.

Lu Jing. 2015. Control Strategies of Large Wind Farms Integration Through AC/DC Parallel Transmission System Based on VSC-HVDC. *Power System Technology*, 39(3): 639–646.

Li, Haifeng. 2009. Performance of directional protection based on variation of power-frequencey components in HVDC/AC interconnected system: Part One DC-system impedance of power-frequnncy component variation. *Automation of Electric Power Systems*, 33(9): 41–46.

Shen, Hongming. 2015. Effect Analysis of AC/DC Interconnected Network on Distance Protection Performance and Countermeaures. *Automation of Electric Power Systems*, 39(11): 58–63.

Xu Zheng. 2012. *Flexible HVDC transmission system*. Beijing: China Machines Press.

Xu, Feng. 2012. A control and protection scheme of multi-terminal DC power system for DC line fault[J]. *Automation of Electric Power Systems*, 36(6): 74–78.

Zhang, Pu. 2012. Performance of Distance Protection for Transmission Lines in an HVDC /AC Interconnected Power System. *Automation of Electric Power Systems*, 36(6): 56–62.

Zhang, Baohui. 2010. *Power system relay protection*. Beijing: China Electric Power Press.

Zhang, Pu. 2010. Performance of current differential protection for transmission lines in HVDC /AC interconnected system. *Power System Protection and Control*, 38(10): 1–5.

Research on the impact of generator tripping considering wind power participating in frequency regulation

Jianfeng Shi, Guanchao Zhao & Cui Wang
Qujing Power Supply Bureau, Qujing, Yunnan Province, China

ABSTRACT: The asynchronous interconnection between Yunnan Power Grid and Southern Power Grid is achieved by High Voltage Direct Current (HVDC) transmission lines in Yunnan-Guangxi via Luxi back-to-back, YongFu, JinZhong, and Yunnan-Guangdong via ChuSui, PuQiao, NiuCong bipolar. After the realization of asynchronous interconnection, the frequency of the supply network can increase when unipolar/bipolar blocking appears in any HVDC line, with which the probability of problems accompanied will increase. Besides, the permeability of the renewable energy sources, such as the small hydropower and wind power, in the power grid will further improve. The total amount of renewable energy has become comparable with that of the traditional energy. Therefore, the application of wind power participating in frequency regulation is incorporated into engineering. In addition, different properties, such as primary frequency characteristics, structure, and characteristics of generators of prime movers, cause different tripping impacts of different generators. Models, to explore and compare tripping impacts of different generators before and after wind power participating in frequency regulation, are built with PSD-BPA on the basis of the analysis of the operating principle of thermal power units, large hydropower units, low-capacity hydropower units, and wind turbine generators, after which a simulation was carried out. Comparison and analysis indicated that when wind power units participate in frequency regulation, the system achieves a better characteristic of frequency recovery.

1 INTRODUCTION

The total installed capacity of Yunnan Power Grid is 9516.952 MW, in which the proportions of medium-capacity or high-capacity hydropower, low-capacity hydropower, thermal power, and wind power are 13.34%, 8.32%, 69.77%, and 10.31%, respectively. With the large-scale connection of renewable energy sources, the permeability gradually increases, which will reach 14.43% by the end of 2016 and be higher in practice when considering large-scale thermal power off the line. In addition, the utilization hours of wind power in Yunnan Power Grid is also higher than that in other regions of China, of which the average value reached 2919 in 2013, first in the country, while the utilization hours is 2080 in the whole country. In pace with the increasing permeability of renewable energy sources, especially wind power, the demand for the application of wind power participating in frequency regulation becomes urgent. By 2018, Yunnan power grid will accomplish the application of wind power participating in frequency regulation, which will bring significant changes to frequency characteristics of Yunnan Power Grid. Simultaneously, because of the further improvement of permeability, the wind power and other renewable energy sources should get more atten-

tion when formulating generator tripping schemes for high-frequency situation and stability control aims.

In recent years, on the basis of the actual requirement of engineering, a great amount of research has been carried out on the fault of AC tie lines, high-frequency generator tripping, and stability control generator tripping after DC blocking by researchers, and valuable results have been obtained. In Zhang Dan et al. (2013) a hybrid scheme on stability control of regional power network and high-frequency generator tripping was proposed, combining simulation and the actual situation of power grid operation, to analyze the adaptability of the program. The coordination of high-frequency generator tripping and the under-frequency load shedding measures were investigated by Li Zheng et al. (2012). In 2016, Yunnan Power Grid will realize asynchronous interconnection with Southern Power Grid. In Ma Zhiheng (2015), aiming at DC fault, which might be caused by frequency off-limit, the measures on stability control generator tripping, and DC additional frequency control are proposed, and the adaptability of the scheme on high-frequency generator tripping was analyzed. In Dong Zhe (2016), the transient energy generator tripping control effect index and system oscillation recovery effect index was defined. The difference of changes when improving first-swing in transient progress and damping changes caused by the different wind generator tripping measures was analyzed, and further study on the influence of oscillation in the system recovery process was carried out.

With the permeability increase of wind energy and other renewable energy, frequency regulation with renewable energy is gradually becoming a demand in the field of engineering practice. On the basis of the analysis of the response characteristics of ordinary asynchronous wind turbine and Doubly Fed Induction Generator (DFIG) in frequency fluctuation, it was found that the output characteristics of asynchronous wind turbine would change, but the frequency fluctuation had less effect on DFIG (G. Lalor et al. 2005). In Guang Hongliang (2007), the different response characteristics of fixed-speed wind generator and DFIG in grid frequency fluctuation are analyzed, and the active power and reactive power decoupling control in response to the change of power grid frequency was extremely limited, and a modified frequency control component providing short-term power to the system was also proposed. According to the wind power grid system and real-time change of frequency response characteristics, the frequency control strategy for wind energy and storage system was proposed by Jiang Wang (2015). The strategy can effectively stabilize the fluctuations of wind power grid, provide active power support for frequency control for the wind farm in the system, and effectively improve the reliability of wind farm in the system of frequency adjustment.

On the basis of the analysis of working principle and operation characteristics of thermal power units, large hydropower units, and small hydropower and wind generator units, the generator tripping impact of different types units before and after the wind turbines participating in primary frequency regulation was simulated. This comparative analysis showed that the grid system would have a better frequency regulating characteristic when wind turbines were involved in generator tripping.

2 ANALYSIS OF INFLUENCING FACTORS OF FREQUENCY CHARACTERISTICS OF SYSTEM AND CONVENTIONAL GENERATOR UNIT

2.1 Frequency characteristics of the electric power system

The power–frequency characteristic of the electric power system is the dynamic characteristic of the system frequency when the active power is unbalanced. It results due to factors such as load frequency characteristics, generator frequency characteristics, and system voltage. Overall, when the load active power is greater than the output, the system frequency will decrease; conversely, the system frequency will rise. Any bus and unit in the system shares the same-frequency dynamic changing process, which can be calculated and analyzed by the equivalent single-machine model. The equation of the model can be expressed as (He Yong 1992):

$$
\begin{cases}
T_S \dfrac{d\Delta f}{dt} = -\Delta P \\
T_G \dfrac{d\Delta P_G}{dt} + \Delta P_G = -k_G \Delta f \\
\Delta P_L = -k_L \Delta f \\
\Delta P = \Delta P_L - \Delta P_G + \Delta P_0
\end{cases}
\tag{1}
$$

where k_G and k_L are the frequency adjustment modulus of the generating units and loads, respectively; ΔP_0 is the initial unbalance power of the system; ΔP is the power deviation; ΔP_L and ΔP_G are the active power change of the load and the generator due to the frequency variation, respectively; T_G is the generator inertia time constant; and T_S is the system inertia time constant. The model of equivalent single-machine system is shown in Figure 1.

The frequency response equation of the system can be expressed as:

$$
\begin{cases}
\Delta f(t) = \dfrac{\Delta P_0}{k_L} \left[1 - 2A_m e^{-\frac{1}{2}\left(\frac{1}{T_G}+\frac{1}{T_S}\right)t} \cos(ft+k) \right] \\
f = \sqrt{\dfrac{k_D}{T_S T_G} - \dfrac{1}{4}\left(\dfrac{1}{T_G}+\dfrac{1}{T_S}\right)^2} \\
A_m = \dfrac{1}{2fT_S} \sqrt{k_D k_G} \\
k = \arg\tan\left[\dfrac{k_D}{fT_S} - \dfrac{1}{2f}\left(\dfrac{1}{T_G}+\dfrac{1}{T_S}\right)\right]
\end{cases}
\tag{2}
$$

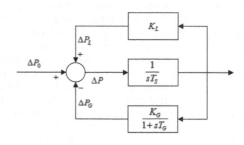

Figure 1. Model graph of single-machine system.

From formulas (1) and (2), we can get the following conclusion: if speed control system is taken into account, frequency dynamic process of the system is a curve decaying at a certain time constant, which mainly depends on the power deviation of the power grid, the frequency characteristics of the load, and the inertia time constant of all generators in the system. In addition, it can be seen that the initial changing rates of frequency are directly proportional to the power deviation and inversely proportional to the system's overall inertial time constant; the steady-state frequency deviation is determined by the power deviation and the power regulation effect coefficient.

2.2 *Frequency characteristics of the conventional units*

When the load quantity changes, the load phase angle will change, but the generator power angle will not change immediately. The power assumed by each generator set is in fact closely related to the unbalanced power, load disturbance, and the electrical distance between each generator set. Therefore, the unit power shortage can be expressed as:

$$\Delta P_i = \frac{E_i U_k B_{ik} \cos \delta_{ik0}}{\sum\limits_{j=1}^{m} E_j U_k B_{jk} \cos \delta_{jk0}} \Delta P_L(0+) \tag{3}$$

where E_i is the transient EMF of the generator bus, U_k is the voltage of load bus k, B_{ik} is the transferring admittance between bus i and bus k, and δ_{ik0} is the phase angle difference between bus i and bus k before the unbalance power occurs.

In 0.5–2 s after the disturbance occurs, the distribution principle of the disturbance power is proportional to the rotary inertia of online unit, and it is irrelevant to the location of generator sets. At this time, ΔP_i can be represented as:

$$\Delta P_i = (\frac{J_i}{\sum\limits_{j=1}^{m} J_j}) \Delta P_L(0^+) \tag{4}$$

where J_i is the rotary inertia of unit i.

After the disturbance occurs at about 3 s, the speed control system of the online generator units starts to adjust the output power of the generator so as to participate in the primary adjustment of the system frequency. The change of electromagnetic power of generator set and its changing rate are very important to the speed control system and rotating reserve of the generator set. The generator power shortage is closely related to the unit capacity, speed control system characteristics, and spinning reserve.

At present, the adjustment coefficient of thermal power units in Yunnan Power Grid is generally 0.05, and the adjustment coefficient of hydropower unit is generally between 0.04 and 0.05. In terms of the moderating effect of frequency deviation, hydroelectric generating units are generally better than thermal power units.

3 BASIC PRINCIPLE AND FREQUENCY CHARACTERISTICS OF THE WIND TURBINE

As the wind power units in certain wind farms of Yunnan Power Grid are DFIGs, the electro-mechanical transient model based on the widely applied wind turbine model is established to analyze the basic principle and frequency characteristic of doubly fed wind turbine.

3.1 *Operation control strategy of wind turbine generator system connected to the grid*

The VSCF wind turbine, operating according to the maximum wind power tracking strategy, can regulate the speed of fans with the change of wind speed, thereby significantly enhancing the efficiency of DFIG wind power system. Most VSCF DFIG wind power systems consist of power electronic devices and a motor, and can be divided into three categories: full power converter-type VSCF wind power system with no gear box, full power converter-type VSCF wind power system with gear box, and AC-excited VSCF doubly fed induction wind power system. VSCF wind power system is connected to the grid through the PWM converter, with easier operation and convenient control, and it can track the optimal tip speed ratio under different wind speeds, thereby improving the efficiency of the wind energy utilization (Jiang Wang. & Lu Jiping. 2014). However, the Maximum Power Point Tracking (MPPT) control method is not involved in the system frequency control. Doubly fed induction wind turbine can work in four quadrants under the control of converter, which realize decoupling of system frequency and rotating speed of the rotor, but it does not make contribution to the system inertia, and has no ability to change the frequency response of the system, either. The relationship between the frequencies of the stator and rotor windings is as follows:

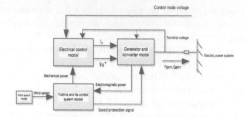

Figure 2. Fan control structure principle diagram.

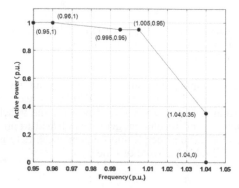

Figure 3. Frequency response curve of APC.

$$f_1 = \frac{pn}{60} \pm f_2 \qquad (5)$$

where f_1 is the stator current frequency, p is half of the number of generator poles, n is the rotor speed, and f_2 is the rotor current frequency. Formula (5) shows that, when the generator speed changes are caused by the change of wind speed, the constant output f_1 can be achieved by adjusting the f_2; thus, operate by the way of VSCF (Cao Jun et al. 2009).

3.2 Control principle of the fans participating in frequency modulation

It is evident from Fig. 2 that, according to the order of control system, the generator can inject active power and reactive power into the system through the inverter; meanwhile, the low-voltage and over-voltage protection functions can be simulated. The electric control model WindCONTROL includes the reactive power control of the wind farm. The output signal of WindCONTROL is the reactive power control signal. The prime mover and its control system reflect the control of the mechanical part, including pitch control and active power control.

Table 1. Comparison of wind power frequency modulation technology.

Frequency modulation demand	Frequency modulation mode	Scope of application	Advantage	Disadvantage
Inertial response	Simulation inertia control	Full wind speed condition	1. Provide the inertial response and contribute to the dynamic stability of the system: 2. Fast response.	1. Shorter duration; 2. Rotor speed recovery, results in secondary reduction in frequency: 3. When frequency and wind speed are too low, or totally opposite, it is difficult to provide effective inertia.
Primary frequency modulation	Rotor speed control	Medium and low wind speed conditions	1. Respond quickly and contribute to the dynamic stability of the system: 2. Provide a standby for primary frequency modulation.	1. When wind speed is high, it is difficult to supply the spare capacity of the system: 2. Fluctuation of wind speed influences the reliability of reserve capacity: 3. Less power generation, less benefit.
	Pitch control	Full wind speed condition, mainly used for high wind speed	1. Provide a standby for primary frequency modulation in full wind conditions: 2. Competitive regulating ability and wide range of power regulation.	1. Limited by mechanical characteristics, the response speed is low, and the contribution to the dynamic stability of the system is small: 2. Blades mechanical loss increases, which reduces the operating life of the units: 3. The random fluctuation of wind speed affects the reliability of reserve capacity.

Active control includes active power control command module and additional control module (such as power control active module, including the system frequency response function, limiting the maximum fan power output function and others). When APC is enabled, the fan retains a certain spinning reserve in the normal operating range of frequency (typically 5%). When the power shortage leads to frequency changes in power grid, fans will have certain frequency regulation capability by changing the fan output through this control module, which can boost the stability of power grid. The frequency response curve of APC is shown in Fig. 3.

Fig. 3 shows that adjustment coefficient of fan is 0.075 when frequency is in the range of 1.005–1.040, and it is 0.7 when frequency is in the range of 0.96–0.995. There is a significant difference between fans and traditional unit in terms of adjustment coefficient.

At present, the main control strategies for wind power frequency modulation are simulation inertia control, rotor speed control, pitch control, and droop control. Their characteristics are shown in Table 1.

Typical inertial time constant of a synchronous generator in a large power plant is usually in the range of 2–9 s, while the typical inertia time constant of wind turbine is between 2 and 6 s. This illustrates that the wind turbine power system does not reduce the moment of inertia of the whole electric power system after connecting to the grid. It is appropriate to take full advantage of the inertia of the wind turbine, taking into account that the "implied inertia" of the wind turbine participates in the frequency regulation of power system. In addition, in order to make the wind turbine characteristics closer to the conventional synchronous generator unit, some fans are also equipped the Wind Inertia function. When a large disturbance occurs in the power system, it makes the wind power generator provide inertia similar to the traditional generator, and the transient stability of the system is also improved.

4 UNIT OPERATING FREQUENCY LIMIT AND HIGH-FREQUENCY GENERATOR TRIPPING STRATEGY

Power system security control device is a safe and automatic device to maintain the stable operation of power system, which focuses on transient stability control of power system. It is an important part of the second and third lines of defense that is essential to maintain the security, stability, and reliability of power system. According to the characteristics of different types of power supply, the allowable operating range of frequency is shown

Table 2. Allowable operating time of the abnormal frequency of turbine generator (GBT 31464-2015).

Frequency range/Hz	Cumulative allowable operating time/min	Allowable operating time at a time/s
51.0–51.5	>30	>30
50.5–51.0	>180	>180
48.5–50.5	Continuous operation	Continuous operation
48.0–48.5	>300	>300
47.5–48.0	>60	>60
47.0–47.5	>10	>20
46.5–47.0	>2	>5

Table 3. Operating rules of wind farm within the frequency range (GBT-19963-2011).

Power system frequency range/Hz	Requirements
<48	According to the lowest frequency of wind turbines within the allowable operating time
48–49.5	The wind farm is required to have the ability to run at least 30 min when the frequency is below 49.5 Hz
49.5–50.2	Continuous operation
>50.2 Hz	When the frequency is higher than 50.2 Hz, the wind farm is required to have the ability to run at least 5 min and execute the order of reducing the output power or high-frequency generator tripping strategy given by power system dispatching agencies, and does not allow the shutdown state of wind turbine.

by the relevant departments. Water turbine generator is usually better than turbine generator and meets the requirements of system scheduling. The operating frequency limit of a typical turbine generator and wind fan is shown in Tables 2 and 3.

5 INFLUENCE OF GENERATOR TRIPPING OF DIFFERENT TYPES OF UNITS ON FREQUENCY

In order to compare the effects of the generator tripping of different types of units after the participation of wind turbine in the frequency regulation, we take a city power grid in Yunnan as the research objective to construct a simulation example. The transmission channel losses 310 MW power transmission when DC blocking occurs and the equivalent capacity of thermal power units, medium and large hydropower generating units, small-capacity hydropower units and wind turbines have been tripped in order to maintain the power balance of the system. Study the influence on the frequency in the four generator tripping conditions.

The wind turbine generator uses GE model of PSD-BPA in the simulation. The model mainly includes the generator converter model, reactive power control model, electric control model, motor control model, rotor mechanical model, wind power model, active control and active rate limiting model, and Wind INERTIA function module. The participation of fans in frequency regulation can be realized by setting Active Power Control (APC) function, the Wind INERTIA function module. The function of the APC mainly includes: limiting the maximum active power output of the wind farm, retaining the active margin of the wind farm and responding the system frequency change and other functions. In the normal circumstances, the active power output of the wind farm can be limited by setting the frequency-active power curve and retaining the active margin. And the output power of wind power generator is adjusted according to the frequency change. When the wind power generator does not participate in the frequency regulation, the recovery process of the power grid frequency after tripping equivalent capacity of hydropower units, wind power units, thermal power units, and low-capacity hydropower units is presented in Fig. 4.

The simulation results in Fig. 4 show that minimum frequency deviation is achieved by tripping the wind turbine generator, and it recovers with the highest speed. The deviation of cutting small-capacity hydropower units, thermal power units, and hydropower units is gradually increasing.

When the wind power generator participates in the frequency regulation, the frequency recovery

process after cutting equivalent capacity units is presented in Fig. 5.

The simulation results show that the minimum frequency deviation is achieved by tripping the wind turbine generator, and it recovers with the highest speed. In addition, the recovery process is better than wind power, which does not participate in the frequency regulation because of the involvement of the wind power units in frequency regulation. The effects of generator tripping of different units with the participation of wind power generation in the regulation and without the participation of wind power generation are presented in Fig. 6.

The simulation results show that the participation of wind power can make the frequency recovery effect better than that when the wind power

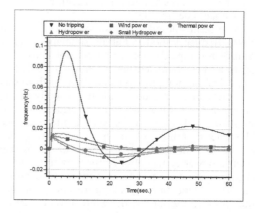

Figure 5. Frequency recovery process after generator tripping when the wind power participates in frequency regulation.

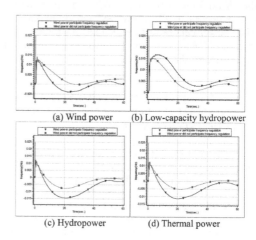

(a) Wind power   (b) Low-capacity hydropower

(c) Hydropower   (d) Thermal power

Figure 6. Effect of generator tripping of different units.

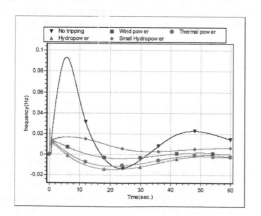

Figure 4. Frequency recovery process when wind power is not involved in frequency modulation.

does not participate in the regulation after cutting the wind power units, hydropower units, or thermal power units.

6 CONCLUSION

The participation of the wind power makes the recovery effect better than when the wind power does not participate in the regulation, regardless of the type of generator tripping.

The sequence of the generator tripping does not change either with the wind power participating in the frequency regulation or without the wind power. In view of the frequency recovery effect, the sequence of generator tripping is wind power, low-capacity hydropower, thermal power, and hydropower.

Because of the complexity of the startup and shutdown operation of the thermal power units, and the additional fuel cost and the vulnerability of the equipment during the startup and shutdown process, the final optimal generator tripping sequence is wind power, low-capacity hydropower, thermal power, and hydropower.

REFERENCES

Cao Jun. 2009. Frequency Control Strategy of Variable-speed Constant-frequency Doubly-fed Induction Generator Wind Turbines[J]. *Automation of Electric Power Systems* 33(13): 78–82.

Dong Zhe. 2016. A Transient Tripping Decision-Making Model for Sending-Out Wind Power System[J]. *Power System Technology* 5(40): 1348–1354.

GBT 31464-2015. 2007. Operation Criterion of Power Network[S]. Beijing: China Electric Power Press.

GBT-19963-2011. 2009. Technical Rule for Connecting Wind Farm to Power System.

Guan Longliang. 2007. Simulation on Frequency Control of Doubly Fed Induction Generator Based Wind Turbine[J]. *Automation of Electric Power Systems* 31(7): 61–65.

He yong. 1992. Analysis of Power-Frequency Dynamics in Large Scale Multi-Machine Power Systems[J]. *Automation of Electric Power Systems*: 28–33.

Jiang Wang, Lu Jiping. 2014. Research on Probabilistic Model of Droop Control Coefficient of Grid-Connected Wind Farm[J]. *Power System Technology* 38(12): 3431–3435.

Jiang Wang. 2015. Research on Frequency Response and Control Strategy of Wind Farm Integrated Power Systems[D]. Chongqing: Chongqing University.

Lalor, G., A. Mullane, M. O'Malley. 2005. Frequency Control and Wind Turbine Technologies[J]. *IEEE Transactions on Power System* 20(4): 1905–1913.

Li Zheng. 2012. Overview of Isolated Power System Frequency Stability Control[J]. *East China Electric Power* 7(40): 1173–1177.

Ma Zhiheng. 2015. Analysis and Strategy Research of Frequency Stability of Asynchronous Interconnection[D]. Zhejiang University.

Zhang Dan. 2013. The Optimization of Regional Power Grid Stability Control and High Frequency Generator-tripping[J]. *Yunan Electric Power* 3(41): 19–21.

Research on the credit risk of direct trade and countermeasures of large power users

Chao Ma & Lin Guo
Chongqing Electric Power Trade Center, Chongqing, China

Zhicheng Yu
Electric Power Research Institute, State Grid Chongqing Electric Power Corporation, Chongqing, China

Lei Wang
State Grid Electric Power Research Institute, Nanjing, China
Beijing Kedong Power Control System Co. Ltd., Beijing, China

Zhiqiang Zhao
Chongqing Electric Power Trade Center, Chongqing, China

Guojun He
Electric Power Research Institute, State Grid Chongqing Electric Power Corporation, Chongqing, China

ABSTRACT: With the step-by-step development of the unified national electricity market, direct transactions between generation enterprises and large users have been more and more frequent. In this paper, direct trade credit system, evaluation index system, evaluation method, and the corresponding countermeasures aiming to large users are introduced. These can clear the two responsibilities, supervise contract execution, and ensure a smooth operation of the grid. It is highly meaningful to promote the transaction according to the research presented in this paper.

1 INTRODUCTION

With the development of direct transactions, the large power users will no longer be charged by power grid. Thus, the risk of arrears will reduce. And at the same time, the large power users have to sign "agreement about the sale and purchase of electricity" with the generation enterprises. This agreement, on the basis of their predictable credit, can restrict and regulate both sides. However, from another perspective, it is difficult for the bad credit users to find transaction object. Thus, power grid company can only supply power to small users or bad credit users, making it more difficult to collect the electricity fees. With constant descending of the credit of user group, it will be more difficult to operate power grid company. The large users' credit evaluation system, aiming to motivate users to improve their credit, will evaluate their credit and then adopt different service strategies according to their different credit conditions. This paper deals with the large users' credit evaluation index system, evaluation methods, and countermeasures.

2 CONSTRUCTION OF EVALUATION SYSTEM ON DIRECT TRANSACTIONS CREDIT RISK

Specifications such as electricity fees, quantity of electricity, and fund are the key consideration of customer credit status. In countries with higher level of credit management, enterprise and individual credit records, collected by the national credit system, can be queried by their collaborators. It is evident that this cooperation based on information symmetry greatly reduces the credit risk of the seller. However, our national credit system is short of unified standard and not much mature at present. Some organizations like the tax authorities and the financial sector set up user credit record index on the basis of their own assessment need. Indicators such as the current actual contribution rate, the cumulative contribution rate, the actual power rate, electricity consumption proportion, electricity charge return, and current fund occupied rate are based on the actual situation of power grid company. The current stage can be defined by appraising personnel The default is usually the period of January 1st to the evaluating day. The large users' credit management index system is shown in Figure 1.

2.1 Technical index

2.1.1 Current actual contribution rate
The current actual contribution rate = current real electricity charge/current receivable electricity fees

It counts monthly in the current period. This index is quantitative and can reflect the user payment. In general, the higher this index value, the higher the score and better the credibility.

2.1.2 Cumulative contribution rate
The cumulative contributionrate =
The real cumulative electrcity charge / The receivable cumulative electricity fees

This index counts monthly and reflects the contribution situation, including the part of default. The receivable cumulative electricity fees are equivalent to the receivable electricity fees in this stage and last year. This stage represents the period from the beginning of the year to the statistical month. Similarly, the higher this index value, the higher the score and better the credibility.

2.1.3 Actual power rate
The actual power rate =
$$\frac{\sum_{i=1}^{n} \frac{\text{The actual electricity consumption in month i}}{\text{The plan electricity consumption in month i}}}{n}$$

This index reflects the ability of user's electricity consumption. Similarly, the higher this index value, the higher the score and better the credibility.

2.1.4 Electricity consumption proportion
Electricity consumption proportion =
The actual electricity consumption in this stage / All users' actual electricity consumption in this stage

This index reflects the proportion of users' electricity consumption. Similarly, the higher this index value, the higher the score and better the credibility.

2.1.5 Electricity charge return
Electricity charge return =
$$\frac{\text{The receivable electricity fees} - \text{The purchase price} \times \text{Electricity consumption}}{\text{The purchase price} \times \text{Electricity consumption}}$$

This index reflects users' contribution rate for the profit of power supply bureau. The higher this index value, the higher the score and better the credibility.

2.1.6 Current fund occupied rate
This stage fund occupied rate
$$= \frac{\sum_{i=1}^{n} \frac{\text{Fund occupied ratethe in ith day}}{\text{Receivable electricity fees in ith day}}}{n}$$

This index reflects the loss of power supply bureau due to users' default. Similarly, the higher this index value, the higher the score and better the credibility.

2.2 Deviation index

The large users' actual electricity consumption usually deviates from monthly or yearly planning trading electricity. This is the meaning of deviation. Those deviation indexes can reflect the implementation of the direct transaction contract.

2.2.1 Electricity deviation ratio
Elextricity deviation ratio =
$$\frac{\text{Large users' actual power consumption} - \text{Plan trading power consumption}}{\text{Plan trading power consumption}}$$

This index reflects the deviation between large users' actual power consumption and planning trading power consumption. The higher this index absolute value, the higher the score and better the credibility.

2.2.2 Days of over 3% electrical deviation
Days of over 3% electrical deviation =
Days of over 103% plan consumption daily + Days of below 97% plan power consumption daily

This index reflects the number of days that large users' power consumption greatly deviate from the plan consumption. The less the number of days, the better the credibility in a statistical period.

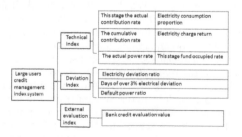

Figure 1. Large users' credit management index system.

Table 1. Index system reflecting the financial credit.

Quantitative indexes (X1 score)		Qualitative indexes (X2 score)
Index category	Basic index	Evaluation index
Recovery of electricity price (X11)	Recovery rate of electricity price (a11 score) The number of accumulated default (a12 score) The amount of default (a13 score)	1. The recent result of corporate credit level (X21 score)
Debt paying ability (X12)	Asset-liability ratio (b11 score) Cash liquidity ratio (b12 score) Liquidity ratio (b13 score)	2. Corporate reputation index (X22 score) 3. The share of product market (X23 score)
Profitability (X13)	Total assets net profit margin (c11 score) Net Profit Margin on Sales (c12 score)	4. The basic qualities of leadership (X24 score)
The growth ability (X14)	Sales growth rate (d11 score) Net profit growth rate (d12 score) 3 years' average profit growth rate (d13 score)	5. Management and development strategy (X25 score)

2.2.3 *Default power ratio*

Default power ratio means the proportion that default power consumption accounted for total direct transaction electricity in a period. Default power consumption is the sum of over 103% and below 97% trading power consumption. This index reflects the proportion of default power consumption and the lower this index value, the higher the score and better the credibility.

2.3 *External evaluation index*

Bank credit evaluation value: This index, users' credit evaluated by banks, refers to banks' evaluation and reflects users' finance. The higher this index value, the higher the score and better the credibility. The index system reflecting the financial credit is shown in Table 1.

3 THE METHOD OF EVALUATING DIRECT TRADING CREDIT

The Analytic Hierarchy Process (AHP), based on resolving elements related to the decision making into objectives, standards, and programs, is an effective method of qualitative and quantitative analysis. It was introduced by professor Thomas L. Saaty from the University of Pittsburgh in the 1970s when he applied network system theory and the multiple objective comprehensive evaluation method to study the distribution of electric power according to each industrial sector's contribution. AHP has high reliability because it helps to capture both subjective and objective aspects of a decision and is especially suitable for lesser indexes just as this paper.

4 APPLICATION OF EVALUATING USERS' CREDIT

There are four levels: AAA, AA, A, and B. It is necessary to make the levels keep flexibility due to the difference of users' specification, development potential, and external evaluation. Therefore, a Hunan Power Grid Corp staff was invited to set up the adjustable proportion that the four levels take up respectively. The evaluation value of two levels at the critical point needs comparison to make the levels more rational. If the value of the first user in the next level equals the value of the last user at the next higher level, then classify the first user in the next level as the next higher level. For example, if the value of the first user belonging to AA equals the value of the last user belonging to AAA, then classify the first user in AA level as the next higher AAA level. Finally, users in different levels will have different services.

4.1 *Management recommendations for AAA*

AAA means high-quantity users who can conscientiously carry out the contract on power supply and demand, pay electricity fees timely, handle the relevant procedures according to the regulations, and actively cooperate with the power supply department.

The purpose of serving these users is to keep the relationship with power supply and improve their satisfaction.

4.2 *Management recommendations for AA*

These users have both value and credit issues. When treating with them, the principle is that pushing them to pay electricity fees timely, making a supplementary payment for default, and improving their credit.

4.3 *Management recommendations for A*

These users are the enterprises banned or restricted by country. They hinder the development of power industry and occupy the fund to some extent. They have poor prospects for development and always come out of the default. However, it is an opportunity to take back electricity fees after exhortation. The purpose of treating these users is to call back the default as soon as possible for reducing the loss of power plants and regulate their electricity consumption to safeguard the interests of power enterprises.

4.4 *Management recommendations for B*

These users have the bad credit because they not only have the massive malicious default but also have difficulty in operation and turnover of capital. It is a serious threat to the production operation of the power supply company due to the large default even though they are just a small part of all users. Therefore, the power plants should punish them when incentive measures fail to work.

5 CONCLUSION

In this paper, the factors that may influence a large user credit, large user credit index system, and corresponding application method are proposed. The purpose of setting up different service strategies for users with different credit ratings is to retrieve the default situation, reduce the loss of power enterprises, motivate users to enhance their credit, and then gradually improve the industry's credit consciousness.

REFERENCES

Deng Xue, Li Jiaming, Zeng Haojian, Chen Junyang, Zhao Junfeng. Weight calculation method and application research based on Score analysis method. [J]. Practice and understanding of Mathematics, 2012, 07:93–100.

Sun Bo, Shi Quansheng. Analysis of the price design model of direct purchase of large consumers under the profit balance[J]. East China Electric Power. 2013, 41(1): 47–49.

Tan Zhongfu, Ju Liwei, Chen Zhihong, Xing Tong, Chen Kun, Zhao baozhu. Optimal model of electricity purchasing decision for power grid enterprises based on conditional risk value model[J]. East China Electric Power, 2013, 06:1296–1301 K. Elissa, "Title of paper if known," unpublished.

Wang Beibei, LIU Xiaocong, LI Yang. A study on the daily scheduling and operation simulation of large capacity wind power access considering user side interaction[J]. Proceedings of The Chinese Society for Electrical Engineering, 2013, 33(22): 35–44.

Xia Qing, Bai Yang, Zhong Haiwang, Chen Qixin. The system design and suggestions for the promotion of direct purchase of electricity by large consumers in China[J]. Automation of Electric power systems, 2013, 20, 1–7.

Zhang Zongyi, Kang Yali, Guo Xinglei. Score analysis of combined model of direct purchase of electricity by large consumers based on spectral risk measure[J]. Transactions of China Electrotechnical Society, 2013, 01: 2013, 01:266–270+284.

Zhou Ming, Zheng Yanan, Li Gengyin. Integrated allocation method for fixed cost of large customer transfer cost considering time sharing price and power quality[J]. Proceedings of The Chinese Society for Electrical Engineering, 2008, 28(19): 125–130.

Electromechanical Control Technology and Transportation – Jia & Wu (Eds)
© 2017 Taylor & Francis Group, London, ISBN 978-1-138-06752-3

Application and simulation of active disturbance rejection technology in distributed power supply control

Fei Ou, Hejin Xiong & Deming Lei
School of Automation, Wuhan University of Technology, Wuhan, China

ABSTRACT: The control strategy of active disturbance rejection control which is used in the automatic control system is practical and effective. The technology of active disturbance rejection control is shown to have many advantages, such as a simpler implementation, higher precision, and better dynamic performance than traditional control technology through a large number of theoretical studies and practical engineering applications. The structure and working principles of the Active Disturbance Rejection Controller (ADRC) are introduced in this paper. The active disturbance rejection controller is used in the distributed power supply, and we made a simulation in MATLAB/Simulink in order to compare it to the conventional control method. The simulation results showed that this control strategy could well control the system of distributed power supply.

1 INTRODUCTION

A distributed power supply (DG) is a set of low-capacity generator units configured near the user's site in order to meet the special needs of individual customers or to support large-scale economic operation of the distribution network. The technology of distributed power grid has attracted more and more attention worldwide as a useful supplement in the system-centralized power generation, which is mainly determined by its economy, environmental protection, flexibility, and unique peak shaving effect.

The control strategy of active disturbance rejection control is based on the classical control theory and the modern control theory proposed by Han Jingqing in the 1990s. It is a new nonlinear robust control technology, which does not depend on the precise mathematical model of the controlled objects. The core of the ADRC is the extended-state observer, which can be used to observe the system uncertainties, including system modeling error and the system's internal and external disturbance. The extended-state observer compensates the uncertainties of the system by the error-state feedback, and it has strong robustness and model adaptability.

In this paper, the active disturbance rejection controller is applied to the system of distributed power control and the simulation is carried out.

2 ACTIVE DISTURBANCE REJECTION CONTROL

The ADRC consists of three parts: Tracking Differentiator (TD), Extended-State Observer (ESO),

and Nonlinear State Error Feedback (NLSEF). Z11 is the transition process of the system input; z12 is the differential of z11; z21, z22, and z23 are state observations; u is the controlled quantity; y is system actual output; and μ is the synthesis of all system disturbances, as shown in Figure 1.

TD has good differential characteristics that help arrange the transition process and track the input signal quickly and without overshoot to avoid the mutation of the set value. The drastic change of the controlled variables and the overshoot of the system output quantity greatly solve the contradiction between the quick response and overshoot of the system. It is subject to certain restrictions when ADRC is used in the occasions with high rapidity. All types of internal or external factors of the system disturbance down to the system disturbance by ESO are core part of ADRC. The internal and external perturbations and the state variables of the system are estimated by ESO to give the corresponding compensation so as to realize the dynamic feedback linearization of the system. The output of TD is compared with the

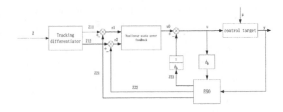

Figure 1. ADRC control block diagram of the principle.

ESO observation to obtain the error of the system state variable. The error amount is entered into the NLSEF to sum the compensation variable from the ESO so as to obtain the controlled variable of the controlled object. The design of the ADRC controller can be greatly simplified without considering the linearity or nonlinearity, time-variance or time-invariant in the controller design because the ADRC divides the object according to the time scale of the system.

The following is the second-order ADRC equation:

TD equation:

$$\begin{cases} \dot{Z}_{11} = Z_{12} \\ \dot{Z}_{12} = f(Z_{11}, Z_{12}, r, h) \end{cases} \quad (1)$$

ESO equation:

$$\begin{cases} \xi = Z_{11} - i \\ \dot{Z}_{21} = Z_{22} - \beta_1 f(\xi, \alpha_1, \delta_1) + b_0 u \\ \dot{Z}_{22} = -\beta_2 f(\xi, \alpha_1, \delta_1) \end{cases} \quad (2)$$

NLSEF equation:

$$\begin{cases} \xi_1 = Z_{11} - Z_{12} \\ u_0 = \beta_3 f(\xi, \alpha_2, \beta_2) \\ u = u_0 - Z_{22}/b_0 \end{cases} \quad (3)$$

$$f(Z_{11}, Z_{12}, r, h) = -rsng\left(Z_{11} - h + \frac{Z_{12}|Z_{12}|}{2r}\right) \quad (4)$$

$$f(\xi, \alpha, \beta) = \begin{cases} |\xi|^a sng(\xi) & |\xi| \geq \delta \\ \xi/\delta^{1-\alpha} & |\xi| < \delta \end{cases} \quad (5)$$

where α1, 2, δ, δ1, δ2, β1, β2, β3, ξ, and b0 are adjustable parameters. The nonlinear function (f) is used to arrange the transition process, and r is the speed factor; if r is larger, the tracking speed is higher; and h is the step size. The control performance of ADRC mainly depends on the rational selection of parameters, and the adjustment of parameter mainly depends on the designer's engineering experience and the continuous use of simulation. At present, the method of ADRC parameters adjustment can be divided into two steps: the TD/ESO/NLSEF were regarded as three independent parts. First, the parameters of TD and ESO were configured until the two parts are adjusted to satisfactory results and the system parameters can be set up finally. Second, the whole system parameters of ADRC are set up combined with NLSEF. In this paper, a new tuning method is proposed to use ADRC in the distributed power

supply; the ADRC is applied to the tracking control of current, and ESO is used to observe the unknown load disturbance. The estimation and compensation toward the timely and accurate change of load can effectively suppress the impact of various disturbances.

3 DISTRIBUTED POWER CONTROL MODEL

DG distributed power supply topology was composed of four parts, including the DC capacitor, three-phase full-bridge, the output filter, and a controller. As shown in Figure 2, the DC voltage is replaced with a DC voltage source. LC filter can suppress the harmonic current, which is configured between the full-bridge output and the machine side.

Figure 3 shows a detailed block diagram of the controller section shown in Figure 2. The controller can track the grid-connected current through the ADRC control under the synchronous rotating coordinate system by sampling the inductor current and the terminal voltage. In addition, this is the idea of using no phase-locked loops. It is not necessary to position the d-axis on the a-axis of the voltage vector as long as the coordinate transformation used in the controller uses the same rotation system, so that we can remove the PLL. This is a grid-based power-tracking reference current generation algorithm based on phase-locked loop current detection technology.

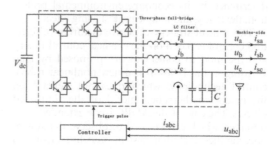

Figure 2. Topological structure of DG.

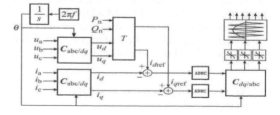

Figure 3. Control block diagram of DG in synchronous rotating frame.

The following is the generation of the tracking reference current:

$$u = \begin{bmatrix} u_d \\ u_q \\ u_0 \end{bmatrix} = \begin{bmatrix} \dfrac{\sqrt{6}}{2} V_m \cos(\varphi_u - \theta_0) \\ \dfrac{\sqrt{6}}{2} V_m \sin(\varphi_u - \theta_0) \\ 0 \end{bmatrix} \qquad (6)$$

Similarly, for the transformed current vector i:

$$i_t = \begin{bmatrix} i_d \\ i_q \\ i_0 \end{bmatrix} = \begin{bmatrix} \dfrac{\sqrt{6}}{2} i_m \cos(\varphi_u - \theta_0) \\ \dfrac{\sqrt{6}}{2} i_m \sin(\varphi_u - \theta_0) \\ 0 \end{bmatrix} \qquad (7)$$

Suppose DG's rated operating powers are Pn and Qn:

$$\begin{cases} P_n = 1.5 V_m I_m \cos(\varphi_u - \varphi_i) = u_d i_d + u_q i_q \\ Q_n = 1.5 V_m I_m \sin(\varphi_u - \varphi_i) = u_d i_d - u_q i_q \end{cases} \qquad (8)$$

From formula (8), the reference current is written as:

$$\begin{cases} i_{dref} = \dfrac{u_d P_n + u_q Q_n}{u_d^2 + u_q^2} \\ i_{qref} = \dfrac{u_q P_n - u_d Q_n}{u_d^2 + u_q^2} \end{cases} \qquad (9)$$

4 SIMULATION AND EXPERIMENTAL RESULTS

In the following, we will use Simulink in MATLAB to simulate the above control method and compare the simulation results of the PI control method to those of the ADRC control method. Finally, we can reach a conclusion according to the simulation results.

The parameters of the auto-disturbance rejection controller are determined, and we obtain a set of better controller parameters.

For the Tracking Differentiator (TD): $a_{01} = 0.23$, $a_{02} = 370$, $a_{03} = 5.2$, r = 3000;

For the Extended-State Observer (ESO): $\beta_1 = 34$, $\beta_2 = 530$, $\beta_3 = 170$, $\beta_4 = 45$, d = 0.0025, b = 18, $\delta_1 = 0.5$, $\delta_2 = 0.25$, $\delta_3 = 0.125$;

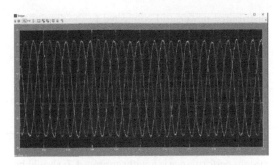

Figure 4.   PI controls the current output of the distributed power supply.

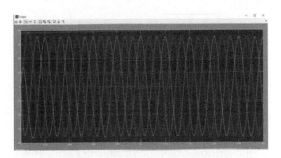

Figure 5.   Active disturbance rejection control of current output of distributed power supply.

For the nonlinear state error feedback (ESLEF): $a_1 = 1$, $a_2 = 43$, $a_3 = 1$, $k_p = 0.55$, $k_i = 0.16$, $k_d = 0.1$, d = 5000

A comparison of the use of ADRC control in distributed power and the use of PI control in distributed power supply from simulation results is shown below.

Comparing Figures 4 and 5, we know that the distributed power supply grid-connected current is symmetrical, the harmonic content is very small, and the output is relatively stable. It can be seen that the distributed power supply under the control of the self-disturbance rejection method can effectively improve the power quality of the microgrid operation after the grid is connected.

5 CONCLUSION

In this paper, the use of an active disturbance rejection controller in distributed power supply is described, which can effectively control the harmonics and convergence of the distributed power supply. Distributed power generation technology is an important part of the power industry in implementing energy transformation in order to promote today's sustainable development. The research of the different topologies and control strategies of the distributed power grid connected to the distributed generation system has great practical significance in that it cannot only strengthen the harmonic control and optimize the power quality, but also improve the reliability and stability of the system, which is conducive to its healthy development. The simulation results show that the active disturbance rejection controller significantly improved the performance of the system compared to the conventional PI controller; the stability time is shorter, the steady-state accuracy is higher, the robustness is higher, and the system performance is greatly improved. It should be noted that the results of the simulation experiments are not directly applicable to those of the actual test procedure, and further work is still on the way.

ACKNOWLEDGMENT

This paper was supported by the National Natural Science Foundation of China (No. 61573264).

REFERENCES

Jinlan Cui and Tianqi Liu. Distributed generation and its grid interconnection issue[J]. Modem Electric Power, 2007, 24(3):53–57.

Kim KS and Rew KH. Reduced order disturbance observer for discrete-time linear systems[J]. Automatica, 2013, 49(4): 968–975.

Kim Seul-Ki, Jeon, Jin-Hong, Cho and Chang-Hee. Modeling and simulation of a grid-connected PV generation system for electromagnetic transient analysis[J]. Solar Energy, 2009, 83(5):664–678.

Lasseter RH. Microgrids distributed generation[J]. Journal of Energy Engineering, 2007, 133(3): 144–149.

Ming Ding and Min Wang. Distributed generation technology[J]. Electric Power Automation Equipment, 2004, 24(7): 31–36.

Sawant RR and Chandorkar MCA. multi-functional four-leg grid connected compensator[J]. IEEE Trans on Industry Applications, 2009, 45(1): 249–259.

Wang Chengshan, Xiao Zhaoxia and Wang Shouxiang. Synthetical control and analysis of microgrid[J]. Automation of Electric Power Systems, 2008, 32(7): 98–103.

Yang Huan, Zhao Rongxiang and Cheng Fangbin. Delay compensation for harmonics detection based on synchronous reference frame without phase lock loop[J]. Proceedings of the CSEE, 2008, 28(27): 78–83.

Zhao C and Huang Y. ADRC based input disturbance rejection for minimum phase plants with unknown orders and/or uncertain relative degrees[J]. Journal of Systems Science and Complexity, 2012, 25(4): 625–640.

Zheng zeng, Huan Yang and Rongxiang Zhao. Multi—function Grid—connected Inverter and Its Application in Microgrid. Automation of Electric Power Systems, 2012, 36.

Zhiqun Wang, Shouzhen Zhu and Shuangxi Zhou et al. Study on location and penetration of distributed generations[J], Proceedings of the CSU-EPSA, 2005, 17(1): 53–58.

Research on vector control system of SMC material transverse-flux PMSM

Bo Cui & Yanliang Xu
Department of Electrical Engineering, Shandong University, Jinan, China

Yun Zhang & Wei Yin
Automation Research Institute, Shandong Academy of Sciences, Jinan, China

ABSTRACT: This paper designed a vector control system of a soft magnetic composite transverse-flux Permanent Magnet Synchronous Motor (PMSM) based on TMS320F28335. First of all, this paper analyzed the mathematical models of PMSM, the principle of vector control and the design method of digital PID controllers. Then the hardware circuit design scheme and the software flow charts of the vector control system were presented. Finally, the experimental research of the control system was carried out on the experimental platform. All the experimental results verified the feasibility and practicability of the design scheme.

1 INTRODUCTION

The Permanent Magnet Synchronous Motor (PMSM) has advantages of simple structure, small volume, high power density, small torque ripple, high efficiency, etc. [1]. At present, PMSM control system has been widely used in many high accuracy control field, such as aerospace, CNC machine tool, electric vehicle and so on.

The structure of PMSM is similar to that of the traditional winding type synchronous motor. The difference between the two kinds of synchronous motor lies in that the rotor excitation of PMSM is provided by permanent magnet instead of the field winding. Since PMSM is a kind of AC motor, it is a multi-variable, nonlinear and strong coupled system which results in the complexity and difficulty of its precise control [2]. Benefited from the vector control theory developed in recent years, the control performance of the PMSM has become comparable with that of DC motor.

This paper designs a vector control system of a SMC transverse flux permanent magnet synchronous motor based on TMS320F28335. The mathematical model of PMSM, the principle of vector control and the design method of digital PI controller are firstly analyzed. Then the hardware circuit design scheme and software flow chart of the vector control system are presented. Finally, the experimental research is carried out and the

feasibility and practicability of design scheme are verified by experimental results.

2 VECTOR CONTROL OF PMSM

2.1 *Mathematical model of PMSM*

As mentioned, PMSM is a multi-variable, nonlinear and strong coupled system and its voltage and flux-linkage equations in the three phase stationary coordinate system are complex. In order to simplify the mathematical model of PMSM, the following assumptions are usually made [3]: (1) The three-phase winding of the stator are completely symmetrical and Y connected; (2) The permanent magnet of the rotor generates the main magnetic field and there is no damper winding on the rotor; (3) Hysteresis loss and eddy current loss of the core are all ignored; (4) The resistance and inductance of motor winding are constant.

There is no electromagnetic coupling between the two current components of i_d and i_q due to the orthogonal distribution of them after coordinate transformation. The decoupling control of PMSM can be achieved by controlling the magnetizing current component i_d and the torque current component i_q respectively.

The position of different coordinate systems is shown in Figure 1.

The mathematical models of PMSM in $dq0$ synchronous rotating coordinate system are as follows:

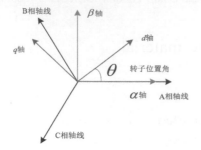

Figure 1. Relative position of different coordinate systems.

1. The voltage equations can be expressed as:

$$\begin{cases} u_d = R_s i_d + L_d p i_d - \omega L_q i_q \\ u_q = R_s i_q + L_q p i_q + \omega(L_d i_d + \psi_f) \end{cases} \quad (1)$$

where u_d = d-axis stator voltage; u_q = q-axis stator voltage; i_d = d-axis stator current; i_q = q-axis stator current; R_s = phase resistance of stator; ω = electrical angular velocity; L_d = d-axis equivalent resistance; L_q = q-axis equivalent resistance; ψ_f = the rotor coupling flux linkage on the stator; P = differential operator.

2. The flux linkage equations can be expressed as:

$$\begin{cases} \psi_d = L_d i_d + \psi_f \\ \psi_q = L_q i_q \end{cases} \quad (2)$$

where ψ_d = d-axis flux linkage of stator; ψ_q = q-axis flux linkage of stator.

3. The electromagnetic torque is expressed as:

$$\begin{aligned} T_e &= 1.5 p_p(\psi_d i_q - \psi_q i_d) \\ &= 1.5 p_p[\psi_f i_q + (L_d - L_q)i_d i_q] \end{aligned} \quad (3)$$

where T_e = electromagnetic torque; P_p = pole pairs of the stator.

4. The motion equation can be expressed as:

$$T_e - T_l = J\frac{d\omega_r}{dt} + R_\Omega \omega_r \quad (4)$$

where ω_r = mechanical angular velocity; T_l = load toque; J = moment of inertia; and R_Ω = the damping coefficient.

Tnhe $i_d = 0$ current control method is adopted in this paper and the electromagnetic torque is simplified as:

$$T_e = 1.5 p_p \psi_f i_q \quad (5)$$

It can be found from (5) that ψ_d becomes constant and T_e is proportional to i_q taking into account of the $i_d = 0$ current control method, as a result, the control performance of PMSM can be comparable with that of DC motor.

2.2 *Principle of vector control*

The schematic diagram of PMSM vector control is shown in Figure 2. The system is composed of the following functional modules: PMSM, three phase inverter, PID controller, coordinate transformations, etc. [4].

The working principle is as follows: as shown in Figure 2, it is a double-loop system with the speed loop and current loop being the outer loop and inner loop, respectively. The feedback speed ω and rotor position angle θ are obtained by resolver for speed PID controlling and coordinate transformation. The input signal of speed PID controller is the deviation of the given speed ω^* and the feedback speed ω and its output is set as the given current i^*_q of the current i_q PID controller. Meanwhile, in order to achieve the $i_d = 0$ current control strategy, the given current i_d of the current i_d PID controller is set as 0. The feedback current i_A, i_B and i_C are detected by current sensors and can be transformed into i_q, i_d through Clarke and Park transformation. The output of the two current PID controllers u^*_d and u^*_q are transformed into u^*_α and u^*_β which can be set as the reference voltage of SVPWM functional module. The six PWM signals generated by the SVPWM functional module can control the switching status of the inverter.

2.3 *Design of digital PID controller*

PID controller is widely used in a lot of applications [5] due to its excellent characteristics, such as simple algorithm, good robustness as well as strong anti—interference ability.

The three PID controllers of this control system are all discrete. The control strategy is as follows:

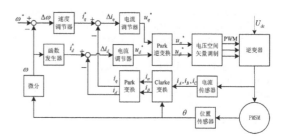

Figure 2. Schematic diagram of PMSM vector control.

$$u(k) = K_p e(k) + K_i \sum_{j=0}^{k} e(j) \qquad (6)$$

where K_p = proportional coefficient; K_i = the integral coefficient; $e(k)$ = deviation at time of k; and $u(k)$ = the output of PID controller at the time of k.

This paper adopts the PID control algorithm that can prevent the integral saturation phenomenon. The current PID controller is taken into account as an example to elaborate on the control principle:

$$e(k) = i_k^* - i_k$$
$$U_k = X_{k-1} + K_p e(k)$$

$$U_{out} = \begin{cases} U_{max} & U_k > U_{max} \\ U_k & \\ U_{min} & U_k < U_{min} \end{cases} \qquad (7)$$

$$X_k = X_{k-1} + K_i e(k) + K_c (U_{out} - U_k)$$

where i_k^* = given current at time k; i_k = the feedback current at time k; X_{k-1} = actual output of PID controller at time $(k - 1)$; X_k = actual output of PID controller at time k; U_{max} = positive amplitude limiter of integrator; U_{min} = negative amplitude limiter of integrator; U_{out} = the theoretical output of integrator; $K_c = K_i / K_p$ = the integral amending coefficient which can compensate the accumulated integral value using limit deviation and limit integral saturation time.

3 HARDWARE DESIGN

The control system is composed of the hardware circuit and software system. The hardware circuit is the running platform of motor control algorithm and also has functions of signal detection and hardware protection. Figure 3 shows the hardware

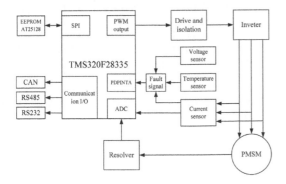

Figure 3. Hardware structure of the control system.

structure of this vector control system. The hardware circuit is composed of the following functional modules: main control circuit, power circuit, signal conditioning circuit, communication circuit and protection circuit.

3.1 Main control circuit

The system uses the DSP TMS320F28335 of Texas Instruments as the main control chip. The chip is a high-performance 32 bit digital signal processor with 6.67 ns instruction cycles and 150 MHz dominant frequency. It can fully meet the requirements of SMC motor's high-frequency operation and can realize the stability and rapidity of the control system.

In addition, there exist several key parameters or variables needed to be saved in non-volatile memory in the running process of the system. This paper uses EEPROMAT25128 of ATMEL Company to save the parameters and variables.

3.2 Power circuit

The power circuit mainly includes the power supply circuit of inverter, the power-driven circuit and the inverter circuit.

1. In order to simplify the design of the system and reduce the electromagnetic interference, the 144 V regulated power supply is directly used as the power supply device of the inverter.
2. The functions of the power-driven circuit lie in the following two points: on the one hand, it can amplify the PWM driving signal and strengthen its driving ability. On the other hand, it can also isolate the weak electricity system on the DSP side from the strong electrical system on the inverter side in order to reduce the EMF and protect the main control circuit. The driver IC IR2101 of IR company is used in this system and it takes great importance in PWM signal driving and isolation.
3. The inverter circuit has a three-phase full-bridge topology which is shown in Figure 4. After all the comprehensive consideration of voltage level, drain current, conducting resistance, gate capacitance and other factors, the MOSFET IRFP4568PbF is used in the inverter circuit.

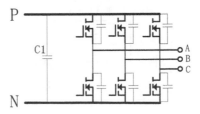

Figure 4. The structure topology of inverter.

And the diodes integrated inside the MOSFET are used as the freewheeling device in order to reduce the cost. In Figure 4, C_1 represents the electrolytic capacitor and the high frequency non-inductive capacitance that can stabilize the voltage and absorb the surge voltage during the switching process of MOSFET.

For the purpose of reducing the surge voltage and extending the service life of switching devices, the absorption circuit is designed, which can be divided into three types: type C, type RC and type RCD. In which, type C absorption circuit is the simplest one. Therefore, this system uses the type C absorption circuit in which the parallel capacitor can slow the rising speed of the surge voltage and achieve zero voltage turning-off.

The requirements for MOSFET absorption circuit are as follows: (1) The wiring inductance of the main circuit is small enough; (2) Adopt the low-inductance absorption capacity and make absorption capacitor parallel with MOSFET.

3.3 Signal detecting and conditioning circuits

Signal detecting and conditioning circuits are composed of temperature detecting circuit, phase current detecting and conditioning circuit, DC bus voltage detecting and conditioning circuit as well as resolver excitation and output signal conditioning circuit.

1. Hall temperature sensor is used in the temperature detecting circuit as shown in Figure 5.
   As shown in Figure 5, the output signal of hall temperature sensor is pull up to 3.3 V by VCC, R_1 and R_2 and then limited to a range of (0–3.3) V by diode MMBD1505A.
2. Three high-precision hall current sensors are used as the sampling device of the three phases current to avoid the sampling error caused by current phase deviation. The system uses CSM300LTA closed-loop hall current sensor with rated input current of 300A and rated output current of 100mA.
   The requirement for bus current measurement accuracy is relatively low since the detect-

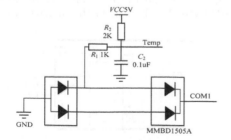

Figure 5. Temperature detecting circuit.

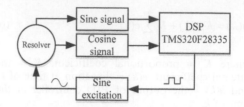

Figure 6. Topology of the decoding circuit.

ing is used in the over-current protection. As a result, the open-looped hall current sensor CS300N is used.
   The main function of current conditioning circuit is to filter out the high frequency interference in the sampled signal and modulate the voltage amplitude of the analog signal to a range of (0–3) V.
3. The function of DC bus voltage circuit is to realize over-voltage and under-voltage protections and hall voltage sensor is used for this purpose.
4. Compared with encoder, resolver has advantages of low cost, high reliability, easy maintenance and strong anti-interference ability. Based on which, the system uses sine-cosine revolver BRT as the rotor position detecting device. Sine-cosine resolver is a kind of special rotor position detection device and its output signals are two analog voltages which have sine and cosine relationship with rotor position angle. The rotor position angle can be obtained after decoding from the analog voltages.
   At present, the most common decoding method is to use special decoding IC, such as AD2S1200 with high integration and high reliability, but the cost is high. In order to reduce the cost of production, this paper designed a new decoding method using DSP and signal conditioning circuit. The topology of the decoding circuit is shown in Figure 6.
   The basic working principle is as follows: The PWM signals with duty cycle 50% and frequency 10 kHz generated by TIME0 of DSP is transformed into sine signal to excite the resolver. Then the output analog signals of the resolver are limited to a range of (0–3) V and sampled by ADC port of DSP.
5. CAN bus communication circuit, RS232 circuit and RS485 circuit are designed in this system so as to ensure the reliable and quick communication.

4   SOFTWARE SYSTEM DESIGN

Since the vector control algorithm is realized in the software system. In order to meet the requirements of real time, reliability and easy maintainability, the following optimization methods are adopted during the designing process of the software system [6]:

1. The software system is programmed in terms of C language and modular design method.
2. All variables are transformed into Per-unit value in order to promote the precision and portability of the software.
3. All the floating point operations are transformed into the fixed-point ones which the DSP can deal with. In this system, Q15 format is adopted to realize the above transformation.
4. The method of multiple sampling and averaging is adopted on the purpose of enhancing the precision of the system.

The whole software system is composed of the main program and the interrupt service subroutines, which will be described in detail in the following part.

4.1 Main program

The flow chart of main program is shown in Figure 7. The functions of main program include: system initialization (watchdog, timer clock, interrupts vector table, etc.), peripheral initialization (EV manager, AD, SCI, etc.) and enters the endless loop waiting for interrupt response after initialization.

4.2 Interrupt service subroutines

Interrupt service subroutines are the core parts of the software system. There are three types of interrupt service subroutines, namely, the Timer 1 underflow interrupt subroutine, the fault protection interrupt subroutine as well as the communication interrupt subroutine.

1. Timer 1 underflow interrupt subroutine
The vector control algorithm is mainly completed in the Timer 1 underflow interrupt subroutine. Its task basically includes phase current, rotor position angle and rotating speed acquisition, PID regulation, SVPWM algorithm, etc. The flow chart of Timer 1 underflow interrupt subroutine is shown in Figure 8. The control system is double-looped.

The software system has a double closed-loop structure of current and speed.

The current loop is the inner loop in which the sampling period of current is 0.1 ms the same as the switching period of MOSFET. The speed loop is the outer loop in which the PID adjustment period is 1ms which means the speed loop PID regulation time is ten times of the current loop PID regulation time.

2. Fault protection interrupt subroutine
The software protection program is designed in order to avoid fault damage to the hard circuit and achieve the double protection of hardware and software. The flow chart of fault protection interrupt subroutine is shown in Figure 9. The functions of fault protection interrupt subroutine mainly includes overheat protection, overcurrent protection, under voltage-protection and over-voltage protection.

The procedure of realization of the protection functions is as follows: when fault occurs, the

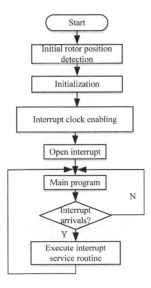

Figure 7.   Flow chart of the main program.

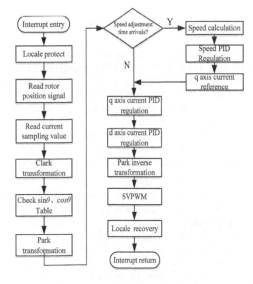

Figure 8.   Flow chart of Time1 underflow interrupt subroutine.

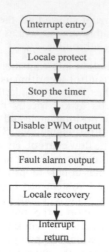

Figure 9. Flow chart of fault protection interrupts subroutine.

TZ1 port of TMS320F28335 will capture the low level signal generated by the hardware protection circuit. According to the special characteristics of DSP, the six PWM signals will be blocked when the electrical level transforms from high to low, so that the fault protection function is realized.

3. Communications interrupt subroutine
   Multiple communication interrupt subroutines, such as the CAN bus communication, the RS232 and RS485 communication interrupt subroutines are designed in order to meet the communication needs with host computer and peripherals.

5 EXPERIMENTAL RESULTS

The parameters of the experimental motor are shown in Table 1. A novel SMC material is chose as the motor core material in order to decrease the iron loss and the difficulty of manufacturing.

Figure 10 shows the experimental motor and the control system. In order to test the characteristics of the control system comprehensively, no-load starting experiment at a speed of 1000 rpm as well as starting and breaking experiments were both carried out.

1. No-load starting test
   The experimental results above show that this control system has good start-up performance with small overshoot and excellent tracking ability, as shown in Figure 11 and Figure 12.
2. Experiment of starting and braking
   It can be seen from the experimental results that the adjusting time of starting and braking is short and the impulse current is small, as shown in Figure 13 and Figure 14.

Table 1. Parameters of experimental motor.

Parameters	Value
Rated power(kW)	2
Rated voltage	220
Rated speed (rpm)	3000
Rated frequency(Hz)	250
Pole pairs(p)	5
Mechanical loss ratio	0.03
Rated operating temperature(°C)	75
Core material	SMC

Figure 10. Experimental platform of control system.

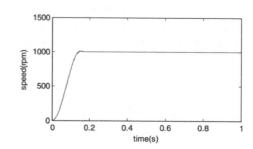

Figure 11. Speed waveform of no-load starting.

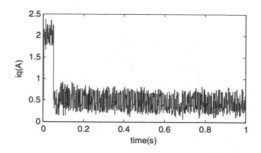

Figure 12. Current i_q waveform of no-load starting.

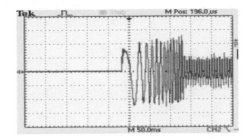

Figure 13. Phase current when motor starts.

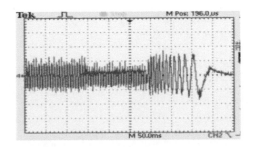

Figure 14. Phase current when motor brakes.

6 CONCLUSIONS

This paper designed a vector control system of SMC transverse flux PMSM based onT-MS320F28335.The experimental results show that this control system has good dynamic and static performances and verifies the correctness of the theoretical analysis and the feasibility of the design scheme. The research work of this paper has great reference value to the research of the high-precision PMSM servo system.

ACKNOWLEDGEMENT

This work was supported by the Key Research and Development Plan Project of Shandong Province (2015ZDXX0601B01; 2016GGC01023).

REFERENCES

[1] Li Xuchuan (2002). Magnetic field oriented control of PMSM based on DSP. *J. Medium and small motors*. 29:17–20.
[2] Shigeo Morimoto, Yoji Takeda & Takao Hirasa (1990). Current Control Method for Permanent Magnet Synchronous motor. *J. IEEE Transactions on Power Electronics*. 5:1–3.
[3] Liu Tingting, Tan Yu & Wu Gang (2009). Simulation of PMSM Vector Control System Based on Matlab/Simulink. *C. 2009 International Conference on Measuring Technology and Mechatronics Automation*. 16: 343–346.
[4] Marco Tursi & Chiricozzi (2010). Surface Mounted Permanent Magnet Synchronous Motors Accounting for Resistive Voltage Drop. *J. IEEE Transaction on Industry Electronics*. 57:440–448.
[5] Li G & Liu Z G (2009). Adaptive Speed Control for Permanent Magnet Synchronous Motor System with Variations of Load Inertia. *J. Transactions on Industrial Electronics*. 56: 3050–3059.
[6] Xu Dong & H Wei (2011). A Digital High Performance PMSM Servo System Based on DSP and FPGA. *J. Proceedings of International Conference on Industrial Electronics and Application*. 22: 2742–2746.

Current control of LCL grid-connected inverter based on ADRC

Jian Sun, Hejin Xiong & Deming Lei
School of Automation, Wuhan University of Technology, Hubei, P.R. China

ABSTRACT: LCL filter has received wide attention as grid-connected inverter; however, it produces a large harmonic current. PI controller was generally used to control the current in the system of the traditional LCL grid-connected inverter; but it is difficult to obtain a very good result even though the control method is simple. In this paper, the structure and working principle of active disturbance rejection controller is introduced, and the active disturbance rejection controller is used in the LCL filter system and we simulated this using MATLAB/Simulink. Compared with the conventional PI control method, the simulation results show that this control strategy can well control the single-phase LCL filter and significantly reduce the harmonic current.

1 INTRODUCTION

With the rapid development of new energy grid-connected power generation technology, the grid-connected inverter has become important. LCL filters are widely studied because of their merits, including small volume, high quality of grid-connected current, and lower requirement of inductance than single inductor L filter.

High-power nonlinear load is widely used in power system, which brings a great harmonic pollution to the power grid. The performance of existing microprocessors is high and hence some mature control algorithms, such as adaptive control, sliding mode control, neural networks, fuzzy control, and other intelligent control strategies, can be applied to power electronic system to solve the problem due to the rapid development of the power electronics technology. The performance of the power filter will be greatly improved by the combination of the power electronic technology and the intelligent control algorithm.

Active Disturbance Rejection Controller (ADRC) is a kind of feedback linearization controller proposed by Han Jingqing in the 1990s (Han J Q, 2009). The control method is evolved by a Nonlinear PID control (NLPID) strategy. It is a practical digital control technology, which does not depend on the precise model of the controlled objects. ADRC is a type of practical digital control technology that is derived from the summary and induction of the computer simulation experiment results.

The method of the ADRC can automatically compensate the mismatch of the object model and the disturbance from the outside world to realize the dynamic feedback linearization of the dynamic

system and then use the nonlinear configuration to construct the nonlinear state error feedback control law to improve the control performances of the closed-loop system. Therefore, the active disturbance rejection controller has been widely used in the field of estimation and control in uncertain systems.

Harmonic signals will be produced in the processes of sampling in LCL inverter, which seriously affect the quality of the grid current. PI controller is used in the control of the LCL grid-connected inverter (Zhang Binfeng, 2016). The control effect of the PI controller is not very high in spite of the simple structure. In this paper, the active disturbance rejection controller is applied to the LCL grid-connected inverter and the simulation research is carried out.

2 SINGLE-PHASE LCL FILTER MODEL

The LCL filter and inverter control block diagram is shown in Figure 1, where L1 is the inductance in the side of inverter, L2 is the inductance in the side of power grid, C is the filter capacitor, Udc is the DC voltage, and Ug is the grid voltage.

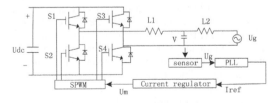

Figure 1. LCL inverter control block diagram.

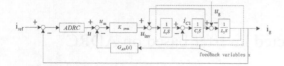

Figure 2. LCL filter and inverter control structure of ADRC.

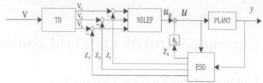

Figure 3. Structure diagram of third-order active disturbance rejection controller.

The control structure of the LCL filter grid-connected inverter is shown in Figure 2.

The open-loop transfer function of the system is shown in Formula (1). The active disturbance rejection control does not need to detect the load current and supply voltage, but it only regards it as an unknown interference of the system by using the Nonlinear State Error Feedback (NSLEF) to compensate it:

$$G_S = \frac{KK_{pwm}}{S^3 a_0 + S^2 a_1 + S a_2 + KK_{pwm}} \tag{1}$$

$$a_0 = L_1 L_2 C \tag{2}$$

$$a_1 = C(L_1 R_2 + L_2 R_1 + L_2 KK_{pwm}) \tag{3}$$

$$a_2 = R_1 R_2 + CR_2 KK_{pwm} + L_1 + L_2 + R_1 + R_2 \tag{4}$$

3 ACTIVE DISTURBANCE REJECTION CONTROLLER

3.1 *Structure of the ADRC*

A third-order active disturbance rejection controller is designed in this paper, whose structure diagram is shown in Figure 3. ADRC consists of three parts: the Tracking Differentiator (TD), the Nonlinear State Error Feedback (ESLEF), and the Extended State Observer (ESO). The tracking differentiator is able to schedule transitions for the inputs of the system.

Tracking Differentiator (TD) was designed for arranging a transition process for the input of system and it can get a smooth input signal to solve the contradiction between the system speed and overshoot. v1 is a process transition of v, v2 is a differential signal of v1, and v3 is a differential signal of v2. ESO is the most important part of the ADRC, which is used to estimate the controlled object and can not only estimate z1, z2, and z3 of each state variable, but also get the right end estimate of the object equation, which means compensation factor. The Nonlinear State Error Feedback (NSLEF) combines the tracking signal output of the differentiator with the error signal estimated by the differential signal and the Extended State Observer (ESO) through the state variable to

obtain the u0 signal, which is combined with the error signal to obtain the control signal, u.

3.2 *Algorithm of the ADRC*

The algorithm of the Tracking Differentiator (TD):

$$TD: \begin{cases} \dot{V}1 = V2 \\ \dot{V}2 = V3 \\ \dot{V}3 = -a_{01} \cdot r^3 \cdot V1 - a_{02} \cdot r^2 \cdot V2 \\ \quad - a_{03} \cdot r \cdot V3 + a_{03} \cdot r^3 \cdot V0 \end{cases} \tag{5}$$

The algorithm of the Extended State Observer (ESO):

$$ESO: \begin{cases} e = Z1 - y \\ \dot{Z}1 = Z2 + (Z2 - \beta_1 \cdot e) \\ \dot{Z}2 = Z3 - \beta_2 \cdot fal(e, \delta_1, d) \\ \dot{Z}3 = Z4 - \beta_3 \cdot fal(e, \delta_2, d) + bu \\ \dot{Z}4 = Z3 - \beta_4 \cdot fal(e, \delta_3, d) \\ fal(e, \delta, d) = \begin{cases} e \cdot d^{(\delta-1)}; |e| >= d \\ |e|^\delta \cdot sign(e); |e| >= d \end{cases} \end{cases} \tag{6}$$

The algorithm of the Nonlinear State Error Feedback (ESLEF):

$$NLSEF: \begin{cases} \varepsilon_1 = V1 - Z1 \\ \varepsilon_2 = V2 - Z2 \\ \varepsilon_3 = V3 - Z3 \\ u_0 = k_p \cdot fal(\varepsilon_1, a_1, d) + k_d \cdot fal(\varepsilon_2, a_2, d) \\ \quad + k_i \cdot fal(\varepsilon_3, a_3, d) \end{cases}$$

$$\tag{7}$$

The parameters of the ADRC are shown in Formulas (5), (6), and (7). The parameters of the Tracking Differentiator (TD) include a01, a02, a03, and r; the parameters of the Extended State Observer (ESO) include β1, β2, β3, β4, d, b, δ1, δ2, and δ3; the parameters of the Error Feedback (NSLEF) include a1, a2, a3, kp, ki, kd, and d; the parameter r is determined by the speed of the transition process and the affordability of the system; and the parameters β1, β2, β3, and β4 are determined by the sampling step used by the system.

4 SIMULATION ANALYSIS

Simulink in MATLAB is used for the control method of simulation in this paper.

4.1 Parameter selection of the controller

The Tracking Differentiator (TD): $a01 = 0.12$, $a02 = 490$, $a03 = 4.8$, $r = 3000$.

The Extended State Observer (ESO): $\beta1 = 37$, $\beta2 = 55$, $\beta3 = 210$, $\beta4 = 45$, $d = 0.0025$, $b = 18$, $\delta1 = 0.5$, $\delta2 = 0.25$, $\delta3 = 0.125$.

The Nonlinear State Error Feedback (ESLEF): $a1 = 1$, $a2 = 43$, $a3 = 1$, $kp = 0.55$, $ki = 0.16$, $kd = 0.1$, $d = 5000$.

4.2 Simulation results

The controller has been generally chosen to use the PI controller in the current control in the traditional LCL grid-connected inverter. The simulation diagram of the LCL filter and inverter control structure of A-DRC is shown in Figure 4.

In order to verify the correctness of the theoretical analysis and the stability and dynamic performance of the control system, simulation of the single-phase LCL inverter is carried out in MATLAB R2013a/Simulink simulation environment. The control effect of the grid-connected current is shown in Figure 5.

The ADRC controller was used in the current control of the LCL grid-connected inverter in this paper, and Figure 6 shows the current waveform.

In order to further analyze the quality of grid-connected current, a Fourier analysis of the simulation waveforms for the first five cycles of the grid-connected current is performed.

The Total Harmonic Distortion (THD) of current in the system of LCL filter is up to 5.80%, which is shown in Figure 7, whereas the use of the ADRC controller can reduce the THD of current, which is only up to 5.80% that is shown in Figure 8.

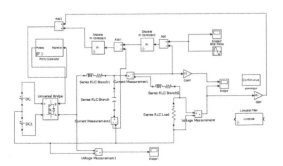

Figure 4. LCL filter and inverter control structure of ADRC.

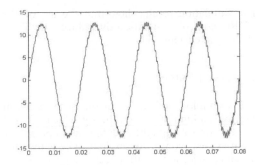

Figure 5. Current waveform of the single-phase LCL inverter with PI controller.

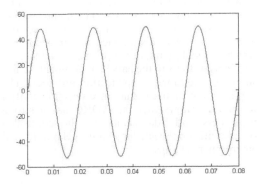

Figure 6. Current waveform of the single-phase LCL inverter with ADRC controller.

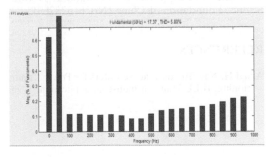

Figure 7. Current of Fourier analysis with PI controller.

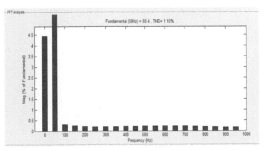

Figure 8. Current of Fourier analysis with ADRC controller.

The simulation results show that the use of the ADRC controller significantly reduces the THD of the grid current.

5 CONCLUSIONS

The technology of active disturbance rejection control is a kind of control theory, which is based on the control thought and the development of the nonlinear control technology. The active disturbance rejection control technology does not comply with the classification of disturbances, but unifies the perturbation and expands a state by its own internal expansion state observer to characterize the unified disturbance. The compensation of this disturbance reflects the excellent control idea of the active disturbance rejection control, which can still maintain the precision and show a strong robustness to the harsh environment. It also shows a wide range of application prospects. The application of active disturbance rejection control technology shows that the ADRC has a strong robustness and adaptability to the controlled object with nonlinear, large time delay and bad environment change.

ACKNOWLEDGMENT

This study was supported by the National Natural Science Foundation of China (No. 61573264).

REFERENCES

Akagi H. New Trends in active Filters for Power Conditioning. IEEE Trans on Industrial application, 1996, 32 (6).

Han, J.Q. "From PID technique to active disturbances rejection control technique," vol.56, no.3, Mar. 2009, pp. 900–906.

Han J Q. Active disturbance rejection control technique—the technique for estimating and compensating the uncertainties [M]. Beijing: National Defense Industry Press. 2009.5.

Han J Q. Auto disturbance rejection control technique [J]. Frontier Science, 2007, 1(1): 24–31.

Hu Xuefeng, Wei Zheng, Chen Yihan, Gong Chunying, Zhang Jiayan. A Control Strategy for Grid-connected Inverters with LCL Filters. Proceedings of the CSEE. Vol.2 No.27 Sep. 25, 2012.

Lesnicar A, Marquardt R. A new modular voltage source inverter topology [C]//Proc. 10th IEEE European Conference on Power Electronics and Applications. Toulouse, 2003:1–10.

Liu, J.L. "Advanced PID control MATLAB simulation," Electronic Industry Press, China, Mar. 2011, pp. 228–231.

Peng Shuangjian, Luo An, Rong Fei, Wu Jingbing, Lu Wenkun. Single-phase Photovoltaic Grid-connected Control Strategy with LCL Filter. Proceedings of the CSEE. Vol.31 No.21. Jul.25, 2011.

Sun Wei, Wu Xiaojie, Dai Peng, Zhou Juan. An overview of current control strategy for three-phase voltage source rectifier with LCL. Filter [J]. Transactions of China Electro technical Society, 2008, 23(1):90–96 (in Chinese).

Wang Dong, Xue Shilong, Zong Yanling. The Simulation of Double–Loop Grid Current Control Technique for Grid-Connected Inverter Using Single LCL Filter. Mar. 25, 2012, Vol 29 No.2.

Xia D, Heydt GT. Harmonic power flow studies Part I Formulation and solution. IEEE Transaction on Power Apparatus and Systems, 1982, 101(6).

Zhang Binfeng, Xu Jinming, Xie Shaojun. Analysis and Suppression of the Aliasing in Real-Time Sampling for Grid-connected LCL-filtered Inverters. Proceedings of the CSEE, Vol.36 No.15 Aug. 5, 2016.

Electromechanical Control Technology and Transportation – Jia & Wu (Eds)
© 2017 Taylor & Francis Group, London, ISBN 978-1-138-06752-3

Design and application of cold water storage technique for a foreign-related oil fuel power plant project

Dichun Yan, Qian Xu & Zhaoyang Gu
Henan Electric Power Survey and Design Institute, Zhengzhou, China

ABSTRACT: In this paper, we present the design scheme and design difficulties of the combined cold water storage and air-conditioning system as well as effective measures in the selection of AHUs and cold water storage system of a foreign-related oil fuel power plant in a hot and dry desert climate.

1 INTRODUCTION

In recent years, foreign-related power plants have been widely applied in various districts with various environments, and some of the meteorological conditions are extremely poor and hence the HVAC design for those projects faced many challenges.

The first principle of foreign-related HVAC design is that the contract should be strictly complied with, but in most cases, the design scheme announced in the contract is not the most appropriate scheme for the local meteorological characteristics. And, different projects have different special requirements because of various districts and environments. For example, power plants located in Africa emit a huge amount of heat from the generator and ancillary facilities in the power house because of the hot and dry weather. If there is not enough ventilating and cooling measures while operating, the operating condition and machine life of main facilities, electric equipment as well as the working environment of staff will be obviously affected. For normal projects, mechanical ventilation is enough to remove waste heat. However, for some hot and dry districts, the ventilation design temperature can even reach 50–55°C, and the indoor temperature can reach 60–70°C or even higher when the emitted heat from the equipment is considered. For such circumstances, mere ventilation cannot meet the indoor temperature requirements and thus the cooling technique must be adopted, that is, the fresh air should be cooled before being supplied to the indoor. Conventional electric cooling or evaporating cooling is not suitable for the hot, dry, and lack-of-water environment. Therefore, a set of cost-effective combined cold water storage and air-conditioning system shall be applied, and its reasonable cooling and regulating strategies can help to shift peak loads to off-peak and realize energy-saving.

The cold water storage technique applied in such project mainly takes the advantage of large temperature difference between day and night, and the system will produce and store cold at night and release it to terminal AHUs at daytime so as to remove the heat from the building.

A real project in Niamey will be introduced in this paper, and according to the local meteorological characteristics, some issues about the cold storage system and the combined ventilation and air-conditioning system will be discussed.

2 PROJECT PROFILE

In this paper, we will use the Projet de la Centrale Thermique Diesel 100 MW de GOROU BANDA de Niamey as an example to introduce the technique application. The project is located in GOROU BANDA district, south of Niamey, capital of the Republic of Niger. The planning capacity is five sets of 20 MW diesel generator, of which four sets will be built in the first phase of the project.

2.1 *Indoor/outdoor design condition and main equipment capacity and parameters*

2.1.1 *Local weather parameter*
According to the contract, the meteorological parameters are shown as below:

Highest design temperature:	Shade 50°C/ outside 55°C
Annual average temperature:	30°C
Daily mean temperature:	35°C
Annual minimum temperature:	10°C
Humidity in rainy season:	60–90%
Humidity in dry season:	10–30%
Average wind speed:	7.8 m/s (28.08 km/h)
Maximum wind speed:	46 m/s (165.6 km/h)
Highest attitude:	300 m

2.1.2 Power house indoor design condition

Maximum temperature in power house: <55°C
Maximum temperature around the generator: <58°C

2.1.3 Main equipment's capacity and parameter in the generator house (hereinafter referred to as power house)

Single diesel generator unit's heat radiation data are provided by the manufacturer as below:

Total: 1260 kW (ambient temperature: 55°C)

Engine 750 kW, generator 430 kW, auxiliaries 80 kW.

2.2 Main technical characteristics

2.2.1 Design particularity caused by meteorological characteristics

The local hourly ambient temperature in 24 h of typical day during the hottest season is shown in Table 1, which is provided by the employer.

Fig. 1 shows the hourly air temperature and relative humidity of a typical day in April.

The meteorological parameters indicate that the project carried out in windy and sandy areas with large temperature difference between day and

Table 1.   Hourly ambient air temperature.

Time	0:00	1:00	2:00	3:00	4:00	5:00
Jan.	31.70	31.03	30.37	27.50	27.17	26.83
Feb.	35.20	34.53	33.87	31.00	30.67	30.33
Mar.	34.50	33.83	33.17	30.30	29.97	29.63
Apr.	**36.00**	**35.33**	**34.67**	**34.00**	**33.67**	**33.33**
May	34.50	33.83	33.17	32.50	32.17	31.83
June	35.60	34.93	34.27	31.40	31.07	30.73
July	31.50	30.83	30.17	27.30	26.97	26.63
Aug.	29.90	29.23	28.57	25.70	25.37	25.03
Sep.	32.00	31.33	30.67	27.80	27.47	27.13
Oct.	32.90	32.23	31.57	28.70	28.37	28.03
Nov.	32.20	31.53	30.87	28.00	27.67	27.33
Dec.	31.20	30.53	29.87	27.00	26.67	26.33

Time	6:00	7:00	8:00	9:00	10:00	11:00
Jan.	26.50	27.67	28.83	30.00	31.87	33.73
Feb.	30.00	31.17	32.33	33.50	35.37	37.23
Mar.	29.30	30.47	31.63	32.80	34.67	36.53
Apr.	**33.00**	**34.17**	**35.33**	**36.50**	**38.37**	**40.23**
May	31.50	32.67	33.83	35.00	36.87	38.73
June	30.40	31.57	32.73	33.90	35.77	37.63
July	26.30	27.47	28.63	29.80	31.67	33.53
Aug.	24.70	25.87	27.03	28.20	30.07	31.93
Sep.	26.80	27.97	29.13	30.30	32.17	34.03
Oct.	27.70	28.87	30.03	31.20	33.07	34.93
Nov.	27.00	28.17	29.33	30.50	32.37	34.23
Dec.	26.00	27.17	28.33	29.50	31.37	33.23

(Continued)

Table 1.   (Continued).

Time	12:00	13:00	14:00	15:00	16:00	17:00
Jan.	35.60	36.73	37.87	39.00	38.47	37.93
Feb.	39.10	40.23	41.37	42.50	41.97	41.43
Mar.	38.40	39.53	40.67	41.80	41.27	40.73
Apr.	**42.10**	**43.23**	**44.37**	**45.50**	**44.97**	**44.43**
May	40.60	41.73	42.87	44.00	43.47	42.93
June	39.50	40.63	41.77	42.90	42.37	41.83
July	35.40	36.53	37.67	38.80	38.27	37.73
Aug.	33.80	34.93	36.07	37.20	36.67	36.13
Sep.	35.90	37.03	38.17	39.30	38.77	38.23
Oct.	36.80	37.93	39.07	40.20	39.67	39.13
Nov.	36.10	37.23	38.37	39.50	38.97	38.43
Dec.	35.10	36.23	37.37	38.50	37.97	37.43

Time	18:00	19:00	20:00	21:00	22:00	23:00
Jan.	37.40	36.87	36.33	35.80	34.43	33.07
Feb.	40.90	40.37	39.83	39.30	37.93	36.57
Mar.	40.20	39.67	39.13	38.60	37.23	35.87
Apr.	**43.90**	**42.63**	**41.37**	**40.10**	**38.73**	**37.37**
May	42.40	41.13	39.87	38.60	37.23	35.87
June	41.30	40.77	40.23	39.70	38.33	36.97
July	37.20	36.67	36.13	35.60	34.23	32.87
Aug.	35.60	35.07	34.53	34.00	32.63	31.27
Sep.	37.70	37.17	36.63	36.10	34.73	33.37
Oct.	38.60	38.07	37.53	37.00	35.63	34.27
Nov.	37.90	37.37	36.83	36.30	34.93	33.57
Dec.	36.90	36.37	35.83	35.30	33.93	32.57

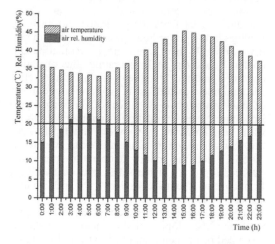

Figure 1.   Hourly air temperature and relative humidity of a typical day in April.

night or the different seasons, and the outdoor air temperature is extremely high and last a long time, especially in summer. Therefore, it is reasonable to set the highest outdoor air temperature (shade 50°C, outside 55°C) as the design temperature.

Meanwhile, the effects shall be taken into account in design not only from the high ambient temperature, but also from local arid, windy, and sandy climate. Through preliminary research, rainy season last only 4 months each year, and there is no underground water resource; thus, the water supply for the power plant is only provided by a seasonal river outside the site, which is drought almost the whole year except the rainy season.

2.2.2 *Design particularity from the contract limitation*

Fully enclosed structure is adopted for the power house in order to prevent sand damage, so the ventilation system shall be a mechanical blow and exhaust type, fresh air system shall have a reasonable precooling process during hot season, and PNN must be guaranteed in any case (PNN: continuous power that the power station can provide on the power input point, after being subtracted by all power loss and all auxiliary equipment's energy consumption).

According to contract terms, it is obvious that conventional A.C. scheme (using electric cooling + fresh air handling unit) cannot meet requirements of the contract, because when outdoor temperature is 55°C, the power consumption of the HVAC system in direct-cooling mode will lead PNN to be lower than the contract limitation. Therefore, HVAC system must be in a reasonable cooling mode with lower running cost, heat-exchanging

between large temp difference, low-temperature air supply, favorable quality, and reliability. Above all, combined system is a better selection to meet both contract requirements and the climate, for its function of peak load shifting of electrical demand.

2.2.3 *Particularity in equipment selection*

For the particular local meteorological characteristics, measures of sand and dust prevention shall be taken into account in the selection of HVAC equipment.

3 SYSTEM DESIGN

3.1 *Power house ventilation and cooling scheme*

With the analysis of meteorological parameters, operating environment of main facilities, and contract's limitation, the final HVAC system for power house is shown in Fig. 2, which is mainly composed of the following subsystems: thermal insulation cold storage pool for cold water storage, modular air-handling units for supplying fresh air, roof fans for exhausting waste air, and water chiller for producing cold water.

The main facilities include: air-cooled chillers, thermal insulation cold-storage water pool, plate heat exchanger, cold water storage pumps, cold water release pumps, terminal system cycling pumps, roof fans, modular air-handling units, valves and meters, and the control system.

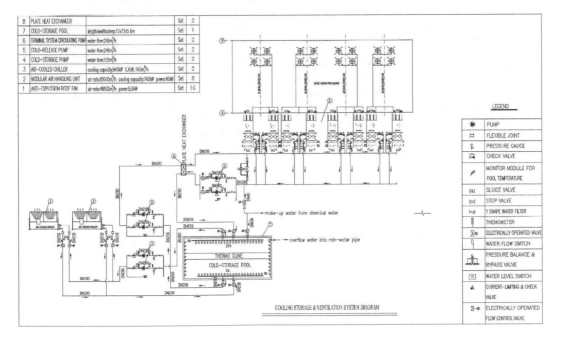

Figure 2.   Power house HVAC system diagram.

95

3.2 Design essentials

3.2.1 Design process and main principles

According to contract terms and the April hourly air temperature and relative humidity, the typical day with lowest outdoor air temp 33°C and its corresponding relative humidity 18.8% is selected for design, and 55°C is used as the exhausting air temperature. Then, the system calculation is performed on the basis of those design parameters.

It was confirmed by the manufacturer that the airflow capacity of air supply units can be increased with the constant dimension. Considering energy-saving and water shortage, the maximum level of airflow capacity is provided by the manufacturer under the constant dimension. With the determined airflow capacity, we rechecked the temperature of supply and exhausted air, and 38°C is confirmed as the critical temperature point for the normal ventilation mode. (When the temperature is no higher than 38°C, normal ventilation working mode is enough. On the contrary, the cooling ventilation working mode should be started.)

Therefore, power house ventilation system is classified into two working modes:

a. Mere ventilation working mode

When outdoor air temperature is lower than 38°C, mere ventilation working mode is selected. Fresh air is directly supplied after filtration, and waste heat and moisture will be exhausted by roof fans to guarantee the highest exhausting air temperature lower than 55°C.

b. Cooling ventilation working mode

When the outdoor air temperature exceeded 38°C, cooling ventilation working mode is started. Besides terminal ventilation system (modular air-handling unit and roof fan), the cold storage and release system will be put into operation to supply cold water for the modular air-handling units to guarantee the highest air supply temperature remaining on 38°C.

3.2.2 Main design principles of cold water storage and release system

3.2.2.1 Cold water storage medium selection

Two commonly used cold storage mediums are water and ice. Referring to related studies and data in Practical HVAC handbook, the chilled water storage system has much advantage than the ice storage system in the aspects of energy consumption, initial cost, and operation fees. Hence, the chilled water storage system is selected for this project.

3.2.2.2 Cold storage and release system components

The project, located in Niamey, Africa, has tropical desert climate with the rainy season merely between June and September. In the early period of operation, the water used for cold storage system can only be delivered from a seasonal river near the power plant, stored after treated. On the basis of the dry, windy, and sandy climate character of this district, fully-underground and thermal insulation structure is adopted for the cold storage pool in order to minimize daily evaporation loss and heat loss.

The cold storage and release system is mainly composed of:

a. Air-cooled water chiller

By taking advantage of relative low temperature and high cooling efficiency during nighttime, the primary-side hot water, which is refluxed after heat-exchanged during daytime, can be circularly cooled with air-cooling method by the air-cooled screw water chiller. The produced 4–5°C cold water would be pumped into a pool for storage.

b. Thermal insulation-type sealed cold storage pool

The thermal insulation-type cold storage pool is a kind of sunken cube near the cold storage station, fully buried underground to accommodate primary-side circulating water. The cold water is produced and stored in the pool during nighttime, then delivered to the primary-side heat exchanger at daytime. The primary-side hot water would be refluxed into the pool at the meantime after heat-exchanged. Naturally stratified water storage technique is applied to the pool with hot, buoyant water on top and cold, dense water below. As the outer wall is made of concrete and the inner wall is made of anti-corrosion and thermal insulation material, there is no heat transfer between water and soil. The temperature difference in the pool can reach 15°C. Strict water proofing measures shall be taken at the pipeline crossing point when the cold storage pipe and cold release pipe pass through the side wall. Before the initial operation, the pool shall be fully filled with normal temperature water at one time, and circularly used after operation. An equal partition is constructed at the center of the pool to ensure that when half of the pool is under maintenance, the other half can keep working, that is, the pool is not out of service completely. In addition, the relief valve and pipe are prepared for drainage when repairing the access manhole, monitoring system of water temperature and water level.

c. Water distribution system

The water distribution system is composed of pipe network, diffusers, valves, and accessories. The main function is to import the hot or cold water into the pool without agitation, forming a horizontal distributed gravity flow along the bottom or top of the pool depending on the density difference instead of inertia force and a thin thermocline, which could minimize the mixing of hot and cold water.

d. Water–water plate-type heat exchanger
The main functions of the water–water plate-type heat exchanger are:

1. When the plant needs to be cooled during daytime, the cold owned by the water of cold storage station (primary side) would be transferred to the water of secondary side. Then, the secondary-side water would be delivered into the air-conditioning terminals after cooled, and the heated primary-side water would refluxed to the cold storage pool waiting for the next cold charging cycle during nighttime.
2. Isolate the primary-side and secondary-side water to ensure the water quality.
3. Prevent the system from back flowing at the earlier stage of operation.

e. Circulating pumps for cold storage
Circulating pumps for cold storage are used for pumping the low-temperature cold water produced by water chiller into the cold storage pool. The middle temperature water, which cannot meet the design temperature, would circulate to the chiller and continuously cooled until the water quantity and temperature in the pool can reach the design value. Pumps are variable-frequency regulated.

f. Circulating pumps for cold release
Circulating pumps for cold release are used for pumping the 4–5°C cold water of primary side into the heat exchanger to complete cold transferring from primary side. Then, the heated primary-side water would be refluxed to the cold storage pool waiting for the next cold charging cycle during nighttime. Pumps are variable-frequency regulated.

g. Circulating pumps for terminal cooling water system
Circulating pumps for terminal cooling water system are used for circulating secondary water. Secondary water cooled by the heat exchanger will be pumped into the terminals—modular air-handling units for cold releasing, the heated water will return to the heat exchanger and circularly cooled at the same time.

h. Modular air-handling units
Modular air-handling units are used for transferring cold provided by the secondary side to the fresh air through internal cooling coil. The cooled air will be conveyed to the power house by the draught fan when the supply temperature reached the design value.

AHUs mainly comprise the following function sections: air inlet section, primary and medium-efficiency filter, cooling coil section, supply air fan, flow equalization section, muffler, air supply section, and so on. In addition, if the outside temperature is low enough, the fresh air need not to be cooled, and AHUs can be simply used as ventilation system.

3.2.3 *Main design principles and function of the control system*

The HVAC system of the power house is configured with a set of auto-control system. It can realize auto-control of all the related equipment and system of the cold storage station and power house by the PLC and host computer method. Communication terminals and hardware interface are also reserved for central control room of the plant.

Control principles and objectives:

Ensure safe and reliable operation and switch of all the equipment and motor valves of the system under the design parameters to maintain the temperature of power house among the design scope.

Adjust the operation mode of the system, sets of operating machine, devices, and valves of the monitoring and control system according to different cooling load to provide stable-temperature water to the terminal system.

According to the control system, the whole HVAC system can be divided into three kinds of operation mode:

a. When the outdoor temperature is higher than the critical value, the cold storage system and terminal water cooling system shall be started to realize cooling ventilation.
b. When the outdoor temperature is among a small range of the critical value, the cold release circulation of the primary side shall be closed, and circulating pumps for terminal cooling water system are started to use the remained cold of cooling water in the secondary side for cooling ventilation.
c. When the outdoor temperature is lower than the critical value, both water circulation of primary side and secondary side shall be closed, simply using the AHUs as mere ventilation system.

According to such logic, the maximum level of energy-saving can be committed by detailing the operation mode.

The control and monitor scope and design requirements of the control system mainly include the following items:

1. Remote start and stop, condition report, malfunction alarm of the water chiller as well as sequencing control, sets control, automatic reset, parameter modification, etc.
2. Start and stop, condition report, sets control and malfunction alarm of roof exhaust fans, and modular air-handling units.

3. Display and control of flux, temperature and pressure of cold water in main supply and return pipe.
4. Measure and display of the temperature and flux of the main cold water supply pipe.
5. Display and control of the supply/return water temperature of the chiller.
6. Display and control of the supply/return water temperature of the water–water plate-type heat exchanger.
7. Display and control of the supply/return water temperature of the cold storage pool.
8. Display of the open and close states, regulation state of motor valves.
9. Local start and stop, self-lock of the water pump as well as the running state display and malfunction alarm.
10. Measure and display of the stored cold of the pool.
11. Display and record of main devices' electrical parameters.
12. Display of the temperature layer of the cold storage pool.
13. Telemetry and display of the liquid level of the pool as well as the control and regulation of makeup water.
14. Flexible manual and automatic switch of the control system.
15. System compatibility: realize the communication with central control system according to the remained communication interface.
16. Computer intelligent analysis and optimal control:
   Computer mathematical model is established to select the best operation mode according to outdoor temperature and record by taking the full advantage of cold storage system in meeting the terminal cooling load thus saving the operating cost.
17. Dynamic monitoring of the temperature field of the pool.
18. Intercom system with functions of remind, help, set, malfunction inquiring, historic records, etc.
19. Measure, analyze, and publish.
20. Optimal control software could not only optimize the load distribution between water chiller and cold storage system, but also set the operation mode of the system and running number of the cooling fan on the basis of all-day hourly cooling load prediction and correction according to the local meteorological and historical load records.
21. Auto-control devices could control sequenced start-up/shutdown of the equipment and opening/closing/regulating of related valves as well as running condition checking, overload protection, malfunction diagnosis, alarm, etc.
22. Display of outdoor dry and wet temperature.
23. Auto power-off when the water level and temperature in pool reached design value.
24. Control the opening state of water supply valve and by-pass valve to provide stable-temperature water according to the outlet water temperature of the pool.
25. When the water chiller and the pool are cooperated to supply cold water, the hourly water temperature supplied to the water chiller can be automatically set and the proportion undertook by the water chiller and the pool can be automatically adjusted to reach the purpose of optical control.

3.2.4 *Selection and calculation of the main equipment and system*

Selection and calculation of HVAC equipment and system of the power house in this project mainly include the following items:

Flux calculation of roof fan and modular AHUs in power house, cooling coil selection of modular AHUs, water chiller selection, water quantity calculation for cold storage, water–water plate-type heat exchanger and pumps selection, hydraulic calculation of the system, etc.

Because of limited space, only cooling coil calculation and cold release duration are introduced in detail.

3.2.4.1 Cooling coil calculation and selection

Original calculation conditions for cooling coil's row number and heat exchange area are:

- Outdoor air temperature:
  $38°C < t_{in} < 55°C$
- Inlet cooling water temperature of the coil:
  $7°C < t_{ws} < 9°C$

Calculation and selection principles:

Cooling coil's row number and heat exchange area per set are shown in Table 2 and are decided on the basis of extreme outdoor temperature of 55°C and supply air temperature of 38°C.

Table 2. Cooling coil's row number and heat-exchange area per set.

Dr Bulb temperature at fan inlet (°C DB)	Wet Bulb temperature at fan inlet (°C WB)	Dry Bulb temperature at fan outlet (°C DB)	Wet Bulb temperature at fan outlet (°C WB)	Inlet cooling water temperature of coil (°C)
55	25.5	37.9	21	7
45.5	20.7	34.7	17.3	7

Table 3. Hourly cooling water flow rate.

Time	11:00	12:00	13:00	14:00	15:00	16:00	17:00	18:00	19:00	20:00
Hourly air temperature in April (MAX)	40.23	42.10	43.23	44.37	45.50	44.97	44.43	43.90	42.63	41.37
Hourly air relative humidity in April (%)	11.5	10.5	9	9	9	9	10.5	12	13	14.5
Supply air temperature of fan (T_{DB}/T_{WB})	31.45/ 17.05	32.6/ 17.3	33.56/ 17.16	34.14/ 18.44	34.68/ 17.28	34.14/ 19.44	34.14/ 18.44	34.16/ 19.16	33.6/ 19.12	32.49/ 18.99
Cooling water flow rate of coil (t/h)	15.084	14.98	14.58	15.516	16	15.816	15.516	15.516	14.508	14.796
Cooling load capacity provided by coil (kW)	367.5	382.3	389.45	403.57	406.72	405.57	403.57	403.77	386.97	377.24
Inlet and outlet water temperatures of coil (°C)	7.0/ 28.0	7.0/ 29.0	7.0/ 30.0	7.0/ 30.0	7.0/ 29.0	7.0/ 29.0	7.0/ 30.0	7.0/ 30.0	7.0/ 30.0	7.0/ 28.0

Hourly cooling water flow rate is decided according to the hourly calculation temperature and relative humidity of the hottest typical day (hottest 45°C) in conditions of ensuring the outlet temperature of the air fan to be 38°C.

The coil is calculated on the basis of the ambient air condition of 55°C/9% (yearly highest temperature in the hottest month). The lowest supply air temperature can only be calculated when the water in coil has a nonlaminar flow (i.e., the water flow rate in coil should not be less than 0.55 m/s).

The difference between hourly calculation temperature and extreme design temperature is quite large, but the supply air temperature should always be 38°C. Thus, there is no one single coil that could meet all of these needs for the quiet wide temperature range.

The hourly calculated results from 11:00 to 21:00 are provided in Table 3. For other times, the difference between hourly air temperature and 38°C is so small that self-circulation of reserved water in the terminal cooling pipeline is enough to cool the air.

The cooling load requirements can be met by regulating the cooling water flow in the terminal pipeline. When the reserved water is not enough to take away the heat, the cold release system on the primary side shall be restarted. That is also a way of energy-saving.

3.2.4.2 Determination of cold release duration in typical summer day by area-method

The calculation outdoor ventilation temperature and humidity is determined on the basis of the hourly temperature of typical day in April where the local highest average temperature of all year occurred. As Figure 1 shows, the area between the temperature curve and 38°C line is the area that needs to be cooled with the amount of 32.3°C • h.

When the fresh air temperature is supposed to be lowered 17°C (55–38°C = 17°C) by the ventilation system, the cold release duration should be 32.3[°C • h] / 17 [°C] = 1.9 [h].

The subsequent calculation and selection of water chiller, water quantity of the pool, water–water plate-type heat exchanger and pumps are based on the above calculated water quantity and cold release duration.

4 CONCLUSION

For the design stage of a foreign-related power plant, the key point is to capture the professional related contract terms, implement the corresponding specifications into the preliminary design and construction drawing design, and guarantee that the finished design could meet the requirements of both employer and the contract. Meanwhile, the particularity of the design and equipment selection caused by the contract and the local meteorology shall be paid much attention to for the HVAC.

Comprehensive energy-saving strategies that are applicable for this project have been proposed on the basis of meeting the contract and corresponding to the design specifications and standards, combining the cold storage system with the cooling and ventilating system to provide reasonable cooling and regulating strategies, and to realize energy saving by shifting peak loads to off-peak. Personalized control methods were used to guarantee that all the equipment of the system could be safely and reliably operated and switched under the design parameters, and various indoor temperature control strategies could be realized flexibly according to various cooling loads caused by the changeable outside temperatures.

The feasibility of using a cold storage system for cooling and ventilating the power house of a foreign-related oil fuel power plant in a hot and dry desert climate is demonstrated. It turns out to be an effective means of cooling for such kinds of hot plants with its favorable energy-saving effect.

REFERENCES

Lu Yaoqing. 2008. Practical HVAC handbook (second edition). China building industry press.

Lu, Y.Z., R.Z. Wang, M. Zhang and S. Jiangzhou. 2013. Adsorption cold storage system with zeolite–water working pair used for locomotive air conditioning. Energy Conversion and Management. 44(10), 1733–1743.

Wang Xiaolin, Dennis Mike. 2013. Influencing factors on the energy saving performance of battery storage and phase change cold storage in a PV cooling system.

Xie Wangdu. 2012. Ventilating and air conditioning system design for projects concerning foreign countries in particular climatic conditions, Journal Heating Ventilating and Airconditioning. HV & AC. 42(3), 23–26, 113.

Zhu Deming. 2002. HVAC design of a foreign fossil fuel power plant project, Journal Heating Ventilating and Airconditioning. HV & AC. 32(3), 63–70.

Electromechanical Control Technology and Transportation – Jia & Wu (Eds)
© 2017 Taylor & Francis Group, London, ISBN 978-1-138-06752-3

Research on software requirements dependencies

Yuqing Yan
School of Finance, Guangdong University of Foreign Studies, Guangzhou, China

ABSTRACT: The research of requirements dependencies is task-based from different points of views in software development. For the usage of the concepts and terminologies in this field there is not yet consensus, which has caused confusions for new researchers. This paper takes into account the relationships that exist among software requirements, points out some problems in the research of requirements dependencies including the problems of accurate definitions and useful classifications of dependencies, discusses the reasons why these problems existed, and contributes some suggestions for solving them.

1 INTRODUCTION

Requirements may affect or depend on each other in some ways. The relationships among requirements are called requirements dependencies (Carlshamre P, 2001 & Dahlstedt S G, 2003). Investigating such relationships is significant for making right decisions in requirements acquirement and analysis, and also necessary for effective management of requirements change (Morkos B, 2014). Many researchers have contributed to this field including Pohl (Klaus Pohl, 1996) who regarded requirements dependencies as a traceability mechanism. Yet, the research done in this field had some problems in defining terminology or classification of dependencies, which have caused some divergences and confusion for readers and new researchers.

Here, the problems which can be found in current literature will be discussed and solutions will be proposed. The structure of this paper is as follows: In Section 2 different concepts and objects for expressing the relationships among requirements are enumerated, the problems in usage of these terminologies are analyzed and suggestions to solve these problems are presented. In Section 3 the problems in classifying dependencies are pointed out and some proposals for solving them are provided. Section 4 proposes the further research for requirements dependencies.

2 THE PROBLEMS IN DEFINING REQUIREMENTS DEPENDENCIES AND SOLUTIONS

2.1 *Related works — the concepts of and objects connected by dependencies*

The research about requirements dependencies is considered as a critical area of requirements engineering (Morkos B, 2014). It is important to understand the terminology and paradigm in this area for clearly identifying dependencies. There are different descriptions about dependencies based on different tasks and requirements objects. For example, (Easterbrook et al, 1996) focused on the relationships between viewpoints in requirements specification. Viewpoints are requirements elicited from multiple perspectives which overlap, complement and contradict each other. A dependencies model as a part of requirements traceability was developed by (Pohl, 1996). This was a milestone in the field and focused on any type and granularity of trace objects used in the requirements engineering process. (Dahlstedt et al, 2003) discussed the relationships between specific types of trace objects-explicitly stated requirements. They used the term interdependencies to express the relationships between the same type of trace objects. (Carlshamre et al, 2001) denoted the relationships between requirements as interdependencies, the framework of interdependencies was used to plan requirements release. (Robinson et al, 2003) reviewed the state of the arts of requirements dependencies that were called requirements interactions. (Spanoudakis et al, 1999) addressed the relation between requirements specifications called overlap which was used to detect the inconsistency between requirements specifications. (Ramesh et al, 2001) aimed at building requirements traceability models to ensure continued alignment between stakeholder requirements and various software outputs in the process of the system development. They presented four types of traceability links and four types of dependencies. The objects they were concerned with included requirements, design objects and development resources (e.g., money, electricity, data). (Knethen et al, 2002) investigated requirements recycling in a systematic way, discussed the relations between documentation entities (e.g. functional requirements, design classes), and defined one type of relationship called Horizontal relationship. Each documentation entity represented certain logical entities. By using the technique of NLP, (Chitchyan et al, 1996) classified predicate verbs of requirements sentences described by natural language into semantically related groups from the requirements dependencies perspective. (Zhang et al, 2005) paid attention to feature dependencies of requirements. A feature was a set of tight-related requirements from user/customer views.

2.2 The problems

It can be seen by the discussion above that there are different concepts and terms to denote the relationship between requirements. Different synonymous words were given, which has brought some confusion to new readers. The main problems can be concluded as follows:

1. Non specific concepts
   Firstly, the concept of software objects was not specific. For example, the model of Pohl (Klaus Pohl, 1996) dealt with any type of trace objects used in the requirements engineering process. Here, "trace objects" were not specific, so a further clarification was needed. Secondly, the relationship between objects was not specific and not bound for any certain development phrase such as (Ramesh et al, 2001) and (Knethen et al, 2002).
2. Non specific scopes of requirements
   While talking about the requirements relationship, the scope of requirements objects was normally unspecific. For example, (Dahlstedt et al, 2003) focused on "explicitly stated requirements" which were not defined explicitly. It is acknowledged that requirements are satisfied with such characteristics as complete, correct, feasible, necessary, prioritized, unambiguous, verifiable (Karl E, 2003). But it is difficult to judge if "explicitly stated requirements" had such characteristics or were stable or volatile? Requirements easily change, "stated" states may not be always kept. When change proposals to requirements are raised, the "stated" requirements will possibly not be stated/confirmed. After change proposals to requirements are implemented through a series of change steps, the states of requirements will evolve. They defined "interdependencies" as the relationship that existed between trace objects of the same type. But they did not explicitly point out "the same type" objects belonged to which development phrase.
3. Less comparable classifications
   Different researchers presented different classification models of requirements dependencies based on different issues. And even for the same

issue, the classification types were completely not the same, without any intersection on them.

2.3 The reasons

The problems have indicated that the discipline of the research to requirements dependencies is not yet built, and there are no explicit concepts, methods and frameworks to follow. What caused this phenomenon?

IEEE 87 defined a requirements as: 1) a condition of capability needed by a user to solve a problem or archieve; 2) a condition or a capability that must be met or possessed by a system to satisfy a contract, standard, specification or other formally imposed document [IEE87]. And IEEE 98 also gave similar definition of a requirement. But it was not followed by all software development organizations (Glinz M, 2007). And the ways of expressing requirements are diversified based on the methods of developing software. For example, in object-oriented software development, use cases are requirements, features are requirements, business rules are requirements, etc. The granularity of requirements are different, or abstract, or concrete, some represent single requirements, others represent specifications. For accurate requirement definition is unavailable, which has caused the arbitrary categories of requirements, so do the requirements dependencies (William N, 2003 & Glinz M, 2007).

2.4 The solutions

The confusion of defining requirements dependencies has bothered many researchers. So Dahlstedt et al suggested [2] that they would not distinguish between dependencies and relationship or other kind of the terms that are used to express the relationships among requirements, they just wanted to find out the different ways by which requirements could be related to each other as well as how they affected each other. The term they used was interdependencies. But this attitude could not be helpful to solve such problems. Hence, two solutions are provided as follows.

1. Further standardizing requirement definition
   Normally, the term of software requirement is related to a development paradigm such as structural or object-oriented or agile method. Different development methods define different attributes, types or structures of requirement. It is necessary for international standard organizations such as ISO to redefine the definition of requirement in more details by different levels, types, structures, terms, properties for compliance with the type of projects, paradigms used in software engineering and methods used in

requirements engineering. It will be beneficial to have standardization of requirements engineering in practice, and lead the various research of requirements engineering more disciplined and normalization, and so does the research of requirements dependencies. And when researchers investigate this field, they should explicitly state the requirements objects that fit the standard of software engineering.

2. Developing formal requirement definition
In order to accurately expressing requirements, it is necessary to develop the formal expressions of requirements including formally defining attributes, granularity, level, structure of a requirement, then formal discipline of requirement acquirement and an analysis can be formed, so do requirements dependencies.

3 THE PROBLEMS OF CLASSIFICATION MODELS IN REQUIREMENTS DEPENDENCIES AND A SOLUTION

3.1 Typical requirements dependencies

Classifying requirements dependencies could provide viewpoints and perspectives for better observing, understanding, acquiring and organizing requirements. Table 1 lists some typical classifications of requirements dependencies from literature.

3.2 The problems

In Table 1, different classifications were established based on different and specific issues, even for the same authors the classifications of dependencies they built were not the same such as (Dahlstedt S G, 2003) and (A. G. Dahlstedt, 2004) of Dahlstedt et al. From the References in the end of each paper, it can be seen that research in this field has formed a reference system of relevant works, but the discipline has not yet been formed. Less research pointed out real problems in this field and compared with other classification models. Most of them were subsets of Pohl's models (Klaus Pohl, 1996) or (Carlshamre et al, 2002). Here, some specific problems are found below:

1. The same terms express different dependencies or repeated definition
For example, "Refinement" dependency was used in (Zhang W, 2005) and (Jason P, 2003), but the semantic was not completely the same; Two different dependencies in (A. G. Dahlstedt, 2004) had the same sub-dependency "Requires".

2. Different words express the same types of dependencies
For example, the meaning of "Abstraction" dependency in (Klaus Pohl, 1996) was the same

Table 1. Typical requirements dependencies.

Types	Sub-types
Structural, Cost/Value (interdependencies) [2]	(Requires, Explains, Similar_to, Conflicts_with, Influences), (Increases/Decreases_cost_of, Increases/Decreasese_value_of)
Not Specified [3]	Any relationship between requirements
Condition, Content, Documents, Evolution, Abstraction (dependencies) [4]	(Contraints, Precondition), (Similar, Compares, Contradicts, Conflicts), (Example_for, Test_case_for, Purpose, Background, Comments), (Replaces, Satisfies, Based_on, Formalizes, laborates), (Generalizes, Refines)
Affect, Move, Rest, Communicate, Mental Actions, General Actions (relationship) [5]	Semantically there are many sub-types of dependencies
Horizontal (relationship) [9]	Refinement, Dependencies
Refinement, Constraint, Influence, Interaction (dependencies) [10]	(Decomposition, Characterization, Specialization), (Binary constraints, Group constraints, Complex constraints), (Inform, Resource-configure, Meta-level-configure, Flow)
Structural, Constrain, Cost/Value (interdependencies) [13]	(Refined_to, Change_to, Similar_to, Requires), (Requires, Conflicts-with;), (Increase/Decrease cost, Increase/Decrease value)
Structure, Resources, Task, Causality, Time (refined types of interactions) [6]	
Total Overlap, Partial Overlap, Inclusive Overlap, No Overlap (overlap) [7]	
Depends-On, Evolves-To, Satisfies, Rationale (traceability link), Goal, Task, Resource, Temporal (dependencies) [8]	

as "Structural" dependency in (Dahlstedt S G, 2003), "Temporal" dependency in meaning in (William N, 2003) was the same as "Time" dependency in (Carlshamre P, 2002), etc.

3. Scarce of consistency and compatibility
Requirements are expectations of stakeholders which will be analyzed, decomposed into subgoals, and designed as well implemented in later phrases. No matter how they are expressed by any languages, intrinsically they are the same, hence the types of requirements dependencies should follow some internal modes and principles, independent of stakeholders' thoughts. But the different classification models of dependencies in current literature were not built by following some clear general principles, or the principles were too vague.

4. Scarce of clarity and comprehensibility
Usually the definition of dependencies was informal, so the semantics were not clear and understandable. For example, in (Klaus Pohl, 1996), "Contradicts" dependency and "Conflicts" dependency were hard to be distinguished semantically. The dependencies of "Documents", "Compares" and "Elaborates" were hard to understand.

5. Scarce of accuracy and completeness
Less accuracy of the dependency classifications is due to informal definitions of dependencies and no guidance in establishing them. And completeness is related to accuracy. Requirements dependencies and its classifications are objective, which should serve industry practices. Before explicitly defining dependencies, researchers need to carefully observe practical phenomena in industry and discover some rules in development process by the principles of software engineering and domain knowledge, to acquire rational awareness, to form abstract concepts, so as to establish theoretical models of dependencies at the same time when it is desirable in practices.

6. Arbitrary Derivation
Moreover, Ruzanna et al (Steve Easterbrook, 1996) pointed out a problem in classification that the types of dependencies were arbitrarily derived by software reqirements analysts who used their limited knowledge.

3.3 *The idea to solve the problems*

Requirements dependencies are conceptual frameworks of relationship in requirements space. For a specific dependency, it constructs a concrete relationship structure among requirements. In this structure, related requirements compose a requirements model. As requirements versions evolve, the structure will change. By observing the process of its evolution, requirements would be understood better (Jason P, 2003), and it will be beneficial to evaluate requirements risks and reuse requirements including requirements relationships (Von Knethen A, 2002).

The problems existed in current research of this field are neither good to communicate and learn from each other among researchers, nor to study and use existing research results. Indistinct concepts will result in cognitive difference, and in reverse affect correct understanding and formation of concepts. These problems have resulted in difficulties in clearly classifying requirements dependencies and effectively identifying and managing them, as well as using them in practice and profoundly investigating them in theory (Dahlstedt S G, 2003).

The key point to solving these problems is to establish some mechanisms to integrate different dependencies that they are currently scattering, incompatible and non-exchangeable, to study requirements dependencies upon a height of theory and methodology, to build theoretical models of requirements dependencies in the context of requirements engineering. And it is necessary to introduce classification guidelines, to gradually establish disciplines in this field.

Here five guidelines are established to classify requirements dependencies based on how requirements may affect each other and the principle of reducing the complexity of a system by dividing it into parts—abstraction, view partitioning, and modularization or abstractions, views, and components (Joachim Karlsson, 1997). They are: Abstract, Design and Implementation Oriented, Cost Controlled, Risk Controlled, and Evolution Controlled. The research (Yan Yuqing, 2011) has shown that these five guidelines are adaptable to the goal of software engineering (Roger S, 2010): Producing the proper, practicable and reasonable cost products.

Except introducing guidelines, the naming system about requirements dependencies deserves to be addressed. The name should be of contextual meaning according to its function, and is easily commanded and learnt. The names of some dependencies are global, which are universal, can be used in any phrases of the whole process of software development such as "Similar" or "Similar_to" dependency (Dahlstedt S G, 2003 & Klaus Pohl, 1996), whatever in any phrase of software development, whatever in requirements engineering, design, coding or testing phrase, it exists in different ways such as functional similarity, non-functional similarity, structural similarity, semantic similarity of verb/action (Steve Easterbrook, 1996), etc. But some dependencies are local, only can be used in certain abstract level, they must be some sub-types of dependencies. Sub-types of requirements dependencies can form a hierarchical structure composed by multiple levels, which depends on practical needs. Thus, the naming system of requirements dependencies should be very flexible under the five guidelines of classification.

4 CONCLUSION

This paper analyzes the current state of the research of requirements dependencies, points out the problems in this field and the reasons why these problems existed, and presents the solutions. The further work is to do more case studies in industry, to find more and more practical usages of requirements dependencies in practical development, and to verify the significance and value of the suggestions in this paper.

ACKNOWLEDGEMENT

I very much appreciate peer reviewers who pointed out some spelling or grammar mistakes in the original paper, and thank my college WANG Xianjun who helped me in correcting the mistakes that reviews pointed out.

REFERENCES

Balasubramaniam Ramesh, and Matthias Jarke. Toward Reference Models for Requirements Traceability [J]. IEEE Transactions on Software Engineering, 2001, 27 (1): 58–93.

Carlshamre P, Sandahl K, Lindvall M, et al. An Industrial Survey of Requirements Interdependencies in Software Product Release Plannin [C]. IEEE International Symposium on Requirements Engineering, 2001. Proceedings. 2002: 84–91.

Dahlstedt A.G. Requirements Interdependencies Towards an Understanding of their Nature and Context of Use (Technique Report). 2004.

Dahlstedt S G, Persson A. Requirements Interdependencies—Moulding the State of Research into a Research Agenda [J]. Ninth International Workshop on Requirements Engineering Foundation for Software Quality, 2003: 71–80.

George Spanoudakis, Anthony Finkelstein, David Till. Overlaps in Requirements Engineering [J]. Automated Software Engineering, 1999, 6: 171–198.

Glinz M. On Non-Functional Requirements [C]. IEEE International Requirements Engineering Conference. IEEE, 2007: 21–26.

Jason P. Charvat. Determine user requirements now to avoid problems later [J]. 2003.

Joachim Karlsson, Stefan Oisson, And Kevin Ryan. Improved Practical Support for Large-scale Requirements Prioritisation [J]. Requirements Engineering Journal, 1997, 2(1): 51–60.

Karl E. Wiegers. Software Requirements (Second Edition) [M]. Redmond, Washington: Microsoft Press, 2003.

Klaus Pohl. Process-Centered Requirements Engineering [M]. New York, NY, USA: John Wiley & Sons Inc., 1996.

Morkos B, Mathieson J, Summers J D. Comparative analysis of requirements change prediction models: manual, linguistic, and neural network [J]. Research in Engineering Design, 2014, 25(2): 139–156.

Roger S. Pressman. Software Engineering: A Practitioner's Approach [M] (7th). China Machine Press, 2010.

Steve Easterbrook, Bashar Nuseibe. Using ViewPoints for Inconsistency Management [J]. Software Engineering Journal, 1996, 11(1): 31–43.

Von Knethen A, Paech B, Kiedaisch F, et al. Systematic requirements recycling through abstraction and traceability [J]. 2002: 273–281.

William N. Bohinson, Suzanne D. Pawlowski, Vecheslav Volkov. Requirements Interaction Management [J]. ACM Computing Surveys (CSUR), 2003, 35(2): 132–190.

Yan Yuqing. The Study of Requirements Evolution and Dependency [D]. Guangzhou: Sun Yatsen University, 128pages. 2011.

Zhang W, Mei H, Zhao H. A Feature-Oriented Approach to Modeling Requirements Dependencies [C]. IEEE International Conference on Requirements Engineering. IEEE, 2005: 273–282.

Electromechanical Control Technology and Transportation – Jia & Wu (Eds)
© 2017 Taylor & Francis Group, London, ISBN 978-1-138-06752-3

LQR based optimal voltage control of rural electric power grid with high penetration of PVs

C. Zhang, Z.Q. Deng, J. Zhou & H.F. Bian
State Grid Electric Power Research Institute, Beijing, China

A.N. Xiao
State Grid Anhui Electric Power Company, Hefei, China

W.X. Zhang
Anhui Electric Power Research Institute, Hefei, China

ABSTRACT: Confronted with the high penetration of the photovoltaic generations in the rural electric power grid and the correspondingly voltage problem, the subject addressed in this paper is the rural distribution network with photovoltaic generations. The optimal voltage control issue is studied in this paper. LQR based optimal voltage control method is proposed to improve the voltage profile. Simulations are accomplished to verify the effectiveness of the proposed method in different operation conditions, and the simulation results show that the proposed control strategy is effective in voltage improvement.

1 INTRODUCTION

According to the "13th Five-Year" development plan for the utilization of solar energy, issued by the China National Energy Board, Rural Electric Power Grid (REPG) will be faced with 15 GW Photovoltaic generations (PVs) integration. However, the China rural electric power grid has some instincts such as long line distance, low load rate, large peal-valley load difference, and large seasonal load difference, and thus its operation and control are tough issues and not yet completed solved (X. L. Dai, 1999). The integration of PVs, in one hand, makes the voltage control of the rural electric power grid much more complicated; and in the other hand, brings control capability to the rural electric power grid.

Therefore, this paper studies the optimal voltage control issue of rural electric power grid with high penetration of PVs. The chapters of this paper are organized as follows. Firstly, considering the instincts of the rural electric power grid, the voltage effect of high penetration of PVs is analyzed and the voltage problems of the rural electric power grid are put forward in chapter two. Secondly, taking the various control capabilities into account and with the objective of minimal control effort, the LQR based optimal voltage control model is built in chapter three. Thirdly, some typical operation cases of rural electric power grid with PVs are studied in chapter four to validate the effeteness of the optimal control method proposed in this paper. And the chapter five gives some results and conclusions.

2 PVS EFFECT AND VOLTAGE CONTROL ISSUES OF THE REPG WITH PVS

2.1 *The negative effect of PVs*

As for a long distance rural electric power grid, when the loads are small and the active power output of the PVs are large, which usually happens in the daytime, the nodal voltages of the PV connection points might exceed the high limit of the voltage permission range (0.93~1.07 p.u.), especially for the PVs in the end of the distribution line. That is, the active power output of the PV might cause over-voltage problem of the rural distribution lines, especially when the loads are too small so that the inverse power flow happens in the distribution lines.

2.2 *The positive effect of PVs*

When the loads are large, which usually happens in the nighttime, the nodal voltages of the loads might exceed the low limit of the voltage permission range, especially for the loads in the end of the distribution line. In this condition, though PV is not generating active power into grid, PV inverter can generate some reactive power to improve voltage profile of the distribution line. That is, the reactive power generation capability of the PV inverter has positive effect in dealing with low voltage problem of the rural electric power grid.

And in the above condition mentioned in section 1.1, if the PV inverter can absorb some reactive power from the distribution network, the voltage can be reduced a bit so that the voltage profile is improved. The voltage improvement relies on the reactive power generation capability of the PVs, which is limited by the capacity of the PV inverter.

2.3 The voltage control issues of the REPG with PVs

As analyzed in the section 2.1 and 2.2, the PVs may have positive or negative effect on rural electric power grid, and thus different control strategies are needed in different operation conditions. And in reality, the operation condition of the rural electric power grid varies during the whole day and between different season days. In addition, the control capabilities of the various PVs should be taken into account simultaneously because of the interactive effect among them, and thus coordinated control shall be realized in the REPG with PVs.

Considering the above aspects integrated, the control strategies of the REPG with PVs is complicated and coordinated control strategies of the PVs should be obtained under different operation conditions.

3 LQR BASED OPTIMAL VOLTAGE CONTROL MODELING

A Linear Quadratic Regulator (LQR) (G. Z. Zhao, 2010) based optimal voltage control model for REPG with PVs will be proposed in this part taking the various control capabilities into account and with the objective of minimal control effort.

The control capabilities mentioned here are the reactive power generation capabilities of the PVs, including injecting reactive power into the grid and absorbing the reactive power from the grid.

In the LQR based optimal voltage control model of the REPG with PVs, the objective is formulated as follow:

$$\min J = \int_0^\infty \left(X^T Q X + U^T R u \right) dt \tag{1}$$

where

1. X is the state matrix of the studied system, that is, the REPG with PVs. Here X can be expressed as follow:

$$X[1] = \begin{cases} \Delta P_{loss} = P_{loss}^a - P_{loss}^b, if\ P_{loss}^a > P_{loss}^b \\ 0, if\ P_{loss}^a \le P_{loss}^b \end{cases} \tag{2}$$
$$X[2:N] = \Delta V[2:N] = V[2:N] - V_n$$

where

N is the number of the nodes of the studied system;

P_{loss}^b is the active power losses of the studied system under the current control strategy;

P_{loss}^a is the active power losses of the studied system after a new control strategy applied;

ΔP_{loss} is the addition active power losses of the studied system due to a new control strategy applied;

$\Delta V_{N \times 1}$ is the voltage deviation matrix of the nodes of the studied system;

$V_{N \times 1}$ is the voltage matrix of the nodes of the studied system;

V_n is the objective voltage of the nodes of the studied system, which is 1 in p.u. value.

The first node of the studied system is taken as the slack bus in this model.

2. u is the control strategy applied in the system. Here, u can be expressed as follow:

$$u_{M \times 1} = \Delta Q_{M \times 1} \tag{3}$$

where

M is the number of the nodes with reactive power generation capability;

$\Delta Q_{M \times 1}$ is the reactive power generation increment matrix of the nodes with reactive power generation capability, which should be solved;

3. J is total cost which should be minimized;
4. $Q \ge 0$ and $R > 0$ are symmetric matrices; they are weighted matrices here for state deviations and control efforts.

The linearized model in a specific operation condition of the studied system in the form of continuous-time linear time-invariant system can be expressed as follow:

$$\dot{x} = Ax + Bu \tag{4}$$

Then, the system can be stabilized by static state feedback control:

$$u = -Kx \tag{5}$$

According to Bellman's optimally principle, the solution of this optimal control problem is given as:

$$K = R^{-1} B^T P \tag{6}$$

where P is the unique positive definite solution of the Algebraic Riccati Equation (ARE):

$$A^T P + PA - PBR^{-1}B^T P + Q = 0 \tag{7}$$

4 CASE STUDY

Some cases of typical rural electric power grid with PVs are studied in this part to validate the effeteness of the optimal voltage control method proposed in this paper.

The studied system is a substation area with a main transformer, a 10 kV bus, three 10 kV feeders, 23 load nodes, and 10 PVs at different load nodes. The branch parameters are listed in Table 1, and the node parameters in Table 2. The voltage of the 10 kV bus is assumed to be kept at 1.0 p.u. which is adjusted by the tap changer of the main transformer and capacitors at bus 1.

Therefore, the task is to solve the optimal reactive power output of the PVs so that the quadratic sum of the voltage deviations of all the nodes is minimized as well as the quadratic sum of the additional reactive power outputs and additional active power losses. However, to avoid the controller to react to the voltage deviations too frequently, a threshold is set as 0.02 for the voltage deviations. That is, there are no need to take a control action if the voltages are between 0.98 and 1.02.

To solve the optimal voltage control strategy, the parameters of LQR based method are defined in Table 3.

4.1 Case 1: The loads are minimum and the PVs largest

In this case, the loads are all in their minimal value (20% of the values shown in Table 2), while the PVs are generating largest active power (200% of the

Table 1. The branch parameters.

Branch no.	Start bus	End bus	R	X	B
1	1	2	0.001168	0.001790	0
2	2	3	0.005915	0.006499	0
3	3	4	0.005915	0.006499	0
4	4	5	0.005915	0.006499	0
5	5	6	0.005915	0.006499	0
6	6	7	0.005915	0.006499	0
7	7	8	0.005915	0.006499	0
8	1	9	0.005380	0.005911	0
9	9	10	0.005380	0.005911	0
10	10	11	0.005380	0.005911	0
11	11	12	0.005380	0.005911	0
12	12	13	0.005380	0.005911	0
13	13	14	0.005380	0.005911	0
14	12	15	0.005380	0.005911	0
15	15	16	0.005380	0.005911	0
16	16	17	0.005380	0.005911	0
17	1	18	0.000964	0.001477	0
18	18	19	0.000964	0.001477	0
19	19	20	0.000964	0.001477	0
20	20	21	0.004665	0.005940	0
21	21	22	0.004665	0.005940	0
22	22	23	0.004665	0.005940	0
23	23	24	0.004665	0.005940	0

Table 2. The node parameters.

Bus no.	Pd	Qd	Pg	Qg	Qgmax	Qgmin	Qc
1	Slack bus				100	−100	4.32
2	0.0962	0.0466	0	0	0	0	0
3	0.0962	0.0466	0	0	0	0	0
4	0.0962	0.0466	0.4	0	2	−2	0
5	0.0962	0.0466	0	0	0	0	0
6	0.0962	0.0466	0.4	0	2	−2	0
7	0.0962	0.0466	0	0	0	0	0
8	0.0962	0.0466	0.4	0	2	−2	0
9	0.1479	0.0716	0	0	0	0	0
10	0.1479	0.0716	0.4	0	2	−2	0
11	0.1479	0.0716	0	0	0	0	0
12	0.1479	0.0716	0.4	0	2	−2	0
13	0.1479	0.0716	0	0	0	0	0
14	0.1479	0.0716	0	0	0	0	0
15	0.1479	0.0716	0.4	0	2	−2	0
16	0.1479	0.0716	0.4	0	2	−2	0
17	0.1479	0.0716	0	0	0	0	0
18	0.4859	0.2353	0	0	0	0	0
19	0.4859	0.2353	0	0	0	0	0
20	0.4859	0.2353	0.4	0	2	−2	0
21	0.4859	0.2353	0	0	0	0	0
22	0.4859	0.2353	0	0	0	0	0
23	0.4859	0.2353	0.4	0	2	−2	0
24	0.4859	0.2353	0.4	0	2	−2	0

Table 3. The parameters of the LQR based method.

Parameter	Value	Parameter	Value
Q	diag(100)	R	diag(1)
N	24	M	10

Table 4. The optimal strategy of case 1.

Node no.	8	16	24
ΔQ_g	−0.85	−1.21	−0.11

values shown in Table 2) but not generating reactive power.

By the LQR based method, the optimal strategy is obtained as shown in Table 4.

As shown in Table 4, PVs should absorb some reactive power from the grid to decrease the voltages.

The voltage profiles before and after the control strategy is executed are compared and the results are shown in Figure 1.

4.2 Case 2: The loads are minimum and the PVs zeroes

In this case, the loads are the same as in case 1 and the PVs are not generating active power or reactive power.

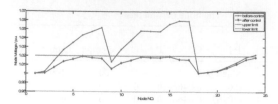

Figure 1. The voltage profile improvement of case 1.

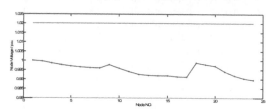

Figure 2. The voltage profile of case 2.

By the LQR based method, it's not necessary for the PVs to generate reactive power. It's obvious because the voltages in this case are beyond the limits as shown in Figure 2.

4.3 Case 3: The loads are maximum and the PVs largest

In this case, the loads are all in their maximal value (120% of the values shown in Table 2), and the PVs are generating largest active power but not generating reactive power as the same as in case 1.

By the LQR based method, the optimal strategy is obtained as shown in Table 5.

As shown in Table 5, the PV at node 8 should absorb some reactive power from the grid since the voltages of the nodes 5~8 exceed the upper limit, as similar as the PV at node 16, but the ΔQ_g at node 16 is much smaller than that at node 8 because the voltage of the node 8 is just a little higher than the upper limit. And the PV at node 24 should generate some reactive power into the grid to improve the voltage of the nodes at the terminal of the distribution line.

The voltage profiles before and after the control strategy is executed are compared and the results are shown in Figure 3.

4.4 Case 4: The loads are maximum and the PVs zeroes

In this case, the loads are the same as in case 3, while the PVs are not generating active power or reactive power as the same as in case 2.

By the LQR based method, the optimal strategy is obtained as shown in Table 6.

As shown in Table 6, PVs should generate some reactive power at different nodes into the grid to improve the voltage of the nodes along the distri-

Table 5. The optimal strategy of case 3.

Node no.	8	16	24
ΔQ_g	−0.36	−0.018	0.43

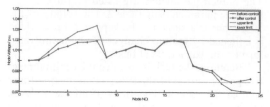

Figure 3. The voltage profile improvement of case 3.

Table 6. The optimal strategy of case 4.

Node no.	8	12	16	23	24
ΔQ_g	0.1	0.44	0.84	1.45	0.42

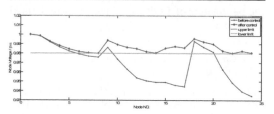

Figure 4. The voltage profile improvement of case 4.

bution line. The PVs work as reactive power compensator in this case.

The voltage profiles before and after the control strategy is executed are compared and the results are shown in Figure 4.

5 CONCLUSIONS

In this paper, the optimal voltage control issue of the rural electric power grid with PVs is studied. The optimal voltage control model is established based on LQR theory. The feasibility of the proposed method is illustrated by several typical operation cases of rural electric power grid with PVs. The proposed method provides a practical technique for voltage improvement optimization.

REFERENCES

China National Energy Board. "13th Five-Year" development plan for the utilization of solar energy. 2016.

Dai. X. L., Application of reactive compensation technology for distribution system [J]. Power System Technology, 1999, 23(6): 11–14.

Zhao. G. Z., Modern Control Theory. China Machine Press, 2010.

Electromechanical Control Technology and Transportation – Jia & Wu (Eds)
© *2017 Taylor & Francis Group, London, ISBN 978-1-138-06752-3*

Overvoltage analysis and suppression in the double-neutral section electrical sectioning device

Yuxin Liu & Guanghui Liu
Department of Electrical Engineering, Zhengzhou Railway Vocational and Technical College, Zhengzhou, China

ABSTRACT: Overvoltage is easily caused when an electric locomotive goes through an electrical sectioning device, thereby leading to the breakdown of the equipment's insulation. A bilateral passive RC suppression method is proposed to restrain overvoltage using an eight-span overlapping electrical sectioning device. The combined simulation of an electrical locomotive–electrical sectioning–traction network is established, which is based on the seven-span single-neutral section and eight-span double-neutral section model. Overvoltage simulation is done in three cases, where all transition processes of electrical sectioning are included. Results show that combined simulation results are consistent with the theoretical values, and the overvoltage caused by the seven-span electrical sectioning is equivalent to the experimental data, which verify the correctness of the chain circuit model and the effectiveness of the combined simulation. The maximum overvoltage on the bilateral passive RC circuit of the eight-span electrical sectioning device over 60 kV and suppressive effect is obvious.

1 INTRODUCTION

The 25 kV single-phase AC power system is applied to electric railway, which supplies it to electric locomotives. The negative sequence current of single-phase traction load will be reduced through the method of cyclic phase. At present, numerous electrical sectioning devices transit from five span to seven span, which are widely used in high-speed railway (Zhao Xiaole, 2010; Zhou Zhigang, 2004 & Sui Yanmin, 2010). Operating experience and experiments show that overvoltage is caused by electric locomotive going through seven-span electrical sectioning device in a short time. It is also affected by random factors, which may cause the breakdown of equipment's insulation and threaten safety of electric locomotive.

Eight-span double-neutral section device is applied to Rome–Naples high-speed railway as well as Shi-tai & Wu-guang passenger line. Compared with the general application of seven-span electric section (Gu Yinan, 2009), electric locomotive going through eight-span double-neutral section will be powered off during the transition process. However, the analysis and improvement effect of overvoltage (FU Qiang, 2005) shocking insulation system in the transition process are studied and described in the literature.

The traction network can be described as a composite chain circuit model, which consisted of longitudinal series components or transverse parallel components. A new method that RC (Liu Mingguang, 2007) circuit applied to double-neutral section has been proposed.

2 OVERLAPPING ELECTRICAL SECTIONING DEVICE

2.1 Seven-span overlapping electrical sectioning

Functions of mechanical segmentation and electric split phase are included in overlapping electrical sectioning device. Add a seven-span length neutral line to seven-span overlapping electrical sectioning device, which ensures the middle five-span insulation, and two four-span insulated overlapping sections are overlapped. Each span is about 55 m; neutral line length is about 300 m; transition zone length is about 60 m; electricity area length is about 70 m; the parallel distance between the neutral conductor and catenary is about 0.5 m; and hyphenation sign is about 120 m away from no electricity area, as shown in Fig. 1.

2.2 Eight-span double-neutral segment electrical sectioning overlap

On the double-neutral segment electrical sectioning overlap, an air gap is added between two neutral lines; on a neutral line, the mutual

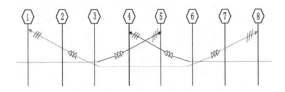

Figure 1.   Seven-span overlapping sectioning device.

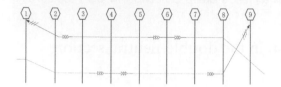

Figure 2. Double-neutral segments electrical sectioning device.

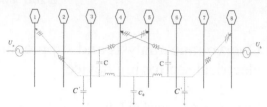

Figure 3. Seven-span single-neutral section.

inductance of two-phase voltage can be reduced, and the induced voltage distribution of neutral line is improved, which overcomes the limitation of single-neutral section. Eight-span overlapping electrical sectioning consisted of two four-span insulated overlapping sections, as shown in Fig. 2.

3   OVERVOLTAGE SUPPRESSION SIMULATIONS

3.1   *Simulation of single-neutral section*

By using the electromagnetic transient tool, the locomotive network electric phase model diagram and the diagram of seven-span/eight-span and eight-span with bilateral RC circuits are established; when the locomotive goes into the electric split phase, over-voltage of locomotive can be simulated by regulating the initial phase angle. The dynamic simulation model of electric split phase is established, and the electrical locomotive going through overlapping electrical sectioning device is simulated. Single-neutral section is shown in Fig. 3.

In Figure 3, C_{12} is the capacitance between neutral section and catenary; C' is the capacitance of neutral line to ground; and C_0 is the capacitance of neutral line with no electricity zone to ground.

Before electrical locomotive going into the split phase, the pantograph connects with catenary only; after entering the electric split phase, it connects the catenary and the neutral line at the same time; finally, it enters the area without electricity. When the electrical locomotive goes out of the electric split phase, the pantograph contacts with the catenary and neutral line at the same time; then it connects with catenary only; finally, the locomotive goes out of the electric split phase. When the electrical locomotive goes through the area without electricity, the pantograph contacts with neutral line only; when the electrical locomotive goes through the area without electricity, the induced capacitance that is in the locomotive and neutral line forms the oscillation circuit with the influence of the residual pressure, which generates overvoltage because of the distributed capacitance and inductance of neutral line. When the locomotive goes into the split phase, initial phase angle of traction transformer secondary side voltage (Φ)

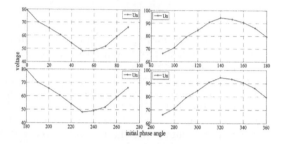

Figure 4. Data of single-neutral section.

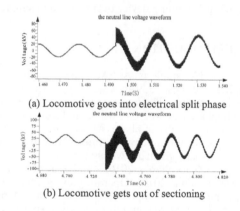

(a) Locomotive goes into electrical split phase

(b) Locomotive gets out of sectioning

Figure 5. $\Phi = 140°$ voltage waveform of neutral line.

is different. If Φ_1 is the initial phase of traction power supply, then $\Delta\Phi = \Phi - \Phi_1$ is fixed. When the locomotive goes into electricity phase, Φ_1 is in the dynamic change, the phase angle Φ of voltage on pantograph will be changed, and overvoltage of neutral line is shown in Fig. 4.

The simulation waveform is shown as follows:

When $\Phi = 140°$, the maximum overvoltage is 95.46 kV on the neutral line; when $\Phi = 320°$, the maximum overvoltage is 95.57kV. From Fig. 5, when the electrical locomotive passes through the electric split phase, the maximum overvoltage reaches 95.57kV. When electrical locomotive is traveling through electrical sectioning, the randomness of overvoltage is very large, and the maximum overvoltage cannot be captured easily. On the basis of the electromagnetic transient model, when the electrical locomotive goes through seven-span overlapping sectioning, the maximum overvoltage reaches 95.57 kV, and the overvoltage stimulation results are the same as the measured experimental data. On the basis of this model, further analyses are conducted in Section 3.2.

3.2 *Effect of double-neural section*

When passenger and freight electric locomotives are traveling in the railway, in order to solve restrictions of pantographs, the eight-span overlapping sectioning of double-neutral line and three ports appear. It is shown in Fig. 6.

In Figure 6, C_1 is the capacitance between neutral section 1 and catenary; C_{10} is the capacitance of neutral line parallel to ground; C_1' is the capacitance of neutral line with no electricity zone to ground; C_{12} is the capacitance between neutral line 1 to neutral line 2; C_2 is the capacitance between neutral section 1 and catenary; C_{20} is the capacitance of neutral line parallel to ground; and C_2' is the capacitance of neutral line with no electricity zone to ground.

When Φ is different, overvoltage on the neutral line is shown in Fig.7.

When electric locomotive goes through overlapping electrical sectioning, simulations of neutral line voltage are shown in Figs. 8 and 9.

From Fig. 9, overvoltage can be restrained when electric locomotive goes through overlapping electrical sectioning to some degree by making use of the double-neutral segment. However, when electric locomotive goes through double-neural section electrical sectioning overlap, threefold overvoltage

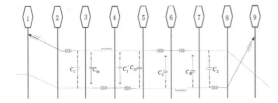

Figure 6. Double-neutral line equivalence.

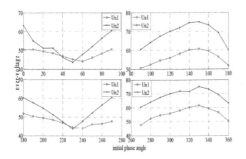

Figure 7. Overvoltage on double-neutral section.

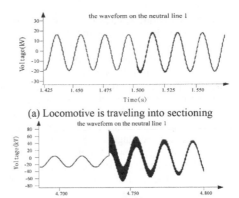

(a) Locomotive is traveling into sectioning

(b) Locomotive gets out of sectioning

Figure 8. $\phi = 330°$ waveform on neutral line 1.

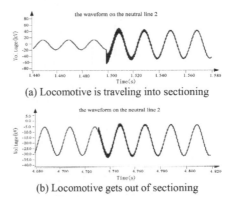

(a) Locomotive is traveling into sectioning

(b) Locomotive gets out of sectioning

Figure 9. $\Phi = 330°$ waveform on neutral line 2.

still may be produced, which is close to the value of overvoltage of locomotive roof insulation. Thus, in order to study overvoltage protection, bilateral suppression circuit is adopted in the double-neutral section.

3.3 *Suppression simulation of double-neutral section with RC circuit*

Former researches pointed out that low-frequency oscillation voltage of neutral line has a great effect on overvoltage. In order to decrease oscillation, protection device is equipped in the neutral line. When RC protection circuit is installed in the whole circuit, the parameters of the whole circuit will change. If the capacitance of RC protection circuit becomes greater, the oscillation frequency of the circuit will be reduced; and because of resistance, the circuit will transform from underdamped state to critically damped or overdamped state, causing overvoltage of the whole circuit become smaller or overvoltage not generated. It is shown in Fig. 10.

Here, Φ is the initial phase when electric locomotive travels into electric sectioning. Overvoltage on neutral line when Φ is different is shown in Fig. 11.

When $\Phi = 150°$, simulation of overvoltage on neutral line is shown in Figs. 12 and 13.

When RC protection circuit is applied to double-neutral section, suppressive effect of the maximum over-voltage on neutral line is obvious. Compared with seven-span single neutral line, when $\Phi = 320°$, the maximum overvoltage is 75.57 kV; when $\Phi = 330°$, in the eight-span double-neutral section (without RC protection device), the maximum overvoltage on the neutral line 2 is 75.15 kV; while RC protection device has been applied to double-

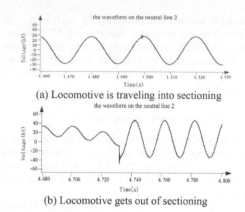

(a) Locomotive is traveling into sectioning

(b) Locomotive gets out of sectioning

Figure 13. $\Phi = 150°$ waveform on neutral line 2.

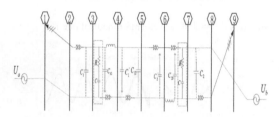

Figure 10.   Double-neutral section bilateral with RC.

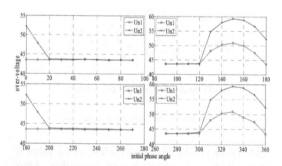

Figure 11.   Double-neutral section of both sides with RC circuit.

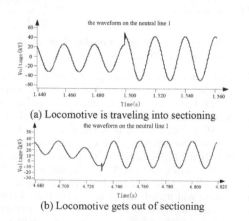

(a) Locomotive is traveling into sectioning

(b) Locomotive gets out of sectioning

Figure 12. $\Phi = 150°$ waveform on neutral line 1.

neutral section, when $\Phi = 330°$, the maximum overvoltage on the neutral line 2 decreases to 59.28 kV.

4   CONCLUSIONS

1. Using electromagnetic transient tools and on the basis of the seven-span and eight-span double-neutral segment electrical sectioning overlap model, a simulation model of a traction power supply system has been built.
2. The maximum overvoltage in seven-span electrical sectioning overlap is 95.57 kV. At the same time, the maximum overvoltage of the eight-span double-neutral section is 75.10 kV. When an RC protection device is applied to the double-neutral section, the maximum overvoltage on the neutral line 2 decreases to 59.28 kV, and the suppressive effect is obvious.

REFERENCES

Fu Qiang. Overlapping Electrical Sectioning Device of OCS [J]. Electric Railway, 2005(2): 30–32.

Gu Yinan, Wang Yi, Wang Hong. Over-voltage Analysis during Electric Locomotive's Auto-passing the Phase Division [J]. Electric Railway, 2009(3):1–3.

Liu Mingguang, Lu Yan'an, WEI Hongwei, et al. Study on Articulated Neutral Section Over-voltage Test [J]. Electric Railway, 2007(4):15–17.

Sui Yanmin. Key Problems of Electrified Railway Catenary Electric Split Phase Technology. Chinese Railways, 2010(11):30–34.

Zhao Xiaole, Zhang Xian, Li Xiaolei. Preliminary Discussion on the Technology about Catenary on Three Fracture Overlapping Electrical Sectioning Device, Railway Construction Technology, 2010(7):22–29.

Zhou Zhigang, Zhang Baoqi. A Kind of Catenary Electric Split Phase Technology Suitable for High-speed Electrified Railway. Chinese Railways, 2004(11):(33–35).

Electromechanical Control Technology and Transportation – Jia & Wu (Eds)
© *2017 Taylor & Francis Group, London, ISBN 978-1-138-06752-3*

Control strategy of a car component stiffness test based on RBF neural network PID

Zhiming Wang & Yihui Yan
Department of Mechatronics Engineering and Automation, Shanghai University, Shanghai, China

ABSTRACT: The fact that different car components have different material properties makes the control strategy very complex in a components stiffness test. With the traditional PID control algorithm it is difficult to achieve stable control in a stiffness test, because its three parameters are fixed. A PID control method based on RBF neural network is put forward, which can make use of neural network learning to adjust the three parameters of PID according to the actual situation. RBF neural network PID can get its best three PID parameters when testing each kind of different car component. In this study, we use MATLAB to simulate the RBF neural network PID and traditional PID and compare the effects of two control methods. It proves that RBF neural network PID can achieve stable control with a small overshoot, fast response, high stability, and short time.

1 INTRODUCTION

The body structure design of car mainly determines its reliability, safety, comfort, and beauty. The body of car directly bears complex loads from outside in the traffic accident. High-quality car components can reduce casualties. Therefore, the study about stiffness of car components is significant.

In this paper, our main work is to design a stiffness test machine for different car components, build a mathematical model, and study the relevant control strategy. We can prove that the RBF neural network PID method can achieve stable control with a small overshoot, fast response, high stability, and short time.

2 MECHANICAL STRUCTURE OF STIFFNESS TEST

Mechanical structure of car component testing machine is composed of a base structure (No. 1), a tower (No. 2), and a principal axis (No. 5), as shown in Figure 1. The base structure uses a motor and synchronous belt to drive two large wire rods to realize the horizontal motion of the tower, which is the shelf of the testing machine that move the principal axis up and down. Servo cylinder is fixed in the principal axis, which can realize six-axis movement with the movement of the tower and the principal axis.

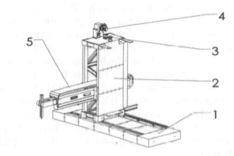

Figure 1. Mechanical structure.

3 CONTROL SYSTEM

Taking into account the accuracy and real-time control, the test machine adopts electric servo drive control mode. The hardware platform includes the PLC controller, drive, load sensor, servo electric cylinder, and some signal sensor components, as shown in Figure 2.

The working principle of the control system is that the PLC controller outputs pulse signal to the servo driver, and the servo driver controls the servo cylinder with position control mode or speed control mode. The servo electric cylinder is provided with a load sensor. The PLC controller reads real-time load value sent by the AD converter module; then, the PLC controller sends load value to PC terminal with serial communication of Modbus protocol. PC terminal reads the digital value and

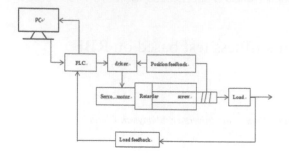

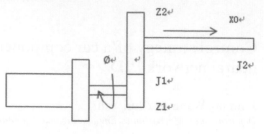

Figure 2. Control system.

Figure 3. Servo cylinder.

converts it to analog value. Finally, PC terminal controls the whole system with relevant control strategy.

4 DESIGN OF CONTROL STRATEGY

4.1 Mathematical model

PC sends the target value to the PLC controller, which in fact is a proportional component process, and the ratio is $K1$. PLC sends a number of pulses to the servo drive according to the real-time value of the load sensor, which is also a proportion component process, and the ratio is $K2$. The servo driver receives the pulse signal to control the angle of the servo motor, which is a cumulative process. The number of accumulated pulses continue to drive the motor, which is an integral part, whose transfer function is K_θ/S, and K_θ is the integral coefficient.

The implementation mechanism of the testing machine is the servo cylinder. The working principle of the servo cylinder is that the rotation of servo motor is changed into the reciprocating linear motion of screw rod with speed reducer. The internal schematic diagram of the servo cylinder is shown in Figure 3. The general cylinder is composed of a servo motor, a speed reducer, and a screw rod. The input angle of motor is θ, X_0 is the displacement of screw output, and Z_1 and Z_2 are gears of reducer, J_1 and J_2 are the moments of inertia of axes 1 and 2, respectively. Electric cylinder is a complex mechanical system inertia device, which integrates the moment of inertia, the damping coefficient, and the elastic modulus of all components, which can be equivalent to the quality of spring and damping system. J_Σ, D_Σ, and K_Σ are equivalent moments of inertia, damping coefficient, and stiffness coefficient of the whole system, respectively. The principle of differential equations can be obtained as follows according to Newton's law:

$$J_\Sigma \frac{d^2\theta}{dt^2} + D_\Sigma \frac{d\theta}{dt} + K_\Sigma\theta = K_\Sigma\theta_i \tag{1}$$

where θ is the input angle of the motor, and the output angle displacement of screw rod is θ_i. And the relationship between X_0 and θ is:

$$\theta = \frac{2\pi}{P} \frac{Z_2}{Z_1} X_0 \tag{2}$$

The following equation can be obtained from equations (1) and (2):

$$J_\Sigma \frac{d^2 X_0}{dt^2} + D_\Sigma \frac{dX_0}{dt} + K_\Sigma X_0 = \frac{P}{2\pi} \frac{Z1}{Z2} \cdot K_\Sigma\theta_i \tag{3}$$

The transfer function of the cylinder is obtained by Laplace transform from (3):

$$G(s) = \frac{X_0(s)}{G(s)} = \frac{K_\alpha \cdot K_\Sigma\theta_i}{J_\Sigma S^2 + D_\Sigma S + K_\Sigma} \tag{4}$$

Formula (4) can also be expressed as (5):

$$G(s) = \frac{X_0(s)}{G(s)} = \frac{\lambda_n}{S^2 + 2\zeta_n\omega_n S + \omega_n^2} \tag{5}$$

And $\omega_n = \sqrt{\frac{K_\Sigma}{J_\Sigma}}$, $\zeta_n = \frac{D}{2J_\Sigma}\sqrt{\frac{J_\Sigma}{K_\Sigma}}$, $\lambda_n = \frac{Z_1}{Z_2}\frac{P}{2\pi}\frac{K_\Sigma}{J_\Sigma}$ in the equation. Clearly, the system transfer function of the electric cylinder is a typical two time oscillation function.

The force value of the measured object is obtained by the moving of the horizontal displacement of the electric cylinder shaft; the longer the displacement, the greater the value of the load. When the target force value is reached, the motor stops rotating and displacement is not changed. Therefore, the process can be simplified as a transition to a smooth process, which can be seen as an inertial link, so we can know that the transfer function is:

$$G_1(s) = \frac{K_3}{T_2 S + 1} \tag{6}$$

where T_3 is the time constant and K_3 is the proportional constant. The feedback channel is only provided to detect the role of real-time load, which

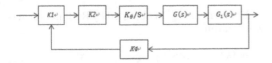

Figure 4.   Transfer function flow diagram.

can be seen as a proportion of links process; however, it does not participate in the process of signal transmission system. As mentioned above, the transfer function block diagram of each flow of the car parts stiffness testing machine is shown in Figure 4. We can conclude that the transfer function of the whole system is:

$$G(s)_n = \frac{K_1 \cdot K_2 \cdot K_\theta \lambda_n \cdot K_3}{S(S^2 + 2\zeta_n \omega_n S + \omega_n^2)(T_2 S + 1)} \tag{7}$$

Through the above research, it is concluded that the system consists of three parts: an integral link, an oscillating link, and an inertial link. The transfer function diagram of the whole system is shown in Figure 4.

4.2   RBF PID controller design

Radial Basis Function (RBF) neural network is a neural network proposed by J. Moody and C. Darken in the late 1980s, RBF neural network simulates the structure of the neural network in the human brain, and it is a kind of local approximation network, which can approximate any continuous function with any precision. The radial basis function network structure is composed of a hidden layer and a linear output layer, and the topology structure is shown as Figure 5.

Radial basis function is used as the activation function of network, and the most commonly used radial basis function is Gauss function, which has some advantages, such as simple expression, good radial symmetry, and good smoothness.

The network structure includes n input node, m hidden layer nodes, and P output nodes. The RBF network of the first j hidden layer nodes of the output is:

$$q_j = \exp\left(-\frac{\|X - C_j^2\|}{2\sigma_j^2}\right) \tag{8}$$

where X is an n-dimensional input vector, C_j is the center vector of the first j hidden layer nodes, and the input vector X has the same dimension:

$$C_j = [C_{j1}, C_{j2}, ..., C_{j3}]^T \quad i = 1, 2, ..., n.$$

σ_j is the basic broadband parameter of the hidden layer node i. The output of the first K node of the network output layer is:

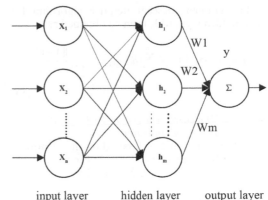

input layer   hidden layer   output layer

Figure 5.   RBF neural network.

$$y_k = \sum_{j=1}^m w_{kj} q_j - \theta_k \tag{9}$$

$k = 1, 2, ..., p$. w_{kj} is the connection weight between neuron K and neuron j and θ_k is the threshold of the first K output node.

Select the target function of network training:

$$j = \frac{1}{2} \sum_{k=1}^p (d_k - y_k)^2 = \frac{1}{2} \sum_{k=1}^p e_k^2 \tag{10}$$

where d_k is the ideal output of the second k output node, and y_k is the actual output of the first k output node.

Partial derivative of the center vector is:

$$\frac{\partial J}{\partial C_{ji}} = \left[\frac{\partial J}{\partial y_k} \cdot \frac{\partial y_k}{\partial q_j}\right] \cdot \frac{\partial q_j}{\partial C_{ji}}$$
$$= -\left[\sum_{k=1}^p (d_k - y_k)\right] \cdot q_j \frac{x_i - c_{ji}}{\sigma_j^2} \tag{11}$$

The gradient descent iterative algorithm for each component of the node center can be obtained as:

$$c_{ji}(k+1) = c_{ji}(k) + \alpha\left[\sum_{k=1}^p (d_k - y_k)w_{kj}\right] \cdot q_j \frac{x_i - c_{ji}}{\sigma_j^2} \tag{12}$$

Partial derivative of width is:

$$\frac{\partial J}{\partial \sigma_j} = \left[\frac{\partial J}{\partial y_k} \cdot \frac{\partial y_k}{\partial q_j}\right] \cdot \frac{\partial q_j}{\partial \sigma_j}$$
$$= -\left[\sum_{k=1}^p (d_k - y_k)w_{kj}\right] \cdot q_j \frac{\|X - C_j\|^2}{\sigma_j^3} \tag{13}$$

The gradient descent iterative algorithm for each hidden layer width can be obtained as:

$$\sigma_j(k+1) = \sigma_j(k) + \beta\left[\sum_{k=1}^{p}(d_k - y_k)w_{kj}\right]q_j\frac{\|X - C_j\|^2}{\sigma_j^3}$$

(14)

4.2.1 PID controller based on RBF neural network identification

The BRF neural network based on the identification of PID controller is shown in Figure 6.

We use incremental PID controller:

$$e(k) = rin(k) - yout(k)$$

(15)

The three inputs of PID are:

$$xc(1) = e(k) - e(k-1)$$
$$xc(2) = e(k)$$
$$xc(3) = e(k) - 2e(k-1) + e(k-2)$$

(16)

The learning performance of neural network is:

$$E(k) = \frac{1}{2}e^2(k)$$

(17)

The three parameters of PID (ki, kp, and kd) are adjusted by the gradient descent method:

$$\Delta k_p = -\eta\frac{\partial E}{\partial k_p} = -\eta\frac{\partial E}{\partial yout}\frac{\partial yout}{\partial u}\frac{\partial u}{\partial k_p}$$
$$= \eta e(k)\frac{\partial yout}{\partial u}xc(1)$$

(18)

$$\Delta k_i = -\eta\frac{\partial E}{\partial k_i} = -\eta\frac{\partial E}{\partial yout}\frac{\partial yout}{\partial u}\frac{\partial u}{\partial k_i}$$
$$= \eta e(k)\frac{\partial yout}{\partial u}xc(2)$$

(19)

$$\Delta k_d = -\eta\frac{\partial E}{\partial k_d} = -\eta\frac{\partial E}{\partial yout}\frac{\partial yout}{\partial u}\frac{\partial u}{\partial k_d}$$
$$= \eta e(k)\frac{\partial yout}{\partial u}xc(3)$$

(20)

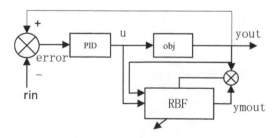

Figure 6.   RBF PID controller.

The formula $\frac{\partial yout}{\partial u}$ is obtained by RBF neural network output:

$$\frac{\partial you(k)}{\partial u(k)} \approx \frac{\partial myou(k)}{\partial u(k)} = \sum_{j=1}^{m}w_jh_j\frac{c_j - x^i}{2b_j^2}$$

(21)

$myou(k)$ is RBF neural network output in k time.

5   EXPERIMENTAL ANALYSIS

We use MATLAB to simulate the RBF neural network PID and traditional PID and compare the effects of two control method. After analyzing the system, the coefficients of each stage are confirmed, and the transfer function of whole system is obtained as:

$$\frac{80}{0.001s^4 + 0.003s^3 + 15S^2 + 25S + 3}.$$

(22)

Because the coefficient of S^4 and S^3 is close to zero, the transfer function is simplified as:

$$\frac{80}{15S^2 + 25S + 3}.$$

(23)

The signal of the simulation input is a step signal of the amplification factor of 500, which represents the target load of input 500 N. The simulation result is shown in Figure 7.

The experimental result shows that the conventional PID has a larger overshoot, while RBF neural network PID can quickly respond with almost no overshoot and become stable with relatively short time.

Then, we input a square wave signal with a frequency of 2 and amplitude of 10 to test the fast response effect of two kinds of control. From Figure 8, we can know that conventional PID cannot track wave signal source, while RBF PID can quickly track wave signal without overshoot.

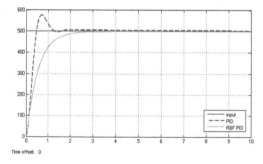

Figure 7.   Simulation results of step signal.

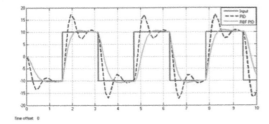

Figure 8. Simulation results of wave signal.

From the simulation results of two signals, we can conclude that RBF PID can achieve stable control with a small overshoot, fast response, high stability, and short time, because RBF neural network can adjust the three parameters of PID through neural network learning.

6 SUMMARY

RBF neural network is a kind of local approximation network, which can approximate any continuous function with any precision and make use of the neural network learning to adjust the three parameters of PID according to the actual situation. We try to use RBF PID in the automobile components stiffness test, because it can get its best three PID parameters depending on neural network online learning when testing each kind of different auto parts. It can prove that RBF neural network PID can achieve stable control with a small overshoot, fast response, high stability, and short time in a car components stiffness test.

REFERENCES

Cheng Longlong, Zhang Guangju, Wan Baikun, Hao Linlin, Qi Hongzhi, Ming Dong. Radial Basis Function Neural Network-based PID model for functional electrical stimulation system control [J]. Conference proceedings: ... Annual International Conference of the IEEE Engineering in Medicine and Biology Society. IEEE Engineering in Medicine and Biology Society. Conference, 2009, 2009.

Helsen, J., L. Cremers, P. Mas, P. Sas. Global static and dynamic car body stiffness based on a single experimental modal analysis test. Proceedings of ISMA, 2010.

Huaizhong Chen. Research of the Electro-hydraulic Servo System Based on RBF Fuzzy Neural Network Controller [J]. Journal of Software, 2012, 79.

Ireneusz Czarnowski, Piotr Jędrzejowicz. Designing RBF Networks Using the Agent-Based Population Learning Algorithm [J]. New Generation Computing, 2014, 323–324.

Ito Kazuhisa, Yamada Tsuyoshi, Ikeo Shigeru, Takahashi Koji. Application of Simple Adaptive Control to Water Hydraulic Servo Cylinder System [J]. Chinese Journal of Mechanical Engineering, 2012, 05: 882–888.

Jie Zhao, Jun Zhong, Jizhuang Fan, Seungik Baek. Position Control of a Pneumatic Muscle Actuator Using RBF Neural Network Tuned PID Controller [J]. Mathematical Problems in Engineering, 2015, 2015.

Khaleel, M. A., K. I. Johnson, J. E. Deibler and R. W. Davies. Effect of Glazing System Parameters on Glazing System Contribution to a Lightweight Vehicle's Torsional Stiffness and Weight [C]. SAE paper, 2000-01-2719, 20.

Lei Yu, Shumin Fei, Jun Huang, Yu Gao. Trajectory Switching Control of Robotic Manipulators Based on RBF Neural Networks [J]. Circuits, Systems, and Signal Processing, 2014, 334.

Pachón-García, F. T., A. Jiménez-Barco, J. M. Paniagua-Sánchez, M. Rufo-Pérez. New approach based on ANN and RBF for analyzing the spatial distribution of electromagnetic field from an exposure standpoint [J]. Neural Computing and Applications, 2014, 256.

Roger Achkar, Souhad Mcheik, Youssef Harkouss. Accurate Wavelet Neural Network for Efficient Controlling of an Active Magnetic Bearing System [J]. Journal of Computer Science, 2010, 612.

Rui-Bin Feng, Chi-Sing Leung, A.G. Constantinides. LCA based RBF training algorithm for the concurrent fault situation [J]. Neurocomputing, 2016.

Sedighizadeh, M., Rezazadeh, A. A modified Adaptive Wavelet PID Control Based on Reinforcement Learning for Wind Energy Conversion System Control [J]. Advances in Electrical and Computer Engineering, 2010, 102.

Shuqing Zhang, Yongtao Hu, Hongyan Bao, Xinxin Li. Parameters determination method of phase-space reconstruction based on differential entropy ratio and RBF neural network [J]. Journal of Electronics (China), 2014, 311.

Xia Rongfei, Huang Jun, Chen Yifei, Feng Yongjian. A study of the method of the thermal conductivity measurement for VIPs with improved RBF neural networks [J]. Measurement, 2016.

Yanhui Xi, Hui Peng, Xiaohong Chen. A sequential learning algorithm based on adaptive particle filtering for RBF networks [J]. Neural Computing and Applications, 2014, 253–254.

Yanjun Li, Yi Liu, Liangsheng Song, Shan Liu. A rapid particle swarm optimization algorithm with convergence analysis [J]. IFAC Proceedings Volumes, 2014, 473.

Zeng Songwei, Hu Haigen, Xu Lihong, Li Guanghui. Nonlinear Adaptive PID Control for Greenhouse Environment Based on RBF Network [J]. Sensors, 2012, 125.

Zhi Huang, Hong Yuan. Research on regional ionospheric TEC modeling using RBF neural network [J]. Science China Technological Sciences, 2014, 576.

Zhong-Qiang, Wu, Wen-Jing Jia, Li-Ru Zhao, Chang-Han Wu. Maximum wind power tracking based on cloud, RBF neural network [J]. Renewable Energy, 2016, 86.

Electromechanical Control Technology and Transportation – Jia & Wu (Eds)
© 2017 Taylor & Francis Group, London, ISBN 978-1-138-06752-3

A new preference-based model to solve the cold start problem in a recommender system

Kun Liu, Wei Liu & Xiaoyun Chen
School of Information Science and Engineering, Lanzhou University, China

ABSTRACT: A Recommender System (RS) aims to provide personalized recommendations to users for specific items (e.g., movies, music, and books). Popular techniques involve the Content-Based (CB) model, Collaborative Filtering (CF), and Knowledge-Based (KB) approaches. The cold start issue is a serious problem in RS and is divided into the cold users' problem and the cold items' problem. In case of new users, RS does not have enough information about their preferences in order to give recommendations. For this situation, we propose a new model, named preference-based recommender, to solve the cold users' problem. Our model first groups the users into clusters according to the attributes of items that have been rated by users. Then, it searches the neighbors of cold users from these clusters with a similarity measure (the neighbors of a cold user are chosen from the cluster to which the cold user belongs). Finally, we use the rating information of the neighbors of a cold user to give recommendations. Our experiments show the performances in different situations adopting the well-known data sets provided by the GroupLens research group. The results reveal the superiority of our recommender model compared to several other methods.

1 INTRODUCTION

With the rapid development of electronic commerce and social media, more and more users have joined platforms such as Amazon or Facebook. The huge amount of available information makes users overwhelmed and indecisive. Users have to spend much time and energy in searching for the information they wanted. Unfortunately, sometimes they cannot get the expected results. In order to tackle this situation, recommender systems have appeared to solve the overload information problem. Recommender systems aim to provide personalized service through analyzing the users' behavior or products information, which makes it easier for them to get the products or information they wanted. Until now there are several widely used approaches in recommender systems.

The Collaborative Filtering (CF) (Breese et al. 1998) is an important and popular method for recommender system. The CF approach is classified into user-based CF (Herlocker et al. 1999) and item-based CF (Wang 2006, Deshpande 2004). The key of the CF method is to find similar users (similar items) of target user (target item). Even though the CF recommender model is widely used in E-commerce, it does not perform well in sparse data. The scarcity of data is because users have few ratings on items. Several research works have been conducted to improve the CF recommendation and performance, and significant progress has been made.

The Content-Based (CB) (Lops et al. 2011) recommender system tries to match users' preferences through items' descriptions. The CB recommender model has three steps: 1) item representation: this step extracts the features that can represent items according to items' descriptions; 2) profile learning: this phase is to gain the users' preferences through the features of items they liked or disliked; and 3) recommendation generation: this stage matches users' preferences with items' features, and the item that has the highest similarity with user's preference will be recommended. The CB method is easy to understand, but it is difficult to extract the preferences of users.

The Knowledge-Based (KB) (Trewin 2000) recommender system is a kind of filtering system. The KB approach includes constraint-based recommendation and instance-based recommendation. The constraint-based recommendation method creates constraint conditions on the basis of the background information of different trades and uses it to lead users to make recommendations. The instance-based recommendation method matches the items that users wanted through items' features. The KB method does not need users rating for items, but the performance of the KB recommender system strongly depends on the system designer's knowledge of the trade.

The hybrid recommender system (Burke 2002) integrates two or more approaches together to make recommendation. The hybrid recommender systems combining collaborative filtering with content-based or knowledge-based information are the most common methods.

In the recommender system, an import issue is the cold start problem (Schein 2002, Lika 2014), which occurs due to the lack of prior information about new users and items. In many cases, the cold start problem leads to low accuracy in recommendations. Therefore, we propose a new RS model to alleviate this problem.

In addition, we note that, using rated items to represent a user, as in conventional collaborative filtering, can only capture the user's preference at a low level. Considering this situation, both Alice and Bob like action movies, but they have not watched same movies. If we use traditional CF methods to measure the similarity between Alice and Bob, they will not be similar at all as there are no correlated items between them. It is not the result we wanted. Even though Alice and Bob do not have any shared items, both are fans of action movies. Thus, the model we proposed has considered the genres of items to solve this problem. Our model first groups all users into clusters according to the information of items that have been rated by users. Then, we obtain similar users of cold users from these clusters using a similarity measure. Finally, we use the rating information of the similar users of cold users to make recommendations. In this paper, Section 2 discusses recent advances related to the cold start problem. Section 3 presents our approach to address the cold start problem. Our proposed method is evaluated in Section 4, and it compares to several techniques. Section 5 reports our main conclusions and provides some directions for our future work.

2 RELATED WORK

Collaborative Filtering (CF), as a kind of personalized recommendation technique, recommends items to users according to their preferences. However, it suffers the cold start problem, and a number of research efforts deal with it to enhance the recommendation accuracy. There are many similarity measures in CF: PIP (Proximity-Impact-Popularity) (Ahn 2008), Pearson Correlation Coefficient (PCC) (Resnick et al. 1994), cosine similarity (Adomavicius & Tuzhilin 2005), and so on. Haifeng Liu (2014) proposed a New Heuristic Similarity Model (NHSM) inspired from PIP model. In this model, he does not only consider the local context for common ratings of each pair users, but also take the global preference of each user's ratings into account. NHSM consists of two parts. The first part combines PSS with the modified Jaccard similarity metric, and the second part is the URP measure considering the rating preference of each user. Yi Cai (2014) proposed a typicality-based collaborative filtering method named Tyco, in which the neighbors of users are found on the basis of users typicality in user groups. In this measure, users have different typicality degrees in different user groups, and items have different typicality degrees in different item groups. Andre Luiz Vizine Pereira (2015) introduced a hybrid approach that combines collaborative filtering recommendation with demographic information. This approach is based on an existing algorithm: SCOAL (Simultaneous Co-Clustering and Learning) (Deodhar & Ghosh 2007).

In summary, much work has been done to address the cold start problem. Some proposed new similarity measures to enhance the recommendation accuracy in CF. Some prediction models typically make use of the information about characteristics of items and users' demographics. There are also many types of hybrid methods that have been introduced in order to make better predictions in the cold start situation.

3 THE NEW RECOMMENDER MODEL

In this section, we introduce a method to solve the cold start problem in detail. The main architecture of our proposed model is depicted in Fig. 1. There are four processes in our method. First, we make a count on attributes of items that have been rated by users, which can get users' preferences. Second, we group all users into clusters based on their preferences. Third, we find the corresponding neighbors of cold users from the clusters to which cold users belong. Fourth, we utilize the rating information of cold users' neighbors to make recommendations. The most popular items of the cold users' neighbors will be recommended to them.

In each data set, all users have rated several items and each item has some attributes. For example, in the movie data set, every movie has genres like action, adventure, comedy, war, and so on. And each movie might have more than one genre. The first step of our model is to count the attributes

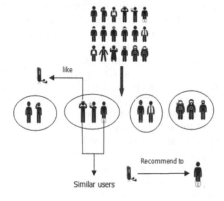

Figure 1. Main architecture of our proposed model.

of items that have been rated by users, so we can get the preferences of users. A user's preference is denoted by vector P_i as follows:

$$P_i = \{p_{i1}, p_{i2}, ..., p_{im}\} \qquad (1)$$

where m is the number of items' attributes and p_{ij} is the number of attribute j of items that have been rated by user i, so p_{ij} indicates the degree of user i liking the items that have attribute j. The preference information of all users is presented by the preference matrix:

$$\text{Preference} = \begin{pmatrix} p_{11} & \cdots & p_{1m} \\ \vdots & \ddots & \vdots \\ p_{n1} & \cdots & p_{nm} \end{pmatrix} \qquad (2)$$

where n is the number of users.

In the second step, we group users into clusters on the basis of their preferences obtained in the first step. We choose k-means algorithm (Jain 2010) to group all users into clusters. The process of this step is formulated as follows:

$$\text{Clusters} = K - \text{means}(\text{Preference}) \qquad (3)$$

In the third step, we aim to get the neighbors of cold users. The neighbors of a cold user are chosen from the cluster to which the cold user belongs (the neighbors of cold users come from training users). And we define a similarity measure to get the neighbors of these cold users as follows:

$$\text{Sim}(u,v) = \sum_{u \in \text{cold}} \sum_{v \in \text{train}} \frac{\sum_{i=1}^{m} u_i \cdot v_i}{m} \qquad (4)$$

where m is the number of items' attributes. If p_i of user u is not less than 1, u_i equals 1; otherwise, it is 0. If p_i of user v is more than 4, v_i equals 1; otherwise, it is 0. The "cold" and "train" represent cold users and training users.

Finally, we utilize the rating information of cold users' neighbors to recommend items. The items that are most popular in cold users' neighbors are recommended to them. We consider popular items are those that have been rated much more times by users.

4 EXPERIMENTS

4.1 Data set

We use two sets of MovieLens in our experiments. MovieLens data sets are collected by the GroupLens Research Project at the University of Minnesota. These two data sets are widely used in recommender system domain. The first data set is called ML-100 K. There are 100,000 ratings with 943 users and 1682 movies in it. The second data set is called ML-1M, which contains 1,000,209 ratings of 3952 movies made by 6040 persons who join the MovieLens in 2000. We cut the ML-1M data set into two parts: ML-1M-1 and ML-1M−2. Each part has 3020 users. In these data sets, each user has rated at least 20 movies, and the ratings follow 1 (bad) to 5 (excellent) numerical scale. The movie's information includes movie id, movie title, and movie genres.

In our experiment, each data set is divided into two parts: 80% users are selected as training users and the remaining are testing users. Moreover, we only choose 5–10 items that have been rated by testing users as training items and the rest as testing items. In this paper, the testing users are cold users.

4.2 Evaluation metrics

Many researchers predict the ratings of users' unrated items. Mean Absolute Error (MAE) and Rooted Mean Squared Error (RMSE) (Willmott & Matsuura 2005) are the most evaluation metrics for them. However, in many cases, better MAE or RMSE is not equal to better recommendation. The precision and recall are more suitable in top-N recommendation (Cremonesi 2010, Bobadilla 2013).

Precision. The precision is the proportion of recommended items that the testing users actually liked in testing items. The higher the precision, the better the recommendation. Let n be the amount of items, which the testing user likes and appear in the recommendation list. The precision is computed as follows:

$$\text{Precision} = \frac{n}{\text{TopN}} \qquad (5)$$

Recall. The recall score is the average proportion of items in recommendation list that appear among the testing items from testing users. This measure should be as high as possible for good performance. Let τ be the number of testing items of one testing user. Hence, the recall is computed as follows:

$$\text{Recall} = \frac{n}{\tau} \qquad (6)$$

4.3 Experimental scenarios

In our experiments, we compare our model with several methods. NHSM [13] and Jaccard are approaches using the traditional CF model. NHSM-P uses the NHSM similarity measure to choose the neighbors, but it recommends the popular items to cold users. Random-C method randomly chooses the cluster to which cold users belong. Random method randomly chooses the users as cold users' neighbors. Three parameters affect the performance of recommendation in our model: the number of clusters setting in K-means algorithm, the number of similar users, and the number of recommendations. The reason that we compare to NHSM-P

Table 1. Performances of precision on different C. where K = 10 and N = 10.

C Dataset	2	3	4	5	6
ML-100K	0.3857	0.4005	0.3889	**0.4132**	0.4062
ML-1M-1	**0.3775**	0.3705	0.3542	0.3339	0.3363
ML-1M-2	0.4222	**0.4227**	0.4066	0.4111	0.4073

Table 2. Performances of recall on different C. where K = 10 and N = 10.

C Dataset	2	3	4	5	6
ML-100K	0.0521	0.0573	0.0558	**0.0766**	0.0748
ML-1M-1	**0.0434**	0.0428	0.0404	0.0388	0.0396
ML-1M-2	0.0419	0.0423	0.0414	**0.0424**	0.0418

method is that we want to illustrate that it is not the key to recommend popular items to users. It is our model to work for better performance. We also choose the random method to be compared because we want to show the importance of the clustering step. Comparing to traditional CF methods, we want to demonstrate the shortages of CF that have been pointed out in Section 1.

Let C denote the number of clusters in the k-means algorithm. Different C will lead to different recommendation results in our model. We first analyze the impact of C on the performance. Tables 1 and 2 show the results with different C, where both K and N are 10. We note that an appointed C does not show the best results in all the three data sets, but when C = 3, the overall performance is not bad. Therefore, in the experiment of comparison, in our model, we determine C = 3.

In our experiments, the preference is our method, and we let K denote the number of cold users' neighbors. Figs. 2–4 show the performances of different methods on evaluation metrics precision and recall. It is clear that precision and recall improve with increasing of K in our method. We also note that our method improves more than 100% compared to traditional CF methods NHSM and Jaccard. Only in Fig. 2, the recall performance of our model is a little worse than that of NHSM-P, the others are the best results. And

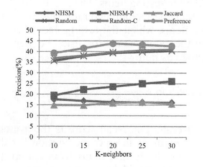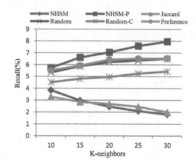

Figure 2. Performances of different methods with different K on ML-100K data set, where N is 10.

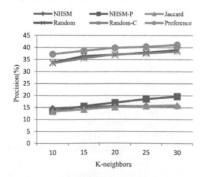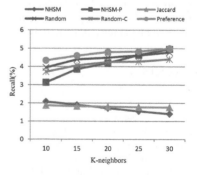

Figure 3. Performances of different methods with different K on ML-1M-1 data set, where N is 10.

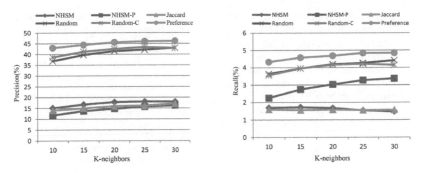

Figure 4.   Performances of different methods with different K on ML-1M-2 data set, where N is 10.

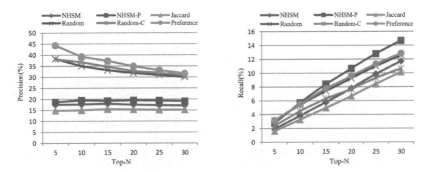

Figure 5.   Performances of different methods with different N on ML-100K data set, where K is 10.

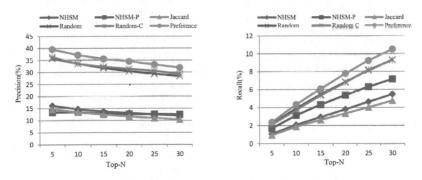

Figure 6.   Performances of different methods with different N on ML-1M-1 data set, where K is 10.

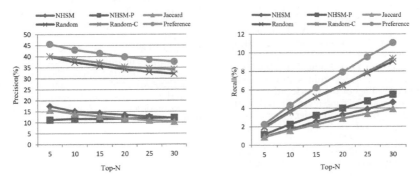

Figure 7.   The performances of different methods with different N on ML-1M-2 dataset where K is 10.

in Haifeng Liu's paper, his novel method NHSM shows better performance in traditional CF model with different similarity measures. In his experiment, he chose 1–10 items as training items.

In top-N recommendation, different N will lead to different recommendation performances. In our experiments, N ranges from 5 to 30. Figs. 5–7 show the performances of different methods. Our method also shows the best results in all conditions except the recall performance in Fig. 5. It is obvious that our measure has remarkable improvement compared to other methods.

5   CONCLUSIONS

In this paper, we presented a new recommender model to alleviate the user cold start problems of RS. We first analyzed the shortages of traditional CF methods. In our proposed model, we first classified all users into groups on the basis of their preferences. Then, we obtained the nearest neighbors of cold users using a similarity measure. Finally, popular items of cold users' neighbors were recommended. For demonstrating the effectiveness of our metric, several experiments were conducted compared to different methods on two widely used data sets. In this paper, we also illustrated why we chose these methods for the comparison. From the experimental results in the end, we can see that our model shows the best performance and it improves more than 100% compared to CF methods. In our method, we alleviate the cold start problem of users with few ratings. When users have no rating, we call this phenomenon the pure cold start problem. In future, we will focus on the pure cold start problem.

REFERENCES

Adomavicius, G., & Tuzhilin, A. (2005). Toward the next generation of recommender systems: A survey of the state-of-the-art and possible extensions. IEEE transactions on knowledge and data engineering, 17(6), 734–749.

Ahn, H. J. (2008). A new similarity measure for collaborative filtering to alleviate the new user cold-starting problem. Information Sciences, 178(1), 37–51.

Bobadilla, J., Ortega, F., Hernando, A., & Gutiérrez, A. (2013). Recommender systems survey. Knowledge-Based Syste-ms, 46, 109–132.

Breese, J. S., Heckerman, D., & Kadie, C. (1998). Empirical analysis of predictive algorithms for collaborative filtering. In Proceedings of the Fourteenth conference on Uncertainty in artificial intelligence (pp. 43–52).

Burke, R. (2002). Hybrid recommender systems: Survey and experiments. User modeling and user-adapted intera-ction, 12(4), 331–370.

Cai, Y., Leung, H. F., Li, Q., Min, H., Tang, J., & Li, J. (2014). Typicality-based collaborative filtering recom-menda-tion. IEEE Transactions on Knowledge and Data Engineering, 26(3), 766–779.

Cremonesi, P., Koren, Y., & Turrin, R. (2010, September). Performance of recommender algorithms on top-n recommendation tasks. In Proceedings of the fourth ACM conference on Recommender systems (pp. 39–46). ACM.

Deodhar, M., & Ghosh, J. (2007, August). A framework for simultaneous co-clustering and learning from complex data. In Proceedings of the 13th ACM SIGKDD international conference on Knowledge discovery and data mining (pp. 250–259). ACM.

Deshpande, M., & Karypis, G. (2004). Item-based top-n recommendation algorithms. ACM Transactions on Information Systems (TOIS), 22(1), 143–177.

Herlocker, J. L., Konstan, J. A., Borchers, A., & Riedl, J. (1999, August). An algorithmic framework for performing collaborative filtering. In Proceedings of the 22nd annual international ACM SIGIR conference on Research and development in information retrieval (pp. 230–237). ACM.

Jain, A. K. (2010). Data clustering: 50 years beyond K-means. Pattern recognition letters, 31(8), 651–666.

Lika, B., Kolomvatsos, K., & Hadjiefthymiades, S. (2014). Facing the cold start problem in recommender systems. Expert Systems with Applications, 41(4), 2065–2073.

Liu, H., Hu, Z., Mian, A., Tian, H., & Zhu, X. (2014). A new user similarity model to improve the accuracy of collaborative filtering. Knowledge-Based Systems, 56, 156–166.

Lops, P., De Gemmis, M., & Semeraro, G. (2011). Content-based recommender systems: State of the art and trends. In Recommender systems handbook (pp. 73–105).

Pereira, A. L. V., & Hruschka, E. R. (2015). Simultaneous co-clustering and learning to address the cold start problem in recommender systems. Knowledge-Based Systems, 82, 11–19.

Resnick, P., Iacovou, N., Suchak, M., Bergstrom, P., & Riedl, J. (1994, October). GroupLens: an open architecture for collaborative filtering of netnews. In Proceedings of the 1994 ACM conference on Computer supported cooperative work (pp. 175–186). ACM.

Schein, A. I., Popescul, A., Ungar, L. H., & Pennock, D. M. (2002, August). Methods and metrics for cold-start recommendations. In Proceedings of the 25th annual international ACM SIGIR conference on Research and development in information retrieval (pp. 253–260). ACM.

Trewin, S. (2000). Knowledge-based recommender systems. Encyclopedia of library and information science.

Wang, J., De Vries, A. P., & Reinders, M. J. (2006, August). Unifying user-based and item-based collaborative filtering approaches by similarity fusion. In Proceedings of the 29th annual international ACM SIGIR conference on Research and development in information retrieval (pp. 501–508).

Willmott, C. J., & Matsuura, K. (2005). Advantages of the mean absolute error (MAE) over the root mean square error (RMSE) in assessing average model performance. Climate research, 30(1), 79–82.

Electromechanical Control Technology and Transportation – Jia & Wu (Eds)
© 2017 Taylor & Francis Group, London, ISBN 978-1-138-06752-3

Low-power single-supply stimulator based on Cortex-M0+

Y. Sun, L. Liu, H. Gao & K. Xu
Qiushi Academy for Advanced Studies (QAAS) and College of Biomedical Engineering,
Zhejiang University, Hangzhou, China

L. Qin
Zhejiang Institute for the Control of Medical Device, Hangzhou, China

S. Ye
College of Biomedical Engineering, Zhejiang University, Hangzhou, China

ABSTRACT: Recently, deep brain stimulation has been increasingly effective in controlling animal robots and treating many patients with brain disorders, and animal models have been used for relative studies. In order to design a low-power and low-cost brain stimulator for small animal models, we propose a constant-current stimulator with single supply on the basis of a Cortex-M0+ processor. Single-power scheme and MCU operation mode transformation reduce the power consumption to 25% of the previous stimulator for each stimulation. The output error is less than 2% in different load cases, and the stable time is less than 25 µs. A biological experiment was carried out on an SD rat with different constant-current stimuli. After the experiment, the circuit simulation was carried out to verify the effectiveness of our stimulator.

1 INTRODUCTION

Recently, Deep Brain Stimulation (DBS) has been widely used to control animal robots (Zeno et al., 2016); treat brain disorders, such as depression (Cook et al., 2014), Parkinson's disease (Connolly et al., 2015, Ferrucci et al., 2016), and epilepsy (Krishna et al., 2016); and restore consciousness inpatients in a coma or vegetative state (Shah and Schiff, 2010). Most of the clinical translational applications of DBS were started from animal models. However, the major commercial stimulators are not flexible for large-scale experiments because of their high cost and high power consumption. Thus, it is mandatory to develop specific stimulators that mimic therapeutic action provided in human patients and adapted to small animals. Constant-current stimulators are favored because of their safety, established methods of charge balancing, and overall facility of implementation (Arfin and Sarpeshkar, 2012, Arfin et al., 2009). In the last decades, many types of constant-current stimulators have been designed for different requirements with their hardware framework sharing common constraints in size, power, and functionality (Bashirullah, 2010). S. Zanos et al. designed a portable system allowing electrical stimulation with programmable system-on-chip devices (CY8C29466) and duplicate supply, whose size is $63 \times 63 \times 30$ mm^3 and power consumption is relatively high (Zanos et al., 2011). The Activa System proposed by Medtronic for DBS has ultralow power consumption (Ondo et al., 2007) and high stability, but a common laboratory cannot afford it due to its high price. Chen et al. proposed a constant-current stimulator with power consumption of 70 mA for each stimulation, which could only work for 2 h with a 3.7V, 80 mAH lithium battery in dual power scheme (Chen et al., 2013), and whose power consumption is too large to be applied to long-term tests. Therefore, we aim to design a low-power-consumption stimulator fulfilling the requirements of small, high precision of stimulation waveform and low cost for laboratory research.

Although dual-power scheme is normally selected in stimulator designing to output biphasic pulse (Kolbl et al., 2016), single-power scheme could reduce the power supply range to reduce the power consumption with simpler circuit. Moreover, since stimuli are inputted discretely in most experiments, the power supply of constant-current module can be closed, and MCU can run in low-power mode to reduce the power consumption further during the nonstimulation stage. In this paper, a low-power single-supply stimulator based on Chen's stimulator (Chen et al., 2013) has been designed. H-bridge circuit is chosen for the stimulation output module to realize biphasic pulse output with single supply. Furthermore, MCU with Cortex-M0+ kernel is chosen to switch operation modes efficiently between run mode and low-power mode and ensure the performance of the system with decreased power consumption.

2 METHOD

2.1 *Overall system architecture*

The conceptual system-level block diagram of the stimulator is shown in Figure 1(A). This system includes four modules: the power module, the constant-current module, the Bluetooth module, and the MCU module. When working, the Bluetooth module receives stimulation start command and mode transformation command from the upper computer and sends the commands to MCU by Universal Synchronous Asynchronous Receiver Transmitter (USART). The MCU decodes the commands, starts the stimulation, and switches different modes for power consumption reduction. According to the stimulation parameters, MCU outputs stable voltage by DAC; then, the constant-current source module transforms the stable voltage into a stable current. Because the single-supply scheme is adopted in our system, H-bridge is used to realize biphasic pulse (Figure 1(B)). The stimulation waveform can be realized by the control of SW1, SW2, SW3, and SW4 in H-bridge with the help of MCU. As shown in Figure 1(B), SW1 and SW2 are paired together and opened at time point a to output positive pulse phase, while SW3 and SW4 are paired together and opened at time point c to output negative pulse phase. At time points b and d, all four switches are closed to make current down to 0 approximately.

2.2 *Hardware design*

The whole system is powered by a single 3.7 V lithium battery. A TPS61040 chip is used to generate 15 V source. And Low-Dropout regulators (LDO, ADP7142) are added to provide ultra-low-noise supplies. Because MCU is used to output constant voltage for constant current module, another LDO (TPS79933) chip is used to ensure that the MCU has a low-noise voltage supply.

Several requirements are considered in the constant-current source module. First, the circuit should have the ability of outputting 5–150 µA current as used in most rodent deep brain stimulation experiments. Second, the input supply voltage range should be wider than 15 V. Third, the slew rate of the current source module should be high,

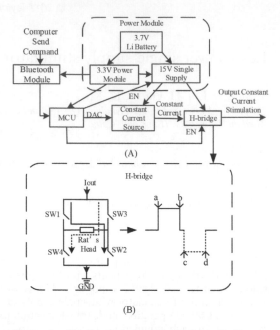

Figure 1. (A) Conceptual system-level block diagram of the whole system. (B) H-bridge circuit.

so that the output can reach stable values in a short time. On the basis of the above requirements, LM8272 is chosen for the circuit. For constant-current module, we complete it with a feedback-type constant-current source.

Two ADG409s are chosen to constitute the H-bridge circuit, which is a dual four-channel multiplexer featuring low resistance (100 Ω Max), low power consumption (1.25 mW Max), low leakage current (5 nA Max), and short conversion time (300 ns Max).

In order to reduce the power consumption more effectively, ultra-low-power STM32 L053C8 single chip based on Cortex-M0+ is chosen for this system, which has multiple power ranges and clock sources, flexible and high-efficient operating modes, and short wake-up time. Different power ranges and clock frequencies are chosen in different stages of system operation. In the nonstimulation stage, low power range (1.2 V) and slow clock frequency (65.5 kHz) are chosen to minimize the MCU power consumption, whereas in the stimulation stage, high-performance power range (1.8 V) and fast clock (16 MHz) frequency are chosen to shorten the span of the stimulation task and then lengthen the low-power run mode time.

2.3 *Software design*

In order to reduce the power consumption efficiently, different operation modes are used in this

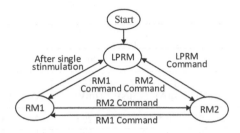

Figure 2. State diagram for mode transformation.

stimulator, including Running Mode1 (RM1), Running Mode2 (RM2), and Low Power Running Mode (LPRM). The state diagram is shown in Figure 2.

In the nonstimulation stage, the MCU enters into LPRM with a low power range (1.2 V) and a low clock frequency (65.5 kHz). The 15 V single supply, unused DAC and clock sources are all closed to minimize the power consumption in the LPRM. In the stimulation stage, the system could automatically switch between RM1 and RM2 depending on the task requirements. High-performance power range (1.8 V) and high clock frequency (16 MHz) are chosen in RM1 and RM2 to shorten the span of the stimulation task and lengthen the low-power running mode time. RM1 is designed for low-frequency stimulation tasks with a long interval between two stimulation pulses. In this mode, the MCU will automatically enter into LPRM after each stimulation to lower power consumption until another stimulation command is received. When the stimulation interval is too short to allow the system to switch modes between two adjacent stimuli, the MCU will enter into RM2 to get ready for every stimulation command until LPRM command is received or no stimulation command is received after a long time (5 s).

3 EXPERIMENT AND RESULT

3.1 Device characteristic test

The whole system power consumption of two power schemes and three operation modes is measured with 4 V input from DC power supply. In order to measure the power consumption of the stimulator, a 10 Ω resistor is connected with the stimulator in series, and an oscilloscope is used to record the voltage of the 10 Ω resistor. The total current of the whole system, representing the whole system power consumption, is obtained by computing the current of the 10 Ω resistor. All the power consumption measurements are made with 50 μA amplitude, 100 μs width stimulation pulse in 100 Hz frequency, and 50 kΩ load.

In order to compare the power consumption difference between single power and dual power, the stimulator is set to keep outputting stimulus in two power schemes. The power consumption is 13.7 mA with dual power and 10.3 mA with single power in the above stimulation conditions. Using single-power scheme, the power consumption is about 20 mA for each stimulation, which is 25% of that of the previous stimulator (80 mA) (Chen et al., 2013).

The power consumption in LPRM is 5.0 mA. In order to evaluate the power consumption of RM1 and RM2, we assume that PC sends stimulation commands at a fixed time interval, and every command will start 10 stimulation pulses, as shown in Figure 3. The average power consumption of different command sending time interval in RM1 and RM2 is shown in Table 1. The power consumption in RM1 is obviously lower than that in RM2. And the shorter the interval of command sending, the more reduced the power consumption.

The feasibility of the stimulator is analyzed by constant-current characteristic tests; 10, 33, 50, and 100 kΩ resistors are chosen for the typical load resistance. An oscilloscope is used to record the voltage waveform of the resistors. The DC and AC characteristics are verified with 20, 50, 70, and 100 μA constant-current outputs. The stable voltage (U) and the time (T) required to reach stable voltage levels are presented in Table 2. As shown in the table, there is no obvious change of the output current (I) with the change of the load resistance, and the error is less than ±2%.

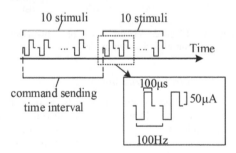

Figure 3. Stimulation waveform diagram.

Table 1. Power consumption in different operation modes.

Time interval (ms)	Mode	
	RM1 (mA)	RM2 (mA)
5000	10.7	20.9
1000	11.2	22.2
200	17.8	23.7

Table 2. Current value under different parameters and loads.

	I_{out} (μA)*											
	100			70			50			20		
R (kΩ)**	U (mV)	I (μA)	T (μs)	U (mV)	I (μA)	T (μs)	U (mV)	I (μA)	T (μs)	U (mV)	I (μA)	T (μs)
9.91	1000	100.9	6	680	68.6	7	500	50.5	7	200	20.2	3
33.85	3350	99.0	6	2350	69.4	7	1680	49.6	7	680	20.1	3
49.70	5000	100.6	12	3450	69.4	12	2500	50.3	12	1000	20.1	12
99.90	10000	100.1	25	7000	70.1	25	5000	50.1	25	2000	20.0	25

*Output current from constant-current module.
**Load impedance.

Figure 4. Schematic diagram of biological experiments.

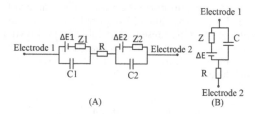

(A)   (B)

Figure 5. Electrical circuit model of a two-electrode system. (B) Simplified circuit model.

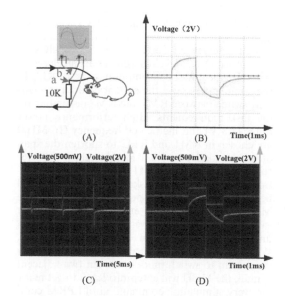

(A)   (B) Time(1ms)

(C)   (D)

Figure 6. (A) Schematic diagram of constant-current stimulation experiments. (B) Simulation pulse of voltage output with 20 μA stimulation. (C) From left to right, the stimulation currents are 20, 30, 100, and 10 μA, respectively. And the pulse width of all stimulations is 100 μs. The upper waveform is recorded on a, and the lower waveform is recorded on b. (D) The stimulation current is 20 μA, and the pulse width is 1 ms. The upper pulse is recorded on a, and the lower pulse is recorded on b.

3.2 Biological experiments

Constant-current stimulation tests were carried out on a Sprague-Dawley (SD) rat through a pair of electrodes implanted into the barrel cortex to validate the stimulator's feasibility on the animal, as is shown in Figure 4.

The impedance of rat's brain tissue is measured and analyzed. As shown in Figure 6(A), a 10 kΩ resistor and the electrodes on the rat's head are connected in series to get the actual current value. The SD rat was given a series of 10, 20, 30 and 100 μA constant-current stimulations, respectively, and the voltage values on a and b were recorded by an oscilloscope. The experimental results are shown in Figure 6(C).

The stimulation results were analyzed by an electrical model of the electrode/tissue. The electrical circuit model of a two-electrode system (Merrill et al., 2005), including the double-layer capacitance (C), Faradaic impedance (Z), equilibrium potential (ΔE), and resolution resistance (R), is shown in Figure 5(A). Assuming that Z1 = Z2 and C1 = C2, the circuit is simplified to Figure 5(B).

A typical 20 μA constant-current stimulation waveform is shown in Figure 6(D), whose positive waveform was simulated by software Multisim 12.0. The simulation result is shown in Figure 6(B), and the parameters are R = 25 kΩ, Z = 57 kΩ, ΔE = 300 mV, C = 10 nF.

4 DISCUSSION AND CONCLUSION

In this paper, a nerve constant-current stimulator is proposed, with an output error of less than ±2%, fulfilling design requirements.

Using a single-power scheme, Cortex-M0+ processor, and low-power chips, the power consumption is reduced to 25% of that of the previous stimulator (Chen et al., 2013) for each stimulation. The utilization of LPRUN reduces the power consumption further with a low stimulation frequency. Therefore, this stimulator is greatly suitable for low-frequency stimulation experiments.

Moreover, the waveform we obtained from biological experiments was simulated well by a circuit model, which proves the effectiveness of our stimulator, and that voltage can be measured in animal experiments to analyze the electrical model of the electrode/tissue during stimulation.

Furthermore, as Cortex-M0+ is a high-performance 32-bit ARM Cortex processor kernel, more system functions can be developed with this MCU, such as the record and analysis of stimulation waveform, which can be considered in the next version of the stimulator.

REFERENCES

Arfin, S.K., Long, M.A., Fee, M.S. & Sarpeshkar, R. 2009. Wireless neural stimulation in freely behaving small animals. *Journal of neurophysiology,* 102, 598–605.

Arfin, S.K. & Sarpeshkar, R. 2012. An energy-efficient, adiabatic electrode stimulator with inductive energy recycling and feedback current regulation. *Biomedical Circuits and Systems, IEEE Transactions on,* 6, 1–14.

Bashirullah, R. 2010. Wireless implants. *Microwave Magazine, IEEE,* 11, S14–S23.

Chen, X., Xu, K., Ye, S., Guo, S. & Zheng, X. A remote constant current stimulator designed for rat-robot navigation. Engineering in Medicine and Biology Society (EMBC), 2013 35th Annual International Conference of the IEEE, 2013. IEEE, 2168–2171.

Connolly, A.T., Kaemmerer, W.F., Dani, S., Stanslaski, S.R., Panken, E., Johnson, M.D. & Denison, T. Guiding deep brain stimulation contact selection using local field potentials sensed by a chronically implanted device in Parkinson's disease patients. 2015 7th International IEEE/EMBS Conference on Neural Engineering (NER), 2015. IEEE, 840–843.

Cook, I.A., Espinoza, R. & Leuchter, A.F. 2014. Neuromodulation for Depression. *Neurosurgery Clinics of North America,* 25, 103–116.

Ferrucci, R., Mameli, F., Ruggiero, F. & Priori, A. 2016. Transcranial direct current stimulation as treatment for Parkinson's disease and other movement disorders. *Basal Ganglia,* 6, 53–61.

Kolbl, F., Nkaoua, G., Naudet, F., Berthier, F., Faggiani, E., Renaud, S., Benazzouz, A. & Lewis, N. 2016. An Embedded Deep Brain Stimulator for Biphasic Chronic Experiments in Freely Moving Rodents. *IEEE Transactions on Biomedical Circuits and Systems,* 10, 72–84.

Krishna, V., Sammartino, F., King, N.K.K., So, R.Q.Y. & Wennberg, R. 2016. Neuromodulation for Epilepsy. *Neurosurgery Clinics of North America,* 27, 123–131.

Merrill, D.R., Bikson, M. & Jefferys, J.G. 2005. Electrical stimulation of excitable tissue: design of efficacious and safe protocols. *Journal of neuroscience methods,* 141, 171–198.

Ondo, W.G., Meilak, C. & Vuong, K.D. 2007. Predictors of battery life for the Activa® Soletra 7426 Neurostimulator. *Parkinsonism & related disorders,* 13, 240–242.

Shah, S.A. & Schiff, N.D. 2010. Central thalamic deep brain stimulation for cognitive neuromodulation—a review of proposed mechanisms and investigational studies. *European Journal of Neuroscience,* 32, 1135–1144.

Zanos, S., Richardson, A.G., Shupe, L., Miles, F.P. & Fetz, E.E. 2011. The Neurochip-2: an autonomous head-fixed computer for recording and stimulating in freely behaving monkeys. *Neural Systems and Rehabilitation Engineering, IEEE Transactions on,* 19, 427–435.

Zeno, P.J., Patel, S. & Sobh, T.M. 2016. Review of Neurobiologically Based Mobile Robot Navigation System Research Performed Since 2000. *Journal of Robotics,* 2016.

Combining carrier-phase and Doppler observations for precise velocity estimation with a stand-alone GPS receiver

Feng Li, Qiang Li & Wei Qi
Beijing Institute of Tracking and Telecommunications Technology, Beijing, China

ABSTRACT: A Comprehensive Method for Velocity Estimation (CMVE) is proposed to combine the carrier-phase and Doppler observations to determine the precise velocity. In the CMVE, the central difference of the carrier phase is used to generate the carrier-phase-derived Doppler, which is fused with the raw Doppler by comparing and selecting. The CMVE can reduce the influence of cycle slips and the receiver clock jumps. Thus, the velocity errors mainly come from the receiver noise and errors of the carrier-phase-derived Doppler, both of which are modeled and analyzed in the paper. Experiments are carried out with a stand-alone GPS receiver, and only the L1 data are used for velocity estimation. In respect of the statistical velocity errors, the CMVE can always generate higher accurate velocity than when only using the raw Doppler.

1 INTRODUCTION

There are three methods of GPS velocity estimation: the method using the raw Doppler observations, the method using the time-differenced positions, and the Time-Differenced Carrier-Phase (TDCP) method. In the method using the time-differenced positions, the velocity precision has a high correlation with the position precision, and usually on the order of decimeter per second or worse. For static or constant velocity conditions, the error is on the order of centimeters per second using the raw Doppler observations, and on the order of millimeters per second using the TDCP method (Ryan et al. 1997; Freda et al. 2015).

Many adjustments have to be implemented on the carrier-phase observations before the TDCP method can be applied for velocity estimation. For example, cycle slips and receiver clock jumps have to be detected and repaired, and both result in step changes in carrier-phase observations. And it is difficult to repair the detected cycle slips exactly.

To obtain precise GPS velocity estimates under various conditions, from constant velocity to high dynamic flights, and to avoid the complex post-processing on the carrier-phase observations, we present a technique that calculates the velocity by combining the carrier-phase and Doppler observations, namely the "Comprehensive Method for Velocity Estimation" (CMVE).

2 DESCRIPTION OF THE CMVE

2.1 Models of the CMVE

For a stand-alone single-frequency GPS receiver, the measurement model of velocity determination can be expressed as (Angrisano et al. 2013; Xie 2014):

$$\dot{\rho}^j = \left(v^j - v_u\right)\cdot l^j + c\cdot\left(\delta\dot{f}_u - \delta\dot{f}^{\,j}\right) + \xi^j \tag{1}$$

where superscript j stands for the index of different satellites, $\dot{\rho}^j$ is the rate of pseudo-range variation, v^j is the velocity of the satellite, v_u is the velocity of the receiver, l^j is the unit line-of-sight vector from the receiver to the satellite, c is the light velocity, $\delta\dot{f}_u$ and $\delta\dot{f}^{\,j}$ are the frequency shifts of the receiver clock and the satellite clock, respectively, and ξ^j is the sum of the common mode errors: ephemeris, ionosphere, troposphere, multipath, and the receiver noise:

$$\dot{\rho}^j = \begin{cases} \lambda f_\phi^j & \left|f_\phi^j - f_D^j\right| \le \delta f \\ \lambda f_D^j & \left|f_\phi^j - f_D^j\right| > \delta f \end{cases} \tag{2}$$

where λ is the wavelength of the carrier, f_ϕ^j is the Carrier-Phase-Derived Doppler (CPD), f_D^j is the raw Doppler observables outputted by the receiver, and δf is the threshold value of the data fusion, which is determined by the data quality of the carrier-phase and raw Doppler observations.

The carrier-phase-derived Doppler is a central difference approximation of the Doppler shift of the carrier, based on the Taylor Series expansion method (Hebert 1995; Cannon et al. 1997; Zhang et al. 2005; Wang and Xu 2011):

$$f_{\phi,k}^j \approx \frac{\phi_{k+1}^j - \phi_{k-1}^j}{2h} \tag{3}$$

where $f_{\phi,k}^j$ is the carrier-phase-derived Doppler at the epoch k, ϕ_{k+1}^j and ϕ_{k-1}^j are the values of carrier phases at epochs $k+1$ and $k-1$, respectively, and h is the sample interval, which is usually no longer than 1 s.

The velocity of the receiver is solved with the Weighed Least Squares (WLS) method (How, Pohlman and Park 2002):

$$x = \left(G^T \cdot W \cdot G\right)^{-1} \cdot G^T \cdot W \cdot y \tag{4}$$

where

$$x = \begin{bmatrix} v_x \\ v_y \\ v_z \\ \delta f_u \end{bmatrix}; G = \begin{bmatrix} -l_x^1 & -l_y^1 & -l_z^1 & 1 \\ -l_x^2 & -l_y^2 & -l_z^2 & 1 \\ & \vdots & & \\ -l_x^n & -l_y^n & -l_z^n & 1 \end{bmatrix}_{n \times 4};$$

$$y = \begin{bmatrix} \dot{\rho}^1 - v^1 \cdot l^1 + c \cdot \delta f^1 \\ \dot{\rho}^2 - v^2 \cdot l^2 + c \cdot \delta f^2 \\ \vdots \\ \dot{\rho}^n - v^n \cdot l^n + c \cdot \delta f^n \end{bmatrix}_{n \times 1};$$

where l_x^j, l_y^j and l_z^j are the components of the unit line-of-sight vector l^j, n is the number of satellites involved in the velocity estimation, and W is the weight matrix:

$$W_{pq} = \begin{cases} \sin(EL^p) & p = q \\ 0 & p \neq q \end{cases} \tag{5}$$

where W_{pq} is the component of the matrix W and EL^p is the elevation of the satellite whose index number is p.

2.2 Robust property of raw Doppler observations

High-quality raw Doppler observables are obtained on the basis of the CMVE. The data quality of the raw Doppler observations can be evaluated with the Doppler residual vectors after velocity estimation using the raw Doppler observables. The Doppler residual vectors can be deduced as follows:

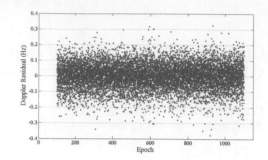

Figure 1. Doppler residuals of all the satellites involved in the velocity estimation in an actual flight test.

$$\hat{D} = \frac{1}{\lambda} \cdot (y - Gx) \tag{6}$$

where $\hat{D}$ is the Doppler residual vectors, the definitions of y, G, and x are the same as those in Equation 4, and only the raw Doppler observations are used in the calculations of all the parameters.

The Doppler residual vectors calculated with GPS receiver data from an actual flight test are shown in Figure 1. The Doppler residuals could not be used for evaluating the Doppler observable accuracy quantitatively. However, the results indicate that the Doppler observables are robust in dynamic scenarios. The Doppler residuals can also be used to reject some satellite observations with problems from the satellites or the receiver channels.

2.3 Determination of the threshold in the CMVE

The threshold value (δf) can be determined by the comparison between the CPD and the raw Doppler:

$$\hat{f}_{d,k}^j = f_{\phi,k}^j - f_{D,k}^j \tag{7}$$

where $f_{\phi,k}^j$ and $f_{D,k}^j$ are the CPD and the raw Doppler of the same satellite at the same epoch, respectively.

The selected threshold should ensure that the carrier phase will be used for velocity estimation in most epochs and that the peaks of the CPD caused by the cycle slips and the receiver clock jumps will not influence the velocity precision. Zhang et al. (2012) found that the receiver clock jumps can enable velocity errors reach up to 0.2 m/s for the horizontal component and 0.5 m/s for the vertical component. Big cycle slips and the receiver clock jumps will result in peaks in the difference between the CPD and the raw Doppler, thus their influence can be reduced by the selection of the threshold.

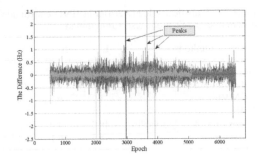

Figure 2. Differences of all satellites between the CPD and the raw Doppler in a static test.

In this mean, the CMVE can process carrier-phase observations with cycle slips directly.

For example, the Doppler difference between the CPD and the raw Doppler calculated with GPS receiver data from a static test is shown in Figure 2. The threshold can be set to 1 Hz in the example.

2.4 Error analysis of the CMVE

The velocity errors have a linear correlation with the errors of the rate of pseudo-range variation, as shown in Equation 8. The linear correlation is related to the satellite geometry configurations relative to the user's position, and can be measured by Dilution of Precision (DOP) terms (Vodhanel 2011):

$$\begin{bmatrix} \Delta v_x \\ \Delta v_y \\ \Delta v_z \\ \Delta \delta f_u \end{bmatrix} = (G^T \cdot W \cdot G)^{-1} \cdot G^T \cdot W \cdot \begin{bmatrix} \Delta \dot{\rho}^1 \\ \Delta \dot{\rho}^2 \\ \vdots \\ \Delta \dot{\rho}^n \end{bmatrix} \quad (8)$$

At epochs using the raw Doppler for velocity estimation, the velocity errors mainly come from the receiver noise. At epochs using the carrier-phase observations for velocity estimation, the velocity errors mainly come from the receiver noise and the errors of the carrier-phase-derived Doppler.

Errors of the CPD include the carrier-phase measurement error and the truncation error. Assuming that the observation error of the carrier phase due to the receiver noise is M_ϕ, the measurement error can be calculated as:

$$E_\phi = \frac{\sqrt{2}}{2} \cdot \frac{M_\phi}{h} \quad (9)$$

where E_ϕ is the measurement error, which is related to the sampling time and the order of the CPD. Along with the increase in N, E_ϕ gets bigger, but the rate of increase of the speed of E_ϕ gets lower.

The truncation error (E_t) can be calculated as:

$$E_t = -\frac{h^2}{6\lambda} \cdot \vec{\varepsilon}_{u,k} \cdot \vec{l}_k \quad (10)$$

where $\vec{\varepsilon}_{u,k}$ is the jerk of the receiver movement with respect to the earth and $\vec{l}_k$ is the unit line-of-sight vector from the receiver to the satellite.

If the jerk of the receiver is too high, there will be a big truncation error in the CPD. In the research of Hebert et al. (1997), the velocity errors are correlated directly with the jerk when using the carrier-phase observations. High jerk scenarios are common for flights, such as the moments that the engine starts to work or turns off, and the moment a gust of wind blows on the aircraft.

3 TESTS AND RESULTS

3.1 Static-mode tests

In the static-mode test, data are collected from a NovAtel ProPak-II-RT2 double-frequency receiver connected to a GPS702 antenna. The data are collected at a 1 Hz rate, and only L1 data are used for velocity estimation. Several methods are used for velocity estimation, such as:

1. using the time-differenced positions;
2. using the raw Doppler observables;
3. using the CMVE.

The first thousand epochs from the results are used for velocity precision analysis in the East-North-Up (ENU) coordinate system, and the theoretical velocity of the receiver is zero. The RMS-Errors (Root Mean Square Error) are shown in Table 1.

The velocity RMS-Errors are on the order of millimeters per second when using the CMVE, much better than that using the raw Doppler observables.

Table 1. Statistical velocity errors in ENU coordinate system in static-mode tests (unit: mm/s).

Different methods	E	N	U
Mean			
Time-differenced positions	−0.5	3.1	5.1
Raw Doppler observables	1.0	−0.4	−5.4
CMVE	0.6	−0.6	−5.7
RMS			
Time-differenced positions	2.9	4.2	10.6
Raw Doppler observables	9.8	11.8	27.8
CMVE	0.7	0.9	2.2

As seen from Table 1, the average of velocity errors is on the order of submillimeters per second in the East (E) and North (N) components, and on the order of millimeter per second in the Up (U) component. This indicates that there is some low-frequency variation in the velocity series due to unmodeled error or ionosphere and troposphere residual effects (Li et al. 2014).

3.2 *Dynamic-mode tests*

In the dynamic-mode tests, a GPS simulator is applied in the experiments. A stand-alone single-frequency GPS receiver is subjected to aircraft dynamic of up to 70 m/s², with flight velocities of up to 7 km/s. The theoretical positions, velocities, and accelerations of the aircraft are recorded by the simulator. The variations of absolute velocity and acceleration are shown in Figures 3 and 4, respectively.

The raw Doppler observables and the CMVE are used to evaluate velocities of the aircraft, and the velocity errors are shown in Table 2. The results show that the CMVE can generate better accuracy velocity than the raw Doppler observables in dynamic situation.

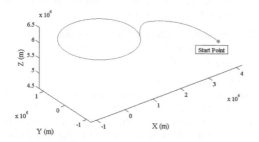

Figure 3. Aircraft trajectory in ECEF coordinate system.

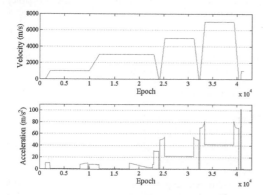

Figure 4. Motion state of the aircraft in the simulation.

Table 2. Statistical velocity errors in ENU coordinate system in dynamic mode tests (unit: mm/s).

Different methods	E	N	U
Mean			
Raw Doppler observables	−0.7	4.0	63.8
CMVE	1.3	−0.5	0.1
RMS			
Raw Doppler observables	48.2	34.0	129
CMVE	15.3	6.1	17.3

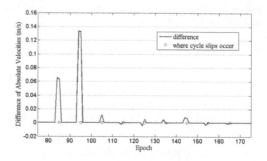

Figure 5. Changes of absolute velocities when adding cycle slips increasing gradually from one cycle to nine cycles, using the CMVE.

3.3 *Reduction in the influence of cycle slips using the CMVE*

The CMVE can reduce the influence of the cycle slips, as an additional function. It is analyzed with the same GPS data as those in the static-mode tests. Signals from eight satellites are received and recorded, whose elevations vary from 11.2° to 57.3° at the start epoch.

In the tests, cycle slips are added to the observations of the satellite PRN 16 (elevation of 39.1°) and the satellite PRN 22 (elevation of 21.6°). Cycle slips are added at nine points with equal time intervals from the 85th epoch to the 195th epoch. The differences in absolute velocities before and after adding cycle slips are analyzed. In the first test, the values of cycle slips at different epochs increase gradually from one cycle to nine cycles, and the results are shown in Figure 5. In the second test, the values of cycle slips at different epochs are all the same at the value of three cycles, and the results are shown in Figure 6. For the purpose of comparison, in the third test, the values of cycle slips at different epochs increase gradually from one cycle to nine cycles, and the TDCP method is used for velocity estimation. The results are shown in Figure 7.

In above tests, the raw Doppler remains unchanged, when cycle slips are added to the carrier phase. When the value of cycle slips is bigger

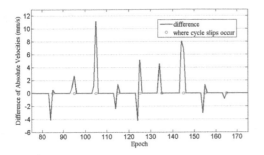

Figure 6. Changes of absolute velocities when adding cycle slips with the value of three cycles, using the CMVE.

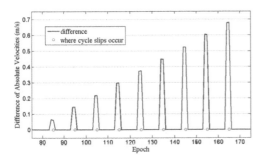

Figure 7. Changes of absolute velocities when adding cycle slips increasing gradually from one cycle to nine cycles, using the TDCP.

than 2 cycles, the influence to velocity precision can be reduced below centimeters per second, by using the CMVE. However, the performance is related to the data quality and the Doppler threshold. Slightly different results are obtained when processing different GPS data.

4 CONCLUSIONS

This investigation proposes a method to combine the carrier-phase and Doppler observations to generate a precise velocity. The CMVE method aims to combine the accurate properties of the carrier-phase observations with the robust property of the Doppler observations simply.

The feasibility of the CMVE has been studied by analysis of the Doppler residuals and the differences between the CPD and the raw Doppler. Through analysis and tests, it is found that the CMVE is applicable for different dynamic situations and performs better than the raw Doppler observables. Different orders of CPD are used in the CMVE, and the error of the CPD is related to the receiver noise and the aircraft dynamic.

In addition, it is found that the CMVE can reduce the influence of the big cycle slips.

For further study, more efforts should be made to reduce the influence of the truncation error on the CPD in a high dynamic situation. Smooth processing method can also be used to eliminate velocity error peaks.

REFERENCES

Angrisano A, Gioia C, Gaglione S, Core G D (2013) GNSS reliability testing in signal-degraded scenario. *International Journal of Navigation and Observation.*

Cannon M E, Lachapelle G, Szarmes M C, Hebert J M, Keith J, Jokerst S (1997) DGPS kinematic carrier phase signal simulation analysis for precise velocity and position determination. *Navigation* 44(2): 231–245.

Ding W, Wang J (2011) Precise velocity estimation with a stand-alone GPS receiver. *Journal of Navigation* 64(02):311–325. doi:10.1017/S0373463310000482.

Freda P, Angrisano A, Gaglione S, Troisi S (2015) Time-differenced carrier phases technique for precise GNSS velocity estimation. *GPS Solut* 19(2): 335–341.

Hebert J M (1995) Velocity determination for an inverted pseudolite navigation reference system. *Air University, U.S.A.*

Hebert J, Keith J, Ryan S, Szarmes M, Lachapelle G, Cannon M E (1997) DGPS kinematic carrier phase signal simulation analysis for precise aircraft velocity determination. *Proceedings of Annual Meeting of the Institute of Navigation* 44(2): 231–245.

How J, Pohlman N, Park C W (2002) GPS Estimation Algorithms for Precise Velocity, Slip and Racetrack Position Measurements. *SAE Technical Paper.*

Li M, Li W, Fang R, Shi C, Zhao Q (2014) Real-time high-precision earthquake monitoring using single-frequency GPS receivers. *GPS Solut* 19(1): 27–35.

Ryan S, Lachapelle G, Cannon M E (1997) DGPS kinematic carrier phase signal simulation analysis in the velocity domain. *Proceedings of International Technical Meeting of the Satellite Division of the Institute of Navigation* 1035–1045.

Vodhanel M T (2011) Problems in GPS accuracy. Dissertations & Theses-Gradworks.

Wang Q X, Xu T H (2011) Combining GPS carrier phase and Doppler observations for precise velocity determination. *Science China Physics Mechanics & Astronomy* 54(6): 1022–1028.

Xie G (2014) Principle of GPS and receiver design (in Chinese). Publishing house of electronics industry, Beijing.

Zhang J (2007) Precise velocity and acceleration determination using a standalone GPS receiver in real time. RMIT, Australia.

Zhang J, Zhang K, Grenfell R, Li Y, Deakin R (2005) Real-time Doppler/Doppler rate derivation for dynamic applications. *Positioning* 4(1&2): 95–105.

Zhang X, Guo B, Guo F, Du C (2012) Influence of clock jump on the velocity and acceleration estimation with a single GPS receiver based on carrier-phase-derived Doppler. *GPS Solut* 17(4): 549–559.

Electromechanical Control Technology and Transportation – Jia & Wu (Eds)
© *2017 Taylor & Francis Group, London, ISBN 978-1-138-06752-3*

Research on the influence of non-scene factors on visual-based loop closure detection

Zihao Zhang & Ruifang Dong

School of Control and Computer Engineering, North China Electric Power University, Beijing, China

ABSTRACT: Loop closure detection is a key step in eliminating the accumulated errors of the Simultaneous Localization and Mapping (SLAM) algorithms to ensure accurate results. This is particularly important for vision-based SLAM systems. For the SLAM system working outdoors, it is susceptible to many non-scene environmental factors (light, weather, pedestrian, vehicle) when using the classical Bag of Visual Words (BoVW) algorithm for image matching which occupies a dominant position in the loop closure detection tasks. This paper lists and analyzes several non-scene environmental factors and their effects on the loop closure detection algorithm based on image matching. In addition, a dataset concerns these non-scene factors are built to test the related performance of the visual-based SLAM algorithm. And two famous closed loop detection algorithm are tested with this dataset.

1 INTRODUCTION

The core problem in the field of autonomous mobile robots is the autonomous localization of robots. How to calculate the pose of robots through various sensors is a prerequisite for the autonomous movement and path planning of robots. To solve this problem, Simultaneous Localization and Mapping (SLAM) system has been proposed (Smith R, 1987). SLAM research focuses on the autonomous mobile robot in the unknown environment obtaining a spatial map, and determining the poses through a variety of sensor collections. In a large-scale and complex environment, the map creation failure and position loss due to accumulated error are always presented. Therefore, it is a very important task to eliminate the accumulated errors generated by the pose estimation in the SLAM operation and realize the correct deviation, optimize the map data, especially in a visual-based SLAM system without depth data. Loop closure detection and closed-loop fusion technique have been proposed to complete this task.

Loop closure detection is the problem of determining whether a mobile robot has returned to a previously visited location, by this way, the SLAM system calculates the deviation between the map coordinate system and the world coordinate system. When the previously visited location is detected, robot will correct its pose estimation and optimize the map data by using the offset been generated by former step, this process is named closed-loop fusion. In this paper we are interested in visual loop closure detection, which formulates

a solution to the problem exploiting visual data, i.e., using images captured by the robot, and it is the key step of the whole accumulated-error-climination task.

To detect a closed-loop, several structure of loop closure detection system been proposed. Clemente et al. (Clemente L A, 2007) proposed a map-to-map matching method by finding similarities between common features of different sub-graphs. (Williams et al., 2008) proposed a image-to-map matching method with relocalization-based techniques. (Cummins et al. 2008) proposed a image-to-image matching method by recognizing the visual appearance of the explored region, which called FAB-MAP. Compared with the first two methods, it is more suitable for large-scale scene in the closed loop detection. This method not only obtains the similarity between the observed positions of the two robots, but also derives the probability that they come from the same region, that is, the probability of occurrence of the loop closure response, by constructing the generative model in the probability frame. The key of this method lies in the modeling of robot vision scene and the quick similarity matching between images. Recent studies show that the most popular and successful loop closure detection method always matches the current view of the robot with those views correspond to previously visited locations in the map. In this case, the problem of loop closure detection is essentially an image matching problem.

Despite significant progress has been obtained in visual loop closure detection, challenges remain especially in dynamic environments experience, for example, weather changes and mobile objects displacement, etc. To test the impact of these factors on visual SLAM system, we build a dataset contains 4600 images with different weather, pedestrian and vehicle conditions, and analyze result of 2 open-sourced loop closure detection system with 6 image descriptors run on it.

The rest of this paper is organized as follows. Section II gives a brief introduction to the related work on visual features for loop closure detection. In Section III we analyze the interference of non-scene factors to visual odometry calculation and image feature extraction. Section IV presents the dataset with various non-scene factors. Section V shows experimental results with our dataset to compare the performance of two popular and successful loop closure detection algorithm. Finally, we conclude the paper in Section VI with a short discussion and future work.

2 RELATED WORK

As mentioned above, our focus in this paper is to investigate the performance of image matching based visual loop closure detection in the presence of non-scene factors. In this section, we give a brief review of 2 common loop closure detection system: FAB-MAP (Cummins M, 2008) and DBOW (Gálvez-López D, 2011 & 2012).

Image matching typically proceeds into two steps: image description and similarity measurement. An image descriptor compresses an image into a one-dimensional vector that is more compact and discriminating than the original image, and it is the most critical step in visual loop closure detection. Many image description techniques exist for visual loop closure detection have enjoyed tremendous success.

FAB-MAP (Cummins M, 2008) use Bag-of-Visual-Words (BoVW) (Sivic J, 2003) to extract image feature. The BoVW descriptors been considered as the most successful image descriptors in visual loop closure, they are local keypoint descriptors. BoVW was originally proposed for image retrieval (Sivic J, 2003), it characterizes an image as a histogram of visual words where visual words are simply vector-quantized versions of the local keypoint descriptors such as SIFT (Lowe D G, 2004) and SURF (Bay H, 2006). To build a BoVW descriptor, a visual vocabulary is first created offline by clustering a large number of keypoint descriptors whose cluster centres form the visual words of the vocabulary. In online operation, the local keypoints of a given image are first detected and described. Each descriptor is then vector-quantized and the histogram of the vector-quantized keypoint descriptors is used as the image descriptor. While the image feature extraction is completed success-

fully, Mahalanobis distance will be used to calculate the similarity of images. BoVW is adopted successfully in the FAB-MAP for computing image similarity and has become one of the most popular techniques in visual loop closure detection algorithms. Because of the invariance properties of local image features such as SIFT and SURF in building the BoVW descriptor, FAB-MAP achieved an excellent performance, becoming one of the standard baseline algorithms in loop closure detection research.

As similar with FAB-MAP, DBOW is a loop closure detection algorithm based on typically bags of visual word. It is also necessary to extract the features of the sample scene images and build the dictionary model by k-means++ offline training. Unlike FAB-MAP, DBOW uses FAST keypoint (Rosten E, 2006) as the feature extraction algorithm and BRIEF (Calonder M, 2010) or ORB (Rublee E, 2012) descriptor as the feature descriptor. Due to BRIEF uses binary coding to extract the descriptors from the surrounding area of the feature points, the descriptor is simpler, and the storage space is smaller than SIFT and SURF. Hamming Distance is used to calculate the similarity of image, which is more faster than the similarity measurement used in FAB-MAP.

SIFT and SURF are popular because they are invariant to lighting, scale and rotation changes and show a good behavior in view of slight perspective changes. However, these features usually require 100 to 700 ms to extract (Lowe D G, 2004 & Bay H, 2006). A BRIEF descriptor encodes much less information than a SURF descriptor, since BRIEF is not scale or rotation invariant, but this does not reduce its accuracy, as reported by the author of DBOW. The ORB descriptor outperforms the SIFT and SURF in the aspects of computation time, maintaining rotation and scale invariance, in theory it is the ideal choice until now.

Obviously, the performance of the above loop closure detection algorithm depends on local keypoint descriptors, which are developed to provide limited understanding of image semantics and do compute every elements in the image regardless of what it is. Therefore, non-scene factor is bound to have a serious impact on the calculation.

3 NON-SCENE FACTOR

In this paper, we define light, weather, pedestrian, vehicle factors of the real world environment as non-scene factors since they are nothing to do with scene-centric image matching which is the key step of loop closure detection. Weather change will definitely result in the change of illumination conditions, and may change the surface of ground and structures within some extreme cases, such as rain and snow. As for mobile objects, such as pedestrian and vehicle, will appear in different positions of image, cover and replace elements of the scene. These unstable non-scene factors will highly affect tracking and image matching in a visual-based SLAM system, even cause loop closure detection failure by false positive image matching.

Since visual-based SLAM system tracks the image feature of the surrounding scene to calculate the translation and rotation of camera pose, and the traditional feature extraction algorithm is to calculate all the elements of the image, there is no filtering of these unnecessary elements. As shown in Figure 1, when moving objects such as pedestrians and vehicles appear in the image, these moving objects will not only block the scene information in the environment, but also its characteristics will be calculated. The representation of the target image is not equivalent to the representation of the scene in which the robot is located. Rotation and translation will be calculated between the two key frames which contain these objects to varying degrees. If the moving object in the image occupies a larger layout of the image, it will seriously affect the visual odometry calculation of visual-based SLAM operation, in further resulting in greater error. So at this stage pose estimation and navigation will be not robust enough while robot in a crowded, traffic complex environment.

The acquisition time interval between two images need to be matched in loop closure detection is longer than the acquisition interval between the two images that need to be calculated in visual odometry calculation, the latter one counts in seconds, and the former one counts in minutes, even in hours or days while in a large-scale environment. As shown in Figure 2, two images from asame location in different time may have different pedestrian, vehicle component which can lead the image matching program into error.

Not only the moving object is a kind of the interfering elements of image feature extraction, when robot works for a long period of time, there may be a sudden change of weather or the shift between day and night, the situation here will also have a great disturb on image matching.

4 TEST DATASET

FAB-MAP and DBOW are 2 most popular and successful loop closure detection algorithms in recent studies. They all have a very good performance on the existing public dataset in Table 1.

To check the performance of these two algorithm, with disturb from non-scene factors, we choose Malaga 2009 Parking 6L and City Centre dataset as target datasets for test, because they are outdoors or urban environment and have a appropriate dynamic environmental change. But this is not enough since the level of their environmental change is regular or lower, we need a dataset with

Figure 1. The interference of pedestrians and vehicles on the visual odometry calculation.

Figure 2. False matching between two images from the same location on a different time.

Table 1. Public datasets for loop closure detection.

Dataset	Description	Total length(m)	Revisited length(m)
New College (Smith M, 2009)	Outdoors, dynamic	2260	1570
Bicocca 2009-02-25b (Bon-arini A, 2009)	Indoors, static	760	113
Ford Campus 2 (Pandey G, 2011)	Urban, slightly dynamic	4004	280
Malaga 2009 Parking 6L (Blanco J L, 2009)	Outdoors, slightly dynamic	1192	162
City Centre (Cummins M, 2008)	Urban, dynamic	2025	801

Table 2. NCEPU campus dataset.

Category	Factor description		
	Weather	Pedestrian flow	Vehicle flow
00530	fine	none	none
01030	light rain	slight	slight
01230	light rain	heavy	heavy
11400	after rain	slight	slight

Figure 3.   Sample image from NCEPU campus dataset.

heavily dynamic environmental change of non-scene factors to test those algorithm's robustness.

In this paper, we build a key frame set named NCEPU campus dataset with the characters of the environment been mentioned above. As shown in Table 2, the dataset contains 4600 images and 4 categories, corresponds to the image information of 126 fixed scenes in different time, different weather, and different flow conditions, so as to simulate the key frame images collected in visual-based SLAM system. The category name is the time of data acquisition, for example, 00530 represents the data collected at 5:30 am on the first day, and the environmental factors are different in different time periods. The four categories of data are used in two combinations to simulate the robot in two different environmental conditions through the same path, and by calculating the image matching degree to determine whether the loop is closed.

Figure 3 shows the sample images of our dataset. It can be seen that the images in the dataset are from real campus scene. The image content is mainly on the roads and nearby trees and buildings, including rainy weather, different pedestrian and vehicle content.

5   EXPERIMENTAL EVALUATION

We evaluate the FAB-MAP and DBOW algorithms with the existing dataset and our dataset in this section, evaluation results are presented in the terms of precision and recall. OpenFAB-MAP (Cummins M, 2011) and DBOW2 are used in the experiment, they are all open-sourced implementation of their algorithms.

First, we use SURF as image descriptor of FAB-MAP, run it on Malaga 2009 Parking 6L and City Centre dataset, the result with a probability threshold 98% of being at the same place than some previous image shows in Table 3.

Then, we use BRIEF as image descriptor of DBOW, and run it on the same dataset with same probability threshold, the result shows in Table 4.

The result data above shows two of these algorithms achieved a high recall without false positive on Malaga 2009 Parking 6L dataset, which have only slight dynamic of environmental change. But when the environmental changes heavily to regular level on City Centre dataset, their performance degrades to half or even lower. To test their limits, we keep testing them with same configuration on our dataset which have heavily dynamic of non-scene factors, and the result shows in Table 5.

In the experiment above, we combine two different categories of the NECPU campus dataset into one sub-dataset to simulate environmental change. For instance, the sub-dataset 01030 & 01140 consist of category-01030 and category-01140 represents an environment with weather change and same pedestrian-vehicle-flow condition. From the Table 5, we can see neither FAB-MAP nor DBOW can do a great job in an environment with non-scene factor change, the recall is extremely low when there is no false positive, it means there is a lot of false negative, and the loop closure detector will consider most of incoming image as a sign

Table 3.   Precision and Recall of FAB-MAP.

Dataset	Precision (%)	Recall (%)
Malaga 2009 Parking 6L	100	68.52
City Centre	100	38.77

Table 4.   Precision and Recall of DBOW.

Dataset	Precision (%)	Recall (%)
Malaga 2009 Parking 6L	100	74.75
City Centre	100	30.61

Table 5.   Precision and Recall of DBOW and FAB-MAP tested on NCEPU campus dataset.

| Sub-Dataset | FAB-MAP | | DBOW | |
	Precision (%)	Recall (%)	Precision (%)	Recall (%)
01030&11400	100	24.06	100	19.83
00530&01030	100	9.39	100	8.61
01030&01230	100	23.64	100	20.14
00530&01230	100	9.50	100	7.65

of new location the robot never arrived, especially with weather changes.

6 CONCLUSIONS

We have presented in this paper a study on some factors affect loop closure detection algorithm and a dataset concerns these factors are built to test their influences. Experiment result shows loop closure detection is not just a simple image matching system, it is based on scene-centric image matching and only need part of the information in image. The non-scene factors of environment, such as weather, pedestrian, vehicle, do have a great negative influence on visual-based loop closure detection. When SLAM system use BoW-based loop closure detection, especially the changes of weather could do a lot more disturbance on matching result. The robustness of these loop closure detection algorithms without image semantic understanding and filtering still need to be improved.

REFERENCES

Bay H, Tuytelaars T, Gool L V. SURF: Speeded Up Robust Features.[J]. Computer Vision & Image Understanding, 2006, 110(3):404–417.

Blanco J L, Moreno F A, Gonzalez J. A collect-ion of outdoor robotic datasets with centimeter-accuracy ground truth[J]. Autonomous Robots, 2009, 27(4):327–351.

Bonarini A, Burgard W, Fontana G, et al. RAWSEEDS: Robotics Advancement through Web-publishing of Sensorial and Elaborated Extensive Data Sets[J]. 2009.

Calonder M, Lepetit V, Strecha C, et al. BRIEF: binary robust independent elementary features[C] Computer Vision—ECCV 2010, European Confere-nce on Computer Vision, Heraklion, Crete, Greece, September 5–11, 2010, Proceedings. 2010:778–792.

Clemente L A, Davison A J, Reid I D, et al. Mapping Large Loops with a Single Hand-Held Camera.[C]//

Robotics: Science and Systems III, June 27–30, 2007, Georgia Institute of Technology, Atlanta, Georgia, USA. 2007.

Cummins M, Newman P. Appearance-only SLAM at large scale with FAB-MAP 2.0[J]. International Journal of Robotics Research, 2011, 30(9):1100–1123.

Cummins M, Newman P. FAB-MAP: Probabilistic Localization and Mapping in the Space of Appearance[J]. International Journal of Robotics Research, 2008, 27(6):647–665.

Galvez-López D, Tardos J D. Bags of Binary Words for Fast Place Recognition in Image Sequen-ces[J]. Robotics IEEE Transactions on, 2012, 28(5):1188–1197.

Gálvez-López D, Tardós J D. Real-time loop detection with bags of binary words[C] IEEE/RSJ International Conference on Intelligent Robots & Systems. 2011:51–58.

Information on https://github.com/dorian3d/D-BoW2

Lowe D G, Lowe D G. Distinctive Image Features from Scale-Invariant Keypoints[J]. International Journal of Computer Vision, 2004, 60(2):91–110.

Pandey G, Mcbride J R, Eustice R M. Ford Campus vision and lidar data set[J]. International Journal of Robotics Research, 2011, 30(13):1543–1552.

Rosten E, Drummond T. Machine Learning for High-Speed Corner Detection[C] European Conference on Computer Vision. Springer-Verlag, 2006:430–443.

Rublee E, Rabaud V, Konolige K, et al. ORB: An efficient alternative to SIFT or SURF[C] IEEE International Conference on Computer Vision. IEEE, 2011:2564–2571.

Sivic J, Zisserman A. Video Google: A Text Retrieval Approach to Object Matching in Videos[C]/ IEEE International Conference on Computer Vision, 2003. Proceedings. 2003:1470.

Smith M, Baldwin I, Churchill W, et al. The New College Vision and Laser Data Set[J]. International Journal of Robotics Research, 2009, 28(5):595–599.

Smith R, Self M, Cheeseman P. A stochastic map for uncertain spatial relationships[C] International Symposium on Robotics Research. MIT Press, 1987:467–474.

Williams B, Cummins M, Neira J, et al. An image-to-map loop closing method for monocular SLAM[C] International Conference on Intelligent Robots and Systems, September 22–26, 2008, Acropolis Convention Center, Nice, France. 2008:2053–2059.

Electromechanical Control Technology and Transportation – Jia & Wu (Eds)
© *2017 Taylor & Francis Group, London, ISBN 978-1-138-06752-3*

Analysis of residential electricity consumption behavior based on the improved Apriori algorithm

Bing Zhang
School of Electrical Engineering and Information, Sichuan University, Chengdu, China

ABSTRACT: As the traditional Apriori algorithm has the problem that the data processing capability is obviously decreased when the data quantity is large and the minimum support degree is low, an improved Apriori algorithm is proposed and used to analyze the residential electricity consumption behavior. The proposed algorithm uses the MapReduce parallel computing model of Hadoop platform to divide the database into a subset of local data and mine frequent itemsets. The experimental results show that the improved Apriori algorithm has a significant improvement in the data processing ability. We use the proposed algorithm to analyze the relationship between the household electricity consumption and six dimensions of family structure, housing area, household income, appliance type, energy-saving awareness, daily maximum temperature, and daily consumption. The experiment uses 10 effective association rules and verifies the feasibility of the proposed algorithm in the analysis of residential electricity consumption behavior.

1 INTRODUCTION

In recent years, big data has become a new concept of widespread concern. Smart grid is one of the most important fields in the application of big data, on which scholars worldwide have been conducting relevant theory and application (Zhang Dongxia; Liu Keyan, 2015; Doug Dorr, 2014). In 2013, China Institute of Electrical Engineering Information Committee issued a document named "China power big data development white paper", which has expounded the connotation and characteristics of electric power big data. This paper analyzed its development prospect and set off a wave of electric power big data research.

With the rapid development of the Chinese economy, people's quality of life generally improved, which gradually increased the residential electricity consumption. Recently, power companies have established many electrical systems, such as electric information acquisition system, power marketing system, and customer service information system. These systems have accumulated a large number of residential electricity information, which can fully exploit the potential value of data by combing with demographic data, economic data, and meteorological data. Therefore, the application of big data technology in the field of residential electricity has broad prospects.

At present, there are few studies on the analysis of residential electricity consumption behavior. Zhang Suxiang (2013) studied the model of analysis of electricity consumption behavior on the basis of the intelligent community data. This paper used k-means clustering algorithm based on cloud computing technology to divide 600 households into five categories, including vacant rooms, office workers, families, the elderly and the old+ office worker families, and business users. By means of a questionnaire survey, He Yongxiu (2012) collected the residential electricity consumption data in different regions of China and analyzed the relationship between household income, housing area, and electricity expenditure from a single dimension. However, because of the small sample size, the sample selection is subjective, and the quality of the survey results cannot be guaranteed. Wang Jiye (2015) proposed the use of association mining technology and analyzed the relationship between the influence factors and electricity consumption behavior from time, space, user types, and other dimensions.

In this paper, an improved Apriori algorithm is proposed by combining the Hadoop platform with the traditional Apriori algorithm. We analyzed the relationship between six dimensions and household electricity consumption on the basis of 9132 households' electricity information from August 2014 to December 2014. The experimental results show that the improved Apriori algorithm can effectively and accurately analyze the household electricity information association and has obvious advantages compared with the traditional Apriori algorithm.

2 TOTAL FRAMEWORK

The electricity consumption of the residential data source comes from the electric power infor-

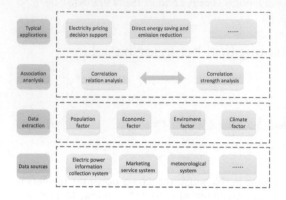

Figure 1. Framework of residential electricity consumption behavior analysis under big data.

mation collection system, the marketing service system, the community management system, the meteorological system, and so on. The independent system has accumulated a large number of residential electricity information. We analyze the association between different dimensions and power consumption by extracting residential consumption data and external data. Mine the association strength among the various factors to give full play to the potential value of the data, which can help improve the residential electricity consumption behavior understanding and awareness. It can also provide favorable guidance for optimal formulation strategy of power use and for the government to make the price subsidy policy and provide decision support. The total framework of residential electricity consumption behavior analysis under big data is shown in Figure 1.

3 IMPROVED APRIORI ALGORITHM

The Apriori algorithm is a very effective method for mining association rules, but when the amount of data is too large and the initial support is low, the efficiency of the algorithm is significantly decreased. The performance of the defect is mainly reflected in: (1) layer search need to scan the database repeatedly, increase the load of I/O; (2) frequent itemset connects itself to generate candidate itemset, the length of the layer is increasing exponentially; even after pruning, the candidate itemset still has a large length (Liu Huating, 2009).

The Hadoop platform is the basic framework of distributed system developed by the Apache foundation, which is suitable for parallel process-

ing of massive data. It uses the distributed storage and computing to overcome the shortcomings of the traditional Apriori algorithm. The core of the Hadoop platform is the distributed file system (Hadoop Distributed File System, HDFS) and MapReduce computing model. HDFS includes a management node (NameNode) and multiple data nodes (DataNode). MapReduce is a parallel computing model to complete the distributed storage and computing tasks with HDFS. Database (Hadoop database, Hbase) and data warehouse Hive based on Hadoop can achieve efficient storage, management, and analysis of big data. The Hadoop platform framework is shown in Figure 2.

In this paper, the Apriori algorithm based on Hadoop is used to analyze the relationship between the composition of family members, the area of housing, the annual income, the type of household appliances, the awareness of energy saving, the maximum temperature, and the daily household electricity consumption. The specific steps of the algorithm are as follows:

1. Divide the database D used to store household electricity information into n subsets of similar data length (D_1, D_2, ... D_n) and sent them to m node ($m \leq n$) to execute the map function.

2. Format n data subset to generate $<key_1$, $value_1>$, specific format for the $<T_{id}$, list$>$, wherein T_{id} represents a family user ID and list represents the influence factors as family member composition and housing area.

3. Use the map function to scan each data subset $<T_{id}$, list$>$, generate the intermediate $<key_2$, $value_2>$ pairs, specific $<itemsets_k$, sup$>$, itemsets k represents K set, sup represents the corresponding support.

4. Use the Combiner function to combine the $<itemsets_k$, sup$>$ on the local map nodes to

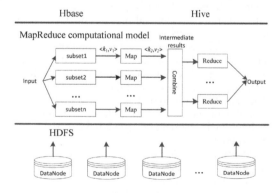

Figure 2. Framework of Hadoop platform.

generate local 1-candidate itemsets. Local 1-frequent itemsets are generated according to local minimum support.

5. Use the traditional Apriori algorithm to generate the local k-frequent itemsets in each node.
6. Use the Reduce function to combine local k-frequent itemsets $<itemsets_k, sup>$, the global candidate itemsets are obtained.
7. Scan the database D again and generate global frequent itemsets based on global minimum support.
8. Divide all the frequent itemsets into p subsets of similar data length and send them to q map nodes $(q \leq p)$.
9. Format p data subset to generate $<T_{id}, freq>$, wherein T_{id} represents the number of frequent itemsets and $freq$ stands for the frequent itemsets.
10. Execute the map function, scan each data subset $<T_{id}, freq>$ to produce $<freq, rule>$, wherein $freq$ represents a specific frequent itemset and $rule$ represents association rule.
11. Use the Reduce function to combine $<freq, rule>$ on each node and output the strong association rules.

The improved Apriori algorithm flowchart is shown in Figure 3.

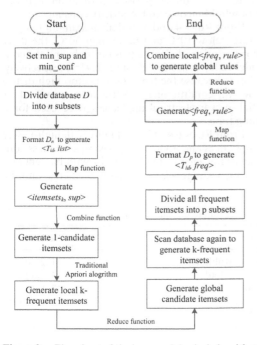

Figure 3. Flowchart of the improved Apriori algorithm.

4 EXPERIMENTAL RESULT

4.1 Experimental environment

Electricity consumption information of 9132 residential households from August 2014 to December 2014 includes family structure, housing area, household annual income, appliance type, energy-saving awareness, daily maximum temperature, and daily consumption.

Experimental environment: Hadoop cluster using Hadoop1.0.4 version, one management node server, and four data node servers. Set up two HDFS copies, the data node can simultaneously perform two Map tasks and two Reduce tasks, the other set as the default configuration.

4.2 Performance analysis of the improved Apriori algorithm

In order to verify whether the improved Apriori algorithm could solve the defects when the performance of large amount of data and the minimum support degree with low data processing capacity decreased significantly, the efficiency of the improved Apriori algorithm is compared with the traditional Apriori algorithm in different data scale and different minimum support conditions.

Experiment 1: Compare the consuming time between traditional Apriori algorithm and improved Apriori algorithm in analyzing residential electricity consumption behavior under different data scale and then calculate the consuming time. MapReduce computing model under the Hadoop platform sets up one management nodes, one data nodes, and four data nodes. The results of the experiment are shown in Table 1.

According to the results of Table 1, with the increasing of the data volume, the consuming time of the traditional Apriori algorithm increases obviously. When the data size is 140 million, the traditional algorithm cannot output the result because of insufficient memory. However, the consuming time of the improved Apriori algorithm increases

Table 1. Consume time contrast between two algorithms under different data scales.

Data vol./million	A/s	B/s (1 node)	B/s (4 nodes)
40	182.6	121.4	77.2
60	215.4	149.2	88.7
80	263.7	187.5	107.3
100	383.2	237.2	124.8
120	630.3	318.3	148.6
140	Error	392.7	178.9

*A represents the traditional Apriori algorithm, B represents the improved Apriori algorithm.

slowly, and the four data nodes take less time than the one data node. This shows that the improved Apriori algorithm is effective in the analysis of the residential electricity consumption behavior under big data, and multiple data nodes are more efficient than single data node.

Experiment 2: Compare the consuming time between the traditional Apriori algorithm and the improved Apriori algorithm in analyzing residential electricity consumption behavior under different minimum support. Set up one management node, one data node, and four data nodes. The results of the experiment are shown in Table 2, where A represents the traditional Apriori algorithm and B represents the improved Apriori algorithm.

As can be seen from Table 2, with the constant reduction of the minimum support, the consuming time of the traditional Apriori algorithm increases significantly, while the improved Apriori algorithm increases slowly. When the minimum support is less than 10%, the traditional algorithm cannot output the result because of insufficient memory. It shows that the improved Apriori algorithm can find strong association rules with low frequency, avoid missing the effective association rules, and make the data mining results more accurate.

4.3 Analysis of residential electricity consumption behavior

We use the improved Apriori algorithm proposed to analyze the association relation and intensity between family structure, housing area, household annual income, appliance type, energy-saving awareness, daily maximum temperature, and daily consumption. The code for the household user information is shown in Table 3. We consider consumption more than 10 kWh as high energy consumption, 5–10 kWh as medium energy consumption, less than 5 kWh as low energy consumption.

Convert 9132 residents' electricity information of each day from August 2014 to December 2014 into type code as the input and the association rules as the output. Set the minimum support as min_sup = 0.1, minimum confidence as min_conf = 0.5, filter out the home user information association rules as shown in Table 4.

Rules interpretation:

According to rule 1, the family of the multiple generations is high energy consumption, and the confidence level is 72.53%. This is because the multigenerations family has a complex constitution, and the number of power users and the range of electrical consumption time are large.

Rule 2 shows that the family with housing area more than 160 m² consumes high energy. The confidence level is 63.66%. This shows that the larger housing area can easily lead to higher consumption.

Rule 3 shows that the family with annual income more than 0.3 million consumes medium energy. The confidence level is 50.42%. This shows that there is no direct positive correlation between household income and electricity consumption.

From Rules 4 and 5, at daily maximum temperature below 10°C and more than 30°C, the daily consumption of electricity is the high, and the confidence levels are 50.65% and 71.41%, respectively. This is because the temperature is too high or too low, air-conditioning, electric heater, and other high-power electrical appliances were used for a long time.

From Rules 6 and 7, the family with high-power appliances and good energy-saving awareness consumes low energy, but the family with bad energy-saving awareness is consumes high energy, and the

Table 2. Consume time contrast between two algorithms under different minimum support.

Data vol./million	A/s	B/s (1 node)	B/s (4 nodes)
10	Error	391.3	182.8
20	610.2	307.2	155.3
30	413.6	260.3	124.4
40	315.2	175.4	105.1
50	261.7	141.3	83.3

*A represents the traditional Apriori algorithm, B represents the improved Apriori algorithm.

Table 3. Consume time contrast between the two algorithms under different minimum support.

Type	Classification	Code
Family structure	{Youth, elder, Multiple}	{1, 2, 3}
Housing area/ m²	{<80, 80–160, ≥ 160}	{4, 5, 6}
Household annual income/million	{<0.1, 0.1–0.3, ≥ 0.3}	{7, 8, 9}
Appliance type	{small, medium, large}	{10, 11, 12}
Energy-saving awareness	{bad, good}	{13, 14}
Daily maximum temperature/°C	{<10, 10–30, ≥ 30}	{15, 16, 17}
Daily consumption	{low, medium, high}	{18, 19, 20}

Table 4. Association rules of electricity consumption information for family subscribers.

No.	Association	Sup	Conf
1	$3 \Rightarrow 20$	10.32%	72.53%
2	$6 \Rightarrow 20$	14.25%	63.66%
3	$9 \Rightarrow 19$	11.26%	50.42%
4	$12 \wedge 15 \Rightarrow 20$	16.24%	50.65%
5	$12 \wedge 17 \Rightarrow 20$	18.73%	71.41%
6	$12 \wedge 14 \Rightarrow 18$	21.52%	53.73%
7	$12 \wedge 13 \Rightarrow 20$	25.75%	74.28%
8	$1 \wedge 4 \wedge 7 \Rightarrow 18$	10.12%	75.14%
9	$1 \wedge 5 \wedge 8 \Rightarrow 20$	10.31%	77.52%
10	$2 \wedge 5 \wedge 8 \Rightarrow 18$	14.67%	68.34%

confidence levels are 53.73% and 74.28%, respectively. It can be seen that a good energy-saving awareness can control the increase in electricity consumption to a certain extent.

According to Rule 8, the young family whose housing area is less than 80 m² and the annual income is less than 0.1 million consumes low energy, and the confidence is 75.14%. This is because many young families have just entered the community, and the economic conditions are relatively tight, mainly to meet the basic living electricity.

From Rules 9 and 10, for the same area of housing and household annual incomes, the young family consumes high energy, the elder family consumes low energy, and the confidence levels are 77.52% and 68.34%, respectively. This is because elderly families pay more attention to thrift than younger families.

5 CONCLUSIONS

This paper first uses association analysis algorithm into residential electricity consumption behavior and presents a calculating method on the basis of the improved Apriori algorithm, which uses MapReduce parallel computing model under Hadoop platform to divide the database into a plurality of local data subsets and mine the frequent itemsets separately.

The experimental results show that the improved Apriori algorithm can effectively improve the data processing ability. Ten effective association rules are put forward in the experiment, which proves that the Apriori algorithm is effective, accurate, and feasible in the application of the analysis of residential electricity consumption behavior under big data.

The rules obtained can improve the understanding of the residential behavior and provide references for the formulation of the pricing policies. In the next step, we can select the important factors influencing the power consumption in the association rules and establish the forecasting model of the household electricity consumption.

REFERENCES

Apache, Apache Hadoop core [EB/OL].2012-08 [2013-02]. http://hadoop.apache.org/core.

C3 Energy analytics platform (TM) [EB/OL]. California, C3 Energy. 2013. http://www.C3energy.com.

Doug Dorr. Data analytics and applications newsletter (DMD and TMD demonstrations) [EB/OL]. California: Electric Power Research Institute. 2014 [2015-02-11]. http://smartgrid.epri.com/.

He Yongxiu, Wang Bing, Xiong Wei, et al. Analysis of residents' smart electricity consumption behavior based on fuzzy synthetic evaluation and the design of interactive mechanism[J]. Power System Technology, 2012, 36(10): 247–252 (in Chinese).

Informatization Committee of the CSEE. White paper of electric power big data of China[M]. Beijing: China Electric Power Press, 2013: 10–15.

Liu Huating, Guo Renxiang, Jiang Hao. Research and Improvement of Apriori Algorithm Forming Association Rules [J]. Computer Applications and Software. 2009, 26(1): 146–149.

Liu Keyan, Sheng Wanxing, Zhang Dongxia, et al. Big data application requirements and scenario analysis in smart distribution network[J]. Proceedings of the CSEE, 2015, 35(2): 287–293.

Wang Jiye, Ji Zhixiang, Shi Mengjie, et al. Scenario Analysisi and Application Research on Big Data in Smart Power Distribution and Consumption Systems[J]. Proceedings of the CSEE, 2015, 35(8): 1829–1836.

Zhang Dongxia, Miao Xin, Liu Liping, et al. Research on development strategy for smart grid bigdata[J]. Proceedings of the CSEE, 2015, 35(1): 2–12.

Zhang Suxiang, Liu Jianming, Zhao Bingzhen, et al. Cloud computing-based analysis on residential electricity consumption behavior[J]. Power System Technology, 2013, 37(6): 1542–1546.

Electromechanical Control Technology and Transportation – Jia & Wu (Eds)
© 2017 Taylor & Francis Group, London, ISBN 978-1-138-06752-3

Triaxial test analysis of flax fiber Xigeda soil

Yuehua Liang & Jie Wang
Architecture and Civil Engineering College, Panzhihua University, Si Chuan, China

Kecai Chen
Architecture and Civil Engineering College, Panzhihua University, Si Chuan, China
Architecture and Civil Engineering College, Southwest University of Science and Technology, Si Chuan, China

Jun Zhang
Architecture and Civil Engineering College, Panzhihua University, Si Chuan, China

ABSTRACT: The flax fiber Xigeda soil is triaxially tested under unconsolidated/undrained conditions. Through the different combinations of flax fiber content, the stress–strain relationship of flax fiber Xigeda soil is found, and the factors affecting the stress–strain relationship and shear strength of flax fiber are analyzed. Through the stress circular envelope curve, cohesion (c) and the internal friction angle φ are calculated. The test results showed that: (1) in comparison with soil, the shear strength of flax fiber Xigeda soil was obviously improved; (2) flax fiber Xigeda soil was better than that in elastic–plastic deformation; (3) for the shear strength index, the internal friction angle of flax fiber Xigeda-reinforced soil changes was lower than that of prime soil (its absolute value of relative rate of change is less than 46%), but the cohesion was much larger than that of prime soil (the maximal increased multiple is nearly 13.5).

1 INTRODUCTION

As the Xigeda formation is relatively new, the pre-consolidation pressure is small and the diagenesis is not completed. It is a weakly cemented semi-diagenetic stratum, which shows the physical and mechanical characteristics of the non-soil and rock-like rock (Ruan, 2013). The parameters of Xigeda soil in the project indicate that the Xigeda formation is a very soft rock close to the soil property, with low natural dry density, high porosity ratio, loose soil structure and low strength. As the Xigeda soil contains a large number of illite-based clay minerals, it has a high sensitivity of water. In the flooding, with the loss of water under repeated action, the shear strength is getting lower (Wang, 2004). In the water-rich sliding surface, its shear strength index value can be reduced to 1.18 kPa, and c value down to 4.22 kPa. The mechanical strength of the whole Xigeda soil layer is high, but when the rock mass is destroyed, the strength of the rock is reduced by about one-third to half (Li, 2008). The unique physical and mechanical characteristics of the Xigeda soil have resulted in its low strength, unstable nature, easy to slip and other engineering geological features. The shear capacity of Xigeda soil has a high practical value. The shear capacity of Xigeda soil can be improved by reinforcement, so a study on Xigeda-reinforced soil is necessary. The results of its research on our practical engineering will also provide a good theoretical knowledge.

The so-called fiber is incorporated into the soil continuous fiber sand or soil in order to achieve a method of strengthening the soil. Fiber soil is non-rigid and brittle. It is suitable for overcoming soil deformation occurring in civil engineering. Meanwhile, in road construction, the technique can be used to solve common problems, such as highway embankment, soft foundation treatment retaining walls and so on. At present, domestic and foreign applications have not yet directly applied flax fiber in the soil.

In this test, Xigeda soil and flax fiber in Panxi area were selected as the main experimental materials. Under different confining pressures ($\sigma 3 = 100$ kPa, $\sigma 3 = 200$ kPa, $\sigma 3 = 300$ kPa), the stress–strain curve of the specimen and the shear strength of the specimen were analyzed to obtain the tensile properties of the fiber. Therefore, it is important to study the engineering properties of flax fiber-reinforced soil.

2 TEST EQUIPMENT, MATERIALS AND TESTING PROGRAM

2.1 Test equipment

This test used the SLB-1 A stress–strain-controlled triaxial shear penetration tester manufactured by Nanjing Road Instrument Co., Ltd. (Fig. 1). Tri-axial test specimens had a diameter of 39.1 mm and a sample height of 80 mm.

2.2 Test material

2.2.1 Xigeda soil
The soil used in this test was Xigeda soil from Panzhihua University. The test parameters are shown in Table 1, and the soil was medium-compressed silty clay.

2.2.2 Fiber material
The test was done on linear linen fiber material with dark brown color and soft texture. The length was 3 cm long according to the size of sample; the fiber content (mass ratio) of the soil samples is 0, 0.5%, 1.0%, 1.5%, 2% and 3%, respectively.

2.3 Sample preparation

The samples were prepared according to the standard size and labeled according to the its fiber content. After 7 days, the sample was taken out and subjected to the saturation test by suction saturation method.

2.4 Test program

The unconfirmed undrained test (UU) method was used for this triaxial test. Considering the effect of incorporation of the fibers on the soil shear strength ratio, six kinds of fiber mixing ratio (0%, 0.5%, 1.0%, 1.5%, 2%, 3%) were set in the flax fiber Xigeda soil test. In order to obtain the complete envelope of the strength, each sample was tested under three kinds of confining pressure (100, 200, and 300 kPa). The strain rate in the test

was 1 mm/min. All specimens were cut to failure or reached a certain axial strain (15%, 12 mm) at the failure point.

3 ANALYSIS OF THE TEST RESULTS

3.1 Effect of flax fiber content on soil strength

When the elastic strain is about 3%, the growth rate of stress Xigeda prime soil was the fastest. Fiber content was relatively stable at a growth curve of 3%.

The stress–strain curves of the prime soils were convex in shape. When the flax fiber was added, the shape of the curve was concave and convex. The stress–strain relationship curve was changed from convex to concave as the amount of flax fiber increases. This indicated that the elastic–plastic deformation capacity of flax fiber soil increases.

When the soil strain is small, the flax fiber in the soil reinforced by it undergoes elastic deformation. At this stage, the fiber is not in the initial stage of the fiber, and the distance between the soils is small, so the curve is gentle. As the strain increases, the distance between the soils is reduced, and the curve becomes steeper and steeper. The axial strain reaches about 5%, and the stress–strain curve is stable in the plastic deformation stage. But the reinforcement effect still comes into play.

3.2 Effect of fiber content on fiber soil C value and the value of φ

According to Mohr–Coulomb strength theory, when the shear stress of any plane in the soil is equal to the shear strength of the soil, the critical state of the point on the brink of destruction is called the "limit equilibrium state" The Mohr Circle and Intensity Envelope Curve of Fiber Xigeda Soil. Figures 2 to 7 show the effect of flax fiber content on the strength of the Xigeda soil. It wasn't difficult to find that the incorporation of fibers had an effect on the internal friction angle and cohesion of the soil compared with the plain soil. The increasing tendency of the cohesion of the soil was more obvious with the increase of the fiber content (see Table 2).

From Table 2, the following can be obtained:
The change of fiber content affects the internal friction angle and cohesion of the fiber.

Figure 1. Picture of the test equipment.

Table 1. The physico-mechanical indexes of Xigeda soil.

Moisture content	Proportion g/cm³	Void ratio	Plastic limits IP%	Liquid limit WL%	Compression factor	Compression modulus
19%	2.77	0.35	16.5	38.6	0.31	4.35

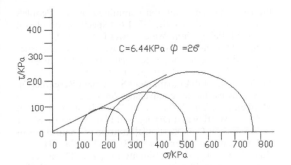

Figure 2. Mohr stress circle of unreinforced soil.

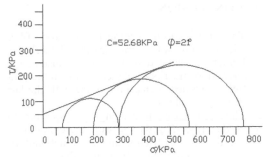

Figure 5. Mohr stress circle of fiber soil (content 1.5%).

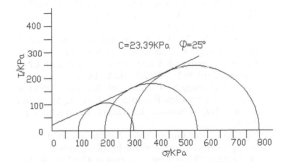

Figure 3. Mohr stress circle of fiber soil (content 0.5%).

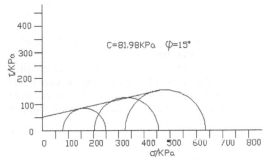

Figure 6. Mohr stress circle of fiber soil (content 2%).

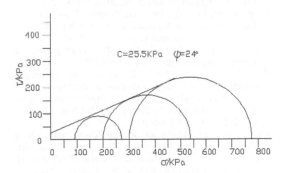

Figure 4. Mohr stress circle of fiber soil (content 1%).

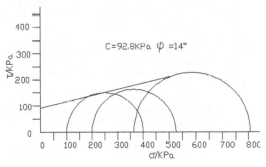

Figure 7. Mohr stress circle of fiber soil (content 3%).

The cohesion c increases significantly with increasing flax fiber content. With the flax amount of 3%, the c value increases by 13.5 times. The friction angle φ changed little with the increase of flax fiber content. When the content of flax fiber was in the range of 0 to 1.5%, φ was almost unchanged. With flax fiber content of 1.5–2%, the φ value decreased, and it wasn't a linear relationship.

In view that above, cohesion c improved the flax fiber Xigeda soil's shear strength. After incorporation of the fiber, the fibers in the soil particles produced a bending mechanism, erosion mecha-

Table 2. Comparison of shear strength indexes.

Amount of linen	0%	0.5%	1%	1.5	2%	3%
c (KPa)	6.4	23.4	25.5	52.7	82	92.8
φ (°)	26	25	24	21	15	14

nism, and intertwined mechanism (Cheng, 2013). A variety of mechanisms greatly enhanced the soil particles and fiber cohesion, so fiber shear strength of soil was significantly enhanced.

4 CONCLUSION

1. The stress–strain relationship of flax fiber Xigeda soil was approximately hyperbolic. Under the same reinforcement condition, the strength of the fiber soil increased with the confining pressure increasing, in line with the law of the general soil.
2. The main factors influencing the shear strength of flax fiber Xigeda soil were the cohesion c, and the internal friction angle φ, which had little effect on the shear strength of fiber.
3. Under the same confining pressure and different fiber content, the increase of shear strength of flax fiber-reinforced soil increased with the increase in the amount of fiber added, but the increase of shear strength was nonlinear with the fiber content increase.
4. Flax fiber could increase the soil ability of elastic-plastic deformation.
5. When the soil was in the elastic deformation stage, the stress of Xigeda soil increased rapidly. When the soil was near the failure state, the fiber soil could bear more pressure than the plain soil.

REFERENCES

Chen Lin, Sun Jinkun, Chen Kecai, Jiao Tao, Luo Qiang, Guo Jian. Experiment Study of physicomechanical Behavior of Xigeda Soil Reinforced by Eupatorium Adenophorum. Industrial Construction, (2016) 05–124, 127.

Chen Lin, Zhu Jian. Deformation Mechanism and Stabilitty Assessment of Xigeda Landslide. Journal of Geological Hazards and Environment Preservation, (2015) 03-11-16.

Chen Wei, Luo Qiang, Sun Jinkun, Jiao Tao, et al. Study on the improvement technology of peasant house wall material of Xi- Geda soil in Panzhihua west area. Sichuan Building Science, (2010) 03–201–04.

Cheng Zh L, Zhang Y M, Ruan D H, et al. The Mechanical Properties of The Fiber Soil Test Results. Low Temperature Architecture Technology, 2013, 35(8): 103–104, 140.

Industry standard editorial committee of the People's Republic of China. SL 237–1999 Specification of soil test[S]. Beijig:China Water Power Press, 1999.

Kong X H, Jiang G L, Wang Z M. Dynamic characteristics of red-mudstone soils under cyclic loads [J], Hydrogeology & Engineering Geology, 2014 (4):80–86.

Li Ch Y. High Fill Embankment Settlement Research[D]. Southwest Jiaotong university, 2008.

Manbeian T. The Influence of Soil Moisture Suction, Cyclic Wetting and Drying, and Plant Roots on Shear Strength of Cohesive Soil[J]. Journal of Central South University of Technology, 1981(1):113, 128, 131.

Ruan Xiao-long, Huang Shuang-hua, Luo Qiang, et al. Study on the Performance of Xigeda Soil with Different Lime Mixture Ratios. Journal of Xihua University (Natural Science Edition), (2014) 01-99-103.

Ruan X L, Research on Modification and Reinforcement Mechanism of Xigeda-soil in High Fill Slope[D]. Xihua university, 2013.

Wang Ya L, Hu Y. Dynamic triaxial test and research on the dynamics characteristics of saturated sandy[J]. Chinese Journal of Underground Space and Engineering, 2010, 6(2): 295–299. (in Chinese).

Wang Y. Extremely Soft Rock Embankment Deformation Characteristics Research-Xi-Pan Highway, for Example[D]. Chengdu University of Technology, 2004.

Xiang Gui-fu, Ren Guang-ming, Nie De- xin, et al. Study on optimal moisture content in mixed filling of Xige-da strata. The Chinese Journal of Geological Hazard and Control, 2004, 2:48–52.

Yang Sh, Li G F, Luo J H, et al. Triaxial Test of PVC Texsol Analysis[J]. Journal of Beijing University of Civil Engineering and Architecture, 2010, 26(4): 12–15.

Zeng Qiang, Liu Wen-lian, Xu Ze-min, Tian Lin. The Influence of Soil Moisture Suction, Characteristics of shear modulus and damping of thesilt and clay of the Xigeda Group in Jingjiu[J]. Hydrogeology & Engineering Geology, 2014 (4): 80–86.

Zhou Hong-fu, Nie De-xin, Zhong Hua-jie. Technique experiment using Xigeda stratum as expressway subgrade materials[J]. Journal of Chengdu University of Technology(Science & Technology Edition), 2013, 6 (3): 295–299. (in Chinese).

Zuo Yong-zhen, Zhang wei Zhang Xiao-chuan, Dang chen.Engineering Properties of Xigeda Strata Siltstone as the Filling Material of Earth-Rock dam[J]. Journal of Yangtze River Scientific Research Institute, 2016, 33(03): 84–88. (in Chinese).

Electromechanical Control Technology and Transportation – Jia & Wu (Eds)
© 2017 Taylor & Francis Group, London, ISBN 978-1-138-06752-3

Study on the control measures of the loose accumulation body slope

Fa Wang
National and Local Joint Engineering Laboratory of Traffic Civil Engineering Materials,
Chongqing Jiaotong University, Chongqing, China

Bingyang Chen
China Merchants Chongqing Communications Technology Research and Design Institute Co. Ltd.,
Chongqing, China

ABSTRACT: Earthquakes can cause serious damage to roads, with serious geological hazards occurring in most sections. On the basis of field investigations, the geological structures of the loose accumulation body slope and types of geological hazards are summarized and analyzed. According to the scale of the side slope and its influence on engineering, three schemes are proposed for landslide treatment, namely avoidance, reinforcement, drainage and monitoring measures. No disturbance and pre-reinforcement treatment schemes are proposed for the unstable slopes to ensure road safety.

1 INTRODUCTION

Landslide is a common adverse geological phenomenon in mountainous areas. The landslide occurring along the highway can negatively affect the normal operation of traffic and transportation of the highway in case of minor landslide, or block transportation, damage highway pavement and subgrade, causing flood damage on the highway in case of a major landslide. Development of landslide along the highway has attracted more and more attention from the transportation authority and researchers for many years, but there is little discussion about the mechanism of its development.

In mountainous regions, the landslide occurring along the highway is of loose mass mostly. Landslide mass consists of typical slope and residual sediments of Quaternary Period, taking on loose structure of rock mass, which is large in porosity, high in compression property, and high in water permeability. The subterrain is often argillaceous mass, which is poor in water permeability, such as argillite, clay shale, etc. The interface between loose media (slope and residual deposit) and subterrain is often the case where the sliding face (or zone) is located.

Therefore, it is of universal significance to study the development mechanism of the landslide. The paper sets out from the fundamental factors affecting the development of landslide, studies in depth how the loose accumulation body slope is formed, and proposes the safety improvement measures to ensure safe transportation of the highway.

2 OVERVIEW OF THE PROJECT

The highway project is located in the southeast of Qinghai-Tibet Plateau, on the east side of Qionglai mountain chain, with Songpan and A'ba massif on its northeast, Jintang arc structural zone on its southwest, and Longmenshan geosynclinal area on its south. The geological structure belongs to S-shaped Xuecheng–Wolong structure on the east side of Jintang arc fold belt in the geosynclinal area in the west of Sichuan Province. The geological structure is complicated. The structural system borders with Maowen Fault on its east and connects obliquely with Jiudingshan Cathaysian structural system. The structural system is complicated by distortion of the mainstay, Xuelongbao Mountain, comprising a series of S-shaped and cambered linear fold, with cambered compression and scissor fault. Its northeast section extends in the 60° direction, only 10–20 km in width. Its middle section, i.e. Lixian County–Xuelongbao zone, approaches the center of torsion, so it is curved into an S-shape. In the section, the cambered compression and scissor fault is developed evidently, so there are especially dense folds. The southwest section extends in the 220° direction, and fans out gradually, with more than 40 km in width and more than 150 km in length. The section extends into Xiaojin–Baoxing Range, meeting with the east side of Jintang cambered structure. Faults disperse mainly in the south and east of S-shaped structure. The surrounding of Xuelongbao Mountain is the area where faults concentrate, which meets obliquely with or is generally parallel to Jiudingshan Cathaysianc tectonic line.

Because of powerful crustal stress in the survey area, the brittle soft rocks break and produce fissures. In addition, the powerful weathering effect destroys the linkage of rocks, and reduces the strength of rock, causing adverse geological phenomenon in engineering, such as landslide, collapse, and falling fragments along the highway.

The outcropping strata in the survey area include metamorphic rocks of Devonian (D), Silurian (S) and Ordovician (O) in the Paleozoic erathem. The lithology includes mainly phyllitizated schists, slates and phyllites. Because of the effect of late tectonic movement, quartz of network structure distributes commonly in the rock masses. The accumulated layer of the Quaternary Period distributes widely in the survey area, which includes mainly alluvium and diluvium deposits, colluvial deposit, gravel soil and broken stone soil of Holocene Series and Pleistocene Series, as well as alluvium and diluvium deposit, slopewash and colluvial deposit of Holocene Series.

The relative elevation difference in the highway is generally 2000 m or higher. Zhagunao River is the main water system in the survey area, which is the basis for local erosion and the centralized drainage zone of ground water. The project site is complex in geological structure, and vulnerable in ecological environment. There are different types of engineering geological disasters in various positions of the existing highway. The geological condition for engineering is evaluated to be moderate.

3 TREATMENT PRINCIPLE OF LANDSLIDE AND UNSTABLE SLOPE AS WELL AS THE DESIGN OVERVIEW

3.1 The design principle

The main principle of subgrade design in the loose accumulation body slope section is to avoid major hazards, eradicate minor hazard and prevent secondary hazards. In accordance with the size, characteristics and influence on engineering of side slope, three schemes, i.e. avoidance, reinforcement and no treatment (drainage + monitoring measures), could be adopted for treatment. Non-disturbance and pre-reinforcement schemes are adopted mainly for the unstable slopes.

3.2 Overview of the design

Landslide of the loose accumulated body occurs in four positions and unstable slope occurs in one position in total. In view of different conditions of landside and unstable slope in different sections, specific engineering measures are adopted respectively. The measures are described as below against engineering sites.

3.2.1 Landslide at Mazha

Mazha Landslide is close to the highway, affecting a 80 m length section. The slide mass consists mainly of crushed stones containing soil. The landslide bed consists mainly of crushed stones containing soil, silt containing breccias, and block stones mixed with soil. The entire plane shape of the landslide is like a Horseshoe. There are steep ridges of 2.00–4.00 height at rear and lateral edges, evident at the boundary. Its shear crack is near the elevation of highway pavement. The landslide is mainly presented by slipping and collapse of soil mass at the front edge. It belongs to shallow retrogressive landslide of small scale.

The landslide is located on a steep slope. A part of toe of the slope is excavated because of construction of the highway, so the free face is increased, providing condition of space for landslide movement. Soil mass softened by rainfall provides the condition of external force for deformation of landslide. A large part of surface water percolates through the soil mass, which not only increases the weight of landslide mass, but also softens the soil containing crushed stones, forming a weak face in the crushed stone-soil layer. Therefore, in rainy season, the volumetric weight of landslide mass increases while value C and Φ decrease, so that the soil mass at the front edge is vulnerable to becoming unstable and extending backward, resulting in extension of the scope of deformation.

The current condition of typical road section is simulated and analyzed. The coefficient of stabilization Fs under storm condition is less than 1, in unstable condition. Such condition is similar with the actual condition, as shown in Figure 1.

It is demonstrated by geological survey report and inverse analysis that values of the parameters of slide-zone soil are taken as below: under natural condition: C, 5 kPa; φ, 32°; under storm condition, C, 4 kPa; φ, 30°. Where the safety coefficient under storm condition is taken to be 1.15, the residual sliding force of landslide is 156.54 kN; where the safety coefficient under storm and

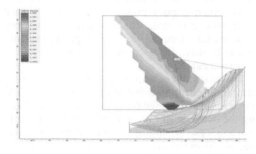

Figure 1. Simulation and analysis of the current condition of the road section.

seismic condition is taken to be 1.05, and the residual sliding force of landslide is 335.67 kN.

The route of the highway passes through the front edge of the landslide by means of low fill and shallow cut. Because of the characteristics mentioned above, the landslide is reinforced by means of pile-plank retaining wall + counter-pressure of the front edge, and treated by means of water cut-off and drainage.

Landslide retaining measures: diameter of anti-slide piles: 1.5 m × 2.0 m; spacing: 5.0 m; concrete: C30; retaining plates made of C30 concrete set between the piles. The cut slope wall is adopted at both sides to connect smoothly to the two ends.

Water cut-off and drainage measures: 60 cm × 60 cm trapezoid cut-off trench, laid with mortar rubble masonry of M7.5, is set on cut slope wall rear edge and both sides of the landslide. In order to drain the ground water inside the slope freely, inclined water drains are set at the front edge of landslide: spacing: 15.0 m; drilling hole: 130, in which soft permeable pipe of diameter 80 is provided.

Supervision of Mazha landslide shall be intensified, so that measures of reinforcement and treatment measures can be adjusted in time according to the condition of landslide and the result can be analyzed and appraised.

3.2.2 *Landslide at Tiejiangpu*

Tiejiangpu Landslide is located on the left side of Section K3+240~K3+420 of the highway, close to the highway, affecting a section that is 160 m in length. The slide mass mainly consists of crushed stones containing soil. The landslide bed mainly consists of crushed stones and pebbles. The entire plane shape of the landslide is like two horseshoes. There is a steep ridge of 1.0–2.0 height at the rear edge, not evident in the lateral boundary. Some portions of its front edge have extended into Zagu'nao River by 1.0–2.0 m, and formed steep ridge of 3.0–5.0 m high under scouring of the river. The rear edge of the landslide is higher than the shear crack by 50 m. The landslide is 78–105 m long, 100–200 m wide, 5°–27° in the main sliding direction. The slide mass is 7.00m in average thickness, 8000 m² in area, 5.6 × 104 m³ in volume. It belongs to shallow retrogressive landslide of intermediate scale.

The landslide is located on a steep slope of 15°–68° angle. Because of scouring of the river, its free face is increased, providing the space condition for landslide movement. The soil mass softened by rainfall provides the condition of external force for the deformation of landslide. A large part of surface water percolates through the soil mass, which not only increases the weight of landslide mass, but also softens the soil or silt containing

crushed stones, forming a weak face in the crushed stone-soil layer. Therefore, during the rainy season, the volumetric weight of landslide mass increases while the values of C and Φ decrease, resulting in the occurrence of the landslide. The landslide is retrogressive.

At present, the landslide is in a limit equilibrium state. It is greatly liable to slide entirely under the influence of rainfall, earthquake and cut slope.

It is demonstrated by geological survey report and inverse analysis that values of the parameters of slide-zone soil are made as below: under natural condition: $C = 5$ kPa; $\varphi = 28°$; under storm condition: $C = 4$ kPa; $\varphi = 26°$. Where safety coefficient under storm condition is made to be 1.15, the residual sliding force of landslide is 513.38–718.78 kN. The stable force under seismic condition is 1353.2–1484.1 kN.

The route of the highway passes through the front edge of the landslide by means of low fill and shallow cut. Because of the characteristics mentioned above, the landslide is reinforced by means of pile-plank retaining wall + counter-pressure of front edge, and treated by means of water cut-off and drainage.

Landslide retaining measures: diameter of anti-slide piles: 2.0 m × 3.0 m; spacing: 6.0 m; concrete: C30; retaining plates made of C30 concrete set between the piles.

Water cut-off and drainage measures: 60 cm × 60 cm trapezoid cutoff trench, laid with mortar rubble masonry of M7.5, is set on the rear edge and both sides of the landslide. In order to drain the ground water inside the slope freely, inclined water drains are set at the front edge of landslide: spacing: 15.0 m; drilling hole: 130, in which soft permeable pipe of diameter 80 is provided.

3.2.3 *Landslide at Kangba*

The entire plane shape of Kangba Landslide is like a horseshoe. There is a steep ridge, which is 30 m high at its rear edge. Its lateral upper edge is evident because of a steep ridge is 122 m high, while its lateral lower edge is not evident because of a steep ridge is only 0.5 m high. The boundary is not evident because of human activities. The front edge of the old landslide has extended into Zagu'nao River. Its shear crack is at Zagu'nao River bed. The landslide is 460 m in length, 320 m in width, 29°–31° in main sliding direction. The slide mass is 25.00 m in average thickness, 82800 m² in area, 1697.4 × 104 m³ in volume, which is retrogressive landslide of superlarge scale and medium thickness.

The entire landslide is in stable state. Because of the free face produced by sand quarrying, secondary columnar fracture can be detected in the affected scope, which results in local deformation in old landslide, generating secondary landslide. The deformation of the secondary landslide is mainly presented by sliding degree, a long-term continuous activity deformation. Horizontal continuous deformation: 1.0–2.0 m/year; annual deformation: 0.1–0.2 m/year; vertical deformation: 0.05–0.1 m/year, evident in deformation. Meanwhile, the tension fracture of slope mass as well as local sliding and collapse are witnessed in the landslide. Local landslide trace is detected on the upper part of slope, but no movement of entire mass is detected. At present, the old landslide is in stable state, and the secondary landslide is relatively stable. However, the unfavorable factors, especially sustained storm will induce easily sliding of entire secondary landslide, endangering the transportation and safety of the national highway under the slope.

The altered highway passes through the front edge of the landslide, 20.0–25.0 m away from its shear crack. There is a platform in the front edge of landslide, which could accommodate the sliding material, serving as a buffer, so the landslide affects the highway less and no special measure is needed for treatment. Since the old landslide is huge in scale, complex in influencing factors, monitoring and early warning are needed after its surrounding is improved by means of water cut-off and drainage. The stable degree of the landslide is still unsatisfactory, so monitoring and early warning are needed after surrounding of the landside is improved by means of water cut-off and drainage, and corresponding treatment measures are taken according to the monitoring and early-warning result.

Water cut-off and drainage measures: trapezoid cutoff trenches are set at the middle part, rear edge and both sides of the landslide. The cutoff trenches are of 60 cm × 60 cm section at the rear edge and both sides, while that is of 60 cm × 60 cm section at the middle part of the landslide. All trenches are laid with mortar rubble masonry of Mu7.5.

3.2.4 Unstable side slope at Yangjiaping

The slope mass is formed mainly with pebble soil. The highway passes through the slope mainly by the means of excavation, which increases the free face of slope. The soil mass softened by rainfall provides the condition of external force for deformation of the landslide. A large part of surface water percolates through the soil mass, which not only increases the weight of landslide mass, but also softens the soil containing breccias, forming weak face. In rainy season, the volumetric weight of landslide mass increases while value C and Φ decrease, resulting in deformation of soil mass of the slope, and endangering the power-transmission iron tower on its top.

Because of the characteristics mentioned above, soil mass of the slope shall be treated by means of pre-reinforcement. The subgrade of the section shall be reinforced by means of pile-plank retaining wall. Based on the data at construction site adjacent to the section, value of C of soil containing pebbles and crushed stones under natural state is taken to be 10 kPa, and that of φ is taken to be 28°. In case of storm state, value of C is taken to be 9 kPa, and that of φ is taken to be 27°.

Landslide retaining measures: diameter of anti-slide piles: 1.5 m × 2.0 m; spacing: 5.0 m; concrete: C30; retaining plates made of C30 concrete set between the piles.

During and after construction, monitoring of the slope shall be intensified, so that the side slope can be reinforced and treated in time according to its condition and the result of reinforcement, and treatment can be analyzed and appraised.

3.2.5 Landslide at Lijiaba

The landslide locates adjacently on the left side of section K9+000~K9+140 (Pile No.) of the highway, and is away from the altered highway. The entire plane state of the landslide is similar to a horseshoe. There are steep ridges of 0.50–1.50 m high at the rear and side edges, evident in boundary. Its shear crack is near the elevation of internal side of the highway, while its rear edge is 21 m higher than the highway. The landslide is 10 m in length, 50–140 m in width, 45° in main sliding direction. The slide mass is 4.00 m in average thickness, 1752.40 m² in area, 0.70 × 104 m³ in volume, taking on sliding and collapse mainly, belonging to shallow retrogressive landslide of small scale.

The landslide locates on a relatively steep slope of 25°–45° grade. A part of slope toe is excavated because of construction of the highway, so the free face is increased, providing the space condition for landslide movement. The soil mass softened by rainfall, agricultural irrigation and domestic water provides the external power for the deformation of the landslide. A large part of surface water percolates through the soil mass, which not only increases the weight of landslide mass, but also softens the soil containing crushed stones, forming a weak face in the layer. Therefore, in the rainy season, the volumetric weight of the soil mass at rear edge of the landslide increases while values of C and Φ decrease, resulting in occurrence of the landslide.

At present, the landslide is in limit equilibrium state. It is greatly liable to slide entirely under the influence of rainfall, earthquake and cut slope. The value of geotechnical parameters is taken on the basis of inversed analysis. Also taking the survey report into consideration, the value of C is taken to be 7.5 KPa and φ to be 25° in case of silt containing crushed stones in natural state, while C assumed to be 6.5 kPa and φ to be 24° in the storm state. The landslide is small in scale. There is a buffer space for accumulation between it and the altered highway, so the landslide can be treated by means of counter-pressure combined with water cut-off and drainage. After such treatment, the coefficient of stabilization is Fs = 1.06 in case of seismic condition and Fs = 1.36 in case of storm condition, meeting the design requirements. Water cut-off and drainage engineering: trapezoid cutoff trenches are set at the rear edge and both sides of the landslide. The trenches are of 40 cm × 40 cm section, laid with mortar rubble masonry of M7.5.

Monitoring shall be intensified during construction so that the reinforcement and treatment measures can be adjusted in time according to the landslide condition, as well as the result of treatment can be analyzed and appraised. The stability of the landslide neither exacerbates nor improves. Therefore, the landslide is still treated by means of counter-pressure after its external environment is improved by water cut-off and drainage, and monitoring and early-warning shall be conducted so that corresponding measures can be taken according to the monitored or warning condition.

4 CONCLUSION AND SUGGESTIONS

In order to ensure the alteration of the highway meeting the requirement of "eradication of minor hazard, enhancement of anti-hazard ability, improvement of safe transportation condition and passable capability of the road", measures shall be taken against following schemes on the basis of the principle and route characteristics mentioned above:

1. If a landslide is far away from the route and has no negative influence on the engineering, it need not be subject to treatment. Only measures of water drainage + monitoring are adopted.
2. For landslides of a medium and small scale that are developing near the highway and may affect the safety of project facilities directly, water drainage systems are set according to the characteristics of the landslide. The rear edge of the landslide shall be cut to reduce the pressure while its front edge shall be loaded to counteract the pressure. The front and middle parts of the landslide are reinforced by anti-slip piles, anti-slip retaining walls, anchored bar type grid beams, pile-plank retaining walls or treated comprehensively.
3. For the super-large landslide that affects the highway seriously and requires a great amount of costs to treat, a measure of avoidance shall be adopted. If it is difficult to avoid the landslide and the highway must pass through the section where there is danger of landslides, in addition to minimization of the disturbance of the landslide, a scheme of temporary treatment + monitoring and early warning shall be adopted.

ACKNOWLEDGMENTS

This study was supported by the open funds of the National and Local Joint Engineering Laboratory of Traffic Civil Engineering Materials of Chongqing Jiaotong University, the Chongqing Key Laboratory of Mountainous Road Structure and Material, and the Hi-tech Laboratory for Mountain Road Construction and Maintenance.

REFERENCES

Kaizong Xia, Congxing Chen, Zude Lu. Graphic analysis of the stability of bedding rock slope under consideration of hydraulic action. Rock and Soil Mechanics. J. 10 (2014).

Sheng Zeng, Tian Hu, Jian Zhao. Study on stability evaluation of accumulation body slope based on Fuzzy-AHP. Journal of Engineering Geology. J. 06 (2012).

Xiumei Ding, Guangshi Liu, Runqiu Huang. The application of shear strain increment in the stability of bulk slope. Advance in Earth Sciences. J. S1 (2004).

Yongfang Ruan, Fei Wang, Xingzhou Huang. Spatial variability of accumulation slope under the influence of rainfall. Mountain Research. J. 01 (2015).

Zhiliang Sun, Lingwwei Kong, Aiguo Guo. Mechanism of slope deformation and instability of accumulation body slope under earthquake action. Rock and Soil Mechanics. J. 12 (2015).

Electromechanical Control Technology and Transportation – Jia & Wu (Eds)
© 2017 Taylor & Francis Group, London, ISBN 978-1-138-06752-3

Analysis of an allocation plan for urban complex air conditioning cold and heat sources in Changchun

Yidan Zhu, Chunqing Wang & Haibin Li
School of Municipal and Environmental Engineering, JiLin Jianzhu University, Changchun, China

ABSTRACT: Changchun, a city located in the cold region, with unique climate characteristics, always has its buildings designed in multiple functions and supported by different architectural energy resources. Within this paper, we described and compared two kinds of air conditioning schemes certain urban complexes in Changchun in terms of their cold and heat supply sources. The dynamic loads of air conditioning in our subjects were tracked and their energy consumption in cold and heat supply via two types of air-conditioning configurations was calculated. It also includes comprehensive analysis in other factors, such as initial investment, operating cost, comfort and stability. The results demonstrate that the system of electrical refrigeration (centrifugal chiller + screw chiller) combined with gas-fired hot water boiler is superior to electrical refrigeration (centrifugal chiller) combined with gas-fired hot water boiler system. Additionally, when the architectural scale is relatively larger, it always needs air conditioning and hot water.

1 ENGINEERING SITUATION

This project is carried out in Changchun city. The chosen building is always used as the commercial office complex, which was designed in following floor/allocation plan: the area of construction as 128662 m^2, the area of above-ground building as 19585 m^2, the area of underground construction as 9077 m^2, the building height as 99.8 m, 24 floors above the ground and 1 floor underground.

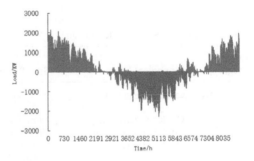

Figure 1. The annual hourly cooling load in the section of shopping malls.

2 HOURLY LOAD

In Changchun, the air-conditioning season usually starts from around June 11th and ends in August 31st, and its heating season is within the period of October 20th to April 6th.

Through setting all kinds of the parameters, we estimated the annual hourly load of the complex as 8760 hours. The hourly cooling load of its shopping malls and office departments within its air-conditioning season are summarized and shown in Figure 1 and Figure 2, respectively.

From Figure1, it shows that the maximum cooling load of shopping malls throughout the year occurs in the early August, which has the peak cooling load of about 2400 kW. In addition, the maximum heat load value of the shopping malls appeared in early January with the peak heat load value is about 2100 kW. From Figure 2, it turns out that the maximum cooling load of office depart-

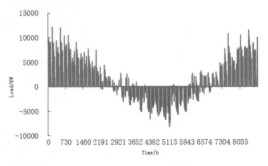

Figure 2. The annual hourly cooling load in the section of office departments.

ment throughout the year also appeared in early August, which has the peak cooling load value as about 7600 kW. Similarly, as summarized above, the maximum heat load value of such office department appeared in early January with the peak heat load as 12500 kW. The statistical quantile intervals of each load rate were summarized in Table 1.

Data show that load rate interval occupied hours of shopping malls and office buildings in the air conditioning and heating reason. The variation of each load rate in hours is related to the local climate characteristics, the designed function and function size of the building, the design parameters of the air conditioning and so on. The data makes it feasible to calculate the annual cost of operating air conditioning as the cold and heat source.

Table 1. The number of hours each load rate interval statistics.

Operating season	Building category	Load rate									
		0~10%	10~20%	20~30%	30~40%	40~50%	50~60%	60~70%	70~80%	80~90%	90~100%
Heating season	Department store	502	288	284	312	348	198	153	68	29	15
	Office building	233	191	233	234	202	117	72	31	11	7
Air conditioning season	Department store	76	78	120	139	227	211	130	62	20	3
	Office building	120	48	59	85	118	83	83	38	17	9

Table 2. Initial investment in scheme one.

Building category	Equipment Name	Specifications	Quantity (Set)	Unit Price (Million)	Total Price (Million)
Department store	Centrifugal chiller	Refrigerating capacity 2461.2 kW, power input 426 kW	1	1.697	1.697
	Chilled water pump	Flow 430 m³/h, head 40 m	2	0.052	0.104
	Cooling water pump	Flow 510 m³/h, head 30 m	2	0.05	0.1
	Cooling tower	Quantity of circulating water 250 m³/h	1	0.1	0.1
	Gas-fired hot water boiler	Heating output 930 kW	1	0.143	0.143
		Heating output 1163 kW	1	0.178	0.178
	Hot water circulating pump	Flow 100 m³/h, head 30 m	2	0.021	0.042
		Flow 120 m³/h, head 30 m	2	0.021	0.042
	Boiler feed pump	Flow 6 m³/h, head 140 m	2	0.0023	0.0046
Office building	Centrifugal chiller	Refrigerating capacity 1758 kW	1	1.022	1.022
		Refrigerating capacity 2988.6 kW	2	1.6	3.2
	Chilled water pump	Flow 320 m³/h, head 40 m	2	0.04	0.08
		Flow 530 m³/h, head 40 m	3	0.042	0.126
	Cooling water pump	Flow 380 m³/h, head 30 m	2	0.042	0.084
		Flow 620 m³/h, head 30 m	3	0.052	0.156
	Cooling tower	Quantity of circulating water 400 m³/h	1	0.08	0.08
		Quantity of circulating water 600 m³/h	2	0.12	0.24
	Gas-fired hot water boiler	Heating output 5600 kW	1	0.86	0.86
		Heating output 7000 kW	1	1.075	1.075
	Hot water circulating pump	Flow 500 m³/h, head 30 m	2	0.042	0.084
		Flow 640 m³/h, head 30 m	2	0.046	0.092
	Boiler feed pump	Flow 10 m³/h, head 140 m	2	0.0023	0.0046
	Aggregate				9.5142

3 ECONOMIC ANALYSIS OF COLD AND HEAT SOURCE PROJECT

According to the characteristics of energy structure and the cooling and heating load of the buildings in Changchun, two kinds of schemes with different composites of cold and heat source are proposed and described as followings:

Scheme one: centrifugal chiller+gas-fired hot water boiler. Specifically, the shopping mall selects a centrifugal chiller which has refrigeration capacity as 2461.2 kW, and two gas-fired hot water boilers with heating capacity as 930 kW and 1163 kW, respectively; as for the office departments, they use two centrifugal chillers which have refrigeration capacity as 1758 kW and 2988.6 kW and a set of gas-fired hot water boiler with heating capacity as 5600 kW and 7000 kW.

Scheme two: centrifugal chiller + screw chiller + gas-fired hot water boiler. Specifically, the shopping mall selects a screw chiller which has refrigeration capacity as 1196.2 kW, and two gas-fired hot water boiler with heating capacity as 930 kW and 1163 kW; as for the office departments, they use a screw chiller with refrigeration capacity

as1702.2 kW, a centrifugal chiller whose refrigeration capacity is 2988.6 kW, and a set of gas-fired hot water boiler with heating capacity as 5600 kW and 7000 kW, respectively.

The tables are provided conditions for calculating the annual operating cost of air conditioning cold and heat source.

3.1 The analysis of initial investment

The preliminary investment details of aforementioned two kinds of schemes are summarized and tabulated in Tables 2 and 3. Analysis based on those two tables indicate that the initial investment of scheme two saves 0.0892 million than scheme one.

3.2 Analysis for operation cost

As for scheme one, store energy consumption calculations of air-conditioning season are tabulated in Table 4, the summary of an annual energy consumption and costs are made in Table 5; for scheme two, the summary of the annual energy consumption and costs is made in Table 6. Within them, the price rates of natural gas and electricity calculated are based on the actual commercial

Table 3. Initial investment in scheme three.

Building category	Equipment Name	Specifications	Quantity (Set)	Unit price (Million)	Total price (Million)
Department store	Screw chiller	Refrigerating capacity 1196.2 kW	2	0.84	1.68
	Chilled water pump	Flow 210 m³/h, head 40 m	3	0.035	0.105
	Cooling water pump	Flow 270 m³/h, head 30 m	3	0.033	0.099
	Cooling tower	Quantity of circulating water 250 m³/h	2	0.05	0.1
	Gas-fired hot water boiler	Heating output 930 kW	1	0.143	0.143
	Gas-fired hot water boiler	Heating output 1163 kW	1	0.178	0.178
	Hot water circulating pump	Flow 100 m³/h, head 30 m	2	0.021	0.042
	Hot water circulating pump	Flow 120 m³/h, head 30 m	2	0.021	0.042
	Boiler feed pump	Flow 6 m³/h, head 140 m	2	0.0023	0.0046
Office building	Screw chiller	Refrigerating capacity 1702.2 kW	1	0.95	0.95
	Chilled water pump	Refrigerating capacity 2988.6 kW	2	1.6	3.2
	Chilled water pump	Flow 320 m³/h, head 40 m	2	0.04	0.08
	Chilled water pump	Flow 530 m³/h, head 40 m	3	0.042	0.126
	Cooling water pump	Flow 380 m³/h, head 30 m	2	0.042	0.084
	Cooling water pump	Flow 620 m³/h, head 30 m	3	0.052	0.156
	Cooling tower	Quantity of circulating wate 400 m³/h	1	0.08	0.08
	Cooling tower	Quantity of circulating wate 600 m³/h	2	0.12	0.24
	Gas-fired hot water boiler	Heating output 5600 kW	1	0.86	0.86
	Gas-fired hot water boiler	Heating output 7000 kW	1	1.075	1.075
	Hot water circulating pump	Flow 500 m³/h, head 30 m	2	0.042	0.084
	Hot water circulating pump	Flow 640 m³/h, head 30 m	2	0.046	0.092
	Boiler feed pump	Flow 10 m³/h, head 140 m	2	0.0023	0.0046
	Aggregate				9.425

Table 4. Energy consumption calculation table in market in air conditioning season in scheme one.

Load rate (%)	<10	10~20	20~30	30~40	40~50	50~60	60~70	70~80	80~90	90~100
Cold load (kW)	240	480	720	960	120	1440	1680	1920	2160	2400
Operation period for air conditioning systems (h)	76	78	120	139	227	211	130	62	20	3
Centrifugal chiller' operating mechanism	1×0.1	1×0.2	1×0.3	1×0.4	1×0.5	1×0.6	1×0.7	1×0.8	1×0.9	1×1.0
Cold water unit' power (kW)	92.4	130.9	156	175.7	193.3	221.2	255	297.4	352.2	425.9
Cooling tower' operating mechanism	1	1	1	1	1	1	1	1	1	1
Cooling tower' power (kW)	20	20	20	20	20	20	20	20	20	20
Operating mechanism of circular pump of frozen water	1	1	1	1	1	1	1	1	1	1
Power of circular pump of frozen water (kW)	75	75	75	75	75	75	75	75	75	75
Operating mechanism of Circular pump of cooling water	1	1	1	1	1	1	1	1	1	1
Power of circular pump of cooling water (kW)	55	55	55	55	55	55	55	55	55	55
Total power (kWh)	18422	21910.2	36720	45272.3	77929.1	78323.2	52650	27738.8	10044	1727.7
Total power consumption of cooling season kwh					370737.7 (kWh)					

prices of the gas and electricity in Changchun city, where the natural gas is 3.92 yuan/m^3 and the electricity is 0.947 yuan/(kW·h). As for the information in Tables 5 and 6, it can be summarized that the scheme two saved 0.065 million of total operation cost (4.241 million) than that by the scheme one (4.306 million).

3.3 Analysis of annual operating cost

There are different methods of economic analysis for the cold and heat source available for the public. Within this paper, we adopt the method in which the annual operating cost is inputted into the analysis, and the minimum annual operating costs of the cold and heat source program is accepted as the optimized design. The approximate estimate formulas of the annual operating cost are listed as follows:

$$YTC = COF+COR \qquad (1)$$

$$COF = C/n+0.5(R+0.8Rs)+INS \qquad (2)$$

where: YTC is the annual operating cost, COF is the fixed cost; COR is the operating cost; INS is the insurance premium; C is the equipment cost; n is the recoupment period; R is the interest rate; Rs is the tax rates. Among them, the compensation period of air conditioning units and gas boiler take 20 years, the deposit rates is set as 4.75%, the tax rate temporarily takes 5.5%; insurance is not included temporarily. In this project, the operating fees of those two schemes are summarized in Table 7. From the analysis of the annual operation, it demonstrates that scheme two is superior to the first scheme.

4 ANALYSIS OF ENERGY CONSUMPTION

The annual consumption of coal in those two kinds of air conditioning programs are summarized in Tables 5 and 6. Clearly, the annual electricity consumption of scheme two is 1030308 kWh, which is lower than the scheme one with consumption value as 69409.1 kWh.

Table 5. Annual energy consumption summary and cost's in scheme one.

Operating season	Building category	Total power consumption (KWh)	Total gas consumption (m³)	Total operation cost (Million)
Heating season	Department store	54021.7	181832.1	0.764
	Office building	141680	650983.2	2.686
Air conditioning season	Department store	370737.7	–	0.351
	Office building	533277.7	–	0.505
Aggregate		1099717.1	832815.3	4.306

Table 6. Annual energy consumption summary and cost's in scheme two.

Operating season	Building category	Total power consumption (KWh)	Total gas consumption (m³)	Total operation cost (Million)
Heating season	Department store	54021.7	181832.1	0.764
	Office building	141680	650983.2	2.686
Air conditioning season	Department store	294429.2	–	0.279
	Office building	540177.1	–	0.512
Aggregate		1030308	832815.3	4.241

Table 7. Comparison table of annual operating cost.

	Scheme one	Scheme two
Initial investment of cold source scheme	9.5142	9.425
COF (Million)	0.911	0.902
Annual operating cost COR (Million)	4.306	4.241
Annual cost YTC (Million)	5.217	5.143

5 CONCLUSION

From the perspective of investment economy, the initial investment amount of scheme two is lower than that of scheme one, and the annual expense of scheme two is also lower than that of scheme one. Hence, the centrifugal chiller + screw chiller + gas hot water boiler system is more economical and cost effective. From the perspective of energy consumption, the power consumption of scheme two is lower than that of scheme one through the whole year. All these support that the system in scheme two is featured as being energy-saving.

Through simulating the energy consumption of the complex building by DEST software and the eco-nomic analysis about the air conditioning heat and cold sources, a final conclusion can be drawn that the best air conditioning heat and cold source scheme for the complex building is the centrifugal chiller + screw water chiller + gas hot water boiler system.

REFERENCES

Guo Lei. The situation of the building energy consumption and saving potential [J]. Journal of Railway Engineering, 4:75–78(2006).

Lu Liaoqing, Practical heating and air conditioning design handbook [M]. Beijing: Chinese Building Industry Press, (2008).

Yan Da, Xie Xiaona, Song Fangting, Building environment design simulation software DeST, The HVAC air conditioning system. 34 (7):48–56(2014).

Zhang Sufang, Hao he. Calculation method of comprehensive energy efficiency ratio of air conditioning water system, The HVAC air conditioning system., 43 (11): 63–66(2013).

Zhou Xin, Yan Da, Hong Tianzhen, Comparative simulation study of building energy simulation software, The HVAC air conditioning system, 44 (4): 113–122(2014).

Electromechanical Control Technology and Transportation – Jia & Wu (Eds)
© 2017 Taylor & Francis Group, London, ISBN 978-1-138-06752-3

Analysis of the effect of the solar cell layout in photovoltaic greenhouses

Zhiwu Ge
Physics Department, Hainan Normal University, Haikou, China

Ye Li
Mechanics Department, Hainan University, Haikou, China

ABSTRACT: In the combination process of photovoltaic electric power generation with agriculture in China, there is currently no uniform standard, and the solar cell layout in photovoltaic greenhouses is in a state of chaos. Many built photovoltaic greenhouses are not able to cultivate plants. To search for solutions to these problems, two typical solar cell panels were constructed to analyze the effect of the different photovoltaic cells' arrangement, and the light intensity and temperature distributions on the gray ceramic tile ground were measured. The light intensity curve showed that the intensity under the panel with slit vents changes obviously. The maximum differences can be 708 w/m², while under the panel with square aperture, it can only be up to 220 w/m². The light intensity changes with time and increased gradually from morning until noon, and then decreased slowly. The change and difference of temperature are very complex, and these are related to material and wind speed, and so on. The solar cells layout on photovoltaic panels can have an important influence on environmental parameters and plants below in the greenhouse. The conclusion is very important for constructing high energy efficiency solar greenhouses.

1 INTRODUCTION

After 10 years of rapid expansion, the monotonous photovoltaic power generation business has just tried hard to obtain the support of local government currently. Combining PhotoVoltaic (pv) with other industries has become a suitable way, and is also the encouraged direction of the policy, one of which is 'agriculture + pv' mode. Because of the common and complementary aspect of PhotoVoltaic (pv) and agriculture on the land use, the industry has been seeking a combination for breakthrough, thereby making it one of the important developing directions currently in China. But there are many problems arising in the process, especially the photovoltaic panels' layout and arrangement, and the corresponding influence on plants below (1–4, 2016).

In this paper, two typical photovoltaic panels are studied in which solar cells are arranged differently, the physical parameters are tested under panels, and the influences and the different effects are analyzed.

2 A TYPICAL SOLAR CELL LAYOUT ON PANELS AND EFFECT ANALYSIS

According to the analysis of the solar cell layout in many photovoltaic greenhouses already built, there are two kinds of typical photovoltaic cells' arrangement or layout on panels.

Figure 1 and 2 show the typical photovoltaic cells' arrangement or layout on panels. Figure 1 shows the panels with slit vents, where each cell has a width of 2 cm, and each slit vent also has a width of 2 cm. Figure 2 shows the panels with square

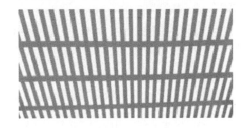

Figure 1. Picture of the panels with slit vents.

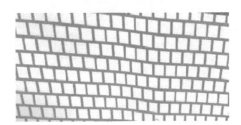

Figure 2. Picture of the panels with square aperture.

aperture, where each cell is 1.3 cm wide, and the dimensions of the square aperture are 2 cm × 2 cm.

The light intensity curve, natural light photo, and infrared temperature distribution image are at 3 m below the panels. The gray ceramic tile ground will be detected next.

2.1 Detection results at 9.20 a.m.

Figures 3–5 show the light intensity distributions under panels, above the ground at 9.20 a.m. Figure 4(a) shows part of the light intensity data in Figure 3, and Figure 4(b) shows part of the data in Figure 5.

From Figure 3, the following equation can be obtained:

$$\Delta L = I_{max} - I_{min} = 830.4 - 187.9 = 642.1 \ (\text{w/m}^2) \quad (1)$$

From Figure 3, the following equation can be obtained:

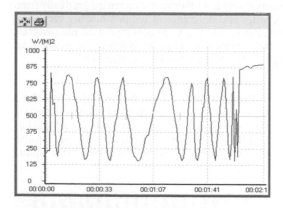

Figure 3. Screenshot showing light intensity distribution under panels with slit vents.

No.s	Time	W/[M]2	
1	00:00:00	214.0	
2	00:00:01	230.6	
3	00:00:02	230.9	
4	00:00:03	830.4	
5	00:00:04	586.7	
6	00:00:05	600.3	
7	00:00:06	263.3	
8	00:00:07	187.9	
9	00:00:08	291.6	
10	00:00:09	324.6	
11	00:00:10	458.6	
12	00:00:11	726.1	
13	00:00:12	769.4	
14	00:00:13	811.9	

(a)

No.s	Time	W/[M]2	
1	00:00:00	304.3	
2	00:00:01	311.9	
3	00:00:02	385.6	
4	00:00:03	339.2	
5	00:00:04	331.4	
6	00:00:05	298.7	
7	00:00:06	261.8	
8	00:00:07	426.4	
9	00:00:08	489.8	
10	00:00:09	524.6	
11	00:00:10	480	
12	00:00:11	446.5	
13	00:00:12	406.5	
14	00:00:13	334.6	

(b)

Figure 4. Screenshots of part of the light intensity data under panels.

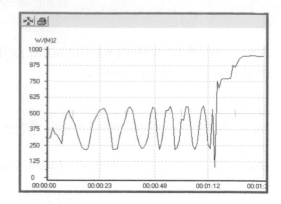

Figure 5. Screenshots of the light intensity distribution under panels with square aperture.

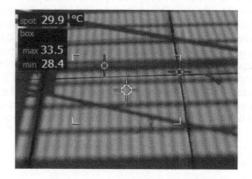

Figure 6. Picture of shadow of panels with slit vents.

$$\Delta L = I_{max} - I_{min} = 524.6 - 304.3 = 220.3 \ (\text{w/m}^2) \quad (2)$$

Where ΔL is the light intensity difference, I_{max} is the maximum value of light intensity, and I_{min} is the minimum value of light intensity. In Figure 10 and Figure 12, the flat part of the curve at the far right shows the light intensity distributions without panels' shade.

Figures 6–9 show the shadow and infrared images on the ground at 9.20 a.m. Photos show the light intensity distribution differently, with a temperature difference of about 2°C on the gray ceramic tile. From figures shown above, it can be observed that the influence factors on temperature are very complex.

2.2 Detection results at 12

Figure 10–12 show the light intensity distributions under panels, above the ground at 12.

Figure 11(a) shows one part of the light intensity data in Figure 10, and Figure 11(b) shows one part of the light intensity data in Figure 12.

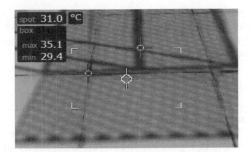

Figure 7. Picture of shadow of panels with square aperture.

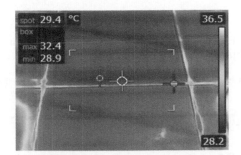

Figure 8. Image of panels with slit vents.

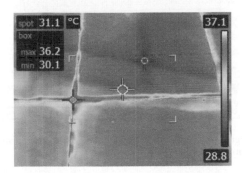

Figure 9. Image of panels with square aperture.

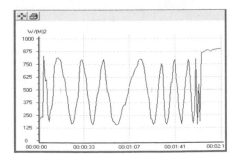

Figure 10. Screenshot of light intensity distribution under panels with slit vents.

No.s	Time	W/[M]2
1	00:00:00	873.8
2	00:00:01	881
3	00:00:02	877.1
4	00:00:03	782
5	00:00:04	592.3
6	00:00:05	283.3
7	00:00:06	173
8	00:00:07	242.2
9	00:00:08	398.1
10	00:00:09	374.6
11	00:00:10	651.7
12	00:00:11	665.8
13	00:00:12	803.2
14	00:00:13	876.5

(a)

No.s	Time	W/[M]2
1	00:00:00	304.3
2	00:00:01	311.9
3	00:00:02	385.6
4	00:00:03	339.2
5	00:00:04	331.4
6	00:00:05	298.7
7	00:00:06	261.8
8	00:00:07	426.4
9	00:00:08	489.8
10	00:00:09	524.6
11	00:00:10	480
12	00:00:11	446.5
13	00:00:12	406.5
14	00:00:13	334.6

(b)

Figure 11. Screenshot of the part of the light intensity data under panels.

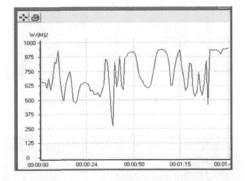

Figure 12. Screenshot of the light intensity under panels with square aperture.

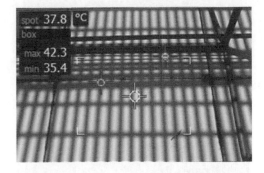

Figure 13. Picture of the shadow of panels with slit vents.

From Figure10, the following equation can be obtained:

$$\Delta L = I_{max} - I_{min} = 881 - 173 = 708 \text{ (w/m}^2) \tag{3}$$

From Figure 12, the following equation can be obtained:

$$\Delta L = I_{max} - I_{min} = 524 - 304 = 220 \ (\text{w/m}^2) \qquad (4)$$

where ΔL is the light intensity difference, I_{max} is the maximum value of light intensity, and I_{min} is the minimum value of light intensity.

Figures 13–16 show the shadow and infrared images on the ground at 12. The light intensity is distributed differently, with a temperature difference of about 3°C on the ceramic tile, not consid-

ering the soil between tiles and the wind change while detecting. From figures above, it can also be observed that the influence factors on temperature are very complex.

2.3 Detection results at 3.20 p.m.

Figures 17–19 show the light intensity distribution under panels, above the ground at 3.20 p.m.

Figure 18(a) shows one part of the light intensity data in Figure 17, and Figure 18(b) shows one part of the light intensity data in Figure 19.

From Figure 17, the following equation can be obtained:

$$\Delta L = I_{max} - I_{min} = 316 - 286 = 30 \ (\text{w/m}^2) \qquad (5)$$

From Figure 19, the following equation can be obtained:

$$\Delta L = I_{max} - I_{min} = 465 - 369 = 96 \ (\text{w/m}^2) \qquad (6)$$

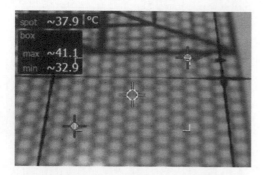

Figure 14.   Shadow of panels with square aperture.

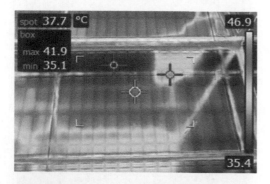

Figure 15.   Image of panels with slit vents.

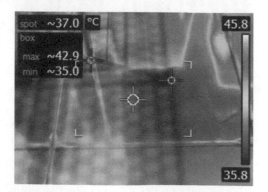

Figure 16.   Image of panels with square aperture.

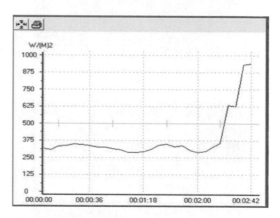

Figure 17.   Screenshot showing the light intensity distribution under panels with slit vents.

No.s	Time	W/(M)2
1	00:00:00	316
2	00:00:06	308.4
3	00:00:12	332.2
4	00:00:18	338
5	00:00:24	348.5
6	00:00:30	343.6
7	00:00:36	336.1
8	00:00:42	325.6
9	00:00:48	324.3
10	00:00:54	313.2
11	00:01:00	306.5
12	00:01:06	288.2
13	00:01:12	286
14	00:01:18	293

(a)

No.s	Time	W/(M)2
1	00:00:00	369.3
2	00:00:06	375.4
3	00:00:12	382.5
4	00:00:18	396.4
5	00:00:24	430.4
6	00:00:30	452.8
7	00:00:36	459.8
8	00:00:42	458.9
9	00:00:48	465.9
10	00:00:54	447.9
11	00:01:00	437.4
12	00:01:06	424.8
13	00:01:12	398.9
14	00:01:18	417.4

(b)

Figure 18.   Screenshot of one part of the light intensity data under panels.

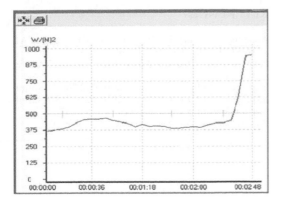

Figure 19. Screenshot of the light intensity distribution under panels with square aperture.

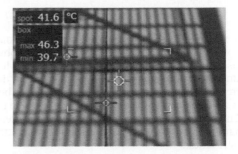

Figure 20. Picture of the shadow of panels with slit vents.

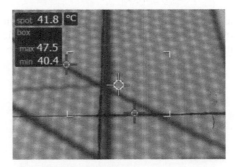

Figure 21. Picture of the shadow of panels with square aperture.

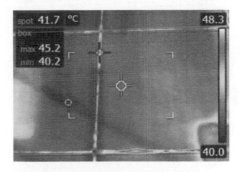

Figure 22. Image of panels with slit vents.

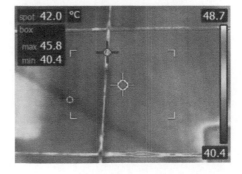

Figure 23. Image of panels with square aperture.

where ΔL is the light intensity difference, I_{max} is the maximum value of light intensity, and I_{min} is the minimum value of light intensity.

Figure 20–23 show the shadow and infrared images on the ground at 3.20 p.m. The light intensity is distributed differently, with a temperature difference of about 2 °C on the ceramic tile. But if the soil between tiles is considered, the temperature difference can be 8 °C, not considering the wind change while detecting. From figures above, it can be observed that the influence factors on temperature are very complex, which need much further study later.

From the natural light photographs shown in Figure 6 and Figure 7, Figure 13 and Figure 14, and Figure 20 and Figure 21, it can be observed that the light intensity distribution under the panels with slit vents changes regularly almost as a sharp sine wave, while the light intensity distribution under panels with square aperture changes in a very complicated manner; the small hole imaging effect appears, and the light intensity distribution is given by using the following equation (Haiping Zhu, et al. 1999):

$$H(x, N_c) = 4J_1(U)[\cos^{-1}(x)-x(1-x^2)^{1/2}]/\pi U \quad (7)$$

$$\text{where } U = 8N_c x [\cos^{-1}(x)-x(1-x^2)^{1/2}] \quad (8)$$

$$x = a/a_0, \ a_0 = 2a/\lambda D_{20} \quad (9)$$

$$N_c = a_2[1/d_1 + 1/d_2]/\lambda \quad (10)$$

where $J_1(U)$ is the first order Bessel function, x is the normalized space frequency, a_0 is the cut-off spatial frequency, d_1 is the object distance, d_2 is the image distance, a is the small hole's radius, λ is the wavelength, and N_c is the Fresnel number.

The influence factors of temperature are very complex, and these will be published in future.

3 CONCLUSIONS

In this paper, three groups of detection results showed that panels with slit vents can cause obvious light intensity differences in greenhouses while panels with square aperture cause much fewer differences when compared with the former. It is better to use panels with square aperture to shade the plants below. Of course, there are a lot of problems that require further research, such as temperature distribution.

ACKNOWLEDGMENT

The authors gratefully acknowledge the Hainan Province Natural Science Foundation for financial support (grant number 20165198).

REFERENCES

Analysis of Pinhole Imaging System, OPTICAL INSTRUMENTS, 21(3):12–14.
Haiping Zhu, Xiaoqing Jiang. 1999. Optical Transfer Function.
http://guangfu.bjx.com.cn/news/20160222/709780.shtml
http://guangfu.bjx.com.cn/news/20160624/745328.shtml
http://guangfu.bjx.com.cn/special/?id = 543117
http://www.solarzoom.com/article-73301–1.html

Electromechanical Control Technology and Transportation – Jia & Wu (Eds)
© 2017 Taylor & Francis Group, London, ISBN 978-1-138-06752-3

Review of the space requirements for environmental protection

AiLing Cai, Run Wang & BingJie Sun
College of Resources and Environment Science, Hubei University, Wuhan, China
Key Laboratory of Regional Development and Environmental Response (Hubei Province), Wuhan, China

ABSTRACT: With the acceleration of urbanization, environmental problems have become increasingly serious. Environmental protection creates explicit space requirements. In recent years, research on environmental protection has drawn more and more attention, but awareness of space requirements that are related to environmental protection lags behind. Based on the previous research, a useful reference is provided in this paper for researchers to carry out related research by systematically summarizing the current achievements, methods, practices, and models of environmental protection. And then, it elaborates on the environmental space planning, eco-red line and an environmental protection plan to implement space requirements. Lastly, the space system of environmental protection is constructed. The overall coordination control, quantification monitor, and innovation technology must be carried out to protect the environment so as to achieve sustainable development.

1 INTRODUCTION

Nowadays, environmental issues and resource exhaustion are becoming more and more serious, and therefore environmental protection is an urgent requirement. Processes driven by human communities are the core causes of transformations occurring in natural systems. While the environment as a whole is composed of natural and man-dominated sub-systems, space is limited and the two compete for it (Petrişor et al. 2013). The proposed "Catch a Total Protection" and "Five Concepts of Innovation, Harmony, Green, Open, Shared" programs indicate that people's awareness of environmental protection has improved. How to implement environmental protection planning regionally? How to evaluate (supervise or inspect) the environmental protection work from space? How to determine the space requirements of environmental protection? How is ecological protection accounted for in spatial protection? These are the questions that must be answered so that environmental protection can be implemented exactly. The only way to truly protect the environment is by carrying out space requirements, such as the eco-red line and green infrastructures. In this article, the research of space requirements of environmental protection is the area of focus. Relevant research ideas, research outcomes, vulnerabilities, and problems that should be noted and put forward about the issues of space requirements of environmental protection are summarized in this article. The framework system of space requirements is explored and a reference is provided to researchers and a means of inspection planning results is provided via this article.

2 RESEARCH REVIEWS

2.1 History of environmental protection

Environmental protection generally refers to all actions, which address real or potential environmental problems, thereby coordinating the relationship between humans and the environment, and safeguarding economic and social sustainable development.

Rachel Carson published a book entitled Silent Spring in 1962, which alerted a large audience to the environmental and human dangers of indiscriminate use of pesticides, such as DDT. Because of the book's warning, the United States Government began to pay attention to the environment and established the Environmental Protection Agency in 1970. After the United Nations Conference on the human environment in 1972, "environmental protection" was widely adopted as an official policy. The Founder of the Environment Improvement Trust is CA Gajendra Kumar Jain who has been working with volunteers since 1988 (Pallangyo 2007).

With respect to industrial wastes, China put forward the "Comprehensive Utilization" policy in 1956. And then, in the late 1960s, "concepts of recycling and 'three wastes'" (waste gas, waste water, and waste residues) treatment was proposed. Until the 1970s, it switched to "Environment Protection". Chinese environmental protection began in 1972. China was one of the first developing countries to implement a sustainable development strategy. At that time, the Guan Ting Reservoir Protection was established in Beijing, and the

Three Waste Dealing Official in Hebei Province was established to study about reservoir pollution and to protect the reservoir. After that, DDT was banned across the country. What is more, China set up an environmental protection office in 1973, which is also called the Ministry of Environmental Protection, to unify the management of national environmental protection. The main responsibilities of the Government's environmental protection sector councils are to implement control of pollutants' discharge policies, to encourage the development of emissions control technology, and to protect the environment. Command and control, economic incentives, voluntary instruments, and public participation are measures used in China.

Developed countries attach great importance to environmental protection at an early stage, while developing countries do so later. Both need to deal with the relationship between economic development and environmental protection. Environmental protection shall exercise unified supervision and management. It is important for people to do everything they can to protect the environment in their daily lives.

2.2 The space questions of environmental protection

The environment became an international issue after 1960. Spatial planning is a powerful tool for protecting the environment and environmental space is helpful for strengthening the efficiency of the space program. Foreign space requirements of environmental protection generally implement spatial planning. 'Sustainable spatial development' needs to find "a territorial balance of satisfying at the same rate the economic, social and environmental needs of present and future generations" (Petrisor 2013). The concept of 'environmental space' has been put forward as a means for providing specific meaning to sustainability (Ton 2007). The concept of environmental space has been put forward as a political tool in an action plan "Sustainable Netherlands". Hanson (Hanson, 1999) argues that the concept has a good chance of becoming quite fashionable. But there are still many questions about it. Spatial planning, in particular, the ecological environmental planning, has become an important measure to prevent the destruction of the environment (Meng, You 2015). The environmental space planning scheme in Germany uses the principles of compensation (Peithmann 2008), prevention, and cooperation to protect the environment. Environmental objectives play a leading role in federal space planning. It is a revelation that environmental issues should be brought into the space program to adapt diversity, integrity, and sustainability. Environmental space has been put forward as a means for 'making concrete' that which is meant by the term "sustainability".

China has established "The National Main Body Function Planning" and "The National Ecological Function Zoning" schemes to divide the space of ecological protection. Environmental space is used to optimize patterns of regional development, and the eco-red line is used to regulate the scale of regional development, and the environmental protection plan should be integrated into "Multiple Planning Integration" to ensure the implementation.

3 SPACE REQUIREMENTS OF THE ENVIRONMENT

3.1 The systems of environmental space planning

Environmental protection needs to begin in defining the regional scale from villages to countries. The environmental protection of villages can be implemented through land-use planning, and by increasing the land for green and environmental protection. Given the scarcity of lands in the city, the urban land is subject to rent theory and this should be intensively used; it is because of this reason that the environmental protection land could not be increased. Not only should the city increase green infrastructures, but it should also increase environmental facilities, coordinate functional partitions, and increase the propaganda of environmental protection. Focusing on different functions of the development, limiting development, and prohibiting development, macro-scale protection needs spatial regulation.

1. Land use planning
Environmental protection should be linked to land use planning. Based on the delimitation of nature conservation, forest park, and landscape, it uses spatial overlay as a core area of ecological protection to construct regional security patterns, including biological corridors and stream buffers (Xiao et al. 2015).
2. Urban planning
Urban development planning places urban construction as the object space, thereby involving in master planning, detailed planning, and so on. Space and technology are used as tools to mainly solve the problems of "coordinates". The ecological space in a city includes GI, greenbelt, and Green Roof. Urban planning should play the role of a connecting link between strategic planning and specialized planning (He 2015).
3. Main function zoning planning
Main function zoning planning explains spatial arrangement, direction, and development

controlling principles at the national and provincial level. It ensures the implementation of environmental protection space at the strategic level.

The government should lay down conditions to speed up the developing area as soon as possible and use the method of fiscal transfer payments, such as ecological compensation mechanisms and development of the local capital to balance development. The environment is a cross-border issue and should be solved at the regional level. A large-scale area is given priority in the space strategy.

3.2 Ecological environment of space under the red line

The ecology-red line includes three aspects, namely eco-service lines, safety barrier lines, and bio-diversity lines, which comprise important ecological function areas, environmentally sensitive areas, and vulnerable areas. Main functional regions are determined against the zone and other important ecological areas. From the view of the 13th Five Year Plan, the eco-red line in Jiangsu province accounted for 40.6%, Sichuan accounted for 23.4%, Chongqing accounted for 37.3%, Jiangxi accounted for 33.7%, and Shanghai accounted for 44.5%. The regional industry and resources affect the range of the ecological red line.

Eco-red line blocks are the most important, ecologically fragile, and sensitive areas. To ensure the eco-red line's operation, a sound ecological compensation mechanism should be established and increased financial transfer paid to efforts in the area which is ecologically relatively vulnerable. And then, policy methods should be used to support green industry development.

3.3 Implementing an environmental protection plan

The problems of environmental pollution, ecological damage, higher environmental risks, and urban environmental function in the position becomes vacant, thereby reflecting the urban environment's mismatch in the layout pattern, environmental resource overload, and so on (Tang et al. 2015). The environment planning should blend with the "Multiple Planning Integration" scheme. MPI combines land use planning; the national, economic, and social planning; and environmental protection planning, which can solve the contradiction between different planning spaces, thereby reducing the number of ecological spaces. Also, people should identify the conservation eco-red line, build an environment controlling system, and implement an environmental protection plan (Zeng 2015).

In recent years, the Environmental and Planning Institute established a technique of urban environmental planning and the work was carried out in more than 10 cities like Yichang. Yichang and Weihai established an air grid simulation model, a 10-square-foot piece of water environment control system, and a GIS platform for the important, sensitive, vulnerable ecological areas, statutory reserves, and ecological protection of important ecological systems. On the platform of "Multiple Planning Integration", Guangzhou carried out the ecological protection of the eco-red line and hierarchical control system.

4 CONSTRUCTION OF AN ENVIRONMENTAL PROTECTION SYSTEM

4.1 Environment space control system

Ecological planning of space comes from the Utopia, Garden City and Broad Acres of the City. There are four basic modes: ecological belt around metropolis, wedge-shaped eco-mode, eco-core model, and axial direction to the eco-mode.

The ecological space can be implemented by the following three kinds of domestic and foreign actions: (1) ecological functional area planning focuses on delimiting ecological functions and controlling of urban areas to guide urban development and protection of the surrounding natural environment. (2) Ecological pattern planning has strategic and systematic characteristics. From the macroscopic angle, it focuses on general layout planning for all kinds of ecological space in the city. (3) By using technological means, like Remote Sensing (RS), GIS, and the quantitative analysis method, ecological control line planning analyzes the layout, gross, function, and morphology of the ecological space by drawing specific boundaries.

Taking Wuhan as an example, an ecological space system is constructed based on the ecological adaptability evaluation in the GIS platform. By using the grid pattern of "Ring-wedge-corridor", the ecological space system in Wuhan differentiates banning of construction areas, restricted areas and suitable areas to complete "two-axis and two-loop, six-wedge and multi-corridor" (He and Wang 2009).

4.2 Ecology space systems

Ecology space systems consist of water space, air space, green infrastructure space, vertical planting space, and eco-environment space.

Water is an important environmental carrier. Environmental protection requires the construction of water space. Firstly, total water resources and efficiency should be controlled in order to reduce water consumption. And then, through facilities establishment and ecological rehabilitation projects, the water ecosystem is protected. Atmospheric environmental control needs to establish a classification system, thereby preventing pollution from fountainhead, and guiding the reasonable industrial layout. The green infrastructure aims at increasing the level of urban greening. Based on a traditional pattern, it is more efficient at constructing the green space. According to the different environmental characteristics of the space, vertical planting can use different green materials and different ways to protect the environment.

In order to solve the problems of increasingly conflicted, space-broken, ecological environment deterioration and low effectiveness in the city, a composite ecological space should be built.

4.3 Construction of a space system for environmental protection

Based on GIS, RS technology, and space platforms, space management and control system of environmental protection contains a partitioned and graded system, thereby controlling size, eco-red line, and multiple gauges to ease environmental burden and optimize the space development pattern at different scales in rural, urban, and regional zones.

There are many challenges to achieve ecological urbanization development in China, such as the absence of environmental features, environment pattern mismatch, environmental resource overload, ecological function degradation, etc. which need theory, such as control and ecological carrying capacity. "Partitioned management-responsibility main body-policy measure" should be completed on the mechanism.

The establishment of environmental control systems should enhance five core tasks in ecosystem services and technical paths.

1. Overall coordination in the environmental space:
Space control uses the allocation of resources as the core, aiming at harmonious development of the economy, society, and ecology. The environment space should coordinate with regional layout, urban and rural construction, traffic development, and land use.
2. Quantify the total amount of the ecological land:
With urbanization, the urban ecological space is eroded, and so it is particularly important to take scientific measures to calculate the total ecological land and hierarchically develop an urban ecological security pattern.

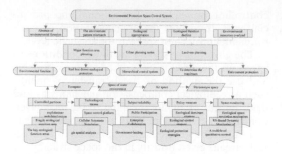

Figure 1. Workflow of the space management and control system of environmental protection.

3. Technological innovations:
With the help of a high-performance computing and technology innovation, banning of construction areas, restricted areas, and suitable areas is easy to conduct and implement. Cellular Automata is used to simulate the environment space dynamic evolution and ecological footprint of the carrying capacity. These two parameters can define an ecologically sensitive region.
4. Flexible strategy in space control:
The space control strategy is important in implementing ecological protection. It needs to adjust measures under local conditions to ensure the dynamic balance of the ecological environment and economic development. The flexible strategy is set aside for urban development and the system should build up ecological space.
5. Research on Ecological Protection Strategies.
Fig. 1 shows the content of environment space. Establishment of an ecological control line system should rise to the level of law enforcement. By using GIS and other technical means to monitor the environment, the basic ecological control line is strengthened and the residents are encouraged to participate. What is more, the communication between the government and the public should be strengthened.

5 CONCLUSIONS

Environmental protection requires the building of ecological space. It stressed building a compound ecological space with ecologic–economic–social function. For example, green infrastructures within the urban area are attached to ecological functions in non-ecological land use. The space development system is based on building space planning and the way of using the control system. The spatial planning system is given priority to space governance, structure optimization, and hierarchical management. Environmental space should form the mechanism of common participation, the cost share, and revenue sharing to realize fairness and harmony in the region. Also, the construction of the space system in environmental protection must be sustained by an equally built enforcement structure.

Only by having clear space requirements and by implementation of environmental space can we protect the environment and achieve sustainable development.

REFERENCES

Hanson, M. (1999) 'Sustainability' rendered usable? The idea of environmental space, Environmental Politics, 8(4), pp. 206–210.

He Mei, Wang Yun. (2009). Construction and Protection Strategies of Urban Eco-spatial System inWuhan City. Planners. J. 25(9):30–34.

He Zizhang. (2015). Study on the "One" of "Multiple Plans Integration": Reflection on The Practices of Xiamen. J. Urban Development Studies. 58–88.

Meng Guangwen, VOGT Joachim. (2005). Spatial Planning as a Tool for Ecological Environmental Protection: the German Experience and its Significance in China. J. Progress in Geography, 06:21–30.

Pallangyo, D.M. (2007). "Environmental Law in Tanzania; How Far Have We Gone?". LEAD: Law, Environment & Development Journal 3 (1). Volume 3/1.

Peithmann O. (2008) One space—one plan: Proposal for reform of the spatial planning system. J. Raumforschung Und Raumordnung (5):429–439.

Petrisor A.-I. (2008), Toward a definition of sustainable spatial development [in Romanian], Amenajarea Teritoriului si Urbanismul, 7, 3–4, 1–5, Alfa Press, Iasi, Romania.

Petrişor, Alexandru-Ionuţ, & Petrişor, Liliana Elza. (2013). The shifting relationship between urban and spatial planning and the protection of the environment: romania as a case study. Present Environment & Sustainable Development. J. Vol. 7, no. 1.

Tang Yanqiu, Liu Deshao, Li Jian, Jiang Hongqiang. (2015). Reflection on the Position of Environmental Planning in Mutiple Planning Integration. Environmental Protection. J. 07:55–59.

Ton Bührs. (2007). Sharing environmental space: the role of law, economics and politics. Journal of Environmental Planning & Management, 47(47), 429–447.

Xiao He, Jin Xianfeng, Chen Jiaquan, Liangtao Li. (2015). Multiple Planning Integration fuse in eco-environmental planning. J. Management Observer, 26:67–68.

Zeng Yu. (2015). Ecological environment and the "Multiple Planning Integration". Development of the Western Region. J. 04:79–83.

Analysis of the light hydrocarbon recovery rate in the negative pressure crude stabilization system

Haisheng Bi & Haixia Wang
College of Electromechanical Engineering, Qingdao University of Science and Technology, Qingdao, China

Jianfei Chen
Shengli Oilfield Technical Test Center, Sinopec, Dongying, China

Xing Zhang
Huangdao Oil Depot, Sinopec Pipeline Storage and Transportation Co. Ltd., Qingdao, China

Dedong Hu
College of Electromechanical Engineering, Qingdao University of Science and Technology, Qingdao, China

ABSTRACT: The negative pressure stabilization technology is the main method used for achieving crude stabilization in the oil field, which has been successfully employed at the combination station in Chinese oilfields in recent years. The process simulation software HYSYS is used to analyze the main factors that have an effect on the light hydrocarbon recovery rate, and then to determine the optimum parameters including the operating temperature, operating pressure, treatment capacity, and water cut in the stabilization system. The final technical project can improve the optimal operation level and economic benefits in the whole oilfield.

1 INTRODUCTION

At the high water cut stage of Chinese oilfields production, the liquid-producing capacity increases sharply, especially the water cut in the crude oil increases substantially, while the oil production and flash gas decrease. High-energy consumption and low efficiency operation have become serious problems in the stabilization system, which affect the efficiency of the whole oil and gas processing system (Haiqin Wang, 2011). The light hydrocarbon recovery rate is one of the crucial factors affecting the efficiency and energy consumption of the stabilization process, which is determined by parameters such as operating temperature, pressure, and water cut in the crude stream from electric dehydrator into a stabilizer. And so, it is very necessary to choose the proper temperature and pressure parameters, control the water cut to ensure the optimal operation of the system, improve the processing efficiency, and reduce energy consumption.

1.1 *Process of negative pressure flash crude stabilization*

The stabilization of the crude oil is very important for reducing the evaporation loss in storage and transportation of crude oil and for recovering light hydrocarbons composed of methane (C1) to pentane (C5). The typical negative pressure stabilization process of the combination station in the oilfield is shown in Fig. 1. The crude stream with water can be turned into a purified crude stream after processing in an electric dehydrator. And then, the purified crude stream with certain pressure is routed to the negative pressure stabilizer to conduct flash separation. The light hydrocarbon from the top of the stabilizer is cooled by using the condenser. The cooled stream is fed to the three-phase separator to conduct gravitational settling and separated into non-condensable gas, light hydrocarbon, and waste water. Meanwhile, the stabilized crude stream from the bottom is exported to the crude oil storage tank by using crude transfer pumps.

1.2 *Optimized object and conditions*

The crude stream flow rate at the outlet of the electric dehydrator is 1134.5 m^3/d (Jing Li, 2006), and the mass flow rate is 4.035×10^4 kg/h. The other parameters are listed in Table1.

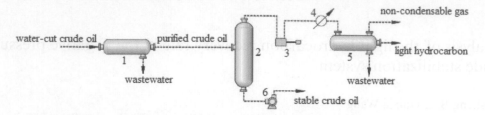

Figure 1. Typical configuration of the crude negative pressure stabilization process.
1-Electric dehydrator; 2-stable vacuum tower; 3-negative pressure compressor; 4-cooled condenser; 5-three-phase separator; and 6-centrifugal pump.

Table1. Simulation conditions.

Parameters	Value
Mass flow, kg/h	4.035×10^4
	0.5
Water cut, %	1.0
	1.5
Density, kg/m³	853.5
Pressure, kPa	40–70
Temperature, °C	55–90

2 RESULTS AND DISCUSSION

2.1 *Recovery rate–pressure–water cut relationship*

According to the gas–liquid two-phase equilibrium principle (Shuchu Feng, 2006), the lower the operating pressure in the stabilizer, the smaller is the light hydrocarbon partial pressure in the gas phase. Therefore, light hydrocarbons can be extracted much easily from the crude oil by flashing, but light hydrocarbon recovery shows distinct laws at different temperatures. The relationship is shown in Fig. 2.

According to Fig. 2(a) and (b), when the temperature is below 70°C, the recovery rate is not sensitive to water cut and therefore it has a slight drop with an increase in the water cut, but its magnitude of decline reduces obviously with an increase in pressure. The biggest decline is 0.76% from 40 kPa to 45 kPa while the smallest is 0.11% from 65 kPa to 70 kPa; Fig. 2(c) shows that, when the temperature is above 70 °C and the pressure is below 60 kPa, the recovery rate changes significantly with an increase in the water cut, and is almost linear when the pressure is below 50 kPa. These show that water cut has a great effect on the recovery rate at a higher temperature and lower pressure.

2.2 *Recovery rate–temperature–water cut relationship*

The recovery rate is more sensitive at a temperature of 60°C–70°C. Similarly, the higher the

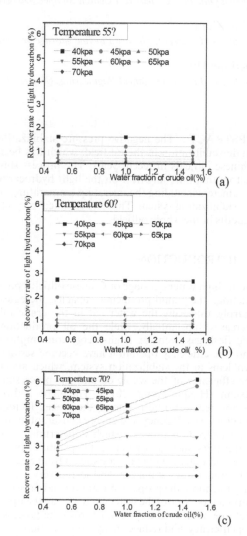

Figure 2. Graphs showing the relationship between recovery rate and water cut and pressure.

temperature is, the more intense the molecular motion of the surface liquid will be. In this way, it is easier to overcome molecular attraction and escape from the phase interface. Other components can also be carried together into the gas phase by the light component molecular interaction. (Diego Fuentes-Cano & Meng Cao, 2013). The different changing rules are shown in Fig. 3.

Fig. 3(a) shows that, when the pressure is 40 kPa, the transition temperature is 65°C; when the temperature is above 65°C, the recovery rate rises almost linearly with an increase in the temperature, and the higher the water cut is, the more obvious the growth becomes. In Fig. 3(b), 50 kPa is the transition pressure and 65°C is the critical temperature. Whenever it is above or below this point, there is an absolutely opposite trend between the recovery rate and water cut i.e. the recovery rate reduces with an increase in the water cut at a temperature below 65°C but rises above 65°C. The

curve approaches a horizontal line at 60 kPa, but the magnitude of the increase of the recovery rate rises with an increase in the temperature. The biggest growth is about 0.3%.

2.3 Recovery rate–pressure–temperature relationship

Temperature and pressure are the two main factors that affect the recovery rate. But it takes on different laws with different water cuts. The relationship is shown in Fig. 4.

By comparing (a), (b), and (c) in Fig. 4, the recovery rate declines with an increase in the pressure under the same temperature and water cut conditions. However, it increases with an

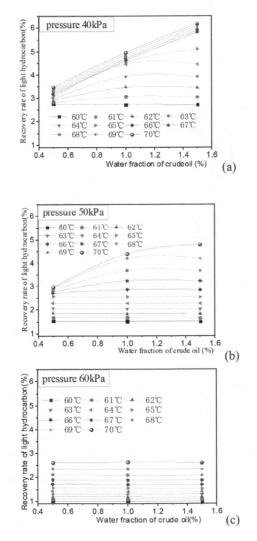

Figure 3. Graphs showing the relationship between recovery rate and water cut and temperature.

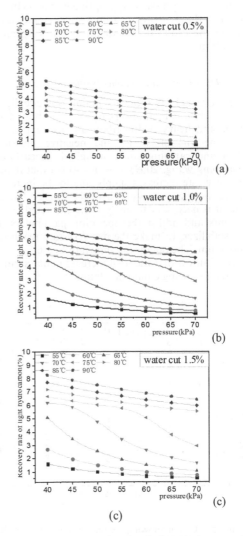

Figure 4. Graphs showing the relationship between the recovery rate and pressure and temperature.

increase in the temperature, while the magnitude of the increase initially rises up substantially and then reduces gradually to approach a constant of about 0.5%. There is a similar special region in the curve at a different water cut, called as the sensitive temperature region or the sensitive pressure region. The sensitive temperature increases with an increase in the water cut. On the contrary, the sensitive pressure decreases. In order to interpret the detail of the special region, the characteristic at 1.0% water cut is shown in Fig. 5.

As shown in Fig. 5, the temperature of the recovery rate abruptly changes at about 66°C, and the pressure at about 50 kPa. The recovery rate changes slightly at a pressure of 40 kPa and a temperature range of 65°C–70°C. The above indirectly shows that, increasing the temperature and reducing the pressure appropriately can help to enhance the light hydrocarbon recovery rate and ensure the crude oil treatment system's optimal operation.

2.4 *The extracted components analysis*

The extracted components mainly include light hydrocarbon, water, and heavy hydrocarbon. The water and heavy hydrocarbon are also extracted into the gas phase for the negative pressure effect and carrying effect of light hydrocarbons (Nianbing wang, 2009). It is not worth that if the recovery rate increases simply due to the increase in the water and heavy hydrocarbon mass. Therefore, it is necessary to analyze the extracted components and adjust the operation parameters in the stabilizer, to make the recovery rate reach its maximum. The details are shown in Fig. 6.

It is known that the total increase in mass satisfies the following equation relation:

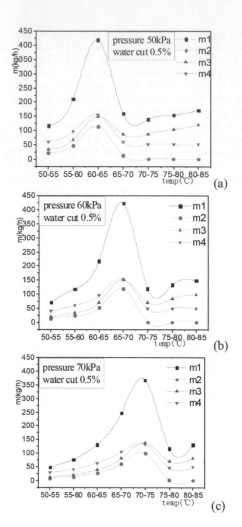

Figure 6.   Mass analysis curve of the extracted component (I).

$$m_1 = m_2 + m_3 + m_4 \tag{1}$$

where

m_1 is the total increased mass; m_2 is the increased mass of water; m_3 is the increased mass of heavy hydrocarbons; and m_4 is the increased mass of light hydrocarbons.

As can be seen from Fig. 6, the increasing amount peaks almost appear in the same temperature range under the same water cut and pressure conditions. m_4 is greater than m_3 and m_2 before the mass peaks, but it changes differently after the peak value i.e. the increase of heavy hydrocarbons is greater than the light hydrocarbon at a higher temperature region. The mass peaks migrate to the high-temperature zone with an increase in pressure. It is discovered that peaks

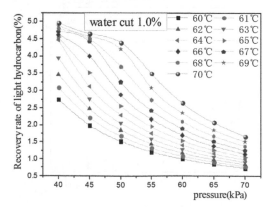

Figure 5.   Graph showing the relationship between the recovery rate and pressure (60°C–70°C).

are located at 50 kPa corresponding to the temperature zone of 60°C–65°C, 60 kPa pressure corresponding to a temperature zone of 65°C–70°C, and 70 kPa corresponding to a temperature zone of 70°C–75°C. The valley points have the similar laws. In addition, the decrease in peaks with an increase in pressure indirectly indicates that the light hydrocarbon is the main extracted component. A similar law also appears in other different water cuts. The details are shown in Fig. 7 and Fig. 8.

In Fig. 7, except for the similar law found in Fig. 6, it can be seen that the increasing mass of water and heavy hydrocarbon components are a little beyond the light components, and the increasing mass of water is obvious for an increasing water cut.

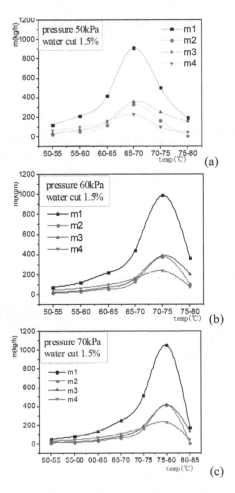

Figure 8.   Mass analysis curve of the extracted component (III).

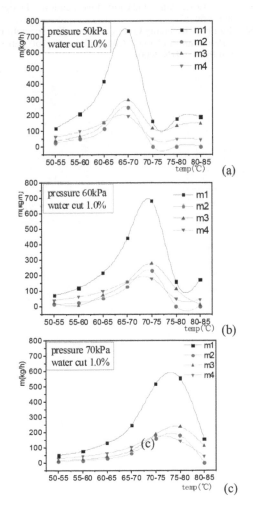

Figure 7.   Mass analysis curve of the extracted component (II).

Similarly, the mass peaks in Fig. 8 follow the similar law as mentioned above. The increasing mass of water is more obvious and almost beyond the heavy hydrocarbon to be the first. The change of the increasing mass of the light component is no longer apparent with an increase in the pressure. Both Fig. 6 and Fig. 7 also show that with an increase in the water cut in crude oil, the effect of water in the process of stabilization has changed. The escape of water molecules plays a leading role in the process of stabilization. The water-carrying effect and water vapor's dilution effect make it easy for the extraction of the light component (A. Jothy, 2011). What is more, one part of the heavy hydrocarbons is also extracted inevitably. Under these conditions, the increase in peaks with an increase in the pressure directly indicates that the heavy hydrocarbon and water are the main extracted components. And so, only under the right temperature and pressure, the effect can be obvious.

3 CONCLUSION

By analyzing the influence of the water cut, the operating temperature, and pressure on the recovery rate, the variation laws of the increase in mass peak migration of each component with changing water cut and pressure are found out. It has shown that, when the crude stream enters into the stabilizer with the water cut of 1.0%, in this case, it is helpful to improve the recovery rate of light hydrocarbons with the water-carrying effect and water vapor's dilution effect.

ACKNOWLEDGMENTS

The authors gratefully acknowledge the Qingdao University of Science & Technology Scientific Research Starting Foundation and Shengli Oilfield and Huangdao oil depot from Sinopec for financial support.

REFERENCES

Diego Fuentes-Cano, Alberto Gómez-Barea, Susanna Nilsson et al. The influence of temperature and steam on the yields of tar and lighthydrocarbon compounds during devolatilization of dried sewage sludgein a fluidized bed. Fuel 2013; 108:341–350.

Haiqin Wang: Energy Saving Technology Research Report of Oil & Gas Processing system in Zhongyuan Oilfield (Survey, Design & Research Institute of ZPEB, 2011. 3).

Jing Li, Juying Su: Journal of Jianghan Petroleum Institute. 2003; 25:124–126.

Jothy, A., B.L. Ooi, W.L. Hum. High recovery crude and condensate stabilization process, Offshore Technology Conference May, 2011.

Meng Cao, Yongan Gu. Temperature effects on the phase behaviour, mutual interactionsand oil recovery of a light crude oil-CO_2 system. Fluid Phase Equilibria 2013; 356:78–89.

Nianbing wang, Dan Song: China Investigation & Design 2009; 3:50–52.

Shuchu Feng, Kuichang Guo: Container Transportation of Petroleum and Gas (China University of Petroleum Press, CHN 2006).

Electromechanical Control Technology and Transportation – Jia & Wu (Eds)
© *2017 Taylor & Francis Group, London, ISBN 978-1-138-06752-3*

Control strategy optimization and experimental study on startup of range extender for extended range electric vehicles

Limian Wang
China Agricultural University, China
Beijing Automotive Industry Advanced Technical School, China

Zhenghe Song & Shumao Wang
China Agricultural University, China

Dihua Yi & Yueyuan Wei
Beijing Electric Vehicle Co. Ltd., China

Zhenyu Zhang
Beijing Automotive Industry Advanced Technical School, China

Jinlong Zhou
Beijing Electric Vehicle Co. Ltd., China

ABSTRACT: Range extender is composed of engine, generator (ISG motor) and the link mechanism. ISG motor is mounted on the exit side of the engine crankshaft to achieve the purposes of starting the engine and generating power. The start-up time of the range extender is related to the battery capacity. The range extender will start up intermittently during vehicle movement, which has great impact on driving comfort, thus it deserves higher attention for start-up performance. This paper starts from optimizing the calibration of engine startup control strategy, then performs evaluation on the start-up comfort, start-up success rate, energy consumption and other characteristics. The range extender selected is composed of an ISG motor, and a naturally aspirated inline four-cylinder gasoline engine. The startup conditions are tuned in the power system bench of the range extender. As a result, the start-up comfort is improved, the start-up success rate is increased and the start energy consumption is decreased by improving the drag target speed, the motor-driven exit time, optimizing the amount of fuel injection and so on.

1 INTRODUCTION

Pure electric vehicle is limited by driving range and long charging time, which have caused setbacks to many electric vehicle manufacturers and consumers over the time. The function of power generation during the running of the extended range electric vehicle can effectively improve the driving range, which has attracted more and more attention in the development of new energy vehicles. The extended range electric vehicle engine is used only for power generation, not to participate in the drive, so that it is not constrained by the speed and drive. Specific generation conditions allow the engine to work at the optimum economic zone, and thus improve the engine efficiency greatly.

In recent years, traditional vehicle engine and motor control technology has been relatively mature (Wang, 2006; Zweiri, 1998; Ma, 2005). More and

more studies are geared towards extended range vehicle engine control strategy optimization while researches on engine start control are rare in the literature. There is an ISG motor (integrated starter generator) in the extended range vehicle, which not only has the function of power generation, but also serves as the role of the engine starter. ISG motor can output large torque, and its speed is flexible and controllable. This characteristic is superior to the engine starting control, and has greater improvement potential. The research on the control method of permanent magnet synchronous motor is more mature. (Grenier, 1997; Li, 2005; Jiang, 2005). According to the starting process of the hybrid electric vehicle engine, the method of using the maximum torque control algorithm to drive the engine to the target speed in the shortest time has been investigated (Zhao, 2013), but systematic and comprehensive studies on the engine start-up reliability, economy, and start-up comfort have not been reported. Unlike traditional vehicles, the start of the extended-vehicle engine is influenced by the running conditions of the vehicle, not the driver. The engine will start up irregularly during vehicle movement, which has a great impact on driving comfort. From this point of view, it is necessary to optimize the engine starting characteristics in order to improve the engine starting performance and the driving comfort.

This paper starts from the optimization of the starting strategy of a range extender, then extends to optimized calibration of the vehicle and the bench to enhance the engine starting performance by improving the start-up efficiency, comfort, economy and reliability etc.

2 STARTUP METHOD INVESTIGATION

The start of the extended range engine is different from the conventional engine, in that extended range engine lacks the traditional starter, which is replaced by the flexible and controllable ISG motor as the starter. The engine start process refers to either the engine status change from static to idle in the ISG drive or the process of achieving the set target speed. Three conditions, e.g., adequate start torque, sufficiently high target rotational speed (Li, 2005), and injection and ignition at the right time, are essential in order for the engine to successfully complete the start.

2.1 Adequate start torque

Engine starting process can be divided into two stages, e.g., the crankshaft is dragged from the static to the target speed, then the crankshaft maintains the target speed until the injection and ignition are started successfully. To complete the above two stages and to ensure a smooth start, the engine must overcome various obstacles including frictions from piston movement, the reaction force of the compressed gas in the cylinder, the inertial force of moving parts, and so on. And the engine must be able to reach the set speed in a very short time.

2.2 Sufficiently high target rotational speed

The engine start speed needs to be high enough to ensure that the engine starts up quickly and reliably. For petrol engine, it must form a combustible gas mixture in the cylinder before the compression stroke is finished and ignited, which in turn requires a high engine speed. ISG motor is a DC motor with concentrated winding permanent magnet and brushless, which features many

characteristics to fully meet start-up requirements such as the rotational speed and torque are controllable in a wide range, and the control scheme is flexible.

2.3 Injection and ignition at the right time

Even after the engine reaches the appropriate speed, the engine might fail to start if the drive exists before starting successfully. If the drive does not exit after starting successfully, it will incur a waste of energy. So selecting both appropriate driver exit time and ignition time plays an important role in the engine starting performance.

Optimized calibration experiments on the range extender are conducted using the different drag target speed, different motor-driven exit time, different amount of fuel injection, ISG start and drive energy consumption etc.

3 HYBRID BENCH BUILDING

The experiment uses a pure serial extended range structure. ISG motor is mounted directly on the crankshaft of the engine. The ISG motor is controlled by the motor controller GCU while the engine is controlled by the engine controller ECU. GCU, ECU, BMS (battery controller) communicate with the vehicle controller VCU via CANbus. The system architecture and communication protocol are illustrated in Figure 1.

Range extender used in this experiment is composed of the ISG motor and the inline four-cylinder engine. Its main parameters are shown in Table 1. Power analyser with model WT 3000 is used to measure the ISG drive energy consumption and FTF2-400 / 50-600 BS battery simulator is used to emulate the power battery and output 350 V high voltage. Engine fuel consumption is measured by AVL_S11C0001 fuel analyser.

The integrated range extender is attached to the laboratory test bench, which is controlled by the

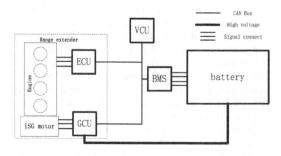

Figure 1. Extended range control system architecture.

186

Table 1. Range extender system engine and ISG parameters.

Name	Item	Parameter
Engine	Drive form	FF
	Manufacturer	BMPC/A151
	Type	Inline four-cylinder
	Arrangement form	State V standard
	Displacement (ml)	1499
	Max Power (kW/(r/min))	84/6000
	Max Torque (N·m/(r/min))	142/4800
	Bore and stroke (mm)	Φ75*84.8
	Compression ratio	10.5
ISG	Peak power (kW)	53
	Peak torque (N·m)	180
	rated power (kW)	35
	rated torque (N·m)	112
	Max speed (r/min)	6000

Figure 2. Test bench and data acquisition equipment.

PC software and calibration tools (installed outdoors). Data acquisition equipment (Fig. 2) is used to record data during the test.

Engine start control logic is shown in Figure 3.

After receiving the engine start signal, the pump starts running, providing the drag target rotation speed. Then, the ISG starts driving the engine. The duration of the driving time and exit time can be characterized by the duration of drag target rotation speed. When the engine speed reaches the target speed, the engine state is changed from shut down to start, which allows injection and ignition at the same time. When the speed reaches a certain level and maintains that level for a pre-specified duration, the state of the engine is changed from start to running. At this point, the successful start flag is set, which concludes the engine start up process.

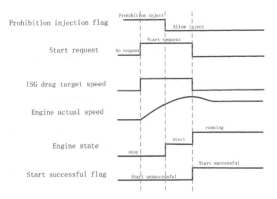

Figure 3. Engine start control logic.

4 EXPERIMENTAL PROGRAM VERIFICATION

The starter of the extended-engine is an ISG motor, which can be controlled to produce speed and torque at different levels. This, in turn, makes it possible to control the engine start more precisely. In principle, the engine is driven by the ISG to a certain target speed. Then injection and ignition are initiated to complete the engine start. Due to uncertainty of the running after the engine ignition, speed variation is tremendous, which potentially causes negative impact to the engine performance. To compensate this uncertainty, the motor drive time is changed by varying the ISG motor drive exit time, and the engine starting conditions. In this paper, two different start control modes are explored to verify the starter performance.

As illustrated in Figure 4, the drive exit time can be characterized by the ISG drag target speed. The different control modes are realized by changing the target rotation speed duration time.

In Mode 1, ISG motor drags the engine to the target speed, of which four (4) different speeds are selected within the range of 800–1400 rpm, then quits work. At same time, ignition is permitted.

Similarly, in Mode 2, ISG motor drags the engine to the four pre-selected target speeds, in the range of 800–1400 rpm. However, in this mode, ignition is allowed only after the target speed is reached. Then ISG quits work after the engine start successful flag is set.

A test bench research on mode 1 is performed. It is observed that some of the attempted starts are unsuccessful with the preset amount of fuel injection at a temperature of 30–45°C. Then the fuel injection quantity is increased. Fifty (50) experiments at different speeds are conducted to evaluate the success rate. The start air-fuel ratio is used as a surrogate for the fuel injection quantity. The test results are shown in Table 2.

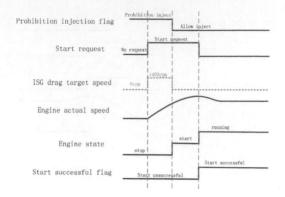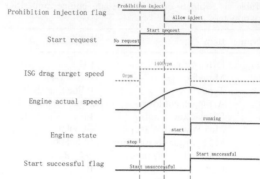

Figure 4. Comparison of start logic in mode 1 and mode 2.

Table 2. Comparison of the start fuel injection quantity and success rate at different speed in mode 1.

Drag target speed/rpm	Start air-fuel ratio	Success percentage
800	13	48%
	7	90%
1000	13	70%
	7	98%
1200	13	82%
	7	100%
1400	13	86%
	7	100%

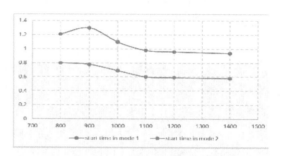

Figure 5. Comparison of start-up time in different modes.

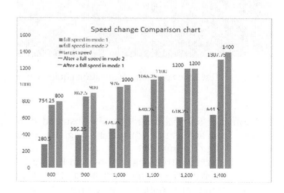

Figure 6. Comparison of speed.

Data in Table 2 indicate there are unsuccessful starts with the default amount of fuel injection if the ISG is allowed to quit driving after the engine was dragged to the target speed. The success rate is low at the low speed, while the success rate has been improved significantly with the enriched fuel injection. Apparently, fuel consumption is higher in mode 1.

The different start effects are explored using the enriched fuel injection in mode 1, and using the default fuel injection quantity in mode 2 (not enriched). Six different target rotational speeds are selected in the test to evaluate the start-up time, start-up comfort. Resulting data are compared in Figure 5.

From the results it can be seen that, the start time becomes shorter as the drag target speed increases. The start time in mode 1 is noticeably above 0.4s and is significantly higher than that in mode 2. As a result, the start-up reliability and start-up speed in mode 2 are much better than that for mode 1.

Through data in Figure 6, it can be seen that, the speed fluctuation reaches 500 rpm during the start-up process in mode 1 while the speed fluctuation

is only 100 rpm in mode 2. Consequently mode 2 renders a better start-up comfort.

From speed curve (Figure 7), for mode 1, it can be seen that the ISG exits driving when the engine speed reaches 1000 rpm at 0.5s, then the engine speed drops significantly, and finally it rises to 1000 rpm to complete the start-up process. From speed curve, in mode 2, ISG motor exits driving after engine start is completed. It is noticeable that the engine

188

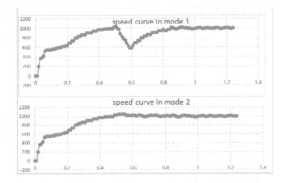

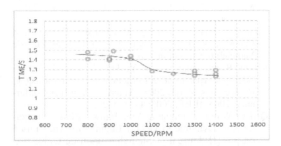

Figure 7. Comparison of the start-up speed fluctuation in different modes.

Figure 8. Start-up time at different speeds in mode 2.

Table 3. Energy statistics at different speeds.

Target speed /rpm	ISG Power consumption /J	Engine consumption /J	Total energy consumption /J
800	4046.15	61790	65836.15
900	4106.67	61637	65743.67
1000	4362.5	61215	65577.5
1100	4709.17	61080	65789.17
1200	5018.29	60715	65733.29
1400	5729.35	60318	66047.35

$$E_{elec}[J] = \int_0^t V_H I_{ins} dt$$

$$E_{fuel}[\mathrm{J}] = \int_0^t Q_{fuel}[\mathrm{kg/s}] dt$$

In the formula, $E_{elec}[J]$ is the motor power consumption in the starting process. The motor current is collected in the test to obtain the motor instantaneous current I_{ins} in the starting process. V_H is the motor voltage, and $E_{fuel}[J]$ is the engine consumption. The engine fuel flow is collected in the test to get the instantaneous flow $Q_{fuel}[kg/s]$.

The total energy consumption in the starting process is calculated by the following formula.

$$E_{total}[J] = E_{elec}[J] + E_{fuel}[J]$$

The energy consumption in the engine starting process at different target speeds is calculated. Six different target speeds are selected, e.g., 800 rpm, 900 rpm, 1000 rpm, 1100 rpm, 1200 rpm, and 1400 rpm respectively. Results are tabulated in Table 3.

Through analysis of ISG power consumption and energy consumption in the starting process, it is found that the power, with which ISG motor needs to drive the engine, requires increasing as the target speed steps up, but the energy consumed in engine starting process decreases. Interestingly, it is found that total energy consumption decreases first, then increases, with the lowest point at the target speed 1000 rpm.

speed does not experience significant fluctuations. The smoother start-up speed in mode 2 delivers a better start-up comfort than that in mode 1.

Ranking by multiple factors, such as, start-up time, start-up economy, start-up comfort and other factors, mode 2 is significantly better than mode 1. As such, mode 2 is selected as the start-up control method.

Under the start control strategy in mode 2, it is attempted to determine the best drag speed for different target speeds. At the temperature of 30–45°C, summary of start-up time at seven different target speeds, is shown in Figure 8.

The response of the motor is very fast that it takes very short time to achieve the target speed. However, the current engine speed has a strong effect on the ignition required for a successful start. The start-up time becomes shorter as the target speed increases.

5 START ENERGY CONSUMPTION EVALUATION

The engine energy consumption and ISG electricity consumption at different rotational speeds in the engine start-up process are calculated using the formula (DeBruin, 2013) shown below.

6 CONCLUSIONS

Two different control modes are analyzed and the best start control mode is determined from the viewpoint of start-up success rate, start-up comfort, and start-up time for different ISG drive exit time during the engine start-up process. Then experimental studies on the energy consumption of different start speeds are performed to determine the best drag target speed. Conclusions are drawn as follows.

During ISG engine start-up, if engine is allowed to quit immediately after the engine speed reaches the target speed, the engine speed will fluctuate considerably. It is challenging to start at a low speed, which requires enriched fuel injection in order to improve the start-up success rate.

After the engine reaches the target speed, if ISG continues to participate in the drive, the start-up success rate will be improved dramatically. When the engine speeds raise fast, ISG motor operation mode changes from drive to generation. This plays a role in regulating speed, thus effectively improves the start-up comfort. Also the start-up time is shorter.

In the engine start-up process, the ISG motor consumes power, and the engine also consumes energy. With the increasing of the drag target speed, ISG motor will need to drag it to a higher engine speed, so the power consumption increases. The engine start energy consumption decreases because the engine is at a higher speed and easier to start. The total energy consumption of the engine and ISG motor in the start-up process decreases first and then increases. The best drag target speed is 1000 rpm for the range extender in this experiment.

ACKNOWLEDGEMENTS

This project was funded by Technology Support Project (Grant No. 2015BAG05B00).

REFERENCES

DeBruin L. Energy and Feasibility Analysis of Gasoline Engine Start/Stop Technology[J]. 2013.

Grenier D, Dessaint L A, Akhrif O, et al. Experimental nonlinear torque control of a permanent-magnet synchronous motor using saliency[J]. IEEE Transactions on Industrial Electronics, 1997, 44(5): 680–687.

Hongpeng Li, Datong Qin, Yang Yang, XU Jia-shu. Dynamic Simulation of Automobile Engine Starting Process[J]. Journal of Chongqing University: Natural Science Edition, 2005, 28(6): 4–8.

Hongpeng Li. Study on Torque Control and Simulation for ISG based on HEV Engine Starting Performance[D]. Mechanical Engineering, Chongqing University, 2005.

Jiang Wang, Jing Wang, Xiangyang Fei. Nonliner PI Speed Control of Permanent Magnetic Synchronous Motor[J]. Proceedings of the CSEE, 2005, 25(7): 125–130.

Jiangling Zhao. Coupling Mechanism Design and the Mode-Switch Control of Full Hybrid Electric Vehicle[D]. Chongqing University, 2013.

Ma Q, Rajagopalan S S V, Yurkovich S, et al. A high fidelity starter model for engine start simulations[C]// Proceedings of the 2005, American Control Conference, 2005. IEEE, 2005: 4423–4427.

Qishan Wang, Chengliang Yin, Jianwu Zhang, Instantaneous Dynamic Model for Engine Governor Based on p-ω.[J]. Journal of Shanghai Jiaotong University, 2006, 40(1): 152–156.

Zweiri Y H, Whidborne J F, Seneviratne L D. A mathematical transient model for the dynamics of a single cylinder diesel engine[C]//Simulation'98. International Conference on (Conf. Publ. No. 457). IET, 1998: 145–151.

Application of metal additive manufacturing in shipping and marine engineering

Chen Chao, Liming Liu & Jiangmin Xu
School of Mechanical Engineering, Jiangsu University of Science and Technology, Zhenjiang, China

ABSTRACT: Beginning with explaining the basic conception of Additive Manufacturing (AM), this paper attempts to discuss the technical principle, application, and development of Selective Laser Melting (SLM) and Laser Solid Forming (LSF), in which AM is applied. It is emphatically analyzed that AM could be applied for manufacturing the metal parts of marine engines and typography LNG carburetor in the area of shipping and marine engineering. Meanwhile, this paper also focuses on analyzing the restoration and surface modification of marine propellers, gears, impellers, etc. Therefore, the results indicate that AM technology has a broad prospect of application and value, which can be widely applied to the processing of various metal materials such as titanium alloy, stainless steel, and aluminum alloy.

1 INTRODUCTION

Additive manufacturing, commonly known as 3D printing technology, is a technology formed by hierarchical processing and printing packing material layer-by-layer to manufacture three-dimensional products according to a designed 3D model. This technology is different from the reducing material manufacturing technology like traditional mechanical processing and material manufacturing technology like casting and forging. The 3D printing technology is known as the core technology of the third industrial revolution on account of synthesizing cutting-edge technology in many other areas, such as digital modeling technology, electromechanical control technology, information technology, material science and chemistry, and so on, which has been widely used in the automotive industry, biomedical, aerospace, and other industries (Zhang, 2016).

Since the mid-1990s, metal additive manufacturing technology, which is now generally named as metal laser cladding forming technology has become one of the hottest research and development directions in the field of Metal Rapid Prototyping (MRP), and has been dubbed in some different names during recent years, such as Laser Solid Forming of powder feeding mode (LSF), Selective Laser Melting (SLM) and so on (Yan, 2006). Two kinds of representative technological principles of metal additive manufacturing are shown in Figure 1.

2 THE APPLICATION OF METAL ADDITIVE MANUFACTURING BOTH AT HOME AND ABROAD

Since 3D printing technology appeared in the mid-1980s, it has entered its mature stage and has the initial framework of a large-scale system, especially in recent years. At the same time, the comprehensive competition situation of 3D printing technology has appeared as a basic pattern of being led by the United States, collaboratively developed in Europe with Japan following and China after it. Born in the late 1980s, the first generation of the 3D printer has gone through three periods: printing the nonmetal models through the technology of powder bonding; giving priority to developing molds; and evolving into rapid manufacturing. Then, the second generation of 3D printer that evolved from printing models has been able to print high-precision functional products in recent years. The third generation of a 3D printer is predicted to be born in the next 10 years. Under the background of intelligent manufacturing, it will become a part of some intelligent manufacturing platforms through combining 3D printer

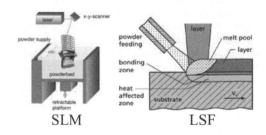

SLM   LSF

Figure 1. The diagram showing the representative technical principles of metal additive manufacturing.

Table 1. Equipment list of domestic and international typical metal additive manufacturing.

Order	Brand	Country	Tech	Model	Maximum size (mm)	Main technical features
1	Concept Laser	German	SLM	X2000R	$800 \times 500 \times 400$	Island exposure technique, two-way powder paving
2	EOS	German	SLM	M400	$400 \times 400 \times 400$	Substrate preheating, unidirectional powder paving
3	Renishaw	UK	SLM	AM250	$250 \times 250 \times 300$	Gravity powder paving
4	3D Systems	America	SLM	DMP320	$275 \times 275 \times 420$	Discrete exposure technique, unidirectional powder paving
5	Bo Lite	China	LSF	LSF-VI	$1500 \times 1500 \times 600$	Coaxial powder feeding
6	Hua Shu	China	SLM	FM271M	$275 \times 275 \times 320$	Powder feeding under single cylinder, unidirectional powder paving

technology with other advanced technologies like Internet of Things technology, big data, cloud computing, robot, intelligent materials, and so on.

Most of the share of the world is at present occupied by German products in the field of metal precision digital additive manufacturing equipment. The foreign supplies of metal precision digital additive manufacturing equipment include German EOS company, German Concept Laser company, German DMG, German Realizer GmbH company, German SLM Solutions GmbH company, UK Renishaw company, America 3D Systems company, and so on (Bao, 2006). The model of X2000R produced by German Concept Laser company can print products of the maximum size of 800 mm × 400 mm × 500 mm through the SLM technology. The LENS technology applying coaxial or paraxial powder feeding has no more limitations and can print small-sized parts, but it is much worse compared to SLM technology equipment in terms of size precision, surface quality, and density.

Since the early 1990s in China, under the support of the National Ministry of Science and Technology and other departments, some universities such as Xi'an Jiaotong University, Huazhong University of Science and Technology, and Tsinghua University have achieved great progress in aspects of typical forming equipment, software, materials, industrialization, and so on. Subsequently, many domestic universities and research institutions also launched a related study. For instance, the Northwest Industrial University, Beijing University of Aeronautics and Astronautics, Nanjing University of Aeronautics and Astronautics, Shanghai Jiaotong University, Jiangsu University of Science and Technology and other units, all carry out some exploration work for research and application. Some domestic companies, such as Xi'an Bo Lite, Changsha Huashu Tech, and Wuhan Binhu Electromechanical Technology Industry, have developed a kind of metal additive manufacturing equipment which has independent intellectual property, making great progress in the aspects of typical forming equipment, software, materials, and industrialization. The equipment industry had not been realized until 2000, which was close to the level of foreign products and changed the early situation of relying on imports. With the support from national and local bodies, many service centers have been established in the whole country. The users of equipment include medical treatment, aerospace, automobile industry, war industry, mold, electronic appliance, shipbuilding industry, and so on, which promotes the development of Chinese manufacturing technology. The equipment list of domestic and international typical metal additive manufacturing is shown in Table 1.

3 APPLICATION OF METAL ADDITIVE MANUFACTURING IN SHIPPING AND MARINE ENGINEERING

In the shipping and marine engineering field, marine engineering equipment is a general term for all kinds of equipment used to develop, exploit, and protect the ocean, and it is also the precondition and foundation for marine economy development. High-tech ships with high complexity and quantity of value are the significant impetus of transformation and upgradation in the shipbuilding industry. In foreign countries, there are many cases of application of metal additive manufacturing in shipping and marine engineering, which have referential values for our country's application of this technology in the area of shipping and marine engineering. The specific applications are shown in Table 2.

Our country always pays attention to the development of shipping and marine engineering equipment industry department in the MIIT launching a series of interpretations of the file *Made in China 2025* at the end of last year. The interpretation claims that shipping industry is a comprehensive

Table 2. The application of metal additive technology in the field of foreign ship and maritime work.

Order	Research unit	Application contents
1	Maersk Line	Using 3D printer to manufacture the container ship's spare parts. The company claims that it will install 3D printers on container ships, and is investigating the possibility and future utilization rate of printing of metal power laser sintering
2	Hyundai Heavy Industries	Hyundai Heavy Industries hopes to print multiple ship components by 3D printers, in order to save time and money in the process of production and promote the industry.
3	Siemens	Using EOS metal printers to print gas turbine components
4	Samsung Heavy Industries, Daewoo Shipbuilding & Marine Engineering (DSME)	The 3D printing center is established by three shipbuilding enterprises and Ulsan University to achieve localized production of 15 kinds of major marine accessories in the center at the end of this year and to achieve localization of as many as 165 different kinds of ship components by the year of 2018
5	Mitsubishi Electric	Using metal additive manufacturing technology to produce spare parts of terminal blade of steam turbines
6	Rotterdam	Setting up 3D printing center. The project involves marine spare parts of 3D printing
7	The United States Navy Department	Trying to install a 3D printer on the Wasp Class amphibious assault ship USS Essex, and hoping the 3D printer will print some spare parts on the ship to reduce the fleet's logistics supply

and strategic industry which provides technology and equipment for water transportation, ocean resource development, and national defense construction, is the essential ingredient of the high-end equipment manufacturing industry in our country, and is the foundation and support of establishing a powerful marine nation strategy. Therefore, *Made in China 2025* accelerates the marine engineering equipment and high-tech ships as one of the top ten key development fields, guiding the developing target in the next 10 years. It also points out the development direction of marine engineering equipment and high-tech ships for our country.

Its significance is as follows: firstly, accelerating the development of marine engineering equipment and high-tech ships is the only way to establish a powerful marine nation; secondly, it is the necessary condition of transforming our country into a great shipbuilding power; thirdly, it is crucial to promote China's industrial transformation and upgradation.

An academic exchange of "Ships and ocean engineering application of 3D printing technology" was held by Beijing Shipbuilding Engineering Society on August 25, 2014. Cao Lin who works in the economic research center of CSIC introduced the applications of 3D printing technology in the shipping industry in a report *Investigation report of applications of 3D printing technology in the shipping industry*. According to his speech, based on what we have done, models and products in which 3D printing technology has been successfully applied and popularized are the shipping industry, especially in the process of research and design. It can be complementary to traditional manufacturing technology. In the medium term, it

is hopeful to be a breakthrough of 3D printing in manufacturing and restoration technology used in complex structural parts, special parts, alloy components, and composite material components in ships. In the long term, 3D printing technology is expected to lead the development direction of the shipping industry, and then promote the revolution of the shipping industry system in the future. The applications of metal additive manufacturing performed so far in the domestic shipbuilding research institute are shown in Table 3.

Lv Delong, a senior engineer in the Promotion and Transformation of Scientific and Technological Achievements Research Center in Science and Technology, Industry of National Defense, considers the shipping industry as an ingredient in the manufacturing industry and is inevitably influenced by 3D printing technology. First of all, this technology will speed up new product developments in the shipping industry and reduce the cost of development. On account of the convenience to shaping up, 3D printing technology can better verify and improve design parameters and ideas so that it increases design level and efficiency. Secondly, it makes the production of auxiliary products more economic and quicker. Due to the small batch of auxiliary products in ships, most of them are customizable products. The 3D printing technology can adjust the parameters at any time and make the cost lower in producing a small batch or even just a single piece of product. The 3D printing technology can manufacture the product which traditional manufacturing methods cannot do and achieve higher accuracy and smoothness. Traditional manufacturing methods are reducing and cutting off

Table 3. Survey of application of metal additive manufacturing in shipbuilding research institute.

Order	Research institute	Situation of application
1	701	Taking up an experiment to shape the impeller in an underwater heat chamber, and if it succeeds, they will provide impeller printing service. Planning a print antenna on the water to solve the problem of complex antenna processing
2	702	Using 3D printing technology to manufacture noise reduction hollow impeller
3	703	Printing stator blade for gas turbine and complex structure
4	704	Manufacturing the parts of metal cold oil pipe
5	705	Newly increased a team—U3 team
6	708	Printing a titanium alloy impeller and trial-manufacture propeller
7	711	Research of a new diesel engine and its key parts
8	717	To reduce the weight of the navigation support frame, gyroscope, shipboard instrument and frame type parts
9	719	Application on radiator, impeller, and honeycomb structure
10	723	Application on radar antenna
11	724	Application on radar antenna

the needless on work blank through multidimensional processing, but it is difficult to manufacture complex products. Any complex product can be quickly and accurately manufactured using three-dimensional design and 3D printing technology, which solves the problem that it is difficult to manufacture complex parts in the past and that high precision leads to the better performance in manufacturing products. The coherence points of metal additive manufacturing and shipping and marine engineering are shown in Figure 2.

Metal additive manufacturing is widely used in the aerospace field. Parts and components used in the aerospace field are high-performance, complicated and unique, with high material cost, which is consistent with metal additive manufacturing, whose characteristics exactly suit the complex parts, small batch, multi-species, and high material utilization. Therefore, metal additive manufacturing is widely used in the aerospace field. Traditional shipping and marine engineering have many differences compared with metal additive manufacturing. Hu Keyi, a chief engineer in Jiangnan Shipbuilding Group considers that there is no big chance for the application of metal additive manufacturing in shipping and marine engineering. Considering material and technology principles at present, the application of 3D printing is not in as wide a range as what we thought with many limitations. However, it deserves expectation for application of metal additive manufacturing in shipping and marine engineering with the development of metal additive manufacturing and breakthrough of new materials and technology.

This paper discusses the application of metal additive manufacturing in shipping and marine engineering by taking an example of several typical parts in shipping and marine engineering. One typography LNG carburetor is manufactured by Marine Equipment Research Institute of Jiangsu University of Science and Technology using SLM technology. LNG is widely used in the world as a kind of clean energy and annual consumption of LNG runs into hundreds of millions of tons. Typography carburetor as a compact heat exchange technology compared with open shelves carburetor only has one-fiftieth volume and one-tenth weight when they deal with the same amount of LNG. At present only Heatrac in the UK has mastered this technology. Marine Equipment Research Institute of Jiangsu University of Science and Technology firstly produced the core parts of heat exchange and tested it successfully at normal temperature and low temperature and high-pressure conditions (−168°C, 120 atmospheres). So the marketization and domestication of this product are quite promising.

In the shipbuilding industry, 3D printing technology is mainly used in small-sized parts with a complicated structure, such as an impeller, engine blades, radiator, and small propeller. But the LSF technology shares a large market in shipping and marine engineering (see Figure 3). The best application of LSF technology is to repair the vulnerable parts in shipping and marine engineering, including internal combustion engine blades, propeller shafts, and gears (see Figure 4 and Figure 5). It can be used for reconstructing the material which vulnerable parts have lost, to maintain structural integrity. This technology is called "laser remanufacturing technology", which is a process of quickly repairing the damaged parts to achieve or even exceed the original performance.

The main application of laser surface modification is laser surface hardening, laser surface alloying, laser surface remelting and solidification, and laser surface cladding. In recent years, there are more applications in shipping and marine engineering (as shown in Figures 6–9). A marine diesel engine works in poor working conditions. The hardware performance of a diesel engine is always tested with conditions of high temperature, heavy

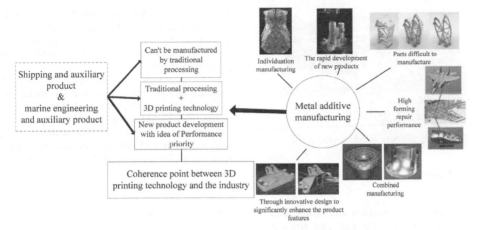

Figure 2.　Coherence points of metal additive manufacturing and shipping and marine engineering.

Figure 3.　3D printing propeller.

Figure 5.　Laser repairing blade of an internal combustion engine.

Figure 4.　Repair of a turbine impeller.

Figure 6.　Marine auxiliary engine crankshaft repaired by laser cladding.

Figure 7.　Laser cladding of flame ring in cylinder liners in diesel engine.

Figure 8.　Laser surface hardening of cylinder liner.

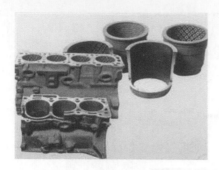

Figure 9. Laser surface hardening of cylinder and cylinder liner.

load, low speed, and vibration and so on, and the ocean environment, etchant gas, and low-quality fuel make it worse. The key parts of a marine diesel engine, such as cylinder liner, piston, connecting rod, and crankshaft, are the heart of a diesel engine and their working quality directly affects its performance. Because of technical reasons, these parts need non-effective strengthening treatment or even no treatment. Therefore, a series of problems show up in a diesel engine, such as low operating life, high oil consumption, severe wear, the difficulty of maintaining and repairing, etc. One of the effective ways to solve these problems is to take surface strengthening of necessary parts. Laser surface modification can enhance the surface with hardness, wear properties, and corrosion resistance, and it is hopeful to overcome defects of traditional techniques.

4 CONCLUSION

Metal additive manufacturing has some technical limitations in its popularization and application. Equipment using SLM technology shows deficiencies such as low efficiency, small printing size, and lack of quality control. Those using LSF technology also have problems in efficiency and accuracy. Therefore, it is still a long way to go for the pioneering and pre-guiding 3D printing technology to be truly applied to shipping and marine engineering. The new breakthrough in the development of metal additive manufacturing not only depends on the technical progress but also needs a group of people with knowledge of metal additive manufacturing. Nowadays, with the continuous development of metal additive manufacturing technology and application under the background of the third industrial revolution and Made in China 2025, these people need to reexamine the design, research and development, manufacture, sales, logistics, maintenance, and even enterprise value chain. Because additive manufacturing is going to fuse with the industry perfectly, this group of people needs to think of metal additive manufacturing in seven aspects: free design thinking, mass customization thinking, integrated part thinking, technology supplement thinking, demand activation thinking, platform service thinking, and collaborative innovation thinking. With the further development of additive manufacturing and the increase in people with the knowledge of additive manufacturing, the application of metal additive manufacturing in shipping and marine engineering is still bright and promising. With the development of digitization, intelligence, 3D printing technology and materials technology, additive manufacturing will eventually raise a revolutionary impact on the traditional shipbuilding industry.

REFERENCES

Bao Guoguang, Zhao Modian. On the Essential Characteristics and Industrialization Measures of 3D Printing[J]. Journal of Northeastern University (Social Science), 2016 18(2):111–117.

Everton, Sarah K., Review of in-situ process monitoring and in-situ metrology for metal additive manufacturing[J]. Materials and Design, 2016 95:431–445.

Hebert, Rainer J, Viewpoint: metallurgical aspects of power bed metal additive manufacturing [J]. Journal of Materials Science, 2016 51(3):1165–1175.

Li Dicheng, He Jiankang, Tian Xiaoyong, etc. Additive Manufacturing:Integrated Fabrication of Macro/ Microstructures [J]. Journal of Mechanical Engineering, 2013 49(6):129–135.

Nguyen, Alex T.T., Hierarchical surface features for improved bonding and fracture toughness of metal-metal and metal-composite bonded joints[J]. International Journal of Adhesion and Adhesives, 2016 66:81–92.

Seifi, Mohsen, Overview of materials qualification needs for metal additive manufacturing[J]. JOM, 2016 68(3):747–764.

Snyder, Jacob C etc, Build direction effects on additively manufactured channels[J]. Journal of Turbomachinery, 2016 136(5).

Stavroulakis, P.I., Invited review article: review of post-process optical form metrology for industrial-grade metal additive manufactured parts[J]. Review of Scientific instruments, 2016 87(4).

Vartanian, Kenneth, Accelerating industrial adoption of metal additive manufacturing technology[J]. 2016 68(3):806–810.

Verlee, B Dornal T etc, Density and porosity control of sintered 316l stainless steel parts produced by additive manufacturing [J]. Power synthesis and processing for controlled microstructure, 2012 55(4):260–267.

Yan Yongnian, Zhang Renjie, Lin Feng. New Development of Laser Rapid Prototyping Technique [J]. New Technology & New Process. 2006(9):7–9.

Yuan Maoqiang, Guo Lijie, Wang Yongqiang. Development and Application of Additive Manufacturing Technologies [J]. MACHINE TOOL & HYDRAULICS, 2016 44(5):183–188.

Zhang Xiaowei. Application of metal additive manufacturing in aero-engine [J]. Journal of Aerospace Power, 2016 31(1):10–16.

Electromechanical Control Technology and Transportation – Jia & Wu (Eds)
© 2017 Taylor & Francis Group, London, ISBN 978-1-138-06752-3

Research of assembly line balance at F company decelerator

Wei Jiang, Yanhua Ma & Xinyu Zhu
Institute of Mechanical Science and Engineering, Jilin University, Changchun, China

ABSTRACT: Currently, the situation of manufacturing market is more and more competitive. The assembly line balancing is an important means to improve the production management level. Based on the drive assembly line of F company for the study, this paper launched the research by using scientific research methods and tools, like witness simulation, dual-population genetic algorithm, MATLAB and Witness. The analysis on parameters of results before and after optimization, including rhythm, balance rate, smoothness index and delay rate, have been provided useful decision support for F company and improved the production efficiency. The research methods and conclusions not only have some reference value, but also have a direct practical fuction on optimizing the other similar production lines.

1 F COMPANY REDUCER ASSEMBLY LINE SIMULATION STUDY

1.1 The overview of F company reducer assembly line

F company reducer assembly system is composed of a series of serial and parallel manufacturing units, which mainly consists of the assembly line and sub-assembly line, with a total of 14 stations. The assembly line and sub-assembly line have 10 stations and 4 stations, respectively.

Each station worker assembles the parts, including the common parts and standard parts supplied by different matching manufacturers, to the reducer in accordance with the production process. The assembly routes and stations layout figure is shown in Figure 1.

Assembly line: G1-Save shell on the line, G2-Loaded bearing, G3-Split bearing cap, G4-Mounted differential, G5-Adjusting backlash, G6-Assembly test, G7-Off cover, G8-Hit cap bolts, G9-Mounted cylinder pump, G10-Assembly offline.

Sub-assembly line: F1-Big wheel assembly, F2-Hit by a bolt tooth, F3-Semi-mounted planetary pinion, F4-Pressure shell bridge.

In this paper, the beat of the reducer assembly line has already been given. The process of assembling follows the beat.

This reducer assembly line has three characteristics:

1. Time of each job element is determined.
2. Each Job element is the least natural operating unit and can not be assigned to two workstations.
3. All the required parts are purchased from the other manufacturers.

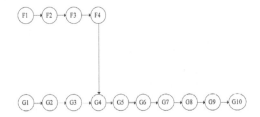

Figure 1. The reducer assembly line of assembly routes and stations layout figure.

1.2 Reducer assembly work stations and time division

This paper used a "triple standard deviation method" to determine the operating time. The operation schedule for each job element is shown in Table 1.

2 OPTIMIZATION BASED ON IMPROVED GENETIC ALGORITHM FOR F COMPANY REDUCER ASSEMBLY LINE BALANCING

2.1 The conclusions of the research on assembly line and sub-assembly line after using simulation and double population genetic algorithm

After using simulation analysis and double population genetic algorithm, the relevant data is obtained. The work load rate before and after Optimization of assembly line and sub-assembly line is shown in Figure 2 and Figure 3.

As it can be seen from Figure 2 and Figure 3, no matter the reducer assembly line or sub-assembly line, their work load rate has reached a better balance after optimization.

Table 1. Reducer assembly line operation schedule for each job element.

No.	Job element	Operation time (s)	No.	Job element	Operation time (s)
1	Pressed into the big wheel	20	25	Mounted bearings	9
2	Loaded bolt	5	26	Add to bridge the difference between the shell	13
3	Pine jig	3	27	Bolts matter	17
4	Pinch big wheel bolts	42	28	Inspected once every 30 preload	7
5	White marker pen	3	29	Record value on the card	3
6	Apply lubrication	6	30	With a chalk mark	7
7	Release bearing	2	31	Write value in three equal portions	4
8	Pressure bearing	20	32	Adjust backlash	40
9	Pressed into the differential housing bridge	23	33	Screw	4
10	Confirm differential assembly	4	34	Detected once every 30	7
11	Save lifted shell	7	35	Painted red lead powder	2
12	In the assembly position mantissa write tail number	2	36	Noise measurements	46
13	To intensify	7	37	Review and tighten the preload	13
14	Check for tightness (beat)	10	38	Take cover	7
15	Gumming	4	39	Straighten cover	6
16	Loading washer	6	40	Confirm complete coupling elements	4
17	Gumming	6	41	Pine jig	5
18	Mounted bearings	5	42	Tighten bolts	55
19	The locating bearing pressed against	12	43	Remove the sealing material	10
20	Loaded bolt	10	44	Mounted cylinder pump	40
21	Tighten bolts	34	45	Downline	14
22	Mark	8	46	Equipment operation	28
23	Check the main differential is the order of the teeth	4	47	Inversion detection	20
24	The sequence number of teeth being zoned Road	3	48	Crane lifting	10

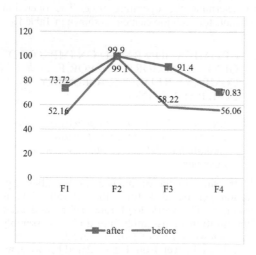

Figure 2. Contrast diagram of work load rate of assembly line of reducer before and after optimization.

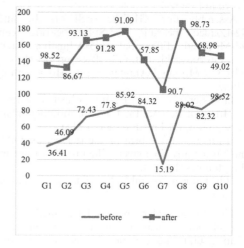

Figure 3. Contrast diagram of work load rate of reducer's sub-assembly line work before and after optimization.

Table 2. Each set of results before and after optimization comparison table.

Optimization project	Assembly line			Sub-assembly line		
	Before optimization	After optimization	Optimization results	Before optimization	After optimization	Optimization results
Equilibrium rate	69	82.8	13.8 ↑	66.7	76.2	9.5 ↑
Balance delay rate	31	16.8	14.2 ↓	33.3	23.8	9.5 ↓
Smoothness index	92	45	47 ↓	37	23	14 ↓

2.2 The content classification and correlation diagrams of assembly line job positions before and after optimization are shown in Figure 4 and Figure 5

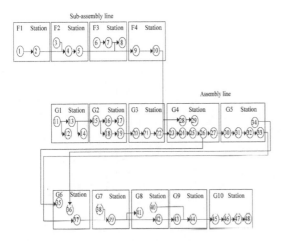

Figure 4. The flow diagram of original assembly process.

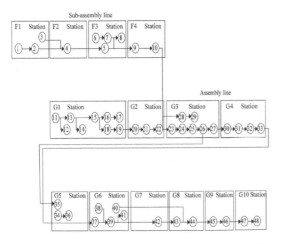

Figure 5. The flow diagram of optimized assembly process.

2.3 The result analysis of F Company reducer assembly line before and after optimization

Based on the data above, this paper calculated the equilibrium rate, smoothness index and balance delay rate. We can find that these three parameters have been optimized. The comparison chart is shown in Table 2.

As shown in Table 2, the equilibrium rate of reducer assembly line optimized from 69% to 82.8%. Meanwhile, the equilibrium rate of reducer sub-assembly line optimized from 66.7% to 76.2%.

The balancing delay rate of optimized reducer assembly line and sub-assembly line has been reduced, respectively, from the original 31% and 33.3% to 16.8% and 23.8%.

The smoothness index of optimized reducer assembly line and sub-assembly line has been reduced, respectively, from the original 92% and 37% to 45% and 23%.

These parameters show that the reducer assembly line at the F company has been greatly improved and in a good balance after using double population genetic algorithm.

3 CONCLUSION

This paper mainly studied the assembly line of F company. Firstly, the tests and inspections of the assembly line found out the original one was seriously imbalanced by using Witness simulation modeling. Secondly, the double population genetic algorithm was adopted to study the assembly line balancing optimization. Finally, after redeploying the assembly line according to the results of optimization, we found that the assembly line productivity has been greatly improved.

This research has achieved the expected effects, which will bring the advantages for the manufacturing enterprise in the intense market competition.

REFERENCES

Gutijahr A. L. & Nemhauser G. L. 1964. An algorithm for the line balancing problem [J]. Management Science, 11: 308–315.

Hoffman T. R. 1963. Assembly line balancing with a precedence matrix [J]. Management Science, 9(4):551–563.

Jaekson J. R. 1956. A computing Procedure for a line balancing problem [J]. Management Science, 2, 261–271.

Sabuneuoglu I. & E. Erel. & M. Tanyer. 2000. Assembly line balancing using genetic algorithms [J]. Journal of Intelligent Manufacturing, 11:295–310.

Salveson M E. 1955. The assembly line balancing problem [J]. Journal of Industrial Engineering. 6(3): 18–25.

Thomospolous N. T. 1970. Mixed-model line balancing with smoothed station assignments [J]. Management Science, 16: 593–603.

Wolfgang Kreutzer. 1986. System simulation: programming styles and languages [J]. Sydney: Reading, Mass: Addison-Wesley.

Yuedong zhan, Ying luo. Journal of research on Modeling and Simulation of [J]. System simulation of logistics automation system based on Petri net 2001, 13(4): 501–504.

Fractal study on the surface topography of shearing marks based on the double spectrum method

B.C. Wang
Shenzhen University, Shenzhen, China

C. Jing & L.T. Li
Guangdong Hengzheng Forensic Clinical Medicine Institute, Guangzhou, China

ABSTRACT: Three kinds of shearing mark samples are made by three kinds of wire broken pliers with different processing patterns on the blade flank. The surface of the shearing mark samples are collected digitally by surface topography tools. The general fractal dimension of three kinds of shearing mark samples is calculated by using the general fractal theory. The characteristics of three kinds of general fractal dimension spectrum and singular spectrum are discussed in detail. It actualizes the extraction and recognition of the surface characteristics of the marks. The study shows that the surface characteristics of the marks can be analyzed and recognized reliably by the combined use of general dimension spectrum and singularity spectrum. It provides an identification method for the surface characteristics of the marks. It also provides the basis for a conclusion of the shearing tools.

1 INTRODUCTION

On the scenes of a crime case, various traces will be left when using tools to destroy objects by a criminal. The striation tool marks are one of the traces. It is usually made by pliers, scissors, and other tools. In the process of cutting objects, the continuous striations that are visible or microscopically visible on the sections will be formed. The visual comparison is the main traditional inspection method of tool marks. It is a comparison of the striation tool marks collected from crime scenes with testing marks made by the suspected tools. The traces left by tools are confirmed based on the striation comparison, which is the "form inspection". The surface topography of the striation tool marks is very complex and irregular in practical situations. It is also random and chaotic. Therefore, the surface topography of striation tool marks cannot be described by digital technologies for a long time. The analysis and inspection method cannot be done digitally.

Fractal theory is one of the important branches of nonlinear science. It studies the scale-invariant object at the structure level. It represents the behavioral characteristics of the nonlinear system by the self-affine characteristic of the study object. (Brown & Savary 1991, Gagnepain & Roques-Carmes 1990, Ganti & Bhushan 1995). Fractal dimension is one of the important parameters that is used to characterize the singularity of a chaotic attractor quantitatively. It is used to represent the digital characteristics of the nonlinear system's behavior. It is used to study the surface topography. Some progress has been made in the recognition of characteristics (Mou & Yang 2010, Su et al. 2014, Wang et al. 2015, Li et al. 2014). Three kinds of wire broken pliers of different processing patterns on the blade flank are used as objects of trace tools in this study. It includes the wire broken pliers with rough milling striation on the blade flank, the wire broken pliers with fine milling striation on the blade flank, and the wire broken pliers with grinding on the blade flank. The lead flake is used as the destruction object. The samples of shearing marks are made. The fractal dimensions of the profile curve are calculated by applying the general fractal dimension. Taking the general fractal dimension spectrum and singular spectrum as the characteristic parameters, the fractal dimension characteristic value of different shearing mark surfaces are found out. It actualizes the extraction and recognition of the surface characteristics of the marks. The surface characteristics of the marks can be analyzed and recognized by the combined use of the general dimension spectrum and singularity spectrum. Therefore, it can also increase the reliability.

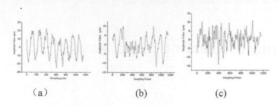

(a)   (b)   (c)

Figure 1.   Three kinds of profile curves.

2 EXPERIMENTAL CONDITIONS AND DATA ACQUISITION

Using the Austrian auto-zoom 3D surface topography device InfiniteFocus, the shearing marks are collected digitally. A three-dimensional shearing mark is stored in the computer. In this experiment, the eyepiece magnification is 10 times and the sampling resolution is 1.1 μm. The stable part of the mark characteristic is marked by the application software. The profile curve which is perpendicular to the surface of the mark is obtained. The profile curve is perpendicular to the mark surface formed by the wire broken pliers with rough milling striation in the blade flank, as shown in Figure 1(a). The profile curve is perpendicular to the mark surface formed by the wire broken pliers with fine milling striation in the blade flank, as shown in Figure 1(b). The profile curve is perpendicular to the mark surface formed by the wire broken pliers with grinding blade flank, as shown in Figure 1(c).

3 GENERAL DIMENSION SPECTRUM AND SINGULAR SPECTRUM OF THE PROFILE CURVES

The fractal theory provides a theoretical basis for the description of the nonlinear behavior system. It is used to quantitatively characterize the singularity of chaotic attractor. The fractal dimension makes up for the deficiency of the traditional analysis model to a certain extent. However, it is found in the study that it is not enough to describe the nonlinear system behavior characteristics only by one fractal dimension. So, it involves the fractal dimension description of various scales and various parameters. The general fractal theory is introduced to calculate the fractal dimension of the profile curves of the surfaces formed by three kinds of wire broken pliers with different processing patterns on the blade flank.

The general fractal is used to describe the non-uniform random probability distribution of the fractal geometry on different layers. The fractal set is divided into N parts, and P_i is the distribution probability when measuring the part i by the

measurement ε_i. The different probabilities of different parts are represented by the scale index α_i. The mathematical expression is given by:

$$P_i(\varepsilon_i) = \varepsilon_i^{\alpha_i} \quad i = 1, 2, 3, \ldots, N \tag{1}$$

Taking the limit of two ends of formula (1), its expression is given by:

$$\alpha = \lim_{\varepsilon \to 0} \frac{\ln P_i(\varepsilon_i)}{\ln \varepsilon_i} = \frac{\ln P(\varepsilon)}{\ln \varepsilon} \tag{2}$$

According to the fractal formula, α is the fractal dimension of some part of the fractal set. Its value reflects the probability measurement of the quality of this part. It controls the singularity of the probability density. It is called singularity index. α is also called the Lipschitz–Holder index. Since there are a number of parts, a spectrum $f(\alpha)$ that is formed by the infinite sequence of different α is obtained. Obviously, $f(\alpha)$ will change with the change in the singularity index α. The curves of α–$f(\alpha)$ is called the singularity spectrum. It is called the α–$f(\alpha)$ spectrum for short. It is an important parameter to describe the general fractal.

There is another set of parameters that can be chosen in the information theory method, namely Q and D_q whose mathematical expression is given by:

$$D_q = \lim_{\varepsilon \to 0} \frac{1}{q-1} \frac{\ln \chi_q(\varepsilon)}{\ln \varepsilon} \tag{3}$$

The general fractal dimension D_q is an important parameter. It has different meanings with different q values. The q and D_q are one set of parameters that describe the general fractal. The curves of q–D_q are called the general dimension spectrum. The relationship between two sets of parameters is given by:

$$f(\alpha) = \alpha q - (q-1)Dq \tag{4}$$

$$\alpha = \frac{d[(q-1)D_q]}{dq} \tag{5}$$

If one of the parameters is known, the other set of parameters can be obtained.

The most important problem in practical applications is how to calculate the general fractal dimension D_q according to experimental results and observation data. How do we seek and find the scaling characteristics according to the general fractal dimension spectrum q–D_q. Calculation measures of D_q spectrum include: covering method, fixed radius method, and fixed quality method.

The general dimension spectrum of profile curves of the surfaces, which are formed by three

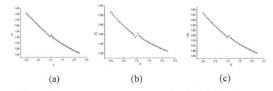

(a)   (b)   (c)

Figure 2. General fractal dimension spectrum of profile curves.

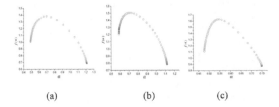

(a)   (b)   (c)

Figure 3. Three kinds of profile curves of the singular spectrum.

Table 1. Characteristic values of the fractal dimension spectrum.

Dq	Sample a	Sample b	Sample c
D_0	1.3827	1.5070	1.6275
$D_{0.95}$	1.3036	1.4532	1.5341
D_1	1.3041	1.4562	1.5353
$D_{1.05}$	1.3073	1.4621	1.5395
D_2	1.2534	1.4289	1.4796

Table 2. Characteristic values of the fractal singular spectrum.

Type	Spectrum width	Spectrum peak value	Singular index at spectrum peak value
Sample a	0.9073	1.5376	0.5103
Sample b	0.6986	1.4124	0.6841
Sample c	0.5262	1.2784	0.6847

kinds of wire broken pliers with different processing patterns on the blade flank, is obtained according to formula (3), as shown in Figure 2.

The general fractal singularity spectrum of the three kinds of profile curves is obtained according to the formula of the general fractal dimension and the relationship between two sets of parameters (4) and (5), as shown in Figure 3.

It can be seen that the dimension spectrum and singularity spectrum are different with different profile curves. The characteristic values of the fractal dimension spectrum and the singular spectrum are also different.

4 THE CHARACTERISTIC ANALYSIS OF THE PROFILE CURVE BASED ON THE GENERAL FRACTAL DIMENSION SPECTRUM

In order to analyze the characteristics of the profile curves, the characteristic values of D_0, $D_{0.95}$, D_1, $D_{1.05}$ and D_2 are extracted from the general fractal dimension spectrum. These values are given in Table 1.

The characteristic values extracted from the singular spectrum are given in Table 2.

In the calculation of the general fractal dimension, taking the iteration order of q = 0.05, a series of fractal dimensions are shown in Figure 2. It can be seen from the fractal dimensions shown in Figure 2 that the dimensional curves will decrease monotonically with the continuous increase in the q value. This means that the dimensional value will decrease with increase in q value. The fractal dimension value in the q = 0.95 will generate

high mutations and the fractal dimension value in the q = 1.05 will generate low mutations. So, the dimensions at the two points and fractal dimension of D_0, D_1 and D_2 are combined together to describe the characteristics of the profile curves. The identification rule of the fractal dimension group is established. The five special points of fractal dimension (D_0, $D_{0.95}$, D_1, $D_{1.05}$ and D_2) at q = 0, 0.95, 1, 1.05, and 2 are taken as a group to distinguish the different profile curves. It can be seen from the results that the three kinds of profile curves are different and their dimension spectra are also different.

Discussion on the experimental results: according to the spectrum curves shown in Figure 3 and the characteristics values are shown in Table 2, the corresponding spectrum width is small and the peak value is big when the marks of the grinding blade flank are left. It shows that the profile curves of the shear mark surface are more precise and more complex. The corresponding spectral width is big and the peak value is small when the marks of rough milling striations on the blade flank are left. It shows that the profile curves of the shear mark surface are less precise and less complex. So, the singular spectrum can be used to divide the surface topography of the shearing marks quantitatively. The geometric characteristics of the shearing marks are expressed scientifically.

In order to use the fractal dimension spectrum to describe the characteristics of different profile curves, the characteristics values of the fractal dimension spectrum are summed. The expression is given by:

Table 3. The sum values of the general dimension characteristic parameters of different profile curves.

Profile curves	Sample a	Sample b	Sample c
Sum of characteristics values	6.5511	7.3074	7.716

Table 4. The sum of three kinds of general fractal singular spectrum characteristic parameters of profile curves.

Profile curves	Sample a	Sample b	Sample c
Sum of characteristics values	2.9552	2.7915	2.4893

$$D_t = \sum_j \beta_j D_j \qquad (6)$$

β_i is the weight coefficient; in this calculation, $\beta_i = 1$.

The characteristic parameter values of the general fractal dimension spectrum of three kinds of different profile curves can be calculated by formula (6). The calculation results are summarized in Table 3.

From Table 2 the sum value of the characteristic parameters of the general fractal singular spectrum is expressed as:

$$f_\alpha = \lambda_1 \Delta\alpha + \lambda_2 f_m + \lambda_3 \alpha_{fm} \qquad (7)$$

λ_i is the weight coefficient; in this calculation, $\lambda_i = 1$.

A set of characteristic parameters of three kinds of profile curves can be calculated according to formula (7). The calculation results are summarized in Table 4.

It can be seen from the calculation results that the characteristic parameters dimension spectrum and singular spectrum are different when the profile curves are different. The distinction is also big.

5 CONCLUSIONS

In this paper, the general dimension spectrum q~D_q and singular spectrum (α-$f(\alpha)$ spectrum for short) of three kinds of profile curves are plotted. The profile curves are different and their dimension spectrums are also different. The dimension value of the different q values and spectrum width, spectrum peak value, and singular index at the spectrum peak value are analyzed. According to the characteristics of the spectrum, the characteristics of profile curves are described by characteristic values of five dimensions (D_0, $D_{0.95}$, D_1, $D_{1.05}$ and D_2). By using the fractal dimension group to identify the profile curves, the recognition rate is more reliable than using a single dimension.

The sum value of the characteristic parameters of the dimension spectrum and singular spectrum are calculated based on this calculation. The sum value of the characteristic parameters is combined with several of the characteristic parameter information so that the characteristics of profile curves can be reflected more comprehensively and accurately. It can greatly improve the reliability of the recognition of the contour curves formed by different tools. The results show that the sum values of the characteristic parameters of three kinds of profile curves are significantly different and the distinction is big. Therefore, the characteristic parameters of the dimension spectrum and the singular spectrum can be used as the characteristic index of different profile curves. It can be used as a criterion for identifying different tools. The study provides a basis for the inference of shearing tools. The study also provides a new method for the quantitative examination of shearing marks.

ACKNOWLEDGMENTS

This work was supported by the Nature Science Fund of China (NSFC), No. 61571307.

REFERENCES

Brown C.A., G. Savary (1991). Describing ground surface texture using contact profilometry and fractal analysis. Wear, 147:211–226.

Gagnepain J.J., C. Roques-Carmes (1990). Fractal approach to characterizations of rubber wear surfaces as a function of load and velocity. Wear, 141: 73–84.

Ganti S., B. Bhushan (1995). Generalized fractal analysis and its applications to engineering surfaces. Wear, 180: 17–34.

Li Y.X., M.Z. Gao, K. Ma (2014). Developments In Calculation Theory of Fractal Dimension of Rough Surface, Advances in Mathematics, Vol. 41, No. 4, P. 397–408, (In Chinese).

Mou L., M. Yang (2010). Research on Criminal Tool Wear based on Multi-scale Fractal Feature of the Texture Image. ICCASM 2010, pp: 685–688, ISBN: 978-1-4244-7236-9.

Su Y.W., W. Chen, A.B. Zhu, et al. (2014). Contact and Wear Simulation Fractal Surfaces, Journal of Xi An Jiaotong University, Vol. 47, No. 7, P. 52–55, (In Chinese).

Wang B.C., Z Y Wang, C. Jing (2015). Study on Examination Method of Striation Marks Based on Fractal Theory. Applied Mechanics and Materials, Vol. 740: 553–556.

Electromechanical Control Technology and Transportation – Jia & Wu (Eds)
© 2017 Taylor & Francis Group, London, ISBN 978-1-138-06752-3

Research on the end effect of the linear phase-shifting transformer

Xin Xiong & Jinghong Zhao
School of Electrical Engineering, Naval University of Engineering, Wuhan, China

Hongbing Ding
Supervisor Office Equipment Repair of Navy Shanghai Area, Shanghai, China

Pan Sun & Wei Peng
School of Electrical Engineering, Naval University of Engineering, Wuhan, China

ABSTRACT: The linear phase-shifting transformer is a new type of transformer based on the structure of the linear motor. Compared to the traditional phase-shifting transformer, the linear phase-shifting transformer can achieve characteristics such as any angle shift, simple structure, and winding production. Due to the opening at the edge of the transformer, the linear phase-shifting transformer has a similar end effect to a linear motor. In this paper, the end effect of the linear phase-shifting transformer end effect is analyzed in theory, and then the optimization scheme was proposed to reduce or eliminate the end effect of the linear phase-shifting transformer's vice edge induction electric potential and the influence of the output. Finally, simulation is used to verify the effectiveness of the optimization scheme.

1 INTRODUCTION

The phase-shifting transformer is an electrical device that is widely used in the electric energy converter technique. It plays an important role in the multi-pulse rectifier and multi-module inverter.

Traditional phase-shifting transformers have structures of a heart column, the phase-shifting transformers' magnetic circuit also comes with the structure of a heart column, and the main magnetic field is a pulsating magnetic field. The output of each phase winding is usually composed of different numbers of turns in the coil and different core columns in a series, by means of changing the internal coil winding, to realize the phase shift between different windings. Moreover, different connection modes can only achieve a phase angle. When the number of phases increases, the winding design and connection becomes very complex and difficult, and the transformer's volume and weight increases.

In order to simplify the winding connection mode and realize the multi-angle phase-shifting, a linear phase-shifting transformer is proposed according to the working principle of the linear motor. The linear phase-shifting transformer has the characteristics such as simple structure, easy fabrication, modularization, and easy expansion. This paper first introduces the basic principle of the linear phase-shifting transformer, the end effect of the linear motor influence on the motor's performance when it is running, examines whether the linear phase-shifting transformer also has the similar end effect, and the influence of the end effect of the linear phase-shifting transformer on its working modes. It also puts forward the optimization measures to reduce or eliminate the end effect of the transformer output. Finally, the effectiveness of this scheme is verified by simulation.

2 BASIC INTRODUCTION OF THE LINEAR PHASE-SHIFTING TRANSFORMER

The linear phase-shifting transformer is a new type of transformer based on the structure of the linear motor. Its working principle is similar to the linear motor. The linear phase-shifting transformer is supplied with three-phase alternating current. When the three-phase current changes with time, the air gap magnetic field according to the A, B, and C phases moves along a straight line, known as traveling wave magnetic field, as shown in Figure 1. The speed of the traveling wave magnetic field is $V_s = 2f\tau$, where f is the current frequency and τ is the pole distance. The auxiliary side coil generates inductive electromotive force undercutting the magnetic induction line and outputs the three-phase alternating current through the coil winding. Since the primary side is fixed, slip $s = (V_s - V)/V_s = (V_s - 0)/V_s = 1$, the linear

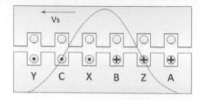

Figure 1. Working principle diagram of the linear phase-shifting transformer.

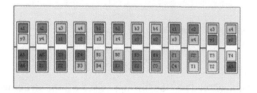

Figure 2. Structure of the linear phase-shifting transformer.

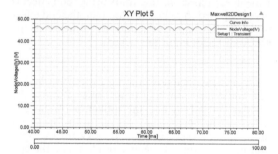

Figure 3. Rectifier output voltage.

phase-shifting transformer can be equivalent to a linear motor if the slip is 1.

The linear phase-shifting transformer can be used for a rectifier or inverter. In this case, the 12-phase linear phase-shifting transformer, for example, the linear phase-shifting transformer structure as shown in Figure 2. When the linear phase-shifting transformer is to be used as a rectifier, the primary with three-phase sinusoidal alternating current, twelve-phase sinusoidal alternating current with 30 degrees apart can be obtained in the auxiliary side, and then the rectifier can obtain 24-pulse direct current, as shown in Figure 3. On the one hand, the DC ripple coefficient of a rectifier is small; on the other hand, the primary alternating current harmonic is small.

When the linear phase-shifting transformer is used as an inverter, the square wave inverter can be directly used. The inverter outputs four groups of three-phase square alternating current access to the primary side, the auxiliary winding generates

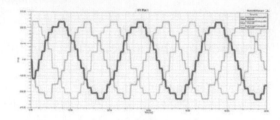

Figure 4. No-load voltage of the inverter output.

inductive three-phase step wave electromotive force, and the no-load voltage is a 24-step wave, as shown in Figure 4. Under the two working modes, the ripple coefficient of the direct current or harmonic of the alternating current can reach less than 5% without adding the filter.

3 END-EFFECT ANALYSIS

This paper studies the end effect of the linear phase-shifting transformer from both ends in the direction of movement, within the calculated energy loss caused by the end effect. It will be the circuit factor into the equivalent circuit and puts forward methods for weakening the end effect according to results of the analytics calculated.

3.1 The cause of the end effect

Due to the longitudinal and lateral two side ends of the linear phase-shifting transformer core, the end effects are caused. The lateral end effect is mainly caused by the unequal width of the primary side and part of the core and conductor overhang. The longitudinal end effect is caused by the opening of both ends of the traveling wave magnetic field. Since the lateral conductor does not cut the magnetic induction line, this paper does not consider the lateral end effect of the linear phase-shifting transformer, and only considers its longitudinal end effect. The end effect refers to the longitudinal end effect in the following section.

As the magnetic field will be enhanced in the corner of the ferromagnetic material, in order to facilitate research, this paper put the flux density space on the auxiliary side of the distribution line as approximate, as shown in Figure 5. This paper neglected the flux ripple component and slot ripple due to asymmetric primary circuit factors, and the increased flux density at the core at both ends as a straight line.

Now setting the air gap flux as shown in Figure 5, in the magnetic flux density changing area at both ends of the secondary core, as shown

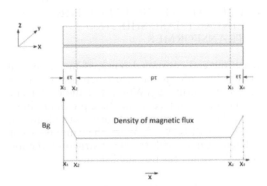

Figure 5.   The cause of the end effect.

in Figure 5, x_1, x_2 and x_3, x_4, the auxiliary side conductor generates the inductive electromotive force. Furthermore, there is extra current which is diffident from the stability current of the auxiliary side load, causing additional consumption.

3.2   End-effect electromotive force and current

According to the Maxwell equation and the constitutive relation, the air gap flux density and current density can be expressed by the formulae (1–2).

$$\mathrm{rot}\, E_e = -\frac{\partial B_e}{\partial t} \tag{1}$$

$$J_e = \sigma E_e \tag{2}$$

In the formulae, E_e is the end effect electromotive force (V/m), σ is the auxiliary side conductor conductivity (S/m), J_e is the end effect current density (A/m^2), and B_e is the air gap flux density of the end effect region (Wb/m^2).

As there is no magnetic body outside the core, the side leakage reactance can be ignored. When calculating the electromotive force and current density at the ends, the following assumptions should be made:

1. The air gap flux density is only the Z-axis component, the induced electromotive force is only the Y-axis component and considers only the fundamental wave. The high harmonics are neglected;
2. The auxiliary conductor is composed of non-magnetic material, the conductivity is isotropic, and there is no skin effect;
3. No end-effect current exists in the auxiliary side conductor other than the end-effect region.

Based on the above assumptions, E_e, J_e, B_e can be expressed as:

$$\begin{cases} E_e = jE_{ey} \\ J_e = jJ_{ey} \\ B_e = KB_{ez} \end{cases} \tag{3}$$

Now, setting B_{ez} and $B_{ez''}$ as the air gap flux density in the end effect regions, according to the relationship curve of rectangular air gap magnetic induction intensity and the size in the last reference, selecting the B_e/B_m value as 4, the middle magnetic field direction and end magnetic field direction exist at an angle of 45 degrees, so $B_{ez}/B_m = 2\sqrt{2}$. B_{ez} and $B_{ez''}$ can be expressed as formulae (4) and (6). $B_{ez'}$ is the air gap flux density for the middle part. With the spatial variation of air gap flux density, the end effect electromotive force E_{ey}, $E_{ey''}$ and the middle part of the electromotive force $E_{ey'}$ can be obtained, such as shown in formulae (8)–(10). Figure 6 shows the magnetic flux density along with the change of location.

$$B_{ez} = \left[\frac{(2\sqrt{2}-1)(\varepsilon\tau - x)}{10} + 1\right] B_m \sin\left(\omega t - \frac{\pi}{\tau}x\right) \tag{4}$$

$$B_{ez'} = B_m \sin\left(\omega t - \frac{\pi}{\tau}x\right) \tag{5}$$

$$B_{ez''} = \left[\frac{(2\sqrt{2}-1)x''}{10} + 1\right] B_m \sin\left(\omega t - \frac{\pi}{\tau}x''\right) \tag{6}$$

In the formulae $x'' = x - (p+\varepsilon)\tau$, the relation between E_{ey} and B_{ez} can be obtained by the formulae (1)–(3):

$$E_{ey} = \int \frac{\partial B_{ez}}{\partial t} dx \tag{7}$$

$$E_{ey} = \omega B_m \left[(a(\varepsilon\tau - x)+1)\frac{\tau}{\pi}\sin\left(\omega t - \frac{\pi}{\tau}x\right) - \frac{a\tau^2}{\pi^2}\cos\left(\omega t - \frac{\pi}{\tau}x\right)\right] \tag{8}$$

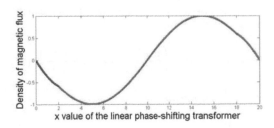

Figure 6.   The magnetic flux density along with the change of location.

$$E_{ey'} = \frac{\omega B_{m\tau}}{\pi} \sin\left(\omega t - \frac{\pi}{\tau}x\right) \quad (9)$$

$$E_{ey''} = \omega B_m \left[(ax''+1)\frac{\tau}{\pi}\sin\left(\omega t - \frac{\pi}{\tau}x''\right) \right.$$
$$\left. - \frac{a\tau^2}{\pi^2}\cos\left(\omega t - \frac{\pi}{\tau}x''\right) \right] \quad (10)$$

In the formulae $a = \frac{2\sqrt{2}-1}{10}$, $\varepsilon \le 1/2$, $x'' = x - (\varepsilon + p)\tau$. The formulae (11)–(13) showed the current density of different regions.

$$J_{ey} = \sigma\omega B_m \left[(a(\varepsilon\tau-x)+1)\frac{\tau}{\pi}\sin\left(\omega t - \frac{\pi}{\tau}x\right) \right.$$
$$\left. - \frac{a\tau^2}{\pi^2}\cos\left(\omega t - \frac{\pi}{\tau}x\right) \right] \quad (11)$$

$$J_{ey'} = \frac{\sigma\omega B_{m\tau}}{\pi} \sin\left(\omega t - \frac{\pi}{\tau}x\right) \quad (12)$$

$$J_{ey''} = \sigma\omega B_m \left[(ax''+1)\frac{\tau}{\pi}\sin\left(\omega t - \frac{\pi}{\tau}x''\right) \right.$$
$$\left. - \frac{a\tau^2}{\pi^2}\cos\left(\omega t - \frac{\pi}{\tau}x''\right) \right] \quad (13)$$

3.3 Consumed power caused by the end effect

The consumed power of different regions can be calculated by formulae (14)–(16).

$$W_e = \frac{\omega}{\pi} l_y h \int_0^{\pi/\omega} \int_0^{\varepsilon\tau} E_{ey} J_{ey} dx dt \quad (14)$$

$$W_e' = \frac{\omega}{\pi} l_y h \int_0^{\pi/\omega} \int_{\varepsilon\tau}^{(p+\varepsilon)\tau} E_{ey'} J_{ey'} dx \quad (15)$$

$$W_e'' = \frac{\omega}{\pi} l_y h \int_0^{\pi/\omega} \int_{(p+\varepsilon)\tau}^{(p+2\varepsilon)\tau} E_{ey''} J_{ey''} dx \quad (16)$$

where l_y is the auxiliary side conductor width (m) and h is the auxiliary side conductor thickness (m).

Substituting E_{ey}, $E_{ey'}$, $E_{ey''}$, J_{ey}, $J_{ey'}$, $J_{ey''}$ into formulae (14)–(16), the consumed power can be calculated as follows:

$$W_e = \frac{1}{2}\sigma l_y h \varepsilon\tau\omega^2 B_m^2 \left(\frac{a^2\varepsilon^2\tau^2}{3} + a\varepsilon\tau + 1 + \frac{a^2\tau^4}{\pi^4} \right) \quad (17)$$

$$W_e' = \frac{1}{2\pi^2}\sigma l_y h \tau^3 p \omega^2 B_m^2 \quad (18)$$

$$W_e'' = W_e \quad (19)$$

$$\frac{W_e}{W_e'} = \frac{\varepsilon^2\pi^2}{p\tau^2}\left(\frac{a^2\varepsilon^2\tau^2}{3} + a\varepsilon\tau + 1 + \frac{a^2\tau^4}{\pi^4} \right) \quad (20)$$

4 EQUIVALENT CIRCUIT OF THE LINEAR PHASE-SHIFTING TRANSFORMER

When the end consumed power is the equivalent circuit, first, the relationship between the magnetic flux density B_m (Wb/m^2) at the center and the induction voltage E_1 in the equivalent circuit is calculated. The number of turns per pole for the stator core set is $N_{1p}/2$. When the windings are in different regions with the same structure, the formulae (21) and (22) are established:

$$e_{1p} = N_{1p}\tau l_y \frac{2}{\pi}\frac{dB_m \sin\omega t}{dt} = N_{1p}\frac{d\phi_{1p}}{dt} \quad (21)$$

$$e_{1e} = N_{1p}\varepsilon\tau l_y \frac{1}{\pi}\frac{dB_m \sin\omega t}{dt} = N_{1p}\frac{d\phi_{1e}}{dt} \quad (22)$$

In the formula, e_{1p} is the primary side center region with each pole distance induction electromotive force instantaneous value (V), e_{1e} is the primary side end region induction electromotive force instantaneous value (V), $\phi_{1p} = (2/\pi)\tau l_y B_m \sin\omega t$ is the center region each pole distance magnetic flux (Wb), and $\phi_{1e} = (1/\pi)\varepsilon\tau l_y B_m \sin\omega t$ is the end-effect region magnetic flux (Wb). Each phase of the original winding has the following relation:

$$e_1 = pe_{1p} + 2e_{1e} \quad (23)$$

$$e_1 = (p+\varepsilon)N_{1p}\tau l_y \frac{2\omega B_m}{\pi}\cos\omega t \quad (24)$$

$$B_m = \frac{\sqrt{2}\pi}{2}\frac{E_1}{(p+\varepsilon)\tau l_y N_{1p}\omega} \quad (25)$$

In the formula, e_1 is the primary side each phase induction electromotive force instantaneous value (V) and E_1 is the primary side of each phase induction electromotive force effective value (V). Substituting formula (25) into formula (17), we obtain:

$$W_e = \frac{\pi^2}{4}\frac{\sigma\varepsilon h E_1^2}{(p+\varepsilon)^2 \tau l_y N_{1p}}\left(\frac{a^2\varepsilon^2\tau^2}{3} + a\varepsilon\tau + 1 + \frac{a^2\tau^4}{\pi^4} \right) \quad (26)$$

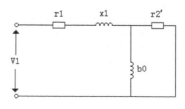

Figure 7. T equivalent circuit.

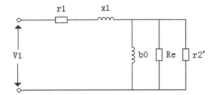

Figure 8. Equivalent circuit when the end effect is included.

Each phase resistance value of the primary edge can be calculated, and as a circuit factor, it can be taken into the equivalent circuit:

$$R_e = \frac{3E_1^2}{W_e + W_e''} = \frac{12(p+\varepsilon)^2 \tau l_y N_{1p}}{\sigma \varepsilon h \pi^2} \frac{1}{1+\beta} = \frac{r_e}{1+\beta} \quad (27)$$

In the formula, $r_e = 12(p+\varepsilon)^2 \tau l_y N_{1p} / \sigma \varepsilon h \pi^2$, and $\beta = \left(\frac{a^2 \varepsilon^2 \tau^2}{3} + a\varepsilon\tau + \frac{a^2 \tau^4}{\pi^4}\right)$. Figure 8 shows equivalent circuit when the end effect is included.

5 OPTIMIZATION SCHEME AND VERIFICATION OF THE END EFFECT

5.1 Optimization scheme of the end effect

Formula (28) can be derived by Formulae (21)–(23):

$$\frac{e_{1e}}{e_1} = \frac{\varepsilon N_{1p} \tau l_y \dfrac{2\omega B_m \cos \omega t}{\pi}}{(p+\varepsilon) N_{1p} \tau l_y \dfrac{2\omega B_m \cos \omega t}{\pi}} = \frac{\varepsilon}{p+\varepsilon} \quad (28)$$

By formula (28), it can be concluded that when the end effect region is much smaller than the middle part, the ratio between end effect induced potential and the total induced potential will be very small. Formula (20) shows that when the values of ε are smaller, the ratio between the end–effect of the consumed power and the middle part of the consumed power is smaller.

According to the above conclusions, the end effect can be reduced by decreasing the end region length value of ε. For the linear phase-shifting transformer, this point can be achieved. The linear phase-shifting transformer is fixed, its air gap can be very small or even zero, and the corresponding end-effect region will be very small.

On the other hand, the conductor of the linear phase-shifting transformer is fixed. If the conductor is in the middle of the region without the end effect, then the force will not be induced by the end effect. Combined with the previous conclusion, we can make the air gap small, while the edge teeth larger, so that the conductor is not at the end-effect

area. Theoretically, the influence of the end effect can be eliminated.

In the beginning of the hypothesis, the primary winding in the air gap flux density distribution is produced as the approximate line, but, in fact, the intermediate region may not be a complete line, as well as the influence of the vortex core, and so on. Therefore, the influence of the end effect cannot be eliminated completely in the optimization scheme, but the influence of the end effect can be greatly weakened.

5.2 Simulation results

The linear phase-shifting transformer was taken and applied to the multi-module inverter as an example, and through the simulation to verify the above conclusions and determine the distribution of magnetic field edge by the magnetic field lines and distribution of the edge tooth part, as shown in Figures 9 and 10. Figure 9 is the magnetic force line distribution with the small air gap and large side teeth. The end magnetic field has some divergence in the air gap and the end effect is very obvious. Figure 10 shows the magnetic force line distribution with a large air gap and small side teeth, the end magnetic field has a significant improvement in the air gap divergence, the end effect is greatly weakened, and conclusion in the fourth part is verified. Figure 11 shows the output voltage of the inverter with a small air gap and large side teeth, and the harmonic component is about 3.23% (the voltage harmonic component in the situation of Figure 9 corresponds to 6.51%). Figures 12 and 13 show the relations between harmonics, efficiency,

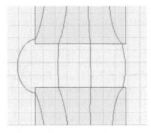

Figure 9. Magnetic force line distribution with the small air gap and large side teeth.

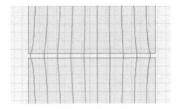

Figure 10. Magnetic force line distribution with the large air gap and small side teeth.

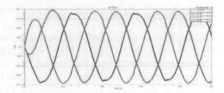

Figure 11. Inverter output voltage.

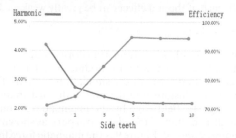

Figure 12. The relationship between side teeth and harmonic and efficiency.

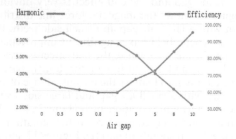

Figure 13. The relationship between air gap and harmonic and efficiency.

and different tooth sizes with air gap sizes in the two cases.

It can be seen from Figure 12, when the edge tooth is below 5 mm, the harmonic decreases with the increase in side teeth and the efficiency is correspondingly reduced, which is mainly caused by the end effect. When the edge tooth is above 5 mm, the harmonic and efficiency are mainly affected by other parameters (e.g. the size of the air gap, the switch performance).

It can be seen from Figure 13, when the air gap of is 1 mm or more, the main factors influencing the harmonic and efficiency is the end effect. When the air gap is above 1 mm, the main factor influencing the harmonic is too close between the cores, causing waveform distortion.

6 CONCLUSIONS

This paper introduced a new kind of phase-shifting transformer based on the linear motor structure. Based on the analysis of the end effect of the linear phase-shifting transformer, the inductive potential, current density, and consumed power of the linear phase-shifting transformer were obtained. Two methods were proposed to weaken the influence of the end effect. Finally, the effectiveness of the two methods was verified by simulation, and the following conclusions were obtained:

1. The end effect of the linear phase-shifting transformer is closely related to the size of the end-effect region. The influence of the end effect is weakened by reducing the size of the air gap to reduce the end-effect region.
2. The end effect of the linear phase-shifting transformer acts on the conductor, increasing the width of the core side tooth, and the conductor is not in the end-effect region, in order to achieve the goal of weakening the influence of the end effect.
3. The analytical results of the end effect of the second and third sections are based on the assumption that the flux density in the air gap is approximated as a broken line. Considering this factor and the influence of the core vortex, the methods of the first two results cannot completely eliminate the influence of the end effect, but the two methods can still greatly weaken the end effect.

REFERENCES

(1978). Linear induction motor (Translation set). *M. Beijing: Science Publishing Company.*

Deng Jiangming, Chen Tefang, Tang Jianxiang (2015). A Compensated Predict Current Strategy for Single-sided Linear Induction Motors Considering Dynamical Eddy-effects and Load Vibrations. *J. Proceedings of the CSEE, 35(15):*3956–3963.

Lin Qiren, Zhao Youmin (1983) Principle of magnetic circuit design. *M. Beijing: China Machine Press.*

Michael I. Levin. (1996) Phase shifting transformer or autotransformer. *P, US. Patent, no.5543771,* 8.

Paice, D. A., et al, (1989) Multi-pulse converter system, *P. US. Patent, no. 4876634,* 10.

Sun Hao, (2014) Research on the Edge Effect of a Linear Motor Based on the Electromagnetic Properties. *D. TianJin: Hebei University of Technology.*

Tang Jianxiang, Jiang Xinhua, Deng Jiangming. (2015) An Improved State Filter for End-effect Cross Control of Single-sided Linear Induction Motors. *J. Proceedings of the CSEE,35(23):*6179–6187.

Tang Yunqiu, Liang Yanping (2010) Motor electromagnetic field analysis and calculation. *M. Beijing: China Machine Press.*

Wang Tiejun, Fang Fang & Jang Xiaoyi (2016). Application of Round-Shaped Transformers in 24 Phase Rectifier Systems *J. Transactions of China Electrotechnical Society, 31(13):* 172–179.

Ye Yunyue (2000). Principle and application of linear motor. *M. Beijing: China Machine Press,*

Zhou D. (2001). Twelve phase transformer configuration, *P. US. Patent, no. 6198647.*

Zhou Ting, Xu Xin, Sun Mingcan (2016). Investigation of Air-gap Field End-effect in Permanent Magnet Axial-flux Motor. *J. Micromotors, 49(3):*14–17.

Electromechanical Control Technology and Transportation – Jia & Wu (Eds)
© 2017 Taylor & Francis Group, London, ISBN 978-1-138-06752-3

A method of building a high-precision landmark library for landmark navigation

Dianlv Zhang & Bo Yang
Beihang University, Beijing, China

ABSTRACT: Satellite navigation based on landmarks information can be used for all kind of satellites which can periodically obtain images of earth surface for its high accuracy and independence. Global landmark library and landmarks are the basis of high precision navigation. A new method for selection of landmark control area and automated establishment of landmark library is proposed. Accordingly, a satellite autonomous navigation scheme based on landmark information is designed. To verify the feasibility of the proposed scheme, a simulation program is designed, and an earth-oriented three-axis stabilized satellite with the altitude of 1100 km is used as an example in this paper. The results show that the position error of the landmark navigation system based on the new landmark library is 100 m, and velocity error is 0.09 m/s.

1 INTRODUCTION

Landmark-based autonomous navigation of aircraft is a kind of navigation method which uses the calibrated landmarks to carry out autonomous positioning. It has the characteristics of high precision and strong autonomy, and its navigation error doesn't accumulate over time. Through this method, all kind of aircrafts and near space vehicle can get enough information during navigation. As a result, landmark navigation can be used to correct inertial navigation errors, or used in geographic calibration of earth observation satellites.

Kau proposed a navigation method using linear landmark in 1975. This is an early interactive method of landmark navigation. This method demands enough manually selected landmarks, which means a lot of repetitive work is involved. At the same time, manual selection of landmarks involves a lot of manual operation lacking of consistency standards. The subjectivity of staff will affect the quality of landmarks selected, a direct result of which is a significant decline in landmark matching accuracy. The error of landmark matching can seriously affect the landmark navigation. Therefore, automatic landmark generation is necessary.

Lots of research have been done about landmark navigation. Emery et al proposed the automatic landmark navigation method based on maximum correlation coefficient (MCC). With its easy implementation, fast computation speed and high accuracy, the MCC algorithm is the most widely used landmark generation method. SSEC designed the automatic landmark navigation system for a geostationary satellite. They believe that the MCC algorithm cannot be applied when the landmark information is not enough. Lu and Zhao applied landmark navigation for geostationary meteorological satellite and polar-orbiting meteorological satellite. But the navigation was still completed through interactive method. Yang et al proposed a method of establishing control point based on ground images to generate landmarks automatically, but only landmarks contain coastline and rivers were selected.

To deal with the limitation in automatic landmark generation, this paper proposes a landmark selection principle and related navigation algorithm. The area of rough texture is used as the bases to generate landmarks and establish the global landmark library. Then automatic landmark matching can be realized through the landmark library, which improves the navigation accuracy effectively.

2 LANDMARK NAVIGATION

Landmark navigation works as the flow chart shown in Figure 1. First, match the landmark with real-time satellite image with MCC algorithm to get several location of calibrated landmarks on real-time image. Then, use these locations to calculate satellite information such as position and speed. Rosborough et al has proven that accurate landmark navigation need at least to match two landmarks.

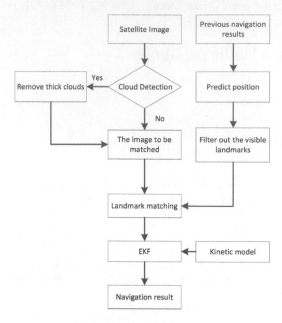

Figure 1. Landmark navigation schematic.

The landmark navigation need accurate attitude measurement or prediction. Errors from attitude measurement will lead to position and velocity errors in navigation. In order to reduce the disturbance from clouds in matching process, strict cloud detection should be executed.

3 FOUNDATION OF GLOBAL LANDMARK

3.1 *Principle of selected landmark*

In landmark navigation scheme, the navigation accuracy will be directly affected by the landmark library. For global landmark library, it is essential to build landmark template in different lighting conditions. And each landmark requires at least eight templates: daytime, night, spring, summer, autumn and winter. These landmarks should have two following characteristics:

1. The landmark image has clear structural features.
2. The landmark image has the same projection pattern as the satellite remote sensing image to be matched.

The principles of landmark selection are followed as:

A. Landmarks are mainly composed of land-water boundary like lakes, rivers, coastlines and islands, as the gray values of water and land are obviously different.

B. To reduce the probability of mismatches, the structural features of landmarks must be unique. Because of the uncertainty of the angle between the satellite image and landmarks, the slope of the land-water boundary cannot be extracted as a feature.
C. High curvature points and inflection points can be used to create of land-water landmarks. These control points should be first calculated by data from the World Bank and the World Coastline Vector Library before be used to generate landmarks.
D. In the non-land-water area, there exists a large amount of irregular area, and high-curvature points which cannot be selected as control points. But the irregular natural coarse textures in these area can provide unique features that can be chosen as the basis of landmarks.

3.2 *Establishment of global ground control area*

The area with enough coarse textures can be defined as the control area and can serve as the basis for landmark. In the process of selecting the control area, there is no necessity to concern the specific characteristics of the ground texture, for the texture reflected by the roughness of the terrain can sufficiently provide the navigation information. Therefore, it is unnecessary to calculate the gray scale co-occurrence matrix intensively to describe the characteristics of image texture.

Let the size of image $f(x, y)$ be $M \times N$, and the size of chosen local window be $m \times n$, the ground roughness can be described with the following gray-scale field characteristic parameters:

1. The standard deviation of grayscale, which describes the degree of grayscale deviating from the mean value. It reflects the fluctuation degree of the grayscale in general:

$$\begin{cases} \sigma = \sqrt{\dfrac{1}{m(n-1)}\displaystyle\sum_{i=1}^{m}\sum_{j=1}^{n}\left[f(i,j)-\bar{f}\right]^2} \\ \bar{f} = \dfrac{1}{mn}\displaystyle\sum_{i=1}^{m}\sum_{j=1}^{n}f(i,j) \end{cases} \qquad (1)$$

2. Grayscale roughness, which describes the degree of grayscale's variation. The roughness positively correlates with grayscale fluctuation. The roughness in the x-axis and y-axis directions is defined as:

$$\begin{cases} r_x = \dfrac{1}{m(n-1)}\displaystyle\sum_{i=1}^{m}\sum_{j=1}^{n-1}\left|f(i,j)-f(i,j+1)\right| \\ r_y = \dfrac{1}{n(m-1)}\displaystyle\sum_{i=1}^{m-1}\sum_{j=1}^{n}\left|f(i,j)-f(i+1,j)\right| \end{cases} \qquad (2)$$

3. The local grayscale correlation coefficient, which describes the relativity of the local grayscale, High autocorrelation will greatly affect the accuracy of the landmark matching. The local gray correlation coefficients in the x- and y-axis directions are:

$$
\begin{cases}
R_x = \dfrac{1}{m(n-1)\sigma^2}\displaystyle\sum_{i=1}^{m}\sum_{j=1}^{n-1}\left[f(i,j)-\overline{f}\right]\left[f(i,j+1)-\overline{f}\right] \\
R_x = \dfrac{1}{n(m-1)\sigma^2}\displaystyle\sum_{i=1}^{m-1}\sum_{j=1}^{n}\left[f(i,j)-\overline{f}\right]\left[f(i+1,j)-\overline{f}\right]
\end{cases}
$$

$$(3)$$

In fact, the roughness and the local correlation coefficient are consistent when they express the local variation of the terrain.

Figure 2 is a remote sensing image of the Tianshan Mountains near the border between China and Kazakhstan. It is a typical remote sensing image with a large number of mixed topographies. Except for land-water boundary which is commonly selected as landmarks, it contains snow mountains, grasslands, forests, deserts and other topographies. Latter analysis is based on this typical image.

When m, n equals 5, the standard deviation and roughness of Figure 2 are shown in Figures 3–5.

The grayscale standard deviation is taken as the measurement of features of local topography with grayscale roughness taken as reference. In other words, once grayscale standard deviation and roughness of a region are greater than certain thresholds, the region can be chosen as control area.

According to Figures 3–5, let the threshold of grayscale standard deviation be 30, and the threshold of grayscale roughness be 6. The control area chosen is shown in Figure 6.

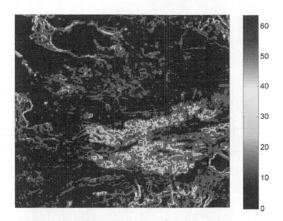

Figure 3. Standard deviation of gray scale.

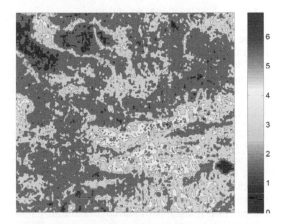

Figure 4. X-axis grayscale roughness.

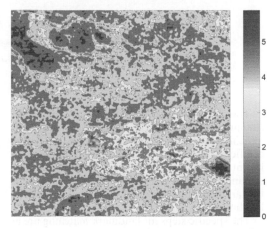

Figure 5. Y-axis grayscale roughness.

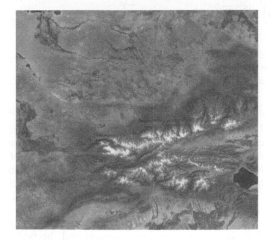

Figure 2. Remote sensing image of mixed landscape.

After the process of calculating control area, the area chosen can be expanded to be landmarks of global landmark library (the size of the landmark

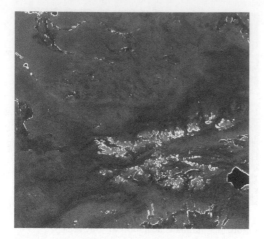

Figure 6. Selected ground control area.

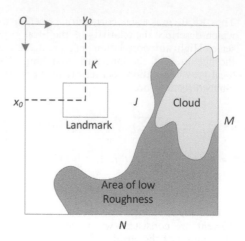

Figure 7. The vector relation in landmark navigation.

image in this paper is 30 × 30). The amount of landmark library using this method increased by 43.97% than the landmark library that contains only land-water landmarks.

3.3 *Landmark matching*

Maximum correlation coefficient is used in landmark matching. Assume the size of satellite image $f(x, y)$ is $M \times N$, and the size of landmark image $w(x, y)$ is $J \times K$, as shown in Figure 7. Calculate the cross correlation coefficient $R(x, y)$ between the landmark image and every possible region in the satellite image. Normalized Maximum Cross Correlation is used to express the correlation of the two, which is expressed as:

remove the image with low standard deviation and roughness. In this step, the threshold of standard deviation can be set as 15, the threshold of roughness can be set as 1.2;

3. Select landmarks from the landmark library that correspond to the lighting conditions (day, night, spring, summer, autumn and winter), and that may appear in the satellite's field of view;

4. Calculate the correlation coefficient between the chosen landmark and each possible location of the remaining image. If the maximum correlation coefficient $R_m(x_m, y_m)$ is greater than 0.72, the landmark image matches the satellite image at location (x_m, y_m).

$$R(x,y) = \frac{\sum_s \sum_t \left[f(s,t) - \overline{f}(s,t) \right] \left[w(x+s, y+t) - \overline{w} \right]}{\sqrt{\sum_s \sum_t \left[f(s,t) - \overline{f}(s,t) \right]^2} \sqrt{\sum_s \sum_t \left[w(x+s, y+t) - \overline{w} \right]^2}} \tag{4}$$

The closer $R(x, y)$ value is to 1, the higher the similarity of the images is, and the higher the matching precision is. The threshold of correlation coefficient $R(x, y)$ is 0.72. If the correlation coefficient $R(x, y)$ reaches the maximum value $R_m(x_m, y_m)$ and the maximum value is greater than 0.72, the landmark image matches the satellite image at location (x_m, y_m).

The specific steps for landmark matching are as follows:

1. Remove the thick clouds after strict cloud detection;
2. Calculate the grayscale standard deviation and grayscale roughness of the remaining image according to Equation (1) and (2). Then,

4 SIMULATION

4.1 *Observation equation for landmark navigation*

For a given time, the real-time photographic image can be obtained by the earth observation equipment. After landmark matching, a number of locations of known landmarks can be given on the real-time image. Let P_i be the center of one of these landmarks located on point (x, y) of the real-time image, showed in Figure 8.

In order to simplify the calculation, assuming that the camera coordinate system and the body-fixed coordinate system coincide. So the look vector should be:

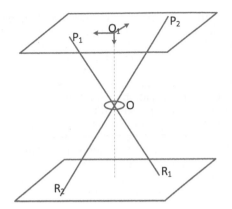

Figure 8. Geometry of camera imaging.

$$u_B = \frac{1}{\sqrt{\left(x_i \cdot d_p\right)^2 + \left(y_i \cdot d_p\right)^2 + f^2}} \begin{bmatrix} x_i \cdot d_p \\ y_i \cdot d_p \\ f \end{bmatrix} \quad (5)$$

where u_B = unit look vector in the body-fixed coordinate system; f = focus of camera; d_p = size of each pixel. Rotate the unit vector into the Earth-fixed system:

$$u_I = TA^T u_B \quad (6)$$

where A = attitude rotation matrix; T = rotation matrix from spacecraft-geodetic system to Earth-fixed system.

In most cases, locations of several landmarks can be found in real-time images. The relationship between the satellite and the landmark is shown in Figure 9.

According to the figure, the relation between satellite and a certain landmark is:

$$R = du + R_V \quad (7)$$

where R_V = displacement vector of satellite; R = displacement vector of landmarks; d = magnitude of the look vector.

The landmark position R can be found in the landmark library; the unit look vector in the Earth-fixed system is known; the magnitude of the look vector can be obtained using the law of cosine:

$$d = R \cdot u + \sqrt{R^2 - R_V^2 + (R \cdot u)^2} \quad (8)$$

which depends on the unknown R_V. But the predicted position of the next step based on the kinetic model can be used as an initial value for R_V. With the predicted position, the magnitude of look vector d can be calculated from (8). So the observation equation is:

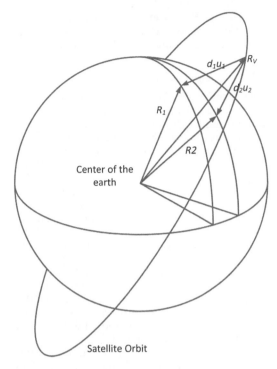

Figure 9. The vector relation in landmark navigation.

$$Z(k) = u + v - \frac{1}{d}(R - R_V) + v \quad (9)$$

4.2 State model

Using gravitational field model with quadratic zonal harmonic term (J_2) perturbation and the other perturbation force being equivalent to Gaussian noise, the state model of satellite is:

$$\begin{cases} \dfrac{dx}{dt} = v_x \\ \dfrac{dy}{dt} = v_y \\ \dfrac{dz}{dt} = v_z \\ \dfrac{dv_x}{dt} = -\mu\dfrac{x}{r^3}\left[1 - J_2\left(\dfrac{R_e}{r}\right)^2\left(7.5\dfrac{z^2}{r^2} - 1.5\right)\right] + \Delta F_x \\ \dfrac{dv_y}{dt} = -\mu\dfrac{y}{r^3}\left[1 - J_2\left(\dfrac{R_e}{r}\right)^2\left(7.5\dfrac{z^2}{r^2} - 1.5\right)\right] + \Delta F_y \\ \dfrac{dv_x}{dt} = -\mu\dfrac{z}{r^3}\left[1 - J_2\left(\dfrac{R_e}{r}\right)^2\left(7.5\dfrac{z^2}{r^2} - 1.5\right)\right] + \Delta F_z \\ r = \sqrt{x^2 + y^2 + z^2} \end{cases}$$

$$(10)$$

where (x, y, z) = position of the satellite; (v_x, v_y, v_z) = velocity of the satellite; μ = the constant of earth gravitation; ΔF = sum of all other perturbation.

4.3 *Simulation result*

The numerical simulation of this paper is based on the observation of three-axis stabilized satellites with an orbital altitude of 1000 km. The parameters of the sample orbit are shown in Table 1.

In the simulation, the satellites are stable in three axes and are oriented to the ground. Referring to FY-3, the resolution of image at satellite bottom point is 1100 m, and the image width is 2900 km. It is also assumed that the satellite can obtain a usable image every 30 seconds, and all images are taken at daytime. The navigation landmark library uses the spring daytime part of the landmark library, and the EKF filtering period is 5 seconds. in each step of filtering, information of 3 matched images is used at most in navigation.

Under the simulation conditions, Figure 10 is the simulation results in five cycles. The average position error is 125.8274 m (3σ) and the average velocity error is 0.1049 m/s (3σ).

If the look vector is calculated from satellite position and landmark position instead of landmark matching, the error introduced by the image's offset at shooting moment can be avoided. Under the same condition, the average position error is 99.1255 m (3σ), the average velocity error is 0.0869 m/s (3σ).

Furtherly, to verify the feasibility of this method, a simulation for satellites at 500–2000 km altitude. The results are shown in Table 2.

On the other hand, the navigation landmarks used in each step of the navigation is restricted to three. There are more than 70% of the images match at least 4 navigation landmarks in the 5 cycles of the simulation for the image width of 2900 km. The number of matched landmarks can also have an effect on navigation results. Since at least two landmarks are required to obtain the specific navigation result, for different amount of landmarks, the navigation result is shown in Table 3.

It can be seen that the amount of landmarks used in navigation cannot improve the accuracy of navigation effectively, but more landmarks can significantly speed up the convergence. The

Table 2. Comparison of simulation results of different orbital heights.

Orbital height m	Position error m	Velocity error m/s
500	62.7134	0.0582
800	87.8328	0.0803
1100	99.1255	0.0869
1400	123.505	0.1094
1700	165.6077	0.1300
2000	192.3637	0.1415

Table 3. Comparison of simulation results of different number of landmarks.

Number of landmarks	Position error m	Velocity error m/s
2	99.9177	0.0882
3	99.1255	0.0869
4	98.8984	0.0852

Table 1. The orbital elements of the sample orbit.

a m	e	i°	ω°	M°
7,478,137.0	0.0	0.0	0.0	0.0

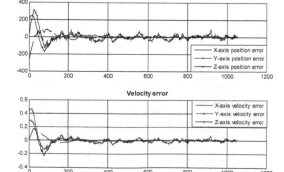

Figure 10. Navigation error.

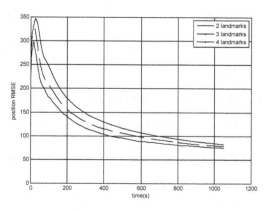

Figure 11. Comparison of RMSE for different amount of landmarks.

RMSE curves of the three simulations are shown in Figure 11. Therefore, when a sufficient amount of navigation landmarks can be obtained for a long time, the amount of landmarks used in the navigation calculation can be appropriately reduced; when the landmark information is not acquired in many steps of filtering, more landmarks should be used to speed up convergence during the navigation.

5 CONCLUSION

In this paper, a new automatic landmark selection method and automatic generation method of landmark library are proposed for the three-axis stabilized satellites using landmark navigation. In order to validate the feasibility of the method, a simulation is carried out by using the established landmark library and the actual satellite data. The results show that the navigation system using this established landmark library has high navigation precision and high reliability.

It is essential to notice that: (1) Landmark selection schemes are extended to all kind of landscape, which are greatly affected by seasons. Therefore, it is necessary to prepare landmarks for different seasons and update them frequently. (2) In all simulations, it is assumed that the camera coordinate system coincides with the satellite body coordinate system. When the earth observation satellite needs to track certain target, the different projections of landmarks are needed in the establishment of the landmark library.

REFERENCES

Bachmann, M. and Bendix, J., 1992. An improved algorithm for NOAA-AVHRR image referencing. International Journal of Remote Sensing, 13(16), pp.3205–3215.

Dai, X. and Khorram, S., 1998. The effects of image misregistration on the accuracy of remotely sensed change detection. IEEE Transactions on Geoscience and Remote sensing, 36(5), pp.1566–1577.

Emery, W.J., Baldwin, D. and Matthews, D., 2003. Maximum cross correlation automatic satellite image navigation and attitude corrections for open-ocean image navigation. IEEE transactions on geoscience and remote sensing, 41(1), pp.33–42.

Guo, Q., Yang, L., Zhao, X., Feng, X., Lin, W., Zhang, Q. and Wei, C., 2013. Research and optimization of landmark matching algorithm for meteorological satellite image navigation. Computer Engineering and Applications, 49(24).

Ho, D. and Asem, A., 1986. NOAA AVHRR image referencing. International Journal of Remote Sensing, 7(7), pp.895–904.

Illera, P., Delgado, J.A. and Calle, A., 1996. A navigation algorithm for satellite images. International Journal of Remote Sensing, 17(3), pp.577–588.

Kau, S.P., 1975, August. Autonomous satellite orbital navigation using known and unknown earth landmarks. In AIAA, Guidance and Control Conference (Vol. 1).

Kim, T., Lee, T.Y. and Choi, H.J., 2005, January. Landmark extraction, matching, and processing for automated image navigation of geostationary weather satellites. In Fourth International Asia-Pacific Environmental Remote Sensing Symposium 2004: Remote Sensing of the Atmosphere, Ocean, Environment, and Space (pp. 30–37). International Society for Optics and Photonics.

Le Moigne, J., Campbell, W.J. and Cromp, R.F., 2002. An automated parallel image registration technique based on the correlation of wavelet features. IEEE Transactions on Geoscience and Remote Sensing, 40(8), pp.1849–1864.

Lu, Y., 1992. Landmark navigation method of geostationary meteorological satellite. Chinese Journal of Computational Physics, 9(4), pp.775–777.

Rosborough, G.W., Baldwin, D.G. and Emery, W.J., 1994. Precise AVHRR image navigation. IEEE Transactions on Geoscience and Remote Sensing, 32(3), pp.644–657.

Townshend, J.R., Justice, C.O., Gurney, C. and McManus, J., 1992. The impact of misregistration on change detection. IEEE Transactions on Geoscience and Remote Sensing, 30(5), pp.1054–1060.

Yang, L. and Yang, Z., 2009. The Automated Landmark Navigation of the Polar Meteorological Satellite. Journal of Applied Meteorological Science, 20(3), pp.329–336.

Yang, L., Feng, X., Guo, Q., Lu, F. and Zhang, X, 2011. Automatic Geometric Precision Correction of Fengyun-2 Meteorological Satellite Imagery. Computer Engineering and Applications, 47(3), pp.202–206.

Electromechanical Control Technology and Transportation – Jia & Wu (Eds)
© 2017 Taylor & Francis Group, London, ISBN 978-1-138-06752-3

Design of a wireless power supply electromagnetic coupler for the underwater unmanned vehicle

Pan Sun, Jinghong Zhao, Xusheng Wu, Wei Gao & Peng Jiang
School of Electrical Engineering, Naval University of Engineering, China

ABSTRACT: The Underwater Unmanned Vehicle (UUV) is an important piece of equipment for ocean exploration. Limited by its own carrying power, the UUV has a weak endurance. For the UUV, the use of wireless power supply technology can effectively solve this problem. In this paper, we first use the ANSOFT simulation software to establish the mutual inductance model and the excitation model of a wireless power transmission system and then we use the finite element method to analyze coupler parameters. From the simulation result, it can be found that the optimized electromagnetic coupler has good coupling characteristics. Therefore, it can take good advantage of the wireless power transmission system and provide sufficient power for underwater electromechanical equipment efficiently.

1 INTRODUCTION

With its own various sensors and weapons, the UUV can complete a series of important military missions such as remote communication relay, anti-submarine alert, underwater reconnaissance and surveillance, and mine submarine. The traditional UUV charging methods are of two main kinds: one is charging by returning to the shore base or desk; the other is carrying out underwater wet plug charging by the cable system. There are some problems in both these charging methods: first, the degrees of automation are low for both methods. The two methods need a manual operation. Besides, they are time-consuming and labor-intensive. Secondly, wet plugging method requires a large actuating force, which results in serious wear and reduces charging times sharply. Thirdly, the hiding performance of the two methods is both bad, which are easy to expose themselves.

The UUV underwater wireless charging technology is a new power transfer method that wirelessly charges the UUV through the seawater medium. This method can overcome a series of problems of the traditional contact charging method, which can fully exploit the technical and tactical performance of the UUV. In this paper, aiming at the problems faced by the traditional UUV charging method, we put forward a scheme that designs a new type of electromagnetic coupler for a wireless charge of the UUV underwater using wireless charging technology.

2 NEW ELECTROMAGNETIC COUPLER STRUCTURE

Aiming at the requirement of streamlined shape and low positioning accuracy of the underwater unmanned vehicle, we design a coupler with the rectangular magnetic core structure as shown in Figure 1. The black part of the circle is a secondary magnetic core, and the outer black part is the primary magnetic core. The silver white part is the pressure hull structure. The yellow part is the winding. The green part is the aluminum plate shield to prevent leakage.

The magnetic core material chosen is Mn-Zn ferrite, which has good electromagnetic shielding

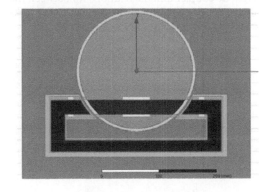

Figure 1. Loosely coupled transformer structure.

properties to reduce the electromagnetic interference outside. Besides, the two coils can obtain a higher coupling coefficient. As a power transmission interface, the electromagnetic coupler can transfer electrical energy without bare conductor contact on both sides with precise positioning. This method avoids the danger of electric shock, sparks, and leakage risk from traditional contact interface. So it is suitable for application in the special environment. Moreover, it reduces the mechanical wear of the interface and extends the service life.

3 MODELING OF THE NEW ELECTROMAGNETIC COUPLER

3.1 *Reluctance model*

Using the reluctance modeling method is an effective method to analyze the electromagnetic characteristics of the electromagnetic coupler. Reluctance is a parameter which describes the magnetic flux transfer capability. The size of reluctance is related to the size of the magnetic core structure and permeability of the material. The magnetic field in the primary coil is distributed in the structure of the electromagnetic coupler and the space near the electromagnetic coupler. Due to the separation structure of the coupler, the magnetic flux near the core gap forms a divergent distribution as shown in Figure 2. On the one hand, this distribution form leads to the coupler magnetic circuit and a part of the magnetic flux does not pass through the secondary coil and becomes the leakage magnetic flux. On the other hand, this kind of distribution makes the reluctance in the main circuit of the two coils increase, which reflects the magnetic coupling relationship between the two coils of the electromagnetic coupler and the relationship between the main magnetic circuit and the magnetic flux leakage circuit.

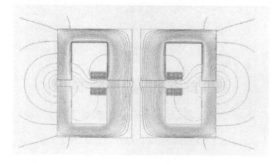

Figure 2. Magnetic field distribution of loosely coupled transformer.

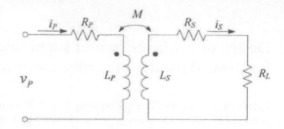

Figure 3. Mutual inductance coupling circuit of the loosely coupled transformer.

3.2 *Mutual inductor circuit model*

Figure 3 shows the mutual inductor circuit model of the electromagnetic coupler, which is respectively the self-inductance of the primary coil and the secondary coil, the mutual inductance between the two coils and load resistance. This is different from the excitation circuit model in that there is no difference between the excitation inductance and leakage inductance. The induction voltage is expressed by the mutual inductance coefficient. The coupling degree between the two coils is expressed by the coupling coefficient. There is only primary current and secondary current in the model current, which is more close to the physical structure of the coupler. So, the mutual inductance model has more advantages than the excitation model in the analysis of the electrical characteristics of the system.

4 FINITE ELEMENT SIMULATION OF THE NEW ELECTROMAGNETIC COUPLER

In this paper, at 100 kHz, the primary and secondary coils are both 10 turns. The input valid value is simulated under the voltage source excitation of 100 V. At first, the secondary coil no-load condition is simulated under the condition that the input valid value of the voltage source excitation is 100 V. From Figure 4, we can understand that the maximum magnetic field strength of the loosely coupled transformer under this condition is 0.156 T. The induction intensity is very ideal, which verifies the coupling degree of the loosely coupled transformer.

We can get the self-inductance coefficient of the two coils by ANSOFT steady state field simulation. From the Figure 5, we can know that with the increase in air gap, the coupling coefficient decreases. In this paper, when the air gap is 8 mm in the system, the coupling coefficient was 0.593. For the loosely coupled transformer, the above parameter is ideal, which verifies that the loosely

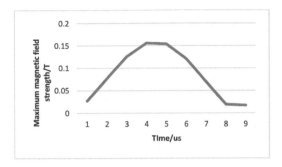

Figure 4. The relationship between the maximum magnetic field strength and time.

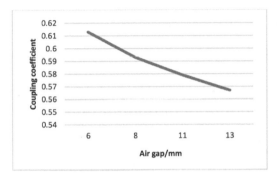

Figure 5. The relationship between the coupling coefficient and the air gap variation.

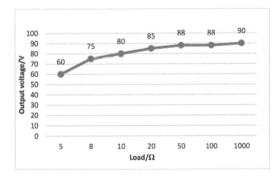

Figure 6. The relationship between output voltage and load variation.

coupled transformer designed in this paper has a strong coupling property.

In this section, we analyze the relationship between parameters of the secondary coil with the load change. In this way, we get the voltage, current, and power values of the primary and secondary coils under different loads to obtain the transmission efficiency. We can get Figure 6 by analyzing voltage with the load change. From the

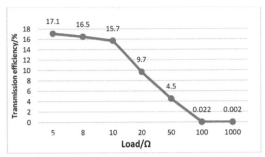

Figure 7. The relationship between transmission efficiency and load variation.

figure, we can know that with the increase in the load, the voltage increases and tends to be gentle. We can get Figure 7 by analyzing the power with load change. From the figure, we can know that with an increase of the load, the output power of the secondary coil declines sharply, and the output efficiency becomes lower and lower. The change of the transmission power is similar to the change of the power.

5 SUMMARY

In this paper, based on the traditional power supply method of analyzing the UUV, we put forward a new underwater electromagnetic coupler and establish a transformer model and a field model of the transmission system. Besides, by using the ANSOFT simulation software, we carry on finite element analysis to the coupling parameter and the result of the simulation illustrates the effectiveness of the design.

REFERENCES

Teng Han, Fang Zhuo, Tao Liu, (2004) Research on Contactless Power Transfer System Using Isolation Transformer. *J. Power Electronic Technology. 38(5).*
Wei Zhang, Qianghong Chen. (2010) The magnetic circuit model of a new contactless transformer and its optimization. *J. Chinese Journal of Electrical Engineering Science. 30(27).*
Xueliang Huang, Linlin Tan, Zhong Chen. (2013) Research and application of wireless transmission technology. *J. Transactions of China Electrotechnical Society. 28(10).*
Yang Li, Qingxin Yang, Zhuo Yan. (2013) Analysis of wireless effective power transmission distance and its influencing factors. *J. Transactions of China Electrotechnical Society, 28(1).*
Zheng Shen, Huiming Chen, Dapeng Jiang. (2007). Modeling analysis and design of contactless transformer. *J. Electric Power System, 26(1).*

Electromechanical Control Technology and Transportation – Jia & Wu (Eds)
© 2017 Taylor & Francis Group, London, ISBN 978-1-138-06752-3

A nonlinear time–frequency analysis method for the low probability of interception

Hao Xu & Fangling Zeng
Electronic Engineering University, Hefei, China

ABSTRACT: The time–frequency analysis of a signal is an important method in signal processing and analysis, and plays an important role in the analysis of low-probability signals. Aiming at the limitation of the Fourier transform in analyzing the time–frequency characteristics of non-stationary signals, a nonlinear time–frequency signal analysis method is introduced and optimized for this method.

1 INTRODUCTION

With the development of electronic technology and information technology, the demand for information technology in various fields is also expanding, and the military field is becoming more and more informational. Information warfare is playing an increasingly important role in the modern war. The radar system is an important access to information channels as well as the threat of war is also increasing. In this case, the low probability of intercept radar came into being. This radar has a strong viability and can effectively avoid the signals emitted by the electronic detection system interception. For the use of an electronic detection system in modern war, with a variety of signals intertwined, a complex electromagnetic environment which is changeable, the need for an electronic detection system with rapid analysis and real-time processing of various signals is required.

Time–frequency analysis method is a good method to analyze radar signals with a low probability of intercept, which can be divided into linear and non-linear transformation. For the low probability of intercept radar signal recognition, we can use time–frequency curve characteristic analysis and through different characteristics of time-frequency curves of different low-probability intercept signals, we can identify several low-probability intercept signals according to different modulation modes. This method can identify the classical low-intercept radar signal accurately under the condition of low SNR.

In this paper, several low-probability intercept signals are introduced and the principle of WVD transform is analyzed. By using this method, the time–frequency analysis of these low-probability signals is carried out, and an improved method for WVD transform is put forward.

2 LOW INTERCEPTION TECHNIQUE

2.1 Low-probability intercept factor

Low-probability intercept factor is a measure of radar low intercept performance parameters (Li, 2002), generally expressed as α, the expression is:

$$\alpha = \frac{R_1}{R_t} \tag{1}$$

where R_1 denotes the maximum detection distance of the intercepted receiver and R_t denotes the maximum distance of radar detection.

From the above formula, if you want to ensure the detection of radar information, you must ensure that the maximum detection distance of the intercepted receiver is greater than the radar maximum detection distance, that is $R_1 > R_t$, then $\alpha > 1$. On the contrary, if the maximum detection range of the radar is larger than the maximum detection distance of the receiver, that is $R_t > R_1$, then $\alpha < 1$, the receiver is difficult to intercept the radar signal and this radar is a low intercept radar.

The lower the probability of intercept is, the stronger it will be and then the smaller the corresponding intercept probability factor will be. Obviously, this low probability of intercept is not absolute as different radars corresponding to different receivers will have different intercept probabilities.

2.2 Low intercepted radar signal characteristics

1. A traditional radar using pulse emission system: this approach is very easy to be intercepted by the peak detection method. Therefore, from the perspective of low-intercept, the signal using the continuous wave transmission system, which will

make the radar signal reach peak power becomes very low and not easily intercepted (Chan, 2007).

2. The use of spread spectrum signal can improve the radar gain, so that not only can it detect the target well, but also will not be easily intercepted by the receiver.

3. On the signal waveform, multi-use large bandwidth, increasing the time width and bandwidth of the signal waveform can reduce the acquisition factor and reduce the probability of acquisition.

3 SEVERAL LOW-PROBABILITY INTERCEPTED RADAR SIGNALS

3.1 Linear frequency modulation signal

Linear Frequency Modulation (LFM) signal is a kind of relatively early pulse compression signal, which is a kind of signal waveform which is often adopted, which is obtained by nonlinear phase modulation and finally obtains a large time-width product. Its advantage is that the matched filter is not sensitive to the Doppler shift of the echo signal and can play a very good pulse pressure.

LFM signal expression is (Lei, 2007):

$$s(t) = a\exp\{j[2\pi(f_0 t + kt^2/2) + \phi]\} \qquad (2)$$

Its autocorrelation function is:

$$R(t,\tau) = a^2 \exp\left[j\left(2\pi f_0\tau + 2\pi k\,\tau t - \pi k\,\tau^2\right)\right] \qquad (3)$$

In the above two formulas, the amplitude of the signal is a. The frequency modulation slope of the signal is k. The initial phase is ϕ. The starting frequency of the signal is f_0 and $f_0 + kt$ is the instantaneous frequency of the signal. It is a good low-intercept radar signal. Its internal structure is more complex. It is not easy to be intercepted and detected by the receiver. It has good anti-jamming ability.

3.2 Phase coded signals

A Phase-coded Signal (PSK) is a large-time wide-bandwidth radar signal that modulates the codeword information in the carrier phase, which reduces the energy per unit of the band making it less susceptible to detection by radar receivers. As for the distance resolution, PSK is also the pulse pressure signal, but also through the time domain nonlinear phase modulation it achieves the purpose of expanding the equivalent bandwidth.

The expression of the phase-coded signal is (Ni, 2007):

$$S(t) = u(t)e^{j2\pi f_0 t} \qquad (4)$$

In addition, $\varphi(t)$ is the phase modulation function:

$$\varphi(t) = \sum_{i=1}^{N} \alpha_i u_0(t - iT_0), \alpha_i \in \left\{\frac{2\pi}{M}1\right\}, \qquad (5)$$

N is the number of symbols. T_0 is the symbol width. M is the binary number. $u_0(t)$ is the unit rectangular pulse.

Phase encoding radar code according to its phase modulation code is different and can be divided into two-phase encoding and multi-phase encoding. At present, a lot of good pseudorandom codes have been obtained, such as Gold code, Barker code, M-sequence, a complementary sequence, and orthogonal sequence, which can extend the radar signal spectrum and can obtain largely matched filter signal processing gain of the anti-detection, interference ability.

3.3 Frequency coding signal

Frequency-encoding (FSK) signal is also a pulse compression signal. The signal uses a pseudo-random sequence where the energy is more dispersed and jump between different frequencies, and has a good low probability of interception performance (Jing, 2001).

The FSK expression is:

$$s(t) = Ae^{j\left[2\pi f_k u_0(t - kT_0)t\right]}e^{j\theta} \qquad (6)$$

The FSK signal does not reduce the peak transmit power and power density, so it is easy to detect, but the advantage is that the pseudo-random sequence is confidential, which makes it difficult for enemy receivers to receive the FSK signal analysis and recognize it. It is easy to be detected, but not easily intercepted, and the same is a good low probability of the intercept signal.

4 WIGNER-VILLE DISTRIBUTION

4.1 Wigner-Ville distribution of the basic principles

The Wigner-Ville distribution was first proposed by Wigner, and Ville introduced it into the signal analysis. It is a quadratic time–frequency characteristic of the signal that not only can describe the frequency of the signal with time, but also describes the signal power at each time–frequency point size, is more intuitive than the linear time–frequency and is a good time–frequency analysis method, which has good time–frequency aggregation (Feng, 2005).

The Wigner-Ville distribution of continuous signals is:

$$W_x(t,w) = \int_{-\infty}^{\infty} x\left(t + \frac{\tau}{2}\right)x^*\left(t - \frac{\tau}{2}\right)e^{-jw\tau}d\tau \qquad (7)$$

The Wigner-Ville distribution of the discrete signal is:

$$W_x(n,w) = 2\sum_{-m}^{m} x(n+k)x^*(n-k)e^{-j2wk} \qquad (8)$$

Discrete signal analysis with the core function $f_l(n)$ of the distribution function is:

$$f_l(n) = x(l+n)x^*(l-n) \qquad (9)$$

Then the Wigner-Ville distribution expression can be rewritten as:

$$W(l,w) = 2\sum_{n=-N}^{N-1} f_l(n)e^{-jwn} \qquad (10)$$

And if:

$$w = \frac{\pi k}{2N}, k = 0,1,2,....,2N-1 \qquad (11)$$

Then:

$$w\left(l,\frac{\pi k}{2N}\right) = 2\sum_{n=-N}^{N-1} f_l(n)e^{-i2\frac{\pi k}{2N}n} \qquad (12)$$

Adjust the value of n so that the expression is rewritten as:

$$w\left(l,\frac{\pi k}{2N}\right) = 2\sum_{n=0}^{2N-1} f_l'(n)e^{-i2\frac{\pi k}{2N}n} \qquad (13)$$

$f_l(n)$ change to:

$$f_l'(n) = \begin{cases} f_l(n), 0 \le n \le N-1 \\ 0, n = N \\ f_l(n-2N), N+1 \le n \le 2N-1 \end{cases} \qquad (14)$$

Finally, we obtain the WVD distribution expression:

$$w(l,k) = 2\sum_{n=0}^{2N-1} f_l'(n)e^{-i2\frac{\pi k}{2N}n} \qquad (15)$$

For example, for amplitude normalized single carrier frequency signals:

$$x(t) = e^{j\left(w_0 t + \frac{1}{2}mt^2\right)} \qquad (16)$$

Then, the kernel function can be calculated as:

$$f_l(t) = x\left(t+\frac{\tau}{2}\right)x^*\left(t-\frac{\tau}{2}\right) = e^{j(w_0+mt)\tau} \qquad (17)$$

We can get the WVD expression as:

$$W(t,w) = \int_{-\infty}^{\infty} f_l(t)e^{-jw\tau}d\tau = \delta\left[w - (w_0 + mt)\right] \qquad (18)$$

The results show that the WVD distribution of the single-carrier low-probability signal is the line spectrum of the continuous impact function along the frequency with the ideal time–frequency aggregation characteristics, but in reality, the signal is not wireless for long, so the distribution map will be different with the ideal, showing a similar dorsal fin.

4.2 Wigner-Ville time–frequency distribution analysis and simulation

For the simulation of the WVD distribution of the LFM signal, the simulation parameters are set as follows: The starting frequency is 20 MHz. The frequency slope is 60 MHz/μs. The sampling frequency is 100 MHz. The simulation results are shown in Figure 1.

From the simulation results, we can get LFM time–frequency information, which shows that WVD distribution for a single carrier frequency LFM signal time–frequency analysis has a good effect.

For 2FSK signal simulation, the simulation parameters are: signal symbol 1111100110101 and the symbol width is set to 1. The simulation results are shown in Figure 2.

As can be seen from the figure, the cross-terms of the signal time–frequency analysis caused a greater interference with some cross-terms more than the signal amplitude making the signal amplitude itself unable to be identified.

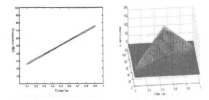

Figure 1. LFM signal WVD distribution and contour.

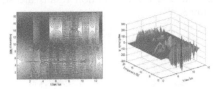

Figure 2. 2FSK signal WVD distribution and contour.

By simulation of the LFM and FSK signals, we can see that the Wigner-Ville distribution has good time–frequency convergence and is a good method for time–frequency analysis, but at the same time because of the existence of cross-terms, it will produce false signals in its distribution, which will have a great influence on time–frequency analysis. In response to this shortcoming, we made the following improvements.

4.3 Improved Wigner-Ville distribution

The generation of the cross-term is the inherent result of the quadratic time–frequency distribution and has a great relationship with the limited support characteristics of the quadratic time–frequency distribution. Therefore, cross-term suppression technology is designed as a kernel function (Zheng, 2013). The most commonly used improvement methods are of three types:

Pseudo-WVD Distribution (PWD), and the expression is:

$$PWD(t,f) = \int_{-\infty}^{\infty} z\left(t + \frac{\tau}{2}\right) z^*\left(t - \frac{\tau}{2}\right) h(\tau) e^{-j2\pi f} d\tau$$
$$= W(t,f) * H(f) \qquad (19)$$

$h(\tau)$ is the window function.

1. Smooth WVD Distribution (SWD), and the expression is as follows:

$$SWD(t,f) = W(t,f) * *G(t,f) \qquad (20)$$

** is time and frequency two-dimensional convolution. $G(t,f)$ is a smooth window function.

2. Smoothed Pseudo WVD Distribution (SPWD), and the expression is:

$$SPWD(t,f)$$
$$= \int_{-\infty}^{\infty} z\left(t - u + \frac{\tau}{2}\right) z^*\left(t - u - \frac{\tau}{2}\right) g(u) h(\tau) e^{-j2\pi f} d\tau$$
$$\qquad (21)$$

and $g(u)h(\tau)$ is a dual real window function.

Next, the PWVD distribution and the SPWVD distribution are respectively performed on the 2FSK signal. The PWVD distribution results are shown in Figure 3.

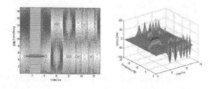

Figure 3.   2FSK contour pseudo WVD distribution.

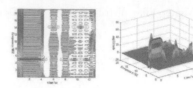

Figure 4.   2FSK signal contour smoothing pseudo WVD distribution.

It can be seen from the above figure that the cross-terms are much more suppressed than those shown in Figure 2. The time–frequency characteristics of the signals are also clear, which indicates that the PWVD distribution improves the time–frequency analysis ability of 2FSK signals.

The simulation results of the SPWVD are shown in Figure 4.

It can be found that the SPWVD distribution has a significant effect on the suppression of cross terms, and almost no effect of cross terms can be seen, and the time–frequency analysis ability of the signal is greatly improved.

5   CONCLUSION

With the development of low-probability interception radar technology, the time–frequency analysis of the Wigner-Ville distribution plays a more and more important role in the signal analysis field. However, with the development of low-probability intercept radar technology, the Wigner-Ville distribution is becoming more and more important in the signal analysis field. The method has a good time–frequency convergence and is a good method for the time-frequency analysis. However, the improvement of WVD distribution, including pseudo-WVD, smooth WVD, and pseudo-smoothing WVD, will be more and more rapid because of the cross-terms and some disturbances in the time–frequency analysis.

REFERENCES

Chan Yunhua LPI radar and its key technology [J] Modern Electronics, 2007, 6 (4): 27–30.
Feng Xiang, Li Jiandong modulation recognition algorithm and performance analysis [J] Radio Science, 2005, 20 (6:): 737–740.
Jing Sheng, Liu Yu, Xi Yi Min linear FM signal parameter estimation algorithm [J] Nanjing University of Aeronautics and Astronautics, 2001, 33 (5): 441–444.
Lei Xuemei LPI signal recognition and parameter estimation [D] Chengdu: University of Electronic Science and Technology, 2007: 22–26.
Li Yingxiang, Zhouxian Min LPI radar signal having characteristics - the new FSK / PSK signal [J] Signal Processing, 2002, 18 (2): 12–15.
Ni Ganfeng LPI radar technology research [D]. Nanjing: Nanjing University of Technology, 2007: 33–38.
Zheng Kai Recognition analysis and parameter estimation of radar signals. LPI [D] Chengdu: University of Electronic Science and Technology, 2013: 35–38.

Electromechanical Control Technology and Transportation – Jia & Wu (Eds)
© 2017 Taylor & Francis Group, London, ISBN 978-1-138-06752-3

Research on the fault diagnosis of the high-voltage circuit breaker based on the feature extraction of the stroke signal

Ke Zhao, Hongtao Li, Jinggang Yang & Yongyong Jia
Jiangsu Electric Power Company Research Institute of State Grid, Nanjing, China

Dege Li, Jianwen Wu, Suliang Ma & Ying Feng
School of Automation Science and Electrical Engineering, Beihang University, Beijing, China

Chuantao Liang & Xinglu Feng
Shandong TaiKai High Voltage Switchgear Co. Ltd., Tai'an, China

ABSTRACT: High-voltage Circuit Breaker (CB) is one of the most important components in a high-voltage circuit. In the process of operation, spring faults such as spring deformation and spring elastic weakening may occur and affect the performance of the CB. The opening/closing springs of the CB undergo different degrees of fatigue during a long running time. When the power system fails, there will be major accidents if the line cannot be cut off in time. In order to detect the types of fault degree of high-voltage CB, we quantified the degree of CB spring failure and determined whether the CB can continue to work. We analyzed the trip signal of the CB in the mechanical vibration signal and proposed a method to process the stroke signal. We processed the stroke signal to obtain the velocity and the variance of volatility distribution to form a feature vector P = [v, σ]. Using the serial Support Vector Machine (SVM) to classify the degree of fault of the spring and define the fitting function with the variance of the spring as the independent variable and the variance of the distribution to express the spring fault diagnosis concretely, it was found that this method can diagnose the spring failure of the high-voltage CB effectively. In addition, experimental results indicate that the fitting function based on wave variance can judge the state of the spring fully, which verifies the practical significance of the prevention of the CB failure.

1 INTRODUCTION

According to the two statistics of an international conference on power grid, mechanical failure is the main fault of high-voltage Circuit Breakers (CB), therefore, it is very important to study the mechanical state of the CB detection. Moreover, research on the mechanical characteristics of high-voltage CB can reflect more than 80% of the operating states (Xu, 2005). Reliable and effective mechanical state online monitoring and analysis technology can help to detect the hidden faults and improve the reliability of the equipment and system operation timely and reliably.

CB is one of the most important electrical equipment of transmission and distribution power system. With the improvement of the transmission and distribution network automation and reliability requirements, the reliability of the CB itself is also increasing (Wu, 2007; Sun, 2011; Wu, 2011). CB in the process of action contains a lot of important state information, the fault information included. Then the fault signals are used to extract the feature vector, which is the key to the analysis of the fault signal.

In recent years, researchers from all over the world research on the CB for different operator agencies, different voltage levels, and different mechanical states. Several methods for dealing with vibration signals and diagnosing mechanical states are given, such as exponential decay oscillation model method (Stokes, 1988), envelope analysis method (Hess, 1992), wavelet packet analysis (Lee, 2003), empirical mode decomposition method (Zhang, 2008; Runde, 1996), dynamic time regularization (Dong, 2011), fractal method (Wu, 2005), and so on. These methods can be used to diagnose one or several high-voltage CB failures. The study of high-voltage CB vibration characteristics has an important significance. Demjanenko (1988) proposed a method based on mathematical statistics. By analyzing the vibration signals measured by the vibration sensor and using the standard deviation as the characteristic parameter, the vibration signal of the mechanical state is quantized. Belief space can be determined by the mean and standard deviation of each state through the comparison for diagnosis of vibration signals in the vibration state of belief space confidence to get the results of the classification of the vibrational state. Bosma (2003) used the state space method to extract four features from the vibration signal, such as standard deviation, mean value, and zero-crossing number. The feature quantity of this method forms a thinking space. The state in space is piled into a "cluster", in order to achieve feature extraction.

Regarding the fault diagnosis of high-voltage CB in recent years, the main research hotspot and emphasis is on vibration sensors to test the vibration signal if it is decomposed, extract the feature vectors to achieve fault signal diagnosis, and make some research results. However, there are only a few studies on the degree of spring failure, the failure of opening/closing the spring, and the method of using the stroke signal to extract the characteristics of vibration signals for fault classification which has not yet been reported. At present, in the switching on process, high-voltage CB spring operating mechanism is used in opening/closing spring energy storage and release to control the CB opening/closing actions. The reliability and stability of the opening/closing action directly affect the safe operation of power systems. So the degree of health of the spring will directly affect the normal operation of the opening/closing spring, which is critical to the running state of the spring. The reliability of the spring operation is an important factor in the reliable operation of the CB. As the nature of the CB spring failure is gradual, and is gradually developed in the operation again and again, the behavior feature of the spring was inevitably changed during the switching process. If the CB spring is operating normally, the change should be within a limited range and should fluctuate up and down at a specific value. If at some point, the change goes well beyond a certain value or if there is a regular migration in one direction, there may be a problem.

For LW30-252 model SF6 high-voltage CB, the typical mechanical faults in the CB include oil buffer leakage fault, spindle jam fault, spring failure, and so on, but this article only considers the opening spring failure based on the study of the characteristics of vibration signal contained in the normal or fault spring of the stroke signal, a new method for fault diagnosis extent of high-voltage CB spring is proposed. At the same time, in order to more effectively guide the production, the paper further determines the extent of the spring failure. The method of curve fitting is used to quantify the degree of spring failure of the high-voltage CB to judge whether the spring can be used or must be replaced immediately, which is of practical significance to prevent the CB failure.

2 DETECTION OF TRAVEL-TIME SIGNALS DURING HIGH-VOLTAGE CB CLOSING AND OPENING SWITCHING

2.1 *Mechanical structure of high-voltage CB*

Figure 1(a) is the structure diagram of a typical column CB of 252 kV, which is equipped with a spring-operated mechanism. Among them, part 1 represents the closing spring and part 8 represents the opening spring. The spring operating mechanism is to ensure that the CB achieves a reliable separation of the main energy storage process, the health of which will directly determine the CB opening and closing process. It is critical for the opening and closing of the response process. The reduction of mechanical reliability of the spring is one of the main causes of CB failure. Therefore, the fatigue working state of the opening spring and the closing spring are the emphasis for research. This article will revolve around spring fault degree. The operating mechanism in this paper is for the SF6 high-voltage CB and is shown in Fig. 1(b).

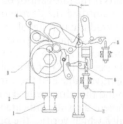

(a) Schematic of the operating mechanism.
1-Closing spring; 2-Oil buffer; 3-Ratchet; 4-Export lever; 5-Opening electromagnet; 6-Closing latch; 7-Closing electromagnet; 8-Opening spring;

(b) Operating mechanism of LW30-252 model SF6 high-voltage CB.

Figure 1. Operating mechanism of a 252 kV porcelain CB.

Equipped with the CT26 spring operating mechanism, it adopts AC electromagnet to control the opening and closing, contact opening distance of 159 mm, over-travel 31 mm. The schematic of the operating mechanism and its corresponding parts is shown in Fig. 1(a).

This article chose the displacement sensor as WDY35D conductive plastic angular displacement sensor, 1 kΩ resistance value, 0.1% linearity. The output voltage of the sensor has a linear relation with the rotation angle. This angular displacement sensor is fixed on the CB operating mechanism of dynamic contact movement characteristics on the relevance of the spindle. It can accurately test the real-time dynamic contact sport characteristic parameters. The position of the angular displacement sensor is shown in position A in Fig. 1(b).

Using the angular displacement sensor to monitor the traveler–time characteristics of a CB, if the opening and closing spring of the CB had mechanical parts that wear and tear, spring fatigue aging, deformation and rust, it will have an impact on the mechanical characteristics of the CB. When the CB is in the opening and closing operation, due to the mechanical vibration of the CB itself (Zhang, 1994; Meng, 2010; Wang, 2008; Ma, 2012), it will lead to a characteristic curve of different degrees of jitter as measured by the angular displacement sensor. The distribution of disturbance of the characteristic curve caused by the vibration of the CB itself is the main content of this paper research.

2.2 The process of opening and closing of high-voltage CB

AC electromagnets are used in the process of opening and closing the high-voltage CB. The main vibration events during the operation of the CB are related to the sequence of actions of the parts (Xu, 2000).

The closing operation of the high-voltage CB starts after receiving the closing command. After the CB receives the closing command, its closing operating coil circuit will be energized and the coil generates enough electromagnetic force to push the core to start moving, then the iron core strike closes the latch. At this point, the closing pawl rotates and the closing latch unlocks the lock to release the energy of the spring, then the closing spring drives the cam and the output shaft to rotate and generates a great force to drive the movable contact to start moving until the moving contact reaches the closing position, so that the oil buffer is compressed in place, and maintains a reliable closing state under the function of the closing spring. During the closing process, the opening spring is compressed to a certain extent so that the energy of the closing spring will reserve a portion of the

energy to the opening spring. Similarly, the opening process of high-voltage CB is mainly making the CB maintain a reliable opening position under the condition that the opening spring can work reliably. If the opening/closing springs have different degrees of fatigue, it cannot cut off the fault when there is a power system failure. This may cause damage to significant equipment and accident of the power grid over a large area, which will cause serious economic losses. At the same time, if the CB cannot reliably close the line when needed, then there will be an undue power failure in the power supply area. Therefore, for the high-voltage CB, to find out the working state of the opening/closing spring in time has a very important significance to improve power supply reliability.

3 DATA ANALYSIS

The main purpose of this paper is to diagnose the fault degree of high-voltage CB spring and the object of the study is the stroke signal. The direction of the study is the extract feature of the stroke signal and the extracted feature is constituted as a feature vector. Then we use artificial intelligence machine to realize fault recognition. Traditional methods to solve these problems have the following several essential steps:

1. Data collection, mainly to provide learning and diagnosis of the input;
2. Data cleaning, mainly dealing with duplication, missing, singular, and other problem data;
3. Data processing, mainly on the cleaned data using data analysis, signal processing, and other methods to extract the signal characteristics of different conditions in order to build the eigenvector for learning machine use;
4. Constructing the learning machine mainly based on the feature vector to achieve fault classification;

The process of fault diagnosis is shown in Fig. 2, in which part A belongs to information collection work, part B belongs to data cleaning, part C and D belong to data processing, and part E belongs to the construction of the learning machine.

3.1 Analysis of the stroke signal of high-voltage CB

The experiment of simulated spring failure was carried out by using the 252 kV pillar CB experimental platform. The pre-compression of the opening spring is 50 mm in normal conditions. In the case of no load CB, the method of adjusting the value of the opening spring pre-compression by the use of tooling to simulate different degrees of breaker

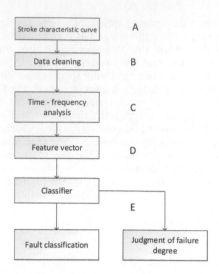

Figure 2. Flow chart of fault diagnosis.

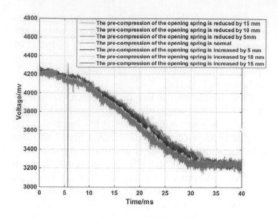

Figure 3. Opening stroke signals under different conditions.

spring failure are: the normal conditions of the opening spring, the pre-compression of the spring reduced by 5, 10, and 15 mm conditions, and the pre-compression of the spring increased by 5, 10, and 15 mm conditions. In each case, 18 sets of trip signal data were collected at a sampling rate of 400 KHz and 126 sets of experimental data were collected.

As the collected stroke signals will inevitably be with interference, resulting in data singularity and missing, when the collected original stroke signals are cleaned, it can obtain 116 sets of valid experimental data. All of the subsequent data are valid data obtained through processing. The original data is shown in Figure 3 where the ordinate for the voltage (unit of mv), of 1 V represents the contact travel of 173 mm.

From Fig. 3, it could be concluded that with the spring pre-compression increasing, the amplitude of fluctuation in stroke signals is increasing. At the same time, when the high-voltage CB opening spring is in a different pre-compression state, there are obvious trend components and high-frequency fluctuating components of the stroke signal. The average velocity of displacement is chosen as the characterization of trend information while the residual information can be understood as disturbance or fluctuation. Time-frequency analysis can be used to extract the intrinsic features of residual information. Therefore, it is possible to characterize the fault by analyzing the trend information and the fluctuation information separately and to identify the type of spring failure by using a certain diagnostic method.

There are many ways to separate the trend. This paper uses the moving average method to smoothen the original stroke signal, which is equivalent to the low-pass filtering of the signal to get the smoothed stroke curve. The basic principle of the moving average method is as follows:

For N nonstationary data $\{y_i\}$, it is nearly smoothed in the interval of every m adjacent data, that is, the mean value is close to the constant. Thus, the value of any one of the m pieces of data can be expressed as an average value of every m pieces of adjacent data and regarded as a measurement result of suppressing a random error or a signal of eliminating noise. The mean value is used to represent the measurement result or signal of the midpoint data or end point data. The general expression is:

$$f_k = \frac{1}{2n+1} \sum_{M=-n}^{n} y_{k+M}, \quad k = n+1, n+2, \ldots, N-n \quad (1)$$

In the formula, m (m = 2n + 1) is the algorithm parameter, and m as the window length directly affects the separation effect. Obviously, the random fluctuation of the result obtained is reduced by the average effect than the original data $\{y_k\}$, that is, it becomes smoother, which can obtain the estimation of random error or noise, i.e., the residual is

$$e_k = y_k - f_k, \quad k = n+1, n+2, \ldots, N-n \quad (2)$$

Through the sliding average, the original data can be filtered out of frequent fluctuations in a number of random fluctuations, which shows a smooth trend, while the random error of the change process also can be obtained so that the statistical characteristics of the amount can be estimated. Smoothing effect under different window lengths (the pre-compression of the spring is a normal condition) as shown in Fig. 4.

It can be seen from Fig. 4 that different m value, that is, different window length directly affects the separation effect. Based on the true representation of the stroke signal, if the value of m is too small, the degree of smoothness is not sufficient resulting in the speed feature representing the trend component being difficult to calculate; at the same time, when the standard deviation of the fluctuation value is too small, the noise will be submerged in the real vibration signal of the stroke signal; if the value of m is too large, smoothing will lead to curve deformation resulting in a larger variance of the standard deviation of the subsequent extraction fluctuation, which will change the characteristic parameters of the actual characteristics of the curve. Therefore, a thorough and appropriate smoothing algorithm without changing the shape of the curve is the key to achieving the separation of trends and fluctuation.

In this paper, after a large number of experiments to deal with the stroke curve, and then through the comparison after processing the waveform effect, the final choice of window length is m = 1000 points, that is 2.5 ms.

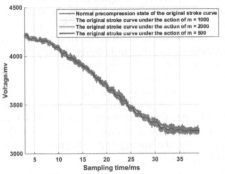

(a) The difference of smoothing waveforms under three windows

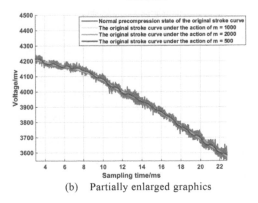

(b) Partially enlarged graphics

Figure 4. Smoothing effect of different M values (window length).

The moving average method is used to process the stroke signal, then the trend component is separated and the fluctuation component of any sample under the seven conditions is obtained by the formula (2). The waveforms of fluctuation of the samples under the seven kinds of operating conditions after separation are shown in Fig. 5.

It can be seen from Fig. 5 that when the pre-compression of the opening spring is in different conditions, the magnitude of the fluctuation in the time domain has obvious distribution characteristics, which can be extracted as the fluctuation feature.

3.2 Feature extraction of trend information of the stroke signal

For obtaining the accurate and smooth trend component, the stroke curve can be used to calculate the opening speed of the contact movement, which can be obtained directly from the numerical differentiation or multi-point solution of the discrete-stroke signal.

Feature extraction of trend components: Supporting that the contact speed is the average speed of the 20–80% stroke during the switching-on process, which indicates the trend information of the displacement curve, according to the contact speed definition, the average speed of contacts is calculated in the seven types of target samples, respectively. The average speed of the contacts is obtained in each sample for every condition and the result is shown in Fig. 6.

Each point in Fig. 6 represents the contact speed of a sample. The velocity is negative because it is the opening process, so the absolute value of speed represents the opening speed. It can be seen from the curve in the figure the trend of the speed of the opening spring in seven different pre-compression states: the speed of the state of the pre-com-

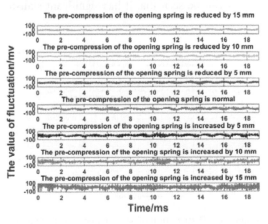

Figure 5. Fluctuation component of the stroke signal.

pression is increased by 15 mm> the speed of the pre-compression of the opening spring increased by 10 mm > the speed of the pre-compression of the opening spring increases by 5 mm > the speed of the normal compression state of the opening spring > the speed of the state of the pre-compression decreased by 5 mm > the speed of the pre-compression of the opening spring decreased by 10 mm > the speed of the pre-compression of the opening spring decreased by 15 mm. The results are consistent with the physical processes of mechanical motion and then describe the contact velocity v characterization and can distinguish spring compression volume changes in compression force changes and spring normal pre-compression force difference. It can be selected as an element in the eigenvector. At the same time, the position marked A in Fig. 6 indicates that the speed of opening is aliased under different working conditions. Only a single feature of speed cannot be used to distinguish between the fault of spring pre-compression increased and the fault of pre-compression reduced. It cannot achieve the purpose of distinguishing the classification of the degree of spring failure, that is, the amount of spring pre-compression increased or reduced cannot be diagnosed. So, under the same conditions, the characteristics of the fluctuation of the spring need to be introduced to achieve the purpose of distinguishing the extent of failure.

3.3 Feature extraction of fluctuation information of the stroke signal

Use the formula (2) to extract fluctuation quantity of the stroke signal, draw the histogram for the amount of fluctuation quantity of any sample under the seven kinds of conditions, and there is a significant difference between them in the distribution of the amplitude of fluctuation quantity and it can be seen that it has significant statisti-

cal characteristics. Using the normal distribution (Mao, 1998) for curve fitting, the random variable x follows a probability distribution with a position parameter σ and a scale parameter σ, and its probability density function is usually expressed by f(x), whose density function is

$$f(x) = \frac{1}{\sqrt{2\pi}\sigma} \exp\left(-\frac{(x-\mu)^2}{2\sigma^2}\right) \qquad (3)$$

The feature extraction of fluctuation: Firstly, considering the time-domain characteristics under seven kinds of operating conditions the difference of amplitude in seven cases will be showing. Therefore, the probability density distribution of fluctuation in the time domain is used to characterize the fluctuation characteristics under seven conditions due to fluctuations in the dispersion in time. Normal distribution fits the corresponding sample drawn probability density distribution curve of seven conditions as shown in Fig. 7.

It can be seen from Fig. 7 that the average of the probability density distributions of the fluctuating components in different spring failure states is approximately equal to zero with little difference. However, there is a significant difference in the shape of the distribution, that is, the standard deviation of the difference is large. So, it can be used to distinguish the fault state. In this paper, it is reasonable to use the probability density fitting function to define the distribution standard deviation as the eigenvector.

From the analysis of the stroke signal, the velocity v and the standard deviation σ of the probability density function were extracted, thereby constituting a vector P = [v, σ] as the eigenvector for the spring fault diagnosis of the high-voltage CB.

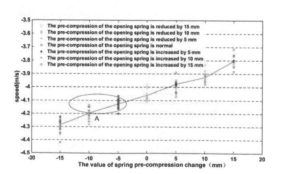

Figure 6. Distribution of sample opening speed at different mechanical conditions.

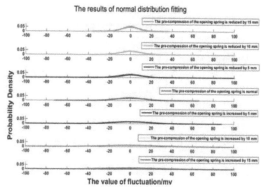

Figure 7. Distribution of fluctuation in the time domain.

4 FAULT DIAGNOSIS BY SVM AND THE ANALYSIS OF DEGREE

4.1 *Fault diagnosis based on SVM*

This paper adopts the Support Vector Machine (SVM) algorithm of sorting which applies to smaller samples, high dimensions, non-linear characteristic, etc. It is suitable for the fault diagnosis extent of the CB opening/closing spring. The basic thought is as shown in Fig. 8.

As shown in Figure 8, the rectangle and the elliptical points represent two kinds of data samples and the line H is the sorting line. The line H_0 and H_1 are parallel to the sorting line, which goes through the nearest samples to the sorting line in the samples. Then the interval between H_0 and H_1 becomes the classification interval. For linearly separable problems, there is always a classification hyperplane $\omega \cdot x + b = 0$, which will divide the training samples into two parts completely. Thus the sample points on H_0 and H_1 are support vectors.

When the optimal separate hyperplane cannot divide the samples into two parts completely, there will be a penalty factor C to control the penalty level for the classification error.

For nonlinear classification problems, if the optimal hyperplane cannot be found to meet the condition of constraint in the original space, the original vector can be mapped into a high-dimensional space by nonlinear transformation for linear classification. Besides, according to the functional theory, when the kernel function $K(x,y)$ meets Mercer condition, the non-linear transform is available and then the classification function can be formed as follows:

$$f(x) = \text{sgn}\left\{ \left(\sum_{i=1}^{n} a_i^* y_i K(x_i, x) + b^* \right) \right\} \qquad (4)$$

As shown in the function, a_i^* is the Lagrangian multiplier, y_i is the classification value, $K(x_i, x)$ is the kernel function, and $K(x_i, x)$ is the classification threshold. There are three kinds of kernel functions usually used: polynomial kernel function, radial basic function, and sigmoid kernel function. So different kernel functions will get different support vector machine classifiers.

The two eigenvectors defined by the travel curve are the trend average speed v and the variance—the standard deviation of the statistical distribution so that there is a binary decision SVM with the serial decision. However, because of the serial structure, the order of the preceding faults determines the correct classification rate. For equal fault types, it is difficult to determine the classification order and because of the spring failure with different degrees of pre-compression reduction or increase, classification data is also difficult to determine. Therefore, it is necessary that the classifier is used to classify the normal pre-compression of the spring, the reduction of the pre-compression of the spring, the increase of the pre-compression of the spring, etc. The classification flow chart is shown in Fig. 9.

It can be seen from Fig. 9 that normal and fault conditions of two can be classified by classifier

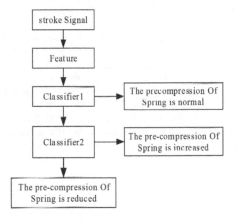

Figure 9. Fault diagnosis model based on serial SVM.

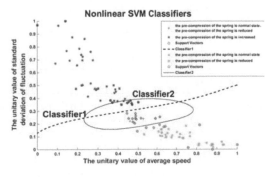

Figure 10. Diagnostic result of nonlinear SVM classifiers.

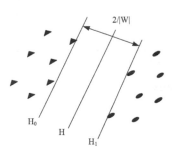

Figure 8. Classification line based on 2D linear separability.

SVM1 and SVM2, which is obtained by analyzing the characteristic vector [mean velocity, standard deviation after fitting the standard normal distribution] and doing the unitary processing.

Fifteen sets of sample data under seven different conditions were trained by the classifier, and the trained SVM was obtained by the nonlinear kernel and the final classification result was obtained as shown in Fig. 10.

Fig. 10 means that the classification range of the normal state of the nonlinear kernel (Gaussian) classification spring is ring-shaped. This method can distinguish the normal state of the spring, the fault state of spring pre-compression increased and the fault state of spring pre-compression reduced. The reliability of the classifier is verified by the experimental data collected under different working conditions and the classification results in Table 1 are obtained.

From Table 1, it can be concluded that the nonlinear nucleus based on the SVM binary tree classification model has a higher diagnostic accuracy for the normal state of the spring, the fault of the spring pre-compression reduced, and the fault of the spring pre-compression increased. This method can better guarantee the reliability of fault diagnosis results.

In practice, when the spring pre-compression is reduced to a very small extent, the spring can continue to be used; but when the spring pre-compression is reduced to a very large extent, at this time, it indicates a severe spring fault. If the CB is operated for a long time, it will result in the closing position that cannot be maintained. At this point, the spring should be replaced. So the calculation of the degree of failure is often more important

than the fault type of judgments in the projects. The degree of fault of the spring should be further studied.

4.2 Determination of spring fault level

Every time when the CB is operated, the detection system processes the travel signal to record and store signal characteristic parameters. Under different conditions, after several operations, the average value of the opening speed (the value of the opening speed is negative) and the mean standard deviation of the fluctuation can be calculated by processing the travel signal as shown in Table 2.

It is shown in Table 2 that based on the normal precondition when the 252 kV column CB pre-compression volume is 50 mm, when the amount of spring pre-compression volume is reduced or increased by a certain degree, the fitting function of the average speed and average standard deviation of the spring pre-compression variation are shown in Figure 11.

It can be seen from Fig. 11 that the changes of opening speed under different working conditions are unobvious, but the mean corresponding standard deviations change greatly. The method of adopting standard deviation to fit the curve of spring pre-compression can well reflect the abrupt change of the mechanical state of the CB. If conduct curve-fitting is based on the data in Table 2 (the mean standard deviation is the dependent variable, the spring pre-compression is the independent variable, the value of the spring pre-compression decrement is negative and the value of the spring pre-compression increment is positive), we can then get the fitted curve function, as follows:

Table 1. Recognition results of the nonlinear kernel SVM model.

Type of failure	Training samples	Measured sample	Correct rate (%)
Pre-compression of the opening spring is normal	15	3	100
Pre-compression of the opening spring decreases	45	4	100
Pre-compression of the opening spring increases	45	4	100
Total	105	11	100

Table 2. Spring failure degree index of the high-voltage CB.

Type of failure	Average speed (v)	Mean standard deviation (σ)
Pre-compression of the opening spring is reduced by 15 mm	−3.8028	7.9752
Pre-compression of the opening spring is reduced by 10 mm	−3.9261	10.1463
Pre-compression of the opening spring is reduced by 5 mm	−3.9776	11.6103
Pre-compression of the opening spring is normal	−4.0604	14.0277
Pre-compression of the opening spring is increased by 5 mm	−4.1246	17.7387
Pre-compression of the opening spring is increased by 10 mm	−4.1962	20.9297
Pre-compression of the opening spring is increased by 15 mm	−4.2864	28.9843

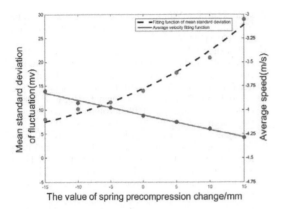

Figure 11.   Results of fitted curves.

$$y = 14.45 * e^{0.04373x} \qquad (5)$$

This paper extracts the mean standard variance characteristic of the fluctuation quantity by collecting the travel-time curve during the normal running of the CB. Then the results of this test are compared with those of the previous operations, which are shown in Fig. 11, from which the changing trend of the parameters can be seen clearly. When the reverse direction changing trend began to accelerate or the parameters changed beyond a certain range, it can be determined that there might be a precursor of the spring failure. At the same time, according to formula (5), the variable quantity of the initial amount of spring compression at this time—x can be obtained. Thus it can be judged that if, at this time, the size of the spring compression volume to test the performance of the spring meets the requirements of the use of the spring, so as to find different levels of hidden trouble, according to the fitting formula, it can be predicted if the spring needs to be replaced or can still be used. Thus the diagnosis of spring failure level has great significance to guide the production of CB and to determine whether the CB is in a reliable working condition.

5   CONCLUSION

This paper studies the LW30–252 SF6 high-voltage CB. The trend signal and fluctuating component of the stroke signal are extracted as the eigenvector of spring fault diagnosis. On the one hand, the design of the serial SVM classifier is to achieve accurate classification of spring failure; on the other hand, in order to make up for the short-comings of the classifier that cannot define the degree of failure, the empirical formula with the standard deviation of the fluctuation is fitted to

achieve the purpose of distinguishing the degree of spring failure. The conclusions are as follows:

1. This paper presents a new method of diagnosing the fault degree of CB spring based on the mechanical vibration signal contained in the stroke signal of high-voltage CB. The purpose of this method is to describe the type of spring fault accurately by introducing the stroke feature quantity, which lays the foundation for the classification of a typical fault of the auxiliary CB.
2. Considering the serial SVM structure, the fault judgment accuracy of the former level directly affects the result of the next level. The serial SVM can effectively judge the fault type, but it is difficult to distinguish the spring degree and refine it.
3. The standard deviation of the fluctuation is used to judge the degree of spring failure. At the same time, the formulae of the spring pre-compression variation and the mean standard deviation are used to achieve the purpose of quantifying the fault degree of the spring, which is of great significance for instructing the actual production.
4. The experimental results show that the new method can correctly identify the state of mechanical failure of the spring and fault classification. The method was confirmed to be reliable and effective without adding special vibration sensors. The data processing is simple, economical, and practical.

REFERENCES

Bosma, A., R.T., Condition Monitoring and Maintenance Strategies for High-Voltage Circuit-Breakers. The 6th International Conference on APSCON, 2003:191–195.
Demjanenko, V., M.L.L., Y.H. Lee, Mechanical Failure Detection of Circuit Breakers. IEEE Trans on Power Delivery, 1988, 3(4):1724–1731.
Dong Wen-zhi, Zhang Chao, Fault Diagnosis of Rolling Bearing Based on EMD Energy Entropy and Support Vector Machine. Mechanical Design and Research, 2011.10. Vol27 No5.
Hess D P, Park S Y, Tangri M K. Noninvasive condition assessment and event timing for power circuit breakers [J]. IEEE Transactions on Power Delivery, 1992, 7(1): 353–360.
Lee D, Lithgow B J, Morrison R E. New fault diagnosis of circuit breakers [J].IEEE Transactions on Power Delivery, 2003, 18(2): 454–459.
Mao Shisong, Wang Jinglong, Pu Xiaolong, et al. Higher Mathematics Statistics [M]. Beijing: Higher Education Press, 1998
Meng Wusheng, Miao Yiwen, Dong Rong. A new differential transformer-type angular displacement

sensor [J]. Small & Special Electrical Machines, 2010, 38(8):32–34.

Runde M, Ottesen G E. Vibration analysis for diagnostic testing of circuit breakers [J]. IEEE Transac—tions on Power Delivery, 1996, 11(4): 1816–1823.

Stokes A D, Timbs L. Diagnostics of circuit breakers [C]. International Conference on Large High voltage Electric Systems, New York: GIGRE, 1988: 1–7.

Sun Yihang, Wu Jianwen, Lian Shijun, et al. Extraction of vibration signal feature vector of circuit breaker based on empirical mod decomposition amount of energy [J]. Transactions of ChinaElectrometricalSociety, 2014, 29(3):228–236.

Sun Yihang, Wu Jianwen. Design on Intelligent Integrated Controller of Circuit Breaker Based on Dual-core CPU.2011 1st International Conference on Electric Power Equipment-Switching Technology [C], 2011, 315–318

Wang Dongsheng, Wang Guimei, Pan Weiwei, et al. Study of fiber grating displacement sensor [J]. Instrument Technique and Sensor, 2008(9):6–8.

Wu Jianwen, Lian Shijun. A kind of fault isolation and lockout features microprocessor-based protection boundary switch [C]. 2007 Anual Conference on Relay Protection and Automation in China.2007.

Wu Jianwen, Sun Yihang Zhang Luming, et al. A range of failure is isolated circuit breaker controller [P]. 201110104084.X.

Wu Zhensheng, Wang Wei, Yang Xuechang, et al. Processing of mechanical vibration signal of high voltage circuit breakers based on fractal theory [J]. High Voltage Engineering, 2005, 31(6): 19–21.

Xu Guozheng, Zhang Jierong, Qian Jia Li, et al. Principle and application of high voltage circuit breaker [M]. Beijing: Tsinghua University Press, 2000.

Xu Jianyuan, Lang, Fucheng, Lin Xin. The present condition and developing tendency of online monitoring technology of mechanical characteristics of High Voltage Circuit Breaker. Huatongjishu, 2005.2.17–21.

Zhang Guogang, Wang Hongwei, Tang Xiang, et al. Vibration signal analysis of hydraulic operating mechanism for high voltage circuit breaker based on EMD method [J]. High Voltage Apparatus, 2008, 44(3): 193–196.

Zhang Yuli, LI Jicheng. High-precision differential transformer displacement sensor [J]. Instrument Technique and Sensor, 1994(2):19–21.

Zijun, MA, Yang Shuanglian. Research on prospects of Measurement principle and application of laser displacement sensor [J]. Gansu Science and Technology, 2012, 28(2):77–78.

Fault diagnosis for thin-walled rolling bearing based on EEMD and TKEO

X.D. He
School of Mechanical Engineering, Northwestern Polytechnical University, Xi'an, China

M. Qiu
School of Mechatronics Engineering, Henan University of Science and Technology, Luoyang, China

ABSTRACT: Thin-walled rolling bearing failures can result in cyclical impulse responses of vibration signals. By processing and analyzing, the fault diagnosis could be detected according to the corresponding response signals. Therefore, this paper proposed a new method combining EEMD (Ensemble Empirical Mode Decomposition) and TKEO (Teager-Kaiser Energy Operator). Firstly, the original signal is demodulated to EEMDs in order to gain the correspondingly IMFs (Intrinsic Mode Function), and then the respective IMF is analyzed by TKEO, finally, energy signals and frequency-domain signals could be gained. It turns out that the effect is obvious and guide ways to control vibrations and noises.

1 INTRODUCTION

Because of their small sectional area, small volume, light weight, long life, high rigidity and low friction torque (Chen, 2000), thin-walled bearings are widely used in the fields of aviation, aerospace, medical devices and robot. The bearing is one of the most widely used and most easily damaged mechanical components in rotating machinery and its running status directly affects the rotating part, even the whole unit. In the actual work of machinery and equipment, the failure rate of rolling bearings is relatively high. If there were a failure, serious consequences could be caused. Therefore, the monitoring and fault diagnosis of rolling bearings are playing an important part (Xu, 2013).

When local injuries or integrated damages happen in rolling bearings, it will inevitably lead to abnormal mechanical operation. Meanwhile, the relative motion between the components will induce periodic pulse signals and then the normal signals will be modulated by the periodic pulse signals in a wide frequency, especially in the early stage of rolling bearing failure. The abnormal response signals just take up a small fraction of the total energy in a wide range, which provides an obstacle to early fault identification and diagnosis. Under the influence of vibration from various components on the shaft, the signal has many disturbances and complex components. Improving the signal-to-noise ratio of fault signals and gaining useful fault information efficiently are becoming the focus of research in the future (Diana, 1998).

Many scholars at home and abroad are committed to the bearing fault diagnosis, and get good results. In 2009, Wu and Huang proposed a method of Ensemble Empirical Mode Decomposition (EEMD) based on Empirical Mode Decomposition (EMD) (Zheng, 2015). In order to achieve signal continuity of white noise, the original signals are decomposed into a number of different modes of the IMF (Intrinsic Mode Function) from high frequency to low frequency, and then effectively solve the mixing problem of the empirical mode decomposition. Shen Changqing (2013) from China University of Science and Technology proposed structural element method to extract impulse response feature combining with morphological filtering. Yang Wangcan(2015) put forward a gear fault diagnosis method based on EEMD multi-scale fuzzy entropy; Wu Xiaotao (2015) from Huazhong University of Science and Technology reconstructs the IMF components according to IMF and SE criteria, and then carries out the filtering method based on steep criteria Treatment. Wang et al. (2012) used the TKEO (Teager Energy Operator) to extract the periodic shocks.

In this paper, Firstly, the original vibration signals are decomposed into several IMFs from high frequency to low frequency according to the EEMD method, so that the continuity of the signal and the mixing of the modal signal can be realized. Secondly, the energy required to generate the signal is tracked according to the TKEO method, and then spectrum analysis are conducted in the frequency domain. The results turn out that the method proposed in this paper can diagnose the fault and prevent the early fault of rolling bearing.

2 BASIC THEORIES

2.1 Basic theory of EEMD

Due to the existence of external disturbance, usually there are mixing problems. Meanwhile, because of the number and distribution of extreme points, the cubic spline curve fitting for EMD method will cause errors. According to the characteristics of the vibration signal itself, the signals are adaptively decomposed into IMFs and a residual from the low-frequency to high-frequency, so that the shortcomings of the EMD method could be overcome. The IMFs gained need to meet the following two conditions:

1. In the entire signal sequence, the number of extreme points and the number of zero-crossing points must be equal or up to a difference of one point.
2. At any point in time, the average values of the upper and lower envelopes determined by the local maximum and minimum values are zeros (Shen, 2013).

The detailed process of EEMD is described in (Shen, 2013).

2.2 Basic theory of TKEO

Open For any continuous signal x(t), the teager energy operator Ψ is defined as

$$\psi\left[x(t)\right] = \left[\dot{x}(t)\right]^2 - x(t)\ddot{x}(t) \tag{1}$$

Where $\dot{x}(t)$ and $\ddot{x}(t)$ represent the first and second order derivative of the signal x(t) versus time t, respectively.

For a discrete signal x(n), since the first order, second order differential does not exist, the difference is taken instead of differential. The Teager energy operator becomes

$$\psi\left[x(n)\right] = \left[x(n)\right]^2 - x(n-1)x(n+1) \tag{2}$$

From equation (2), only three samples are needed to calculate the energy of the signal source n at any time. This method is sensitive and has strong time resolution for the fast changing shock signals. It is especially suitable to monitor instantaneous component in the original signals.

According to reference (Gadivia-Ceballos, 1996; Maragos, 1993), the amplitude $|A(n)|$ and frequency w(n) are as below respectively:

$$|A(n)| = \frac{2\psi\left[x(n)\right]}{\sqrt{\psi\left[x(n+1) - x(n-1)\right]}} \tag{3}$$

$$w(n) \approx \frac{1}{2\pi}\sqrt{\frac{\psi\left[\dot{x}(n)\right]}{\psi\left[x(n)\right]}} \tag{4}$$

In which

$$y(n) = x(n) - x(n-1)$$

3 FAULT DIAGNOSIS BASED ON EEMD AND TKEO

Text First, according to fault characteristic frequency formula of the thin-walled rolling bearing (5), (6), (7) to calculate the outer ring, inner ring, ball fault frequency respectively:

$$f_{outer} = \frac{1}{2}zf_n\left(1 - \frac{D_w}{D_e}\cos\alpha\right) \tag{5}$$

$$f_{inner} = \frac{1}{2}zf_n\left(1 + \frac{D_w}{D_i}\cos\alpha\right) \tag{6}$$

$$f_{ball} = \frac{1}{2}f_n\left(1 - \frac{D_w^2}{D^2}\cos^2\alpha\right)\frac{D}{D_w} \tag{7}$$

where D_w stands for the diameter of running ball, D is the diameter of the bearing raceway or cage, α the contact angle, z the number of balls and f_n is the rotating frequency of the shaft.

Secondly, according to the data collected by the tester and based on the method described in 2.1, the, single-mode component IMFs could be gained after EEMD decomposition of the original signal.

Then, according to the procedure in 2.2, the IMFs at each frequency are decomposed by TKEO operator, and energy and amplitude will be obtained.

Finally, according to the characteristics of frequency and amplitude, the obtained frequency is compared with the calculated frequency. If the peak frequency f_{TKEO} acquired by TKEO operator

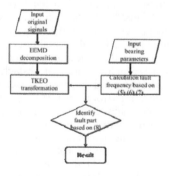

Figure 1. The overall flow chart.

238

and fault frequency have relationship as (8), the damage component could be identified.

$$f_{TKEO} = Nf \qquad (8)$$

where N is a positive integer, f can be f_o, f_i, f_b, or any two of them could coexist.

The overall flow chart is shown as Figure 1:

4 EXPERIMENT

The bearing tester is GM-DXJ-12-1 dynamic performance tester.

The experimental bearing is SKF 6205-2RS JEM series deep groove ball bearings and the bearing parameters are shown in Table 1. According to the rolling bearing parameters, the fault frequencies are shown in Table 2. In order to measure the vibration signals under the fault condition, a 0.5 mm wide and 0.5 mm deep grooves were manufactured by wire cutting in the outer ring of the rolling bearing without affecting the normal running condition of the rolling bearing. During the experiment, the motor rated speed was set to 1770 rpm and NI's multi-channel data acquisition equipment and displacement sensors were used with 12 KHz sampling frequency. The influence of

Table 1. The rolling bearing parameters.

Category (mm)	The inner ring	The outer ring	Ball diameter	The ball number	Pitch diameter
Value	25.00	52.00	7.94	20	39.04

Table 2. The fault frequencies of the experimental bearing.

Fault types	The outer ring	The inner ring	The rolling ball
Fault frequencies	$8.31f_n$	$12.55f_n$	$2.356f_n$

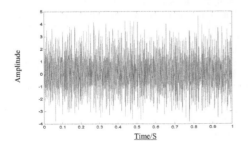

Figure 2. Fault frequency domain signals of the outer ring.

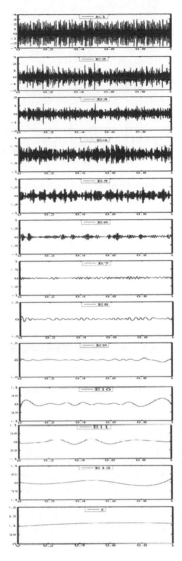

Figure 3. EEMD decomposition of the original signal and residual signal.

external noise should be tried to get rid of as much as possible, so the relative quiet experimental environment should be kept. Time domain signals were acquired as Figure 2.

It was calculated that the fault frequencies of the outer ring, the inner ring and the rolling ball were 245.1 Hz, 370.2 Hz and 69.5 Hz respectively for the thin-walled rolling bearing. Figure 3 shows the EEMD decomposition signals of the original signals (denoted as e1, e2, e3... e12) and the residuals (r) respectively. Based on the IMFs obtained, the first IMF is decomposed by TKEO to get the amplitude-time image and the frequency-time image respectively. The amplitude—time and frequency—time images for first IMF are showed in Figure 4, Figure 5.

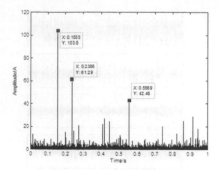

Figure 4.   First IMF amplitude-time waveform.

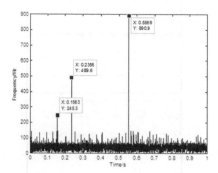

Figure 5.   First IMF frequency-time waveform.

It shows that the thin-walled rolling bearing has sharp energy changes at 0.1555 s, 0.2356 s and 0.5569 s in Figure 4, which imply that the bearing has faults at the corresponding places. It can be seen from Figure 5, the corresponding time of the frequency are 245.1 Hz, 490.3 Hz and 980.5 Hz. The calculated frequencies are one time, two times and three times of natural frequency position, which can determine that the bearing outer ring is in fault.

5   CONCLUSION

In this paper, based on EEMD and Teager energy operator a new method for fault diagnosis is introduced. Through the analysis of the model and the thin-walled bearing experiment, it turns out that:

1. Through EEMD, a series of IMFs can be obtained, and the amplitude and frequency of each intrinsic modal function can be acquired also.
2. Teager energy operator can used to decompose the first IMF, and track the instantaneous amplitude and frequency;
3. Based on of EEMD analysis and the envelope spectrum analysis of instantaneous amplitude, the fault frequency of rolling bearing can be obtained, which can effectively identify the fault location of the rolling bearing.

The method of based on EEMD and Teager energy operator is an effective method to extract bearing fault signals. At the same time, the fault frequency is much obvious in this method, so it can be effectively applied to the fault diagnosis of thin-walled rolling bearings. The method has broad prospects in the field of complex signal fault diagnosis.

ACKNOWLEDGEMENT

The authors are grateful to financial support from the National High-tech R&D Program of China (No. 2015AA043004) and Science and Technology Innovation Outstanding Talent Fund Projects of Henan Province (No. 154200510013). Corresponding author: M.QIU.

REFERENCES

Chen J Q, Zhou H, Xu L L. Review of theoretical research on the load distribute ion in roller bearing [J]. Journal of Beijing Inst itute of Petro—chemical Technology, 2000, 8(1): 47–52.

Diana G, Fossati F, Resta F. High speed railway collecting pantographs active control and overhead lines diagnosis solution [J]. Vehicle System Dynamics, 1998, 30, 69–85.

Gadivia-Ceballos, L, Hansen, Kaiser, JF. Vocal fold pathology assessment using am autocorrelation analysis of the teager energy operator [J]. In: Proceedings of 4th international conferenceon spoken language, 1996 7(21):757–60.

Maragos P, Kaiser JF, Quatieri TF. Energy separation in signals modulations with application to speech analysis[J]. IEEE Transactionson Signal Processing, 1993 41(30): 24–51.

Shen Chang-qing, Peter W. Tse, Zhu Zhong-kui, et al. Rolling element bearing fault diagnosis based on EEMD and improved morphological filtering method [J]. Journal of Vibrations and Shock, 2013, 32(2):39–43.

Wang Tian-jin, Feng Zhi-peng, Hao Ru-jiang, et.al. Fault diagnosis of rolling element bearings based on Teager energy operator [J]. Journal of Vibration and Shock.2012, 31(2):1–5.

Wu Xiao-Tao, Yang Meng, Yuan Xiao-hui, et al. Bearing fault diagnosis using EEMD and improved morphological filtering method based on kurtosis criterion [J]. Journal of Vibtation and Shock, 2015 34(2):38–44.

Xu Ya-jun, Yu De-jie, Sun Yun-song, et al. Roller bearing fault diagnosis using order multi-scale morphology demodulation [J]. Journal of Vibration Engineering, 2013,26(2):252–259.

Yang Wang-can, Zhang Pei-lin, Wang Huai-guang, et al. Gear fault diagnosis based on multiscale fuzzy entropy of EEMD [J]. Journal of Vibration and Shock, 2015, 34(14):163–167.

Zheng Zhi, Jiang Wan-lu, Hu Hao-song, et al. Research on rolling bearings fault diagnosis method based on EEMD morphological spectrum and kernel fuzzy C-means clustering [J]. Journal of Vibration Engineering, 2015,28(2):324–329.

Electromechanical Control Technology and Transportation – Jia & Wu (Eds)
© 2017 Taylor & Francis Group, London, ISBN 978-1-138-06752-3

Equipment technology research on profile controlling and oil displacement in the frigid zone of the Erlian region

Jing Wang, Jing You, Mei Bin & Jianlin Hu
Petroleum Engineering Research Institute of Huabei Oilfield Company, Renqiu, Hebei, China

Ping Li & Shichao Wang
Third Oil Production Plant of Huabei Oilfield Company of Petrochina, Renqiu, Hebei, China

ABSTRACT: It is in the high-watercut and ultra-high watercut stage in Erlian oilfield: the average recovery rate was only 17.3% and the near wellbore processing cannot resolve the extreme heterogeneity problem in deep reservoir. Field test confirms that the profile controlling and oil displacement technique are effective methods to improve recovery factor in the middle-late period of the reservoir development. Erlian oilfield is located in the frigid zone, so the constructions of profile controlling and oil displacement measures in winter require that equipments should possess the higher performance of antifreeze, sand and temperature resistant. This device is mainly used for constructions in cold area in winter such as profile controlling, oil displacement and microbial injection. By using modular design, this device can be combined arbitrarily to different processes. Considering construction characteristics of cold region, all the modules are in heating and insulation design, convenient to antifreezing and unfreezing.

1 INTRODUCTION

The reservoir of Erlian oilfield belongs to the small fan—delta deposit system of cliffs, coarse grains and near source. The types of the reservoir are low permeability sandstone reservoir, medium permeability and conventional heavy oil sandstone reservoir, conglomerate reservoir and so on. The reservoir heterogeneity is strong and the permeability of interwells is $(0.3 \sim 513) \times 10^{-3} \mu m^2$. Recently, the oilfield is in the high-watercut and ultra-high watercut stage and the average recovery rate was only 17.3%. As a result of long—time washing of the injected water, near wellbore processing cannot resolve the problem of extreme heterogeneity in deep reservoir. Weak gel flooding technology can not only adjust the contradictions between layers, but also improve the heterogeneity of deep reservoir, which has broad prospects for the guidance to the Erlian oilfield development, becoming one of the regular stimulation measurements.

Erlian oilfield is located in the frigid zone, so some important topics have become essential which affect the proper working of profile controlling and oil displacement, such as whether the process tech-nology and equipment design for profile controlling and oil displacement are adapt to the needs of injection construction in frigid zone, the needs of the pharmaceutical preparation and the needs of adjusting the conventional well injection parameters and to the implementation of the heating and insulation measures. So, on the basis of summarizing the application practice of profile controlling and oil displacement equipments in recent years, aiming at the problems of winter constructions in cold areas, the equipment technology researches on profile controlling and oil displacement in winter were carried out. This device is mainly used for constructions in cold area in winter such as profile controlling and oil displacement and microbial injection. And this device has applied in 3 wells, achieving good effect.

2 THE DESIGN CONCEPT AND DESIGN BASIS

The excellent anti-freezing function, the quickest way to assemble, advanced working performance, these are the design concepts needed in the deep profile control and flooding. Based on this idea, the module design and heat preservation and heat insulation design were highlighted on the design of the system device.

Modular design mainly meets the site construction requirement, such as convenient transport, quick disassembling and flexible combination. There are six units in the installation: the pump unit 1 set, fluid confection unit 2 sets, water storage unit 1 set, material reserve unit 1 set and duty room 1 set. They work respectively and have their special functions. Through the connection

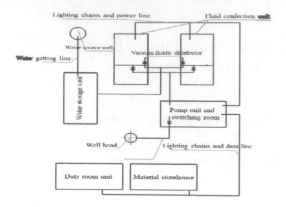

Figure 1. Ideal drawing of the device.

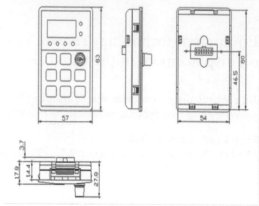

Figure 2. The size diagram of the control panel.

process and pipeline, a complete set of injection system is formed, accomplishing different injection work.

Heat preservation and heat insulation design: all units should strengthen the insulation design, each unit is separate and forms an independent functional module, and each independent module completes a process. According to the requirement of the construction parameters, several function modules can be combined to form a set of units. Anti-corrosion and explosion-proof design: steel materials and anti-corrosive paint are chosen in the liquid pool; explosion-proof design and equipment selection are applied in the electrical equipments. Heat preservation processing is needed outside the process line between functional modules.

3 SCHEME DESIGN

3.1 Optimization design for system process, parameters and allocations

The device adopts modular design and each module constitutes an independent skid mounted unit. When installed, according to the specific situation of the construction site and mutual connection relationship between elements shown in the flow chart, six units are put in right place. The field needs to be flat and level without foundation. Process line is electric heating insulation hose specially designed and manufactured.

The function of oil production LCD and number operation panel (Fig. 2) is for the operating of transverter, the setting of function parameters, condition monitoring and so on. Among them, LED displays the current state of functional parameters digitally; LCD displays annotation and illustration of the state of parameters. Use the union link or quick connector (Fig. 3).

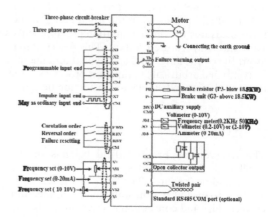

Figure 3. The layout drawing of basic operation.

According to the continuity requirements of the profile controlling, oil displacement and microbial injection work, and different requirements of injection parameters in different blocks, different well location and different measures, the system takes the process of alternate dispensing and continuous injection. In order to adapt to the injection requirements of different blocks and different locations, according to the statistical conditions of injection wells in each block, daily water-injection rate is 30–80 m³, water injection pressure is 7–22 MPa. Considering the safety coefficient, the injection ability of this system is determined as 9.9 m³/h and the injection pressure is 30 MPa.

3.2 The research and application of heat insulation technology and materials

In Erlian cold area, the process of the whole system requires not freezing and blocking in the profile control and oil displacement constructions in winter. Besides, the configuration agent should be well-distributed and cured thoroughly in the dispensing process. Therefore the system was studied further in heat insulation technology and materials.

Take simulation experiments on starting performance in low-temperature and heating temperature rise curve of heating elements.

Take single and composite experiment on the thermal insulation material, waterproof material and coating material. According to the functional requirements of the material, several properties are mainly tested, such as heat resistance and anti-aging characters of various materials, heat insulation characters, waterproof and dampproof characters and comprehensive properties of the composite using.

Based on this, after the selection of performance and parameters and simulation experiment, accord-ing to the needs of technology of profile controlling and oil displacement, the heat insulation technology of Germany was introduced and absorbed, and the pressure heating insulation hose was successfully developed. This technology would process the pressuring, heating, heat insulation, waterproof and insulation of pipeline at an organic whole by using integration design.

Each line can be used alone or in series. Heating temperature can be adjusted from 20°C to 80°C according to the environmental changes, realizing the automatic temperature control. Heat preservation and heat insulation performance is very good and the temperature difference of internal and external pipeline can be up to 40°C–50°C. It not only satisfies the needs of diversity of field process connection, high pressure, waterproof, dampproof, heat insulation and so on, but also solves the problems of the quick destuffing and assembling, convenient storage, easy control of process lines and other issues.

On the basis of the pressure heating insulation hose technology and the compound technology of thermal insulation material, the mode of the system heat insulation is all-round, small temperature difference, multi-stage heat preservation and multi-stage heating. It is convenient to control and can reduce energy consumption.

The wall (six sides) of each unit is with 80 mm glass wool to insulate. At the same time, water storage tank and liquid pool use electricity device to heat. Ensure that all process line uses pressure heat insulation hose to connect.

The fluid confection unit of system is designed to two sets of independent units on the configuration. One unit injects water, at the same time, the other unit deposits chemicals and ripens, increasing the mixing and curing time. Besides, a separate storage unit is set up to ensure water supply and shorten the time of fluid confection. And in the fluid confection unit and water storage unit, there are two sets of 15 KW electric heating device separately which preheat and insulate when injecting water, temperature rising is 5°C to 10°C. Further heating and insulating in the fluid confection work, by rising temperature 5–10°C further, to ensure that the agents dissolve and cure, preventing the liquid frozen.

Considering the convenient connection, process pipings adopt pressure heat insulation hose materials, connected with the union link or quick connector, which is convenient and quick, adapts to the connection of different locations. The heat insulation measures and protective measures of pipes can fully meet demands for the convenient shift, simple control, strong and wear-resisting and higher waterproof.

3.3 Design on reagent preparation technology

According to the characteristics of common materials used in profile control and oil displacement process, vacuum fluidic distributor is designed, which can mix liquid and solid, liquid and liquid. Its characteristic is that the device can draw powder and solid materials into the mixer evenly in the leading of high velocity fluid and the small component powder is fully mixed under the washing of high speed fluid, which can be applied to 0–10000 PPM solid-liquid solution and 0–30000 PPM liquid-liquid solution. The speed of fluid confection is quick and the liquor is uniform.

Through gel monitoring of the field liquid disposing, the results are shown in Table 1, from the

Table 1. Agents gelling table.

Number	Sample time	Sample point	Gelling viscosity mpa·s	Remarks
1	2011–08–11	Upper of liquid pool	2080	Test temperature is 38°C, not open the heating function
		Middle of liquid pool	2100	
		Bottom of liquid pool	2130	
2	2011–10–04	Upper of liquid pool	2080	Outdoor temperature is 0~2°C, open the heating function, temperature of liquid room maintains at 37°C
		Middle of liquid pool	2070	
		Bottom of liquid pool	2110	

Note: The gelling viscosity will be subject to that of monitoring for 24 hours.

gelling viscosity of sampling from the top, middle and lower, the liquid is uniform basically. Under the condition of low temperature, open the heating function, the gelling condition is basically stable compared with before.

4 FIELD TEST AND EVALUATION

After completing adaptability evaluation of equipment control technology according to the requirements of design, field injection test is needed, of which the main purpose is to verify the adaptability to the environment, the liquor quality and reasonality of process parameters configuration of this technology and system device, and the controllability and effect of heat insulation measures through the experiment.

Field test shows that the thermal heating measures is practical and reliable; liquor device has the advantages of simple operation, stable performance, complete dissolution of medicament and uniform solution; displacement is adjustable, pressure is stable; after recovery injecting water, under the condition of keeping the same daily water-injection rate, the injection pressure is rising 1–2 MPa compared with before.

5 CONCLUSIONS

The experiments show that adaptability of the technology and the system is better and the working pressure and injection speed can meet the needs of the field. Liquor quality is higher; solution is uniform and cures thoroughly.

Electric heating device works well and temperature rise rate is appropriate. Piping heat insulation performance is good, completely meeting the need of Erlian area environment of below 40°C.

REFERENCES

Dagao Duan, Yan-song Cui, Zhongliang Deng. Hardware Design for Embedded Wireless Video Surveillance System Based on DM642 [J]. Journal of Electron Devices, dec2005,4(28).

Hinden R, Deering S. Internet Protocol Version 6 (IPv6) Addressing Architecture [S] . RFC 3513. April 2003.

Hongxun Wang, Qi Zhang et al. Principle of oil production technology [M], Bei Jing: Petroleum Industry Press, 1989.

Hongyu Shao, Yantao Tang. The design of efficient profile control liquid injection machine [J]. China Petroleum and Chemical Standard and Quality, 2011(7).

Jianghong Wu, Fuwei Yang. Modification and application effect of profile control pump [J]. Inner Mongolian Petrochemical Industry, Jun 2003.

Jianguo Shen et al. New development of fracturing and acidification technology in Sichun [J], Gas Industries, 2001.

Qi Zhang. Principle and design of oil production engineering [M], China University of Petroleum Press. Dong Ying, Shan Dong, 2006.

Electromechanical Control Technology and Transportation – Jia & Wu (Eds)
© 2017 Taylor & Francis Group, London, ISBN 978-1-138-06752-3

An efficient and reliable meter registration method for the power information collection system

Zhongxing Wu
China Electric Power Research Institute, Beijing, China

Weifeng Wang
State Grid Zhejiang Electric Power Company, Zhejiang, Hangzhou, China

Xingqi Liu & Linan Zhang
China Electric Power Research Institute, Beijing, China

ABSTRACT: In the power collection system, the traditional way of registration has difficulty in meeting the requirements of service development because of the long registration time and reliability issue. In this paper, we present an efficient and reliable meter registration method, which provides multiple parallel registration slots to accelerate the registration process. It also supports multiple-channel access modes as well as effective collision resolution mechanisms. Meanwhile, it introduces an acknowledgement mechanism to increase its reliability. The simulation results indicate that this method significantly reduces the registration time and improves the registration efficiency compared to the existing method. The reliability issue is also solved with respect to the conventional registration mechanism.

1 INTRODUCTION

With the rapid development of the automatic meter reading technology (Tang, 2015; Hu, 2014; Yin, 2015), the efficiency and reliability of the collector are attracting more and more attention. In power collection applications, a collector commonly connects many meters. The collector needs to search for all the downstream meter information and report the newly searched meter information to the master station. The DL/T 645 protocol (1997&2007) provides a wildcard reading operation which supports for abbreviated addressing. However, this scheme has a complex searching algorithm which incurs long average search time and the risk of inadequate searching results.

In this paper, taking all the aforementioned considerations into account, we present a reliable and high-efficiency meter registration method, which is suitable for the power collection system (Xu, 2015). Experiments and analysis show that the proposed method can effectively improve registration efficiency, reduce registration time greatly, and increase reliability (Du, 2015; Gao, 2014) compared with the existing method.

2 MECHANISM DESCRIPTION

The proposed registration mechanism is mainly implemented in the registration process by the collector to collect the information from downstream meters. The network consists of a collector and many meters. In the traditional DL/T 645 protocol, the registration mechanism utilizes wildcard address to finish the collection operations. Briefly, the collector first sends packets containing reading address wildcard AAAAAAAAAAn (n is a fixed value between 0 and 9; A represents a wildcard) to each meter in turn. Then, each meter feeds back whether their own table number matches the wildcard address contained in the received packet. If a collision occurs when the last wildcard is m, the collector will fix the last wildcard to m and select the penultimate wildcard as the polling wildcard, that is, selected packet address wildcard as with AAAAAAAAAnm (n is a fixed value between 0 and 9; m is a fixed value). The collector keeps taking such steps until there is no conflict to complete the whole process of meter reading. To conclude, this method uses polling search table operations and has a large number of invalid search table operations, which incurs a long average search time and the risk of inadequate searching results.

To solve this problem, we propose a method to provide multiple parallel registration slots in a registered command cycle when the collector sends a register command. Meters can select a slot from multiple different slots to send a register request, which significantly improves the efficiency of meter registration and effectively addresses the conflict problem that occurs in conflicting meter

1 Byte	1 Byte	2 Byte	2 Byte	1 Byte	variable	variable
serial number and priority flag (PIID)	registration type	response start time	response time slot length	channel access mode	channel access parameters	payload

Figure 1. Format of registration initiation message.

nodes. Meanwhile, the proposed method supports multiple channel access modes in order for the collector and meters to choose from. In addition, the mechanism has scalability. The complete registration mechanism consists of the meter registration mechanism when the collector powers up or restarts and the keep-alive mechanism after the completion of the meter registration.

2.1 Message structure and function

The registration mechanism defines six kinds of messages, which are registration initiation, registration request, registration response, registration response confirmation, registration keep-alive, and registration keep-alive confirmation.

2.1.1 Registration initiation message format
Figure 1 shows the format of registration initiation message. The main fields are defined as follows:

a. Registration type;
  1. 01, Re-registration,
  2. 02, Registration continuation,
  3. Others, reserved.
b. Response start time, expressed in milliseconds. Recommended values: 200~500.
c. Response time slot length, expressed in milliseconds. Recommended value: 150.
d. Channel access mode;
  1. 01, Address matching access mode (access mode 1),
  2. 02, Hash calculation access mode (access mode 2),
  3. 03, Random access mode (access mode 3),
  4. Others, reserved for extension.
e. Channel access parameter, which is determined by the specific channel access mode, is defined as follows:

1. For address matching access mode:
The destination address, support wildcard format:
It indicates that the collector specifies which wildcard in the meter address is used to match the slot number because each wildcard varies from 0 to 9. The meter sends the registration request message in the matching slot in this access mode.
2. For hash calculation access mode (Bai, 2004; Zhang, 2014):

1. Hash calculation factor, h_coeffs;
2. The maximum value for hash calculation, h_max_val, with no unit;
3. ID number of the meter, addr, with no
The calculation formula is as follows:

$$\left(\sum_i h\_coef fs_i * addr_i \right) \% h\_\mathbf{max}\_val$$

In the above formula, $addr_i$ represents two consecutive decimal numbers out of totally 12 decimal number meter addresses. There are six addresses selected in total. h_coeffs_i is the calculation factor in the hash algorithm. h_max_val is the slot number of hash mapping. Each meter obtains a slot number according to the result of the hash calculation and sends the registration request message in the slot.
3. For random access mode (Mei, 2012):
  1. Size of contention window, expressed in number of back-off units;
  2. Length of back-off unit, expressed in milliseconds;
  3. The total length of the registration cycle, expressed in milliseconds.

The collector will proactively broadcast registration initiation message with registration type set to be a re-registration when the collector powers up or restarts and it starts to invite the meters to register. Then, based on the registration condition, the collector may broadcast the registration initiation message with registration type set to be registration continuation to finish the registration of the individual meter which did not register during re-registration.

2.1.2 Other message formats
The formats of other types of messages are the same, as shown in Figure 2.

2.2 Registration mechanism

2.2.1 Registration process
The registration mechanism is mainly finished by a 4-way handshake. The procedure of registration process is shown in Figure 3. The collector first broadcasts the registration initiation message with registration typesetting to be a re-registration and

1 Byte	variable
serial number and priority flag (PIID)	payload

Figure 2.   Format of other messages.

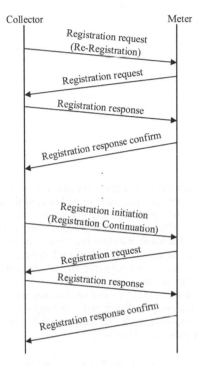

Figure 3.   Registration process.

provides multiple access slots for each meter to send a request in parallel. Depending on the channel access mode, the meter will send a registration request message to the collector at the appropriate access slot after receiving the registration initiation message. After receiving the registration request message, the collector sends the registration response message. In the end, the meter sends the registration response confirmation message to inform the collector that the meter has known that the registration is finished.

After the re-registration process, the condition may exist where some meters are still not registered due to delayed powering up, conflicts, or other reasons. For this condition, the collector broadcasts the registration initiation message with registration type set to be the registration continuation to those unfinished registration meters to continue the registration process. The registration mechanism is under different access modes.

After the collector sends the registration initiation message, the meter will send the registration request message according to the channel access mode. The following examples illustrate three kinds of channel access modes. It is assumed that there are three meters with the meter ID number to be meter 1 (600000000002), meter 2 (700000000035), and meter 3 (800000000065), respectively.

2.2.1.1   Address matching access mode
The collector first initiates registration with the registration type set to be re-registration as shown in Figure 4. The destination address is AAAAAAAAAAAA. So meter 1 (600000000002) transmits registration request in slot 2. The requests of meter 2 (700000000035) and meter 3 (800000000065) collide in slot 5.

Then the collector responds to meter 1 (600000000002). Meter 1 sends a response to confirm to the collector, as shown in Figure 5.

Due to the conflict between meter 2 and meter 3, the collector adjusts the address matching the slot and sends the registration initiation message again. The registration type is registration continuation and the destination address is AAAAAAAAAAA5, as shown in Figure 6. Meter 2 (700000000035) and meter 3 (800000000065) will respond respectively in slot three and slot six.

Then the collector confirms meter 2 and meter 3 followed by the responses from meter 2 and meter 3 respectively, as shown in Figure 7.

Then, the collector continues to send the registration initiation message. If there are no conflicts in the following ten slots, it can be considered that the registration process is completed.

2.2.1.2   Hash calculation access mode
The collector first broadcasts registration initiation message with registration type set to be re-registration, as shown in Figure 8. The result of hash calculations is assumed to be 1, 6, and 9 for three meters, respectively. Then three meters will send requests in the corresponding slots 1, 6, and 9.

After that, the collector responds to the determinate meters in turn and the meters send a confirmation, correspondingly. If a conflict occurs, it can be solved by adjusting the hash calculation factor and the maximum value. Then, the collector changes the registration type to be registration continuation and continues to implement the aforementioned procedure until no conflict occurs.

If no conflicts occurred after the collector sent a registration initiation message, it can be considered that the registration process is finished.

Random access mode
The collector first broadcasts registration initiation message with registration type set to be re-registration as shown in Figure 9. The random back-off length for each meter is between 1 and the contention window. In this example, the random back-off lengths for three meters are selected

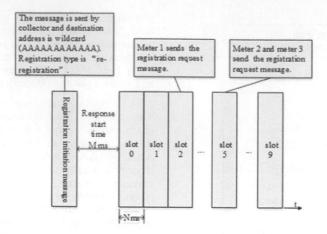

Figure 4.   Re-registration under access mode 1.

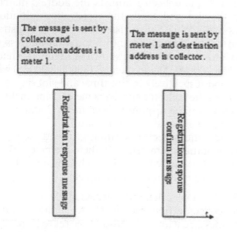

Figure 5.   Registration confirmation under access mode 1.

as 5, 10, and 20 respectively, which are constantly reduced by 1 every A millisecond (i.e., the back-off unit length is A millisecond). When the random back-off value for a certain meter is reduced to 0, it will send a registration request message to the collector while the other meters stop backing off immediately after receiving the registration request message from this meter. The pause will last for N milliseconds (message transmission duration) during which the meter will send the registration request. After that, the other meters start to back off again. The procedure continues to implement until all the random back-off lengths of the meters reach 0 or the total length of the registration cycle is reached. If the registration does not finish within one registration cycle (the length of a registration cycle is L milliseconds), it will register again in the next registration cycle.

Then, registration response and response confirmation messages are exchanged between the collector and corresponding meters as shown in Figure 10.

If a conflict occurs in one registration cycle, the random back-off window length will be modified to solve the conflict problem and the registration type is changed to registration continuation until no conflict occurs.

If the collector does not receive any registration request messages after it sends the registration initiation message, it can be considered that the registration process is finished.

2.2.2   Keep-alive registration mechanism

After the completion of registration, the registration information may change due to changes in the channel environment or device power failures, restart, or other reasons. The keep-alive mechanism realizes the updating of the registration information of the network equipment so that the collector will get the current status of the network.

The collector broadcasts the registration keep-alive message and the registered meters will send the registration keep-alive confirm message. After this process, the collector will delete the registration information of the meters which have left the network according to the received confirm messages.

2.3   Reliability mechanism

In order to guarantee the correct reception of registration response message by a meter, the collector requires the corresponding meter to send the registration confirmation message as a reply to the registration response message. If there is no reception of registration response confirmation message from the corresponding meter, the collector will resend the registration response message.

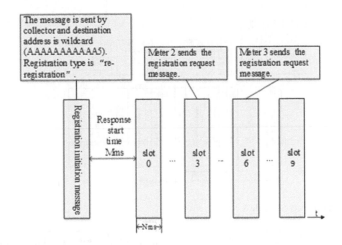

Figure 6.   Contribution of the registration under access mode 1.

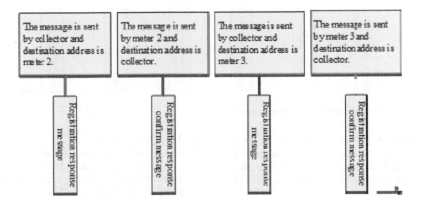

Figure 7.   Registration confirmation.

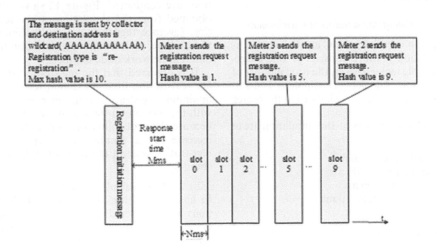

Figure 8.   Re-registration under access mode 2.

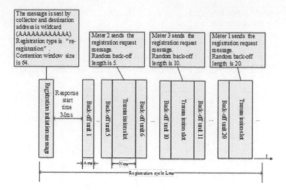

Figure 9.   Re-registration under access mode 3.

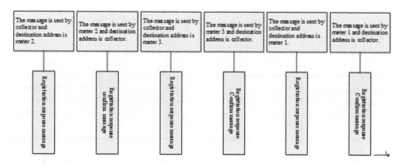

Figure 10.   Registration confirmation under access mode 3.

It will be treated as a registration failure if there is still no reception of registration confirmation message after k (recommended value is 3) times registration response messages are sent by the collector.

3   PERFORMANCE VERIFICATION AND ANALYSIS

3.1   *Introduction of the simulation environment*

The simulation is built by the C environment. The entire simulation network consists of one collector and multiple meter devices, the total number of which is variable. The network topology is a star network where the collector is located at the center, as shown in Figure 11.

The specific parameters of the simulation are as follows:

1. Registration cycle: 3000 ms;
2. Response start time: 200 ms;
3. Reply slot length: 150 ms;
4. A back-off unit length: 10 ms.

3.2   *Performance analysis of the registration mechanism*

The performance analysis is conducted in terms of effectiveness and reliability.

3.2.1   *Effectiveness*

Effectiveness is reflected in the average registration time, in other words, the shorter the average registration time is, the higher the effectiveness of registration mechanism will be.

3.2.1.1   Address matching access mode

In address matching access mode, 10,000 simulations are conducted. Figure 12 shows the results obtained from the simulation of the registration time (average registration time, maximum registration time, and minimum registration time) with different network node numbers.

It is observed that the registration times (average registration time, maximum registration time, minimum registration time) increase accordingly with the increase in the number of meters. This is because the probability of conflict increases with increase in the number of meters. The average registration time in the network with 32 nodes is about 35.6 s.

3.2.1.2   Hash calculation access mode

In hash calculation access mode, 10,000 simulations are conducted. Figure 13 shows the results obtained from the simulation of the registration time with different node numbers.

It can be seen that the average registration time, the maximum registration time, and the minimum registration time, all become longer with the node number increasing. The average registration time of 32 nodes is about 34.7 s.

In order to investigate the influence of the total number of hash slots on effectiveness, the registration time under different total number of hash slots is illustrated in Figure 14 where the number of meter nodes in the network is 32. It can be seen that registration time declines abruptly with the hash slot number until a certain point from where the registration time begins to increase. When the number of meter nodes is relatively small, conflict is the dominant factor which causes a large registration time. Therefore, as the slot number increases, the conflicts decrease and the registration time becomes shorter. However, when the number of hash slots becomes larger, the extra idle slots are the dominant factor which causes the increase of registration time. The registration time increases slowly due to the increment of hash slots.

The average registration time is the minimum when the number of slots is 80.

Next, the average registration time under different meter numbers is evaluated where the number of hash slots is 10 and 80, respectively. Figure 15 shows the impact of different meter numbers on average registration time with different slot numbers. When the meter number is small, the extra idle slots are the dominant factor which causes the increase of registration time. As inferred, the total number of idle slots under 80 slots is much more than that under 10 slots. Therefore, the registration time under 80 slots is longer distinctly than 10 slots. With the increase of meter nodes, the impact of conflicts on registration time becomes more and more significant. So, the less the slot numbers are, the greater the probability of conflicts is, which results in a faster increase of registration time. Finally, the average registration time under 10 slots exceeds that of 80 slots, but the difference is not distinct. So we choose 10 as the hash slot number in the following simulations.

3.2.1.3 Random access mode

In random access mode, 10,000 simulations are conducted. Figure 16 shows the registration time performance with a variation of the node number. The size of the contention window is set to 64. As shown in Figure 16, the average registration time, the maximum registration time, and the minimum registration time become longer with the increase of node numbers.

The registration time varies with the number of meter nodes and the size of the contention window. As shown in the table, with the increase in the number of meter nodes, the registration time becomes

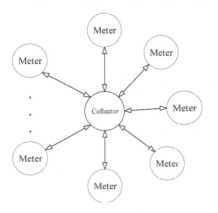

Figure 11. Network topology.

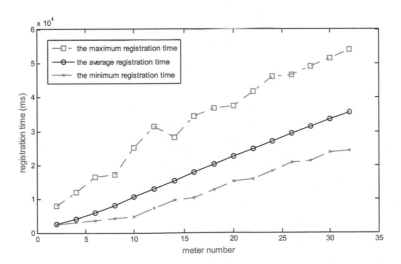

Figure 12. Simulation results of the registration time with different meter numbers in address matching access mode.

251

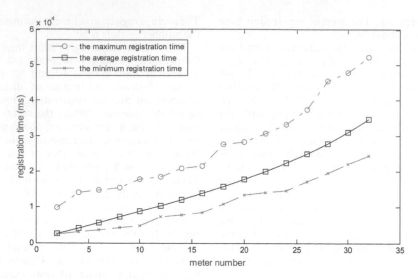

Figure 13. Simulation results of registration time with different meter numbers in hash calculation access mode.

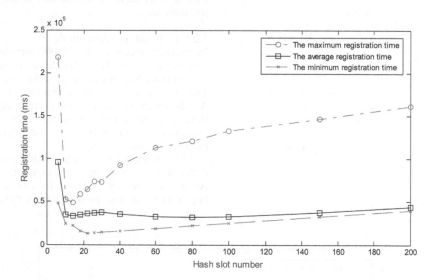

Figure 14. Simulation results of the registration time with different hash slot numbers.

longer. What's more, the selection of the contention window size affects the registration performance to a certain extent. When the contention window size is small, the conflict is the dominant factor in causing an increase of average registration time. However, with the increase of the contention window size, the time spent on backing off is the dominant factor which causes the increase in average registration time. Therefore, with the increasing window size, the overall trend of registration time first declines and then rises again. When the contention window size is 32, 64, and 128, the average registration time changes a little.

3.2.1.4 Summary and comparison

Figure 17 shows the comparison of the average registration time between the registration mechanism of DL/T 645 and the proposed registration method adopting three access modes. It is observed that the average registration time of the four mechanisms become longer with the increase in node numbers. However, the registration time performance of the proposed three access modes is always obviously better than the registration mechanism of the DL/T 645 regardless of the meter nodes, which shows that the average registration time performance of the proposed mechanism is significantly

252

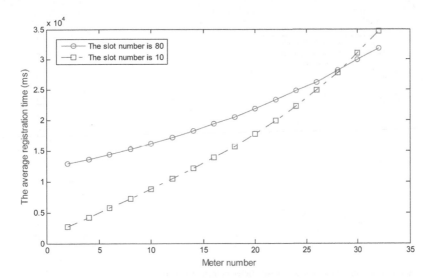

Figure 15. The average registration time with different meter numbers in hash calculation access mode.

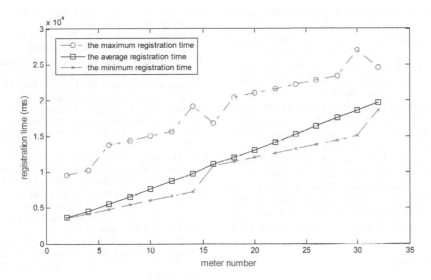

Figure 16. Simulation results of the registration time with different meter numbers in the random access mode.

improved. When the number of nodes is 32, the average registration time of DL/T 645 protocol needs more than 70 s, while the address matching access mode, the Hash calculation access mode and the random access mode only need 35.5 s, 34.7 s, and 19.6 s, respectively, each decreased by 53%, 54%, and 74%, which proves that the performance is significantly improved.

Furthermore, the curves obtained from the proposed mechanism using three different access modes also have differences. We can see from Figure 17 that the average registration time of the random access mode is minimum and hence the

performance is optimal. Hash calculation access mode is in the second place but it can be further improved by adjusting its parameters of the hash calculation. The average registration time of address matching access mode is slightly longer than the other two, but it is still far better than the registration mechanism of the DL/T 645 protocol. In practice, we can select the specific access mechanism according to the actual condition.

3.2.2 Reliability

In the DL/T 645 protocol, when the collector receives two different registration request messages

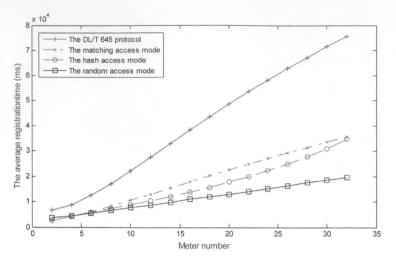

Figure 17. The average registration time comparison with different meter numbers.

from different meters in a single time slot and if these two messages do not collide in some circumstances, the collector will admit the first meter and miss the second meter. In this case, the second meter does not know that it is not registering successfully and thus it will not send the registration request to the collector anymore.

In our proposed registration mechanism, a registration confirmation mechanism is introduced. When the collector receives a registration request message, it will send a registration response message to the meter. As a result, the meters will know whether the registration is successful or not. If not successful, the meter can continue to register using the following registration continuation mechanism, which effectively avoids meter-searching failure and improves the reliability of the registration.

4 CONCLUSION

In this paper, based on the understanding of the registration mechanism of the power collecting system, we propose a more effective and reliable method for meter registration. The simulation results indicate that under three access modes of the proposed method, the registration performance is significantly superior to the traditional one. We also provide a reliability mechanism that effectively solves the problem of invalid searching that exists in the traditional registration mechanism.

REFERENCES

Bai Enjian, Wang Jinggang, Xiao Guozhen. Secure Hash algorithm with variable Hash output [J]. Computer Science, 2004, 31(4): 189–191.

DL/T 645–1997 State Grid Electric Power Research Institute multi-functional electric energy meter communication protocol [S]. People's Republic of China electric power industry standard 1997.

DL/T 645–2007 State Grid Electric Power Research Institute multi-functional electric energy meter communication protocol [S]. People's Republic of China electric power industry standard 2007.

Du Jiang, Guo Ruipeng, Li Chuandong, etc. An important control method for reliability evaluation of power system [J]. Automation of Electric Power Systems, 2015, 39 (5): 69–74.

Gao Huisheng. Analysis on the influence factors of power communication [J]. Power information and communication technology, 2014, 12(11): 1–7.

Hu Jiangyi, Zhu Enguo, Du Xingang, etc. Application status and development trend of electricity information collection system [J]. Automation of Electric Power Systems, 2014(2): 131–135.

Mei Huiping. Research on a realization technology based on the IEEE802.11b random backoff algorithm of MAC layer [J]. Value Engineering, 2012, 31 (28): 211–213.

Tang Yeqiang. Current status and development of automatic meter reading technology using electric power measurement [J]. Digital communication world, 2015(9).

Xu Dongmei, Hu Jingyi, Zhang Cui. Research on power line communication standardization [J]. Information technology and standardization, 2015(5).

Yin Lu, Sun Yun, Li Mingming, etc. The application of automatic meter reading in the calculation of power information collection system [J]. Digital communication world, 2015(9).

Zhang Ying, Wu Hesheng. Comparison and analysis of Hash algorithm for multiple processes [J]. Computer engineering, 2014, 40(9): 71–76.

Electromechanical Control Technology and Transportation – Jia & Wu (Eds)
© 2017 Taylor & Francis Group, London, ISBN 978-1-138-06752-3

Research on the optimal capacity configuration of the energy storage system for an electric vehicle fast-charging station

Ruiming Yuan
State Grid Jibei Electric Power Co. Ltd., Beijing, China
North China Electric Power Research Institute Co. Ltd., Beijing, China

Qinqin Li
School of Electrical Engineering, Beijing Jiaotong University, Beijing, China

Kan Zhong & Zhenyu Jiang
State Grid Jibei Electric Power Co. Ltd., Beijing, China
North China Electric Power Research Institute Co. Ltd., Beijing, China

Qingduo Yin
State Grid Jibei Electric Power Co. Ltd., Beijing, China

Weige Zhang
School of Electrical Engineering, Beijing Jiaotong University, Beijing, China

ABSTRACT: The fast-charging station for electric vehicles has the characteristics of a high charging power and short charging time. However, there are also problems such as severe load fluctuation and a low load rate of the transformer. The addition of the battery energy storage system in the fast-charging station is the most direct and effective way to solve these problems. The energy storage system can reduce the distribution capacity of the charging station and increase the load rate of the transformer by charging during the valley-load period and discharging during the peak-load period. The energy storage system can also use the Time-of-Use (TOU) price by storing energy when the electricity cost is low and releasing it when the electricity cost is high, in order to reduce the electricity cost and stabilize load fluctuations. In this paper, a mathematical model for the daily total cost of the fast-charging station with an energy storage system is established, and a linear programming method is used to solve this problem. Taking an electric bus fast-charging station as an example, the paper also analyzes and studies the charging and discharging rates of the storage battery. The results prove that the established model of the fast-charging station with energy storage system can reduce the cost of the charging station and stabilize load fluctuations.

1 INTRODUCTION

According to statistics, the number of electric vehicles in China has reached 800,000 until September 2016, ranking it first in the world. China's Ministry of Transport has issued the "urban public transport" 13th Five-Year "Development Program." The program made it clear that new energy vehicles will reach 200,000 in urban public transport by 2020 (Qian, 2016). In recent years, with the rapid development of electric vehicles and the extensive construction of high-power direct current (DC) fast-charging station, the charging station has become an important point supplying energy to electric vehicles. The charging time, especially for taxis, buses, and other modes of urban public transport, is short. However, through the data

analysis of the existing fast-charging station, it has been found that the existing fast-charging stations generally have a problem: their transformer capacity selection is too large, which increases the initial investment cost of the charging station and leads to a low utilization of the transformer (Ge, 2013). In addition, a high-power fast DC charging station in the practical application, when load fluctuations are large, can access the grid directly, leading to a greater impact on the power grid. However, it is not conducive to the safe operation of power distribution systems (Yong, 2013; Mastny, 2016). Configuring energy storage system in the fast-charging station is the most direct and effective way to solve the aforementioned problems (Chen, 2016; Negarestani, 2016; García-Triviño, 2016). The energy storage system can reduce the distribution capacity of the charging station and improve the utilization ratio of the transformer through charging in the valley load and discharging in the peak load. The Time-of-Use (TOU) price can also be utilized by storing energy when the cost is low and releasing energy when the cost is high, in order to reduce the electricity cost and stabilize load fluctuations.

At present, there are many studies on the capacity optimization configuration of the energy storage system, but most of them are concerned with wind battery, photovoltaic (PV) battery, wind/PV battery, and smart grid. Energy storage devices are usually installed in order to smooth the power fluctuations of wind farms and PV plants, and reduce the impact on power systems and users (Zheng, 2016; Lin, 2014; Zhou, 2014). Large batteries can be used in smart grids to reduce the average cost of energy supply. The annual energy balance method is proposed for the independent wind power generation system, using batteries to balance the power difference between the power generating system and the power consumption for capacity configuration (Hu, 2012). Wang (2008) proposed the peak integration method, which takes the maximum absolute value of the integration of battery charging and discharging on the entire time axis as the capacity of batteries. Xu (2016) proposed the point-by-point calculation method, by taking charge–discharge times as the objective function, to calculate the corresponding abandoned electric quantity, charge–discharge times of energy storage for obtaining the power and capacity optimal value of energy storage. Codemo (2013) studied the cost saving that can be achieved by some selected energy storage algorithms in smart grids, such as battery capacity, charge/discharge rate, power request process, and cost functions.

However, there is little research on the capacity allocation about the energy storage system of fast-charging stations in the aforementioned studies. In this paper, a mathematical model for the total cost of the fast-charging station with energy storage system is established. The paper aims at minimizing the total cost per day of the system, and considers the power and capacity constraints of the energy storage battery. Then, a linear programming method is applied to obtain the optimal capacity and power of the energy storage system. Besides, the charging and discharging rates of the storage battery are also analyzed and studied.

2 MATHEMATICAL MODEL OF THE FAST-CHARGING STATION WITH ENERGY STORAGE SYSTEM

The model of the fast-charging station with energy storage system is schematically shown in Figure 1.

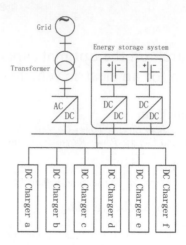

Figure 1. Schematic of the model of the fast-charging station with energy storage system.

The charging station is equipped with a certain number of DC chargers. Establishing the battery energy storage system in the DC bus side of the fast-charging station can stabilize load fluctuations and reduce the impact on the grid. Besides, the energy storage system can also make use of the TOU price to reduce the electricity cost and thus improve the economic benefits of the fast-charging station.

2.1 Optimization objective function

Setting minimizing total cost per day of the fast-charging station with energy storage system as the optimization objective, we have:

$$\min C = \frac{C_1 + C_2}{365} + C_3 \tag{1}$$

$$C_1 = w_1(p_p \cdot P_{bess} + p_e \cdot Q_{bess}) + p_{m1} \cdot Q_{bess} \tag{2}$$

$$C_2 = w_2 \cdot p_t \cdot P_{trans} + p_{m2} \cdot P_{trans} + p_c \cdot P_{trans} \tag{3}$$

$$C_3 = \sum_{t=0}^{24} \left\{ (P_1(t) + P_b(t)) \cdot p_g(t) \right\} \tag{4}$$

$$w_1 = \frac{r_0(1+r_0)^m}{(1+r_0)^m - 1} \tag{5}$$

$$w_2 = \frac{r_0(1+r_0)^n}{(1+r_0)^n - 1} \tag{6}$$

where p_p is the power cost of the energy storage battery (yuan/kW); p_e is the capacity cost of the energy storage battery (yuan/kWh); p_{m1} is the annual Operation and Maintenance (O&M) cost of the energy storage battery (yuan/kWh); P_{bess} is the rated power of the energy storage battery (kW); Q_{bess} is the rated capacity of the energy storage battery (kWh); p_t is the capital cost of the transformer (yuan/kVA); p_{m2} is the annual Operation and Maintenance (O&M) cost of the transformer (yuan/kVA); p_c is the annual basic capacity cost of the transformer (yuan/kVA); P_{trans} is the distribution capacity of the transformer (kVA); P_l (t) is the load demand power (kW); P_b (t) is the energy storage charge and discharge power (kW); p_g (t) is the TOU price of the grid (yuan/kWh); r_0 is the discount rate; m is the lifespan of the energy storage battery (years); n is the lifespan of the transformer (years).

2.2 Constraint conditions

To ensure the rationality of the storage capacity optimization, the limitations of instantaneous power balance, energy storage capacity and power, energy storage State of Charge (SOC), energy storage capacity regression, and the energy storage charge and discharge rates should be considered as constraints:

$$P_g(t)=P_l(t)+P_b(t) \tag{7}$$

$$P_{min} \le P_b(t) \le P_{max} \tag{8}$$

$$\begin{cases} Q(t) = Q(t-1) + P_b(t-1) \\ \quad Q(t) \le Q_{bess} \end{cases} \tag{9}$$

$$\begin{cases} \sum_{t=1}^{24} P_b(t) = 0 \\ \quad SOC(t) = \dfrac{Q(t)}{Q_{bess}} \\ SOC_{min} \le SOC(t) \le SOC\text{max} \end{cases} \tag{10}$$

where P_g (t) is the grid output power (kW); P_{min} and P_{max} are the upper and lower limits of energy storage (kW); Q (t) and Q (t–1) are the energy storage capacity at times t and t–1, respectively (kWh); P_b (t–1) is the energy storage power at time t-1 (kW); SOC (t) is the SOC of energy storage capacity at time t; SOC_{min} and SOC_{max} are the upper and lower limits of the SOC of energy storage.

3 CASE STUDY

3.1 System introduction

Taking a bus fast-charging station as an example, the charging station has eight DC 450 kW charging piles of 58 fully electric buses. The typical daily load curve is shown in Figure 2. The peak load appears at 10:15–11:15 and 18:30–20:00, which are

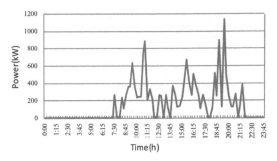

Figure 2. The typical daily load curve of the fast-charging station.

the peak periods of the electricity cost. The valley load appears at 22:00–07:00, which is the valley period of the electricity cost. The peak load of the charging station reaches 1144kW and the average load is 168kW; as a result, the load rate is only 0.147. Normally, the charging station load is lower than the maximum load and the utilization of the transformer is very low.

3.2 Input data and model parameters

The parameters of the energy storage battery are as follows: the capacity cost of the energy storage battery is 1500/kWh, the capacity recovery cost is 300yuan/kWh, the power cost of the energy storage battery is 500yuan/kW, the annual O&M cost of the energy storage battery is 15 yuan/kWh, the lifespan of the energy storage battery is taken as 8 years, and the SOC range of the energy storage battery is chosen as 0.1–0.9.

The parameters of the transformer are as follows: the capital cost of the transformer is 150yuan/kVA, the annual O&M cost of the transformer is 30yuan/kVA, the annual basic capacity cost of the transformer is 384yuan/kVA, and the lifespan of the transformer is taken as 20 years based on industry data.

The discount rate is 6%. The TOU price of the distribution grid is presented in Table 1.

3.3 Results

The daily total cost of the fast-charging station with energy storage system is shown in Figure 3. In order to find the tendency of the total cost towards the change in battery capacities and charging–discharging rates, a case with a wide range of charging and discharging rates (0.5–3C) and energy storage battery capacities (114–800 kWh) was simulated. The optimal results at different charging and discharging rates are summarized in Table 2.

Table 1. The TOU price of the distribution grid.

Division of period	Period	TOU price (yuan/kWh)
Off-peak period	23:00–7:00	0.3946
Semi-peak period	Other time	0.6950
On-peak period	10:00–15:00 and 18:00–21:00	1.0044

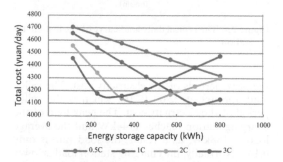

Figure 3. Daily total cost of the fast-charging station with energy storage system.

Table 2. Optimal results at different charging and discharging rates.

	No energy storage	690 kWh/1C	460 kWh/2C	230 kWh/3C
Peak load (kW)	1144	455	374	483
Energy storage cost (yuan)	0	545	465	283
Transformer cost (yuan)	1254	499	453	529
Electricity cost (yuan)	3516	3049	3192	3365
Total cost (yuan)	4770	4093	4110	4177

From Figure 3, it can be seen that the increase in energy storage capacity at the initial range will increase the maximum transferable electricity during the peak period, and therefore the total daily cost will decrease. When the storage capacity increases to a certain value, increased costs of energy storage will be higher than the gains from transferable electricity, and the total cost will increase. Table 2 summarizes the optimal results at different charge and discharge rates. The total cost of the system has different degrees of reduction compared with the situation of no energy storage system. So, it is significant to configure the energy storage in the fast-charging station.

As we can see from Table 2, the total cost of the system is lowest when the energy storage capacity is 690kWh and the charge and discharge rate is 1C. Under this condition, the load of the charging station with and without energy storage can be compared, as shown in Figure 4. The charge and discharge power of energy storage is shown in Figure 5, and the comparison of the results of three situations is given in Table 3.

Figure 4 shows that the peak load of the fast-charging station is reduced considerably after adding the energy storage system. Figure 5 shows that the energy storage system basically charges during the valley price or flat price period, and discharges during the peak electricity price period, contributing to the reduction in the electricity cost. As we can see from Table 3, operating at a large charge and discharge rate contributes to the reduction in the peak load, whereas operating at a small charge and discharge rate contributes to the reduction in the cost of electricity. In general, the daily total cost is the least when the energy storage capacity is 690 kWh and the charge and discharge rate is 1C, but the recovery period is also the longest.

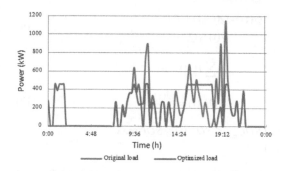

Figure 4. Load comparison of the fast-charging station before and after adding the energy storage system.

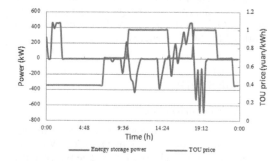

Figure 5. Charge and discharge power of the energy storage system.

Table 3. Comparison of the results.

	Load clipping ratio	Electricity cost ratio	Transformer cost ratio	Rate of return	Recovery period (years)
690 kWh/1C	60%	13%	60%	14%	2.8
460 kWh/2C	67%	9%	64%	13.8%	2.6
230 kWh/3C	58%	4%	58%	12%	2

4 CONCLUSION

This paper established the battery capacity optimization model of the electric vehicle fast-charging station with a battery energy storage system. The following cost factors were considered: battery capacity cost, power cost, and transformer basic capacity cost, as well as the battery capacity constraints, power constraints, charge–discharge rate, SOC constraints, and other operating factors. A linear programming method is adopted to arrive at the optimal value of the capacity of the energy storage system.

Taking a fast-charging station as an example, the result shows that the equipping energy storage system in the fast-charging station can reduce the peak load and charging station distribution capacity, thereby reducing the construction cost of the charging station. The energy storage system can also make use of the TOU electricity price by charging during the valley-price period and discharging during the peak-price period to reduce the operation cost of the charging station.

ACKNOWLEDGMENTS

This paper was supported by the National Key R&D Program of China (2016YFB0900505).

REFERENCES

Chen H, Hu Z, Xu Z, et al. Coordinated charging strategies for electric bus fast charging stations[C]. *Power and Energy Engineering Conference (APPEEC), 2016 IEEE PES Asia-Pacific. IEEE,* 2016: 1174–1179.

Codemo C G, Erseghe T, Zanella A. Energy storage optimization strategies for smart grids[C]. *2013 IEEE International Conference on Communications (ICC).* IEEE, 2013: 4089–4093.

García-Triviño P, Fernández-Ramírez L M, Torreglosa J P, et al. Control of electric vehicles fast charging station supplied by PV/energy storage system/grid[C]. *Energy Conference (ENERGYCON), 2016 IEEE International.* IEEE, 2016: 1–6.

Ge Wenjie, Huang Mei, Zhang Weige. Economic operation analysis of the electric vehicle charging station[J]. *Transactions of China Electrotechnical Society,* 2013, 28(2): 15–21.

Hu Guozhen, Duan Shanxu. Sizing and cost analysis of photovoltaic generation system based on vanadium radix battery[J]. *Transactions of China Electro technical Society.* 2012, 35, 260–267.

Lin Li, Li Liangyu. An optimal capacity configuration method of wind/PV and energy storage co-generation system[C]. *2014 IEEE PES General Meeting| Conference & Exposition.* IEEE, 2014: 1–5.

Mastny P, Moravek J, Vrana M. Concept of fast charging stations with integrated accumulators—Assessment of the impact for operation[C]. *Electric Power Engineering (EPE), 2016 17th International Scientific Conference on.* IEEE, 2016: 1–6.

Negarestani S, Fotuhi-Firuzabad M, Rastegar M, et al. Optimal Sizing of Storage System in a Fast Charging Station for Plug-in Hybrid Electric Vehicles[J]. *IEEE Transactions on Transportation Electrification.* 2016, 4(2): 443–453.

Qian Jin. Ministry of Transport issued "urban public transport" thirteen five "Development Program" [J]. *Engineering construction standardization,* 2016 (8): 21–21.

Wang, X.Y. Determination of battery storage capacity in energy buffer for wind farm[J]. *IEEE Transactions on Energy Conversion,* 2008: 868–878.

Xu Zhenzhou, Liu Bo. Study on the optimization method of energy storage configuration for new energy[C]. *Electricity Distribution (CICED), 2016 China International Conference on.* IEEE, 2016: 1–4.

Yong J Y, Ramachandaramurthy V K, Tan K M, et al. Modeling of electric vehicle fast charging station and impact on network voltage[C]. *Clean Energy and Technology (CEAT), 2013 IEEE Conference on.* IEEE, 2013: 399–404.

Zheng Xiangyu, JiaRong, Xie Yongtao, et al. Energy storage capacity configuration for large-scale wind power accessed to saturated system[C]. *Electricity Distribution (CICED), 2016 China International Conference on.* IEEE, 2016: 1–5.

Zhou Tianpei, Sun Wei. Optimization of battery–supercapacitor hybrid energy storage station in wind/solar generation system[J]. *IEEE Transactions on Sustainable Energy,* 2014, 5(2): 408–415.

Electromechanical Control Technology and Transportation – Jia & Wu (Eds)
© 2017 Taylor & Francis Group, London, ISBN 978-1-138-06752-3

Optimal control for dry dual clutch overlap operations during torque phase in shift

Mingxiang Wu
College of Engineering, Shanghai Normal University TianHua College, Shanghai, P.R. China

ABSTRACT: A mathematical model for dynamic behaviors of the on-coming and off-going clutches during torque phase in shift is proposed. Then, an objective functional counting for friction works of both clutches, shock intensity of the vehicle and engine acceleration is constructed. Further, an optimal control strategy is established by seeking minimum value of the objective functional. The seeking principle is based upon the Pontryagin's minimum principle. As a result, analytically optimal solutions of the engine torque and the dual clutch friction torques with respect to time are derived. Finally, a numerical analysis of transient dynamic responses of first to second gear shift on a ramp road is implemented through the MATLAB/Simulink platform. It is shown by comparison that shock intensity of the vehicle and friction loss between driving and driven plates remain contradictory in the torque phase.

1 INTRODUCTION

1.1 *Research background*

In comparison with Mechanical Transmission (MT), Automatic Transmission (AT) and Continuous Variable Transmission (CVT), Dual Clutch Transmission (DCT) (Kegresse, 2013) has merits in power transmission efficiency and fuel consumption (Rudolph, 2003). For the purpose of comprehensively optimizing shift quality, dual clutch friction wear and fuel efficiency, the dynamics mechanism behind dual clutch engagement behaviors should be investigated in depth at first. Then, proper application of optimal control theory should be implemented to develop dual clutch control technologies.

There have been many literatures published in investigations to dynamics and control performances of dual clutch engagement. Y. Zhang and X. Chen (2003 & 2005) validated their multi-body system dynamic model of wet dual clutch assembly through Modelica/Dymola development platform. Similar works according to Newtonian dynamic modeling methodology are also accomplished by many other researchers (Goetz, 2005; Manish, 2007; Yang, 2007; Liu, 2009; Joshi, 2009; Ahlawat, 2010; Galvagno, 2011; Walker, 2011; Bataus, 2012; Walker, 2013) through Matlab/Simulink simulation platform, with differences in model degrees of freedom. However, as applications of lower order models in driveline control strategy design could dramatically reduce operation time of algorithms, most researchers tended to build driveline models with no more than 4 DOF for very quick calculations with least delay possible in real vehicle operations.

The main control objective is comprehensive optimization of shock intensity, dual clutch friction work and engine acceleration during shift. Kulkarni Manish and Shim Taehyun (2007) investigated influence of different dual clutch overlap operations to engine power transmission qualities during shift. Yang Weibin (2007) and Hu Hongwei (2010) utilized orthogonal experimental design method and genetic algorithm for optimization of all key parameters in the formulation of empirically optimal dual clutch pressures. However, as robustness of the above open loop control techniques is not good in real vehicle operations, many close loop control strategies are proposed and published in recent years. Original close loop control flow charts were conceived by Zhang Yi (2005) and M Goetz (2005), respectively, in which values of optimal engine torque and dual clutch friction torques were timely generated from feedback calculation of measurable state variables, such as speeds of flywheel, clutch driven plates and other driveline components. Subsequent researchers focused on demonstrating validity and robustness of feedback and feedforward close loop control strategies. Main control approaches are fuzzy neural network controls (Qin, 2009; Liu, 2009; Sun, 2011; Sun, 2011) and intelligent PID controls

(Zhang, 2010). In their investigations, throttle opens, throttle open change rates, engine-dual clutch speed differences and their time derivatives were recommended main inputs to calculate values of control variables, such as displacement of dual clutch actuators. Probably due to difficulties in state space-based dynamic modellings and functional analysis skills, it is difficult for these researchers to derive analytically optimal solutions of engine torques and dual clutch friction torques. As a result, the above expert knowledge-based empirical control strategies increase computational burden, limit fields of their applications to few certain types of vehicles, and also are not able to improve shift qualities to a more reasonable fine level. For this reason, Ni (2009) proposed his objective functional regarding friction time and weighted quadratic term of the shock intensity, and derived optimal dual clutch engagement and release laws in shift based on the minimum principle. As engine dynamics was excluded in his dynamic model, the friction work had to be extremely underestimated in an approximate form of friction time for successful solution of his optimal shift control problem. To overcome this problem, researchers (Li, 2010; Chen, 2010; Liu, 2011) established their LQ objective functionals with regard to shock intensity and friction wear of dual clutch, and then derived optimal engagement and release laws of dual clutches in shift. Their derivation approaches rely on the minimum principle. However, as lack of engine maneuvers in objective functionals and approximations of friction loss in linear quadratic form, improvements of shift qualities were also limited. Besides, time spend in offline solution of Riccati equations not only increase computational burden but also restrict applications of these control algorithms in real vehicle operations, especially in the very short torque phase generally accomplished within 0.2 s.

1.2 *Research means*

In view of above imperfections, the present research proposes inclusion of friction work and shock intensity into the proposed objective functional at first. Meanwhile, one feasible engine maneuver in correspondence to dual clutch operations is formulated in terms of square of engine acceleration. Then, engine acceleration is included in the objective functional. Further, the proposed optimal control problem is solved on the basis of the extremum value theorem, and analytical solutions of optimal engine and dual clutch friction torques are derived for ease of engineering applications. Finally, a typical 1st to 2nd gear shift on a ramp road is numerically simulated in Matlab/Sim-

ulink platform, and validity as well as efficiency of the proposed control algorithm are verified.

2 NONLINEAR MATHEMATICAL MODELLING OF THE DCT DYNAMIC BEHAVIORS IN SHIFT

2.1 *Torque transmission characteristic of whole shift process*

The whole shift process can be divided into three main phases in time domain, which has been delimited to shift preparation phase, inertia phase and torque phase, as shown in Fig. 1 and Fig. 2.

In the torque phase, dual clutch overlap operations are characterized by pressure ramping up of the on-coming clutch and pressure ramping down of the off-going clutch. Since dynamic mechanism

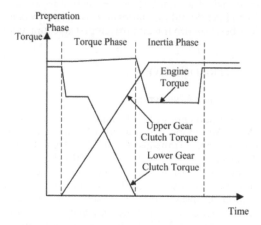

Figure 1. Torque characteristic of engine and clutch in upshift.

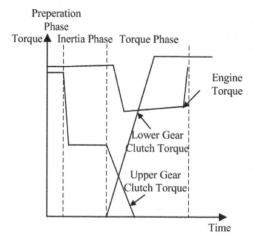

Figure 2. Torque characteristic of engine and clutch in downshift.

behind the inertia phase is similar to that of the launch and shift of AMT as well as the launch of DCT, only the dynamics and control in torque phase are investigated.

2.2 Dynamic modelling of torque phase

Referring to literature (Yang, 2007; Galvagno, 2011; Hu, 2010), an integrated optimal control-oriented DCT driveline dynamic model (As shown in Fig. 3) is built for proposing control strategy aiming at effective compromise of shock intensity, dual clutch friction loss and engine acceleration during shift. The DCT driveline is regarded as an assembly of rigid rotating components, which means only the moment of inertia of each component is considered and the elasticity-damping properties of the DCT driveline are ignored for eliminating unnecessary computational burdens, as indicated in(Yang, 2007; Galvagno, 2011; Hu, 2010).

The dynamic equations of the dual clutch engagement in the torque phase are presented as follows

$$
\begin{cases}
J_e \dot{\omega}_e = T_e - (T_{co} + T_{ce}) \\
J_{co} \dot{\omega}_{co} = T_{co} - T_{mo} \\
J_{ce} \dot{\omega}_{ce} = T_{ce} - T_{me} \\
(J_{oo} + J_{oe} + J_o) \dot{\omega}_o = T_{mo1} + T_{me1} - T_m \\
(J_{out} + J_M) \dot{\omega}_r = T'_m - T_r
\end{cases} \tag{1}
$$

The present research calculates the resisting moment T_r applied on the tire by utilizing the following formulation [27]:

$$
T_r = r_w (F_f + F_a + F_l) \tag{2}
$$

and tire rolling friction resistance F_f, wind resistance F_a and gradient resistance F_i are calculated

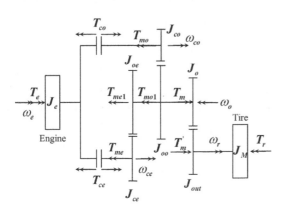

T_{co}
T_{mo}
J_{co}
ω_{co}
J_{oe}
J_o
T_e
J_e
ω_e
T_{me1} T_{mo1} T_m
ω_o
Tire
Engine
T_{me}
J_{oo} T'_m ω_r T_r
J_M
T_{ce}
ω_{ce}
J_{ce}
J_{out}

Figure 3. The simplified dynamic model of the DCT.

through the following formulations (Chen, 2000):

$$
F_f = (M + M_H) \cdot g \cdot f \tag{3}
$$

$$
F_a = \frac{C_D \cdot A \cdot v^2}{21.25} \tag{4}
$$

$$
F_i = (M + M_H) \cdot g \cdot \sin \alpha_i \tag{5}
$$

with $k = \frac{r_w}{i_o}$, $f = f_0 \left(1 + \frac{v^2}{19440}\right)$ [27], where f is rolling resistance coefficient, f_0 is static rolling resistance coefficient, g is acceleration of gravity, v is vehicle velocity, r_w is wheel rolling radius, A is frontal area of the vehicle, C_D is air resistance coefficient, M is total vehicle mass without load, M_H is mass of the load.

To investigate optimal dynamic behaviors of the DCT in torque phase, following state variables, control variables and outputs of the DCT driveline dynamic models are defined:

$$
x_1 = \omega_e, x_2 = \omega_o, x_3 = \dot{\omega}_o \tag{6}
$$

$$
u_1 = T_e, u_2 = T_{co}, u_3 = T_{ce}, u_4 = \dot{u}_2, u_5 = \dot{u}_3 \tag{7}
$$

$$
y_1 = x_1, y_2 = x_2, y_3 = x_3 \tag{8}
$$

Substitution of the above 3 transformations into equation (1) yields the following dynamic equation of the DCT presented in state space.

$$
\dot{x} = f_T(x) + \sum_{i=1}^{5} g_{Ti}(x) u_i \tag{9}
$$

where

$$
f_T(x) = \left[0, -\frac{T_r}{i_o J_{eq}}, 0\right]^T, \quad g_{T1}(x) = \left[\frac{1}{J_e}, 0, 0\right]^T,
$$

$$
g_{T2}(x) = \left[-\frac{1}{J_e}, \frac{i_{go}}{J_{eq}}, 0\right]^T, \quad g_{T3}(x) = \left[-\frac{1}{J_e}, \frac{i_{ge}}{J_{eq}}, 0\right]^T,
$$

$$
g_{T4}(x) = \left[0, 0, \frac{i_{go}}{J_{eq}}\right]^T, \quad g_{T5}(x) = \left[0, 0, \frac{i_{ge}}{J_{eq}}\right]^T.
$$

Naturally, equivalent inertia is calculated in the following form

$$
J_{eq} = \frac{J_{out} + J_M}{i_o^2} + i_{go}^2 J_{co} + i_{ge}^2 J_{ce} + J_{oo} + J_{oe} + J_o \tag{10}
$$

where i_{go} is transmission gear ratio of the odd gear, i_{ge} is transmission gear ratio of the even gear, i_o is final drive gear ratio.

3 OPTIMAL PROCESS CONTROL

3.1 Application of Pontryagin's minimum principle to resolving the integrated optimal engine-dual clutch torque matching problem in torque phase

To effectively compromise shock intensity, friction wear and engine acceleration, a nonlinear objective functional is formulated in terms of friction work, shock intensity and engine acceleration. The functional is presented as follows:

$$J = \int_{t_{Ti}}^{t_{Te}^*} \left[Q_1 u_2 \left(x_1 - i_{go} x_2 \right) + Q_2 u_3 \left(x_1 - i_{ge} x_2 \right) + k^2 Q_3 \left(\frac{i_{go} u_4 + i_{ge} u_5}{J_{eq}} \right)^2 + Q_4 \left(\frac{u_1 - u_2 - u_3}{J_e} \right)^2 \right] dt \quad (11)$$

According to the extremum value theorem, analytical solutions of the optimal engine torque and dual clutch friction torques are derived as following

$$u_1 = U_{11} \cdot \left[e^{\lambda(t-t_{Ti})} - 1 \right] + U_{12} \cdot (t - t_{Ti}) + u_1 \left(t_{Ti} \right) \quad (12)$$

$$u_2 = U_{21} \cdot \left[e^{\lambda(t-t_{Ti})} - 1 \right] + U_{22} \cdot (t - t_{Ti}) + u_2 \left(t_{Ti} \right) \quad (13)$$

$$u_3 = U_{31} \cdot \left[e^{\lambda(t-t_{Ti})} - 1 \right] + U_{32} \cdot (t - t_{Ti}) + u_3 \left(t_{Ti} \right) \quad (14)$$

$$u_4 = U_{41} e^{\lambda(t-t_{Ti})} + U_{42} \quad (15)$$

$$u_5 = U_{51} e^{\lambda(t-t_{Ti})} + U_{52} \quad (16)$$

where

$$\lambda = \frac{Q_1 \left(i_{go} Q_2 - i_{ge} Q_1 \right) J_{eq}}{2 Q_4 \left(Q_1 - Q_2 \right) i_{go}^2 i_{ge}}, \quad k = \frac{r_w}{i_o},$$

$$U_{11} = \left(\frac{i_{ge} - i_{go}}{Q_2 i_{go} - Q_1 i_{ge}} - \frac{J_e}{2 Q_4 \lambda} \right) \times \left(\frac{Q_1 J_{eq}}{2 Q_3 i_{go} k^2 \lambda} \lambda_{3,0} - Q_1 u_2 \left(t_{Ti} \right) - Q_2 u_3 \left(t_{Ti} \right) \right),$$

$$U_{12} = \left(\frac{Q_1 - Q_2}{Q_2 i_{go} - Q_1 i_{ge}} + \frac{Q_1 J_e}{2 Q_4 i_{go} \lambda} \right) \frac{J_{eq} \lambda_{3,0}}{2 Q_3 k^2},$$

$$U_{21} = \frac{1}{Q_2 i_{go} - Q_1 i_{ge}} \left[\frac{Q_1 i_{ge} J_{eq} \lambda_{3,0}}{2 Q_3 i_{go} k^2 \lambda} - i_{ge} \left(Q_1 u_2 \left(t_{Ti} \right) + Q_2 u_3 \left(t_{Ti} \right) \right) \right]$$

$$U_{22} = -\frac{Q_2 J_{eq} \lambda_{3,0}}{2 Q_3 \left(Q_2 i_{go} - Q_1 i_{ge} \right) k^2},$$

$$U_{31} = \frac{1}{Q_2 i_{go} - Q_1 i_{ge}} \left[-\frac{Q_1 J_{eq} \lambda_{3,0}}{2 Q_3 k^2 \lambda} + i_{go} \left(Q_1 u_2 \left(t_{Ti} \right) + Q_2 u_3 \left(t_{Ti} \right) \right) \right],$$

$$U_{32} = \frac{Q_1 J_{eq} \lambda_{3,0}}{2 Q_3 \left(Q_2 i_{go} - Q_1 i_{ge} \right) k^2},$$

$$U_{41} = \frac{1}{Q_2 i_{go} - Q_1 i_{ge}} \left[\frac{Q_1 i_{ge} J_{eq} \lambda_{3,0}}{2 Q_3 i_{go} k^2} - i_{ge} \lambda \left(Q_1 u_2 \left(t_{Ti} \right) + Q_2 u_3 \left(t_{Ti} \right) \right) \right],$$

$$U_{42} = U_{22} = -\frac{Q_2 J_{eq} \lambda_{3,0}}{2 Q_3 \left(Q_2 i_{go} - Q_1 i_{ge} \right) k^2},$$

$$U_{51} = \frac{1}{Q_2 i_{go} - Q_1 i_{ge}} \left[-\frac{Q_1 J_{eq} \lambda_{3,0}}{2 Q_3 k^2} + i_{go} \lambda \left(Q_1 u_2 \left(t_{Ti} \right) + Q_2 u_3 \left(t_{Ti} \right) \right) \right],$$

$$U_{52} = U_{32} = \frac{Q_1 J_{eq} \lambda_{3,0}}{2 Q_3 \left(Q_2 i_{go} - Q_1 i_{ge} \right) k^2}.$$

3.2 Numerical simulation analysis for dual clutch overlap qualities in torque phase

The dynamic model of dual clutch overlap operations in torque phase is built in the Matlab/Simulink/SimDriveline. Values of structure parameters of a midsize passenger car configured with a DCT assembly are listed in Table 1. Since proceeding time of the torque phase is usually within 0.2 s, road resistance torque T_r is regarded to be invariant with time. This hypothesis is verified by some researchers in their investigations into the dynamics and control of the engagement and release processes of DCT dual clutch (Liu, 2011; Liu, 2010).

Numerical predictions obtained for the transient shift responses of the DCT vehicle are shown in Fig. 4. This typical shift condition is a 1st gear to 2nd gear shift on a 6° ramp road. Through careful observations of numerical simulation results, a general conclusion can be derived, which is summarized as: gradual increase in Q_1/Q_3 and Q_2/Q_3

Table 1. Vehicle parameters.

Item	Quantity value	unit	Item	Quantity value	unit
M+M$_H$	1600	kg	i$_{go}$	3.21, 1.42, 0.81	
r$_w$	0.3	m	i$_{ge}$	1.91, 1.05, 0.62	
J$_e$	2.7000	kg*m²	i$_o$	4.2	
J$_{co}$	0.0015	kg*m²	f$_o$	0.02	
J$_{ce}$	0.0010	kg*m²	C$_D$	0.33	
J$_{oo}$	0.0050	kg*m²	αi	0.1	
J$_{oe}$	0.0020	kg*m²			
J$_o$	0.0010	kg*m²			
J$_{out}$	0.0050	kg*m²			
J$_M$	153.80	kg*m²			
A	2.2	m²			

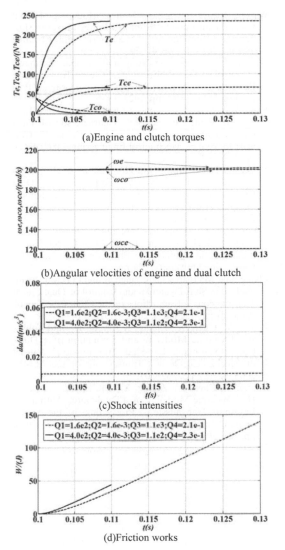

(a)Engine and clutch torques

(b)Angular velocities of engine and dual clutch

(c)Shock intensities

(d)Friction works

Figure 4. Shift qualities of a typical 1st to 2nd gear shift induced by the optimal engine-dual clutch torque transmission in torque phase.

value yields gradual increase of shock intensity and gradual decrease of friction loss. Besides, it should also be noted that the value of Q_4 must be carefully adjusted to prevent occurrence of power circulations, which has been demonstrated in the present numerical simulation work.

4 CONCLUDING REMARKS

A proper dynamic model of dual clutch overlap operations in shift is built with exclusion of gear backlashes and stiffness damping characteristics of all transmission components, just for eliminat-

ing unnecessary computational burdens. Then, an optimal control strategy, featured with proposition of an objective functional including formulations of friction work, shock intensity and engine maneuver, has been established and investigated numerically in depth. Further, shift qualities of a typical 1st to 2nd gear shift on a ramp road are numerically studied in the Matlab/Simulink platform, and a typical phenomenon that friction loss of dual clutch friction discs and vehicle jerk remain contradictory in torque phase is discovered. Finally, it can be concluded that friction loss of dual clutch friction discs and vehicle jerk are still two contradictory indices ought to be primarily considered in torque phase.

Since it can be numerically predicted that power circulation phenomenon is prevented and that excellent compromise of ride comfort, friction loss of friction discs is achieved, it would be eagerly anticipated to testify our proposed control strategy through real vehicle driving tests under typical driving conditions.

REFERENCES

Ahlawat, R., Fathy, H.K., Lee, B., Stein, J.L., Jung, D. "Modeling and Simulation of a Dual-clutch Transmission Vehicle to Analyse the Effect of Pump Selection on Fuel Economy". Vehicle System Dynamics, 2010, 48(7), pp. 851–868.

Bataus, Marius-Valentin., and Vasiliu, Nicolae. "Modeling of A Dual Clutch Transmission for Real-time Simulation". UPB Scientific Bulletin, Series D: Mechanical Engineering, 2012, 74(2), pp. 251–264.

Chen Qinghong. "The Key Control Technology Research of AMT/DCT Automatic Transmissions Universal Development Platform". PhD dissertation, 2010, Chongqing University, Chongqing.

Chen, L. Investigation to dynamics and control of AMT vehicle clutches during engagement (in Chinese). PhD Dissertation, Shanghai Jiaotong University, Shanghai, People's Republic of China, 2000.

Galvagno, E., Velardocchia, M. and Vigliani, A. "Dynamic and Kinematic Model of A Dual Clutch Transmission". Mechanism and Machine Theory, 2011, 46(6), pp. 794–805.

Goetz, M., Levesley, M., and Crolla, D. "Dynamics and Control of Gearshift on Twin-Clutch Transmissions". Proc. IMechE, Part D: J. Automobile Engineering, 2005, 219(8), pp. 951–963.

Hongwei Hu, Xianghong Wang and Yimin Shao. "Optimization of the Shift Quality of Dual Clutch Transmission Using Genetic Algorithm". 2010 Sixth International Conference on Natural Computation (ICNC 2010), pp. 4152–4156.

Joshi, Ajinkya S., Shah, Nirav P., Mi, Chris. "Modeling and Simulation of A Dual Clutch Hybrid Vehicle Powertrain". 5th IEEE Vehicle Power and Propulsion Conference, VPPC '09, 2009, pp. 1666–1673.

Kegresse, A. "Zahn der wechselgetriebe für Kraftfahzeuge". Pat. DE 894 204, 1939.

Li Yuting, Zhao Zhiguo and Zhang Tong. "Research on Optimal Control of Twin Clutch Engagement Pressure for Dual Clutch Transmission". China Mechanical Engineering, 2010, 12(12), pp. 1496–1501.

Liu Hongbo, Lei Yulong, Zhang Jianguo, Jin Lun and Li Youde. "Shift Quality Assessment and Optimization of Dual-clutch Transmission". Journal of Jilin University, 2012, 42(6), pp. 1360–1365.

Liu Xi, Cheng Xiusheng and Feng Wei. "Optimal Control of Gear Shift on Wet Dual-clutch Automatic Transmission (in Chinese)". Transactions of the CSAE, 2011, 27(6), pp. 152–156.

Liu Yonggang, Qin Datong, Jiang Hong and et al. "Shift schedule optimization for dual clutch transmissions". 5th IEEE Vehicle Power and Propulsion Conference, VPPC '09, 2009, pp. 1071–1078.

Manish Kulkarni, Taehyun Shim and Zhang Yi. "Shift dynamics and control of dual-clutch transmissions". Mechanism and Machine Theory, 2007, 42(2), pp. 168–182.

Ni Chunsheng, Lu Tongli and Zhang Jianwu. "Gearshift Control for Dry Dual-clutch Transmissions". WSEAS Transactions on Systems, 2009, 8(11), pp. 1177–1186.

Qin Datong, Zhao Yusheng, Hu Jianjun and Ye Ming. "Analysis of Shifting Control for Dry Dual Clutch System". Journal of Chongqing University, 2009, 32(9), pp. 1016–1023.

Rudolph, F., Steinberg, I., and Günter, F. "Die Doppelkupplung des Direktschaltgetriebes DSG der Volkswagen AG". VDI-Ber., 2003, 1786, pp. 401–411.

Sun Xian'an, and Wu Guangqiang. "Shifting Control Strategy Simulation for Dual Clutch Transmission Vehicle". Journal of Southeast University, 2011, 41(4), pp. 729–733.

Sun Xian'an, and Wu Guangqiang. "Shifting Quality Evaluation System for Dual Clutch Transmission Vehicle". Chinese Journal of Mechanical Engineering, 2011, 47(8), pp. 146–151.

Walker, Paul D. and Zhang Nong. "Modelling of Dual Clutch Transmission Equipped Powertrains for Shift Transient Simulations". Mechanism and Machine Theory, 2013, 60, pp. 47–59.

Walker, Paul D., Zhang Nong and Tamba, Richhard. "Control of Gear Shifts in Dual Clutch Transmission Powertrains". Mechanical Systems and Signal Processing, 2011, 25(6), pp. 1923–1936.

Yang Weibin, Wu Guangqiang, Qin Datong. "Drive Line System Modeling and Shift Characterstic of Dual Clutch Transmission Powertrain (in Chinese)". Chinese Journal of Mechanical Engineering, 2007, 43(7), pp. 188–194.

Yonggang Liu, Datong Qin, Hong Jiang, et al. "A Systematic Model for Dynamics and Control of Dual Clutch Transmissions". Journal of Mechanical Design, Transactions of the ASME, 2009, 131(5), Paper No. 061012.

Yonggang Liu. "Study on Integrated Control of Passenger Vehicles Equipped with Dual Clutch Transmissions (in Chinese)". PhD dissertation, 2010, Chongqing University, Chongqing.

Zhang Jinle, Ma Biao, Zhang Yingfeng and Li Heyan. "Simulation on Shift Dynamics and Control of Dual Clutch Transmissions". Transactions of the Chinese Society of Agricultural Machinery, 2010, 41(5), pp. 6–11.

Zhang, Y., Chen, X., Zhang, X., Jiang, H., and Tobler, W. "Dynamic Modeling and Simulation of a Dual-Clutch Automated Lay-Shaft Transmission". Journal of Mechanical Design, Transactions of the ASME, 2005, 127(2), pp. 302–307.

Zhang, Y., Chen, X., Zhang, X., Tobler, W., and Jiang, H. "Dynamic Modeling of a Dual-clutch Automated Lay-shaft Transmission". Proceedings of the ASME Design Engineering Technical Conference, 2003, pp. 703–708.

Electromechanical Control Technology and Transportation – Jia & Wu (Eds)
© *2017 Taylor & Francis Group, London, ISBN 978-1-138-06752-3*

A nonlinear programming method for solving LCCL resonant structure parameters

Jianxin Gao, Xusheng Wu, Wei Gao, Wei Peng & Xin Xiong
Department of Electrical Engineering, Naval University of Engineering, Wuhan, China

ABSTRACT: To solve the problem that the LCCL resonant structure parameters design process is difficult and heavily constrained, a nonlinear programming method for solving LCCL resonant structure parameters is proposed. According to the equivalent circuit of LCCL resonant structure, the decision variables are selected. Then, the equality constraints and inequality constraints are described by decision variables. Finally, according to the design goal, the objective function is constructed. The complex parameters design problem of LCCL resonance structure is transformed into a general nonlinear programming problem. Simulation and experiment is carried out. The simulation and experimental results show: in the case of using the proposed parameters design method for LCCL resonant structure, when the ICPT system inverter turns off, the power tube enters the anti parallel diode freewheeling state, the instantaneous current value of the power tube is 0, and ZCS of the ICPT inverter is achieved.

1 INTRODUCTION

By using the principle of electromagnetic induction, the Inductive Contactless Power Transfer (ICPT) system can convert electrical energy into magnetic energy. In the form of magnetic field, the ICPT system can realize the contactless transmission of high power electric energy in the larger air gap. However, there is a large air gap between the transmitter and the receiver coil in the ICPT system. The magnetic reluctance in air gap is relatively large, which result that it is difficult to form a concentrated magnetic circuit between transmitter coil and the receiving coil. Thus, the problem of low mutual inductance and large leakage inductance between the transmitting coil and the receiving coil is produced. With the increase of the distance between the transmitting coil and the receiving coil, the mutual inductance between the coils decreases rapidly, and the leakage inductance increases rapidly, which directly affects the power and efficiency of the ICPT system. In order to solve this problem, most of the existing ICPT systems adopt resonance method. Increasing the resonant structure on both sides of the transmitter and receiver, the coils and resonant structures are resonant, that reduces the reactive power caused by the leakage inductance of the coil and improves the power factor of the system. The each stage of the resonant structure can be accurately described by differential equations. According to the circuit structure and the number of independent energy storage devices, the resonant structure can be divided into first-order capacitor series or parallel structure, second-order LC structure, third-order LLC or LCL structure, and the higher-order complex structure.

Most of the researchers have studied the basic structure and the second-order LC structure, however, the research on the third-order and higher-order resonant structures is relatively insufficient. On the research of the third-order resonant structure, most of the researchers use the soft switching method of the resonant converter. The majority of the research on the ICPT system third-order resonant structures is based on the LCL resonant structure.

Because of the constraint conditions are relatively large, if the iterative method is used to solve the parameters of the LCCL resonant structure, the design process needs to rely on the experience of the designer. In addition, the whole design process is complicated, and the iteration is needed. For this reason, this paper proposes a Nonlinear Programming (NLP) method to solve the LCCL resonant structure parameters. Firstly, the decision variables are selected according to the equivalent circuit of LCCL resonant structure. Then, the equality constraints and inequality constraints are described by decision variables. Finally, according to the design goal, the objective function is constructed. In this paper, the complex parameters design problem of LCCL resonance structure is transformed into a general nonlinear programming problem.

2 FORMULATION OF THE PROBLEM

The equivalent circuit of ICPT system LCCL resonant structure is shown in Figure 1.

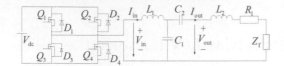

Figure 1. LCCL resonant structure.

The equivalent circuit of ICPT system LCCL resonant structure is shown in Figure 1.

The input of the LCCL resonant structure is connected to the inverter of the ICPT system, and the output is connected with the ICPT system power transmitting coil. High power ICPT inverter system generally uses the full bridge structure as shown in Figure 1. Q_1, Q_2, Q_3, Q_4 is the inverter switch tube, D_1, D_2, D_3, D_4 is the anti parallel diode of the corresponding switch tube. The snubber circuit is omitted in Figure 1. V_{dc} is the DC supply voltage of the ICPT inverter. V_{in} is the input voltage of the LCCL resonant structure. I_{in} is the input current of LCCL resonant structure. V_{out} is the output voltage of the LCCL resonant structure. I_{out} is the output current of LCCL resonant structure. R_1 is the AC resistance of the power transmitting coil (measured at the resonance frequency of the transmitter coil). Z_f is the reflected impedance of the receiver.

2.1 Selection of decision variables and constraint conditions

In order to minimize the number of decision variables, it is easily known that V_{in} and I_{out} can effectively describe the correlation of the LCCL resonant structure parameters based on the analysis of the LCCL resonant structure. Therefore, V_{in} and I_{out} are selected as decision variables. Because V_{in} and I_{out} are continuous values, the optimization model is a continuous optimization model.

Firstly, the resonant frequency of LCCL resonant structure is determined $f_0 = Q_{f0}$. The equality constraint of resonance angular frequency ω_0 can be described as Formula (1–2).

$$f_0 = Q_{f0} \tag{1}$$

$$\omega_0 = 2\pi f_0 \tag{2}$$

At the transmitter coupling coil resonance frequency f_0, the value of the ICPT system transmitter coil inductance L_2 is measured as Q_{L2}, the value of the ICPT system transmitter coil AC resistance R_1 is measured as Q_{R1}, the value of the ICPT system receiver reflection impedance is measured as Z_f. The resistance section of Z_f is noted as Q_{Zf}.

Using Kirchhoff Voltage Law (KVL) and Kirchhoff Current Law (KCL), the equality constraints between the parameters of the LCCL resonant structure can be determined as Formula (3–8). L_2 is

the ICPT system power transmitter coil inductance value, which is not easy to adjust in the actual situation. Therefore, we use the other components cooperating with L_2 to form resonance instead of changing the value of L_2, So, the constraint condition of the Formula (6) is a fixed value constraint.

$$C_1 = \frac{I_{out}}{\omega_0 V_{in}} \tag{3}$$

$$C_2 = \frac{C_1}{-1.25 + \dfrac{L_2}{L_1}} \tag{4}$$

$$L_1 = \frac{V_{in}}{\omega_0 I_{out}} \tag{5}$$

$$L_2 = Q_{L2} \tag{6}$$

$$R_1 = Q_{R1} \tag{7}$$

$$Z_f = Q_{Zf} \tag{8}$$

According to the relations between the parameters of the LCCL resonant structure, the fundamental voltage of each electronic component of the LCCL resonant structure can be analyzed by means of KVL and KCL. We can easily draw the conclusion that the voltage parameters of the LCCL structure electronic components must satisfy the equality constraint as shown in the Formula (9–12).

$$U_{C1} = \sqrt{V_{in}^2 + I_{out}^2(R_1 + Z_f)^2} \tag{9}$$

$$U_{C2} = \frac{I_{out}}{\omega_0 C_2} \tag{10}$$

$$U_{L1} = I_{out}(R_1 + Z_f) \tag{11}$$

$$U_{L2} = \omega_0 I_{out} L_2 \tag{12}$$

Similarly, the current value of the LCCL structure electronic components must satisfy the equality constraints as shown in Formula (13–16).

$$I_{C1} = \sqrt{\left(\frac{I_{out}^2}{V_{in}}(R_1 + Z_f)\right)^2 + I_{out}^2} \tag{13}$$

$$I_{C2} = I_{out} \tag{14}$$

$$I_{L1} = \frac{I_{out}^2}{V_{in}}(R_1 + Z_f) \tag{15}$$

$$I_{L2} = I_{out} \tag{16}$$

Note the rated output power of the LCCL resonant structure as $P_{out, nom}$, so that the output power P_{out} of the LCCL resonant structure must satisfy the inequality constraints shown in the Formula (17).

$$P_{out} = I_{out}^2(Z_f) \ge P_{out, nom} \tag{17}$$

Note the maximum allowable voltage of C_1 component as U_{C1max}, the maximum allowable voltage of C_2 component as U_{C2max}, the maximum allowable voltage of L_1 component as U_{L1max}, the maximum allowable voltage of L_2 component as U_{L2max}. It is easily drawn that the voltage of the LCCL structure electronic components must satisfy the inequality constraint conditions as shown in the Formula (18–21).

$$0 < U_{C1} \leq U_{C1max} \tag{18}$$

$$0 < U_{C2} \leq U_{C2max} \tag{19}$$

$$0 < U_{L1} \leq U_{L1max} \tag{20}$$

$$0 < U_{L2} \leq U_{L2max} \tag{21}$$

Note the maximum allowable current of C_1 component as I_{C1max}, the maximum allowable current of C_2 component as I_{C2max}, the maximum allowable current of L_1 component as I_{L1max}, the maximum allowable current of L_2 component as I_{L2max}. It is easily drawn that the current of the LCCL structure electronic components must satisfy the inequality constraint conditions as shown in the Formula (22–25).

$$0 < I_{C1} \leq I_{C1max} \tag{22}$$

$$0 < I_{C2} \leq I_{C2max} \tag{23}$$

$$0 < I_{L1} \leq I_{L1max} \tag{24}$$

$$0 < I_{L2} \leq I_{L2max} \tag{25}$$

Note the minimum available parameter value of the component C_1, C_2, L_1 in the actual situation as C_{1min}, C_{2min}, L_{1min}, respectively. Note the maximum available parameter value of the component C_1, C_2, L_1 in the actual situation as C_{1max}, C_{2max}, $L_2/1.25$, respectively. Thus, the actual available value of the component parameters of the LCCL resonant structure must be satisfied with the inequality constraint conditions as shown in the Formulas (26–28).

$$C_{1min} \leq C_1 \leq C_{1max} \tag{26}$$

$$C_{2min} \leq C_2 \leq C_{2max} \tag{27}$$

$$L_{1min} \leq L_1 \leq L_2/1.25 \tag{28}$$

Above, the complete mathematical model of the LCCL resonant structure is described by means of the Formula (1–28) in the form of nonlinear programming constraints. The construction method of objective function will be discussed below.

2.2 Constructing objective function

Under the conditions of satisfying the above constraints, the optimal LCCL resonant structure parameters should satisfy the conditions shown in the formula (29), that is, the rated voltage and rated current of the LCCL resonant structure should be as small as possible under the rated conditions, so that the electronic components maximum allowable voltage and current left margin. In the meanwhile, the input voltage value of LCCL resonant structure should be as small as possible.

$$\min\{U_{C1}, U_{C2}, U_{L1}, U_{L2}, I_{C1}, I_{C2}, I_{L1}, I_{L2}, V_{in}\} \tag{29}$$

The analysis formula (29) shows that if the Formula (29) is directly used as the objective function, the problem is a multi-objective optimization problem, each target is constrained by the decision variables, and the optimization of one of the objectives must be based on the cost of other goals. For this reason, this paper does not attempt to find the optimal solution. Instead, we try to find a non inferior solution by constructing a single objective function, so that each target can be optimized as much as possible.

The single objective function is constructed as shown in formula (30). The weight coefficient of U_{C1} is k_1, U_{C2} weight coefficient is k_2, U_{L1} weight coefficient is k_3, U_{L2} weight coefficient is k_4, I_{C1} weight coefficient is k_5, I_{C2} weight coefficient is k_6, I_{L1} weight coefficient is k_7, I_{L2} weight coefficient is k_8, V_{in} weight coefficient is k_9.

$$Z = k_1 U_{C1}^2 + k_2 U_{C2}^2 + k_3 U_{L1}^2 + k_4 U_{L2}^2 + k_5 I_{C1}^2 + k_6 I_{C2}^2 + k_7 I_{L1}^2 + k_8 I_{L2}^2 + k_9 V_{in}^2 \tag{30}$$

In order to achieve the goal that each component of the LCCL resonant structure work in a relatively light load level and the load level difference of each component is as small as possible, the normalization processing of the weight coefficients should be carried out based on the corresponding maximum allowable voltage or current value of the electronic components. Because the value of the objective function has no practical significance, to solve the nonlinear programming problem is in order to obtain the LCCL resonant structure parameters. Therefore, in order to facilitate the computer program to solve the nonlinear programming problem, the weight coefficients can be multiplied by an integer, so that the k_i ($i = 1, 2, \ldots 8$) are in the form of positive integers, only the ratios between k_i ($i = 1, 2, \ldots 8$) needs to be maintained.

In particular, when $k_i = 0$ ($i = 1, 2, \ldots 8$), the objective function is empty, the problem is reduced to a constrained optimization problem with no objective function. It is still meaningful to solve this constrained optimization problem. On the one hand, we can quickly determine whether the nonlinear programming problem is feasible. On the other hand, we can quickly solve a set of feasible solutions to meet the various constraints.

Based on the above constraints and the objective function, the nonlinear programming problem is proposed. V_{in} and I_{out} are taken as decision variables. The optimization target is to minimize the objective function. The constraints are the equality constraints and inequality constraints as shown in the formula (1–28). The optimal V_{in} and I_{out} values can be obtained by solving the nonlinear programming problem, and then using the formulas (3–5), we can get the parameters of the LCCL resonant structure.

Finally, according to the relations between the input fundamental voltage RMS value of the LCCL structure and the inverter input side equivalent DC voltage, the inverter input side equivalent DC voltage V_{dc} can be calculated as shown in formula (31)

$$V_{dc} = \frac{\sqrt{2}\pi}{4}V_{in} \tag{31}$$

Based on the above analysis and mathematical modeling, the design problem of complex LCCL resonance structure is transformed into a general nonlinear programming problem. The number of constraints and decision variables are relatively small. Hence, the trust region method, interior point method and penalty function method can be used to solve the nonlinear programming problem quickly. For solving nonlinear programming problems, this paper does not do further research and discussion.

3 SIMULATION AND EXPERIMENTAL ANALYSIS

Next, the simulation and experimental analysis will be carried out to further analyze and validate the proposed LCCL resonant structure parameter design method.

In this paper, we designed a LCCL resonant structure with a resonant frequency of 40 kHz and an output power of 1 kW. The identified LCCL resonant structural parameters are shown in Table 1. Low tail current IGBT is chosen as the switch tube.

According to the method proposed in this paper, the nonlinear programming problem is as follows:

$$\min Z = U_{C1}{}^2 + U_{C2}{}^2 + U_{L1}{}^2 + U_{L2}{}^2 \\ + 625 \times (I_{c1}{}^2 + I_{c2}{}^2 + I_{L1}{}^2 + I_{L2}{}^2) \tag{32}$$

Subject to

a. The LCCL parameters relations constraints

$$C_1 = \frac{I_{out}}{\omega_0 V_{in}} \tag{33}$$

$$C_2 = \frac{C_1}{\frac{L_2}{L_1} - 1.25} \tag{34}$$

Table 1. Identified parameters of LCCL resonant structure.

Parameter	Value	Parameter	Value
f_0/Hz	40 000	I_{C2max}/A	40
$P_{out,nom}$/W	1000	C_{2min}/μF	0.01
L_2/μH	105.70	C_{2max}/μF	1.32
R_l/Ω	0.05	U_{L1max}/V	2000
R_f/Ω	2.6	I_{L1max}/A	40
U_{C1max}/V	2500	L_{1min}/μH	0
I_{C1max}/A	40	L_{1max}/μH	84.56
C_{1min}/μF	0.01	U_{L2max}/V	1000
C_{1max}/μF	1.32	I_{L2max}/A	40
U_{C2max}/V	2500	—	—

$$L_1 = \frac{V_{in}}{\omega_0 I_{out}} \tag{35}$$

$$L_2 = 105.70 \times 10^{-6} \text{ H} \tag{36}$$

$$R_l = 0.05 \ \Omega \tag{37}$$

$$Z_f = 2.6 \ \Omega \tag{38}$$

b. The resonance frequency constraints

$$f_0 = Q_{f0} = 40000 \text{ Hz} \tag{39}$$

$$\omega_0 = 2\pi f_0 = 2 \times \pi \times 40000 \text{ rad/s} \tag{40}$$

c. The LCCL voltage relations constraints

$$U_{C1} = \sqrt{V_{in}{}^2 + I_{out}{}^2(R_l + Z_f)^2} \tag{41}$$

$$U_{C2} = \frac{I_{out}}{\omega_0 C_2} \tag{42}$$

$$U_{L1} = I_{out}(R_l + Z_f) \tag{43}$$

$$U_{L2} = \omega_0 I_{out} L_2 \tag{44}$$

d. The LCCL current relations constraints

$$I_{C1} = \sqrt{\left(\frac{I_{out}^2}{V_{in}}(R_l + Z_f)\right)^2 + I_{out}^2} \tag{45}$$

$$I_{C2} = I_{out} \tag{46}$$

$$I_{L1} = \frac{I_{out}^2}{V_{in}}(R_l + Z_f) \tag{47}$$

$$I_{L2} = I_{out} \tag{48}$$

e. The LCCL structure output power constraints

$$P_{out} = I_{out}^2(Z_f) \geq 1000 \text{ W} \tag{49}$$

f. The components voltage constraints

$$0 < U_{C1} \leq 2500 \text{ V} \tag{50}$$

270

$$0 < U_{C2} \le 2500 \text{ V} \tag{51}$$

$$0 < U_{L1} \le 2000 \text{ V} \tag{52}$$

$$0 < U_{L2} \le 1000 \text{ V} \tag{53}$$

g. The components maximum allowable current value constraints

$$0 < I_{C1} \le 40 \text{ A} \tag{54}$$

$$0 < I_{C2} \le 40 \text{ A} \tag{55}$$

$$0 < I_{L1} \le 40 \text{ A} \tag{56}$$

$$0 < I_{L2} \le 40 \text{ A} \tag{57}$$

h. The component parameters range constraints

$$0.01 \, \mu\text{F} \le C_1 \le 1.32 \, \mu\text{F} \tag{58}$$

$$0.01 \, \mu\text{F} \le C_2 \le 1.32 \, \mu\text{F} \tag{59}$$

$$0 \, \mu\text{H} \le L_1 \le 84.5572 \, \mu\text{H} \tag{60}$$

In this paper, the Lingo program is used to solve the nonlinear programming problem, and the optimal parameter values L_1, C_1, C_2, and V_{in} of the LCCL resonant structure are obtained. The results are as follows:

$$C_1 \approx 0.3579 \, \mu\text{F} \tag{61}$$

$$C_2 \approx 0.3127 \, \mu\text{F} \tag{62}$$

$$L_1 \sim 44.23 \, \mu\text{H} \tag{63}$$

$$V_{in} \approx 218.01 \text{ V} \tag{64}$$

Calculate the equivalent DC voltage value V_{dc} on the inverter side of ICPT system.

$$V_{dc} = \frac{\sqrt{2}\pi}{4} V_{in} \approx \frac{\sqrt{2}\pi}{4} \times 218.01 \text{ V} \approx 242.15 \text{ V} \tag{65}$$

Figure 2 shows the corresponding relations between the output voltage and current of the LCCL resonant structure and the driving signal of the ICPT system when the proposed parameters design method is used. From top to bottom of Figure 2, the figures are the LCCL resonant structure input voltage waveform, LCCL resonant structure input current waveform, inverter single IGBT drive signal 1, and inverter single IGBT drive signal 2. The dotted line in the Figure 2 corresponds to the switching time of the inverter switch tubes.

As can be seen in Figure 4, when the ICPT system inverter turns off, the power tube enters the anti parallel diode freewheeling state, the instantaneous current value of the power tube is 0, and ZCS of the ICPT inverter is achieved.

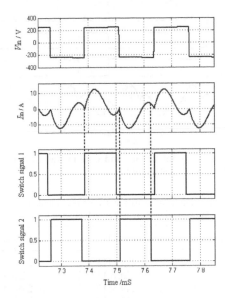

Figure 2. The simulation waveform of LCCL resonant structure using the proposed parameters design method.

In order to verify the parameters design method proposed in this paper, a prototype is built. The schematic of the prototype is shown in Figure 1. The parameters of the LCCL resonant structure are calculated by the proposed method in this paper, as shown in the formula (61–65). Low tail current IGBT is chosen as the switch tube, and in order to prevent full bridge short circuit, 2 μS dead time interval is added to IGBT driver signal.

Figure 3 is the input voltage and current waveform of the LCCL resonant structure designed by the proposed method when the reflection impedance is 2 Ω. The waveform is acquired by using the 200 MB bandwidth probes and Tektronix MSO2024B oscilloscope. The waveform from top to bottom is as follows: the inverter single IGBT driving voltage waveform V_{GE}, the input voltage V_{in} waveform of the LCCL resonant structure, input current I_{in} waveform of the LCCL resonant structure.

As shown in Figure 5, the experimental results are consistent with the simulation results. When the ICPT system inverter turns off, the IGBT enters the anti parallel diode freewheeling state, the instantaneous current value of the IGBT is 0, and ZCS of the ICPT inverter is achieved. The experiment verified that the proposed parameters design method can make ICPT system inverter work in ZCS state, reduce the inverter loss of ICPT system, and enhance the efficiency of the ICPT system.

Table 2 shows the comparison between the fundamental value calculated by the proposed parameters design method and the measured effective value of the prototype. In order to simplify

271

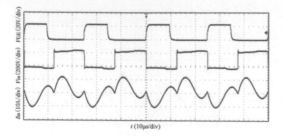

Figure 3. Input voltage and current waveform of the LCCL resonant structure.

Table 2. The comparison between the calculated fundamental value and the measured effective value.

Parameters	Calculated	Measured
U_{C1}/V	224.1171	237.33
I_{C1}/A	20.16117	25.09
U_{C2}/V	249.5512	248.39
I_{C2}/A	19.61161	19.52
U_{L1}/V	51.97078	135.07
I_{L1}/A	4.675197	7.31
U_{L2}/V	520.9713	523.25
I_{L2}/A	19.61161	19.52
V_{in}/V	218.0081	240.58

the expression of the voltage and current of the LCCL resonant structure, the fundamental value is used to approximate. Therefore, a certain error has been produced. Among them, the voltage and current of the C_1 and L_1 elements contain a large number of harmonic, which leads to a relatively large difference between the fundamental value and the measured effective value. At the same time, The C_1 and L_1 components work in the resonant state, they have a strong low-pass filtering effect, higher-order harmonic which C_2, L_2 components contained has been decreased, so the difference between calculated fundamental value and the measured effective value is relatively small.

4 CONCLUSIONS

Aiming at the problem that the LCCL resonant structure parameters design process is difficult and heavily constrained, a nonlinear programming method for solving LCCL resonant structure parameters is proposed. According to the equivalent circuit of LCCL resonant structure, the decision variables are selected. Then, the equality constraints and inequality constraints are described by decision variables. Finally, according to the design goal, the objective function is constructed. The complex parameters design problem

of LCCL resonance structure is transformed into a general nonlinear programming problem. Simulation and experiment is carried out. The simulation and experimental results show: in the case of using the proposed parameters design method for LCCL resonant structure, when the ICPT system inverter turns off, the power tube enters the anti parallel diode freewheeling state, the instantaneous current value of the power tube is 0, and ZCS of the ICPT inverter is achieved.

REFERENCES

Covic G A, Boys J T. 2013. Modern trends in inductive power transfer for transportation applications. *IEEE Journal of Emerging and Selected Topics in Power Electronics* 1(1): 28–41.

Dong Jiqing, Yang Shangping, Huang Tianxiang, et al. 2015. A novel constant current compensation network for magnetically-coupled resonant wireless power transfer system. *Proceedings of the CSEE* 35(17): 4468–4476.

Gao Jianxin, Wu Xusheng, Gao Wei, et al. 2016. Compensation technology of magnetic resonant wireless power transfer transmitter based on LCC. *Transactions of China Electrotechnical Society* 31(S1): 9–15.

Huang C Y, Boys J T, Covic G A, et al. 2010. LCL pick-up circulating current controller for inductive power transfer systems. *IEEE Transactions on Power Electronics* 28(4): 640–646.

Huang Xiaosheng, Chen Wei. 2014. A novel compensation network for ICPT systems. *Proceedings of the CSEE*, 34(18): 3020–3026.

Jiang Yan, Zhou Hong, Hu Wenshan, et al. 2015. Optimal parameter matching based on capacitor array for magnetically-resonant wireless power transfer system. *Electric Power Automation Equipment* 35(11): 129–136.

Jolani F, Yu Y Q, Chen Z Z. 2015. A planar magnetically-coupled resonant wireless power transfer using array of resonators for efficiency enhancement. *IEEE MTT-S International Microwave Symposium.* Phoenix, USA: IEEE, 2015: 1–4.

Liu Chuang, Ge Shukun, Guo Ying, et al. 2016. Double-LCL resonant compensation network for electric vehicles wireless power transfer: experimental study and analysis. *IET Power Electronics* 9(11): 2262–2270.

Nakamura K, Honma K, Ohinata T, et al. 2015. Development of concentric-winding type three-phase variable inductor. *IEEE Transactions on Magnetics* 51(11): 1–4.

Peng Yonglong, Zhu Jinbo, Li Yabin. 2015. Hybrid damping control based on the LC filter inductor voltage feedback and input shaping techniques. *Power system protection and control* 43(2): 103–107.

Richard Z Z, Liu X L, Zhang Z M. 2015. Near-field radiation between grapheme-covered carbon nanotube arrays. *AIP Advances* 5(5): 1–13.

Zhai Yuan, Sun Yue, Su Yugang, et al. 2014. Magnetic resonance wireless power supply system with characteristics of constant voltage. *Transactions of China Electrotechnical Society* 29(9): 12–16.

Electromechanical Control Technology and Transportation – Jia & Wu (Eds)
© 2017 Taylor & Francis Group, London, ISBN 978-1-138-06752-3

Sliding rate analysis of the single loop CDSV transmission system

Baofeng Zhang, Yahui Cui, Mengru Wang & Kai Liu
School of Mechanical and Precision Instrument Engineering, Xi'an University of Technology, Xi'an, China

ABSTRACT: Based on the single loop CDSV transmission system, the relationship was studied between the system sliding rate and the sliding rates of the basic stepless speed change units. Firstly, using the mechanism conversion method, the transmission ratio of the single loop CDSV transmission system was computed. Secondly, according to the value range of the power distribution coefficient, the single loop CDSV transmission system was classified as power distributed (converged) transmission system, power counterclockwise circulation type transmission system and power clockwise circulation type transmission system. Then, the sliding rates were analyzed for the different types of CDSV power transmission system. The influences of the P unit sliding rate on the system sliding rate were gotten, and the influences had three. Namely, the system amplifies the influences of the P unit sliding rate on the system sliding rate, the system reduces the influences of the P unit sliding rate on the system sliding rate, and the P unit sliding rate and the system sliding rate are equal. Analyzing the P unit sliding rate and CDSV transmission system sliding rate contributes to select the different transmission rate, transmission type and P unit sliding rate in order to make the system work efficiently.

1 INTRODUCTION

The single loop system is the system which only has one possible power circuit and DOF = 1[1], as shown in Figure 1. It consists of the differential gear train X unit and the basic closed P unit. In this paper, the basic closed P unit adopts basic stepless transmission and the system can realize stepless speed regulation. Therefore, the system is called the single loop CDSV transmission system (CDSV, short for Closed Differential stepless Variators). The system can realize the better high-power variable speed stepless transmission to adapt to different working conditions. The paper studies the sliding characteristics of the single loop CDSV system in order to obtain the influence of the basic P unit' siding rate on the whole system's sliding rate. Thereby, the system's transmission type, transmission ratio and the P unit's sliding rate are designed reasonably to make the system operate more accurately and efficiently.

2 THE TRANSMISSION RATIO AND POWER FLOW TYPE OF THE SINGLE LOOP CDSV TRANSMISSION SYSTEM

2.1 The transmission ratio

As shown in Figure 1, according to the relative motion principles of theoretical mechanics, the following formulas can be gotten.

$$
\begin{cases}
K = i_{ab}^c = \dfrac{w_a - w_c}{w_b - w_c} \\[2mm]
i_p = i_{\alpha\beta}^\gamma = \dfrac{w_\alpha - w_\gamma}{w_\beta - w_\gamma} \\[2mm]
w_a = w_\beta, w_b = w_\alpha, w_\gamma = 0 \\[2mm]
i_{IO} = \dfrac{w_I}{w_O} = \left(\dfrac{w_a}{w_c}\right)^x
\end{cases}
\tag{1}
$$

The system transmission ratio can be obtained by formula 1.

$$
i_{IO} = \left(\frac{1 - K}{1 - K \cdot i_p}\right)^x
\tag{2}
$$

Thereinto, w_a, w_b, w_c denotes espectively the angular velocity of the three basic building blocks a, b, c for the differential gear train X unit. $w_\alpha, w_\beta, w_\gamma$ expresses respectively the angular velocity of the three basic building blocks α, β, γ for P

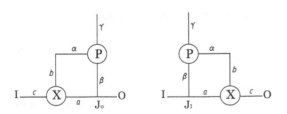

Figure 1. Single loop CDSV transmission system.

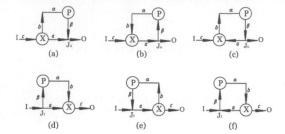

(a) (b) (c)

(d) (e) (f)

Figure 2. The power flow direction of single loop CDSV transmission system.

Table 1. The judgment criterion of single loop CDSV transmission system's power type.

Power distribution coefficient	Power flow transmission type
$q < 0$	Power distribution (collection) system
$0 < q < 1$	Power reverse circulation transmission system
$1 < q$	Power clockwise circulation transmission system

unit. If the single loop transmission system is the type of XP system, $x = -1$. If it's the type of PX system, $x = 1$.

2.2 Power flow type

Due to $\begin{cases} T_a : T_b : T_c = 1 : (-K) : (K-1) \\ P_a : P_b : P_c = T_a \cdot w_a : T_b \cdot w_b : T_c \cdot w_c \end{cases}$, combining the system's transmission ratio, it can be gotten

$$P_a : P_b : P_c = 1 : (-q) : (q-1) \tag{3}$$

Here, $q = K \cdot i_P$ called power distribution coefficient. As shown in Figure 2, by the power distribution coefficient q, the design criteria of transmission type for the single loop system is shown in Table 1 above.

3 THE SLIDING RATE ANALYSIS OF THE SINGLE LOOP CDSV TRANSMISSION SYSTEM

Since the stepless transmission depends on the friction transmission to provide traction to transfer power, there must be slipping. Therefore, in the case of load, the output rotate speed is lower than the no-load output rotate speed, and the speed difference will increase when external load increases. The calculation formula of sliding rate is as following. $\varepsilon = \frac{n_0 - n}{n_0} \times 100\%$ Where, n_0 is the no-load

output rotate speed, n is the output speed with external loads.

In order to facilitate the subsequent calculations, the variables are explained here. In the case of no external load, the P unit's output rotate speed, the system output rotate speed, the P unit's transmission ratio and the system transmission ratio are expressed respectively with $n_P^0, n_O^0, i_P^0, i_{IO}^0$. In the case with external loads, the P unit's output rotate speed, the system output rotate speed, the P unit's transmission ratio and the system transmission ratio are expressed respectively with n_P, n_O, i_P, i_{IO}. The sliding rate of P unit and the sliding rate of system are expressed respectively with $\varepsilon_P, \varepsilon$.

3.1 The sliding rate analysis of the XP type CDSV transmission system

For the XP type CDSV system, by formula 2, formula 4 can be gotten. Then, according to formula 4, formula 5 is gotten.

$$\begin{cases} i_{IO}^0 = \dfrac{n_I}{n_O^0} = \dfrac{1 - K \cdot i_P^0}{1 - K} \\ i_{IO} = \dfrac{n_I}{n_O} = \dfrac{1 - K \cdot i_P}{1 - K} \end{cases} \tag{4}$$

$$\varepsilon = \frac{n_O^0 - n_O}{n_O^0} \times 100\% = \frac{K\left(i_P^0 - i_P\right)}{1 - K \cdot i_P} \tag{5}$$

3.1.1 The power distribution (collection) transmission system $(q < 0)$

As shown in Figure 2(a), the power of P unit is flowing into the unit from α end and flowing out from β end. Then, for the P unit, formula 6 holds. From formula 6, the sliding rate of P unit can be gotten.

$$\begin{cases} i_P^0 = \dfrac{n_\alpha}{n_\beta^0} = \dfrac{n_\alpha}{n_P^0} \\ i_P = \dfrac{n_\alpha}{n_\beta} = \dfrac{n_\alpha}{n_P} \end{cases} \tag{6}$$

$$\varepsilon_P = \frac{n_P^0 - n_P}{n_P^0} \times 100\% = \frac{i_P - i_P^0}{i_P} \tag{7}$$

From formula 6 and 7, it holds.

$$\frac{\varepsilon}{\varepsilon_P} = \frac{K \cdot i_P}{K \cdot i_P - 1} \tag{8}$$

Due to $K \cdot i_P < 0$, then $0 < \varepsilon/\varepsilon_P < 1$. Figure 3 is used to describe the ratio change situation of the system sliding rate and the P unit sliding rate with the change of ip when $K = -0.2$, $i_{pmin} = 3$,

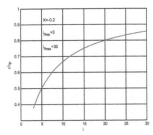

Figure 3. The ratio of system and P unit's sliding rate $(K \cdot i_P < 0)$.

$i_{pmax} = 30$. It can be seen from the graph that the ratio increases as i_p increases, and $\varepsilon/\varepsilon_p$ is more close to 1 when i_p is greater, but always less than 1. Those indicate that the system sliding rate is less than the P unit sliding rate when $K \cdot i_p < 0$. Namely, the system reduces the effect of the P unit sliding rate on the whole system.

3.1.2 The power reverse circulation transmission system (0 < q < 1)

As shown in Figure 2(b), the power of P unit is flowing into the unit from α end and flowing out from β end. Then, for the P unit, it holds. Then by formula 9, formula 10 can be gotten.

$$\begin{cases} i_P^0 = \dfrac{n_\alpha^0}{n_\beta} = \dfrac{n_P^0}{n_\beta} \\ i_\Gamma = \dfrac{n_\alpha}{n_\beta} = \dfrac{n_P}{n_\beta} \end{cases} \tag{9}$$

$$\varepsilon_P = \frac{n_P^0 - n_P}{n_P^0} \times 100\% = \frac{i_P^0 - i_\Gamma}{i_P^0} \tag{10}$$

From formula 9 and 10, it can be gotten.

$$\frac{\varepsilon}{\varepsilon_P} = \frac{K \cdot i_P^0}{1 - K \cdot i_P} \tag{11}$$

Due to $0 < K \cdot i_p < 1$, here, the paper takes $K = 3$, i_p and i_p^0 are both from 0 to 0.3, then $0 < K \cdot i_p < 1$. Figure 4 describes the change conditions of the ratio of the system sliding rate ε and the P unit sliding rate ε_p with the change of i_p and i_p^0 in these parameters range.

It can be seen from the graph that the different values of i_p and i_p^0 are corresponding to the different values of $\varepsilon/\varepsilon_p$. Therefore, it needs to make a concrete analysis on CDSV system sliding rate change.

a. If $\varepsilon/\varepsilon_p < 1$, the sliding rate of system is less than the sliding rate of P unit. According to formula 11, $K \cdot (i_p^0 + i_p) < 1$ can be obtained. $K \cdot (i_p^0 + i_p) < 1$ expresses the lower part of the Figure 2b with the dividing curve of $\varepsilon/\varepsilon_p = 1$. On this, the sliding rate of system is less than the sliding rate of P unit. Namely the system reduces the effect of the sliding rate of P unit on the whole system.

b. If $\varepsilon/\varepsilon_p > 1$, the sliding rate of system is greater than the sliding rate of P unit. According to formula 11, $K \cdot (i_p^0 + i_p) > 1$ can be obtained. $K \cdot (i_p^0 + i_p) > 1$ expresses the upper part of the Figure 2b with the dividing curve of $\varepsilon/\varepsilon_p = 1$. On this, the sliding rate of system is greater than the sliding rate of P unit. Namely the system amplifies the effect of the sliding rate of P unit on the whole system.

c. If $\varepsilon/\varepsilon_p = 1$, the sliding rate of system is equal to the sliding rate of P unit. According to formula 11, $K \cdot (i_p^0 + i_p) = 1$ can be obtained. $K \cdot (i_p^0 + i_p) = 1$ expresses the dividing curve of $\varepsilon/\varepsilon_p = 1$.

The change situation of $\varepsilon/\varepsilon_p$ when 0<q<1 and k=3

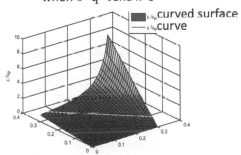

Figure 4. The ratio of system sliding rate and P unit's sliding $(0 < K \cdot i_p < 1)$.

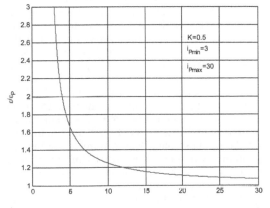

Figure 5. The ratio of system sliding Rate and P unit's sliding $(K \cdot i_p > 1)$.

Table 2. The PX type's ratio of system sliding rate and the P unit's sliding.

PX	The power distribution (collection) transmission system $$(K \cdot i_p < 0)$$ $$\frac{\varepsilon}{\varepsilon_P} = \frac{K \cdot i_P^0}{K \cdot i_P^0 - 1}$$	
	The power reverse circulation transmission system $$(0 < K \cdot i_p < 1)$$ $$\frac{\varepsilon}{\varepsilon_P} = \frac{K \cdot i_P^0}{1 - K \cdot i_p}$$	The change situation of $\varepsilon/\varepsilon_p$ when 0<q<1and k=3
	The power clockwise circulation transmission system $$(K \cdot i_p > 1)$$ $$\frac{\varepsilon}{\varepsilon_P} = \frac{K \cdot i_P^0}{K \cdot i_P^0 - 1}$$	

3.1.3 The power clockwise circulation transmission system (q > 1)

As shown in Figure 2(c), the power of P unit is flowing into the unit from α end and flowing out from β end. So the sliding rate formula of P unit is as formula 7. The ratio of the system sliding rate and the P unit sliding rate is as formula 8.

Due to $K \cdot i_p > 1$, then $\varepsilon/\varepsilon_p = K \cdot i_p/(K \cdot i_p - 1) > 1$. The paper makes K = 0.5, $i_{pmin} = 3$ and $i_{pmax} = 30$ in order to describe the ratio change situation of the system sliding rate and the P unit sliding rate with the change of i_p. As shown in Figure 5, the ratio decreases as i_p increases and $\varepsilon/\varepsilon_p$ is more close to 1 when i_p is greater, but always greater than 1. Those indicate that the system sliding rate is greater than the P unit sliding rate when $K \cdot i_p > 1$. Namely, the system amplifies the effect of the P unit sliding rate on the whole system. What's more, it can be seen from the figure that $\varepsilon/\varepsilon_p$ is close to infinity when $K \cdot i_p$ is close to 1. Namely the sliding rate of system tends to infinity. In other words, although the XP type CDSV system is running when Ki_p is near to 1, there is no output rotational speed.

3.2 Sliding rate analysis of the PX type CDSV transmission system

According to the XP type CDSV system's sliding rate analysis methods, the system sliding rate are analyzed under various forms of power flow circumstances, shown in Table 2.

4 CONCLUSION

In the paper, using mechanisms transformation method, the transmission ratio of the single loop transmission system is calculated. Then according to the power distribution coefficient q of the X unit, the system can be divided into the power distribution (collection) transmission system, the power reverse circulation transmission system and the Power clockwise circulation transmission system. Aiming at different types of transmission system, the sliding rate of P unit and the sliding rate of system are analyzed and calculated. Next, the changing curve of the ratio of the system sliding rate and the P unit sliding rate along with the P unit transmission ratio are drawn using Matlab software. Through the curve, the relationship between the P unit sliding rate and the system sliding rate are analyzed to design system transmission type, transmission ratio and the P unit sliding rate reasonably, and to make the system operate more accurately and efficiently.

REFERENCES

Hailiang Xu. The theory and experimental study on the single loop planetary transmission [D]. Xi'an: Xi'an University of Science and Technology, 2010.

Houxin Li. Energy analysis and active design research on the single loop planetary transmission [D]. Xi'an: Xi'an University of Science and Technology, 2013.

Min Jie. The dynamic characteristics theory of The single loop CVT planetary transmission. Xi'an: Xi'an University of Science and Technology, 2012.

Shichang Chen. Transmission and visual design research on the single loop system [D]. Xi'an: Xi'an University of Science and Technology, 2004.

Yahui Cui. The study on Power distribution (collection) planetary transmission: [P.h.D. Thesis] [D]. Xi'an: China University of Technology, 1998.

Zhongtang Ruan etc. The design and selection guide of mechanical stepless transmission [M]. Beijing: Chemical industry press. 1999. 9.

Research on the power conversion circuit of a single-phase AC input PMSM control system

Yun Zhang
Key Laboratory of Automotive Electronics, Automation Research Institute,
Shandong Academy of Sciences, Jinan, China

Bo Cui
Department of Electrical Engineering, Shandong University, Jinan, China

Zhixue Wang
Key Laboratory of Automotive Electronics, Automation Research Institute,
Shandong Academy of Sciences, Jinan, China

Xu Huang
Department of Electrical Engineering, Shandong University, Jinan, China

ABSTRACT: The power conversion circuit of a single-phase AC input PMSM control system was deeply analyzed and designed in this paper. Firstly, the basic functional units of the motor control system were introduced; then we put emphasis on the power conversion circuit. The capacity of the filter capacitor, the on-state current of the rectifier and the current capacity of the inverter were analyzed in detail and then the calculation formulas of the above key parameters were derived. Finally, the correctness and feasibility of the analysis and the calculation method were verified by the experimental results.

1 INTRODUCTION

Permanent Magnet Synchronous Motor (PMSM) is widely used in many high-performance servo control system due to its advantages of high efficiency, high power density, simple structure, etc. (Xu, 2011; Li, 2011). The calculation of the key parameters in the power conversion circuit is one of the key steps during the designing of the control system. But at present, there is no unified calculation formulas for the key parameters of the power conversion circuit. We usually select device relying on practical experience and experiments which will reduce the development efficiency and increase the cost (Liu, 2010).

Considering the disadvantages of the traditional designing process, the power conversion circuit of a single-phase AC input PMSM control system was deeply analyzed and designed and the calculation formulas of the key parameters were derived in this paper. Finally, experiments were carried out on the established experimental platform.

2 MOTOR CONTROL SYSTEM AND POWER CONVERSION CIRCUIT

The structure of the single-phase AC input PMSM control system is shown in Figure 1. The functions of the different modules are as follows: As shown in Figure 1, the single-phase AC input is converted into DC voltage by the rectifier and filter circuit; The driven circuit of the system plays the role of amplifying the PWM driving signal and isolating the main control circuit from the power circuit (Bai, 2002); The detecting circuit can acquire the rotor position and current signals.

Figure 2 shows the power conversion circuit composed of the rectifier, the filter capacitor and the inverter. AC power source V_{AC} is transformed into sinusoidal half-wave U_D through the rectifier BR and then transformed into DC voltage U_C through capacitor C. The effect of the inductance L_2 is to suppress the instantaneous output current of the rectifier bridge BR and reduce the harmonic component of the DC side voltage of the inverter (Ding, 2010).

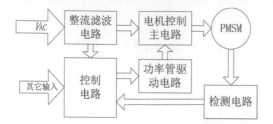

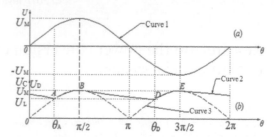

Figure 1. Structure of single-phase AC input PMSM control system.

Figure 3. Voltage waveform when motor is working.

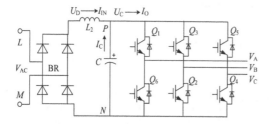

Figure 2. Power conversion circuit.

Three phase full-bridge topology is usually used in the inverter circuit of the PMSM control system.

It is one of the key procedures to design the key parameters of the power conversion circuit including the on-state current of the rectifier bridge BR, the capacity of the filter capacitor C and the current capacity of the inverter circuit.

The concrete analysis and design to each function module was presented as follows.

3 DESIGN OF POWER CONVERSION CIRCUIT

3.1 *Analysis and calculation of capacitor C*

Filter capacitor C has a variety of functions (Xiang, 2004) in the AC power-supplied system, such as: (1) Filting out the harmonic components generated in the chopping process; (2) Smoothing the sinusoidal half-wave voltage; (3) Providing energy to the main circuit of the control system, etc.

The input and output voltage of the rectifier bridge BR and the output voltage of the capacitor C when the control system is working under the rated power are shown in Figure 3.

As shown in Figure 3, curve 1 is the input voltage V_{AC} of the rectifier BR, curve 2 is the output voltage U_C of the capacitor C and curve 3 is the output sinusoidal half-wave voltage U_D of the rectifier BR. During AB and DE period, U_C and U_D are nearly coincident. But in fact, U_D is slightly larger than U_C and the rectifier keeps providing energy to the main circuit and charging the capacitor C

at the same time. During OA and BD period, U_D is becoming smaller than U_C. The rectifier is been cut off and the capacitor C begins discharging to the main circuit.

The above working process cycles during the operation of the motor control system.

The output voltage U_C of capacitor C has the minimum value U_L at A and D and the maximum value U_M at B and E.

The quantity of electric charge of capacitor C Q_B at B point can be expressed as:

$$Q_B = CU_M \tag{1}$$

The quantity of electric charge of capacitor C Q_D at D point can be expressed as:

$$Q_D = CU_L \tag{2}$$

Set the average of the output current of capacitor C as I_{AV} and set the discharging time from B to D as Δt, the following formula can be obtained:

$$I_{AV} = \frac{Q_B - Q_D}{\Delta t} \tag{3}$$

Then formula (4) can be deduced from (1)–(3):

$$I_{AV} = \frac{(U_M - U_L)C}{\Delta t} \tag{4}$$

Meanwhile, transform Δt into $\Delta \theta$:

$$\Delta \theta = 2\pi \cdot f \cdot \Delta t \tag{5}$$

where f is the frequency of the AC input voltage.

The capacitor voltage at D can be expressed as:

$$U_L = U_M \cdot \sin\left(\Delta \theta - \frac{\pi}{2}\right) \tag{6}$$

Then $\Delta \theta$, Δt and I_{AV} can be get from the above formula:

$$\Delta\theta = \frac{\pi}{2} + \arcsin\left(\frac{U_L}{U_M}\right) \tag{7}$$

$$\Delta t = \frac{\frac{\pi}{2} + \arcsin\left(\frac{U_L}{U_M}\right)}{2\pi f} \tag{8}$$

$$I_{AV} = \frac{2\pi f(U_M - U_L)C}{\frac{\pi}{2} + \arcsin\left(\frac{U_L}{U_M}\right)} \tag{9}$$

The output power of capacitor C at D point can be expressed as:

$$P_D = U_L I_{AV} = \frac{2\pi f(U_M - U_L)CU_L}{\frac{\pi}{2} + \arcsin\left(\frac{U_L}{U_M}\right)} \tag{10}$$

Then the calculation formula of capacitor C can be expressed as:

$$C = \frac{P_D \cdot \left(\frac{\pi}{2} + \arcsin\left(\frac{U_L}{U_M}\right)\right)}{2\pi f(U_M - U_L)U_L} \tag{11}$$

The minimum value U_L of U_C at D point shown in Figure 3 is the smallest voltage that capacitor C can provide for the main power circuit. During the designing of the filter, the control requirement can be met once U_L is not smaller than the peak value of the motor's rated line voltage U_N under the rated input power P_N.

The capacity of C can be calculated by replacing U_L with U_N in formula (11). In practical application, considering the voltage fluctuation and the overload ratio, we always select the capacity value larger than the calculated results.

3.2 Calculation of the on-state current of rectifier BR

The voltage and current waveform of the power conversion circuit is shown in Figure 4 in which curve 2 is the capacitor voltage U_C, curve 3 is the output sinusoidal half-wave voltage U_D of rectifier BR without filter capacitor, curve 4 is the charging and discharging current waveform of the filter capacitor C and curve 5 is the waveform of the AC input current.

As analyzed in 3.1, the filter capacitor C is in discharging state during OA and BD period because the sinusoidal half-wave voltage U_D is smaller than capacitor voltage U_D. The discharging current is I_O which is under the horizontal axis as shown in curve 4. There is no input current during the discharging process as shown in curve 5.

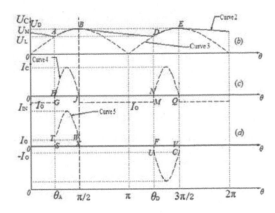

Figure 4. Current waveform when motor is working.

The filter capacitor C is in charging state during AB and DE period while U_D is slightly larger than U_C. The charging current is shown in curve 4 that is above the horizontal axis. As shown in curve 5, there is a AC input current I_{IN} which can provide capacitor charging current I_C and the main circuit current I_O. The current relationship can be expressed as:

$$I_{IN} = I_O + I_C \tag{12}$$

The corresponding electric angle θ_A of voltage U_L at point A can be expressed as:

$$\theta_A = \arcsin\left(\frac{U_L}{U_M}\right) \tag{13}$$

According to the energy conservation theorem, the energy increased in the charging process must be equal to the energy released during the discharging process (Chikh, 2011). The relationship can be expressed as:

$$I_{CA}\left(\frac{\pi}{2} - \theta_A\right) = I_O\left(\frac{\pi}{2} + \theta_A\right) \tag{14}$$

$$I_{CA} = \frac{I_O\left(\frac{\pi}{2} + \arcsin\left(\frac{U_L}{U_M}\right)\right)}{\left(\frac{\pi}{2} - \arcsin\left(\frac{U_L}{U_M}\right)\right)} \tag{15}$$

where I_{CA} is the average of the charging current.

Then the average current of the rectifier I_{INA} can be expressed as:

$$I_{INA} = \frac{\pi \cdot I_O}{\left(\frac{\pi}{2} - \arcsin\left(\frac{U_L}{U_M}\right)\right)} \tag{16}$$

where I_O is the average of the AC input current that is calculated under the rated input power P_N and the rated voltage U_N.

In practical application, the overload capacity is usually take into account by multiplying an overload factor about 1.5–2.

3.3 Current capacity calculation of the inverter

The three-phase stator windings of AC motor are usually Y-connected. The rated parameters of the motor include rated line voltage U, rated power P_N, rated speed n_N, etc.

The RMS of the rated line voltage can be expressed as:

$$U = \frac{\sqrt{2}}{2} \cdot U_N \qquad (17)$$

where U_N is the peak value of the rated line voltage.

If the phase current is I_N and the power factor angle is φ while the PMSM is under rated operation, then there is the following relationship (Tang, 2008):

$$\sqrt{3} \cdot \frac{\sqrt{2}}{2} U_N \cdot I_N \cdot \cos \varphi = P_N \qquad (18)$$

$$I_N = \frac{2P_N}{\sqrt{6}U_N \cos \varphi} \qquad (19)$$

The rated on-state current of the power transistor must be larger than the peak value of the rated motor current. Considering the overload ratio v and the weak magnetic coefficient (the Maximum speed under weak magnetic/the rated speed) a, the selection principle of the on-state current I_{IGBT} can be expressed as:

$$I_{IGBT} > \frac{2v\alpha P_N}{\sqrt{3}U_N \cos \varphi} \qquad (20)$$

4 EXPERIMENTAL RESULTS

On the basis of the analysis and the design methods above, a PMSM control system is completed. The key parameters of the experimental motor are shown in Table 1.

The rated frequency of the AC input is 50 Hz. If we set U_L as 260 V the same as the peak value of the rated line voltage, then the value of the capacitor C will be about 472 uF calculated by formula 11. After considering the overload ratio and the standard sizes of capacitor, a capacitor of 560 uF is selected finally.

Table 1. Parameters of the experimental motor.

Parameters	Value
Rated power (W)	750
Rated voltage (V)	185
Rated speed (rpm)	3000
Rated frequency (Hz)	50
Pole pairs (p)	1
Mechanical loss ratio	0.03
Rated operating temperature (°C)	75

Under the rated operation state, current I_O is about 4A (750/185) which represents that the average input current and the thermal loss are not too large.

The average current of the rectifier I_{INA} is 22A calculated by formula 16. According to the standard specification of the rectifier, the maximum on-state current is selected as 25A.

Considering the overload ratio $v = 2$, the weak magnetic coefficient $a = 2$ and the rated power factor $\cos \varphi = 0.95$, the on-state current I_{IGBT} will be about 14A calculated by formula 19. And the IGBTs with on-state current 15A were selected.

Inductance L_2 has a certain inhibitory effect on the peak current.

During the experiment, the input power of the motor controller is adjusted to 750 W, and the test results are shown in Figure 5. The first curve is the waveform of the AC input voltage V_{AC} and the next one is the capacitor voltage U_C. According to the former analysis, the filter capacitor C is in charging state during DE which corresponds to BC of curve V_{AC}. Set the electric angle at B point on the curve of V_{AC} as θ and the corresponding voltage is 260 V. Then the electric angle at C point is $\pi/2$ which corresponds to a voltage of 310 V.

According to the trigonometric position, the formula below can be obtained:

$$\sin \theta = \frac{U_B}{U_C} = \frac{260}{310} = 0.77 \qquad (21)$$

$$\frac{\theta}{\pi/2} = \frac{1}{3.14/2} = 0.64 \qquad (22)$$

The calculation results are consistent with the former analysis and design.

The third curve is the waveform of the AC input current whose peak value is about 20A.

The last one is the waveform of the charging and discharging current I_C of filter capacitor C. As shown in the figure, the charging current has the same waveform as the AC input current and the discharging current is stable at about 4A.

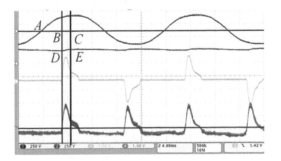

Figure 5. Test curve under input power 750 W.

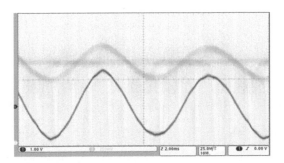

Figure 6. Phase current of motor under 750 W input power.

Figure 6 shows the phase current of the PMSM. The peak value is about 6A which is consistent with the results calculated by the calculation method in 2.3.

5 CONCLUSIONS

This paper deeply analyzed the operating principle of the power conversion circuit of a single-phase AC input PMSM control system. The key parameters were analyzed theoretically and the corresponding calculation method was given. The correctness and feasibility were verified by the experimental results. This research work has certain reference value for the designing of motor control system.

ACKNOWLEDGEMENT

This work was supported by the Key Research and Development Planning Project of Shandong Province (2016GGC01023; 2015ZDXX0601B01).

REFERENCES

Chikh HK, et al (2011). PMSM Vector Control Performance Improvement by Using Pulse with Modulation and Anti-Windup PI Controller. *C. Proceedings of International Conference on Multimedia Computing and Systems.*2011: 1–7.

Liu HY, et al (2010). Design of Controlling System about the High-Power PMSM Based on STM32.*C. Proceedings of International Conference on Compute, Mechatronics, Control and Electronic Engineering.* 2010: 374–377.

Peijiang Li, Ting Ting (2011). Simulation research of permanent magnet synchronous motor control system. *J. Computer simulation.* 28: 255–258.

Tao Bai, Xuchun Li, et al (2002). Field oriented PMSM control system based on DSP. *J. S&M Electric Machines.*29:17–20.

Wen Ding, Deliang Liang, et al (2010). Design and implementation of PMSM vector control system based on D-SP. *J. Micromotors.* 43:72–77.

Wen Xiang, Guo Hai, Jinghua Ji (2004). Full digital vector control speed regulation system based on DSP. *J. Electric Machines and Control.* 8: 175–178.

Xu D, T Wang and H Wei (2011). A Digital High Performance PMSM Servo System Based on DSP and FPG-A. *C. Proceedings of International Conference on Ind-ustrial Electronics and Application.* 2011: 2742–2746.

Yunliao Tang, Yingli Luo (2008). Electrical Machinery. *M. Mechanical Industry Press.* 2008:342–343.

Signal processing and computer science

Electromechanical Control Technology and Transportation – Jia & Wu (Eds)
© *2017 Taylor & Francis Group, London, ISBN 978-1-138-06752-3*

Optimization study on the coal powder conveying system of utility boilers

Fushuang Liu, Jun Zhao & Shougen Hu
School of Energy and Power Engineering, University of Shanghai for Science and Technology, Shang Hai, China

ABSTRACT: Uneven distribution of air and pulverized coal is a major problem in the coal powder conveying system of utility boilers, which may pose pulsating combustion in the hearth. This combustion instability will further lead to some bad consequences, such as steam temperature deviation, superheater bursting, heat transfer deterioration, slagging in the furnace, and even compelling shutdown. In this paper, we analyze and solve the equilibrium problem in the direct blowing pulverizing system of high-capacity utility boilers. By trimming the resistance of powder pipeline and adopting pulverized coal distributor, the pulverized coal conveying system can be operated evenly and accurately, thus improving the efficiency and reliability of the power plant boiler operation.

1 BACKGROUND AND SIGNIFICANCE

Energy is the material base of human activity, so energy shortage and environmental pollution have become two major problems facing the development of China. In response to the proposition of energy saving, environmental protection, and reducing emissions, countries all over the world including China are accelerating the development of nuclear energy. As China is rich in coal but lacking in oil and gas, thermal power will still be the main mode of power generation in China for a long period of time. In recent years, environmental protection and energy conservation has become an important direction of China's power industry structure adjustment in the thermal power industry. Under the guidance of China's policy, we should actively promote optimization and upgrade the industrial system. Meanwhile, with the closing of a large number of small thermal power units with low energy efficiency and heavy pollution, the power plants will gradually develop with high quality and high index.

As one of the three largest host equipment of thermal power plant, utility boiler is vital to improve the efficiency of boiler combustion and reduce emissions. Especially for corner tangential combustion, it can effectively perfect the above two aspects by improving the control accuracy of air-pulverized coal parameters corresponding to furnace burner as well as reducing each burner deviation. At the same time, it can create good furnace combustion conditions.

Uneven distribution of air-pulverized coal refers to the following two conditions: different air-pulverized coal assigned to each nozzle and uneven distribution of pulverized coal due to the cross section granularity in a single nozzle. These can cause local hypoxia, fire difficulties, combustion instability, nozzle local burn, local slagging in furnace boiler, reduced efficiency, and so on. With the increasing of utility boiler capacity, burner quantity, and air leakage and explosion safety problems of the coal-pulverizing storage system (Anguo Zhang, 2010), the direct blowing pulverizing system is widely applied to power plants with high capacity, that is, more than 300 MW. Consequently, the question of wind powder pulverizing inequality becomes more prominent.

In direct blowing pulverizing system of large power plant boilers, unequal distribution is mainly attributable to two aspects: uneven distribution of wind and powder that is sent to every corner of the wind powder pipe and different factors such as powder pipeline length and elbow number leading to different comprehensive resistance coefficients of powder tube (Kefa Cen, 2003). The following two methods are adopted, respectively, to solve the above two problems. The first way aims to balance the pulverized coal concentration, namely, to disperse particles evenly before primary air gets into the pipe. This can be done by further reforming the classifier at coal mill outlet or adding some discrete devices such as pulverized coal distributor. The second way aims at the balance control of parallel air powder pipeline. This kind of method tries to adjust the flow of pulverized coal in each pipeline by adding the throttle element in the powder feeding pipe, thus achieving the purpose of resistance leveling.

2 BALANCING THE POWDER FEEDING PIPELINE RESISTANCE

2.1 Aerodynamic computation

Usually, burners of same layer get pulverized coal by a coal grinding machine. As the quantity of elbow and distance of each branch pipeline to each burner are different, the pipeline resistance of each pipe is not identical, leading to the uneven distribution of wind and powder in branch pipe. The resistance of pulverized coal pipe includes friction resistance, local resistance, increased loss of pulverized coal, and so on. Because pulverized coal in various pipe stress distribution is different, the pipeline internal friction resistance coefficient is also different.

The coal powder conveying system of utility boiler generally calculates the resistance on the basis of *DL/T 5145-2002 Thermal Power Plant Pulverizing System Design and Computing Regulations*. First, set air velocity and pulverized coal concentration of pipe initial section. Then according to the empirical formula, calculate the total resistance of each pipe and install throttling element on accounting. Finally, conduct cold and hot experiment, respectively, and make adjustment to achieve the purpose of leveling.

2.1.1 Pipe friction resistance ($\Delta p_{f\mu}$) calculation

Pipe friction resistance generally refers to the resistance due to the viscosity of the gas and the pipe wall friction during powder airflow through straight pipe, concerning with flow state, relative roughness of pipe inner wall, powder concentration, and pipeline rout. The general formula for calculation is:

$$\Delta p_{f\mu} = \lambda_\mu \frac{L}{D_e} \frac{\rho w^2}{2} \ (Pa) \tag{1}$$

The friction coefficient of powder gas is determined according to the following general formula:

$$\lambda_\mu = \lambda_0 (1 + k_\mu \mu) \tag{2}$$

where

λ_0, λ_μ—The frictional resistance coefficient of pure air and wind powder flow, respectively (λ_0 obtained by experiment, generally taken 0.015);
D_e—The equivalent diameter of the pipe, *m*;
ρ—Average density of wind and powder flow, *kg/m³*;
L—The pipe length, *m*;
μ—The concentration of the powder flow, *kg/kg*;
K—The correction factor (usually obtained by experiment);
w—The average flow velocity of gas, *m/s*.

2.1.2 Head loss of promotion ($\Delta p_{t\mu}$) calculation

In flow of vertical pulverized coal pipeline, pressure loss caused by promote pulverized coal should be taken into account. The following is the calculation of the resistance in the direct blowing pulverizing system:

$$\Delta p_{t\mu} = Z \mu \rho g \ (Pa) \tag{3}$$

Among them:

Z—The elevate height of vertical pipeline, *m*;
g—The gravitational acceleration, *m/s²*.

2.1.3 Local resistance ($\Delta p_{\zeta\mu}$) calculation

Local resistance refers to the resistance that is caused by the change of the section size or the direction of the powder airflow, such as a variety of elbow, the size of the joint, orifice plate, and T-pipe. Local resistance calculation of pipeline elements is shown in (4):

$$\Delta p_{\zeta\mu} = \zeta_\mu \frac{\rho w^2}{2} (Pa) \tag{4}$$

where

ζ_μ—containing powder airflow through the local resistance coefficient of (obtained by experiment).

This standard is applied to the design and calculation of pulverizing system of 65–2000 t/h coal-fired boiler. Because of the different flow field disturbance, artificial factors, limitation of measurement accuracy and so on, the experience formula and the reference coefficient in the calculation rules have some errors. According to the different conditions to calculate, the reference coefficient comes from table or chart; however, this chart of regulation is some trend lines drawn by limited operating point and portability that is poorer in some nonparticular conditions. During the actual power station construction, The certain error of installing location in field pipeline and equipment or some other uncertain factors lead to random piping arrangement such as increasing the elbow to achieve the ideal installation position and so on. It will also be faced with tedious calculation, even be designed and calculated again, which requires a lot of labor and time. With the development of science and technology, upgrading a port of piping components and equipment, such as inventing the dual-core adjustable shrinkage to eliminate the eccentric flow of single-core adjustable shrinkage, shows that the regulation cannot calculate new equipment resistance. Recently, according to the production and application of manufacturer's feedback, we have found that excessive deviation of powder in the actual application affects the normal use of the boiler.

2.2 Numerical simulation

Recently, with the improving of the computer performance, Computational Fluid Dynamics (CFD) based on computer numerical simulation has developed gradually and achieved a very good application. This method is established on the basis of conservation equation under different conditions and combines with the corresponding initial and boundary conditions by using the theory and method of numerical calculation to gas–solid two-phase flow field to building model and simulation. We finally obtain the flow characteristics of the two-phase pulverized coal. It improves the theoretical and experimental research of coal conveying system (Zheng Wen, 2013).

Numerical simulation has great flexibility and saves time, money, and labor. It can simulate flow models with special size, complex shape, high temperature and pressure, toxic flammability, and so on. It draws detailed flow field information, even the information that the experiment could not provide, and further predicts the experimental and explains the experimental phenomena. Although the application of the CFD to study two-phase flow has been successful, there are still many problems to be solved. These problems are mainly manifested in the following two aspects. The first is due to the nature of the flow process of complex real system, which is not enough, and completely accurate mechanism of the model, which is also difficult to establish. Therefore, the imperfect mechanism model often leads to the simulation results that cannot correctly forecast the distribution of all kinds of flow field. We developed and established a complete model for in-depth numerical experiment research. The other is due to the limitation of the calculation conditions (such as computer calculation speed and the use of numerical methods) so that people have to use simplified processing, such as dense-phase flow, ignoring the coupling and interaction, leading to certain errors in simulation results.

For simulation of the coal powder conveying system of utility boiler, solid–gas ratio of 0.3–0.7 generally belongs to the dilute-phase pneumatic conveying. After years of development of two-phase flow of gas–solid research, many achievements have been used in engineering at present. Commercial software such as GAMBIT and ICEM CFD have been able to meet the computing simulation of two-phase flow of dilute phase. At the same time, it uses specific experiments to validate the simulation results so as to improve the simulation. Simulation calculation is very convenient, especially when dealing with the complex pipeline. For new piping components and equipment, it also can undertake numerical simulation to get ideal flow law and resistance calculation results so as to provide a good guide for the practical application.

3 APPLICATION OF PULVERIZED COAL DISTRIBUTOR

From outlet of coal mill to the inlet of parallel transmission pipe, this uneven pulverized coal distribution does not contribute to resistance unevenness but blames the inherent uneven milling equipment, which often needs to install a pulverized coal distributor before the flow branch. In general, pulverized coal distributor utilizes the process of concentration, diffusion, guidance, or mix when powder airflow goes through it, thus achieving the purpose of even powder gas flow distribution. The application of the pulverized coal distributor worldwide mainly has finned diversion type, diffusion type, grille type, and double-adjustable type.

3.1 Ribbed diversion pulverized coal distributor

Finned diversion distributor installs in the 90° corner of pipeline, and the elbow has a disk with a rotating diversion fin. To adjust the direction of the fin, the rotating disk can adjust the distribution of pulverized coal in the downstream branch for the even distribution of the wind and powder. Design air volume range is 17500–65000 m^3/h, and relative deviation of pulverized coal quantity is about 5%. However, the bisecting condition is limited, which in fact is used rarely (Haoliang Jiang, 1976).

3.2 Diffusion-pulverized coal distributor

The distributor is directly installed at the outlet of medium-speed mill, controlling powder flow acceleration, deceleration, and residual rotation during the process of contraction and expansion. By strengthening the disturbance and mixing, uniform distribution can be achieved, and finally pulverized coal can be separated from the side duct. Besides, the structure is relatively simple. It can lead to any number of branch pipes; however, with the increasing of branch number, the distribution uniformity will be reduced greatly. At present, four branches are commonly used. Because of its inherent nonuniform distribution factors, its distribution deviation is in the range of 15–20% (branch number, 4). On taking additional deviation by the difference of pipe length into account, the total distribution deviation can reach 30–40%.

3.3 Grille-pulverized coal distributor

Grille-pulverized coal distributor is the most widely used distributor. For this distributor, uneven pulverized coal airflow in pipe will be divided evenly into many slit airflow by the grid, and through some guidance it gets into the side channels alternately. When the number is larger enough, the powder and air of both sides' channels can achieve uniform distribution at the same time. The primary grille will be evenly split into two, and the secondary grille will be evenly split into four. In order to get a good effect on distribution, the grilles must be very slender and closely arranged. In addition, as high flow rate in the narrow channel often results in a more complex structure and that the grids are easy to wear, the shell material is generally ordinary carbon steel plate, and the lining is abraded with high-chromium cast steel. The density deviation and airflow deviation control are in the range of 10–15% (branch number, 4), and if the additional deflection is taken into account, the total distribution deviations is up to 25–35%, and the airflow resistance is generally less than 900 Pa (Wu Jiang, 2004).

3.4 Double-adjustable pulverized coal distributor

To solve the pulverized coal bias problem caused by the uneven distribution of the pipeline resistance, resistance components such as orifice plate are applied to balance the resistance deviation of pipe. In fact, this is the only possible way of adjustment, but the adjustment of the positive effect is very small and often causes adverse consequences. For example, the volume of air in powder pipe with a higher amount of pulverized coal is always smaller. Thus, to control the flow of pulverized coal, the pipeline resistance must be further increased artificially. This is bound to make a smaller air volume, resulting in the difficulty of coal powder transportation, posing the deposition of pulverized coal, and plugging accidents. Some resistance adjustment will also affect the pulverized coal flow and airflow, and the synthetic effect is orthokinetic. Thus, the uneven distribution problem cannot be effectively solved. As a result, the problem of uneven distribution of pulverized coal is unable to adjust effectively via throttling element, which has been the industry consensus.

Double-adjustable pulverized coal distributor is a new generation of patented products invented by the National Engineering Research Center of Power Plant Boiler Coal Clean Combustion, which mainly sets up a pulverized coal shade air separation unit in the shell. This unit separates pulverized coal into dense-phase space and dilute-phase space. The dense-phase space connects mixing tube by the dense-phase adjusting device, and the dilute-phase space connects mixing tube by the dilute-phase branch pipe resistance adjusting device. Because of the separation unit, airflow can be divided into dense and dilute phases, and can be respectively adjusted. Then, the dense-phase airflow and the dilute-phase airflow can be mixed to realize even distribution of pulverized coal and air in pulverized coal pipe (after mixed tube). The distributor can adjust the pulverized coal flow and airflow separately so as to achieve uniform distribution of gas powder completely. It also can reduce the pulverized coal flow. At the same time, it can enhance airflow in the pipeline so as to overcome the disadvantages of the throttling element (Yueming Wang, 1997; Kefa Cen, 2003).

Double-adjustable pulverized coal distributor in dilute-phase space sets up a throttling damper, which can respectively adjust resistance characteristics of each pulverized coal pipe. Therefore, adjustable shrinkage cavity can be dismantled. Because of the rectangular shell, the internal flow rate is low. The shell is directly made of 16 MN steel without a wear lining board. The cost of the distributor is 50% more than that of the grille-pulverized coal distributor, and the accurate dynamic adjustment depends on accurate on-line measurement technology. As a result, current coal-fired power plants have not achieved a satisfactory performance without grille type. By the practical engineering application, the main technical indicators are as follows: concentration deviation and airflow deviation control is within 8–10%, and wind flow resistance is generally less than 500 Pa (Wu Jiang, 2004).

4 CONCLUSION

1. Grille-pulverized coal distributor is the most widely used distributor type, but not without disadvantages such as a relatively more complex structure, larger flow resistance, and being adjustable in a static situation but not in a dynamic condition. Therefore, we should deal with the optimization and reduce the flow resistance, which can further reduce the wind powder quantity deviation.

2. Double-adjustable pulverized distributor can be arranged randomly, and the amount of air and powder can be adjusted respectively. Therefore, this kind of coal distributor can be widely applied to gain more experience.

3. The accurate regulation depends on accurate on-line measurement technology, so we should continue to improve the on-line measurement precision of the pulverized coal concentration and provide a more accurate basis for the system balance adjustment.

4. Under the guidance of the two-phase flow theory and with the help of numerical analysis method, the pulverized coal conveying system in gas-solid two-phase flow is studied. Through experiment and comparison, the resistance balancing scheme of coal powder conveying system is achieved by numerical calculation so as to explore how to improve the feasibility of the calculation method in "DL/T 5145-5145".

REFERENCES

Anguo Zhang, hui Liang. Power Plant Boiler Coal Powder Preparation and Calculation. *Beijing: China Power Press, 2010. 08*, 50 ~ 61.

DL/T 5145–2002, Thermal Power Plant Pulverizing System Design Technical Regulation (s).

Editorial department of "China Power Encyclopedia". China Power Encyclopedia, Thermal Power Volume. *Beijing: China Power Press, 1995. 05*, 425.

Haoliang Jiang. Introduce the Distribution Effect of Two Kinds of pulverized coal distributor. *J. Journal of Thermal Power Generation, 1976, (2)*, 20.

Kefa Cen. Large Power Station Boiler Safe and Optimal Operation Technology. *Beijing: China Power Press, 2003. 02*, 403~404.

Kefa Cen. Large Power Station Boiler Safe and Optimal Operation Technology. *Beijing: China Power Press, 2003. 02*, 416.

Wu Jiang. Present Situation and Application Prospect to Pulverized Coal Distributor. *J. Journal of Heat Engine Technology, 2004, (1)*, 35~36.

Yueming Wang. Double Adjustable Pulverized Coal Distributor: China, 97218197.0. 1997-06-05.

Zheng Wen. Fluent Fuid Computing Applications Tutorial (2). *Beijing: Tsinghua University Press, 2013. 01*, 2.

Electromechanical Control Technology and Transportation – Jia & Wu (Eds)
© 2017 Taylor & Francis Group, London, ISBN 978-1-138-06752-3

Attack detection technology based on depth feature representation

Jiadong Liu

School of Earth and Space Sciences, Peking University, Beijing, China

ABSTRACT: In traditional attack detection, the popular method is to establish the normal behavior pattern of the network data stream. Then, the detection result is achieved through comparing the network behavior with the normal behavior pattern. However, it is difficult to generate a normal network behavior pattern. In this paper, we use Recurrent Neural Networks (RNN) to analyze and represent the original attack features and depth features and then use Support Vector Machine (SVM) to detect an attack with the depth features. Experimental results indicate that this method can improve the performance of attack detection as well as detect a new, unknown attack in the network.

1 INTRODUCTION

Network space is open and makes life convenient. However, from many types of network security event reports, security vulnerabilities and risks are emerging in an endless stream. In order to defend a network threat, there are two types of methods: static passive protection technology and dynamic active protection technology. In the static active protection technology, a firewall is commonly used. In the dynamic active protection technology, network dynamic states and users' behaviors are monitored. As following, once network threats are detected, the link between hosts and the Internet can be disconnected. Therefore, attack detection technique has been an important technology to keep network security (Davis & Clark 2011).

There are many types of attack detection methods, such as feature matching, expert system, state transition, fuzzy reasoning, data statistics, pattern recognition, machine learning, and deep learning (Liao et al. 2013). Deep learning is viewed as an effective detection method, which can be self-learning to mine much known and unknown information in big and complex network data. In the deep learning method, high-dimension feature space can be mapped to a low-dimension space, and hidden information can be mined (Fiore et al. 2013).

The motivation of deep learning is to establish and simulate the human brain to analyze the data with multilayer nonlinear information processing method. It is important to use deep learning algorithm to build model for expressing and analyzing the features of the network data; then, the network data can be re-expressed (Fiore et al. 2013). First, in forward process, the input data are deep represented by calculating layer by layer. On the contrary, the input data can be reconstruction represented according to the deep represented data.

Second, weight is adjusted on the basis of the difference between the original input data and reconstructed data, which can make the reconstructed data approximate to the original input data. Classification and feature visualization are beneficial for deep learning (Sutskever et al. 2009).

In this paper, we use deep learning method to mine the depth feature representation of the network data, and then the attack behaviors are detected with the depth feature representation. Our work focuses on how to represent depth features with RNN and choose a suitable machine learning method to detect the attack. Compared with the traditional attack detection methods, our method can take full advantages of feature information with RNN, which is an artificial intelligence method. Our method can improve the performance of attack detection.

2 METHODOLOGY

In this paper, attack detection structure based on data feature deep representing is proposed, as shown in Figure 1, which contains three parts: feature processing, feature deep representing based on RNN, and attack detection based on SVM. First, network attack features are extracted from the original network data stream. These features are from the header of the package, content of the network protocol, and packet load of the network. Second, RNN is used to represent and analyze network attack features layer by layer to get a new reconstruction feature space. Third, SVM is used to detect network attack with the reconstruction features.

2.1 *Recurrent Neural Networks*

In RNN, the nodes in the hidden layer are mutually connected. The output of nodes in the hidden layer

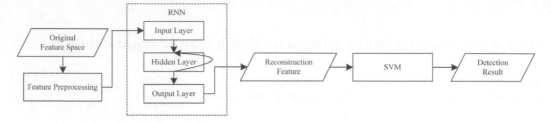

Figure 1. Attack detection structure based on depth feature representation.

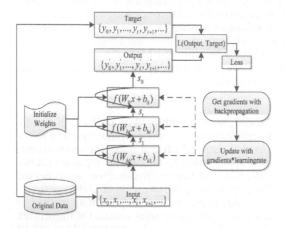

Figure 2. Architecture of RNN representing original feature space.

acts not only on the current layer but also on the input of the front node. Similarly, the input of nodes in the hidden layer depends not only on the input of the current layer but also on the output of the front node (Graves et al. 2013; Gregor et al. 2015).

In the RNN algorithm, there are input units, hidden units, and output units. The input data of input units are $\{x_0, x_1, ..., x_t, x_{t+1}, ...\}$, and the true label of the input data is $\{y_0, y_1, ..., y_t, y_{t+1}, ...\}$. The output data of output units are $\{y_0', y_1', ..., y_t', y_{t+1}', ...\}$, and the output unit of hidden units is $\{s_0, s_1, ..., s_t, s_{t+1}, ...\}$.

In the architecture, as shown in Figure 2, there are several parameters and functions, such as the hidden layer number, the neurons of each hidden layer, the activation function, the learning rate, the loss functions, and the iterations (Lipton et al. 2015). In this paper, let activation function of hidden layer be $ReLu()$ and the activation function of output layer be $Linear()$. Let the loss functions be Mean Square Error (MSE): $mean((output - target)^2)$. The lower the value of MSE, the better is the result. Therefore, the process is to decrease the value of MSE. Meanwhile, the initial values of the weights

$\{W_0, W_1, ... W_t, W_{t+1}, ...\}$ are null, which make the RNN work start. Commonly, the initial weights are set to the uniform distribution.

2.2 Support Vector Machine

Support Vector Machine (SVM) aims to construct a hyperplane to be a classifier (Meyer & Wien 2015). The function of the hyperplane is $w \cdot x + b = 0$, where w is the normal vector and b is the intercept. The hyperplane divides the sample space into two parts: positive part and negative part. The direction of normal vector points to the positive part. In general, there is infinite number of hyperplane, but the optimal one is achieved by maximum margin distance to the nearest training data point of any class. The functional margin of hyperplane to sample (x_i, y_i) is defined as:

$$\hat{\gamma}_i = y_i(w \cdot x_i + b) \tag{1}$$

In general, the larger the margin, the lower the generalization error of the classifier. Therefore, the goal of SVM is to find an optimal hyperplane that has the largest margin distance and gets a high classification accuracy of testing samples.

On the contrary, there are some noise data in samples far from the normal position. We call these noise data as outliers, which usually have a great impact on the correct classification. In order to solve this problem, slack variable ζ_i and penalty factor C are presented. Finally, the original optimal problem is translated into:

$$\min_{w,b,\zeta} \frac{1}{2}\|w\| + C\sum_{i=1}^{n}\zeta_i \tag{2}$$

$$s.t. \quad y_i(w \cdot x_i + b) \geq 1 - \zeta_i, \quad i = 1, 2, ..., n$$

$$\zeta_i \geq 0, \quad i = 1, 2, ..., n$$

Its dual is:

$$\min_{\alpha} \frac{1}{2}\sum_{i=1}^{N}\sum_{j=1}^{N}\alpha_i\alpha_j y_i y_j K(x_i, x_j) - \sum_{j=1}^{N}\alpha_i \tag{3}$$

$$s.t. \quad \sum_{i=1}^{N} \alpha_i y_i = 0 \tag{4}$$

$$0 \le \alpha_i \le C, \quad i = 1, 2, ..., N \tag{5}$$

The decision function is:

$$f(x) = sign\left(\sum_{i=1}^{N} \alpha_i^* y_i K(x \cdot x_j) + b^* \right) \tag{6}$$

where α^* and b^* are optimum solutions of the problem.

3 EXPERIMENTAL EVALUATION

The classifiers are evaluated by a confusion matrix tabulated in Table 1 (for a two-class problem). True Positive (TP) is the number of positive examples correctly classified (True Positives), and similarly for the others.

The confusion matrix provides a simple way of describing the performance of the classifiers on a given data set. There are many other evaluation metrics to measure the learning model, such as *accuracy, recall, precision, F1, ROC,* and *AUC* (L et al. 2015), which are defined as:

$$accuracy = \frac{TP + TN}{TP + FN + FP + TN} \tag{7}$$

$$recall = \frac{TP}{TP + FN} \tag{8}$$

$$precision = \frac{TP}{TP + FP} \tag{9}$$

$$F1 = \frac{(1 + \beta^2) \times precision \times recall}{\beta^2 \times precision + recall}$$
$$= \frac{2 \times TP}{2 \times TP + FN + FP} (\beta = 1) \tag{10}$$

where *accuracy* is the overall accuracy of all data sets, *precision* is the percentage of predicted positive applications actually being benign ones, and *recall* is the percentage of the correctly classified positive examples. In addition, *F1* integrates *recall* and *precision* as a measure of the effectiveness of the classification models. When both *recall* and *precision* are high, the value of *F1* will be high.

Table 1. Confusion matrix.

	Positive	Negative
True	True Positive (TP)	False Negative (FN)
False	False Positive (FP)	True Negative (TN)

4 EXPERIMENTS AND DISCUSSION

In this part, we present the results of the experiments carried out with the proposed model, by calculating the accuracy and F1. The aim is to illustrate that the depth feature mining by deep learning methods can improve the performance of the attract detection.

The environment is set as Intel(R) Xeon(R) CPU E5-2620 v3 @ 2.40 GHz with 32.0 GB RAM and 64 bit Ubuntu 16.04 operating system.

4.1 Data set for experiments

In the experiments, the data set is adopted from Cooperative Association for Internet Data Analysis (CAIDA) (http://www.caida.org/home/). The network traffic data are collected from the backbone network using the high-speed Internet. After initial processing, the number of the instances in the data set is about 103194, containing test data set and training data set. There are four types of attack packages in the data set, containing probe, DOS, R2 L, and U2 L. The feature dimension is 28, which can be divided into three categories: (1) network-type features associated with the overall information of the network link, such as protocol, duration, and service; (2) network traffic statistics features, such as the number of service requests in a period of time; and (3) behavior features in the network traffic, such as the number of landing attempts.

4.2 Experiments on RNN architecture

In this part, we analyze the result of RNN representing the feature with optional configured architecture. We set the number of neurons in each hidden layer to 10. The number of hidden layers and the number of iterations are the parameters, which we set them in the experiment. Otherwise, both the number of input layers and output layers are set to 2. Then, the initial weights are set as uniform in the number of hidden layers and the number of output layers, and the learning rate is set to 0.03. The test loss is recorded in Table 2.

Table 2. Comparison of parameters in RNN architecture.

	20	40	60	80	100
1	0.14	0.13	0.12	0.12	0.12
2	0.051	0.043	0.030	0.024	0.019
3	0.033	0.020	0.019	0.018	0.017
4	0.070	0.051	0.030	0.019	0.017
5	0.096	0.062	0.038	0.025	0.023

As shown in Table 2, we conclude that the result is affected by the number of hidden layers and the number of iterations. First, when the number of hidden layer is fixed, the loss decreases with the increasing of iterations. Secondly, when the number of iterations is fixed, the loss first decreases and then increases with the increasing of the hidden layer. The reason may be that when the number of hidden layer is less, the train data are under-fitting, which causes a high loss. While the number of hidden layers is more, the train data are over-fitting, which also causes high loss. Anyhow, it is a complicated problem to adjust parameters for RNN. In order to achieve better results, we should repeat experiments many times.

4.3 Experiments on depth features with RNN

The depth feature space is obtained with RNN, in which the number of hidden layers is set to 3, and the number of iterations is set to 100. In this part, we compare the depth features with the original features. SVM is used as the classifier. The results are shown in Table 3. According to the results, we find that the depth features outperform the original features. Therefore, the result can further prove RNN improving performance of attack detection.

4.4 Comparison with machine learning methods

Decision Trees (DT), Naive Bayes (NB), SVM, and AdaBoost are commonly applied as classifiers for attract detection. In this section, we will compare their performances.

Table 4 shows the performance of different machine learning classifiers with the depth features

Table 3. Comparison of parameters in RNN architecture.

	Depth features with RNN	Original features
Accuracy	0.9839	0.9558
Recall	0.9931	0.9695
Precision	0.9902	0.9715
F1	0.9916	0.9705

Table 4. Comparison of machine learning methods with features.

		NB	SVM	DT	Ada-Boost
Depth features with RNN	Accuracy	0.8752	0.9839	0.9271	0.9425
	F1	0.9483	0.9916	0.9550	0.9625
Original features	Accuracy	0.9226	0.9558	0.9432	0.9426
	F1	0.9547	0.9705	0.9720	0.9715

with RNN and the original features. First, comparing different features with the same classifier, the results illustrate that the depth features outperform the original features. Second, comparing different machine learning classification models, the differences of the performance are obvious. SVM is the best because it uses not only kernel function to map low-dimension space into high-dimension space to separate features easily but also slack variables and penalty factor to deal with noise data. Therefore, SVM has a high accuracy.

5 CONCLUSIONS

At present, attack detection has a high practical significance, and depth feature representation is a hot research in feature processing. In this paper, we use RNN to analyze and represent original attack features to get depth features and then use SVM to detect attack with the depth features. The results indicate that this method can improve the performance of attack detection. However, there are many parameters to adjust, which is a complicated problem. In future, we would pay more attention to explore a method to automatically select an optimal value for the parameters for RNN.

REFERENCES

Cen, L., C.S. Gates, L. Si, et al. 2015. "A probabilistic discriminative model for Android malware detection with decompiled source code," IEEE Transactions on Dependable and Secure Computing, vol.12, no.4, pp. 400–412.

Davis J.J, Clark A.J. 2011. Data preprocessing for anomaly based network intrusion detection: A review[J]. Computers & Security, 30(6): 353–375.

Fiore U., Palmieri F., Castiglione A., et al. 2013. Network anomaly detection with the restricted Boltzmann machine[J]. Neurocomputing, 122: 13–23.

Graves A., Mohamed A., Hinton G. 2013. Speech recognition with deep recurrent neural networks[C]//2013 IEEE international conference on acoustics, speech and signal processing. IEEE: 6645–6649.

Gregor K., Danihelka I., Graves A., et al. 2015. DRAW: A recurrent neural network for image generation[J]. arXiv preprint arXiv:1502.04623.

http://www.caida.org/home/

Liao H.J, Lin C.H.R., Lin Y.C., et al. 2013. Intrusion detection system: A comprehensive review[J]. Journal of Network and Computer Applications, 36(1): 16–24.

Lipton Z.C., Berkowitz J., Elkan C. 2015. A critical review of recurrent neural networks for sequence learning[J]. arXiv preprint arXiv:1506.00019.

Meyer D., Wien F.H.T. 2015. Support vector machines[J]. The Interface to libsvm in package e1071.

Sutskever I., Hinton G.E., Taylor G.W. 2009. The recurrent temporal restricted boltzmann machine[C] //Advances in Neural Information Processing Systems: 1601–1608.

Electromechanical Control Technology and Transportation – Jia & Wu (Eds)
© 2017 Taylor & Francis Group, London, ISBN 978-1-138-06752-3

Research on classification and recognition of Peking opera facial images based on SIFT features and support vector machine

Peng Zhang, Qing Zhu & Zhi-Qiang Wang
Beijing University of Technology, Beijing, China

ABSTRACT: Peking opera is one of the important representatives of Chinese traditional culture. Especially, its unique facial art has high cultural value, research value, and application value. In order to carry forward the essence of Chinese traditional culture in modern life and make the public have access to and understand the important connotation of the Peking opera facial images, in this paper, we use the classical Peking opera facial image as the image database. According to the art characteristics of the images, the images are classified into four categories. The main differences between the categories are the shape and distribution of the ornamentation. On the basis of the SIFT feature of the image texture feature, the feature vectors are reduced dimensionally and normalized to the illumination factors. By using the support vector machine classifier, the recognition rate can be significantly improved, and at the same time guarantee the original accuracy.

1 INTRODUCTION

The process of recognizing image is step by step initially to extract the features of the image and then use the extracted image features to train the classifiers. Therefore, the image feature extraction algorithm and the classification learning algorithm have become the main research object in recent years. The quality of feature extraction has a fundamental effect on the result of image recognition. At the same time, the classification learning algorithm is a type of algorithm, which can automatically classify the image by learning the feature of the extracted image, and it has a direct influence on the recognition effect.

In this paper, the SIFT feature is used to describe the facial images of Peking opera. The SIFT feature vectors are extracted and dimensions are reduced and normalized for illumination factors. SVM classifiers are used to classify the facial images of Peking opera. We designed a comparative test to analyze and evaluate the results of this paper.

2 RELATED WORK

The basic feature extraction methods for image are focused on color, texture, and shape. The methods of color feature extraction mainly include color histograms, color moments, and color entropy. Banu. M showed how to improve the pathological image retrieval system by studying the expres-sive method of the color characteristics; analyzing the HSV, Lab CIE, and LUV CIE color space; and measuring and matching according to the similarity. Texture feature extraction methods mainly include gray-level co-occurrence matrix, LBP texture features, and texture feature of wavelet transform. Yuan-Zheng LI applied the LBP algorithm to facial expression analysis in face recognition. Shape feature extraction methods mainly include Hu invariant moment, generalized Fourier description, and edge histogram. Xian-quan ZHANG used the invariant moments to classify the paper-cut patterns. Classification learning algorithm mainly include the K-nearest neighbor algorithm, support vector machine, and the neural network learning algorithm. Dai-Yu ZHANG used the support vector machine algorithm to identify the categories of traffic signs.

3 METHODOLOGY

3.1 Scale invariant feature transform

SIFT (Scale Invariant Feature Transform) is a scale-invariant feature transformation algorithm proposed by David G. Lowe in 1999 and refined in 2004. The main calculation steps of generating the SIFT feature vector set are divided into four stages.

3.1.1 Extreme value detection in scale space
The first step in the calculation is to search all the scale and image positions and use Gauss difference formula to efficiently detect potential feature

points. First, we should establish an image Gaussian pyramid, then create the image Gaussian difference pyramid (DOG), and finally finish the extreme value detection on the basis of DOG. In order to detect the extreme point of the DOG space, the 26 pixels (8 in the same layer, and 9 in both the upper and lower layers) of the image are required to be compared with each pixel in the image. When the result shows the maximum or minimum value, the point is the key point of the candidate.

3.1.2 Location of key points

For each candidate key, its spatial location and scale is determined by fitting. Before determining the position and scale of key points, candidate key points should be selected to remove low-contrast candidate key points and edge candidate key points so as to improve the stability of the operator and the repeatability of the descriptors.

3.1.3 Direction of key points

In order to ensure the invariance of the transformation, we need to assign a direction to each key point. All the operations of extracting the image features are based on the image data that have been adjusted according to the given direction, scale, and position. SIFT uses the gradient direction feature of the neighborhood pixels of the feature points to determine the direction of the feature points. In particular, the feature point's direction is determined by the main peak direction of the gradient direction histogram of the feature points and the subpeak direction.

3.1.4 Generation of feature descriptor

The feature point descriptor is generated by calculating and counting the gradient of the pixels around the current scale of the key point.

3.2 Support vector machine

Among the methods of classification and recognition, Cortes and Vapnik proposed the Support Vector Machine (SVM) algorithm as an effective classifier, which shows many advantages in solving small sample, nonlinear, and high-dimensional pattern recognition. The support vector machine is based on the statistical machine learning theory and thus it has a strict theoretical and mathematical foundation. By transforming the problem into convex quadratic optimization problem, not only the optimal solution could be obtained, but also the local optimal could become the global optimal solution, by which way the global optimality of the algorithm can be guaranteed.

SVM is developed from the optimal classification surface in the linear separable case, and the classification function is:

$$f(x) = \text{sign}(w^T x + b) \tag{1}$$

where w is the weight vector of the classification and b is the bias of the classification surface.

For nonlinear problems, it can be transformed into a linear problem in a high-dimensional space by nonlinear transformation, and the optimal classification surface could be obtained in the transformation space. In the optimal classification surface, the linear classification of a certain nonlinear transformation can be achieved by using the appropriate inner product function K (x_i, x_j); at the same time, the computational complexity is not increased. In this case, the classification function is:

$$f(x) = \text{sign}\left(\sum_i y_i \alpha_i K(x, x_i) + b\right) \tag{2}$$

where x_i is the ith dimension of the training space vector X, y_i represents the category of x_i, the value is either 1 or −1, α is the Lagrange multiplier; for most i, αi is zero, and K is a convolution kernel, which transforms the inner product of a high-dimensional feature space into a simple function operation on a low-dimensional input space. Frequently-used kernel functions are polynomial kernel functions, radial basis function, and so on.

Linear kernels are chosen in this paper because they are selected in the original space and can avoid mapping the input data into high-dimensional space. Therefore, the linear kernel function is much faster than other kernel functions.

4 EXPERIMENTS

4.1 Image database

There are 396 classic facial images of Peking opera collected in this paper, which are divided into two image databases: training set and test set. The number of images in the two databases are 224 and 172, respectively. According to the characteristics of the face mask art, they are divided into four categories: "usual face", "three-tile face", "colorful-tile face", and "broken face". Part of the images shown in Figure 1.

The following experiments were carried out according to the following four schemes: HOG + SVM, SURF + SVM, SIFT + SVM, and OptSIFT + SVM. The validity of the algorithm is verified by comparing the classification accuracy and single-pattern recognition speed and model file size. The classification accuracy formula is:

$$CA(\%) = \frac{a}{b} * 100\% \tag{3}$$

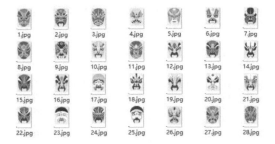

1.jpg	2.jpg	3.jpg	4.jpg	5.jpg	6.jpg	7.jpg
8.jpg	9.jpg	10.jpg	11.jpg	12.jpg	13.jpg	14.jpg
15.jpg	16.jpg	17.jpg	18.jpg	19.jpg	20.jpg	21.jpg
22.jpg	23.jpg	24.jpg	25.jpg	26.jpg	27.jpg	28.jpg

Figure 1.   Part of the images.

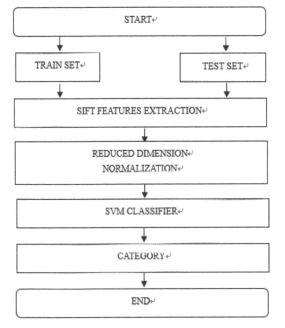

Figure 2.   System flowchart.

where a is the number of correctly classified samples and b is the total number of test samples.

4.2   *System flowchart*

The system flowchart is shown in Figure 2.

4.3   *OptSIFT*

The original SIFT operator is designed as a 128-dimensional feature vector, which allows the SIFT algorithm to maintain the stability of the brightness, scale, rotation, and other factors. In order to accelerate the SIFT algorithm, the SIFT operator is simplified to be a 32-dimensional feature vector. In this paper, on the basis of PCA dimension reduction, it reduces the dimension

Figure 3.   Software interface prototype.

Table 1.   Results of the experiment.

Method	Accuracy %	Speed ms	Size kB
HOG + SVM	86.04	506	1864
SURF + SVM	88.95	1200	540
SIFT + SVM	95.93	1835	471
OptSIFT + SVM	91.28	134	93

of the original matrix by calculating the sample covariance matrix, obtaining feature vectors corresponding to the first 32 features according to the characteristic value to form the feature matrix and making the sample matrix multiply the feature matrix.

In addition, the value of the eigenvector of SIFT itself is proportional to the light intensity. It is found that the SIFT eigenvector has a modulus value greater than 0.4 when the Peking opera images are overexposed. Therefore, it cannot distinguish the texture feature from the exposure region. In order to better adapt to the change of the intensity of the image, in this paper, the eigenvector value greater than 0.4 is normalized to 1.

4.4   *Software interface prototype*

The software interface prototype is shown in Figure 3.

5   RESULT

Table 1 shows the experimental results of the four schemes in terms of accuracy, recognition speed, and model file size.

6 DISCUSSION AND FURTHER RESEARCH

In this paper, SIFT and SVM based on image texture features are selected for learning recognition. The contrastive experiment is designed and contrasted with HOG model based on structural features and SURF based on texture features. In general, the recognition accuracy of texture features is higher than that of structural features. However, the slower recognition and the relatively large model still pose problems. In this paper, the SIFT feature vector is optimized, and the 128-dimensional feature vector is reduced to a 32-dimensional one. At the same time, factors such as illumination are normalized. The algorithm can greatly improve the recognition speed and reduce the model size.

For future research, the color feature is also one of the important manifestations of the Peking opera facial images. As the research continues, it will accurately identify the characters represented by the Peking opera facial images.

REFERENCES

Banu, M. & Nallaperumal, K. 2010. Analysis of Color Feature Extraction Techniques for Pathology Image Retrieval System [C]. *Computational Intelligence and Computing Research*: 1–7.

Dai-Yu, Zhang. 2011. Research and Implementation of Traffic Sign Recognition Technology Based on Color Information and SVM. Northeastern University.

David, G. Lowe. 2004. Distinctive Image Features from Scale-Invariant Keypoints [C]. *International Journal of Computer Vision*: 91–110.

Guo-Xiang, L.I. & Xian-quan, Zhang. 2010. Paper cut-out patterns recognition based on moment invariants and BP neural network. *Computer Engineering and Applications* 46(29):158–160.

Yuan-Zheng, L.I. 2013. Research on the Key Technologies of Object Tracking and Facial Expression Recognition [D]. Xidian University.

Sparse fast Fourier transform implementation on multicore DSP

Yifei Liu & Huihong Gong
School of Biomedical Engineering, Hubei University of Science and Technology, Xianning, Hubei, China

Yanhong Zhou
School of Medicine, Hubei University of Science and Technology, Xianning, Hubei, China

ABSTRACT: Sparse Fast Fourier Transform (SFFT) is a recently proposed highly efficient Fourier transform algorithm. It can also be utilized to extract the K largest frequency coefficients from the N-point sequence, which is sparse in the frequency domain. Although SFFT has a higher algorithm efficiency than FFTW (Frigo & Johnson 2005), the SFFT algorithm still has higher computational requirements because of its complexity. In this paper, we present a parallel implementation of SFFT on the Texas Instruments' TMS320C6678 multicore Digital Signal Processor (DSP). The result showed that the SFFT implementation on multicore DSP could obtain an acceleration of up to 1.9 times that of a single-threaded PC. With the advantages of low power and low cost, DSP-based SFFT implementation is particularly suitable for embedded systems.

1 INTRODUCTION

SFFT (Hassanieh et al. 2012) is a class of sublinear time algorithms. SFFT can be applied in computing the Discrete Fourier Transform (DFT) of a time domain signal, which is sparse in the frequency domain, and SFFT is faster than FFTW for discrete Fourier transform. Besides, similar to compression sensing (Donoho 2006), SFFT can recover the original signal from a down-sampled frequency sparse signal. This feature renders SFFT a wide application in the field of signal compression.

However, SFFT is a type of complex signal processing algorithms because it has random signal permuting sampling and irregular memory access. Because of the high computational requirement of SFFT, Agarwal et al. (2014) implemented a million-point SFFT on Field-Programmable Gate Arrays (FPGAs), and Hu et al. (2012) presented a parallel implementation of SFFT on Graphics Processing Unit (GPU) using a human voice signal as a case.

Traditionally, the processor performance improvements rely on the higher operating frequency. However, due to power constraints, the operating frequency of embedded processors is difficult to improve. With the development of modern multicore processors, it is also possible for multicore DSP to support a real-time implementation of computation-intensive SFFT algorithm at a lower power and a lower clock frequency. In this paper, we study a parallel implementation of SFFT on a multicore DSP for SFFT, which can obtain a broad application in the embedded system.

2 SFFT ALGORITHM

2.1 *Introduction to SFFT algorithm*

SFFT algorithm is described below:

For a time-domain N (N is an integer power of 2) dimension signal x, let x_i denote the ith element of the signal and x° denote its DFT. The sparsity parameter, K, is defined as the number of nonzero Fourier coefficients of signal x. SFFT algorithms include two stages: the location loop and estimation loop stage and the estimate values

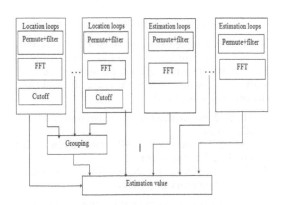

Figure 1. Flow diagram of SFFT algorithms.

stage. Figure 1 shows the flow diagram of the SFFT algorithms:

The first step is a random permutation of the signal x in the time domain. This object of permutation is to reorder a signal's frequency domain x^ and break apart the adjacent nonzero frequency coefficients. After random permutation, the nearby and continuous frequency domain coefficients became discontinuities, and the input signal x discontinuities in the frequency domain will produce some additional frequency components known as spectral leakage. Dolph–Chebyshev flat window functions filter is used to smooth the effects of spectral leakage because its frequency response is nearly flat inside the pass region and has an exponential tail outside it.

In step 2, the permuted and filtered input y is hashed and subsampled into a set of buckets with size B; then, a B-size FFT is performed on $y_i (n/B)$, and $z^ = y^_i (n/B)$.

Step 3, which is called cutoff: let I keep the d*k coordinates of maximum magnitude in z^, which are called bins. The above three steps will be repeated L times to guarantee that the significant coordinates are acquired and are considered as one of the candidate coordinates, and all the loop return L sets of coordinates $I_0 \ldots I_{L-1}$.

Then, grouping step includes a reversed hash and vote process: the reversed hash function is utilized to obtain the set of coordinates mapping to one of the bins. The voting process is to keep the locations that acquire more than half of L votes that are moved into the final location set I'.

The estimate value stage only computes the frequency coefficients of the final candidate locations. More details about SFFT can be found in Hassanieh et al. (2012).

2.2 *Performance evaluation*

The original implementation of SFFT v1.0 is written in C++ and utilizes the C++ standard template library. We rewrite the code on single-core DSP with the following strategies that are different from the original source code:

1. Static array instead of the dynamic memory allocation;
2. Use FFT function in DSP software library for fast Fourier transform;
3. Single precision implementation for reducing on-chip memory requirements.

Several parameters may affect the performance of the SFFT algorithm. We use the follow two

characteristic parameters to evaluate the runtime of SFFT on single-core DSP:

Parameter 1: -N 65536 -K 50 -B 4 -E 2 -L 8 -l 5 -r 4 -t 1e-8 -e 1e-8

Parameter 2: -N 4194304 -K 200 -B 4 -E 0.5 -L 10 -l 3 -r 2 -t 1.e-6 -e 0.5e-8

From the profiling results in Figure 2, we observe the permuting and filter step is the most time-consuming part in SFFT when using parameter 1. With large sparsity K, SFFT commonly has a large number of candidate coordinates in the estimation value stage, and the time taken by the estimation stage will increase. As we have estimated, it can be seen in Figure 3 that when parameter 2 is used, the percentage of time taken by the estimation stage becomes dominant.

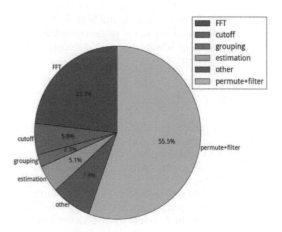

Figure 2. Profile runtime of SFFT on single-core DSP with small sparsity K.

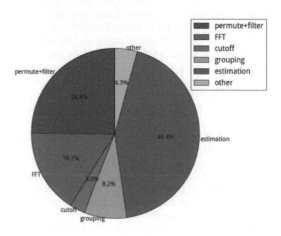

Figure 3. Profile runtime of SFFT on single-core DSP with large sparsity (K).

3 PARALLEL SFFT AND DSP OPTIMIZATION

3.1 *Parallel implementation for SFFT*

As shown in Figure 1, the SFFT algorithm has many round location loops and estimation loops. In the estimation value stage, each loop completes the estimation of the frequency coefficients of a candidate location. TMS320C6678 DSP has eight C66x Fixed/Floating-Point CPU cores (Texas Instruments 2012), and all the cores can compute in parallel. The main idea of the parallel implementation of SFFT algorithm is to map the loop index to DSP core id.

The flow of parallel SFFT algorithm is shown in Figure 4. We use core 0 as the master core; permutation, filter, FFT, and cutoff parallelly run on the core1–core7, which are the slave cores.

When the master core receives the interrupt messages from all the slave cores, it indicates all the slave cores have achieved the location loops and estimation loops. Then, the grouping step votes for the final candidate locations from L sets of coordinates that have been put into the shared memory of 8 DSP cores. After the master core completes the grouping, it will notice all the slave cores to obtain the candidate locations from the shared memory and start the estimation value. When the estimation value stage ends, the master core receives the interrupt messages again from all slave cores and then the SFFT algorithm finishes.

3.2 *DSP optimization for SFFT*

DSP is a microprocessor with single-cycle hardware multiply and accumulate units, deep pipelines, separate program, and data memories (Harvard architecture). The above architectural features make it suitable for real-time and high-performance digital signal process applications. In most cases, compiler optimization for DSP can be very effective. However, there are some cases when the compiler alone is not sufficient for implementing complex operations. In such situations, we use intrinsic operations and optimized C6000 DSP libraries in the SFFT.

Because the permutation and filter of SFFT include complex multiply and accumulate from randomly permuted samples, the original C++ codes are replaced by the intrinsic operations as follows:

result = _complex_mpysp(x_re_im,filter_re_im);
        _daddsp(sampt_re_im,result);

In addition, each location loop and estimation loop in SFFT includes a B-size FFT. TI provides some highly optimized DSP software libraries, such as DSPLIB. The function DSPF_sp_fft-SPxSP of DSPLIB is used in this paper to improve the FFT performance.

4 RESULTS

4.1 *Experimental setup*

The operating system in this paper is Ubuntu 16.04 that runs on a standard PC with AMD A10-7850K processor and with 1600 MHz, 16GB DDR3 memory. Code Composer Studio for Linux 5.5 is used for TMS320C6678 DSP simulation. The optimization option of DSP compiler is enabled by default at optimization levels -O2 in this experiment. The clock frequency of each TMS320C6678 CPU core is set to 1.0 GHz. For keeping high performance and low memory occupancy, SFFT is implemented through the Chip Support Library (CSL) without DSP/BIOS real-time operating system.

4.2 *Results and discussion*

4.2.1 *Execution time versus input signal size N*

In this simulation experiment, the size of the input signal varies from 2^{16} to 2^{22}, and the sparsity K is fixed at 50. Figure 5 shows the SFFT execution times on the single-threaded PC and the single-core, four-core, six-core, and eight-core DSP. The result indicates that the SFFT can be accelerated by a factor of 1.7 for large signal size; the performance improvement of large signal size SFFT is benefited from the location loops and the estimation loops parallelization. However, when the signal size is small, the performance of SFFT on multicore DSP has no distinct advantages over that on the single-threaded PC. There are several reasons to explain the result. First, when the signal size is small, the total run time is relatively short. Furthermore, the grouping step is closely related

Core0   Core1-Core7

Figure 4. Flow diagram of the parallel SFFT algorithm.

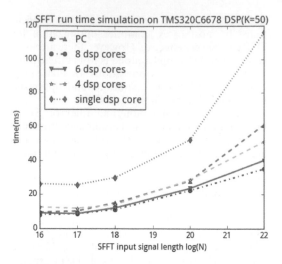

Figure 5. Profiling results for parallel SFFT with various input signal sizes.

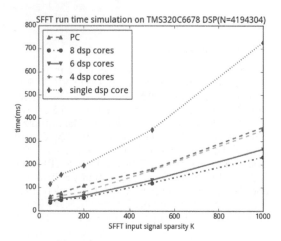

Figure 6. Profiling results for parallel SFFT with various sparsity (K).

to the sparsity factor, K, rather than the signal size, N. In addition, intercore communication and some other operations occupy a larger proportion because of the shorter total execution time in the small signal size parallel SFFT.

4.2.2 *Execution time versus sparsity K*

In this section, the signal size (N) is fixed at 2^{22} and sparsity K varies from 100 to 500. We discuss the effect of the sparsity factor (K) on the execution time of SFFT on multicore DSP. In the SFFT algorithm, the running time of the estimation value and grouping is closely associated with the sparsity (K). Usually, when k increases, the candidate locations for estimation value also increases and more loops are necessary for estimation value. The

parallel implementation for estimation value on multicore DSP is a simple task. However, because there are large vector sum and accumulation of vote for the coordinates in the grouping stage of SFFT, the stage is inconvenient for parallelization. As Figure 3 shows, with a large sparsity (K), the estimation value takes much more time than the grouping stage in SFFT. The performance of SFFT can still benefit from the estimation value parallelization on multicore DSP. The result is shown in Figure 6. We can gain almost the same performance with the PC by using four-core DSP. The speed of SFFT on an eight-core DSP can improve by 90% compared to that of the PC.

5 CONCLUSIONS

In this paper, we implemented a parallel SFFT on the TI TMS320C6678 multicore DSP, utilized the intrinsic operations, and optimized C6000 DSP libraries for further improving of the SFFT performance. The simulation result shows that the DSP with eight CPU cores can yield a 1.9 times speedup over PC for SFFT, and it presents the feasibility of high-performance SFFT on multicore DSP.

The multicore DSP provides a low-cost, low-power, and high-performance solution for SFFT in the embedded application. The future work includes the SFFT implementation on the multicore DSP chip, where we will focus on the memory access and cache optimization.

REFERENCES

Agarwal, A., Hassanieh, H., Abari, O., Hamed, E., Katabi, D. & Arvind. 2014. *High-throughput implementation of a million-point sparse Fourier* Transform. 24th International Conference on Field Programmable Logic and Applications, Munich, Germany, September. 2–4, 2014. pp. 1–6.

Donoho D.L. 2006. *Compressive sensing*. IEEE Transactions on Information Theory 52(4):1289–1306.

Frigo, M. & Johnson, S. 2005. *The Design and Implementation of FFTW3*. Proceedings of the IEEE 93 (2):216–231.

Hassanieh, H., Indyk, P., Katabi, D., & Price, E. 2012. *Simple and practical algorithm for sparse fourier transform*. Proceedings of 23rd Symposium on Discrete Algorithms, Kyoto, Japan, January 17–19, 2012. pp. 1183–1194.

Hu, J., Wang, Z., Qiu, Q., Xiao, W. & Lilja D.J. 2012. *Sparse fast fourier transform on gpus and multi-core cpus*. Computer Architecture and High Performance Computing (SBAC-PAD); IEEE 24th International Symposium on, Washington, DC, USA, October 24–26, 2012. pp. 83–91.

Texas Instruments. 2012. TMS320C6678 Multicore Fixed and Floating-Point Digital Signal Processor Data Manual. http://www.ti.com/lit/ds/sprs691c/sprs691c.pdf.

Electromechanical Control Technology and Transportation – Jia & Wu (Eds)
© *2017 Taylor & Francis Group, London, ISBN 978-1-138-06752-3*

A hybrid level-of-detail modeling and rendering approach for the visualization of large-scale urban scenes

Zhihua Zhang, Haiyin Wang, Zhenbiao Hu & Jiuyan Zhang
Qingdao Geotechnical Investigation and Surveying Research Institute, Qingdao, China

Shengchuan Zhou
Qingdao West-coast Geographical Information and Remote Sensing Centre, Qingdao, China
Qingdao Geotechnical Investigation and Surveying Research Institute, Qingdao, China

ABSTRACT: In this paper, we introduce a user-assisted modeling approach that allows the user to exploit the repetition pattern of a building façade, thus creating a splitting grammar representation for image texture compressing. A novel hybrid LOD rendering method is then proposed for the visualization of large-scale urban scenes. We extract progressive point cloud as well as model border lines from the input models and provide their simplifications encoding in a data structure that allows fast LOD selection. Screen space projected area is used as an LOD selector to switch the rendering between the progressive geometric model and the polygon model that contains procedural textures. The experimental results show a significant performance enhancement compared to that of ground truth rendering and geometric LOD, and the quality of the result is indistinguishable from the original models.

Keywords: inverse procedural modeling, level-of-detail, blue noise sampling, massive urban models

1 INTRODUCTION

Despite recent progress, representing and rendering large-scale urban data remains challenging. Urban scenes are often an important background; they are usually not the main focus, but they provide an important ambiance, without which the scene will lack realism.

Currently, the most-used large-scale 3D model rendering approaches are level-of-detail (Hoppe H, 1996; Sander P V, 2001 & Peng C, 2012), out-of-core rendering (Yoon S E, 2005), and image-based rendering (Gobbetti E, 2005). Urban building usually contains a regular boundary, which results in geometric distortion after geometric level-of-detail processing. Image-based urban rendering also has been addressed by introduced impostor (Sillion F, 1997), texture cluster (Ma Chunyong, 2010), and block maps representation (Cignoni P, 2007), which allow for a more compact simplification and show better performance for long-distance views, but are prone to parallax error and loss of individual properties.

Recently, 3D point cloud modeling has been widely used in modeling city-scale scenes. Point-based rendering is suitable to visualize data model with complex geometric structure and topology (Kobbelt L & Sainz M, 2004) but rarely used in rendering urban model with large-scale spatial distribution pattern.

In this paper, we propose a new user-assistant inverse procedural modeling approach for texture compression, which indicates symmetries and repetitions of building facades through procedural rules. The compressed texture can be directly rendered in GPU and considerable saving of the GPU memory while maintaining comparable efficiency of rendering.

Our secondary contribution is a novel LOD rendering model. We use recursive Wang-tile to create a point-based progressive model and use projected area to determine which primitives (point or polygon model containing procedural compressed textures) and the number of point should be rendered. Our approach shows a significant improvement in data compression and rendering performance, which can represent city-scale 3D senses in real time while obtain indistinguishable visual effect compared with the ground truth.

2 REVERSE PROCEDURAL MODELING

2.1 *Grammar definition*

Our approach describes an input texture image (G) as a context-free attribute grammar. The grammar is an attribute split grammar including terminal and nonterminal symbols, which can be defined by Backus Normal Form as:

$$\langle G \rangle ::= \langle N \rangle \{ \langle N \rangle \} \qquad (1)$$

where N is the nonterminal symbol for representing image transformations and intermediated steps and can be thought of as groups or structures. The terminal symbol T represents final submages that compose the input images. The recursive definition of the nonterminal symbol N is:

$$\langle N \rangle ::= \left[O_p \right] (\langle N \rangle \mid \langle T \rangle) \{ \langle N \rangle \mid \langle T \rangle \} \qquad (2)$$

The operation symbol O includes *spliteX* (split horizontally), *spilteY* (split vertically), *subdivX* (subdivide horizontally), *subdivY* (subdivide vertically), *flipX* (flip horizontally), *flipY* (flip vertically), and *rotateX* (rotate). The parameter p represents divided proportion for the subdivide operation and represents rotation angle for the reverser and rotation operations.

Every terminal and nonterminal symbol is associated to a rectangular image. Terminal symbols represent a certain part of the input texture, whereas nonterminal symbols represent areas that are composed of other nonterminal or terminal symbols. As shown in Figure 1, the example image obviously contains a repetition pattern, containing characters "A" and "B".

Then, the input image G can be represented by two terminal symbols $T_1 = A$, $T_2 = B$ and two nonterminal symbols N_2, N_3. The splitting grammar is:

$$N_1 = spliteX_{0.5}(T_1, N_2)$$
$$N_2 = spliteY_{(1/3)}(N_3(0), N_3(90), N_3(180)) \qquad (3)$$
$$N_3(p) = rotate_p(T_2)$$

According to the procedural grammar representation, we define the image compression factor (C) of the input image as:

$$C = 1 - \sum_{i=1}^{|T|} \frac{R_i}{R_G} \qquad (4)$$

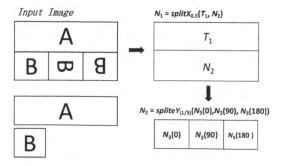

Figure 1. Example image and its procedural grammar representation.

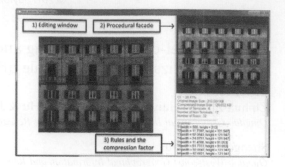

Figure 2. Using interactive modeling tools to represent the original facade in window (1) as a procedural representation. Window (2) shows the results, while window (3) shows the procedural rules and compression ratio.

where R_i is the image resolution of terminal symbol T_i and R_G is the resolution of the input image. Although we could also add the size of the grammar to the compression, the size is usually significantly smaller than the size of the image, which can be safely ignored. In Figure 1, the compression factor is $C = 1/3$, which means we can save almost one-third memory space through our grammar expression.

2.2 *Interactive modeling*

On the basis of the procedural split grammar, we implement a modeling application that allows users to compress a facade in an interactive way. The application contains three operational windows (as shown in Figure 2). The first is the modeling window, which shows the input texture image and allows users to exploit repetition and symmetry pattern of the image. The second window displays the procedural modeling result of the original image in real-time manner, and the third window shows the grammar, rules, and compression statistics.

The application uses dark blue for selected symbols, light blue for terminal symbols, and shows nonterminal symbols with a pattern. This allows a direct visual evaluation for a user to make better estimate of the compression ratio.

3 GEOMETRIC LOD GENERATION

Through inverse procedural modeling, we simplified the input model by reducing texture complexity. For large-scale urban scenes rendering, the input model should also be simplified in a geometric way.

Our strategy is based on the perspective projection because our key observation is that human visual acuity is extremely sensitive to edges (Burr D C, 1989). Urban scenes are usually composed of building with clearly defined boundary. And the common LOD approaches, such as triangle mesh simplification, and point-based rendering, will often generate noise for building edges. At the same time, we hypothesize that simplified urban scenes can be rendered by point and lines, which will support an efficient rendering and will be also visual plausible in distance view.

The input 3D building model is converted into LOD representation consisting of point and line. In this section, we describe how this process can be done.

3.1 Point-based model generation

Point-based modeling for urban scene rendering should follow several important characteristics in order to produce a high-quality visual lossless rendering result.

1. Point samples should be evenly distributed over the building surface.
2. Regular point distribution usually provides regular patterns that is visually disturbing. A good sampling model could remove the regularity by transferring low-frequency artifact to high-frequency noise.
3. The sampling should be progressive, thereby changing the point density without conditions 1 and 2.
4. The sampling should be fast enough for processing large-scale urban scenes.

A method that has the above four characteristics is the multijittering sampling.

As shown in Figure 3, we divide the model surface into multilevel grid and proceed the point cloud sampling. The sampling progress should meet the *N-rooks* condition that the sampling points are randomly selected in the grid and each pair of point should not appear in the same column or row. Point color is sampled by texture mapping, and the sampling level is saved for each point and then used as LOD index during rendering.

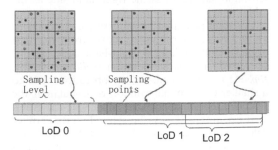

Sampling Level

Sampling points

LoD 0   LoD 1   LoD 2

Figure 3.   Multijittered sampling for point-based level-of-detail modeling generation.

3.2 Line-based model generation

Line generation is based on the angle of the two jointed mesh faces. If the jointed angle is less than 170°, the adjacent edge will be extracted for representing a building structure.

In order to get a higher visual quality, we use the prevailing color from the input as a single color of the line to avoid using texture. In most cases, edges of two joints share the same color. Otherwise, a brighter color should be used because human vision is more sensitive to bright color (Jameson D, 1964).

3.3 Hybrid LOD model ranking

After creating point and line models, we designed a data structure to store primitives and allow for adaptive LOD selection.

The point ranking creates a point sequence for an input model, then any prefix of the sequence is a simplified point-based model that can progressively add details.

In the sampling step, we assigned each point a level index value. We can get a progressive LOD representation by sorting the entire point set according to the index number and reordering the subsequence that has the same index values by the Fish–Yates algorithm, then storing the various points in increasing order.

Line ranking creates a line-based LOD model for progressively rendering any prefix of the sequence as a simplified line-based model. We use the length of line as sorting key and sort lines in decreasing order. Then, we can render visible lines by selecting a sequence prefix through the projected length of line.

As shown in Figure 4, the ranked point and line sequences can combine with the procedural textured polygon model to create a hybrid LOD representation. We use point and line representation in distance view and use procedural textured polygon model in close-ups. We compile the ranked point and line sequences and the polygon model as vertex buffer object, and then the LOD model selection and rendering can be done by a single OpenGL function call.

Figure 4.   Hybrid LOD model. From left to right are input model, different detail levels with different colors, 25% point samples and 25% line samples, 75% point samples and 75% lines, and procedural rendered texture models, respectively.

4 HYBRID RENDERING

The ranked sequences structure provides a flexible and fast way to render the LOD model. The number of point and line is calculated during the rendering from ranked sequences.

Our objective is to select the procedural textured model or the point- and line-based LOD model for visual lossless rendering and minimizing rendered primitives. It is important to calculate proper number of LOD points and lines to guarantee the rendered model is hole-less and visually continuous. In this paper, we use screen-space projector to achieve this.

Let us denote the input model consisting of triangle set by $M = \{M_1 \ldots M_{|M|}\}$ with surface area A_i. The projected raster area $\left(A'_m\right)$ in screen space under a given perspective projection can be approximated by:

$$A'_m \approx A'_u \sum_{1}^{|M|} A_i \tag{5}$$

where A'_u is the raster area of a screen-aligned unit square in model position. Let W denote the resolution of the screen in pixel. Given the total field of view θ and the view distance d, the unit raster area $\left(A'_u\right)$ and unit raster length $\left(L'_u\right)$ can be calculated as:

$$A'_u = L'_u \cdot L'_u \approx \left(\frac{W}{2d \cdot \tan \theta / 2}\right)^2 \tag{6}$$

Let us assume that the input model can be approximated by a point sequence and line sequence as shown in Figure 4. The point sequence has a raster area A'_p, and the line sequence has a raster area A'_l. To get a faithful rendering result, especially to avoid holes, we need:

$$A'_m - A'_l - A'_p \leq 0 \tag{7}$$

4.1 Line-based model rendering

During runtime, we select a suffix from the ranked line sequence. The number of lines is controlled by the projected length. A line is rendered if and only if its projected length is greater than 1 pixel.

Let us assume that the input model is approximated by a line set, and the length of line L_i in the data set is l_i. The raster area covered by the lines is:

$$A'_l \approx L'_u \sum_{i=1}^{|L|} l_i \tag{8}$$

If the raster area has $A'_m - A'_l \leq 0$, we render the model only by lines. Otherwise, we compute the raster area that need to be covered by point to fill the holes.

4.2 Point-based model rendering

In our approach, points are used for filling gaps between line structures and providing façade details. Each point is represented as a point space using OpenGL with size W'_p. The number of rendered points depends on the difference between the approximated raster area of the triangle model and the approximated raster area of the line model. The number of points is then determined by:

$$n_p = j_p (A'_m - A'_l) / W'_p \tag{9}$$

The factor j_p slightly adjusts the point density to avoid gaps caused by sampling and hardware antialiasing, and we use $j_p \in [1.2, 1.5]$ in our experiment. The experimental results show that the value of j_p should increase by 10% after the hardware antialiasing increased to a higher level.

4.3 Procedural rendering

During runtime, we present a new approach for direct GPU-based facade rendering. The texture images correspond to the terminal symbol, and the encoded grammar rules are stored in GPU memory. Then, the (u, v) texture coordinates are evaluated directly from the grammar. The encoded texture image could be significantly smaller than the input image. However, as grammar phrasing is required, there is an overhead imposed by the texture look-up.

In our implementation, we want to avoid branching and loops because of their low performance in modern GPU architectures (Open GL 4.3 and Shader Model 5.0). Our GPU program stores each grammar rules encoded in a data structure that stores the operation, parameters, and the set of the terminal and nonterminal symbols on the right-hand side. Each rule has an ID, and the rules are stored in a table that is indexed by the left-hand side and has the right-hand side as the table elements.

For each pixel that has a texture coordinate (u, v), the rules interpretation starts from the first nonterminal symbol. A right-hand side scan is processed, and each operation (*split, subdivided, flip,* etc.) is evaluated. As the operations lead to a smaller nonterminal symbol, we calculate if the searched texture coordinate falls inside the evaluated operation. If yes, the symbol starts another rule that is found and should be further interpreted. If the founded symbol is a nonterminal one, the interpretation continues with scan operation like above. The scanning stops if a terminal symbol is found. The transformed texture coordinate is then used for texture lookup.

5 RESULTS AND EVALUATION

We created five test scenes showing different urban areas with the aim of converting various scenarios. Each scene contains thousands of architecture models. We use inverse procedural modeling tool to create procedural textured model for data compression and analysis of the space complexity and then introduce our LOD method to test the performance of large-scale urban scenes rendering. The input city model was obtained from Digital Qingdao Urban Data Set, which contains 5.2 GB of geometric models and 42.3 GB of textures.

5.1 Inverse procedural modeling analysis

We introduced 10 participants to test our procedural modeling approach with ages 20–30 years and having some knowledge of 3D modeling.

The objective of the test was to encode and compress building texture, allowing memory use reduction and fast rendering, while maintaining plausible visual quality.

Figure 5 shows a typical texture modeling result. The overall compression was 54%, with the standard deviation around 10%. The maximum compression result was 99%, and the minimum compression result was 12%. The input can be better compressed if it has repeated window or similar repetitive instance, and the occlusion noise such as trees can be rapidly removed in the modeling progress (Figure 5b).

5.2 Rendering performance

We tested the rendering performance on a desktop computer with Intel Xeon E5-2620 at 2.0 GHz with 16 GB memory and an NVidia GTX 960 GPU with 1.5 GB video memory.

We first use five urban scenes that with different building styles to compare the performance with triangle mesh rendering and GLOD (Cohen

Figure 6. City-scale 3D scene rendering at 31–45 frames per second.

Table 1. Rendering time per frame (ms).

Scene	1	2	3	4	5
Ground truth	243	248	323	341	235
GLOD	121	112	168	161	104
Hybrid rendering	23	21	36	31	21
Hybrid rendering (compressed)	33	35	47	41	32

J, 2003) rendering and then introduce a large-scale urban scene for city-scale data visualization (Figure 6).

Table 1 shows the performance differences among triangle mesh rendering, GLOD rendering, hybrid rendering, and hybrid rendering on the basis of the procedural textured model. The result indicated that our approach is approximately five times faster than the GLOD and 10 times faster than the ground truth model rendering. The procedural textured model significantly reduced the space complexity of building texture, thereby improving the rendering performance by approximately 30%. As shown in Figure 6, our approach is capable for massive city model visualization, which can get interactive rendering frame rate while maintaining a plausible visual quality.

6 CONCLUSION

We presented a novel method for large-scale urban scenes modeling and rendering. The main idea is to use a user-assisted approach for inverse procedural modeling of facade images and introduce a hybrid representation of lines, points, and procedural textured building model. Our method efficiently compressed images, obtaining an archived average compression of 54%, and the hybrid LOD representation makes it possible to render large-scale urban scenes in an interactive frame rate.

Figure 5. Image with occluded parts (a) can be quickly resolved by interactive modeling. The compression factor is 73%, which provides considerable memory savings while archiving plausible visual quality.

309

Our method focuses on the building modeling and rendering for urban scenes. Future works include optimization of trees and plants. We also introduce the LIDAR point cloud and an oblique photography model for supporting more data sources of urban modeling.

REFERENCES

Burr D.C, Morrone M.C, Spinelli D. Evidence for edge and bar detectors in human vision[J]. Vision Research, 1989, 29(4): 419–431.

Cignoni P., Di Benedetto M., Ganovelli F., et al. Ray-casted blockmaps for large urban models visualization [J]. Computer Graphics Forum, 2007, 26(3): 405–413.

Cohen J., Luebke D., Duca N., et al. GLOD: A driver-level interface for geometric level of detail[C] //Proceedings of the SIGGRAPH '03 Sketches & Applications. New York: ACM Press, 2003: 1–1.

Gobbetti E., Marton F. Far voxels: a multi-resolution framework for interactive rendering of huge complex 3d models on commodity graphics platforms[J]. ACM Transactions on Graphics, 2005, 24(3): 878–885.

Hoppe H. Progressive meshes[C] //Proceedings of the 23rd Annual Conference on Computer Graphics and Interactive Techniques. New York: ACM Press, 1996: 99–108.

Jameson D., Hurvich L.M. Theory of brightness and color contrast in human vision[J]. Vision Research, 1964, 4(1–2): 135–154.

Kobbelt L., Botsch M. A survey of point-based techniques in computer graphics[J]. Computer and Graphics, 2004, 28(6): 801–814.

Ma Chunyong, Chen Yong, Han Yong, et al. A GPU-based rendering acceleration algorithm for urban simulation[J]. Periodical of Ocean University of China. 2010, 40(7), 141–144 (in Chinese).

Peng C., Cao Y. A GPU-based approach for massive model rendering with frame-to-frame coherence[J]. Computer Graphics Forum, 2012, 31(2pt2): 393–402.

Sainz M., Pajarola R. Point-based rendering techniques[J]. Computer and Graphics, 2004, 28(6): 869–879.

Sander P.V., Snyder J., Gortler S.J., et al. Texture mapping progressive meshes[C] //Proceedings of the 28th Annual Conference on Computer Graphics and Interactive Techniques. New York: ACM Press, 2001: 409–416.

Sillion F., Drettakis G., Bodelet B. Efficient impostor manipulation for real-time visualization of urban scenery[J]. Computer Graphics Forum, 1997, 16(3): 207–218.

Yoon S.E., Salomon B., Gayle R., et al. Quick-VDR: out-of-core view-dependent rendering of gigantic models[J]. IEEE Transactions on Visualization and Computer Graphics, 2005, 11(4): 369–382.

Study on quantitative terrain analysis based on DEM

Man Jiang, Haitao Wei, Wentong Sun & Jiantao Wang
School of Resources and Civil Engineering, Shandong University of Science and Technology,
Tai'an, Shandong, P.R. China

Junzhe Liu
Shandong Experimental High School, Jinan, Shandong, P.R. China

ABSTRACT: This study is based on ArcGIS platform, with the 30 meters resolution DEM image data of Taishan District. The authors studied the relationship between Relief Amplitude and the spatial distribution of contour frequency by using the technology combination of Similar Thought of DBSCAN (Density-Based Spatial Clustering of Applications with Noise) and moving windows; and then presents a method which is: divide the contour; and then aggregating the contour to realize the quantitative grades of the relief. The algorithm steps are as follows: make the closed curve become segments (simplify the contour, and then divide the simplified contour into segments); determine the size of the moving windows (the moving window's size is three times the shortest segment's Minimum Bounding Rectangle (MBR)); the determination of the segments frequency (determine the frequency of segments in each moving window according to its coverage; get the spatial distribution figure of contour frequency); view the spatial distribution; adopt raster data as a display; use the segments frequency as grid values; and get the contour frequency raster graphic of the Mount Tai area. Based on the curve graph in which the x-axis is the moving window position, and the y-axis is the contour frequency, the authors set thresholds for the frequency to help give quantitative grades to the relief. With field investigations and political boundaries data, the contour frequencies of 0~3, 3~6, 6~9, 9~12, 12~15 respectively corresponds to flat slope, gentle slope, slope, steep slope, cliffs. With the relief grades and field investigations, the authors discuss several interesting points of spatial distribution concerning the figure of contour frequency. And the results show that the method is suitable for analyzing the salient relief, and can work with existing terrain analysis methods to provide references for rescue work. It has significant values both in theory and practice.

1   INTRODUCTION

With the rapid development of computers, remote sensing, and GIS technology, people began to try to study new ideas and methods of digital mapping of various landform types. In 1964, Hammond first proposed the automatic classification method of landform types, pointing out it can be determined by the slope and relative fluctuation degree in the statistical area unit (Hammond E H, 1964). This idea was achieved by Dikau in 1991 (Dachau R, 1991), and his method was revised and developed by Baryon, Morgan, et al (Baryon L, 1998 & Morgan J M, 2005). Draught extracted elevation, slope gradient, profile curvature, plane curvature and other terrain factors from the DEM, and used the image segmentation method to complete the classification (Prima O D A, 2006). Since then, many experts and scholars at home and abroad have carried out in-depth research on the automatic classification of landform in the regular statistical unit (Iwahashi J, 2007; R. T., McIvor, 2000; Liu Ail, 2006; Lang Ling Ling, 2007 & Cheng Weiming, 2009). These studies have a common characteristic in using the regular statistical unit to compute various types of terrain factors (such as shape degree of V, table roughness, high process variable coefficients, slope change rate, the average high process and profile curvature and plane curvature rate), to achieve the rapid classification of geomorphologic types. The disadvantage, using the statistical unit, is the rules used to extract the geomorphologic boundary are usually different from traditional artificial boundary topography and particularly in undulating mountain. And the classification criteria and methods are mainly proposed for the whole of China, so it is not completely applicable to the Taishan District, and other small areas. Therefore, this study uses the example of the Taishan District to propose a new algorithm based on DEM combined with the moving window technique, using a new algorithm proposed for the quantitative classification of the topographic relief degree by using the clustering algorithm of DBSCAN clustering algorithm.

2 CONTOUR DENSITY ANALYSIS ALGORITHM

DBSCAN is typical of the cluster algorithm based on density. The main idea of the algorithm is that a point in the spatial database continues clustering if the density of points in its closest region is above a certain threshold value. This paper makes use of an idea similar to DBSCAN to quantitatively analyze the contour density. Relief amplitude, which is the important indicator for describing landform quantitatively and identifying the landform type, reflects the degree of terrain fluctuation. We can divide relief amplitude into several ranks by analyzing the density of the contour. And we can generate the contour by DEM image maps and analyze the density of the contour through the line segment clustering thought based on "dividing-aggregating". The thinking framework is as shown in Figure 1.

First, simplify and break the contour. Divide the simplified contour into segments according to the characteristic point; cluster the segments based on density. Then, expand the segments data on the DBSCAN clustering way using the conventional aiming point procedure. Define the size of the moving window by the shortest segment's MBR; then, divide the space into several groups of moving windows parallel to the horizontal coordinate axis. Determine the frequency of contour line segments in each moving window, using the window position as the horizontal axis, and using the frequency of contour line segments as the vertical axis; get the spatial distribution figure of contour frequency. When we combine the figure with the field survey, we can classify the frequency of the contour by the auxiliary setting frequency threshold value., Different levels of frequency represent different terrain, such as flat slope, gentle slope, slope, steep slope, cliffs. Using the way raster data are displayed, and using the frequency of contour line segments as grid values, we obtain the Taishan District contour frequency distribution grid map that can intuitively display the range of terrain relief.

3 TAISHAN DISTRICT GEOGRAPHIC BACKGROUND

For the Taishan District, geographical coordinates are 36°05′ to 36°20′N, 117°03′ to 117°13′E, 28 kilometers from north to south, 24 kilometers wide from east to west, with a total area of 336.86 square kilometers. The Taishan District is on the western edge of the Tailai Synclinal Basins. The Taishan mountain range is north of Taishan City, south of the Dawen River, with a topography from north to south. The northern part of the Taishan District is the Taishan Scenic Area. The Taishan Scenic Area is at least 200 meters above sea level. The highest elevation is 1532.7 meters, and is the highest peak in Shandong Province. Topographic relief in the southwest is low hills, 170 to 200 meters above sea level. Southeast of Taishan is the Dawen River alluvial plain where the terrain is relatively flat, and about 130 meters above sea level. In the middle of the Taishan District is the Taishan Piedmont Alluvial Plain, with an elevation of about 200 meters above sea level.

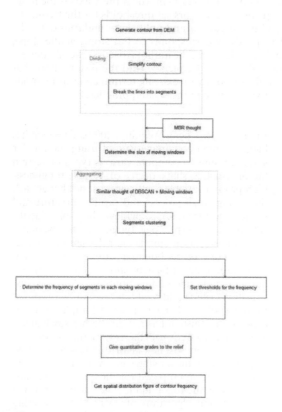

Figure 1. Thinking framework.

4 EXPERIMENTS AND RESULTS

This study uses GDEMDEM (30-meter resolution) ASTGTM2_N36E117, which intercepts near the Taishan Mountain Range of 117.006° to 117.231° longitude, and using 36.201° to 36.330° latitude imagery, with the GCS_WGS_1984 coordinate system, and a 50-meter contour interval. After testing, when the moving window is three times as much as the length of the MBR of the shortest line segment, the contour frequency spatial distribution can be obtained. The topography of the Taishan District can be carefully and accurately reflected. Using the window position as the horizontal axis and the contour frequency as the vertical axis, the spatial distribution figure of contour frequency was obtained. The maximum frequency of the statistical window segment is 15, and the minimum is 0. According to the distribution of frequency and the auxiliary field investigation, setting the contour frequency threshold of 3, 6, 9, 12, the terrain can be divided into 5 grades, namely 0~3, 3~6, 6~9, 9~12, 12~15, reflecting flat slope, gentle slope, slope, steep slope, cliffs.

Selecting undulating terrain changes of typical regions of Dui song shan, and Shi ba pan, where slopes are steep, and doing the contour frequency distribution curve, Graphs 1 and 2 can respectively be obtained. Graph 1's contour frequency curve changes dramatically showing that topography

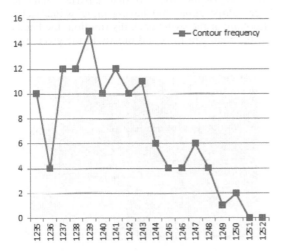

Graph 1. Dui Song Shan contour frequency curve.

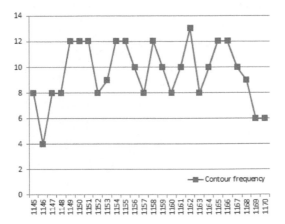

Graph 2. The Shi Ba Pan contour frequency curve.

changes are great. The lowest contour frequency was 0, the maximum was 15. The 9~15 high frequency segment is mainly distributed in the 1237 to 1243 moving window position. The position is Dui song shan's cliffs and steep hills. The moving window position for residential areas of Song shan District have a relatively flat terrain; the 0~6 low frequency segment is mainly distributed in the 1244 to1252 position. Graph 2 for Shi ba pan's curve shows the frequency value is mainly concentrated in the 8 to 13 range, and the frequency range is more focused and concentrated in the high frequency range. The difference of topographic relief is small, mainly because of the steep slope.

Figure 2 is the Taishan District contour frequency distribution grid map, with the white and yellow are-as' contour frequency 0~3 and 3~6 respectively. In detail, white areas represent altitude difference which are less than 150 meters within 1 km; the yellow color areas represent altitude difference less than 300 meters within 1 km. The degree of relief is bigger from green to blue to red, with the contour frequency 6~9, 9~12, 12~15 respectively. The green area is slope and the altitude difference of the green area is between 300~450 meters within 500 meters; the blue area is steep slope and the altitude difference of the blue area is between 450~600 meters within 500 meters; the red area is cliff and the altitude difference of the red area is more than 600 meters within 50 meters. We find from Figure 2 that the green areas are mainly concentrated in and matching with Mount Tai's range, and the slope increases gradually starting from the green. Around the mountain to the south is the city, which is white and yellow with low contour frequency. Examining the lines between green and yellow, we can easily identify an outline of Mount Tai.

In the mountains, the steep slope has an obvious tendency of aggregation according to its blue

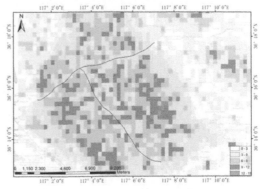

Figure 2. Taishan district contour frequency distribution.

steep zones. Mount Tai can be roughly divided into northern mountains, a southwest mountainous area, and southeast mountains of three parts for its characteristics of clustering. The southeast mountains are the most obvious area of steep slope, and this area has the most famous places of interest such as Nan tian men, Shi ba pan and Tian jie.

The most dangerous section of Mount Tai is Shi ba pan that is located in the northern. Dui song shan. As shown in Figure 3, the black curve represents Shi ba pan. Beginning in Dui song shan to the southeast, and the Nan tian men to the northwest, the Shi ba pan area is mainly a dangerous blue area. In the extreme north of Shi ba pan is Nan tian men, in the flat yellow colored area shown in the Figure 3. In Figure 4, Tian jie starts from Nan tian men, to the Bi xia Temple by a distance of 0.6 km. There are many visitor lodging places and restaurants. Its slope is more flat. As shown in Figure 4, this area is mainly yellow and green.

Figure 5 shows Dui song shan known as the "turquoise bold cliff", whose terrain is dangerous and steep with many cliffs; its counter line frequency mainly includes red and blue. Blue counter lines mainly focus on massif which stands for steep slopes, and reds mainly focus around the massif, which stands for cliff. As opposed to the DEM image base map and analysis, the red zone range of coverage, DEM shows color change areas, which is from light color areas to darker, where the elevation changes significantly, and this is the distinct characteristic of cliffs.

In Figure 6, the white curve is for the Tian Wai Cun Round Hill Road, which starts from Tian Wai Cun, along the Huang xi River, go along Round Hill Road for 13 kilometers, reaching the Zhong tian men. As can be seen from Figure 6, the Round Hill

Road area mainly appears in yellow and green, as far as possible, to avoid road construction in the steep slope zone, and having more switchbacks. Looking at the switchback distribution, the bends in the road can be found mostly in the green and

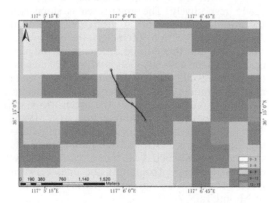

Figure 3. Taishan eighteen contour plate area frequency distribution.

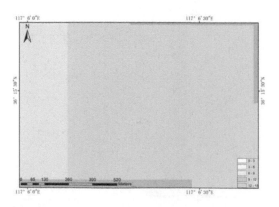

Figure 4. Tian Jie plateau area contour frequency distribution.

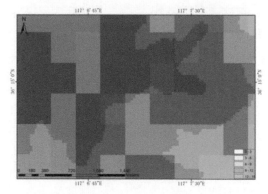

Figure 5. Dui Song Shan DEM and contour frequency contrast.

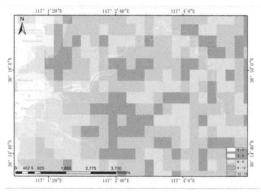

Figure 6. Tian Wai Cun Round Hill Road.

blue areas which are relatively steep. Surrounded by mountains, the road was built in the gentle slope area as much as possible, but limited by the lack of gentle slopes and hindered by steep mountain conditions, inevitably some roads were built in the steep slope area. In order to solve the problem as the slope is steep, switchbacks are used to achieve the purpose of buffer gradient; bends also appear more often in the blue and green areas, as shown in Figure 6.

5 SUMMARY AND OUTLOOK

This article proposes a method for quantitative analysis of topographic relief in small areas as in the case of Mount Tai, extending DBSCAN thought to the line data, and introducing the MBR technique to build moving windows. According to the statistical analysis of the frequency of contour lines in the moving windows, the result of contour frequency space distribution is obtained. This result plays a supplementary role in the analysis of topographic relief in mountainous areas, especially suitable for the study of relative slope comparison, and it can provide necessary decisive information for the construction of road works as well as rescue operations. The results' accuracy of contour frequency space distribution depends greatly on the DEM basic data's accuracy, the size setting of the moving window, and topographic changes in the field. Therefore, improving the DEM's resolution, setting the proper size of the moving window and fully integrating the topographic changes can significantly improve the accuracy of contour frequency space distribution.

In future work, we can store the result of high precision contour frequency-space distribution in a database; analyze the frequency distribution of the contour lines in the next area by Case-Based Reasoning technology (CBR); look for similar historical cases; apply them to new problems; obtain matching cases similar to the current condition of characteristic parameters; revise the solution of the matching cases according to the specific situation; and then apply to the current project status.

ACKNOWLEDGMENT

This work was supported by surveying and mapping scientific research innovation team of Shandong University of Science and Technology. We thank and greatly appreciate Dr. Wei's help.

REFERENCES

Baryon L. Gis analysis of macro landform [M], // SIRC, The 10th Colloquium of the Spatial Information Research Centre. New Zealand: University of Otego, 1998, 35–48.

Cheng Weiming, Zhou Chenghu, Chai Huixia, etc. China's land physiognomy type base wood form quantitative extraction and analysis [J]. Geo-information science, 2009, 11 (6): 725–736.

Dachau R, Barb E, Mark R M. Landform classification of New Mexico by computer [M]. U.S. Geological Survey, 1991.

Ester, Martin, Hans-Peter Krieger, Jorge Sander, Xiaowei Xu, A Density-Based Algorithm for Discovering Clusters in Large Spatial Databases with Noise [C], Proceedings of the 2nd International Conference on Knowledge Discovery and Data Mining (KDD 96), Portland, Oregon, USA, 1996, 226–231.

Hammond E H. Analysis of Properties in Landform Geography: An Application to Broad-scale Landform Mapping [J]. Annals of the Association of American Geographers, 1964 (54): 11–19.

Han J, Kamber M, Pei J. Data Mining Concepts and Techniques [M]. (3nd edition). China Machine Press, 2012: 288–293.

http://baike.baidu.com/link?url=dSNiDiKMTxN2 mmN_OB1a7RH_zvSN4_qsIyghhTB3lyZunisPwJi-QiH5cN6hnchlvmpghAmkoyZFEv0WoKGiwHav8X-OBY-46 HzgslKyqnK

Iwahashi J, Pike R J. Automated classifications of topography from OEMs by an unsupervised nested-means algorithm and a three-part geometric signature [J]. Geomorphology, 2007(86): 409–440.

Lang Ling Ling Cheng Weiming, Zhu Qijiang, etc. Multi-scale DEM extraction relief degree of comparative analysis, the low hilly land in fujian as an example [J]. Geo-information science, 2007, 9 (6): 1–8.

Liu Ail, Tang guoan. Chinese landscape wooden form DEM automatic segmentation study [J]. Geo-Information Science, 2006 (4): 8–14.

McIvor, R.T., P.K., Humphreys. A cage-based reasoning approach to the make or buy decision [J]. Integrated Manufacturing Systems, 2000, 11 (5): 295–310.

Morgan J M, Lash A M. Developing landform maps using ESRI's Model Builder[C]. ESRI User Conference 2005 Proceedings. http://gis.esri.com/library/user-cone/froc05/papers/pap2206.Pdf, 2005.

Prima O D A, Chigoe A, Yokoyama R, etal. Supervised landform classification of Northeast Honshu from DEM-derived thematic maps [J]. Geo morphology, 2006 (78): 373–386.

Electromechanical Control Technology and Transportation – Jia & Wu (Eds)
© 2017 Taylor & Francis Group, London, ISBN 978-1-138-06752-3

A novel interest point detector based on convolutional features with unsupervised feature learning

Qi Jia, Xiaodan Wang & Laien Zhou
Air and Missile Defense College, Air Force Engineering University, Xi'an, China

ABSTRACT: Interest point detection is of great significance in computer vision applications. In this letter, we present a convolutional approach for interest point detection, which detects invariant 2D-features form images under different view conditions of an object or scene. In contrast to other convolution networks that are trained to represent data or solve a classification task, our network is trained for detecting the interest points of an image and the unsupervised feature learning is applied to ensure the algorithm with accuracy and distinctiveness. Also, we present a comparison evaluation on benchmark datasets of affine covariant features, yielding the algorithm of competitive results compared with the traditional approaches on interest point detection.

1 INTRODUCTION

Interest points, where the image gray value changes sharply, are useful low-level features that can provide informative representation for digital images. So interest point detection algorithms are the key techniques in computer vision applications such as image matching (Brown & Lowe 2007), image retrieve (Mikolajczyk & Schmid 2001) and 3d scene reconstruction (Agarwal et al. 2009). An ideal interest point detector should find salient image regions so that they can be repetitively detected despite the challenges as the same scene may exhibit large variations in images under arbitrary viewing conditions such as blur, zoom, rotation, light change and other kinds of disturbance (Tuytelaars & Mikolajczyk 2007).

In this letter, we propose a novel interest point detection algorithm while the inspiration comes from the expressive features representations output by the layers of Convolutional Neural Networks (CNNs) (Cun et al. 1990) due to their high performance in vision applications. Our work is based on the architecture of the first layer in CNNs and in order to apply the feature maps for interest point detection, we adapt a significantly different approach than traditional one.

Our main contribution in this letter is a new approach for interest point detector based on the CNN, and in training procedure, an unsupervised feature learning algorithm is applied to obtain filter parameters for feature map computation.

We review the related work in section 2; the new approach to feature map computation and the detector are introduced in section 3; in section 4, we describe how to train the network with unsupervised feature learning algorithm, and the experimental results are presented in section 5.

2 RELATED WORK

With the development of the computer vision, hundreds of methods have already been proposed with inspiration and perspiration, here, we only review the ones which are close to ours in this section.

2.1 Detection approaches

So far, the most popular detector is Harris detector (Harris 1988), based on the measurement of the Sum of Squared Differences (SSD) in a patch. The resulted corner-like image structure measurement ensures isotropy and is widely used in practice. Therefore, a large number of detectors are proposed based on that idea, which boils down the detection of interest points to the analysis of the eigenvalues of SSD matrix (Kovesi 2003, Kenney et al. 2005).

Another kind of approaches focus on the detection of local image structures by templates, which find interest points by comparing the intensity of surrounding pixels with that of center pixels. The most representative one is the traditional Smallest Univalue Segment Assimilating Nucleus (SUSAN) algorithm (Smith & Brady 1997). An improvement of this approach is Features from Accelerated Segment Test (FAST) (Rosten & Drummond 2006) detector and its variants (Rosten & Drummond 2010), which determines a point as interest point by the circle template surrounding the proposal point. Another improvement of the FAST detector is the building of the decision tree by learning algorithm so that it can accelerate the process of determining. Also, template-based method for junction detection is widely applied since junctions reveal important occlusion relationships between objects (Sinzinger 2008, Dimiccoli & Salembier 2009), and they trend to detect more various features than corner detectors.

For the purpose of scale invariance during detection, the detectors based on multi-scale image processing are proposed. The most popular one is SIFT (Lowe 2004) detector and is variants (Chen et al. 2015), which detects interest points in the multiscale pyramid built by the Gaussian filter. And for better approximating the accuracy and distinctiveness, the wavelet transform and nonlinear scale space is introduced in SURF (Bay et al. 2006) and KAZE (Alcantarilla et al. 2012) detector, and the BRISK (Leutenegger & Siegwart 2011) detector uses a different sampling strategy determining interest point. Furthermore, the multiscale processing can be applied combined with the original approaches, such as the Harris-Affine (Mikolajczyk & Schmid 2004) detector and the junction detector in Xia et al. (2014).

2.2 *Convolutional neural network*

Since the CNNs have recently advanced the state-of-the-art in computer vision applications, it is natural to apply CNN for local features. Many approaches are proposed by the inspiration that the intermediate layers of CNN could output image representations (Oquab et al. 2014), especially convolution kernel network (CKN) (Mairal et al. 2014). However, CNN structure applied for interest points detection is still lack of studies. In this letter, the feature map in the first layer of CNN is applied for interest point detection as we treat the interest point as a kind of low-level feature. Furthermore, in order to better approximate the interest point detector, our approach has a similar but more complex architecture in feature map computation.

2.3 *Unsupervised feature learning*

In the context of computer vision, the Unsupervised Feature Learning (UFL) algorithm (Lam & Wunsch 2013) is applied with the purpose of getting invariant features from a limited amount of labeled data. Several works have shown the advantages of the neural networks which is layer-wised trained in an unsupervised mode (Hinton & Salakhutdinov 2006, Hinton et al. 2006, Ranzato et al. 2007). In this letter, our unsupervised training approach is mainly inspired by those methods.

3 DETECTOR WITH CONVOLUTIONAL FEATURES

In this letter, we apply the feature map generated by the first layer of CNN for interest point detection. Since the CNN shows anisotropy, before detection, rotate invariance should be ensured.

3.1 *Feature map with isotropy*

The architecture of the first layer in CNN is shown in Figure 1a. In this letter, with the purpose of getting rotation invariance, we compute the feature map in a feature augmentation style instead of direct convolutional computation. As shown in Figure 2, by rotating the filters to various angle, a set of augmented filters W is obtained, and a set of feature maps C is computed by the filters W, for each position in C, we choose the max value as the output feature map $Cr(x, y)$ value:

$$Cr(x, y) = \max\left(c_1(x, y), c_2(x, y), ..., c_i(x, y)\right), c_i \in C \tag{1}$$

$$R(p) = \sum_{j=1}^{k} Cr_j(x_{img}, y_{img}) \tag{2}$$

where k is total number of the filters. And we can get different amount of interest points by applying

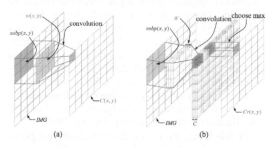

Figure 1. Illustration of feature map computation. (a) Feature map computation in original CNN. (b) Feature map computation with feature augmentation in this letter. W is the set of rotated-augmented filters whose original filter is $w(x, y)$ in (a), C is the set of feature maps computed by W, $Cr(x, y)$ is the output feature map with rotate invariance.

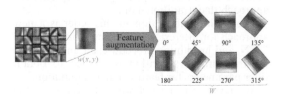

Figure 2. The feature augmentation step in feature map computation, as illustrated in this figure, the original feature is augmented to 8 different orientations.

318

low or high threshold on response value R and different size on the nomaximum suppression window.

4 TRAINING NETWORKS

In the convolution neural network, the feature representation of the intermediate layers relies on filters which are defined in a data-dependent manner. While training, in comparison to CNN, our approach involves the same set of hyper-parameters: number of the filters and the sizes of feature patches, but different from the supervised training in CNN, in this letter, the sparse auto-encoder is applied.

4.1 Auto-encoder

Auto-encoder (Rumelhart et al. 1986) was early used for getting features from non-labeled data by a neural network. An auto-encoder neural network is an unsupervised learning approach that applies backpropagation, setting the output data to be equal to the inputs. By placing constraints to the network cost function, we can extract specific features from the data.

For an image patch vector $subp$, which contains $n \times n$ pixels, an auto-encoder with k hidden units is constructed to extract image feature. After the encoding, the encoder is forced to learn a compressed representation $f_{enc}(subp)$ of the input $subp$, and the decoder try to reconstruct the encoded representation $f_{enc}(subp)$ to the output data $h_{w,b}(subp) = f_{dec}(f_{enc}(subp))$. Note that with the purpose of learning the general representation of the image, we choose large amount of patches that are randomly selected from the image to train the auto-encoder.

The activation function of hidden layer and output layer is:

$$F = S(subp; w, b) \tag{3}$$

$$\hat{subp} = S(F; w', b') \tag{4}$$

where $S(x; w, b)$ is the sigmoid function, x means input, w, b means parameter of the network:

In order to train the network to fit the input data, the backpropagation algorithm has been applied, so the cost function of the network should be established. Given a training set of m image patches vectors $\{subp(1), \ldots, subp(m)\}$, the overall cost function of the network is:

$$J(w, b) = \frac{1}{m} \sum_{i=1}^{m} \left(\frac{1}{2} \left\| subp^{(i)} - \hat{subp}^{(i)} \right\|^2 \right)$$
$$+ \frac{\lambda_1}{2} \sum_{l=1}^{L-1} \sum_{i=1}^{s_l} \sum_{j=1}^{s_{l+1}} (w_{ji}^{(l)})^2 \tag{5}$$

where k means the number of hidden units, L = 3 means the number of the layers, sl means the number of units in l-th layer, s1 = n × n, s2 = k, s3 = n × n mean number of units, w(l) means the weight parameter between layer l and layer l + 1.

The second term in equation 7 is a weight decay term that tends to decrease the magnitude of weights, and helps prevent overfitting (Hinton 1987). The weight decay parameter λ_1 controls the relative importance of the two terms.

Note that the output range of the sigmoid function is [0, 1], so the value of training samples should be scaled to ensure that they lie in the [0, 1] range.

4.2 Sparse regular term

When using auto-encoder to extract feature from image patches, one problem is that the convergence of the network parameter is not well-posed because of the complexity of the image, redundant construct of the network and the noise disturbance. Based on that issue, some regularization terms representing the prior information of the image patch should be added to cost function. One approach to solve this issue is sparse regularization term (Olshausen & Field 2004). Sparsity means that an image patch can be represented by few base patches as show in Figure 3, in which those base patches contain high correlation information of the training patches, for an image, the base patches usually imply edges, lines or surfaces.

One approach for constructing the sparse regularization term $\varphi_2(w, b)$ is to measure the KL distance between a sparse constant ρ that is closed to zero and the average activation $\bar{\rho}_j$ of hidden unit j ($j \in [1, k]$), where:

$$\bar{\rho}_j = \frac{1}{m} \sum_{i=1}^{m} S(subp^{(i)}; w_{ji}^1, b_{ji}^1) \tag{6}$$

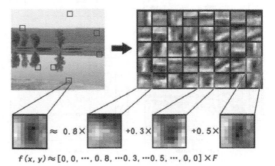

$f(x, y) \approx [0, 0, \cdots, 0.8, \cdots 0.3, \cdots 0.5, \cdots, 0, 0] \times F$

Figure 3. The sparse features of an image, the listed patch is represent by only three sparse features.

So the expression of the sparse regularization term $\varphi(w, b)$ is:

$$\varphi(w,b) = \sum_{j=1}^{s_2} KL(\rho \| \bar{\rho}_j)$$

$$= \sum_{j=1}^{s_2} \rho \log \frac{\rho}{\bar{\rho}_j} + (1+\rho) \log\left(\frac{1-\rho}{1-\bar{\rho}_j}\right) \quad (7)$$

After the construction of the sparse regular term $\varphi(w, b)$, we add it to the network cost function:

$$J_{regularized}(w,b) = \frac{1}{m} \sum_{i=1}^{m} \left(\frac{1}{2} \left\| h_{w,b}(subp^{(i)}) - subp^{(i)} \right\|^2 \right)$$

$$+ \frac{\lambda_1}{2} \sum_{l=1}^{L-1} \sum_{i=1}^{s_l} \sum_{j=1}^{s_{l+1}} \left(w_{ji}^{(l)} \right)^2$$

$$+ \lambda_2 \sum_{j=1}^{s_2} \left(\rho \log \frac{\rho}{\bar{\rho}_j} + (1+\rho) \log\left(\frac{1-\rho}{1-\bar{\rho}_j}\right) \right)$$

$$(8)$$

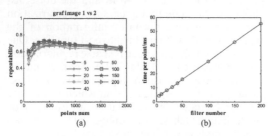

Figure 4. (a) The repeatability graph of the interest points detection algorithm for the "graf" dataset as number of filters is varied from 5 to 200. (b) Timing results as the number of the filters is varied. Best view in color.

With the regularized cost function $J_{regularized}(w, b)$, the backpropagation algorithm is applied to train the network parameters and we get the filters for the convolution network.

5 EXPERIMENTAL RESULTS AND ANALYSIS

In this section, we conduct experiments for detecting affine invariant features in arbitrary images on the benchmark datasets. Evaluation criterion for detector performance is repeatability, which signifies that detection is independent of changes in the imaging conditions (Schmid et al. 2000). Furthermore, our approach is compared against the state-of-the-art interest point detection algorithms include FAST, SIFT, Harris and SUSAN in terms of both algorithm performance and execution time.

In the experiment, the auto-encoder was trained with 5000 random-selected natural image patches of size 11×11, and how the number of the filters affect algorithm performance will be discussed later. During the computation of feature maps, we augment each filter for 12 different orientations.

5.1 Filter number influences

We first consider how the number of the filters influences the performance of our approach. As shown in Figure 4a, with the increase of the filter number from 4, the repeatability is rising, but after reaching 20, the influence of different feature number trends to be insignificant. The conjecture of this phenomenon is that an interest point is detected as a kind of low-level feature, which can be represented

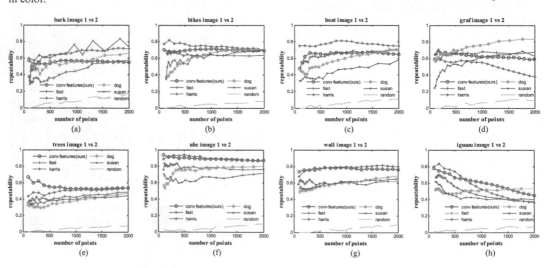

Figure 5. Comparisons of the different detectors as the number of interest points is varied. Note that in our approach, the number of the filters $k = 20$.

Table 1. Timing results for the detectors on "graf" dataset. In our approach, the number of filters is 20.

Convolution features (ours)	3.6474 s
SIFT	4.7669 s
FAST9	3.0557 s
Harris	3.5229 s
SUSAN	7.9376 s
Image Resolution	600×800
Number of Interest points	500

by only a few bases especially when we apply the feature augmentation. Also, as it can be observed in Figure 4b, with the increase of filter number, algorithm execution time also suffers increase.

5.2 Comparison of different detctors

Figure 5 depicts the repeatability result on the benchmark datasets for different detectors include FAST, SIFT, SUSAN, Harris, and convolution features (our approach). As it can be observed form the result, our approach outperform SIFT, SUSAN and FAST detectors in some conditions especially Gaussian noise, viewpoints change and blur image, and outperforms the Harris detectors in zooming and rotation.

Also, we conduct timing tests which are performed on a desktop computer whose CPU is Inter(R) Core(TM) i7–4790 @3.6 GHz with MATLAB R2014a, while the "graf" dataset is used. The timing results for various detectors against convolution detector are presented in Table 1. Note that the training of the filters is not included in the timing test as the filters learned only once from the first image in the image dataset sequence.

The computations in timing test are performed on CPU only and a parallelized method by GPU computing could be applied for real-time performance.

6 CONCLUSIONS

We purpose the convolutional features with unsupervised feature learning for interest point detection. In contrast to previous approaches, our approach presents a different detection approach based on CNN and the computation of feature maps exhibited a feature augmentation style. For training operations, we present an unsupervised feature learning algorithm for learning filters. In the experimental tests, we achieve near state-of-the-art performance on the dataset of affine covariant features.

Further investigations will be taken into the convolution features for interest point detection.

The next one is the building of multilayer convolution network while only the first layer of convolution network is applied in this letter. In addition, the training of the network filters can be better approximated for the interest point detection applications.

REFERENCES

Agarwal, S., Snavely, N., Simon, I., Seitz, S.M., & Szeliski, R. (2009). Building rome in a day. Communications of the Acm, 54(10), 105–112.

Alcantarilla, P. F., Bartoli A., Davison A. J. (2012). KAZE features. Computer Vision–ECCV 2012. Springer Berlin Heidelberg, 2012: 214–227.

Bay, H., Tuytelaars, T., & Van Gool, L. (2006). SURF: speeded up robust features. European Conference on Computer Vision (Vol. 110, pp. 404–417). Springer-Verlag.

Brown, M., & Lowe, D.G. (2007). Automatic panoramic image stitching using invariant features. International Journal of Computer Vision, 74(1), 59–73.

Chen, J., Duan, L.Y., Gao, F., & Cai, J. (2015). A low complexity interest point detector. IEEE Signal Processing Letters, 22(22), 172–176.

Cun, Y.L., Boser, B., Denker, J.S., Howard, R.E., Habbard, W., & Jackel, L.D., et al. (1990). Handwritten digit recognition with a back-propagation network. Advances in Neural Information Processing Systems (Vol. 88, pp. 465). Morgan Kaufmann Publishers Inc.

Dimiccoli, M., & Salembier, P. (2009). Exploiting T-junctions for depth segregation in single images. IEEE International Conference on Acoustics, Speech and Signal Processing (pp. 1229–1232). IEEE Computer Society.

Harris, C. (1988). A combined corner and edge detector. Proc. 4th Alvey Vision Conf., 1988(3), 147–151.

Hinton, G.E. (1987). Connectionist learning procedures. Artificial Intelligence, 40(1–3), 185–234.

Hinton, G.E., & Salakhutdinov, R.R. (2006). Reducing the dimensionality of data with neural networks. Science, 313(5786), 504.

Hinton, G.E., Osindero, S., & Teh, Y.W. (2006). A fast learning algorithm for deep belief nets. Neural Computation, 18(7), 1527–1554.

Kenney, C.S., Zuliani, M., & Manjunath, B.S. (2005). An Axiomatic Approach to Corner Detection. IEEE Computer Society Conference on Computer Vision and Pattern Recognition (Vol. 1, pp. 191–197). IEEE Computer Society.

Kovesi, P. (2003). Phase Congruency Detects Corners and Edges. DICTA (pp. 309–318).

Lam, D., & Wunsch, D. (2013). Unsupervised feature learning classification using an extreme learning machine. International Joint Conference on Neural Networks (pp. 1–5).

Leutenegger, S., Chli, M., & Siegwart, R.Y. (2011). Brisk: binary robust invariant scalable keypoints. ICCV, 58(11), 2548–2555.

Lowe, D.G. (2004). Distinctive image features from scale-invariant keypoints. International Journal of Computer Vision, 60(2), 91–110.

Mairal, J., Koniusz, P., Harchaoui, Z., & Schmid, C. (2014). Convolutional kernel networks. Advances in Neural Information Processing Systems, 2627–2635.

Mikolajczyk, K., & Schmid, C. (2001). Indexing based on scale invariant interest points. Computer Vision, 2001. ICCV 2001. Proceedings. Eighth IEEE International Conference on (Vol. 1, pp. 525–531).

Mikolajczyk, K., & Schmid, C. (2004). Scale & affine invariant interest point detectors. International Journal of Computer Vision, 60(1), 63–86.

Oquab, M., Bottou, L., Laptev, I., & Sivic, J. (2014). Learning and Transferring Mid-level Image Representations Using Convolutional Neural Networks. Computer Vision and Pattern Recognition (pp.1717–1724). IEEE.

Ranzato, M., Huang, F.J., Boureau, Y.L., & Lecun, Y. (2007). Unsupervised Learning of Invariant Feature Hierarchies with Applications to Object Recognition. Computer Vision and Pattern Recognition, 2007. CVPR '07. IEEE Conference on (pp.1–8). IEEE Xplore.

Rosten, E., & Drummond, T. (2006). Machine learning for high-speed corner detection. European Conference on Computer Vision (Vol. 3951, pp. 430–443). Springer-Verlag.

Rosten, E., Porter, R., & Drummond, T. (2010). Faster and better: a machine learning approach to corner detection. IEEE Transactions on Pattern Analysis & Machine Intelligence, 32(1), 105–119.

Rumelhart, David, E., Hinton, Geoffrey, E., Williams, & Ronald, J. (1986). Learning representations by back-propagating errors. Nature, 323(6088), 533–536.

Schmid, C., Mohr, R., & Bauckhage, C. (2000). Evaluation of interest point detectors. International Journal of Computer Vision, 37(2), 151–172.

Sinzinger, E.D. (2008). A model-based approach to junction detection using radial energy. Pattern Recognition, 41(2), 494–505.

Smith, S.M., & Brady, J.M. (1997). Susan—a new approach to low level image processing. International Journal of Computer Vision, 23(1), 45–78.

Tuytelaars, T., & Mikolajczyk, K. (2007). Local invariant feature detectors: a survey. Foundations & Trends® in Computer Graphics & Vision, 3(3), 177–280.

Xia, G.S., Delon, J., & Gousseau, Y. (2014). Accurate junction detection and characterization in natural images. International Journal of Computer Vision, 106(1), 31–56.

Electromechanical Control Technology and Transportation – Jia & Wu (Eds)
© 2017 Taylor & Francis Group, London, ISBN 978-1-138-06752-3

Frame extraction method for indoor scene images

Yang Liu
College of Software Engineering, Chongqing University of Posts and Telecommunications, Chongqing, China

Qingjie Wei
College of Computer Science, Chongqing University of Posts and Telecommunications, Chongqing, China

Ying Qian
College of Software Engineering, Chongqing University of Posts and Telecommunications, Chongqing, China

ABSTRACT: In recent years, the reconstruction of indoor scenes of buildings has become the focus of 3D reconstruction. Accurate, fast, and anti-jamming extraction of frame are the primary tasks of the indoor scene reconstruction. The existing frame extraction methods are complex, and the extraction results are in low fitting degree with the real indoor frame. Therefore, a frame extraction method based on the Manhattan assumption is proposed in this paper. The method uses the clustering algorithm to detect vanishing points and feature segments in the three mutually orthogonal main directions to obtain candidate corner vertices. By deeply understanding the distribution of the candidate corner vertices, the accurate corner vertices are detected and the frame is obtained. The validity of the method is proved by stepwise experiments, and the result of comparative experiments show that the proposed method is superior to the existing methods under poor illumination and occlusion conditions.

1 INTRODUCTION

Extracting frame structure from the indoor scene is the primary task of indoor scene's understanding. It has a great value in dangerous places that people cannot reach, for example, disaster relief and archaeological excavations (Chu 2013). For the human eye, it is easy to identify the frame structure of the indoor scene. However, the illumination, occlusion, and other interference make the computer automatically identify the framework full of challenges. In recent years, more and more methods of indoor scene frame detection have been proposed. Lee et al. (2009) proposed 12 corner models, fitting the frame by binding component's combination with models. This method is based on geometric reasoning and can recover scene structure using only line segments. However, this method cannot work well when abundant, missing, or incorrect extraction existed in the detected line segments. Hedau et al. (2009) generated candidate frames by shooting rays from vanishing points and then selected the best candidate via ranking Support Vector Machines (SVMs). The theory of the method is simple and easy to understand, but manual marking of the furniture was needed as the clues for training classifier of ranking SVM. Guo Lu et al. (2014) proposed the method based on line refinement using cross-ratio constraint and depth

constraint to obtain the indoor scene frame. The method has some advantages in complex indoor scene frame detection task, but the detection process depends strongly on the accuracy of the three mutually orthogonal vanishing points and does not apply to the case of missing vanishing points.

In this paper, on the basis of the former research on indoor frame extraction, a new frame extraction method is proposed to solve the problems of complex theory, vanishing point missing, high dependency of line segment, and so on. The method combines J-linkage vanishing point extraction algorithms and Manhattan assumption, which indoor scenes satisfy together without detecting the vanishing point in all three directions. It has less demand of indoor scene conditions, and it is suitable for five-sided indoor scene images.

2 INDOOR SCENES THEORIES

2.1 Vanishing point

In the perspective projection plane under ideal pinhole imaging model, the lines parallel to the projection space maintain parallelism, while the lines that are not parallel gather to a point. This point is called the vanishing point (Rother 2002).

For the single image, the depth information can be obtained only through some features and prior knowledge of the scene. The vanishing point is determined by a set of lines that are parallel in three-dimensional space and intersect in two-dimensional space. Therefore, it is feasible to use the vanishing point to recover the depth information of the indoor scene.

2.2 Manhattan assumption

The concept of the Manhattan structure was presented by Zhang (2012) and Coughlan (2003). The Manhattan structure means that the lines in the scene are either parallel to each other or perpendicular to each other. That is to say, if we find any three straight lines from the Manhattan structure randomly, the relationship between these three lines in space is certainly one of the following three conditions: 1) three lines perpendicular to each other; 2) two of the lines are parallel and perpendicular to the other; and 3) three lines are parallel to each other.

Many lines in the indoor scene are parallel, and boundary lines in the scene are perpendicular to each other, such as corridor boundaries, doors, and murals. On the basis of these lines, if the three vanishing points in the indoor scene are found, then the vanishing point must be perpendicular to each other, that is, the Manhattan direction. In fact, a two-dimensional indoor scene image has at least one Manhattan orientation. There are many parallel lines in each Manhattan direction from the three-dimensional space, so each Manhattan direction is determined by a set of intersecting lines or a set of parallel lines when projected to two-dimensional images. Therefore, it is very important to find three mutually orthogonal Manhattan directions for indoor scene structure recovery (Li 2014).

For the five-sided indoor scene image, although the vanishing points in three directions are not all necessarily detected, there must be a vanishing point located inside the frame intersected by feature lines of other two directions. As shown in Figure 1, this feature provides us an important clue for the recovery of the indoor frame structure.

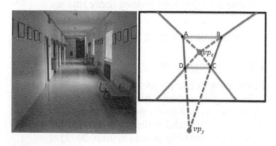

Figure 1. Five-sided indoor scene image, the frame structure, and vanishing point model.

2.3 Corner vertices

In the frame structure shown in Figure 1, *A*, *B*, *C*, *D*, the intersections of frame in three directions are defined as corner vertices. If the location of the vanishing point and four corner vertices are determined, then the frame is naturally determined.

3 VANISHING POINT DETECTION

We use the J-linkage vanishing point algorithm proposed by Tardif et al. (2009) for the vanishing point detection, and the feature line segments are divided into three mutually orthogonal categories on the basis of Manhattan directions. The following sections describe the vanishing point detection algorithm step by step.

3.1 Line segment detection

First, the edge detection is performed using Canny operator, which has good continuity when detecting edges. Canny edge detection method uses the Gaussian function of the first derivative, and it has a good balance between noise suppression and edge detection. Lin (2003) summarized that as it has "non-maximum value suppression" and morphological connection operation, the extracted edge is complete, and the continuity of the edge is better than that of other operators.

Then, the line segments are extracted using MATLAB open-source toolbox provided by Kovesi (2003). Kovesi's algorithm uses the phase coherence to extract various features of the image, such as the step edges, line edges, and edges that integrate both features. Phase coherence is a low-level image processing method, which accords with human's vision mechanism. As phase coherence extracts the frequency feature of the image and is not affected by the change of the contrast, the algorithm can obtain complete image features in case of change in luminance (Li 2011). The set of line segments obtained by the phase coherence algorithm is shown in Figure 2.

3.2 J-linkage vanishing point detection algorithm

For the indoor scene frame structure detection method proposed in this paper, the feature lines in the direction of three vanishing points are important clue and foundation. Therefore, we choose J-linkage clustering algorithm to obtain vanishing point and feature line classification result simultaneously.

The basic idea of J-linkage algorithm proposed by Tardif et al. (2009) is: the data points are put into their similar conceptual space for analysis, and the data point that belongs to the same model instance

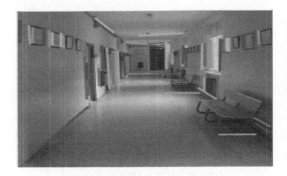

Figure 2. Extraction result of line segments presented using random colors.

Table 1. Summary of the J-linkage clustering algorithm.

J-linkage clustering algorithm

INPUT
  $E_{1...N}$: set of N edges and associated entities
  φ: consensus threshold (2 pixels)
  M: number of vanishing point hypotheses
OUTPUT
  v: set of vanishing points
ALGORITHM
  Build the preference matrix $P \in \{0, 1\}^{N \times M}$
For: $m \leftarrow 1$ to M
  randomly select a subset of 2 edges S from E
  $v_m \leftarrow V(S)$: calculate vanishing point from S
For: $n \leftarrow 1$ to N
  $P_{nm} \leftarrow D(\varepsilon_n, v_m) \leq \varphi$
  Classify edges using J-Linkage clustering on
  P compute vanishing points for each cluster
  $v \leftarrow$ Re-compute vanishing point for each cluster

---

will be gathered in a similar concept space. Table 1 describes the main steps of J-linkage algorithm. The first step is to randomly choose M minimal sample sets of two edges to compute an intersection, that is, an assumed vanishing point. The second step is constructing the preference matrix, P, which is a Boolean matrix. According to whether the location of the line and the assumed vanishing point satisfy the consistency measure function, the value of the matrix is assigned. Finally, it compares P between two clusters. If the similarity is high, they should belong to the same cluster. The expected vanishing points are the assumed vanishing point clustering lines.

Where $D(\varepsilon_n, v_m)$ is the consistency measure function of each line segment with vanishing point, namely the distance between the connection of vanishing point and line segment's midpoint and any endpoint of the line segment.

By using the J-linkage vanishing point detection algorithm, the feature line segment clustering results can be obtained, as shown in Figure 3.

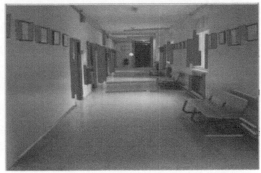

Figure 3. Result of line segments clustering and vanishing point detection. The line segments of different directions are presented by different colors. The vanishing point is shown in yellow.

4   INDOOR SCENES FRAME DETECTION PROCESS

This paper presents a method for frame detection of a single indoor scene image. The method is as follows:

Step 1. According to J-linkage vanishing point detection and linear clustering results, the intersections of line in three mutually orthogonal directions are obtained as the candidate corner vertices.

Step 2. The candidate corner vertices obtained in Step 1 are filtered to determine the final four corner vertices.

Step 3. The vanishing point and corner vertices are connected to obtain the indoor scene frame.

The following section describes the work of each step.

4.1   Candidate corner vertices detection

Assuming that the feature line sets in three directions are E_x, E_y, and E_z, we choose a line from E_x, E_y, and E_z randomly, which are denoted as l_x, l_y, and l_z, respectively. On the basis of Manhattan's assumption, the corner vertices of indoor scene frame must be in the intersection points set of l_x, l_y, and l_z.

With the vanishing point determined, if the locations of four corners are determined, then the frame is naturally determined. Therefore, the main task of the first step is to detect candidate vertices. However, because of the existence of error when detecting a line segment, three feature lines l_x, l_y, and l_z do not necessarily intersect at one point, but the intersection occurs to be a triangle.

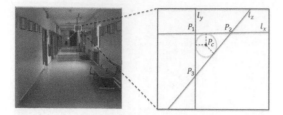

Figure 4. Three feature lines intersecting as a triangle caused by error.

As shown in Figure 4, suppose l_x, l_y, and l_z intersected at $P_1(x_1, y_1)$, $P_2(x_2, y_2)$, and $P_3(x_3, y_3)$, respectively. Because the intersection is due to the error of line extraction, it should lie inside $\Delta P_1 P_2 P_3$. We use inner point of the triangle as the candidate corner vertex, $P_c(x_c, y_c)$, which is calculated as follows:

$$P_c(x_c, y_c) = \begin{pmatrix} \dfrac{|P_2 P_3| x_1 + |P_1 P_3| x_2 + |P_1 P_2| x_3}{|P_2 P_3| + |P_1 P_3| + |P_1 P_2|}, \\ \dfrac{|P_2 P_3| y_1 + |P_1 P_3| y_2 + |P_1 P_2| y_3}{|P_2 P_3| + |P_1 P_3| + |P_1 P_2|} \end{pmatrix} \quad (1)$$

Thus, the candidate vertices are calculated as:

$$P_c(x_c, y_c) = \begin{cases} l_x \times l_y \times l_z & l_x, l_y, l_z \text{ intersect at one point} \\ P_c \text{ given by (1)} & \text{otherwise} \end{cases} \quad (2)$$

In the actual experiment, we only calculate the inner point for triangle, whose length of three sides are all less than the setting threshold (ε) in case of high time complexity. Then, we can obtain the candidate corner vertices set C.

4.2 Corner vertices determination and frame connection

According to the candidate corner vertices detected in the previous step, corner vertices of the frame will be determined. According to the Manhattan assumption, the candidate corner vertices have the following properties: If the results of the line segment and candidate corner vertices detection are accurate, then the point in C should be radially distributed along the depth direction by the vertices of the corners. On the basis of this property, the steps to determine the corner vertices and frames are as follows:

Step 1. Judge the existence of three direction vanishing points, vp_x, vp_y, and vp_z. If they exist, then we connect them, or if vp_x or vp_y does not exist, draw a line parallel to x, y through vp_z.

Step 2. The two lines in Step 1 can divide the image into four quadrants, and the origin is vp_z. The closest point to the origin in the candidate corner vertices of each quadrant is extracted. Then, we obtain four corner vertices of the frame.

Step 3. The origin and four corner vertices are connected and extended outward until it intersects the edges of the image. We remove the part of the line inside the frame and then the scene frame is finally obtained.

5 EXPERIMENTAL RESULTS AND ANALYSIS

The experimental conditions are as follows:

1. Programming environment: We deploy the method on Intel i5 2.20 GHz CPU, 8G RAM, Windows 10 PC. All algorithms, including ours, are performed with MATLAB (2015b).
2. The experimental data set: The test images are indoor scene image sets provided by Denis P. (2008), Hedau (2009), or downloaded from the Internet.
3. Experimental scene: Indoor scenes are usually under illumination variations, occlusions, reflections, and so on. Therefore, we divide the 150 indoor scene images into four representative categories:
   ① In good condition,
   ② Occlusion on the ground,
   ③ Occlusion on the wall,
   ④ Reflection on the ground.

For the process of the method, the parameters were set as:

The thresh value in Canny detector is [0.0000 0.1500], sigma = 2;

The minimum value of Kovesi's line segment detection algorithm is 50 pixels.

The consistency threshold in the J-linkage algorithm is constantly 2 pixels.

The side length threshold (ε) of the triangle when detecting candidate corner vertices is 5 pixels.

5.1 Experimental results

As shown in Figure 5, we selected the most representative image of each category for experiments. The proposed method is compared with the state-of-the-art approaches by Lee et al. (2009) and Hedau et al. (2009).

The results of stepwise experiments show that the proposed algorithm is accurate and can detect most of the feature lines. The frame detected

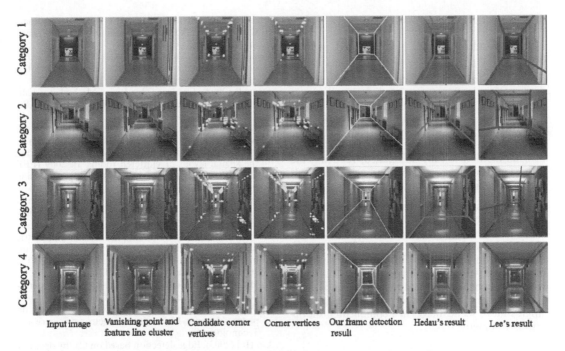

Category 1						
Category 2						
Category 3						
Category 4						
Input image	Vanishing point and feature line cluster	Candidate corner vertices	Corner vertices	Our frame detection result	Hedau's result	Lee's result

Figure 5. Stepwise result of the proposed frame detection algorithm and comparison result.

Table 2. Comparison of the error between the existing algorithms and our method.

	Category ①	Category ②	Category ③	Category ④	Average error
Hedau	0.2907	0.3345	0.6597	0.4993	0.4461
Lee	0.2094	0.5375	0.2506	0.2632	0.3152
Our method	**0.1721**	**0.1063**	**0.0637**	**0.1313**	**0.1184**

in this paper can be accurately fitted to the real frame. In a variety of shooting conditions, it can accurately detect the vertices and frame boundary. In comparison, although Hedau's algorithm can detect the basic frame shape, it shows poor test results under complex shooting conditions (such as the reflection because of illumination). While Lee's algorithm was not affected under the reflection condition, it has poor ability to understand the depth of the direction.

In order to analyze our method, Hedau's algorithm, and Lee's algorithm quantitatively, we use corner error function to evaluate the accuracy of the extracted indoor scene frame. While the vanishing point is determined, if the location of the four corner vertices is determined, then the frame is naturally determined. C_{err} is calculated as:

$$C_{err} = \left(\frac{|AA_g| + |BB_g| + |CC_g| + |DD_g|}{l} \right)^2 \quad (3)$$

Where A, B, C, and D are the detected coordinates of corner vertices, while A_g, B_g, C_g, and D_g are the real coordinates of corner vertices, and l is the diagonal length of the image. Table 2 compares our method and the other two algorithms.

It is evident from the table that the error of corner vertices of our method is significantly lower than that of Hedau's algorithm and Lee's algorithm.

5.2 Limitations analysis

Although the algorithm proposed in this paper is superior to the other two algorithms in detection results, there are some limitations. Since the assumed scene structure model proposed in this paper assumes that the vanishing point must be located inside the middle frame, the algorithm proposed in this paper is only applicable to the five-sided indoor scene structure and not to the three-sided and four-sided scenes.

6 CONCLUSION

The main advantages of the proposed method are:

1. It determines the indoor frame structure by detecting the four corner vertices. The theory is simple and easy to understand.
2. It has less demand in illumination or occlusion conditions, so it is widely used in a variety of scenarios.
3. This method is fully automated, reducing the cost of manual operations and achieving real-time intelligence.

Experimental results show the effectiveness of the proposed method. Furthermore, its performance is superior to that of the existing two algorithms. However, the method proposed in this paper only applies to five-sided indoor scene images. We will add adaptation in the follow-up work. In conclusion, our method is simple and effective, and it is suitable for frame extraction of single indoor scene images in a complicated shooting environment.

ACKNOWLEDGMENT

This work was supported in part by the National Natural Science Foundation of China under Grant No. 61171060.

REFERENCES

Chu, J., GuoLu, A., Wang, L., Pan, C., & Xiang, S. (2013). Indoor frame recovering via line segments refinement and voting. In 2013 IEEE International Conference on Acoustics, Speech and Signal Processing (pp. 1996–2000).

Coughlan, J.M., & Yuille, A.L. (2003). Manhattan world: Orientation and outlier detection by bayesian inference. Neural Computation, 15(5), 1063–1088.

Denis P. (2008). Efficient Edge-Based Methods for Estimating Manhattan Frames in Urban Imagery. M.Sc. Thesis, York University, Canada.

Guo Lu, Anzheng (2014). Key issues of indoor scene reconstruction. Nanchang Aeronautical University.

Hedau, V., Hoiem, D., & Forsyth, D. (2009). Recovering the spatial layout of cluttered rooms. In 2009 IEEE 12th international conference on computer vision (pp. 1849–1856).

Kovesi, P. (2003). Phase congruency detects corners and edges. In The Australian pattern recognition society conference: DICTA 2003.

Lee, D.C., Hebert, M., & Kanade, T. (2009). Geometric reasoning for single image structure recovery. In Computer Vision and Pattern Recognition, 2009. CVPR 2009. IEEE Conference on (pp. 2136–2143).

Li Hua (2014). A study line and vanishing point detection on indoor scene photo algorithm [D].

Li Xuefei (2011). Image outline extraction based on phase coherence. University of Electronic Science and Technology.

Lin Hui (2003). Edge detection based on Canny operator and evaluation [J], Heilongjiang Institute of Technology, 17 (2): 3–6.

Rother, C. (2002). A new approach to vanishing point detection in architectural environments. Image and Vision Computing, 20(9), 647–655.

Tardif, J.P. (2009). Non-iterative approach for fast and accurate vanishing point detection. In 2009 IEEE 12th International Conference on Computer Vision (pp. 1250–1257).

Zhang, L., & Koch, R. (2012). Vanishing points estimation and line classification in a Manhattan world. In Asian Conference on Computer Vision (pp. 38–51).

Electromechanical Control Technology and Transportation – Jia & Wu (Eds)
© *2017 Taylor & Francis Group, London, ISBN 978-1-138-06752-3*

Design of a modified particle analyzer for the automatic identification of the globular foam's boundary

J.J. Xia, X.C. Fu, Z.M. Bao, T. Chen, X.Z. Zhang, Y. Chen, R.J. Wang, C. Hu & L.S. Jing
Tianjin Fire Research Institute of MPS, Tianjin, China

ABSTRACT: The purpose of this study is to identify the globular foam's boundary automatically. First, the foam's microstructure evolution was explained from the geometric viewpoint. The shape of foam just after creation was globular. Second, a particle analyzer was designed to identify the boundary of this globular foam. There were three parts in the analyzing process, including gray part, superimposing part, and linking part. Third, an example of foam graph was identified using this modified particle analyzer automatically. The micro structure parameters included circumference, foam number per unit area, and diameter. The results showed that this modified particle analyzer could be used to identify globular foam's boundaries and analyze the size distribution statistically.

1 INTRODUCTION

Foams were formed by trapping pockets of gas in a liquid or solid. The films of liquid membrane could separate the regions of gas one by one (Bureiko A, 2015). Firefighting foams were produced by liquid foam solution using a generating device (Fu X, 2012). The foam solution would be pushed into the mixing part of generating device. Then, air would be pushed into the moving foam solution at a certain angle. These two phases would be mixed fully in the mixing part. The firefighting foam could be further stable when passing through short or long pipeline. Finally, the firefighting foam would be created, as shown in Figure 1.

The surfactant in the foam's film would keep a relative balance of the foam (Uehara T, 2014). The film in the foam shortly after creation was thick enough. The shape of the foam shortly after creation was globular. Over time, the liquid with some surfactant in the film would flow downward through Plateau border among regions of gas (Le Merrer M, 2012). This flow was the drainage caused by the effect of gravity. Then, the film became thinner and thinner. Over time, the surfactant left in the liquid film could not keep balance between gas and liquid. Some foam bubbles were broken soon. Then, some foam bubbles were combined together. Over time, the different air pressure between neighboring foams might lead to coarsening of some foams. The left foam bubble became bigger and bigger, and its number became smaller and smaller. Therefore, the foam's structure was changing over time, as shown in Figure 2.

Traditional macroscopic parameters on firefighting foam included foam expansion and drain-

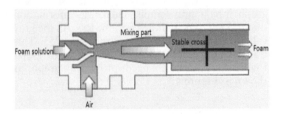

Figure 1. Firefighting foam generating device.

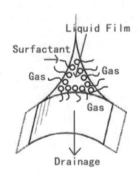

Figure 2. Liquid film with surfactant in Plateau border among regions of gas.

age time. Foam expansion meant the ratio of foam's volume to foam solution's volume. This parameter could reflect the foaming performance of the foam solution. Drainage time (25%) indicates the time required to obtain enough drainage liquid after generation, whose weight could reach 25% of the original foam solution. This parameter could reflect the foam's stability. The foams with the same foam expansion and drainage time might lead to different extinguishing effects. Then, the microscopic parameters on firefighting foam should be studied (Xia J J, 2013 & Laundess A J, 2011).

The foam's structure graphs could be captured by a microscope device. The shape of the globular foam would turn into a polygon after creation over time. There should be some relationship between changing microscopic parameters and the fire-extinguishing result. The qualification of initial foam's structure should be studied, including microscopic circumference, foam number per unit area, and diameter. Therefore, identification of foam's micro boundary was quite essential, and a special analyzer should be designed for globular foam.

2 PARTICLE ANALYZER'S DESIGN

2.1 Gray part

There were three steps in the whole particle analyzer's design. The first step was gray part (Sivaramakrishnan A, 2013). This step could turn a color graph into a gray graph. Then, the information that needed to be handled became less. All the gray value in this graph was between 0 and 255. This transformation was linear. Then, the color graph Figure 3 would be changed into a gray graph, as shown in Figure 4.

2.2 Superimposing part

The second step was the superimposing of two graphs. A copy of Figure 4 and two gray graphs were obtained (Hattori H, 2015). Critical binarization and detection by operator were treated, respectively. Then, the two treated graphs would be combined together through the superimposing part.

One of the two gray graphs should be treated by critical binarization. Then, a black-and-white graph could be obtained as shown in Figure 5.

Figure 3.   Original globular foam's micrograph.

Figure 4.   Globular foam's micrograph after gray part.

Figure 5.   Globular foam's micrograph after binary processing.

The binary threshold was calculated automatically. This critical binarization could reduce the graph's noise. Then, more correct information would be saved for further process.

The other gray graph should be treated using detection by introducing Canny operator. The Canny edge detector used a multistage algorithm to detect a wide range of edges in graphs. This detection could catch as many edges as possible with low error rate. The edge points detected using Canny operator were accurately located on the center of the edge. The edge was marked once and the graph noise did not create false edges generally. The detected graph is as shown in Figure 6. To bold the detected edge, coarsening was carried out. This coarsening could enhance the foam's boundary for superimposing part.

These two graphs were obtained in Figures 5 and 7. To get more information, Figure 7 was superimposed over Figure 5, and the superimpos-

Figure 6. Globular foam's micrograph detected by Canny operator.

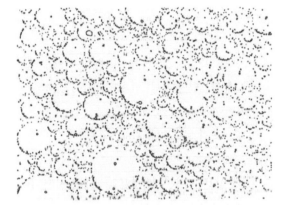

Figure 7. Globular foam's micrograph after expansion.

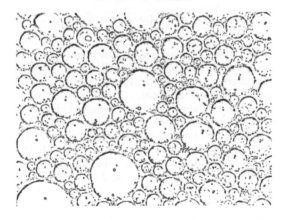

Figure 8. Globular foam's micrograph after superimposing.

ing graph is shown in Figure 8, which shows that the foam's boundary was enhanced. Some noise information were also enhanced, which would be addressed or omitted later.

2.3 Linking part

Linking part could reduce the noisy information and get more accurate boundary from Figure 8 (Li E S, 2009). Then, three types of line were defined as below including Bus Line, Ridge Line, and Arc Line. To the long line with a large central angle, the Bus Line could be used to generate globular foam's boundary actively. To the long line with a small central angle, the Ridge Line could be used to link to Bus Line passively. To the left short line, the Arc Line could not be used to link the Bus Line or Ridge Line. Therefore, the Arc Line should be omitted as noisy information.

Before the formal linking, the Bus Line's globular boundary would be compared with the globular boundaries of the neighboring Bus Line or Ridge Line. If the globular boundary of the Bus Line was located in a similar place to the neighboring Bus Lines or Ridge Lines, the fitting line would connect the original Bus Lines with the neighboring Bus Lines or old Ridge Lines. Then, the number of Bus Lines would decrease. The number of longer Bus Lines would increase. This would increase the further formal linking efficiency, as shown in Figure 9.

Then, the formal linking would be carried out. The linking line would start from the Bus Line's both ends from Figure 10. The linking line's length was calculated automatically. If this line could link to the neighboring Bus Lines or Ridge Lines, then the new linking line would start from this neighboring Bus Lines or Ridge Lines as shown in Figure 11. This process would be carried out until the new linking line from the original Bus Line's one end meets the original Bus Line's other end. Then, the complete globular boundary could be identified. Otherwise, this linking process should be canceled.

After all the linking processes were finished, the final graph is drawn, as shown in Figure 12. In this

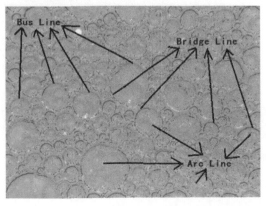

Figure 9. Three types of lines in the globular foam's graph.

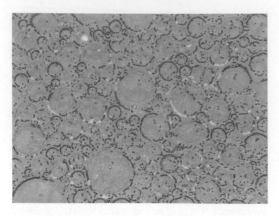

Figure 10. Comparison of neighboring fitting globular boundaries.

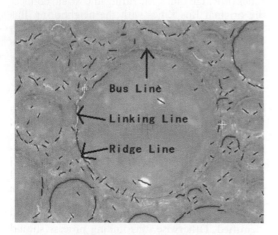

Figure 11. Linking line between Bus Line and Ridge Line.

Figure 12. Result of automatic identification after the linking processes.

Figure 13. Result of automatically identification after the last examination.

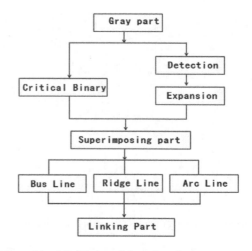

Figure 14. Modified particle analyzer's strategy.

graph, most of the foam's globular boundaries were identified automatically. The left foams' globular boundaries could be examined automatically and linked as shown in Figure 13. If there were some nonglobular foams' boundaries, manual identification should be carried out as a complement.

Therefore, a whole modified particle analyzer for globular foam's boundary was designed. The strategy included gray part, superimposing part, and linking part, as shown in Figure 14 (Xia J, 2014).

3 APPLICATION OF MODIFIED PARTICLE ANALYZER

3.1 Microstructure parameters

The globular foam's boundary could be identified by modified particle analyzer. Then, microstructure parameters could be calculated automatically, including circumference, foam number per unit area, and diameter.

Circumference of a circular or globular object was the linear distance around the foam's edge. This parameter changes with foam's coarsening or breakup. Foam number per unit area was expressed in square millimeter. This number changes with foam's breakup. Diameter was any straight line segment that passes through the center of the globular and whose endpoints lie on the globular. This parameter was changing as foam's coarsening or breakup. Particle size distribution of granular foams was a list of foam's size. This parameter would be changing during the whole drainage process (Gao H, 2016).

3.2 Example

The foams' graph captured by stereo microscope at 129 s is shown in Figure 15. The identi-

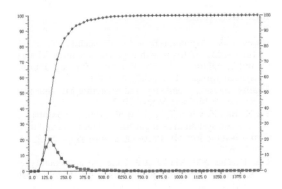

Figure 17. Foam particle distribution of Class A foam graph at 129 s (X-axis: average diameter; Y-axis: left distribution percentage and right cumulative percentage).

Table 1. Parameters of foam graphs in 129 s.

Parameters	Result
Circumference (average)	511.2 μm
Foam number per unit area	25.665/mm²
Diameter (average)	160.32 μm

fied foams graph with bold boundaries is shown in Figure 16. Most of the globular foams were identified automatically. Some of the nonglobular foams were identified manually. Then, the particle size distribution of these foams is shown in Figure 17. The relative parameters and distribution curve were calculated automatically as shown in Table 1.

Figure 15. Class A foam graph at 129 s after creation.

4 CONCLUSION

In summary, the modified particle analyzer of firefighting foam was designed and applied to study globular firefighting foam. The boundary could be identified and relative parameters could be calculated automatically, including circumference, foam number per unit area, and diameter. This analyzer would be applied in the characterization of firefighting foam in different times and from different generating devices in the future.

ACKNOWLEDGMENTS

The study was supported by 2014 Tianjin Postdoctoral Science Foundation, National Science Technology Support Program No. 2014BAK17B03, and CECS Standard Program 2016 79.

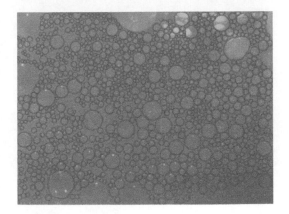

Figure 16. The identified Class A foam graph at 129 s after creation including some nonglobular foams.

REFERENCES

Bureiko A, Arjmandi-Tash O, Kovalchuk N, et al. Interaction of foam with a porous medium: Theory and calculations [J]. The European Physical Journal Special Topics, 2015, 224(2): 459–471.

Foams: physics, chemistry and structure[M]. Springer Science & Business Media, 2013.

Fu X, Bao Z, Chen T, et al. Application of compressed air foam system in extinguishing oil tank fire and middle layer effect [J]. Procedia Engineering, 2012, 45: 669–673.

Gao H, Zhang M, Xia J J, et al. Time and surfactant types dependent model of foams based on the Herschel-Bulkley model [J]. Colloids and Surfaces A: Physicochemical and Engineering Aspects, 2016, 509: 203–213.

Hattori H, Yamamoto K, Kumai H, et al. Stereoscopic image generation device, stereoscopic image display device, stereoscopic image adjustment method, program for causing computer to execute stereoscopic image adjustment method, and recording medium on which the program is recorded: U.S. Patent 9,224,232 [P]. 2015-12-29.

Laundess A J, Rayson M S, Dlugogorski B Z, et al. Small-scale test protocol for firefighting foams DEF (AUST) 5706: effect of bubble size distribution and expansion ratio [J]. Fire technology, 2011, 47(1): 149–162.

Le Merrer M, Cohen-Addad S, Höhler R. Bubble rearrangement duration in foams near the jamming point [J]. Physical review letters, 2012, 108(18): 188301.

Li E S, Zhu S L, Zhu B S, et al. An Adaptive Edge-detection Method Based on the Canny Operator [C]// International Conference on Environmental Science and Information Application Technology, Esiat 2009, Wuhan, China, 4–5 July 2009, 3 Volumes. 2009:465–469.

Sivaramakrishnan A, Karnan M. A novel based approach for extraction of brain tumor in MRI images using soft computing techniques [J]. Journal of Advanced Research in Computer and Communication Engineering, 2013, 2(4): 1845–1848.

Uehara T. Numerical Simulation of Foam Structure Formation and Destruction Process Using Phase-Field Model [C]//Advanced Materials Research. Trans Tech Publications, 2014, 1042: 65–69.

Xia J J, Fu X C, Zhang X Z, et al. Design and Application of Endoscope with Aided Lens for Observation of Fire-Fighting Foam [C]//Applied Mechanics and Materials. Trans Tech Publications, 2013, 271: 823–828.

Xia J. Canny operator-based foam boundary recognition and grain size analysis method: C.N.A Invention Patent ZL201210195198.4[P]. 2014-02-26.

Electromechanical Control Technology and Transportation – Jia & Wu (Eds)
© 2017 Taylor & Francis Group, London, ISBN 978-1-138-06752-3

Application of wireless detection technology in the evaluation of the load-carrying capacity of offshore workover derrick

Yubao Qian
School of Mechanical and Transportation Engineering, China University of Petroleum, Beijing, China
School of Mechanical Engineering, Yangtze University, Jingzhou, Hubei Province, China

Hongwu Zhu
School of Mechanical and Transportation Engineering, China University of Petroleum, Beijing, China

Wenxiu Wu, Jian Hua & Ding Feng
School of Mechanical Engineering, Yangtze University, Jingzhou, Hubei Province, China

ABSTRACT: The derrick on the offshore oil platform is a very important piece of equipment for the workover job. The harsh environment injures derrick's performance, which reduces the load-carrying capacity of the derrick. For security reasons, the carrying capacity of the derrick must be determined. The safety evaluation of workover derrick is based on the strain gauge, which is difficult to paste and has a low accuracy. Wireless measurement technology has more advantages than the strain gauge. Wireless measurement BDI system can collect data without cable transmission, and BDI is easy to paste and has a higher accuracy. The BDI system can obtain an accurate carrying capacity of the derrick, which can ensure the safety of the offshore oil platform. The successful application of the carrying capacity evaluation by BDI system can be used as a reference for relevant personnel, which ensures the security of workover by giving an allowable hook load.

1 INTRODUCTION

The derrick on the offshore oil platform is a very important equipment for drilling and workover. It is used for lifting and lowering the pipe string, placing crane, and suspending traveling block hook (GAO Qingmin, 2013). There are dozens of oil wells on every offshore platform. All workover have been done by a single derrick. The utilization rate of offshore workover derrick is higher, which reduces the life and increases the unsafe factor.

Besides the higher utilization rate, the working environment is very poor. There is increasing significance of fatigue and corrosion effect, aged material, damaged structure, and declined load capacity of derrick, which could be caused by high-salinity sea breeze, working load, wind load, and wave load (An Ziliang, 2011). It is important to avoid hidden dangers during the workover job and to make detection period to obtain the accurate load capacity and correct evaluation of derrick's load capacity and find the structural defect of derrick in service (Gangjun Zhai, 2011).

The strain is proportional to the load within the elastic range of the material, which is part of the theory of derrick's load capacity evaluation. For the hook load of test that is usually less than half

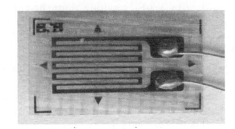

Figure 1. Strain gauge.

of the design load, the load capacity is obtained by linear extrapolation (He Guangtian & Han Dongying, 2011). The derrick load carrying capacity is measured and evaluated according to the standards as follows: Specification for Structural Steel Buildings ANSI/AISC 360-2010), the Chinese National Standard (GB) 25428-2010 edition "Specification for Drilling, Workover Derrick, Base", petroleum industry standard (SY, SY/T, "T"-recommendation) 6326-2012 edition "Method for Testing as well as Evaluating the Load Carrying Capacity of Derrick, Foundation of Oil Drilling Rigs and the Workover Rigs", as shown in Figure 1.

The traditional detection method is as follows. Paste the strain gauge on the main legs of derrick[1], provide a set of hook loads, which is less than design load to hook, collecting strain value by wire strain gauge, linear extrapolation, assessment grades according to *ANSI/AISC 360-2010, GB25428-2010,* and *SY6326-2012*. However, there are many limitations in the traditional detection method. (Hu Jun, 2013).

(1) The location of strain gauge is limited for the long cables. Too many cables make confusion in the scene and increase security risk.
(2) It is inconvenient to carry too many equipment on monkey board to overcome high labor intensity, lower detection efficiency, and lower accuracy.
(3) The dimensions of strain gauge are small and the pasted directions of the strain gauge have great influence on the test of load capacity of workover derricks. The small dimensions require workers to have a higher technology.

There are many problems in traditional load capacity detection methods of workover derrick. Therefore, BDI Wireless Structural Testing System (STS-Wi-Fi) emerged over times.

2 WIRELESS DETECTION EQUIPMENT (BDI SYSTEM)

BDI wireless detection equipment system collects testing data by Wi-Fi, resulting from the tedious work of cable arrangement and connection. It also improves the test efficiency and the safety of testers. Its specific features are as follows:

(1) Wireless: Wireless data transmission, no longer need cable, be replaced by Wi-Fi.
(2) Efficiency: 3–5 min to install a BDI sensor; 32 channels system; and 2–3 h only needed for derrick test.
(3) Higher precision: Accurate sensor.
(4) Portability: Lightweight, efficient, durable, strong anti-interference.

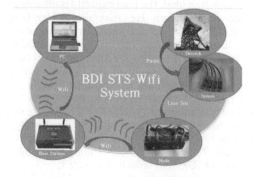

Figure 2. BDI wireless structural testing system equipment components.

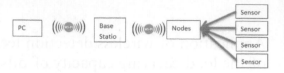

Figure 3. Working principle of BDI wireless structural testing system.

(5) Intelligence: Automatic identification of sensors and data to avoid the chance of manual recording errors.
(6) Compatible with a wide range of sensors and analysis software, a wide range of applications.
(7) The sensors can automatically search the channel, so which node that sensor connected to is less important. However, it is important to pay attention to where is the sensor is pasted.

BDI Wireless Structural Testing System (STS-Wi-Fi) equipment components are shown in Figure 2.

The working principle of BDI Wireless Structural Testing System is shown in Figure 3.

The sensors are arranged on the measuring points of the workover derrick. Every node has four channels and each channel is connected with a sensor by a 5 m cable. The testing data of this part are transferred through the cable. Nodes are fixed on the monkey board. Between the node and base station, there is no cable, and its testing data are transmitted via Wi-Fi. Therefore, it is not needed to mount the cable between the drilling deck and the monkey board; however, it is necessary in the traditional test. Wi-Fi-transmitting makes test work more stable, more accurate, and less intense as well as it minimizes the workload and the work risk. The base station is planting on the drilling deck. Between base station and computer (PC), collecting data are also transmitted via Wi-Fi. Drilling deck is no more cluttered with cables, and clean environment improves work safety.

3 DETECTION AND EVALUATION PROCESS

3.1 *Evaluation principle of workover derrick*

The strain is proportional to the load within the elastic range of the material. By providing a set of loads to hook, a series of strain values can be collected by the BDI STS–Wi-Fi system. Dealing with data, we achieved the fitting linear function. With the linear function, the stress under the design loading can also be obtained. Comparing stress with the allowable stress, the derrick load carrying capacity is clear. Evaluation and rank of derrick can be gained easily according to the standards, such as *ANSI/AISC 360-2010, GB25428-2010,* and *SY/Y 6326-2012*.

3.2 Paste sensors

For the accurate testing data, according to the standard *SY/T 6326-2012*, sensors measuring point were arranged as shown in Figure 4.

As shown in the figure, there are eight measuring points, and each point has four sensors placed in different directions, as shown in Figure 5. The offshore workover derrick is almost a "K"-shaped derrick. Many types of section steels were used on offshore workover derrick. Different types of section steel have involved different patterns to the arrangement.

The most common section steel used on offshore workover derrick is rectangular steel. According to the standard *SY/T 6326-2012*, the arrangement of the sensor pattern should be as shown in Figure 6(a), and all the sensors should be mounted at the same height. The distance edge is 5 mm, as shown in Figure 6 (b).

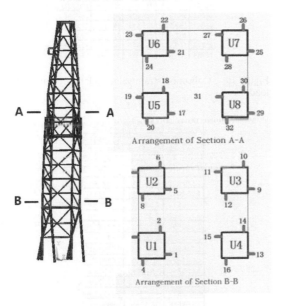

Arrangement of Section A-A

Arrangement of Section B-B

Figure 4.   Pattern of sensor arrangement.

Figure 5.   Pasted sensors of derrick's leg.

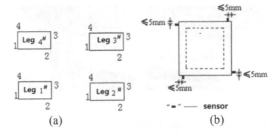

Figure 6.   Arrangement pattern of rectangular steel.

Figure 7.   Connection between sensors and node on field.

The nodes are fixed on the monkey board and connected to sensors with the cable, as shown in Figure 7. It is necessary to remember where the sensors are pasted or the channel to which the sensor was connected. It is worth noting that the antennas should be parallel, which has a strong influence on the measurement signal.

3.3 Measurement process

Take HXJ180 derrick as an example. According to the standards as *ANSI/AISC 360-2010, GB 25428-2010*, and *SY/T 6326-2012*, the minimum hook load is no less than 15% of the design load and the maximum hook load is no less than 25% of the design load. More than three sets of hook load levels are needed.

$$180T \times 15\% = 27T$$

$$180T \times 25\% = 45T$$

The minimum hook load must be greater than 27T, and the maximum hook load must be greater than 45T. The four preset load levels are 27T, 35T, 45T, and 55T.

Loading followed from small to large and steady increase in load. The same steps were done four times with different preset loads.

As shown in Figure 9, collecting data of diffident measure point is a two-dimensional curve.

(a)   (b)

Figure 8.   Collecting strain value.

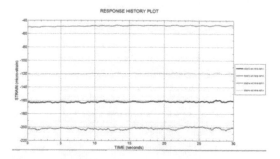

Figure 9.   Number 13 to 16 of strain curves under load 58Ts.

传感器	58	46	35	28
B3857	150.21	118.644	84.1546	63.0061
B3858	183.791	147.805	119.568	96.3874
B3859	147.117	110.244	75.3838	53.4892
B3860	157.589	118.495	80.5353	57.183
B3861	59.1436	48.9094	36.9982	30.5485
B3862	77.551	66.853	52.1235	43.8214
B3864	39.9021	29.3046	17.199	11.0613
B3866	37.0247	27.754	18.7487	13.6911
B3867	55.4372	44.8053	26.889	19.8854
B3869	83.9608	75.5452	58.3948	38.3978
B3870	51.4509	35.1651	23.4063	14.5018
B3871	55.4551	40.8494	28.4669	20.043
B3872	162.032	118.886	93.7592	71.8944
B3873	201.881	162.367	116.19	91.5516
B3874	48.3698	21.5711	4.4047	-2.9266
B3875	119.928	86.6065	55.5578	36.7442
B3876	175.374	136.895	95.8744	71.9291
B3877	161.767	130.434	91.8117	72.1753
B3878	97.4155	74.0099	53.0668	40.2396
B3879	152.15	117.483	81.6476	60.017
B3880	69.026	56.9056	42.772	34.8364
B3881	40.9177	34.6739	31.1352	25.6483
B3882	-12.2467	-15.7719	-16.5363	-17.9174
B3884	36.2583	26.0115	18.3707	12.8203
B3885	58.6041	45.9419	33.3099	25.6297
B3886	54.4475	42.8635	30.6997	23.3363
B3887	40.7329	31.4281	23.1566	17.8273
B3888	58.7993	46.7228	33.3236	26.0295
B3889	160.826	125.511	89.6139	67.695
B3890	162.849	132.411	99.2593	79.1217
B3893	120.131	89.5519	59.5188	42.3439
B3894	159.822	120.335	79.1746	56.8484

Figure 10.   Collection strain value under different hook load.

3.4 Data process

WinGRF, special software provided by BDI Company, was used to deal with collection data. The data were extracted and imported to EXCEL. The result is shown in Figure 10.

By fitting the strain value under different hook loads, the coefficient of fitting formula is as shown in Figure 11. All the goodness of fit are larger than 0.99, and the result shown in Figure 11 is credible. The relationship between hook load and strain is shown in Figure 12, which is the same as obtained using the fitting formula.

Using the fitting formula, we calculate all the linear extrapolation strain and stress under the design load. We further find the maximum stress and compare it with the allowable stress.

According to the fitting formula, we calculate the maximum strain and stress under the design load of derrick.

3.5 Evaluation of carrying capacity

According to the standard SY/T 6326-2012, the strength of the derrick shall satisfy Formulas 1 and 2:

$$\frac{f_a}{F_a} + \frac{C_{mx}f_{br}}{\left(1-\dfrac{f_a}{F'_{ex}}\right)F_{bx}} + \frac{C_{my}f_{by}}{\left(1-\dfrac{f_a}{F'_{ey}}\right)F_{by}} \le 1.0 \qquad (1)$$

拟合X系数	拟合常系数	决定系数
2.695971	-6.53121	0.995581
2.646646	28.37549	0.996597
2.878026	-21.4405	0.994591
3.091344	-23.2944	0.994512
0.889806	7.417840	0.992646
1.054040	16.87156	0.991911
0.898818	-12.4848	0.990753
0.720528	-5.23703	0.991797
1.133594	-9.72316	0.978611
1.386676	7.220910	0.956234
1.118851	-14.7418	0.991161
1.081673	-8.14502	0.994119
2.699856	0.948890	0.983513
5.435387	2.146502	0.992256
4.568517	-46.4542	0.957122
2.558656	-30.1959	0.99323
3.200857	-11.2170	0.994267
2.799556	-0.73489	0.990856
1.755575	-5.79563	0.992505
2.844563	-13.8027	0.99486
1.062633	7.317046	0.993563
0.448492	14.70558	0.991157
0.163106	-22.3054	0.92021
0.711385	-5.80160	0.991283
1.016575	-0.80818	0.994618
0.960966	-1.56289	0.995074
0.702018	-0.49654	0.993883
1.017802	-0.51111	0.992651
2.872581	-6.86453	0.994974
2.589087	12.25776	0.995359
2.401608	-20.5794	0.992587
3.191452	-26.8046	0.991946

Figure 11.   Coefficient of fitting formula.

338

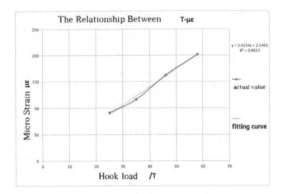

Figure 12. Relationship between hook load and strain of sensor B3873.

Table 1. Maximum of strength coefficient of diffident sections.

Section	A—A	B—B 1	B—B 2
Section average of stress	104.64 MPa	106.041 MPa	98.92 MPa
X direction bending stress	3.97 MPa	9.23883 MPa	14.51 MPa
Z direction bending stress	11.26 MPa	4.32787 MPa	1.585 MPa
Maximum stress	123.86 MPa	119.426 MPa	126.94 MPa
Strength coefficient	0.6530	0.8044	0.7756
Load-carrying capacity		1800 kN	

$$\frac{f_a}{0.60F_y}+\frac{f_{bx}}{F_{bx}}+\frac{f_{by}}{F_{by}} \leq 1.0 \qquad (2)$$

If $f_a/F_a \leq 0.15$, Formula 3 can be used instead of Formula 1 and Formula 2.

$$\frac{f_a}{F_a}+\frac{f_{bx}}{F_{bx}}+\frac{f_{by}}{F_{by}} \leq 1.0 \qquad (3)$$

where

f_a—Axial tension and compression stress of test bar, which are under the design hook load (MPa);

F_a—The allowed axial tension and compression stress is only when the axial tensile stress exists (MPa);

f_b—Compression bending stress of test members under the design hook load (MPa);

F_b—The allowed axial tension and compression stress is only when the bending moment exists (MPa);

F_e—Euler stress divided by a safety factor.

According to the standards as *ANSI/AISC 360-2010, GB25428-2010*, and *SY 6326-2012*, calculate the strength coefficient of the derrick. The strength coefficient of the derrick of all sections is shown in Table 1.

In Table 1, the maximum of strength coefficient is 0.8044, less than 1. According to *SY 6442-2010* and *SY/T 6326-2012*, the load-carrying capacity of derrick meets the requirements of the design hook load (1800 kN), and the derrick was rated at level A.

In the case where the strength coefficient exceeds 1, we calculate the maximum allowable hook load by inverse calculation. The steps are: (1) Assume that the maximum of strength coef-ficient is 1 or the maximum linear extrapolation is equal to allowance stress; (2) Calculate the allowance strain; (3) Calculate the hook load by the fitting formula. Note that the newly obtained hook load is the allowance-carrying capacity of the derrick.

4 CONCLUSIONS

The application of the BDI STS–Wi-Fi system in the evaluation of the load-carrying capacity of the offshore oil derrick has successfully improved the working efficiency, the measurement accuracy as well as the accuracy of the derrick load-carrying capacity. In addition, this application has reduced the labor intensity of the measuring staff, reduced the strict requirements of gauge pasting and avoided scene confusion for wireless equipment.

The accuracy of the evaluation of the derrick load-carrying capacity increases field work safety. And the load-carrying capacity of the offshore derrick can provide guidance for field workover.

ACKNOWLEDGMENTS

The authors thank all members of the assessment team of derrick load-carrying capacity. The authors are also thankful to the members of the maintenance team of a certain offshore platform for their support.

This work was supported by Yangtze University research fund (Study on vibration and test method of heavy drilling rig derrick) and Petro China Innovation Fund Program (2014D-5006-0310).

REFERENCES

An Ziliang, Xiao Lizhong, Kong Lingchao. 2011. Derrick stress monitoring system based on Modbus RTU communication protocol [J]. *Chinese Test,* 06:72–75.

ANSI/AISC 360-2010. Specification for Structural Steel Buildings [S].

Chinese National Standard GB 25428-2010. Specification for Drilling and Workover Derrick, Base [S].

Gangjun Zhai, Zhe Ma, Hang Zhu, 2012. The wind tunnel tests of wind pressure acting on the derrick of deep-water semi-submersible drilling platform. *Energy Procedia* 14:1267–1272.

Gao Qingmin, Li Qiang. 2013. Stress test and bearing evaluation of offshore drilling (repair) derrick [J]. *Shanxi youth*, 06:116.

Han Dongying, Shi Peimin, Zhou Guoqiang, 2011. Safety Evaluation of Marine Derrick Steel Structures Based on Dynamic Measurement and Updated Finite Element Model. *Procedia Engineering* 26:1891–1900.

He Guangtian. 2011. Research on reliability of JJ160/41-K-shap oil derrick [D]. *Daqing petroleum institute*. http://www.vertinfo.com/p-40154.htm

Hu Jun, Tang Yougang, Li Shixi, 2013. Vibration Test and Assessment for an Ocean drilling rig derrick: Taking the ZJ50/3150DB Drilling rig as an Example. *Petrol. Explor. Develop.* 40(1):126–129.

Petroleum industry standard SY/T 6326-2012. Method for Testing and Evaluating on the Load Carrying Capacity of Derrick and Foundation of Oil Drilling Rigs and Workover Rigs [S].

Electromechanical Control Technology and Transportation – Jia & Wu (Eds)
© 2017 Taylor & Francis Group, London, ISBN 978-1-138-06752-3

Visualization of the trade information of agricultural products

Jiaqi Hou & Fucheng Wan
Department of Computer Technology, University of Northwest Nationalities, Lanzhou, China

ABSTRACT: The progress of science and technology has brought a lot of convenience to our lives, especially for data acquisition and processing, and the requirements continue to increase. A single text data expression and analysis has not been fast and efficient for data processing and using when facing massive data during the Internet+ era, and new technology research is imminent. In this case, the technology of data visualization is presented, which can express complex data vividly by using computer graphics and a variety of graphical methods. In this paper, the existing classic visualization model is analyzed, and a new data visualization analysis component model is proposed to deal with the data more efficiently.

1 INTRODUCTION

With the development of data visualization technology, a large number of visualization tools have been developed. The birth of Echarts and Highcharts provides a rich graphics library for the visual development of Web front end. In this paper, a data visualization application is designed and implemented for the problem of agricultural products' trading information visualization on E-commerce platform, which is used to display the geographic information of agricultural products sales point and sales (Zhongfang Ren, 2004). On the basis of MVC design idea, the design of the front page is based on the map of Echarts graphics library while the back end uses the flask framework based on Python language and adopts ORM technology to solve the problem of matching object and data.

2 BASIC CONCEPT

2.1 *Web development*

Web development is based on B/S structure, which is divided into frontend and backend development. The frontend development, development of the page, is based on the technology of HTML, Javascript, and so on (Xiaoting Ma, 2015). HTML is used for the development of static pages, whereas Javascript is for dynamic pages. The backend development, server-side development, is for the realization of the business logic of the site, and the development language of backend mainly include ASP, PHP, Java, and Python.

2.2 *MVC design ideas*

MVC (Model-View-Controller) is introduced by Small-talk-80, an object-oriented design patterns

for the creation of reusable UI procedures (Jun Hu, 2009). MVC divides Web development into three layers: model layer, view layer, and controller. The model layer is responsible for the entity-relation logic of the database, the view layer is responsible for the design of the page, and the control layer is responsible for the interaction of the page with the server data. The key to MVC is that the view layer and the model layer should not communicate with each other (Lei Lei, 2014). The controller will accept interactive commands from the view and manipulate the model and determine whether changes in the model need to change the view

2.3 *Flask framework*

Flask is a lightweight Web development framework based on the Python language, which has rich third-party libraries and can be developed in a very short period into a stable Web application using web2.0 and web3.0 era (Jianye Yang, 2015). Its WSGI toolkit uses Werkzeug, and the template engine uses Jinja2. Compared with the traditional javaEE framework and Microsoft. Net framework, the flask has a shorter development cycle, more concise code, and a wealth of third-party libraries for developers and other advantages.

2.4 *About Echarts*

Ehcarts, a library of javascript, which has now been used in business activities, is developed by the Baidu frontend data visualization team and is a very promising project (Chunling Chen, 2006). By using the lightweight Canvas library ZRender, it can create data visualization charts simply and clearly. It not only has interactive function, but also can be customed personly. We need to import the script tag, initialize with echarts.init, and set option to get the initial histogram in the specific implementation. It is a data visualization technology and can display a lot of data intuitively through a simple graphical form or a dynamic figure so that users can gain access to information from the data more conveniently.

2.5 *ORM*

Through the method of early embedded SQL and the later JDBC/ODBC, the programmer needs to write a lot of code of data access layer during developing an application in which the business logic layer and the user interface layer must be separated (Ziyi Wang, 2016). It is a great waste of resources and manpower to write repetitive code with the same pattern in each project. In this case, the emergence of the Object-Relational Mapping (ORM) technology solves the problem effectively. It generates an automatic mapping between the relational database and the object to avoid the need to deal with complex SQL statements in the specific database operation so that the software designers can only focus on the business logic of the object architecture rather than the underlying code of the database SQL and JDBC. According to statistics, using ORM can reduce software development time and cost by 40%, greatly improve data readability, and simplify the code tuning and testing because of the separation between the business layer and the actual data storage. Many developers and experts have predicted that ORM will become the mainstream model of the object-oriented development.

3 DEMAND ANALYSIS

3.1 *About the user*

The system involves two types of users: system administrator and normal user. The normal users can be divided into customers and sellers. Of course, one user can have two identities at the same time. Customers are more concerned about the quality of agricultural products, the sales of agricultural products, the prices, and the geographical factors. On the contrary, sellers are concerned about the consumption of various agricultural products in different regions during different periods. The system administrator needs to ensure the integrity and correctness of information.

3.2 *Business logic*

The business logic of the system is relatively simple, which mainly presents the corresponding agricultural product data to the user. The system administrator is mainly concerned with the man-

agement of the database table; the customer's main concern is the sale of agricultural products under the name of sales, price, and geographical information; vendor users tend to visit certain agricultural products sales information.

4 SYSTEM DESIGN AND IMPLEMENTATION

4.1 *Architecture design*

According to the demand analysis, the system uses three layers of development model: presentation layer, business logic layer, and data layer. The frontend uses Echarts and Jquery for page development, the server selects Python's Flask framework, and the database uses MySQL, as shown in Figure 1:

4.2 *Database design*

According to the analysis of the business process of the system, it can be concluded that the system consists of three entities: user, agricultural product, and region. User entity can be divided into seller and customer, and the relation between user and agricultural product is one-to-many. The business entity contains product, user, and order. In addition to the basic product name, region, type, price, and geographical area, the product should also have the field about the seller and the

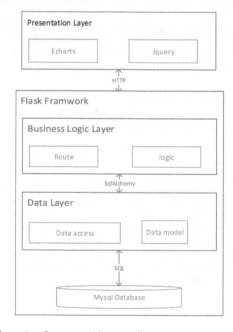

Figure 1. System architecture diagram.

customer in an order. The order entity should contain the field of customer and the date. User is divided into two types: customer and seller, so user must also have the attributes of the role in addition to the basic user name, login password, and mailbox. Taking MySQL database for example, the specific database design is shown in Figure 2.

4.3 Visual interface design

For the visualization design, the first step is the data acquisition and formatting. Data acquisition is the crucial point in the model and the base of future works. In the interactive type of model, the data from front end will reach the back end by a specification process. In the process, the classic ajax request is often required for assistance to get JSON format data results ultimately (Wang Jau-Hwang, 2003). The second step is about the rules for generating data at different levels. These hierarchical data are practical data, which are selected in terms of requirements and associations, and retain attributes associated with the chart dimension (George Lawton, 2002). Take multilevel drilling of pie chart and transforming to column chart from pie chart for example, the former should only be paid attention to common attributes, whereas the latter should also be paid attention to the chart dimension-related attributes.

In the agricultural product visualization system, the visual interface mainly displays the agricultural product category, region and seller information, and so on. Here, an example of a single kind of agricultural product is used to introduce the design (use simulation data). The specific design is shown in Figure 3.

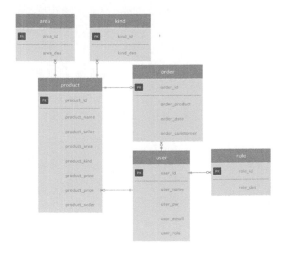

Figure 2.   Database table design diagram.

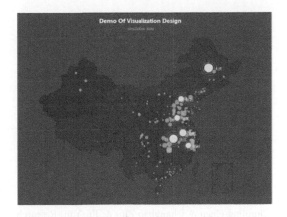

Figure 3.   Interface design diagram.

5   KEY TECHNOLOGIES

According to the requirement analysis, the key of the system is to access data from database. In this page, ORM (Object Relation Mapping) is used to access the database (Paul K Harmer, 2002). ORM, a programming technique, is used to achieve the transformation of the data between different types of systems in an object-oriented programming language. The flask framework implements the ORM mechanism through a third-party library named flask_sqlachemy. The specific code is as follows:

```
from flask_sqlalchemy import SQLAlchemy/
#import the library
app = Flask(__name__) # create an app
app.config['SQLALCHEMY_DATABASE_URI/
'] = /
'mysql://<username>:<password>@<hostname>/
<database>/' # configure the database info
app.config['SQLALCHEMY_COMMIT_
ON_TE/ARDOWN'] = Truc
#commit to the database when code is accessed
db = SQLAlchemy(app) # load sqlalchemy into/
the app extension
```

6   CONCLUSION

With the rapid development of the Internet in the past 20 years, Web apps based on B/S structure gradually replace the old generation (C/S), and now a new generation of Web development models is also developing rapidly. For a Web application, the first principle is convenience. Living in the era of information explosion, text data cannot meet the people's demand for efficient information acquisition. Visualization technology has become an important way for us to obtain data information efficiently. This paper introduces the lightweight Web development framework and a frontend visualization library Echart and presents the application of visualization on the transaction platform.

ACKNOWLEDGMENTS

This study was supported by the National Science-technology Support Plan Projects (No. 2015BAD29B01).

REFERENCES

Chunling Chen & Changbao Zhu & Jin Yan. Research on Software Development Method Based on ORM Technology [J]. Computer and Modernization, 2006, 6(31):57–58.

George Lawton. Virus wars: Fewer attacks, newthreats [J]. IEEE Computer, 2002, 35(12):22–24.

Jianye Yang & Jianping Geng. Real-time Web Data Monitoring System Based on HTML5 [J]. Journal of Guilin University of Electronic Technology, 2015(2): 136–141.

Jun Hu. Data Mining Visualization Model Machine Application [D]. Beijing: Beijing Jiaotong University, 2009.

Lei Lei. Analysis of Common Data Visualization Technology [J]. Modern Television Technology, 2014(9): 137–139.

Paul K Harmer & Paul D Williams & Gregg H Gunsch. An artificial immune system architecture for computer security applications [J]. IEEE Transtraction on Evolutionary Computation, 2002(6):252–280.

Wang Jau-Hwang & Deng P S, Fan Yi-Shen, et al. Virus detection using data mining techinques [C]. Proceedings IEEE 37th Annual 2003 International Carnahan Conference on Digital Object Identifier, 2003:71–76.

Xiaoting Ma. Design and Implementation of Library Data Visualization Analysis System [J]. Journal of Library Science, 2015, 19(12):129–131.

Zhongfang Ren & Hua Zhang. Review of MVC Model [J]. Journal of Computer Applications, 2004, 21(10):1–4.

Ziyi Wang & Chunhai Zhang. Design and Realization of Data Visualization Analysis Component Based on ECharts [J]. Microcomputer and Its Applications, 2016, 16(31):228–229.

Electromechanical Control Technology and Transportation – Jia & Wu (Eds)
© 2017 Taylor & Francis Group, London, ISBN 978-1-138-06752-3

Research and design of multilingual agricultural e-commerce logistics information management system

Sheng Dai, Ning Ma, Ya-ru Cao & Jia-jia Li
National Languages Information Technology Research Institute, Northwest University for Nationalities,
Lanzhou City, Gansu Province, China

ABSTRACT: In this paper, a multilingual e-commerce logistics information management system is designed for the logistics conditions and human customs in the ethnic regions of Northwest China, using the mature Internet technology, so as to improve the logistics situation in minority areas and promote the study of logistics technology. First, according to the actual demand, the main problems are put out, the functions of the logistics system are analyzed, and the whole structure of the system is designed using B/S structure containing five main modules. Finally, combined with the actual development conditions, we select the appropriate technology to achieve the whole system.

1 INTRODUCTION

With the development of Information Technology (IT) and the continuous innovation of the Internet technology, new things continue to emerge. E-commerce as a new business mode with the progress of technology that has deeply penetrated our life. It has changed people's shopping behavior and even the way of life and plays an important role in our daily life.

The logistics is not only an important part of the e-commerce process, but also the basis for the steady development of e-commerce. The so-called logistics refers to the entire process of commodity circulation of goods from the customers and businesses in the network to reach a trading agreement, starting from the merchant through a variety of transport routes to reach the end user. From the beginning of the transport of goods to the final reach of customers and their confirmation and evaluation, the whole complex process belongs to the scope of logistics management. Because of the complexity of the logistics process, there is a problem of information disconnection between the logistics platform and the e-commerce platform, which seriously affects the development of e-commerce.

At present, the logistics industry is generally using the way including freight transport and manual distribution. The specific process is that goods are moved through a variety of transport routes into various subsites in the first stage and then by couriers the goods are sent to customers manually. In the middle of the delivery of goods, customers can check the progress of the delivery of goods through their own documents in the corresponding logistics or e-commerce business website (Zhang Duo, 2000).

The current Chinese logistics industry and e-commerce are still at a relatively low level of development because of the late start. There are still problems such as slow development, lack of standardization in enterprises, shortage of talent, and lack of supervision, which seriously affect the healthy development of the logistics industry and have become the bottleneck restricting the rapid development of e-commerce economy. How to implement electronic commerce and logistics distribution seamless and interactive development, speed up the logistics enterprise informatization, industrialization, intelligence, and humanization construction, enhance the core competitiveness of logistics enterprises; how to build a logistics distribution network keeping up with the development of the times, covering various industries, increase the efficiency of the modern logistics information system, which has higher responsiveness and distribution efficiency, are the two objective requirements of the development of China's information society and the urgent task of developing e-commerce economy.

This paper is aimed at the study of the logistics situation of northwest minority areas in China so as to design the multilingual agricultural products electricity supplier logistics information management system. Facing the complicated language situation in minority areas, the system must realize the multilingual function and multilingual information display and effectively integrate the information of the suppliers, different platforms, logistic providers, and customers so as to build a quick, effective, and stable logistics information system.

2 THE MAIN RESEARCH CONTENT

Combining with the actual research projects and then through theoretical analysis and practical summary, this paper systematically studies and explores the problems in the construction of e-commerce logistics information system, the system's overall solution, and the implementation of safety measures. The main research contents include the following two points:

1. The core functions of logistics information management system:
   a. To achieve the basic logistics distribution tracking function;
   b. To achieve the transfer of information between different objects;
   c. To provide intelligent decision support;
   d. To complete the realization of minority languages.
2. The structure and design of the system:
   The logistics management system implements the following functions:
   a. Publication and query of information;
   b. Business management subsystem;
   c. Cargo-tracking and inquiries;
   d. Logistics order processing;
   e. Platform management.

Information release and query module: To issue logistics information (including freight, vehicles, and routes), business needs, industry news, policy rules, and so on. At the same time, users can query the transportation routes, logistics outlets, and other information, as shown in Figure 1.

Logistics transaction module: To provide consultation service for suppliers and logistics providers, provide transaction tracking service; intelligent decision function provides users with logistics route selection, cost calculation, and other services.

Logistics tracking module: To provide users with real-time monitoring and query of logistics information and set up service evaluation and service information feedback and other functions.

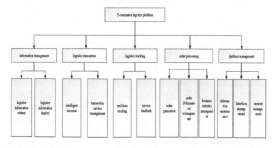

Figure 1. Functional structure of logistics management system.

Order processing module: the users can set the order generation style and manage historical orders and unprocessing orders, while generating reports and conducting business statistics.

Platform management module: The module is mainly used to maintain and manage user information, platform interface, and so on. At the same time, this module also needs to deal with network information security.

3 KEY ISSUES TO BE RESOLVED

3.1 Realization of order processing and report display

Through the database/application, we obtain the original information, as well as provide information content and architecture. Then, the system displays information in accordance with the requirements of the form.

3.2 Online tracking algorithm and implementation

This function enables the upper-level decision maker to store information in a fast, consistent, and interactive manner in a variety of perspectives thus to thoroughly understand the information and make decision.

3.3 Data encryption technology

Data encryption technology is the main security measures used in e-commerce, which can be used in the exchange of information in accordance with the requirements by the both sides of the transaction. It transforms some important information and data from an understandable plaintext into a complicated and incomprehensible ciphertext through some code or password and then propagates it for the receiver to decrypt it in a known way (Wang Shao-rong, 2008).

3.4 Minority language display

In order to meet the needs of customers of different nationalities, to help them on the platform for information query and transaction processing, the system must be displayed in multiple languages.

3.5 Exchange and transmission of different types of data

The system connects different logistics suppliers, customers, public information service platform, and other objects to deal with a variety of different data, to solve the problem of exchange between different data.

4 SYSTEM IMPLEMENTATION

Through analysis and research on the logistics information management system worldwide, we determine the general design principle and system function and support different minority languages so as to realize the multilingual design.

The system uses B/S structure and is divided into three directions in the direction of technology: the front-end services to provide and display, background data, and information security management. There are three levels in software architecture: presentation layer, business layer, and data layer, as shown in Figure 2.

Specific technical application is reflected in the following four aspects:

- Web server implementation. Java EE is a new specification for Java technology, which implements the reorganization of the programming paradigm. In addition, due to the independence of platform and the function of rapid convenient deployment and robust code, the system uses PHP as the front-end code development tools, supplemented by other auxiliary tools to complete. This is a good solution. Using Web service technology to realize the integration of logistics data and services is a good choice.
- Exchange of intermediate data and business processing. The system uses Active MQ as JMS (Java Message Service, Java Message Service) message middleware, Tomcat as the application server and J2EE as the deployment of applications. The business logic is managed using Spring containers. Database access operations are based on MyBatis ORM (Object Relation Mapping, object relational mapping) framework. The bottom layer uses C3P0 connection pool.
- The realization of the background database. We use the relational database system MySQL as a back-end database. This is due to the cross platform of the database. Other reasons are that it has a full network and strong function and is easy to use, easy to manage, fast, safe, and reliable.
- Realization of network information security. E-commerce is based on the electronic information technology, whose core content is online transaction. The security issue in the transaction process is one of the most important concerns to enterprises. The security problem in this system mainly refers to the security of data transmission in the system. In the data security design of the system, data encryption technology is adopted to realize the data transmission security, and the knowledge and technology involved are worth studying.

5 CONCLUSIONS

According to the characteristics of logistics, the logistics system must be optimized and reconstructed. Enterprises must integrate logistics resources and improve the structure of the logistics network. Through the construction of logistics information platform and the integration of new technologies, the data can be sustained simple, efficient, real-time, and accurate so as to enable enterprises to shorten the delivery time and reduce costs and improve customer satisfaction, provide better logistics services for e-commerce, and promote the development of e-commerce. According to the current technological development and system function requirements, this system selects the required technology, puts forward the goal and guiding ideology of the logistics information management system, and lists the main modules involved in the logistics information management system and the adopted technology methods to realize the multilingual function. This paper is of great significance to the construction of logistics information management system based on e-commerce.

ACKNOWLEDGMENT

This study was supported by the Fundamental Research Funds for the Central Universities (No. Yxm2015187).

REFERENCES

Chen Jun edited. *J2EE build enterprise-class application solutions* [M]. Beijing: People's Posts and Telecommunications Press, 2002 (in Chinese).

David A. Chappell, Tyler Jewell. JAVA Web Sevrice [J]. *O, Reilly*, 2002, 11(6):10–14.

Jayanth Jayaram, Keah-Choon Tan. Supply chain integration with third-party logistics providers [J]. *International Journal of Production Economics*. 2010 (2).

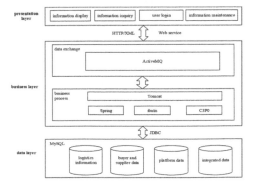

Figure 2. Software structure of logistics information platform.

Liao X.H., Yan Y.S., Wang L., et al. The Design and Implementation of Logistics Monitoring System on Food Safety [J]. *Logistics—the Emerging Frontiers of Transportation and Development in China*, 2008(03): 2206–2212.

Martin Bichler, Simon Field, Hannes Werthner. Introduction: Theory and Application of Electronic Market Design [J]. *Electronic Commerce Research*. 2001 (3).

Robert. C. Martin. Agile Software Development Principle, Patterns, and Practices. 2008.

Wang Shao-rong, Wen De-hua. Development of Data Encryption in E-commerce Security [J]. *Communications Technology*, 2008, 41 (12): 322–324. (in Chinese).

Wen Yu. Design of Software Architecture [M]. (2nd ed.), Beijing: *Electronics Industry Press*, 2012 (in Chinese).

Zhang Duo. *Electronic commerce and logistics* [M]. Beijing: Tsinghua University press, 2000 (in Chinese).

Zheng Guan-wei. *J2EE Technology Insider* [M]. Mechanical Industry Press, 2002 (in Chinese).

Electromechanical Control Technology and Transportation – Jia & Wu (Eds)
© *2017 Taylor & Francis Group, London, ISBN 978-1-138-06752-3*

Codebook design based on the particle swarm algorithm

Bingyu Qu & Ying Qian
Graphic Image and Multimedia Lab, Chongqing University of Posts and Telecommunications, Chongqing, China

ABSTRACT: The initial codebook selection of the LBG algorithm will affect the performance of the final codebook, and the design of the vector quantization codebook with basic particle swarm optimization algorithm is prone to local optimal solution problems. To solve these problems, a new vector quantization codebook design method is proposed by using the influence of inertia weight on particle search ability, the influence of learning factor on particle trajectory, and the relationship between codebook. First, the code words are sorted according to the relationship between the feature quantities of the code words in the codebook, and then the codebook was obtained by randomly selecting a certain number of the sorted code words so that each codebook has a certain relationship. Finally, the inertia weight and learning factor are used to improve the particle swarm algorithm, which can ensure the global convergence of the improved algorithm. Experimental results show that compared with LBG, the Improved Particle Swarm Optimization (IPSO) codebook design algorithm can improve the codebook performance significantly.

1 INTRODUCTION

Vector quantization is an important data compression technology. Vector quantization can use the relationship between adjacent data to eliminate redundancy, with a larger compression ratio, simple codecs, and other advantages. Its application has spread to the voice recognition, image compression, and other fields. Vector quantization includes three major techniques: codebook design, code word search, and code word index allocation. Among these three techniques, codebook design is the most critical one. In the 1980s, Linde, Buzo, and Gray (1980) proposed a classic algorithm for codebook design, which is simple and intuitive; however, it had two major problems. First, the initial codebook selection has a greater impact on the algorithm. The improper initial codebook will not only reduce the convergence rate of the later algorithm, but also reduce the performance of the resulting codebook. Second, the time required for the optimal division of vectors is longer, and the training time of the whole LBG algorithm is longer. In order to solve these problems, scholars have put forward a variety of optimization algorithms (Guo Yanju 2013, Thepade D S D 2013, Sudee P D 2013, 2015), which facilitated further development of the technology. So far, there are genetic algorithm (Delport V 1995), neural network (Krishnamurthy A K 1990), Particle Swarm Optimization (PSO) (Shi Y 1999), and colony optimization algorithm (Li Xia 2004). These algorithms are able to obtain a high-quality codebook. Although the PSO algorithm can overcome the influence of the

initial codebook on the performance of the LBG algorithm, its operation process is more prone to "premature" phenomenon. In order to overcome this shortcoming, scholars have proposed a variety of algorithms for the inertia weight and learning factors (Guo Changyou 2011, Li Li 2010, Kennedy J 1995, Ratnaweera A 2004, Ping Yi C 2013). For the linear weighting algorithm (Shi Y 1998), the global searching ability of the early particles is very strong. If the suitable points cannot be found at this time, the local search will be strengthened with the decrease of the inertia weight so that the algorithm will fall into local optimal. Therefore, scholars have proposed a variety of nonlinear inertia weight improvement algorithms so that the global search capability and local search capability have a good balance.

2 PARTICLE SWARM OPTIMIZATION ALGORITHM

2.1 *Principle of the particle swarm optimization algorithm*

In the 1990s, two scholars in the United States proposed a Particle Swarm Optimization (PSO) algorithm on the basis of the observation of food behavior in birds. Particle Swarm Optimization (PSO) is a population-based stochastic search technique. First, the algorithm initializes a group of particles and randomly obtains the position and velocity information of the particles. Then, it obtains the optimal solution by successive iterations. In the d-dimensional search space, N represents the number of particles in the entire search space, and the i-th particle can be expressed by $\mathbf{X}_i$, the speed can be expressed by $\mathbf{V}_i$, and the maximum speed can be expressed by $\mathbf{V}_{max}$. During each iteration, each particle's position and velocity are calculated, respectively, as:

$$\mathbf{v}_{id}^{k+1} = \mathbf{v}_{id}^k + c_1 r_1(\mathbf{p}_{id} - \mathbf{x}_{id}^k) + c_2 r_2(\mathbf{p}_{gd} - \mathbf{x}_{id}^k) \tag{1}$$

$$\mathbf{x}_{id}^{k+1} = \mathbf{x}_{id}^k + \mathbf{v}_{id}^{k+1} \tag{2}$$

where $1 \leq i \leq N$, $1 \leq d \leq D$ and c_1 and c_2 are two positive constants; r_1 and r_2 are independent random numbers in [0,1]; $\mathbf{P}_{id}$ is the value of "individual extreme value" of the i-th particle in the d-dimension; and $\mathbf{P}_{gd}$ is the value of the "global extremum" of the particle swarm at the d-dimension. Equation (1) comprises three parts on the right-hand side. The first part is the previous particle velocity, which provides necessary momentum that particles can remain constant across the entire search space. The second part is "cognitive" portion of the particles, which represents the particle itself, which can move each particle toward its own best position. The third part is the "social" part, which indicates the mutual cooperation among the particles in the particle swarm, which can move the particles to the global optimum position.

2.2 The standard particle swarm optimization algorithm

When the value of first part in equation (1) is large, it will keep the particle searching for the whole space at all times, and it is easy to miss the optimal solution. When the value of the first part is small, maintaining a low speed in the entire search space flight, the particle will only get local optimal solution. Taking into account these factors, and in order to balance the global and local search capabilities of the PSO, Shi Y (1998) introduced the inertia weight (w) into the particle velocity calculation formula, whose value determines the degree of particle retention to the previous velocity. At this time, equation (1) is modified to:

$$\mathbf{v}_{id}^{k+1} = w\mathbf{v}_{id}^k + c_1 r_1(\mathbf{p}_{id} - \mathbf{x}_{id}^k) + c_2 r_2(\mathbf{p}_{gd} - \mathbf{x}_{id}^k) \tag{3}$$

and the particle position remains unchanged. Usually, once the inertia weight (w) is added to a particle swarm algorithm, it is called the standard particle swarm optimization algorithm. And the concrete realization steps of the algorithm are as follows:

1. Initialization. The size of the population is first selected, and then the start position, start velocity,

individual extremum, and global extremum of the particle are set.

2. First, select the fitness function according to the need and then calculate the value of each particle corresponding to the function.

3. For each particle, the fitness function value is compared with the previous individual extreme value. If it is better than the previous one, then the fitness function value of this time is taken as the latest individual extreme value. Otherwise, the individual extremum remains unchanged.

4. Compare the fitness function values of all current particles, select the particle with the largest fitness function value in this iteration, and then compare the corresponding fitness function values of the particle to the global maximum values. If the comparison result is better than the previous one, then the global extreme value will be regarded as the latest global extreme value. Otherwise, the global extreme value will remain unchanged.

2.3 Particle swarm optimization algorithm parameters

2.3.1 Inertia weight

The inertia weight determines the range in which the particle can search in the entire search space. If the inertia weight is set larger, the particle search range is more extensive. Otherwise, the particle search range is narrow. In general, in order to increase the probability of finding the optimal solution, people always hope that the particle can cover the whole search space as early as possible, and in the latter part, it is expected that the particle search can be limited to a smaller range to accelerate particle convergence. In the 1990 s, Shi Y and Eberhant R C (1998) proved through experiments that the inertia weight (w) is a decreasing function with time, which is expressed as follows:

$$w(\mathrm{t}) = w_{start} - \frac{w_{start} - w_{end}}{t_{max}} \times t \tag{4}$$

where w_{start} and w_{end} are the initial and final values of inertia weight, which are the maximum and minimum values of the inertia weight, respectively, t is the current number of iterations, and t_{max} is the maximum allowed number of iterations.

Although the inertia weight function with decreasing linearity with time can better balance the global search ability and the local search ability, once the particle cannot select a better optimal position in the early stage of the search, then the local search ability is enhanced, and eventually it is easy to achieve the local optimal situation. To this end, scholars have done a series of research on inertia weight and then proposed a variety of optimization algorithms based on inertia weight. Li Li (2010) also proposed a nolinear inertia weight optimization algorithm, which is expressed as follows:

$$w(\mathbf{t}) = w_{end} + (w_{start} - w_{end})\exp\left(-k \times \left(\frac{t}{t_{max}}\right)^2\right) \quad (5)$$

where w_{start} and w_{end} are the initial and final values of inertia weight, which are the maximum and minimum values of the inertia weight, respectively, t is the current number of iterations, t_{max} is the maximum allowed number of iterations, and k is the control factor, controlling the inertia weight with the change of the number of iterations t and the function value of the speed of change.

2.3.2 Learning factor

In the PSO algorithm, although the selected inertia weight can balance the global and local search abilities well, if the selection of the learning factor is not appropriate, it will make the particle fall into the local optimum situation. Kennedy and Eberhart (1995) pointed out that if the "cognitive" part is higher than the "social" part, it will lead to the excessive individual search of the particle in the search space so as to neglect its relationship with the whole. Otherwise, it will lead to premature particles toward the local optimal solution. Therefore, in order to make the two learning factors average, that is, the time that the particle can fly is only half of the total search space time, Kennedy and Eberhart set the value of the two learning factors to two. In the population-based optimization method, it is usually expected that individuals can search in the whole search space in order to avoid local optimization, and later hope to be able to converge globally to achieve global optimization. To this end, Asanga Ratnaweera and Saman K. Halgamuge (2004) mentioned that by changing the learning factor with time, the cognitive part and the social part are reduced. That is, during pre-search, particles will be mainly subject to their own historical information, while the late major will be subject to social information:

$$c_1 = c_{1s} + \frac{(c_{1e} - c_{1s})t}{t_{max}} \quad (6)$$

$$c_2 = c_{2s} + \frac{(c_{2e} - c_{2s})t}{t_{max}} \quad (7)$$

where c_{1s} and c_{2s} are the initial and c_{1e} and c_{2e} are final values, respectively, t is the current number of iterations, and t_{max} is the maximum allowed number of iterations. Asana Ratnaweera found that the

optimal range of learning factors c_1 and c_2 are [2.5, 0.5] and [0.5, 2.5], respectively, by a large number of simulation experiments.

3 DESIGN OF CODEBOOK BASED ON PARTICLE SWARM OPTIMIZATION

As the standard PSO algorithm can achieve better global search in the whole search space, it can be considered to be applied to vector quantization codebook design. At different iterations, each particle is able to search and cluster the training vectors in the training set in different ranges to obtain a new codebook and update the centroid of the new cell. If the termination condition is reached, the obtained optimal codebook is output. Otherwise, the searching and clustering are continued until the codebook with better performance is generated.

3.1 Encoding method

The image is divided into a number of image blocks without any overlap: the training vector and its corresponding dimension, k, and codebook size, N. Although the Particle Swarm Optimization (PSO) algorithm uses real numbers to encode vectors, however, because the objects manipulated in vector quantization are all vectors, the codebook design based on PSO is based on the clustering center coding method. That is to say, each particle is composed of a codebook, which has N code words. Because the position of the particle is a vector of N × K, the velocity should also be a vector of N × K, and the expressions of the two are as follows:

$$\mathbf{Y}_i = \begin{bmatrix} y_{i11} & y_{i12} & \cdots & y_{i1k} \\ y_{i21} & y_{i22} & \cdots & y_{i2k} \\ \vdots & \vdots & \cdots & \vdots \\ y_{iN1} & y_{iN2} & \cdots & y_{iNk} \end{bmatrix} \quad (8)$$

$$\mathbf{V}_i = \begin{bmatrix} v_{i11} & v_{i12} & \cdots & v_{i1k} \\ v_{i21} & v_{i22} & \cdots & v_{i2k} \\ \vdots & \vdots & \cdots & \vdots \\ v_{iN1} & v_{iN2} & \cdots & v_{iNk} \end{bmatrix} \quad (9)$$

where $\mathbf{Y}_i$ and $\mathbf{V}_i$ are the position and velocity of the i-th particle in the particle swarm, respectively.

3.2 Fitness function

In general, PSNR is used to evaluate the compression performance of PSNR, so PSNR is used as the fitness function in PSO. Suppose that the training vector set corresponding to an image is X = {$\mathbf{x}_1$, $\mathbf{x}_2$,

351

$\mathbf{x}_3,..., \mathbf{x}_M\}$, the training vector dimension is k, the codebook size is N, the codebook can be expressed as $Y = \{\mathbf{y}_1, \mathbf{y}_2, \mathbf{y}_3,..., \mathbf{y}_N\}$, then the PSNR corresponding to the image can be shown below:

$$PSNR = 10\lg\frac{255^2 \times k}{D} \qquad (10)$$

$$D = \frac{1}{M}\sum_{i=1}^{M}\min_{1\leq j\leq N}d(\mathbf{x}_i,\mathbf{y}_j) \qquad (11)$$

It can be seen from the above equation that the PSNR corresponding to the image is larger when the mean-squared error between the training vector and the nearest code word in all the codebooks is smaller. Otherwise, the peak signal-to-noise ratio is smaller.

3.3 Description of the codebook design algorithm based on particle swarm optimization

1. Initialization operation. An image is divided into M nonoverlapping image blocks, which is represented by $X = \{\mathbf{x}_1, \mathbf{x}_2, \mathbf{x}_3,..., \mathbf{x}_M\}$, where each image block is used to generate a corresponding training vector of dimension k. N training vectors randomly selected from X are combined into an initial codebook $Y = \{\mathbf{y}_1, \mathbf{y}_2, \mathbf{y}_3,..., \mathbf{y}_N\}$, that is, the starting position of the particle. After repeating L times of randomly chosen operations, a particle swarm $Y^i(t) = \{\mathbf{y}^i_1(t), \mathbf{y}^i_2(t), \mathbf{y}^i_3(t),..., \mathbf{y}^i_N(t)\}$ of size L is generated, and i = 1, 2,...,L. The initial velocity, individual extremum (**P**), and global extrema (**P**$_g$) corresponding to the particle are set to be $[0]_{N\times K}$, and the number of iterations is set to t = 1.
2. The fitness function value. Calculate the PSNR of each particle and compare the calculated value with the individual extremum (**P**). If the calculated value is larger than **P**, update **P**, otherwise do not make any changes:

$$\mathbf{P}^i(t) = \begin{cases} \mathbf{P}^i(t-1) & PSNR(\mathbf{Y}^i(t)) < PSNR(\mathbf{P}^i(t-1)) \\ \mathbf{Y}^i(t) & PSNR(\mathbf{Y}^i(t)) \geq PSNR(\mathbf{P}^i(t-1)) \end{cases}$$
$$(12)$$

where $\mathbf{P}^i(t)$ is the extreme value of the i-th particle corresponding to the iteration number t.
3. Compare the PSNR of the L particles with the PSNR of the particle, which has the global extremum. If a particle with a peak signal-to-noise ratio larger than the global extrema appears in the L particles, the global extrema is updated; otherwise, no changes are made.

$$\mathbf{P}_g(t) = \arg\max(PSNR(\mathbf{Y}^i(t), i=1,2,\cdots,k)) \quad (13)$$

4. Update the particle speed and position using the following corresponding formulas:

$$V^i_j(t+1) = w \cdot \mathbf{V}^i_j(t) + c_1 \cdot r_1 \cdot (\mathbf{P}^i_j(t) - \mathbf{Y}^i_j(t))$$
$$+ c_2 \cdot r_2 \cdot (\mathbf{P}^i_{gj}(t) - \mathbf{Y}^i_j(t)) \qquad (14)$$

$$\mathbf{Y}^i_j(t+1) = \mathbf{Y}^i_j(t) + \mathbf{V}^i_j(t+1) \qquad (15)$$

where i is the i particle, j is the first dimension component j, j = 1,2,3,...,N, c_1 and c_2 are learning factors, r_1 and r_2 are uniformly distributed in the interval [0,1] random number, and w is the inertia weight.
5. When the algorithm iteration is completed or the global extreme value reaches the set value, the algorithm ends and outputs the optimal codebook. Otherwise, let t = t+1, then go to step (2).

4 IMPROVED CODEBOOK DESIGN ALGORITHM

4.1 Improve ideas

Zhang Xubing (2007) mentioned the problem of particle coherency operation and gave some conclusions. In this paper, it is pointed out that the particle swarm optimization problem usually refers to the fact that each component of each particle contains its own meanings, and their order in the particle is also constant. Thus, different particles in the particle group have a consistent internal structure. However, in the design of the codebook using the particle swarm optimization algorithm, since the initial codebook is randomly selected from the training vector set and the code words in the codebook do not have any association, there is no problem of consistency between particles in the particle group and between the particles. In the literature, all the code words in the initial codebook are arranged in ascending order on the basis of the mean of all the code words in the initial codebook, and hence the internal structure of each initial codebook is made uniform, and the diversity of the code words in different codebooks can be said to maintain the diversity of the particle swarm. However, this paper and the paper proposed by Ping Yi C (2013) all use the idea of particle-consistent operation to increase the probability of global convergence of particle swarm. In order to ensure a good initial position of the initial iteration, all the training vectors are sorted in ascending order on the basis of the mean of all the training vectors. Then, the training vector sets are divided evenly according to the number of code words, and the new codebook is obtained by using the nearest-neighbor partitioning principle. The updated codebook is the starting position of the particle.

Ratnaweera A (2004) mentioned that the cognitive part and the social part are reduced by changing the learning factor with time. That is, at the beginning of the search, the particles will be subject to its own historical information, and later mainly social information shall prevail. In order to make full use of particle swarm in the early stage and to converge the particles as soon as possible, this paper takes the intersection of inertia weight of linear decreasing mentioned by Shi Y (1998) and inertia weight of nonlinear decreasing mentioned by Li Li (2010) as the benchmark. When the inertia weight is less than or equal to the intersection point, the learning factors c_1 and c_2 are 2.5 and 1.5, respectively, and when the inertia weight is greater than the intersection point, the learning factors c_1 and c_2 are 1.5 and 2.5, respectively. As the nonlinear inertia weight function given in Li Li (2010) can better balance the global and local search capabilities of particles, it is used in this paper.

4.2 Simulation experiment and result analysis

In this paper, the standard Lena test image with the size of 256 is used as the test image to verify the performance of the proposed algorithm, and the experiment is divided into image blocks with M = 16384. The algorithms adopted in the experiment are the LBG algorithm, Fixed Inertia Weight (PSO-FIW), and the improved algorithm. The corresponding parameter setting of each algorithm is as follows:

1. LBG algorithm, the number of concentrated training vector M = 16384, t = 20;
2. PSO–FIW algorithm, the inertia weight w = 0.792, learning factor $c_1 = c_2 = 2$, L = 10, t = 20;
3. Improved algorithm, the inertia weight

$$w(t) = w_{end} + (w_{start} - w_{end})\exp\left(-3 \times \left(\frac{t}{t_{max}}\right)^2\right)$$

learning factor $c_1 = 0.5$ or 2.5, $c_2 = 2.5$ or 0.5, and $w_{start} = 0.9$, $w_{end} = 0.4$, L = 10, t = 20;

Table 1. Comparison of three algorithms.

Image	Algorithm	PSNR/dB			
		128	256	512	1024
Lena	LBG	30.3309	31.1591	31.8754	32.5606
	PSO-FIW	30.4226	31.1938	31.9560	32.6120
	Improved algorithm	30.6889	31.6460	32.5437	33.4597

Table 1 shows that when compression ratio is different, the three algorithms for standard Lena test image processing were used to get the average PSNR value. It can be seen from the table that the PSNR obtained by the proposed algorithm is higher than that obtained from the LBG algorithm and the PSO-FIW algorithm.

5 CONCLUSION

In this paper, the idea of consistency is used to obtain a good position in the early stage of particle search, which increases the probability of getting the global optimal value. The combination of learning factor and inertia weight can balance the global optimal and local so that the codebook obtained by the improved algorithm is better. The validity of this algorithm is also verified by a series of experiments and comparison with different algorithms.

REFERENCES

Delport V. & Koschorreck M. (1995). Genetic algorithm for codebook design in vector quantization. Eleltronics Letters, 31(2): 84–85.
Guo Changyou (2011). A particles swarm optimisation with adaptive inertia weight. Computer Applications and Software, 28(6), 289–292.
Guo Yanju, Chen Lei & Chen Guoying (2013). Codebook design algorithm for image vector quantization based on improved artificial bee colony. Journal of Computer Applications, 33(9): 2573–2576.
Kennedy J. & Eberhart R. (1995). Particle swarm optimization. IN Proc. IEEE Int. Conf. Neural Networks, 1942–1948.
Krishnamurthy A.K., Ahalt S.C. & Melton D.E., et al. (1990). Neural networks for vector quantization of speech and images. IEEE Journal on Selected Areas inCommunications, 8(8):1449–1457.
Li Li & Niu Ben. Particle Swarm Optimization Algorithm (2010). Beijing: Metallurgical Press, 38–44.
Li Xia, Luo Xuehui & ZhangJihong (2004). Codebook design for image vector quantization with ant colony optimization. Acta Electronica Sinica, 32(7), 1082–1085.
Linde Y., Buzo A. & Gray R. (1980). An algorithm for vector quantizer design. IEEE Transaction on Communication, 28(1): 84–95.
Ping Yi C., Tsai J.T. & Chou J.H., et al. (2013). Improved PSO-LBG to design VQ codebook. SICE Annual Conference (SICE), 2013 Proceedings of. IEEE, 876–879.
Ratnaweera A., Halgamuge S.K. & Watson H.C. (2004). Self-organizing hierarchical particle swarm optimizer with time-varying acceleration coefficients. IEEE Transactions on evolutionary computation, 8(3): 240–255.

Shi Y. & Eberhart R.C. (1999). Empirical study of particle swarm optimization. Evolutionary Computation, 1999. CEC 99. Proceedings of the 1999 Congress on. IEEE, 3.

Shi Y., Eberhart R. (1998). A modified particle swarm optimizer. Evolutionary Computation Proceedings, 1998. IEEE World Congress on Computational Intelligence., The 1998 IEEE International Conference on. IEEE, 69–73.

Sudeep D., Thepade & Vandana Mhaske (2013). New Clustering Algorithm for Vector Quantization using Haar sequence. IEEE 2013 International Conference on Information & Communcation Technologies (ICT), 1144–1149.

Sudeep D., Thepade & Vandana Mhaske (2015). New Clustering Algorithm for Vector Quantization using Hybrid Haar Slant Vector. IEEE 2015 International Conference on Computing Communication Control and Automation (ICCUBEA), 634–640.

Sudeep D., Thepade & Vandana Mhaske, et al. (2013). New Clustering Algorithm for Vector Quantization using Slant Transform. IEEE 2013 1st International Conference on Emerging Trends and Applications in Computer Science (ICETACS), 161–166.

Thepade D.S.D., Mhaske V. & Kurhade V. (2013). New Clustering Algorithm TCEVR for Vector Quantization Using Cosine Transform. In Fifth International Conference on Advances in Recent Technologies in Communication and Computing (ARTCom 2013), 187–194.

Zhang Xubing, Guan Zequn & Xu Jingzhong (2007). Codebook design of image vector quantization based on particle swarm optimization. Computer Applications, 27(12):3051–3054.

Electromechanical Control Technology and Transportation – Jia & Wu (Eds)
© *2017 Taylor & Francis Group, London, ISBN 978-1-138-06752-3*

New lossless compression method for BMP true color images

Z.Z. Ma & Y. Wan
School of Information Science and Engineering, Lanzhou University, Lanzhou, China

ABSTRACT: BMP is a common image file format of image processing under Windows environment. The standard introduces two gray-scale image compression methods, i.e., BI_RLE4 and BI_RLE8, but does not provide any corresponding compression algorithms for the color image, thus it limits the application of BMP color images in practice. In order to solve the problem of excessive occupation of storage space by BMP images, we propose a novel lossless compression algorithm for the 24 bit true color images in this paper. The key idea of our proposed algorithm is pretreating the pixel values in an image firstly, then the correlation image will be reduced by using least square method and the lossless compression of this image is realized by the entropy coding technique. Analysis and experimental results demonstrate that, in comparison with other common ZIP and PNG lossless compression methods, our proposed BMP method is simple, and can significantly improve the compression ratio for the nearly same processing time.

1 INTRODUCTION

BMP (Bitmap) is the standard image file format on Microsoft Windows operating system and it is a bitmap image file format. Typical color depths (the number of bits per pixel) of the BMP image include 1, 4, 8 and 24, which make the BMP image have abundant picture information (Song and Ye 2011), so the compression effect of BMP is not obvious. Therefore, BMP is limitedly used for the gray-scale image and the commonly used method is Run Length Encoding (RLE), which also has its own limitations. RLE can achieve ideal compression efficiency when the color information in the image has very high repetition. On the contrary, this compression method will only increase the data volume while any two adjacent pixels are not same. BMP never provides any compression method for the true color image to save the storage space. Therefore, designing a fast and efficient lossless compression method to extend the application range of the BMP file is of great interest.

In general case, lossy compression can obtain better compression ratio. However, in some practical applications (e.g., aerospace, telemetry, medical treatment, safety protection and picture archiving), recovering the original image needs no image information missed. Obviously, these applications call for the compression method lossless.

In this paper, we propose a novel lossless compression method. The correlation (Jun and Giarra 2016) of an image will be reduced by using the difference in adjacent pixel values and least square method, then the data after being decorrelated will be coded by Huffman. Simulation results demonstrate that,

compared to other two common lossless compression methods (i.e., ZIP and PNG), our proposed method can improve the compression ratio by 36% at average.

The rest of this paper is organized as follows. Section 2 reviews the BMP. Section 3 presents our proposed lossless compression method. Section 4 gives the simulation results.

2 BMP FILE FORMAT

BMP is one of commonly used image file formats on Windows operating system (Murad and Banerjee 2013). Most of the image processing software under the Windows environment supports this format. Internal structures of the BMP file are shown in Table 1.

2.1 *BITMAPFILEHEADER*

The length of BITMAPFILEHEADER is fixed to 14 bytes. The header mainly consists of identification code, size of BMP file and offset byte number from file header to bitmap data.

Table 1. Bitmap file structure.

BITMAPFILEHEADER
BITMAPINFOHEADER RGBQUAD IMAGEDATA

2.2 BITMAPINFOHEADER

The length of **BITMAPINFOHEADER** is fixed to 40 bytes. The bitmap information header mainly records the length of this structure, the width and height of image, the bit number of each pixel and the compressed format.

2.3 RGBQUAD

Each color in RGBQUAD is denoted by a data structure, which is 4-byte and records the relative intensities of R, G and B (red, green and blue). One integer (R, G, B) can describe a pixel of the bitmap, where R, G and B can be any integer between 0 and 255. The total number of the recoded colors in RGBQUAD depends on biBitCount in BITMAP-INFOHEADER (Zu 2000.). E.g., biBitCount = 4 and the total number of the colors in RGB-QUAD is 16, then the length of the RGBQUAD is $2^{\wedge biBitCount} \times 4$ bytes. It is worth mentioned that the true color bitmap has no RGBQUAD.

2.4 IMAGEDATA

BMP stores the pixel data by starting from the lower left corner, and row by row from below to above. Before stored into the IMAGEDATA, byte number in each row of the uncompressed image date must be the multiples of 4 (padding zero if necessary) (Liu 2011.). Likewise, the byte number of every row in the compressed image must be the multiples of 2 (padding zero if necessary). In gray-scale image, the image data is the index in RGBQUAD of this pixel. But for true color bitmap, the image data is described by (R, G, B). For binary image, 1 bit can describe the color of one pixel ("0" and "1" represent "black" and "white", respectively), so 1 byte can describe 8 pixels. But for the true color image, one pixel needs 3-byte storage space.

3 PROPOSED LOSSLESS COMPRESSION METHOD

3.1 *Principe of lossless compression*

There exists strong correlation (which can lead to redundancy) between the image pixels, rows or columns, so we can apply compression processing to the image. Compression schemes can be broadly classified into two categories: lossy compression and lossless compression (Li and Liang 2016). Lossy compression permits reconstruction only of an approximation for the original image, though such method aims to improve the compression ratio. By contrast, lossless compression is a class of data compression algorithms that requires the

original data to be perfectly recovered from the compressed data.

According to the source encoding theory in information theory (Cao and Zhang 2009.), we denote the compression efficiency Z by

$$Z = \frac{H}{L} \times 100\% \qquad (1)$$

where, H is the source entropy, L is the average code length and no more than H. Designing the encoding algorithm aims to let L→H and then achieves optimum coding (Shen and Wei 1999). Entropy encoding is a lossless data compression scheme that is independent of the specific characteristics of the medium. The more frequently utilized entropy coding algorithms include RLE, Huffman coding, arithmetic coding and LZW coding, all of which are based on the statistics of images.

3.2 *Proposed method*

Commonly, the coding efficiency is very low while performing entropy coding on the image data with strong correlation. The basic theoretical framework of image lossless compression coding is predictive coding method, common lossless compression methods are all based on these ideas (Wu 1997). Such framework divides the compression process into two parts, i.e., decorrelation and coding. Decorrelation is used to wipe off the redundant information in the image as much as possible by utilizing correlation within the image data. Accordingly, coding is performing lossless compression by choosing suitable entropy coding according to the characteristic with the image data after being decorrelated.

This paper realized images decorrelation by pretreatment on the difference in adjacent pixel values and minimizing the color value in one pixel by using least square method. Then, the Huffman coding will be used to code the date after being decorrelated.

3.2.1 *Decorrelation*

Each pixel in the BMP true color image is made up of RGB (Red, Green and Blue) components and each component need a byte to present its color information. Now we select *Lena* image to histogram statistics. In common circumstance, the original image histogram is shown in Figure 1. It can be seen that the statistical distribution is very chaotic and smaller values of the probability. Figure 2 and Figure 3 show the histograms after performing different correlation treatment on these five images.

In these figures, the horizontal axis is the probability of the data values' emergency after treatment, while vertical axis is the pixel values after treatment.

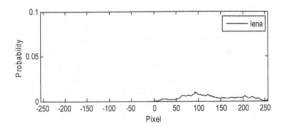

Figure 1. Statistical result of the pixel value.

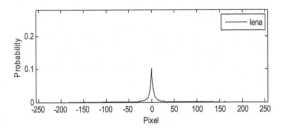

Figure 2. Statistical result of neighboring pixels difference method.

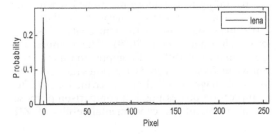

Figure 3. Statistical result of same pixel difference method.

In order to reduce the correlation of the neighboring pixels, we accept the treatment method of calculating the difference between adjacent pixel values and obtain the results shown in Figure 2. It can be seen that the data values after being treated are mainly in the range of −50 and 50, with a smaller fluctuation, thus a regulation have emerged. Therefore, we can use 5 bits to denote the pixels between −50 and 50, and 8 bits to denote other pixels. Besides, we use one extra bit to distinguish positive and negative pixels in Figure 1. It is obvious that the saved storage space with above-mentioned method is limited.

By separating RGB components of the true color image we find that G (green) component can more detailedly reflect some information (e.g., contour). To reduce the necessary coding bit number, on the basis of previous processing, we can calculate the color value of R and B by imposing least square method on current color values of G and R with B and obtain the data distribution shown in Figure 3. The data shown in this figure converge on narrow low frequency range, but with a high probability.

According to character shown in the true color imaging histogram, to a single pixel, we have following treatment formulas.

$$G_i' = G_i - G_{i-1}$$
$$R_i' = G_i' - \alpha(R_i - R_{i-1}) - a \quad (2)$$
$$B_i' = G_i' - \beta(B_i - B_{i-1}) - b$$

where (G_i, R_i, B_i) denotes the current pixel, $(G_{i-1}, R_{i-1}, B_{i-1})$ denotes the previous pixel, (G_i', R_i', B_i') is the pixel obtained by processing (G_i, R_i, B_i), a and b are constants. In order to minimize the value of R_i' and B_i', the coefficients α and β are introduced, which can be calculated by using least square method.

$$\sum_{i=1}^{n} |R_i'|^2 = \sum_{i=1}^{n} \left[G_i' - \alpha(R_i - R_{i-1}) - a \right]^2$$
$$\sum_{i=1}^{n} |B_i'|^2 = \sum_{i=1}^{n} \left[G_i' - \beta(B_i - B_{i-1}) - b \right]^2 \quad (3)$$

These four coefficients in forlmula 3 can be calculated according to the data shown in these images. It can be seen that the values of α and β approximate closely to 1, whereas a and b is smaller than 10^{-3}. To realize imaging lossless compression more conveniently, the approximate values of α and β are set to 1 and the approximate values of a and b are set to 0.

If the absolute value of G_i' is larger than 3, this pixel keeps invariant and is set to (G_i, R_i, B_i). If the pixels less than 4 are denoted by 4 bits: the first bit denotes the decoding control bit, the second one is used for the sign (positive or negative), remaining 2 bits can be used to denote a value less than 4. If the pixels are no less than 4, we use 9 bits to denote these pixels: the first bit still denotes the decoding control bits, remaining 8 bits denote the value of such pixel. To save storage space, we store each 0–1 sequence by use its own relevant decimal digit between 0 and 255. The imaging histogram obtained by using treatment method mentioned above is shown in Figure 3. It can be seen that most pixels are very low, and the probability is significantly different. So we use lossless method by choosing Huffman coding to compress the decorrelated data.

3.2.2 Huffman encoding

Huffman coding is a lossless method and one example of well-known Variable Length Coding (VLC) strategies. This coding method can perfectly recover original image from the compressed data. Huffman coding was developed by David A. Huffman in 1952, and is one of entropy encoding (Ji 2004). The basic idea of Huffman coding is that within a group of source nodes with different possibilities, nodes with higher possibility are assigned with shorter code words, and nodes with lower possibility are assigned with longer code words. Since possibility of each sign within different source data remains essentially unchanged, a table of Huffman coding can be used for data coding.

In the practical application, compression speed and reconstruction speed of Huffman coding are unsatisfactory, but because of its simplicity and efficiency, Huffman coding has still been widely used in the field of data compression. The concrete operating procedures of Huffman coding can be seen in (He 2008).

3.3 Algorithm description of proposed method

Usually the neighboring pixels/rows have strong correlation, the original pixels can be limited into a certain range by taking minus between neighboring pixels and same pixel. Our proposed method is shown as follows:

1. For a given pixel, exchanging this pixel value with the calculated result obtained by taking this pixel minus previous pixel;
2. Calculate the values of R_i' and B_i' by using least square method;
3. If $|R_i'| < 3$, $|G_i'| < 3$ or $|B_i'| < 3$, exchanging the value of (R_i', G_i', B_i') with that of original (R_i, G_i, B_i);
4. Preprocess data for the convenience of coding: the processed pixels less than 4 are denoted by 4 bits while other pixels by 9 bits;
5. Perform Huffman encoding.

4 SIMULATION RESULTS AND ANALYSIS

In this section, we present simulation and compare the compression ratio performance of our method to other two lossless compression methods (i.e., PNG and ZIP).

PNG is an image file format and simultaneously a bitmap image file format. This compression format is an LZ77 derivative algorithm, and can achieve higher compression ratio and perfect reconstruction without any data loss (Sun 2004), which mainly benefits from the used algorithm that can marker the repeated data. To minimize the storage space, PNG usually utilizes different treatment methods for the images with various characteristics.

ZIP is a compression file format that supports the lossless data compression. ZIP file format permits many compression algorithms, of which DEFLATE are the most common one. This format was originally created in 1989 by Phil Katz (Guo 2015), and was first used in PKWARE, Inc.'s PKZIP utility, as a replacement for the previous ARC compression format by Thom Henderson. ZIP is widely used in the fields of compression program and data application. Each byte must be recovered accurately to ensure the integrity and reliability of the information while decompressing.

In order to compare the compression performance of different compression algorithms, we choose data compression ratio as the metric of data compression quality. In the field of computer science, data compression ratio is usually used to measure the decreasing amount of the data compression algorithm. Then we have:

$$R = \frac{U}{C} \qquad (4)$$

where, R denotes data compression ratio, U and C are the sizes of original image and compressed image, respectively.

According to the above-mentioned scheme, we use 15 true color testing images from the USC-SIPI image database of the University of Southern California (Wang 2009). The compression performance of different compression methods is shown in Table 2, where Size denotes the size of the original images and BMP is the saving format after compression processing. For comparison purpose, we also present the experimental results of ZIP

Table 2. Compression ratio for different compression methods.

Image(.bmp)	Resolution	PNG	ZIP	Proposed
Airplane	512×512	1.582	1.585	1.861
Bridge	1280×713	1.579	1.579	1.958
Lena	512×512	1.198	1.198	1.536
Leaf	1018×762	1.269	1.265	1.472
Flower	512×480	1.136	1.135	2.151
Tower	800×600	1.797	1.786	2.249
Water	1024×747	1.115	1.113	1.532
Barbara	720×576	1.145	1.143	1.467
Boats color	787×576	1.306	1.305	1.792
Fruits	512×480	1.133	1.131	1.525
Yacht	512×480	1.157	1.157	1.487
Pepper	512×512	1.258	1.255	1.278
Pens	512×480	1.109	1.111	1.44
Girl	720×576	1.276	1.272	1.573
4.1.04	256×256	1.176	1.169	1.391

and PNG compression methods in Table 2, where PNG format is obtained by transferring the original BMP file format with ImageMagick software. ImageMagick is free software that can be used to create, edit and compose image. ImageMagick also can read, transfer and write different format images. ZIP format is achieved via the ZIP compression software.

By comparing the compression ratio of different methods in Table 2, it is clear that, compared to ZIP and PNG methods, our proposed method can achieve better lossless compression performance improvement for the BMP true color images, i.e., by processing 15 true color images and then saving, the necessary storage space corresponding to our method is significantly less than that of ZIP and PNG methods. By making statistical analysis on the data in Table 2, we conclude that: compared with ZIP method, the compression ratio of our method increases by 36.7% at average; likewise, compared to PNG method, the compression ratios of our method increases by 36.5% at average.

5 CONCLUSION

BMP format file compression is usually limited on the gray-scale images, and Run Length Encoding (RLE) has its own limitations. In general, processing true color images with RLE is very difficult to obtain good compression ratio. In this paper, we reduce the correlation in image by using least square method to calculate the difference between adjacent pixel values and further utilize Huffman coding (one of entropy coding) method to encode the de correlated data. Compared to other compression methods, our proposed method is simple and easy to implement. What's more, it can ensure the entire process lossless with help of the parameters obtained by using least square method. Experimental results for BMP color images show that, compared to other two common lossless compression methods (i.e., ZIP and PNG), our

proposed method can achieve better compression performance.

REFERENCES

Cao, X.H. & Zhang, Z.C. 2009. Information theory and coding. Beijing: TSINGHUA UNIVERSITY PRESS 2009:85–111.

Guo, R.Y. 2015. Analysis and application of compressed file format. China: Beijing.

He, Z.Q. 2008. New data compression algorithm based on Huffman coding. Science Technology and Engineering 8(16):45 31–45 32.

Ji, J.W. 2004. Research and application of the image lossless compression technology. Beijing University of Chemical Technology 8.

Jun, B.H. & Giarra. 2016. Nanoparticle flow velocimetry with image phase correlation for confocal laser scanning microscopy. MEASUREMENT SCIENCE AND TECHNOLOGY 27(10).

Li, Y.M. & Liang, Y. 2016. Temporal Lossless and Lossy Compression in Wireless Sensor Networks. ACM TRANSACTIONS ON SENSOR NETWORKS 12(4).

Liu, H.M. 2011. Conversion of original image data to BMP bitmap document. Techniques of Automation & Applications 30(7):27.

Murad, F.M. & Banerjee. 2013. Image management systems. ASGE Technology Committee Gastrointestinal endoscopy 79(1):15–22.

Shen, L.S. & Wei, H. 1999. Research of image lossless compression. Journal of Data Acquisition & Processing 14(4):486.

Song, Y.W. & Ye. J.F. 2011. Analysis of BMP format file and display algorithm. Modern Electronics Technique 34(20):5.

Sun, H. 2004. Portable Network Graphics. Publishing & Printing (01):19.

Wang, X.Y. 2009. Research on fractal image compression methods. Dalian: Dalian University of Technology.

Wu, X. 1997. Lossless compression of continuous-tone images via context selection, quantization, and modeling. Image Processing, IEEE Transactions on 6(5): 656–664.

Zu, B.F. 2000. Compression and decompression methods for BMP file. Journal of Fushun Petroleum Institute 20(1):67–69.

Electromechanical Control Technology and Transportation – Jia & Wu (Eds)
© 2017 Taylor & Francis Group, London, ISBN 978-1-138-06752-3

Research and implementation of an image compression scheme based on the BTC algorithm

Yu-qi Zhang
Graphics, Images and Multimedia Labs, Chongqing University of Posts and Telecommunications, Chongqing, China

Ying Qian
College of Software Engineering, Chongqing University of Posts and Telecommunications, Chongqing, China

ABSTRACT: Block Truncation Coding (BTC) is an efficient lossy image compression technique. Its compression quality is good but the compression ratio is low, only 2 bits/pixel. To further improve the image compression ratio, a hybrid compression scheme of BTC–DPCM is proposed, which takes the types of image blocks into account and uses different coding algorithms to encode them according to the block type. Image block also has a certain correlation while taking into account the similarity between the image blocks. Experimental results show that a high compression ratio and a good compression quality can be obtained for many types of images. The advantage is obvious especially for smooth images. For images with complex contents, the compression ratio should be increased as much as possible to ensure image quality.

1 INTRODUCTION

With the rapid development of the social network, a huge amount of digital information data needs to be transmitted over the network, especially images and videos. In order to ensure data transmission efficiency and save network bandwidth, the data are compressed and then transmitted. In the past three decades, image compression technology has been continuously improving. The current image compression methods are the spatial domain compression methods, such as the Vector Quantization technology (VQ) proposed by Gray, R. (1984), the BTC algorithm, and the transform domain compression methods, such as the Discrete Cosine Transform (DCT) and the Discrete Wavelet Transform (DWT) compression algorithm.

The BTC image compression algorithm was proposed by Delp and Mitchell (1979), whose basic theory is that the original image is divided into nonoverlapping blocks, with each block quantized into two representative gray values and bitmaps so as to get a compression rate of 2 bits/pixel. The BTC algorithm is simple and easy to implement. It can realize fast encoding and decoding and result in good compression image quality. However, the traditional BTC algorithm has a high pixel bit rate, which greatly limits the BTC algorithm in the current image compression applications. Therefore, in the following research, people focus on how to reduce the image's pixel bit rate.

Lema, M. (1984) proposed the AMBTC algorithm, which could further reduce the encoding time and get better reconstruction image quality at 2 bits/pixel. Yuen, H. (1996) proposed the BTC algorithm on the basis of the human visual model, which uses sixteen 4×4 visual pattern matrices to replace all bitmap matrix and encodes the number of visual pattern matrix. The method not only increased the compression ratio significantly, but also increased the error of reconstruction image. Hu, Y. C. (2003) takes into account the similarity between the image blocks. It searches the blocks around the 12 image blocks that it is similar to the current block. The reconstructed image block is the same as the adjacent block reconstructed image block. Masoudnia, A. (2003) proposed a hybrid coding scheme based on BTC-DWT. It uses BTC for boundary area as it can greatly reduce the distortion of constructed image. Domnic, S. (2006) proposed a scheme that uses Variable Length Coding (VLC) and predictive coding to obtain a compression rate of up to 1.5 bits/pixel. Belgassem, F. (2008) used the interpolation method in the image block bitmap, which only transmits 8 bits of 16 bits, and the other 8 bits are obtained by interpolating. The method obtains a compression rate of 1.5 bits/pixel.

In this paper, Section 2 reviews the BTC algorithm. Section 3 gives an improved algorithm. Section 4 presents the performance evaluation of the proposed algorithm. Section 5 concludes the paper and describes the scope of future work.

2 THE BTC ALGORITHM

Let a digital image be divided into blocks of $n \times n$ pixels. If we set $m = n^2$, $x_1, x_2, ..., x_m$ representing the pixels in the block, its first moment, second moment, and variance are as follows:

$$\bar{x} = \frac{1}{m}\sum_{i=1}^{m} x_i, \quad \bar{x}^2 = \frac{1}{m}\sum_{i=1}^{m} x_i^2, \quad \bar{\sigma}^2 = \bar{x}^2 - (\bar{x})^2 \tag{1}$$

Using the one-bit quantizer, let $\bar{x}$ be the threshold, and two output levels are a and b, such that:

$$\begin{cases} x_i \geq \bar{x} \rightarrow output = b \ \& \ bitmap_{x_i} = 1 \\ x_i < \bar{x} \rightarrow output = a \ \& \ bitmap_{x_i} = 0 \end{cases} \tag{2}$$

where $i = 1,2,3,...m$, a and b denote low and high pixel values, respectively, and the bitmap is a binary image block composed of 0 and 1. The $bitmap_{xi}$ is an element in the bitmap. The central idea of the BTC algorithm is that when the pixel value is replaced by the low level and the high level, the first-order moment and the second-order moment can be maintained. Thus, we can get the value of a and b according to formula (3):

$$\begin{cases} m\bar{x} = (m-q)a + qb \\ m\bar{x}^2 = (m-q)a^2 + qb^2 \end{cases}$$

$$a = \bar{x} - \bar{\sigma}\sqrt{\frac{q}{m-q}} \tag{3}$$

$$b = \bar{x} + \bar{\sigma}\sqrt{\frac{m-q}{q}}$$

where q denotes the number of pixels larger than the threshold in the block. Each image block is compressed into a, b, and an $n \times n$ bitmap. Decoding bitmap in the 1 replaced by b, 0 replaced by a. a, b are respectively expressed in 8 bits, bitmap needs n^2 bits. When the block size is 4×4, we can get a compression rate of 2 bits/pixel.

3 ADAPTIVE COMPRESSION SCHEME BASED ON BTC

3.1 *Image block classification*

There are a large number of uniform parts in an image, such as the background area. And there are also many complex texture areaa, such as a wallpaper. Different image blocks using different encoding methods are necessary. Here, we divide the image block into uniform block, common block, and complex block. The uniform block refers to

the block with little difference between pixels, and the image block is encoded by BTC. If the reconstructed pixel value and original pixel value have little difference, it is a common block, otherwise it is a complex block. Block classification conditions are as follows:

1. Uniform block

$$x_{max} - x_{min} \leq Th1 \tag{4}$$

where x_{max} and x_{min} represent the maximum and minimum values in a block and Th1 is a condition threshold.

2. Common block
The image blocks are coded by BTC to obtain low-level a as well as high-level b, where Th2 is a condition threshold.

$$\begin{cases} x_{max} - b \leq Th2 \\ a - x_{min} \leq Th2 \end{cases} \tag{5}$$

3. Complex block
A complex block does not satisfy formulas (4) and (5).

3.2 *Proposed scheme*

Step1: Let a digital image be divided into blocks of 4×4 pixels.

Step2: To determine the type of an image block, that is to check whether it is a uniform block, reconstruct the block by means of the block. If not, use BTC to encode the current block, then according to (5) to determine whether it is a common block. If it is a common block, the encoded data are transmitted directly. If it is a complex block, it is encoded by the DPCM algorithm proposed by Goodman, D. (1980), using linear prediction, uniform quantization, and quantization step, and the quantization value is reserved only for 2 bits.

Step3: The receiver can decode the data by the number of bits. If it receives 8 bits, use the mean to reconstruct the block. If it receives 32 bits, it is a common block, and use the two levels and bitmap for decoding. If it receives 40 bits, it is a complex block that is decoded by the predictive coding.

Irrelevant block	Above block $X_{i1,j}$
Left block $X_{i,j-1}$	Current block X

Figure 1.   Position relation.

3.3 Coding scheme considering interblock correlation

There are a lot of correlations between the image blocks, and the image can be further compressed. We only considered the similarity between the current block and the image blocks in Figure 1. The special case is the first row and the first column.

$$ratio = \frac{(a_1 - a_2) + (b_1 - b_2)*\left(2 + \frac{n*n}{8}\right) + (c_1 - c_2)*\left(1 + 2*\frac{n*n}{8}\right) + a_2 + b_2 + c_2}{M \times N} \tag{9}$$

In the first row, we only considered the similarity between the image block and its left block. In the first column, we only considered the similarity between the image block and its above block. The similarity conditions of the image blocks are as follows:

1. The similarity condition of the uniform block

$$|\bar{x}_i - \bar{x}_j| \leq Th3 \tag{6}$$

where $\bar{x}_i$ and $\bar{x}_j$ are the mean values of the current block and the adjacent block, respectively.

2. The similarity condition of the common block

$$\begin{cases} |a_i - a_j| \leq Th4 \\ |b_i - b_j| \leq Th4 \\ bitmap_i \oplus bitmap_j \leq Th5 \end{cases} \tag{7}$$

where a_i and a_j are the low levels, b_i and b_j are the high levels, $bitmap_i$ and $bitmap_j$ are the bitmaps of the two adjacent blocks, and Th4 and Th5 are set as boundary values.

3. The similarity condition of the complex block

Similar conditions for complex blocks need to meet the texture of similar blocks. The texture can be represented by a bitmap. It also needs to meet the condition that the average pixel difference of adjacent blocks is less than Th6.

$$\begin{cases} bitmap_i \oplus bitmap_j \leq Th5 \\ \frac{1}{m}\sum_{k=1}^{m} |x_{i(k)} - x_{j(k)}| \leq Th6 \end{cases} \tag{8}$$

4. Decoding method

If the current block meets the similar conditions, the one-bit identifier is transmitted to the decoder, and the decoder uses the before block's reconstructed block as the current reconstructed block.

4 EXPERIMENTAL EVALUATION

4.1 Experimental evaluation index

1. Compression ratio

We calculate the compression ratio according to the number of each type of block $T1 = [a1\ b1\ c1]$ and the number of similar block $T2 = [a2\ b2\ c2]$.

where M, N denote the original image size and n is the size of the block.

2. Compression error

The Mean Square Error (MSE) is used to measure the difference between the compressed image and the original image:

$$MSE = \frac{1}{m}\sum_{i=1}^{n}\sum_{j=1}^{n}(I_{i,j} - I_{i,j}')^2 \tag{10}$$

4.2 Experimental image set

Ten standard images with the size of 512×512, as shown in Figure 2, are used in the experiment.

4.3 Determination of boundary value of relevant conditions

First, we set the Th1, Th2, Th3, Th4, Th5, and Th6 initial values as shown in Table 1.

The standard Lena image of 512×512 was used for the experiment. The test was performed by changing only one of the values, and the values of the other five items in Table 1 were unchanged. For example, the influence of the change of Th1 on MSE and ratio was tested. Th1 = 5, 10, 15, 20, and so on, and Th2 = 30, Th3 = 2, Th4 = 5, Th5 = 4, Th6 = 5 to keep the values in Table 1 unchanged. The results are as follows:

In Figure 3, the MSE and ratio curves in Th1 intersect at Th1 = 30. The intersection can guarantee the compression quality and ensure the compression ratio, so we take Th1 = 30. In the curve of Th2, as the Th2 value increased, we can see that MSE decreased and ratio increased. It no longer changes obviously after Th2 = 70, so we take the Th2 = 70. In the curve of Th3, the MSE increases with the increasing of Th3, the ratio is also increased. The intersection is between 2 and 3. The ratio changes little, so the choice of MSE is small. We set Th3 = 2. In the curve of Th4, the two

Figure 2. Standard image set used in the experiment (in the order of the image name for the image1,image2,... image10).

Table 1. Conditional boundary initial values.

Th1	Th2	Th3	Th4	Th5	Th6
15	30	2	5	4	5

curves intersect at Th4 = 9, so we take Th4 = 9. In the curve of Th5, the two curves intersect at the position between 3 and 4, the MSE of 3 and 4 are only different by 2, and the ratio is increased. We take Th5 = 4. In the curve of Th6, the two curves intersect at position 15, so we take Th6 = 15. We get all the boundary values, as shown in Table 2.

4.4 Experimental results

The experiment uses the size of 4×4 to block. It compares with the mainstream image compression scheme JPEG's core algorithm DCT. DCT coefficient quantization table uses JPEG's standard quantization table. The results are shown in Table 3.

In Table 3, the BTC algorithm and the AMBTC algorithm have stable compression ratio and high reconstructed image quality, but their compression ratio is relatively low. The proposed method does not have a fixed compression ratio. For a smooth image, a large compression ratio can be achieved and the reconstructed image quality is better. The average compression ratio of this method is 9.3, and the MSE is 116.26. It compares with the traditional BTC compression ratio, which has been greatly improved, and the error is only slightly increased. Comparing with the current mainstream DCT compression algorithm, we can see that the proposed method can obtain a higher compression ratio. The error is only slightly increased, such as image 7. This compression ratio using this method is 13.2, and MSE is 49.95. The compression ratio of the DCT compression method is 12.8, and the MSE is 40.29. For the texture-rich image, the compression ratio of this method is slightly smaller than that of the DCT method, but the MSE value is much less, such

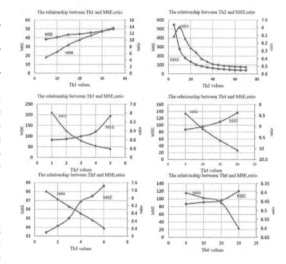

Figure 3. Influence of several boundary conditions on the compression performance.

Table 2. Finalized condition values.

Th1	Th2	Th3	Th4	Th5	Th6
30	70	2	9	4	15

as image 10. The compression ratio and MSE of the proposed method are 7.1 and 254.48, respectively. The MSEs of DCT (10 coefficient) method and DCT (15 coefficient) method are 634.63 and 507.39, respectively. This method is applied for 10 images, whose compression ratios are higher than that of the DCT (15 coefficient) method's compression ratio, and the mean MSE of this method is reduced by nearly 60 compared with the DCT (15 coefficient) method. Figure 4 shows image7 and image10 using the proposed method and the DCT method.

Table 3. Experimental results of several compression algorithms.

Methods	BTC		AMBTC		Proposed method		DCT (10 coefficient)		DCT (15 coefficient)	
Images	MSE	ratio	MSE	ratio	MSE	ratio	MSE	ratio	MSE	ratio
Image1	74.93	4	72.88	4	114.44	8.7	222.69	12.8	185.21	8.53
Image2	39.12	4	36.18	4	56.68	13.5	51.11	12.8	28.69	8.53
Image3	164.47	4	159.10	4	208.92	9.8	592.81	12.8	442.57	8.53
Image4	33.82	4	33.65	4	58.42	9.8	55.90	12.8	38.80	8.53
Image5	130.49	4	127.25	4	146.35	5.3	338.20	12.8	253.74	8.53
Image6	102.59	4	101.26	4	115.65	6.5	177.72	12.8	110.36	8.53
Image7	31.09	4	30.62	4	49.95	13.2	40.29	12.8	23.94	8.53
Image8	31.20	4	31.12	4	48.78	13.2	45.71	12.8	29.67	8.53
Image9	90.21	4	87.03	4	108.93	5.6	177.71	12.8	127.87	8.53
Image10	209.30	4	205.50	4	254.48	7.1	634.63	12.8	507.39	8.53
Average value	90.72	4	88.46	4	116.26	9.3	233.68	12.8	174.82	8.53

a.(original image)

b.(proposed method)

c.(DCT 10 coefficient)

d.(original image)

e.(proposed method)

f.(DCT 15 coefficient)

Figure 4. Comparison of experimental results.

5 CONCLUSION

In this paper, we proposed a hybrid encoding scheme based on BTC–DPCM. Considering the different contents of image blocks, we first classify the image blocks into uniform blocks, common blocks, and complex blocks according to the setting conditions. These three types of image blocks are encoded by different compression algorithms. The BTC algorithm is used to guarantee the compression ratio, the DPCM method is used to ensure the quality of the image with a complex region, while the similarity between image blocks is considered simultaneously. Therefore, this scheme can guarantee the quality of the image. According to the experimental results, this scheme has a higher compression ratio for smooth images compared with the DCT compression scheme. As to rich texture image, the compression ratio of our method is slightly lower than that of the DCT method, but the image quality is far better than that of the DCT compressed image.

REFERENCES

Belgassem, F., Rhoma, E., & Dziech, A. (2008, September). Performance evaluation of interpolative BTC image coding algorithms. In Signals and Electronic Systems, 2008. ICSES'08. International Conference on (pp. 189–192). IEEE.

Delp, E., & Mitchell, O. (1979). Image compression using block truncation coding. IEEE transactions on Communications, 27(9), 1335–1342.

Domnic, S. (2006). BTC image compression with variable length integer codes.

Goodman, D. (1980). Embedded DPCM for variable bit rate transmission. IEEE Transactions on Communications, 28(7), 1040–1046.

Gray, R. (1984). Vector quantization. IEEE Assp Magazine, 1(2), 4–29.

Hu, Y.C. (2003). Improved moment preserving block truncation coding for image compression. Electron. Lett, 39(19), 1377–1379.

Kamel, M., Sun, C.T., & Guan, L. (1991). Image compression by variable block truncation coding with optimal threshold. IEEE Transactions on Signal Processing, 39(1), 208–212.

Lema, M., & Mitchell, O. (1984). Absolute moment block truncation coding and its application to color images. IEEE Transactions on communications, 32(10), 1148–1157.

Masoudnia, A., Sarbazi-Azad, H., & Boussakta, S. (2003, December). A BTC-based technique for improving image compression. In Electronics, Circuits and Systems, 2003. ICECS 2003. Proceedings of the 2003 10th IEEE International Conference on (Vol. 1, pp. 108–111). IEEE.

Mathews, J., Nair, M.S., & Jo, L. (2013, March). Modified BTC algorithm for gray scale images using max-min quantizer. In Automation, Computing, Communication, Control and Compressed Sensing (iMac4 s), 2013 International Multi-Conference on (pp. 377–382). IEEE.

Pennebaker, W.B., & Mitchell, J.L. (1992). JPEG: Still image data compression standard. Springer Science & Business Media.

Yuen, H., & Li, C.K. (1996, May). Quadtree segmented two-dimensional predictive visual pattern BTC image coding. In Circuits and Systems, 1996. ISCAS'96., Connecting the World., 1996 IEEE International Symposium on (Vol. 2, pp. 680–683). IEEE.

Electromechanical Control Technology and Transportation – Jia & Wu (Eds)
© *2017 Taylor & Francis Group, London, ISBN 978-1-138-06752-3*

Research on a plan of making suburb leisure tourism suitable for characteristics of the western Sichuan forest district—taking Sichuan Xinjin Huayuan Town as an example

Pengcheng Xiang
Faculty of Construction Management and Real Estate, Chongqing University, China
Construction Economics and Management Research Center, Chongqing University, China

Tongyun Du
Construction Economics and Management Research Center, Chongqing University, China

Hengyu Gu
Faculty of Government Management, Peking University, China

ABSTRACT: Leisure tourism is developed in full swing in China at present, but the phenomenon of homogenization is widespread. The special landscape plot in western Sichuan forest district is selected in the paper. A suburb tourism development plan strategy suitable for characteristics in western Sichuan foreign disc is discussed through summarizing characteristics of western Sichuan forest district, actual plot and market situation. Suburb leisure tourism development in Sichuan Xinjin Huayuan Town is applied as an example for description, thereby providing reference for suburb leisure tourism development in western Sichuan forest disc and even suburb tourism development in the whole country.

1 INTRODUCTION

Suburb leisure tourism is developed vigorously in China in recent years, which mainly depends on people's ecological consciousness and leisure values. More and more people begin to think that tourism with leisure holiday is for relaxation. Meanwhile, long journey sight-seeing tourism is not suitable for the two-day weekend mode in China; therefore, suburb leisure tourism is characterized by suitable destination, beautiful environment and rich types. The novel self-service or half self-service tourism form with stronger ecological consciousness is widely popular. Development of suburb leisure tourism in China should be constructed and perfected currently. Meanwhile, certain characteristic and distinct themes are required for developing and constructing suburb leisure tourism, and repeating should be avoided. Novel development mode of suburb tourism can change the situation in western Sichuan forest disc. It is regarded as new extension of ecologic tourism connotation in China, which provides direction for studying suburb leisure tourism approach. The approach is suitable for regional characteristics in western Sichuan.

2 WESTERN SICHUAN FOREST DISC CHARACTERISTICS AND DEVELOPMENT STRATEGIES

2.1 *Ecological intelligence- nature combination in design*

Major layout of Chengdu Plain for forest disc settlement shows that western Sichuan forest disc is located in Sichuan Basin, which is surrounded by mountains. The whole Minjiang River system which runs through the region conforms to the principle of traditional settlement site 'on the back of mountain and in front of river, negative to Yin and hugging Yang'. Forest disc is a compound ecosystem, which is in line with the theory in landscape ecology 'patch- corridor- matrix' from the perspective of forest disc landscape unit. Farmland is regarded as the matrix. Forest land has typical functions of flood regulation, pollution purification, climate regulation, etc. from the perspective of the constituent elements of forest disc. Farmland plays the role of artificial wetland. Forest disc buildings are scattered in farmland, which are beneficial for forming good circulation of rural ecosystem (Jiang, 2009). Ecological intelligence in

Western Sichuan forest disc should be used in sub-urb leisure tourism fully. The relationship between people and nature as well as between design and site should be coordinated in design. Therefore, plan design can be combined with natural condition of the site organically. Recycling local materials and waste materials left in the site should be selected as far as possible. Original processing mode should be changed, and their functions and aesthetic appreciation can be innovated.

2.2 Humanities connotation-inheriting historical context

Ecological intelligence contained in Western Sichuan forest disc not only belongs to organic combination of natural environment and living environment, but also is regarded as coordinated coexistence of natural environment and local culture. Western Sichuan forest disc has culture connotation for unifying multiple elements. It is produced mainly because natural and humanity environment is diversified in Sichuan due to influence of history and geographical environment. Taoism culture and immigration culture are included on the basis of inheriting Sichuan culture. Culture features of immigration from different regions are accepted in western Sichuan since western Sichuan undergoes immigration for many times in the history. Immigration culture is produced. Ideas of pursuing nature respect, nature compliance, no constraints and compatibility are embodied in western Sichuan forest disc culture sufficiently. Residential environment with sustainable development, adapting to surrounding environment, should be constructed in suburb leisure tourism development. Artificial environment construction should not be considered from the perspective of coordinating material and environment only. The coordination between culture given by artificial environment and historical context of the site should be further considered. The formation of western Sichuan forest disc should be explored before planning. The association in the aspects of history, local nature, humanities, etc. should be developed. Rich humanities connotation thereof should be disclosed, which should be utilized in the plan design.

2.3 Landscape image- creation of landscape image

Lynch Kevin believes that landscape images belong to a group of images, which are interacted between the observer and the environment (He, 2009). Huang Yuanxiang (2013) believes that the rural landscape image is a personal characteristic in the process of cognizing of rural landscape. It is characterized by the formation of a personal characteristic of mental-map. Forest disc settle-

ment landscape in Western Sichuan has strong integrity, and the relationship between Gestalt psychology graphics and background is eradicated. A landscape patch with stable and harmonious aesthetic feeling is formed in combination of Western Sichuan forest disc, wherein the farmland is regarded as background, house is regarded as figure, and bamboo forest between them is regarded as hedge. A shocking earth landscape is formed from the perspective of landscape aesthetics. It is composed of point (forest disc settlement), line (river and road) and face (fields). Suburb leisure tourism planning has the core of controlling landscape image. Rural landscape image and local context features (Chen, 2011) of the project lot are summarized through analysis. The micro-earth landscape of 'point-line-face' is repaired in western Sichuan forest disc. Visual image of space landscape mode of 'house-forest-farmland' is protected. Natural and quite landscape images of countryside leisure are restored. Concretely speaking, external landscape (formed through landscape distribution and enclosure), landscape skeleton (formed by water road network), and multi-core public space in the settlement are protected.

3 CASE ANALYSIS-DEVELOPMENT STRATEGY OF SUBURB LEISURE TOURISM IN SICHUAN XINJIN HUAYUAN TOWN

3.1 Project plan background

Chengdu Construction Committee proposed to protect and utilize forest disc groups in 2012. Xinjin County was selected as one of pilot zones. Suburb leisure tourism project of Sichuan Xinjin Huayuan Town is located in the Northeast of Xinjin County. It is about 30 km away from Chengdu downtown area, and 15 km away from Xinjin County. The base occupies an area of about 16.3457 hectares primarily. The plot has better natural conditions with long and deep folk culture as well as stronger development value.

3.2 Ideas of village tourism development plan in Sichuan Xinjin Huayuan Town

The principles of 'protection priority and restoration first' should be followed to establish the development mode between forest disc and urban area. Site memory is awakened. Landscape of original forest disc is protected. A forest disc tourism destination with harmonious development between human beings and nature is constructed. Main body of current forest disc should be reserved. Huang forest disc is regarded as the core of forest disc group. Zhonghuang forest disc, Tang forest disc, Luo forest disc and Hou ancestral hall are set around it. Taxus chinensis industry and farming industry are inserted. An interactive mode of driving industry with forest disc and industry connection with forest disc is formed.

3.2.1 Strategy I: city heart—ecological pattern of tourism health preservation, and inclusion into city ecosystem

General layout of water system is determined on the basis of existing water system. Four major hydrological ecological infrastructure systems are constructed. First of all, rainwater conservation collection area is designed in the source. Distribution water system is shown in the form of ecological ditch. Organic form of the whole ecological ditch is obtained from 'capillaries'. Its purpose lies in further exchanging materials with external environment, and water systems can be further purified. A part of water systems can be collected and utilized as landscape water or domestic water. Other purified water is discharged into rivers. Wooden footway, hydrophilic platform, leisure corridor and other landscape facilities are set according to landscape creation demand. Small wetland landscape is set locally. Micro-climate is adjusted when surrounding environment is purified.

Ecological context of original site can be protected well through ecological system perfection and construction. Animal habitats are protected, local energy balance and micro-climate can be changed, and the site can be fully self-updated smoothly, as shown in Figure 1.

3.2.2 Strategy II: forest disc heritage- inheriting traditional culture and extending historical memory of the site

1. Application of regional cultural elements in design

Western Sichuan forest disc should be designed as a whole through extracting regional culture elements. Elements are reconstructed for different times and spaces to form western Sichuan forest disc image, which is rich in regional culture connotation and characteristics, as shown in Figure 2.

2. Vigorous development of customs and cultural activities in western Sichuan forest disc

Comprehensive cultural undertakings should be vigorously developed at the grass-roots level in the aspect of suburb leisure tourism development in response to the national call, thereby promoting culture creative industry development in Southwest China, and driving local people to voluntarily organize customs culture activities, enrich amateur life, and solve employment problems. Meanwhile, culture industry market and consumption can be also expanded through the activities. Reputation and influence of the whole region can be improved.

Figure 1. Schematic diagram of hydrological and ecological infrastructure system.

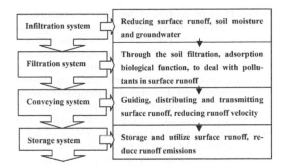

Figure 2. Wall landscape processing of Sichuan Xinjin Huayuan Town.

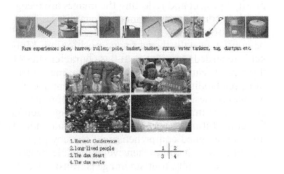

Figure 3. Art and cultural activities of 'farming customs in western Sichuan forest disc'.

Therefore, culture activities based on 'western Si-chuan forest disc farming practices' are planned. History context of western Sichuan forest disc is protected. Meanwhile, folk regional culture is promoted vigorously; connotation and charm of folk cultures are exhibited all over the world, thereby promoting folk culture exchanges and cooperation among different regions, as shown in Figure 3.

3.2.3 Strategy III: forest disc image—self-innovation of western Sichuan forest disc and construction of forest disc protection model

Current problems of western Sichuan forest disc style are analyzed and studied. Regional culture is studied. The results show that western Sichuan forest disc has prominent landscape characteristics.

It has image perception of 'landscape countryside'. The present situation of the regional landscape suffers from the core problem that there are not significant image characteristics in western Sichuan forest disc. Failure phenomena are produced in local regions especially, therefore local area in western Sichuan forest disc begins to disappear slowly, or the local area is eaten by surrounding area, and the image of western Sichuan forest disc is not prominent as a whole. Therefore, creation of landscape image of western Sichuan forest disc should be started with protection of forest disc firstly.

Protection of western Sichuan forest disc landscape image lies in protection of western Sichuan forest disc on one hand. It also lies in creating tourism atmosphere in western Sichuan forest disc on the other hand. Regional rural, leisure and other features become prominent. Image creation in western Sichuan forest disc has one of purposes to create attractive suburb leisure tourism theme image, thereby improving visibility and reputation, and creating active signboard of suburb tourism in western Sichuan forest disc.

Theme image should be determined firstly. Theme image covering area, core area and tourism product project and reflecting the images are recognized. Secondly, tourism image slogan is designed, and country images are distilled. Its design should be consistent with theme image positioning. Market is regarded as guidance. It has characteristics of distinctive features, clearness and easy memorization, readability, etc. Secondly, tourism image elements are designed. It is an internal approach to create rural image. Five major sense image factors for constituting image of western Sichuan forest disc are designed comprehensively, thereby providing tourists with comprehensive and whole-process experience. Finally, tourism image is transmitted. It is external approach to create rural image. A variety of image communication strategies are adopted for shaping and transmitting landscape image in western Sichuan forest disc externally.

3.2.4 *Strategy IV: children world- rural experience tourism product planning based on children world*

Suburb leisure tourism product feature lies in experience. Experience is divided into four major categories according to whether consumer participation belongs to active participation or passive participation as well as whether consumer association or environmental relationship belongs to integration into the situation or information absorption (Duan, 2004). Suburb leisure experience project is designed for Huayuan Town according to five steps of design experience proposed by Pine and Gail (1999) (setting theme, shaping impression, removing negative cue, cooperating to join souvenirs, and mobilizing stimulation by five senses) (Li, 1999). Experience product series are formed, including rural leisure health maintenance, rural cultural experience, ecological tourism, rural visit touring, agriculture specialty shopping, farming life experience, children financial education, fun activities, taxus chinensis industry, etc.

Different from ordinary children amusement and education places, Children science education eco-tourism base in the case is a characteristic in children farming and traditional culture education. Taxus chinensis health maintenance, organic fruits and vegetables, etc. are regarded as highlights. Children are liberated from traditional classroom. Rural landscape is relied for creating the first brand-new science education practice base with own distinct characteristics in western Sichuan forest disc, which integrates children science and education, ecological tourism, organic vegetables, etc, as shown in Figure 4.

3.2.5 *Strategy IV: internal security—unifying group strength and perfecting management mechanism*

Xinjin Huayuan Town Co., Ltd. is constructed. Office, reception department, management department, agricultural product sales department and

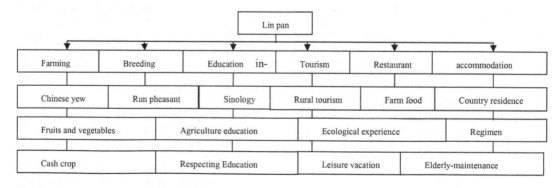

Figure 4.   Rural experience tourism product planning.

other institutions are established. The mode of 'government + company + peasant household' is adopted. The government is responsible for active organization and coordination, and formulation of preferential policies. Good social environment is created for suburb tourism development; Company is mainly responsible for various concrete affairs in the company, which acts as a bridge of communication and exchange between government and peasant households, thereby ensuring profit maximization among all participants, and coordinating interest distribution to all parties; all operators should exert individual motivation fully. Individual operation with farm characteristics is created, and Internet thinking is utilized for developing and promoting individual business fully.

The company is invested by developers and village collectives jointly, which is held according to shares. Public service facilities are invested and constructed by the company. The company constructs a purchased house site into food and beverage department or stores, etc. which can be rented by local farmers. Rent and other income are used as company management fee. It can be used for paying relevant taxes and fees, development bonus and other purposes.

4 CONCLUSION

Western Sichuan forest disc is a suburb leisure tourism mode with regional features. It has three core characteristics- ecological intelligence, cultural connotation and landscape image. It belongs to deep humanity connotation and intelligence achievement which are accumulated by western Sichuan people for utilizing land and adapting to landscape pattern for thousands of years rationally. Therefore, we should constantly consider, reflect, and deeply study how to protect culture connotation and ecological pattern in the project site well, and how to exert characteristics biological tourism resources in western Sichuan forest disc, drive sustainable development of suburb leisure tourism, and improve life level of local residents, etc. when the land in west Sichuan forest disc is developed. Protection of site ecological context and culture tone should be further considered when maximum development profits are guaranteed. Maximum of development operation economic benefits, social benefits and ecological benefits in the whole site can be treated from the perspectives of integrity, long-term and forward-looking at a brand-new level.

REFERENCES

Chen Yulu. Symbiotic harmony with the environment-discussion on ecological significance from the perspective of spatial characteristics. Research on Sichuan Building Science, 2011, 37 (2): 235–237.

Duan Zhaolin. Experience of agricultural and ecological resource conservation. Tang Jianjun. Urban and rural ecological environment construction. Beijing: China Environmental Science Press, 2004:229–233.

He Junhua. Cao Wei. Imaginable city- interpretation of 'City Image'. Chinese and Foreign Architecture, 2009, 10 (2):48–50.

Huang Yuanxiang. Wang Lina, Li Mingcai, Wang Shouhong, He Wei, Sun Dajiang. Influence of western Sichuan forest disc on construction of 'Garden City' landscape image in Chengdu. Chinese Gardening Abstract, 2013, 9 (3): 103–105.

Jiang Tao. Survival intelligence of western Sichuan forest disc and modern enlightenment thereof. Proceedings of Urban Development and Planning Conference, 2009, (8): 126–132.

Li Leilei. Tourism destination image planning: theory and practice. Guangzhou: Guangdong Tourism Publishing House, 1999:151.

A novel distributed control strategy for VSC-MTDC

Guoqiang Sun & Yuping Zheng
NARI Group Corporation (State Grid Electric Power Research Institute), Nanjing, China

Zhinong Wei, Yang Yuan & Zijie Lin
College of Energy and Electrical Engineering, Hohai University, Nanjing, China

ABSTRACT: Voltage Source Converter (VSC)-based Multiterminal HVDC (MTDC) has been considered to be a promising solution to integrate large Off-shore Wind Farms (OWFs) into the power grid. In this paper, we propose a distributed coordinated control strategy. In order to provide reference voltage for wind turbine generators, WF-side VSCs output a sinusoidal voltage with defined frequency and amplitude. On the contrary, a simplified state space equation of the whole MTDC is built, and a control strategy for on-shore VSCs is proposed using Interior Point Method (IPD) to solve the optimization problem. Communication is needed to transmit the output power of WFs and DC voltage of VSCs, and a reduced-order state observer is implemented to reduce the effect of time delay and packet losses. Finally, on the basis of a VSC-MTDC with four terminals, the simulation results on PSCAD/EMTDC have verified that the proposed control strategy can reduce power loss and DC voltage fluctuation.

1 INTRODUCTION

Because of the exhaustion of fossil fuel, Large-Scale Renewable Energy Generation (LSREG), particularly Offshore WFs (OWFs), play an increasingly important role in modern power systems. On the contrary, as the capacity of Voltage Source Converter (VSC) has significantly increased to hundreds of MW, using Mutiterminal HVDC (MTDC) based on VSC to integrate LSREG to power system is considered a promising substitution for thermal power plants (Zhang et al., 2010, Yu et al., 2015, Raza et al., 2015). However, it also presents a great challenge to design a control strategy for VSC-MTDC to meet the grid code as well as enhance the stability of the power system.

Numerous studies have analyzed coordinated control strategies for VSC-MTDC. Most of them can be categorized into the master–slave control and the droop control. In the master–slave control, one master VSC acts as a constant DC voltage source, while the others operate in a power control mode (Chen et al., 2006). Meanwhile, the whole system will collapse when the master VSC fails. Therefore, Rault et al. (2012) proposed a voltage margin method, in which master VSC can be shifted and the shift operation would inevitably cause DC voltage oscillations.

In the droop control, several VSCs with power–voltage (or current–voltage) droop character are responsible for regulating DC voltage instead of master VSC. Thus, the whole system can maintain stability while one or more droop VSCs could fail. However, small droop coefficients will lead to power oscillation, and large droop coefficients will cause DC voltage deviation (Rouzbehi et al., 2015, Chaudhuri and Chaudhuri, 2013). On the basis of the droop control, many modifications have been made. Abdelwahed and Elsaadany (2015) proposed an adaptive droop control strategy, in which the reference voltage and droop gain are changed dynamically by a central controller to fulfill power flow optimization. To prevent overvoltage failure, a coordinated control without communication considering voltage–current characteristics of VSCs in MTDC was proposed by Liang et al. (2012). Two types of droop control were implemented in WF-side VSCs and grid-side VSCs, respectively. Consequently, DC voltage was maintained by grid-side VSCs under normal condition, while when malfunction occurred, WF-side VSCs would take over control.

Employing VSC-HVDC with conventional controllers to integrate large-scale WFs negatively affects frequency stability of on-shore AC grid. Thus, some studies were proposed to use WFs to provide frequency support for onshore AC grid. Silva et al. (2012) deployed an additional frequency–Vdc droop control into on-shore VSCs to create a DC voltage deviation according to AC grid frequency, which would change the frequency of off-shore AC grid and further lead to adjust output powers of wind turbines equipped with deloading controllers (De Almeida and Lopes, 2007, Morren et al., 2006) to provide frequency support. On this basis, Zhu et al. (2013) took advantage of the energy stored on the DC line of VSC-HVDC to increase AC grid inertia. To endow MTDC with inertia mimicry capability, Zhang et al. (2016) proposed a hierarchical control strategy, including OPF program and inertia mimicry component, based on droop control. The small signal model analysis and parameter selection method were also proposed to ensure the whole system under-damped.

In order to enhance the system stability and reduce power losses, a distributed control strategy for VSC-MTDC integrating offshore wind farm is proposed in this paper. The wind farm-side VSCs function as AC voltage sources with defined frequency. At the same time, a power flow optimize element is implemented in grid-side VSCs. Furthermore, a four-terminal VSC-MTDC model was built in PSCAD/EMTD, and the simulated result demonstrated that the proposed control strategy can improve system performance and reduce power losses as well.

2 STATE SPACE MODEL

2.1 State space model of onshore VSC

In general, to provide reference sinusoidal voltage for wind turbine generator, the off-shore VSCs function as an AC voltage source with constant frequency and amplitude. On the contrary, the on-shore VSCs maintain the voltage of DC network and induce power into AC network.

This paper mainly focuses on the active power and DC voltage regulation. The state space model of controller on d-axis and output filter of on-shore VSCs are included. The off-shore VSCs function as a constant-voltage source and instantly injects the output power of the WF. As such, they are regarded as a measureable and uncontrollable power source in this paper.

In on-shore VSCs, the voltage loop and current loop are typically used. Neglecting the high-order harmonic component and the sampling time delay, the block diagram of the inner current loop and LC filter is shown in Figure 1.

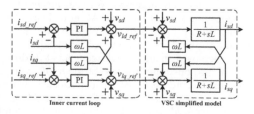

Figure 1. Simplified model of inner current loop and LC filter.

where L and R are the total inductance and resistance of the output filter and AC network, respectively, i_{sd_ref} and i_{sq_ref} are the reference currents on d-axis and q-axis, respectively, and v_{sd} and v_{sq} represent the d-component and q-component of the PCC voltage, respectively.

This block diagram can be decoupled into two simplified current models on d-axis and q-axis, as shown in Figure 2.

The transfer function of both simplified current loops model can be expressed as follows:

$$G_C(s) = \frac{i_{sd}(s)}{i_{sd\_ref}(s)} = \frac{i_{sq}(s)}{i_{sq\_ref}(s)}$$
$$= \frac{sK_{pc} + K_{ic}}{s^2L + s(R + K_{pc}) + K_{ic}} \quad (1)$$

where K_{pc} and K_{ic} represent the proportional and integral constants of the inner current loop, respectively.

Usually, the on-shore VSCs regulate DC voltage, adopting PI controller as outer voltage loop, as shown in Figure 3.

The transfer function of the on-shore VSCi including DC voltage controller and simplified current loop model is obtained as:

$$v_{dc,i}(s) = \frac{G_{1V}(s)\dfrac{1}{sC_{eq,i}}}{1 + G_{1V}(s)\dfrac{1}{sC_{eq,i}}} v_{dc\_ref,i}(s) + \frac{1}{sC_{eq,i}} i_{dc,i}(s)$$

$$(2)$$

$$G_{1V}(s) = \left(K_{pv} + \frac{K_{iv}}{s}\right)G_c(s)\frac{3U_s}{2V_{dc\_base}} \quad (3)$$

where K_{pv} and K_{iv} are the proportional and integral constants of the DC voltage controller, respectively, and $C_{eq,i}$ is the equivalent capacitor of the VSCi.

To implement the proposed control strategy into digital controller, the model should be discretized with sampling period T_s:

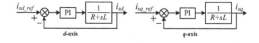

Figure 2. Simplified current model on d-axis and q-axis.

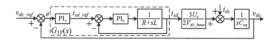

Figure 3. Simplified structure of VSC DC voltage control.

374

$$\begin{cases} x_i(k+1) = A_i x_i(k) + B_{w,i} i_{dc,i}(k) + B_{u,i} v_{dc\_ref,i}(k) \\ y_i(k) = C_i x_i(k) \end{cases} \tag{4}$$

where $x_i(k)$ is the state vector of VSCi. The details of A_i $B_{w,i}$ $B_{u,i}$ C_i can be seen in Appendix A1.

2.2 State space model of VSC-MTDC

According to the nodal voltage method, for a VSC-MTDC integrated with n_w wind farm-side VSC, n_g grid-side VSC, and n_l liaison buses, we have:

$$I = GE \tag{5}$$

where I is the current vector consisting of each DC line $I = [I_{w1}, \ldots, I_{wnw}, I_{g1}, \ldots, I_{gng}, I_{l1}, \ldots, I_{lnl}]$. The inject current of liaison buses should be zero $I_{l1}, \ldots, I_{lnl} = 0$. G represents the admittance matrix of DC network, and E represents the voltage vector of each DC bus $E = [E_{w1}, \ldots, E_{wnw}, E_{g1}, \ldots, E_{gng}, E_{l1}, \ldots, E_{lnl}]$.

The input power of wind farm-side VSCs is equal to the output power of the wind farm to which it is connected:

$$E_{wi} I_{wi} = P_{wi}, i = 1 \ldots n_w \tag{6}$$

According to (3) and (4), the equation for grid-side VSCi can be obtained as:

$$\begin{cases} x_i(k+1) = A_i x_i(k) + B_{u,i} v_{dc\_ref,i}(k) + B_{w,i} e_i GE \\ y_i(k) = C_i x_i(k), \quad i = 1, \ldots, n_g \end{cases} \tag{7}$$

The equations of VSC-MTDC including VSCs and DC network are:

$$\hat{x}(k+1) = A' \hat{x}(k) + B_{uo} \hat{u}(k) + B_{wo} E_e (G_1 \hat{w}(k) + G_2 C' \hat{x}(k)) \tag{8}$$

where

$$\hat{x}(k) = \begin{bmatrix} x_1(k) \\ \vdots \\ x_{n_g}(k) \end{bmatrix}, \hat{u}(k) = \begin{bmatrix} v_{dc\_ref,1}(k) \\ \vdots \\ v_{dc\_ref,n_g}(k) \end{bmatrix}, \hat{w}(k) = \begin{bmatrix} E_{w1} \\ \vdots \\ E_{wn_w} \end{bmatrix},$$

and the details of A' B_{uo} B_{wo} E_e G_1 G_2 C' can be seen in Appendix A2.

3 MODEL PREDICTIVE CONTROL

The power loss of the DC grid is:

$$P_{loss} = \sum_{i=1}^{n_w} P_{wi} + \sum_{j=1}^{n_g} P_{wj} = E^T GE \tag{9}$$

In order to reduce the power loss of DC networks, the optimization problem in the model predictive control is given as follows:

$$\min J = E^T(k+1) GE(k+1)$$
$$\text{s.t.} \begin{cases} \hat{x}(k+1) = A' \hat{x}(k) + B_{uo} \hat{u}(k) \\ \quad + B_{wo} E_e (G_1 \hat{w}(k) + G_2 C' \hat{x}(k)) \\ E_{wi}(k+1) I_{wi}(k+1) = P_{wi}(k), i = 1 \ldots n_w \\ I(k+1) = GE(k+1) \\ E_{min,i} \leq E_i \leq E_{max,i} \\ I_{min,i} \leq I_i \leq I_{max,i}, i = 1 \ldots n_w + n_g + n_l \end{cases} \tag{10}$$

The optimization problem can be regarded as a nonlinear programming problem, including nonlinear equality and inequality constraints, which can be formulated as:

$$\min_x J = f(x) = x^T Hx$$
$$\text{s.t.} \begin{cases} h_i(x) = 0, i = 1, \ldots, n_w \\ g(x) \leq 0 \end{cases} \tag{11}$$

Incorporate the inequality constraints to a logarithmic barrier function:

$$\min_z \varphi_\mu(z) = f(x) - \mu \sum_{i=1}^{4(n_w+n_g)} \ln s_i$$
$$\text{s.t.} \begin{cases} h_i(x) = 0, i = 1, \ldots, n_w \\ g(x) + s = 0 \end{cases} \tag{12}$$

where μ is the barrier parameter and s is a $4(n_w + n_g)$ dimension vector consisting of slack variables. $\mu > 0$ and $s > 0$.

The Lagrange function for (12) can be formulated as follows:

$$\mathcal{L}(z, \lambda; \mu) = f(x) - \mu \sum_{i=1}^{4(n_w+n_g)} \ln s_i + \lambda_h^T h(x) + \lambda_g^T (g(x) + s) \tag{13}$$

where λ_h and λ_g represent the Lagrange multiplier of equality constraints and inequality constraints, respectively, $\lambda_h > 0$, and $\lambda_g > 0$.

The first-order necessary optimality conditions for (13) can be obtained as:

$$\begin{bmatrix} \nabla_x \mathcal{L}(z, \lambda; \mu) \\ \nabla_s \mathcal{L}(z, \lambda; \mu) \end{bmatrix} = \begin{bmatrix} \nabla x^T Hx + A_h(x)^T \lambda_h + A_g(x)^T \lambda_g \\ S\Lambda_g e - \mu e \end{bmatrix}$$
$$= \begin{bmatrix} 0 \\ 0 \end{bmatrix} \tag{14}$$

where S and Λ_g are diagonal matrices, whose diagonal elements are s and λ_g, respectively, $A_h(x)$ and

$A_g(x)$ are the Jacobian matrices of $h(x)$ and $g(x)$, respectively, and e is a $4(n_w + n_g)$ dimension vector, all of whose values are 1.

Define $z = (x, s)$, and a new search direction can be generated using Newton's method:

$$z^+ = z + \alpha_z d_z, \quad \lambda^+ = \lambda + \alpha_\lambda d_\lambda \tag{15}$$

where

$$\begin{cases} \alpha_z \in (0, \max\{\alpha \in (0,1]: s + \alpha d_s \geq (1-\tau)s\}] \\ \alpha_\lambda \in (0, \max\{\alpha \in (0,1]: \lambda_g + \alpha d_g \geq (1-\tau)\lambda_s\}] \end{cases}, \text{ and}$$

$$\begin{bmatrix} d_z \\ d_\lambda \end{bmatrix} = - \begin{bmatrix} \nabla W(z, \lambda; \mu) & A(x)^T \\ A(x) & 0 \end{bmatrix}^{-1} \begin{bmatrix} \nabla_z \mathcal{L}(z, \lambda; \mu) \\ c(z) \end{bmatrix}.$$

Details of d_z, d_λ, $A(x)$, $c(z)$ and $W(z, \lambda; \mu)$ can be found in Appendix A3. The procedure of the interior point method is shown in Figure 4.

Figure 5 presents the detailed implementation of the proposed control strategy.

It should be noted that the output power of each wind farm and the DC voltage of each VSCs should be transferred to each on-shore VSCs through long-distance communication. In general, the wind speed of off-shore wind farm changes slowly, which leads to a limited output power fluctuation. Therefore, the time delay and packet loss during long-distance communication are acceptable.

On the contrary, the time constant of DC network is small, and the time delay and packet loss will cause a huge error on the calculation of objective function, if the DC voltage data are directly used. To solve this problem, a state observer is implemented to estimate the state variables of

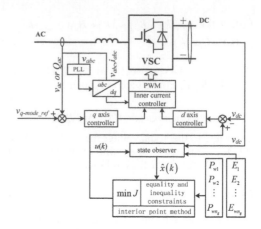

Figure 5. Block diagram of the proposed control strategy.

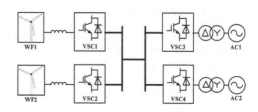

Figure 6. Four-terminal VSC-MTDC testbed.

the whole system, and the estimated values are adjusted with reference to the DC voltage from communication.

4 SIMULATION RESULT

To verify the effectiveness of the proposed control strategy, a VSC-MTDC with four terminal integrating two offshore wind farm was built in PSCAD/EMTDC, as shown in Figure 6. VSC1 and VSC2 employed constant AC voltage control connecting to a wind farm separately, whereas VSC3 and VSC4 adopted the proposed control strategy. $E_{max} = 420$ kV.

In the simulation, at $t = 0–5$ s, WF1 and WF2 generated 193 MW power initially. At $t = 5$ s, the generated power of WF1 and WF2 reduced to 120 and 155 MW, respectively. Finally, at $t = 15$ s, WF1 and WF2 recovered to their original state.

Figures 7–9 present the simulation results. These results demonstrate that the proposed control strategy can maintain at least one of the off-shore VSC DC voltages at E_{max}, effectively reducing the power loss of the DC network. Meanwhile, the DC voltage fluctuation during wind speed variance is reduced.

Initial: $z_0 = (x_0, s_0)$, $\mu_0 > 0$, $k \leftarrow 0$

set the trust-region radius $\Delta_0 > 0$

1: Build maximal clique, according to (12)

2: **while** $\mu_k \neq 0$ **do**

3:   **while** stopping criteria for the barrier problem not matched **do**

4:     **if** $neig \leq 2n+1$ **then**

        compute the search direction $d = (d_z, d_\lambda)$;

        else based on (15), calculate (z_{k+1}, λ_{k+1}) and

          Δ_{k+1};

      end if

5:     **if** the most recent barrier problem was solved in less than three iterations

      then $\mu_{k+1} \leftarrow \mu_k / 100$;

      else $\mu_{k+1} \leftarrow \mu_k / 5$;

      end if

6:     $k \leftarrow k+1$;

    end while

  end while

Figure 4. Flowchart of the interior point method.

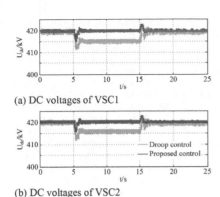

(a) DC voltages of VSC1

(b) DC voltages of VSC2

Figure 7. DC voltages of VSC1 and VSC2.

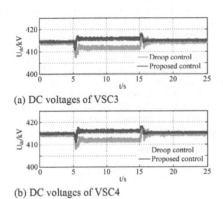

(a) DC voltages of VSC3

(b) DC voltages of VSC4

Figure 8. DC voltages of VSC3 and VSC4.

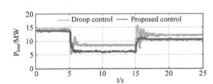

Figure 9. Power loss of DC grid.

5 CONCLUSION

In this paper, a coordinated control strategy for VSC-MTDC based on MPC has been proposed. A power flow optimization framework is implemented to ensure that at least one of the DC voltages of all wind farm-side VSCs can reach the maximum voltage.

Simulation results indicate that, compared with conventional DC voltage droop control, the proposed control strategy can effectively reduce the power loss of DC grid and DC voltage fluctuation.

Appendix A1

$$
A_i = \begin{bmatrix} 1 & 0 & 0 & -b_3 T_s \\ T_s & 1 & -a_2 T_s & -b_2 T_s \\ 0 & T_s & 1-a_1 T_s & -b_1 T_s \\ 0 & 0 & k_{isd} T_s & 1-T_s \end{bmatrix}, B_{w,i} = \begin{bmatrix} 0 \\ 0 \\ 0 \\ -T_s/C_{eq} \end{bmatrix},
$$

$$
B_{u,i} = \begin{bmatrix} b_3 T_s \\ b_2 T_s \\ b_1 T_s \\ T_s \end{bmatrix}, C_i = \begin{bmatrix} 0 \\ 0 \\ 0 \\ 1 \end{bmatrix}^T, a_1 = \frac{R+K_{pc}}{L}, a_2 = \frac{K_{ic}}{L},
$$

$$
b_1 = \frac{K_{pv} K_{pc}}{L}, b_2 = \frac{K_{iv} K_{pc} + K_{pv} K_{ic}}{L}, b_3 = \frac{K_{iv} K_{ic}}{L}
$$

Appendix A2

$$
A_o = \begin{bmatrix} A_1 & & \\ & \ddots & \\ & & A_{n_g} \end{bmatrix}, B_{uo} = \begin{bmatrix} B_{u,1} & & \\ & \ddots & \\ & & B_{u,n_g} \end{bmatrix},
$$

$$
B_{wo} = \begin{bmatrix} B_{w,1} & & \\ & \ddots & \\ & & B_{w,n_g} \end{bmatrix}, E_e = \begin{bmatrix} e_1 & & \\ & \ddots & \\ & & e_{n_g} \end{bmatrix},
$$

$$
G = [G_1 \vdots G_2], C' = \begin{bmatrix} C_1 & & \\ & \ddots & \\ & & C_{n_g} \end{bmatrix}
$$

Appendix A3

$$
d_z = \begin{bmatrix} d_x \\ d_s \end{bmatrix}, d_\lambda = \begin{bmatrix} d_h \\ d_g \end{bmatrix}, A(x) = \begin{bmatrix} A_h(x) & 0 \\ A_g(x) & I_e \end{bmatrix},
$$

$$
c(z) = \begin{bmatrix} h(x) \\ g(x)+s \end{bmatrix},
$$

$$
W(z, \lambda; \mu) = \begin{bmatrix} \nabla_{xx} \mathcal{L}(z, \lambda; \mu) & 0 \\ 0 & S^{-1} \Lambda_g \end{bmatrix}
$$

ACKNOWLEDGMENT

This study was supported by NARI Group Corporation (State Grid Electric Power Research Institute) Postdoctoral Project (Research on Control and Protection of VSC-MTDC for off-shore wind farm integration).

REFERENCES

Abdelwahed, M. A. & Elsaadany, E. Adaptive droop based power sharing control algorithm for offshore multi-terminal VSC-HVDC transmission. Electrical Power and Energy Conference (EPEC), 2015 IEEE, 2015. IEEE, 67–72.

Chaudhuri, N. R. & Chaudhuri, B. 2013. Adaptive droop control for effective power sharing in Multi-Terminal DC (MTDC) grids. *IEEE Transactions on Power Systems,* 28, 21–29.

Chen, H., Wang, C., Zhang, F. & Pan, W. Control strategy research of VSC based multiterminal HVDC system. 2006 IEEE PES Power Systems Conference and Exposition, 2006. IEEE, 1986–1990.

De Almeida, R.G. & Lopes, J.P. 2007. Participation of doubly fed induction wind generators in system frequency regulation. *IEEE transactions on power systems,* 22, 944–950.

Liang, J., Gomis-Bellmunt, O., Ekanayake, J., Jenkins, N. & An, W. 2012. A multi-terminal HVDC transmission system for offshore wind farms with induction generators. *International Journal of Electrical Power & Energy Systems,* 43, 54–62.

Morren, J., De Haan, S.W., Kling, W.L. & Ferreira, J. 2006. Wind turbines emulating inertia and supporting primary frequency control. *IEEE Transactions on Power Systems,* 21, 433–434.

Rault, P., Colas, F., Guillaud, X. & Nguefeu, S. Method for small signal stability analysis of VSC-MTDC grids. 2012 IEEE Power and Energy Society General Meeting, 2012. IEEE, 1–7.

Raza, A.A., Diangou, B.X., Xunwen, C.S. & Weixing, D.L. Appraisal of VSC based MTDC system topologies for offshore wind farms. 2015 9th International Conference on Power Electronics and ECCE Asia (ICPE-ECCE Asia), 2015. IEEE, 2141–2147.

Rouzbehi, K., Miranian, A., Candela, J.I., Luna, A. & Rodriguez, P. 2015. A generalized voltage droop strategy for control of multiterminal DC grids. *IEEE Transactions on Industry Applications,* 51, 607–618.

Silva, B., Moreira, C., Seca, L., Phulpin, Y. & Lopes, J.P. 2012. Provision of inertial and primary frequency control services using offshore multiterminal HVDC networks. *IEEE Transactions on sustainable Energy,* 3, 800–808.

Yu, Y., Feng, Y., Jiang, H. & Qiu, Y. Research on VSC-MTDC for grid integration of wind farm. International Conference on Renewable Power Generation (RPG 2015), 2015. IET, 1–5.

Zhang, L., Ye, T., Xin, Y., Han, F. & Fan, G. 2010. Problems and Measures of Power Grid Accommodating Large Scale Wind Power [J]. *Proceedings of the CSEE,* 25.

Zhang, W., Rouzbehi, K., Luna, A., Gharehpetian, G.B. & Rodriguez, P. 2016. Multi-terminal HVDC grids with inertia mimicry capability. *IET Renewable Power Generation.*

Zhu, J., Booth, C.D., Adam, G.P., Roscoe, A.J. & Bright, C.G. 2013. Inertia emulation control strategy for VSC-HVDC transmission systems. *IEEE Transactions on Power Systems,* 28, 1277–1287.

Electromechanical Control Technology and Transportation – Jia & Wu (Eds)
© 2017 Taylor & Francis Group, London, ISBN 978-1-138-06752-3

Cyber-attack feature processing approach based on MA-LSSVM

Yuanyuan Ma & Gaofeng He
Global Energy Interconnection Research Institute, Nanjing, China

ABSTRACT: Amongst all types of the increasingly mature biological intelligent evolutionary algorithms, feature selection methods based on the evolutionary technology and its hybrid algorithm are emerging. According to the feature selection problem of the high-dimensional small sample safety data, this paper combines the Memetic Algorithm (MA) and Least-Squares Support Vector Machines (LS-SVM) to design a type of wrapper feature selection method (MA-LSSVM). The proposed method utilizes the specialty of least-squares support vector machine to ease the search of an optimal solution to construct a classifier and then regards the classification accuracy as the main component of the memetic algorithm fitness function in the optimization process. The experimental results demonstrate that MA-LSSVM can be more efficiently and stably used to obtain features that largely contribute to the classification precision and then reduce the data dimension and improve the classification efficiency.

1 INTRODUCTION

With the continuous development of computer technology and network, as well as rapid expansion of various types of information, how to effectively search for effective information from huge data has become an inevitable reality. In the aspect of information security, information expansion will cause a wide variety of high-dimensional small sample data. High-dimensional small sample is subject to the following two cases: The first one is that the number of all types of training samples is far less than the dimension of feature space. The second one is that although the quantity of training sample is greater than dimension of the feature space, the two belong to the same order of magnitude (Belhumeur P N, 1997). A large number of redundant features and noise characteristics exist on the characteristic space of most high-dimensional small sample data. On the one hand, such characteristics can reduce data prediction precision of classifier. On the other hand, time and space complexity of training can be significantly reduced which will lead to well-known "curse of dimensionality". In order to effectively explore useful information from a huge amount of high-dimensional small sample data, feature selection has become critical for the analysis and processing of information security high-dimensional data.

At present, the feature selection method presents diversified and comprehensive research tendency. The feature selection method is classified into three types: filter, wrapper, and embedded (Saeys Y, 2007). Filter is unrelated to subsequent sorting algorithm. Perform characteristic evaluation and selection of training sample data on the aspect of statistics directly with training sample data. The wrapper method adopts classification performance of classifier as character subset evaluation standard or part of the standard or part of the standard. The embedded method adopts embedded sorting algorithm as a constituent part. The filtering feature selection algorithm is independent of subsequent machine learning algorithm, and the computational cost is relatively low. Therefore, the operating speed is high. However, the effect of feature selection is relatively general. The MA-LSSVM depends on certain or a variety of learning algorithms at feature selection process. Compared to filter algorithm, the computation cost is relatively higher and efficiency is lower; however, the effect of feature selection is relatively better. For example, in Li L (2001), Li et al. selected a character subset combining with GA (Genetic Algorithm) and kNN (k-Nearest Neighbor) classifier. The SVM (Support Vector Machine) structural risk minimization principle has strong learning and generalization ability. For example, Li K (2013) and Li Y (2009) combined BPSO and support vector machine and put forward the BPSO-SVM feature selection algorithm. In addition, the Least-Squares Support Vector Machine (LS-SVM) was put forward by Suykens et al. It was improved on the basis of standard support vector machine (Li K, 2013). The optimal margin hyperplane was obtained through solving a set of linear equations, which can effectively avoid complex quadratic programming and significantly promote solution speed. Therefore, it has become an effective tool for numerous researchers to solve the problem of different fields. For example, Aydogdu et al. (2014) utilized fuzzy clustering and LS-SVM on failure prediction of the water supply network; Langone R (2015) utilized LS-SVM on failure prediction of industrial device.

Moscato P (1989) proposed the memetic algorithm for the first time. In the standard genetic algorithm process, optimal solution was obtained generally through individual mutation, crossover, and selection as well as adaptive iterative evolution to the individual (Zheng Yamin, 2008). However, for the memetic algorithm, genetic operation object is not ordinary individuals in population space, but the excellent individuals elected in each local area. Genetic operation is to select the adaptable outstanding individual. In addition, new individuals can be generated through crossover operation. Such new individuals may belong to some new areas and will be replaced by adjacent superior individuals in the next generation of local search. And then perform a new round of global evolution. The search solution efficiency of the memetic algorithm based on the global and local search mechanisms is several orders of magnitude higher than that of the traditional genetic algorithms on certain issues. It can be applied to a wide range of areas and satisfactory results can be obtained. For example, Karaoglan I (2015) applied the memetic algorithm to solve mixed cargo. Cattaruzza D (2014) applied the memetic algorithm on studies of vehicle path planning.

In this paper, we study MA-LSSVM to reduce the cost of computing. LS-SVM was selected as the classifier using the feature selection method. Combining with search solution mechanism of the memetic algorithm, high-dimensional feature selection algorithm based on MA-LSSVM security data was proposed.

2 RELATED WORK

2.1 LS-SVM

In this paper, the charts must be indicated with Chinese figures and tables, and shall be clarified in the text (e.g., as shown in Figure 1 and Table 1). The drawing number and figure title are placed at the center below the figure. The table title and number are placed at the center above the figure. Adopt three-line table as far as possible and add auxiliary line if necessary.

LS-SVM supports insensitive loss function of support vector machine and changes to quadratic loss function. Change the inequality constraint to equality constraint. Consequently, solving the optimal separating hyperplane has transformed to a set of linear equations, which can effectively avoid solving complex quadratic programming

thus to significantly increase the solving speed. The problem is described below (Sun Yanfeng, 2006 & Liu Gang, 2012).

Set l has $\{x_i, y_i\}_1^l$ training set of samples. The input data i is $x_i \in R^n$, and the output data i is class $y_i \in \{-1, +1\}$. The sample $\varphi(x) = (\varphi(x_1), \varphi(x_2), ..., \varphi(x_n))$ can be mapped from the original space to a feature space via nonlinear mapping $\varphi(\cdot)$. Support vector machine is to build a classifier. The form is as follows:

$$f(x) = sign\left[w^T \varphi(x) + b\right] \tag{1}$$

Sample x can be classified correctly by f(x). Here, w is the weight vector and b is the threshold value.

Optimization issues solved from LS-SVM are as follows:

$$\min_{w,e} J(w, e) = \frac{1}{2} w^T \varphi(x) + b \tag{2}$$

It satisfies equality constraint:

$$y_i\left[w^T \varphi(x_i) + b\right] + e_i = 1, i = 1, ..., l \tag{3}$$

Lagrange polynomial of dual problem is as follows:

$$L(w, b, e, a) = \\ J(w, e) - \sum_{i=1}^{l} \alpha_i \left\{ y_i \left[w^T \varphi(x_i) + b \right] - 1 + e_i \right\} \tag{4}$$

where $\alpha_i \in R$ is the multiplier of Lagrange. According to the equality constraint, the value can be plus or minus. The optimized conditions are:

$$\begin{cases} \dfrac{\partial L}{\partial w} = 0 \to \sum_{i=1}^{l} \alpha_i y_i \varphi(x_i) \\ \dfrac{\partial L}{\partial b} = 0 \to -\sum_{i=1}^{l} \alpha_i y_i = 0 \\ \dfrac{\partial L}{\partial e_i} = 0 \to \alpha_i = \gamma e_i \qquad i = 1, 2, ..., l \\ \dfrac{\partial L}{\partial \alpha_i} = 0 \to y_i \left[w^T \varphi(x_i) + b \right] - 1 + e_i \quad i = 1, 2, ..., l \end{cases} \tag{5}$$

Formula (5) can be written as the system of linear equation as follows:

$$\begin{bmatrix} I & 0 & 0 & -Z^T \\ 0 & 0 & 0 & -y^T \\ 0 & 0 & \gamma^l & -I \\ Z & y & 1 & 0 \end{bmatrix} \begin{bmatrix} w \\ b \\ e \\ \alpha \end{bmatrix} = \begin{bmatrix} 0 \\ 0 \\ 0 \\ I \end{bmatrix} \tag{6}$$

where $Z=[\varphi(x_1)y_1, \varphi(x_2)y_2,..., \varphi(x_l)y_l]^T, e=[e_1, e_2,..., e_l]^T,$ $y=[y_1, y_2, ..., y_l], I=[1, ..., 1]^T,$ and $\alpha=[\alpha_1, ..., \alpha_l]^T$

Eliminate e and w and utilize Mercer conditions:

$$\Omega_{kj} = y_k y_j \varphi(x_k)^T \varphi(x_j) = y_k y_j \psi(x_k, x_j)$$
$$k, j = 1,..., l \tag{7}$$

The equation set obtained is related to b, α. The equation set (6) is transformed to:

$$\begin{bmatrix} 0 & -y^T \\ y & \Omega+\gamma^{-1}I \end{bmatrix}\begin{bmatrix} b \\ \alpha \end{bmatrix}=\begin{bmatrix} 0 \\ I \end{bmatrix} \tag{8}$$

Assume $A = \Omega + \gamma^{-1}I$. As A is a symmetric positive semidefinite matrix, A^{-1} exists. The solution of the system of linear equation (8) is as shown in (9):

$$b = \frac{y^T A^{-1}I}{y^T A^{-1}y} \quad \alpha = A^{-1}(I - yb) \tag{9}$$

The first equation in formula (6) is replaced by w of formula (1). Utilize formula (7) to derive formula (10):

$$f(x) = sign\left[a_i y_i \psi(x, x_j) + b \right] \tag{10}$$

where α_i and b are the solutions to equation set (8). In formula (10), f(x) is the solved classifier.

2.2 Measurement method of stability

For research on feature selection, the stability of measure feature selection algorithm is a key point. The stability measure of the feature selection algorithm is determined through determining the comparability of different feature selection results. The common measurement method is to conduct feature selection on different training sample sets. Then, perform similarity comparison for the obtained feature selection results to evaluate the stability of feature selection algorithm. At present, according to different feature selection results, there are three types of similarity measurement methods: weighting, ranking, and subset (Li Yun, 2012).

Here, we shall use the following frequently used similarity measurement methods on the basis of character subset and shall analyze the MA-LSSVM feature selection algorithm on the basis of the experimental results.

1. Jaccard index

$$sim_{Jaccard}(f_i, f_j) = \frac{|f_i \cap f_j|}{|f_i \cup f_j|} \tag{11}$$

where f_i is the characteristic vector. The data range of Jaccard index is [0, 1], where 0 refers to totally different character subsets of the two feature selection results and 1 refers to identical character subsets. Consequently, the higher the Jaccard index, the higher the comparability of the character subset.

2. Dice's coefficient:

$$sim_{Dice}(f_i, f_j) = \frac{2|f_i \cap f_j|}{|f_i| + |f_j|} \tag{12}$$

The data range of Dice index is [0, 1], where 0 refers to totally different character subsets of the two feature selection results and 1 refers to identical character subsets. Consequently, the higher the Dice index, the higher the comparability of the character subset. In general, Dice index is similar to Jaccard index. It is noteworthy that Dice index can obtain more accurate evaluation in case of intersection of two character subsets. In addition, it can treat various sizes of subsets.

3. Average Normal Hamming Distance (ANHD)

$$sim_{ANHD}(f_i, f_j) = \frac{1}{m}\sum_{k=1}^{m}|f_i^k - f_j^k| \tag{13}$$

It gives the total number of features, and f_i refers to the dimensional value of the characteristic vector. O refers that the feature is not selected. 1 refers to the feature is selected by the algorithm. The value range of ANHD is [0, 1]. The lower the value, the higher is the comparability of the character subsets.

4. KI, Kuncheva index

$$sim_{Kuncheva}(f_i, f_j) = \frac{|f_i \cap f_j| \cdot m - k^2}{k(m - k)} \tag{14}$$

With increasing of the index of the selected character subset, the possibility of overlap will increase correspondingly. KI can help us to avoid overlapping of two feature subsets by accident. The value range of KI is [−1, 1]. The higher the KI value, more similar are the character subsets of the two selected feature selection results.

3 FEATURE SELECTION BASED ON MEMETIC ALGORITHM

A feature set $F = \{f_1, f_2,..., f_i,..., f_n\}$ is given. The character subset of characteristic set F_s can be expressed as $S = \{s_1, s_2,..., s_i,..., s_n\}$.

$s_i = \{0, 1\}i = 1, 2, \ldots, n$, where s_i refers to the feature i if f_i is selected or not. If f_i is selected, $s_i = 1$; otherwise, $s_i = 0$. Feature selection is to detect the character subset, which can realize the optimal classifier effects. Therefore, a quantitative standard is required to measure the classification capacity of the character subset. In the memetic algorithm, "individual" means the character subset. Fitness evaluation is an important index to evaluate the impact of feature subset on classifier performance. That is to say, it is an index to evaluate individual performance. To this end, the proposed fitness function should include two aspects: (1) Accuracy rate of classifier verification is trained by adopting the features of the character subset. The performance of the classifier is evaluated by adopting cross-validation results and guiding the evolution of the population; (2) Select the quality of the features. Each feature subset contains a certain number of characteristics. If the accuracies of the classifier under two character subsets are the same, the one containing less character subsets will be selected preferentially. On the aspect of classification accuracy and the number of characteristics, special attention shall be paid on classification accuracy. Consequently, in the memetic algorithm, the fitness function shall be determined in the following form for each individual of the group:

$$fit(F_i) = \beta * P(F_i) - \delta * N(F_i) \qquad (15)$$

where, $P(F_i)$ classification accuracy of the characteristic structure classifier F_i selected, namely the character subset corresponding to "1" of the individual F_i to train the classification accuracy of LS-SVM classifier. $N(F_i)$ refers to the quantity concluded in "1" (character subset size). The memetic algorithm is to detect the global maximum. The higher the classification accuracy, the smaller the character subset scale, which represents that the higher the fitness value, it is more possible to win the mutual competition of the individual of the character subset. β and δ are weight parameters for classification accuracy and feature number, and shall be adjusted according to the specific situation. The higher the β value represents that the classification accuracy shall be paid of the character subset. Similarly, a higher value of δ shows that the characteristics number can be acquired.

Accurately speaking, the memetic algorithm is the frame of optimized algorithm. A combination of different search strategies can constitute diversified memetic algorithms. Because of the large amount of high-dimensional small sample data features and few training samples, BDE (Binary Differential Evolution) is adopted as global search strategy (Wu Zhifeng, 2008) and TS (Tabu search) as local search strategy (Sun Yanfeng, 2006).

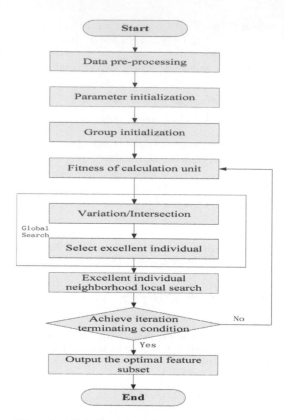

Figure 1. High-dimensional feature selection algorithm process based on security data of MA-LSSVM.

Figure 1 shows a high-dimensional feature selection algorithm process based on the security data of MA-LSSVM. The specific steps are presented as following:

Step 1: Data pre-processing: In order to facilitate data analysis, perform z-score standardization processing on a sample set. Converse original data to dimensionless index assessment value so that the values of each index are of the same order of magnitude.

Step 2: Parameter initialization: In the BDE algorithm, it is necessary to set the value range of initial population M, initial evolution algebra $g = 0$, the largest iterative evolution g_{max}, zoom factor ξ, and crossed factor CR. In the TS algorithm of local search, it is necessary to set the length of tabu list and the maximum iteration $Tstep_{max}$.

Step 3: Population initialization: Solution of randomly generated values within certain limits (individual), $g = 1$.

Step 4: Calculate the fitness value of each individual according to the fitness function.

Step 5: Global search: Constantly update the position of the individuals according to the optimization mechanism of the BDE algorithm through variation, crossover, and selection.

Step 6: Local search: Perform fitness value calculation for excellent individuals generated in global search at neighborhood fields by adopting TS algorithm. Update the optimal solution set if more excellent individuals are searched.

Step 7: Determine termination conditions: There are two termination conditions. The first one is that the number of evolution exceeds the maximum allowable evolution (g_{max}). The second one is that the fitness of the individual exceeds 0.99. If reaching any one of the two conditions, the memetic algorithm process will be terminated. Otherwise, $g = g + 1$, skip to step 4.

Step 8: Output the optimal character subset.

Tao Gong modified the mutation operation of differential evolution algorithm specific to discrete optimization and put forward BDE. The algorithm has significant effects on the aspect of global optimization (Gong T, 2007). In the paper, BDE was adopted to determine the optimal solution and perform global searching of the whole space as well as provide initial solution for local searching.

Appropriate control parameters are extremely important for the BDE algorithm. Therefore, adaptive control parameter adjustment strategy was used in the paper. In the population optimization process, the variance of fitness is:

$$\sigma^2 = \sum_{i=1}^{M} \left(\frac{fit(F_i) - fit_{avg}}{fit_{best}} \right) \tag{16}$$

where $fit(F_i)$ is the fitness of individual i, fit_{avg} is the average fitness value of the current population, fit_{best} is the optimal fitness value of the population, and M is the size of the population.

Considering the distribution of the population, the self-adaption control parameter adjustment strategy of the BDE algorithm is as follows:

$$\xi^g = \xi_{max} - (\xi_{max} - \xi_{min})\left(1 - \frac{\sigma^2 g}{M}\right) \tag{17}$$

$$CR^g = CR_{max} - (CR_{max} - CR_{min})\left(1 - \frac{\sigma^2 g}{M}\right) \tag{18}$$

where g represents the current generation of evolution, σ_g^2 represents the fitness variance of the generation g, ξ^g represents the zoom factor of the current generation of evolution, ξ_{max} and ξ_{min} represent the maximum and minimum values of zoom factor, respectively, CR^g represents the crossed factor of the current generation of evolution, and CR_{max} and CR_{min} represent the maximum and minimum values of the crossover operators, respectively. It can be seen that with diminishing of population evolution-

ary σ^2, the scaling factor decreases, and the interleaving increased gradually (Jiang Liqiang, 2007).

Local search is an important concept of the memetic algorithm. It has an important influence on the performance of the algorithm. The basic concept of local search is based on the greedy algorithm, while constantly looking for a better solution at the current solution neighborhood. It is the process to select the excellent individual at local areas to enhance algorithm performance and individual fitness. TS is an optimization method for heuristic search. It has the following advantages: high search efficiency, memory function, and excellent climbing ability. The TS algorithm is adopted to avoid alternate searching through tabu list and corresponding taboo rules. The aspiration criterion is used to avoid taboo excellent status. However, TS performance relies on the initial solution to a great extent (Sun Yanfeng, 2006).

TS algorithm, as a local search strategy, is used on the BDE algorithm in global search of the evolution of every generation to seek the excellent individuals, by successive iteration, in the local neighborhood to find the best solution in a local neighborhood. The search strategy combining with BDE and TS advantages can reach equilibrium on solution and convergence speed.

4 FEATURE SELECTION BASED ON THE MEMETIC ALGORITHM

Relevant experiment of the paper is performed on PC. The computer adopts Intel Core2 Extreme Q6850 processor, Frequency: 3.00 GHz, 2 GB internal storage, Microsoft Windows 7, and MATLAB 2012b +Weka 3.7.3.

Experimental data adopt three high-dimensional small sample data sets, Colon, Leukemia, and Lung for algorithm performance verification. Such data sets can be loaded from the main page of BRB-AiTayTools: http://levis.tongji.edu.cn/gzli/data/mirror-kentridge.html. Table 1 shows relevant information of data set, including data set, features, instances, and classes.

Compare the classification effects of the two wrapper algorithms, MA-LSSVM and GA-kNN[3], BPSO-SVM[4], proposed in the paper on time efficiency of character subset, stability of feature selection, and character subset on classifier.

Table 1. Description of experimental data set.

Data set	Features	Instances	Classes
Colon	2000	62	2
Leukemia	7129	72	2
Lung	12533	181	2

Set the population size $M = 100$, the largest iterative evolution $g_{max} = 1000$, and the kernel function of the LS-SVM algorithm described in formula (10) $\psi(\cdot)$ adopts radial basis function RBF. In formula (15), $\beta = 100$, $\delta = 0.5$, double exchange and initial interleaving are used in the crossover operation at global search for initial scaling factor $\xi = 0.6$. In local searching, set tabu list length $TL = 20$ and the maximum iteration $Tstep_{max} = 80$. A 10-fold cross-validation is adopted to check the classification precision, namely divide the data set into 10 parts at random. Use nine parts for training. The remaining part is for testing set. The results of the mean value is used as the estimate of the accuracy of algorithm in order to obtain more accurate results, and 10-fold cross-validation is needed to calculate the mean, 30 of each group of algorithm in this paper, and the experiment is performed 10 times for cross validation. The MA-LSSVM algorithm and the average elapsed time of comparing algorithm put forward in the paper are as shown in Table 2. The average classification accuracy of feature selection results on RF (Random Forest) and decision-making tree C4.5 classifiers are shown in Table 3. The two classifiers are integrated in weka. Set confidence factors of pruning of C4.5 confidenceFactor as 0.25. Set the number of Random Forest spanning tree numTrees as 10.

According to Table 2, the iteration speed of MA-LSSVM for feature selection to terminate iteration is significantly higher than that of the feature selection methods, GA-kNN and BPSO-SVM. Combining with Table 3, from feature number-

Table 2. Consuming time (s) of three feature selection methods, MA-LSSVM.

Data set	GA-kNN	BPSO-SVM	MA-LSSVM
Colon	31.55	27.43	18.32
Leukemia	67.43	50.98	36.08
Lung	224.02	198.82	160.91

Table 3. Classification verification results of MA-LSSVM, GA-kNN, and BPSO-SVM on data set.

Data set	Used method	Feature subset size	C4.5 (%)	RF (%)
Colon	MA-LSSVM	4	90.93	91.20
	GA-kNN	9	88.52	89.13
	BPSO-SVM	7	89.67	90.98
Leukemia	MA-LSSVM	14	98.63	99.01
	GA-kNN	15	94.12	95.33
	BPSO-SVM	15	95.53	95.90
Lung	MA-LSSVM	7	99.62	99.77
	GA-kNN	8	98.52	98.42
	BPSO-SVM	8	98.25	98.74

Table 4. Stability comparison of MA-LSSVM, GA-kNN, and BPSO-SVM.

Data set	Used method	Jaccard	Dice	ANHD	KI
Colon	MA-LSSVM	1.0	1.0	0	1.0
	GA-kNN	0.95	0.98	0.05	0.97
	BPSO-SVM	1.0	1.0	0	1.0
Leukemia	MA-LSSVM	1.0	1.0	0	1.0
	GA-kNN	0.9	0.95	0.03	0.9
	BPSO-SVM	1.0	1.0	0	1.0
Lung	MA-LSSVM	1.0	1.0	0	1.0
	GA-kNN	1.0	1.0	0	1.0
	BPSO-SVM	1.0	1.0	0	1.0

ber of the selected feature selection and C4.5 and RF classifiers of character subset, MA-LSSVM is superior to GA-kNN and BPSO-SVM. It illustrates that MA-LSSVM can efficiently find the optimal solution combining with the global search and local search. Furthermore, the LS-SVM algorithm accelerates the iterative evolution process of the population on the aspect of classification effects and speed for nonlinear high-dimensional data. It has obtained improvement on the quality and speed of character subset. In addition, in Table 4, four kinds of similarity measure methods are used for the stability of feature selection results, because each method has to carry out 30 experiments. Therefore, each method has to calculate $30*(30 - 1)/2$ feature selection results for comparability measurement. Therefore, all data in Table 4 are average value of each measurement level. Clause 2.2 illustrates that the MA-LSSVM algorithm put forward in the paper has satisfactory stability on feature selection.

5 SUMMARY

Safety data characteristics processing method (MA-LSSVM) based on the memetic algorithm and LS-SVM was proposed in this paper. Through experimental comparison of three high-dimensional sample data sets and the same type of algorithm, it can be seen that the MA-LSSVM algorithm has high validity and stability on high-dimensional sample data processing. The primary causes for good effects acquired through the algorithm put forward in the paper include: 1) Memetic algorithm adopts an optimization strategy frame according to BDE global search and local search based on TS, which ensures that the algorithm can search for a better solution; 2) Adaptive adjustment for control parameters in global search has accelerated the convergence of the algorithm to avoid local optimum and improved the algorithm performance; 3) LS-SVM is used as the classifier of the optimization process specific to high high-dimensional data latitude and great redundancy, thus improving the execution efficiency of the global algorithm.

REFERENCES

Aydogdu M, Firat M. Estimation of Failure Rate in Water Distribution Network Using Fuzzy Clustering and LS-SVM Methods [J]. Water Resources Management, 2014: 1–16.

Belhumeur P N, Hespanha J P, Kriegman D. Eigenfaces vs. fisherfaces: Recognition using class specific linear projection [J]. Pattern Analysis and Machine Intelligence, IEEE Transactions on, 1997, 19(7): 711–720.

Cattaruzza D, Absi N, Feillet D, et al. A memetic algorithm for the multi trip vehicle routing problem [J]. European Journal of Operational Research, 2014, 236(3): 833–848.

Gong T, Tuson A L. Differential evolution for binary encoding [M]//Soft Computing in Industrial Applications. Springer Berlin Heidelberg, 2007: 251–262.

He Dakuo, Wang Fuli, Zhang Chunmei. Establishment of Parameters of Genetic Algorithm Based on Uniform Design [J]. Journal of Northeastern University(Natural Science), 2003, 24(5): 409–411.

Jiang Liqiang, Guo Zheng, Liu Ouangbin. Study on the strategy of scaling factor in differential evolution algorithm [J]. Chinese journal of Scientific Instrument. 2007: 508–510.

Karaoglan I, Altiparmak F. A memetic algorithm for the capacitated location-routing problem with mixed backhauls [J]. Computers & Operations Research, 2015, 55: 200–216.

Langone R, Alzate C, De Ketelaere B, et al. LS-SVM based spectral clustering and regression for predicting maintenance of industrial machines [J]. Engineering Applications of Artificial Intelligence, 2015, 37: 268–278.

Li K, Gao X, Tian Z, et al. Using the curve moment and the PSO-SVM method to diagnose downhole conditions of a sucker rod pumping unit [J]. Petroleum Science, 2013, 10(1): 73–80.

Li L, Weinberg C R, Darden T A, et al. Gene selection for sample classification based on gene expression data: study of sensitivity to choice of parameters of the GA/KNN method [J]. Bioinformatics, 2001, 17(12): 1131–1142.

Li Y, Zhang N, Li C. Support vector machine forecasting method improved by chaotic particle swarm optimization and its application [J]. Journal of Central South University of Technology, 2009, 16: 478–481.

Li Yun. Research on stable feature selection [J]. Microcomputer & Its Applications, 2012, 31(15): 1–2.

Liu Gang Li Yuan-xiang Zheng Hao. 2-Opt-and-generalized Opposition-based Differential Evolution Algorithm with Reserved Genes [J]. Journal of Chinese Computer Systems, 2012, 33(4): 789–794.

Luo jun, Fan Pengcheng. Improved particle swarm optimization based on genetic hybrid genes [J]. Application Research of Computers, 2009, 26(10): 3716–3717, 3753.

Moscato P. On evolution, search, optimization, genetic algorithms and martial arts: Towards memetic algorithms [J]. Caltech concurrent computation program, C3P Report, 1989, 826: 1989.

Saeys Y, Inza I, Larrañaga P. A review of feature selection techniques in bioinformatics [J]. bioinformatics, 2007, 23(19): 2507–2517.

Sun Yanfeng. A Hybrid Strategy Based on Genetic Algorithm and Tabu Search [J]. Journal of Beijing University of Technology, 2006, 32(3): 258–262.

Suykens J A K, Vandewalle J. Least squares support vector machine classifiers [J]. Neural processing letters, 1999, 9(3): 293–300.

Wu Zhifeng, Huang Houkuan. A Binary-Encoding Differential Evolution Algorithm for Agent Coalition [J]. Journal of Computer Research and Development, 2008, 45(5): 848–852.

Yang Wenlu, Ning Yufu. Integrated cross-factor and metropolis rule group search optimization algorithm [J]. Computer Engineering and Design, 2013, 34(6): 2020–2024.

Zheng Yamin, Improvement of Feature Selection Method Based on Genetic Algorithm [D]. Chongqing University, 2008.

Electromechanical Control Technology and Transportation – Jia & Wu (Eds)
© 2017 Taylor & Francis Group, London, ISBN 978-1-138-06752-3

Design of remote smart home control system based on the internet of things

Yajin Che

Sichuan Vocational College of Information Technology, Sichuan, China

ABSTRACT: With the continuous innovation of related technology and the rapid development of information industry, the Internet of things, as a new product of information technology, has been applied in many fields, especially in the field of residential home. It is of great significance to promote the quality of people's daily life. In view of this, this paper briefly analyzes the collection and transmission of indoor remote control signal. On the basis of this, the design of remote control software is analyzed and discussed. Finally, the practical application test shows that the remote smart home control system designed by this research can realize the effective transmission of the indoor electrical control of information. It can meet the needs of practical use both in controlling accuracy and the delay.

1 INTRODUCTION

Since the advent of the Internet of things, it has been widely used in all aspects of social life, and has brought great changes to people's daily life. The application of Internet of things in the field of home has realized the innovation of design ideas for smart home control system, which brings great convenience to people's daily life and promotes the quality of life. To achieve remote smart home control by using Internet of things, the main work is to use the Internet to achieve the connection of all kinds of household equipment and promote the formation of a more complete internal network environment. Thus it can achieve the automatic control of the equipment and scientific management.

1.1 Design concept of smart home control system

There is a very close relationship between the construction of smart home control system and the gradual improvement of the Internet of things. The application of smart home control system will greatly promote the convenience of people's daily life and improve the quality of life. The design principle of the smart home control system is to implement the remote control and management for the indoor equipment by using the Internet of things technology and the sensor nodes. In essence, the working principle of this system is to achieve the control of equipment based on the external control

information which is obtained from the separation and the compressor delivery. To have a good applicability, it is necessary to ensure that access to the external control signal received by smart home control system has a certain reliability and can be connected with control components. And then the remote intelligent control and management of the furniture equipment can be achieved.

In addition, there are some specific requirements to be met in the design and application of remote smart home control system, such as obtaining control command sensing its model by using the system, achieving real-time attention, controlling and managing different types of control by using completely different signals, and including a subsystem for signal processing.

2 TRANSMISSION AND COLLECTION OF INDOOR REMOTE CONTROL SIGNAL

2.1 Amplification and compensation of remote signal

In order to achieve the effective control of the remote smart home, its precision and distance usually have a great influence on the new generation of smart home system. To ensure the stability of the remote control signal, it is necessary to adopt certain measures to realize the amplification of the circuit, and then achieve the effective control of the signal to ensure that there are no serious mistakes. The energy supply of the amplifying circuit control signal mainly includes two categories: the interior and the exterior. It can amplify the remote control signals and greatly improve the control accuracy. This module mainly has the following functions:

1. High sensitivity. It can not only realize the removal of noise, but also enlarge the difference of the remote signal. It is energy-saving. Generally the circuit current is required to be controlled in: 480 VA; when in standby operation, the voltage is 2.6 VA.
2. In general, the driving voltage will always remain in the range of 2.2 V–3.6 V.
3. The signal value of the voltage change can be obtained in time, and the actual working condition of the electric appliance can be reflected in a timely manner.
4. Low pass filter can be appropriate for the internal signal processing.

If the input control signal is relatively small in the hardware connection, small signal model should be used. FET micro equivalent circuit and small size model of bipolar transistor are basically the same. The field effect pipeline can be regarded as a two port network, and the output interface can be in the range of the drain electrode and the source electrode when the input interface is in the range of the grid electrode and the source electrode. Whatever kind of field effect transistor it is, it can be regarded as zero grid current value. The input interface can be equated to open circuit because of the voltage value between the grid electrode and the source electrode. Thereby it can realize the acquisition of the network and the amplification of the control signal on the bottom, and then convert into a voltage signal.

2.2 Reception of remote signal

In order to ensure that the signal still can be received timely and accurately when there is certain loss in the process of transmission, in this design, the serial interface and the USB interface are used to collect the remote control signal information data, and transmit it to the corresponding electrical control terminal. The designed devices for receiving signal are as follows:

1. Wireless serial receiver
   XL01-232 AP1 is a half-duplex wireless transmission module of UART interface, whose general operating band will be in several Hz bands like 325, 760, 825 and so on. This module strictly meets ETSI and FCC15.247/15.249 certificated standards and the relevant requirements of wireless control. This wireless serial

receiver uses the serial port to debug the software equipment to carry on the output power, the serial port format, the serial port speed, the frequency of work and so on.

2. Discriminator of Multiple Control Command
   In order to be more effective in identifying and dividing the control of the command, WUSB principle needs to be taken. The main connection principle of this part is the network hub and the topology structure. The data information processed by the host machine is sent to each device based on the WUSB host, and at the same time it configures the corresponding address and the loan. Thereby it generates a group relationship between the device and the remote control device. The WUSB host and its corresponding device use the pattern of "Point-to-Point" to achieve directional transmission. WUSB host can achieve the identification of multiple WUSB devices, and at the same time it can be connected to completely different WUSB hosts. When the different types of WUSB devices are in a relatively small space, the most important problem is how to improve the access capability of the equipment and the reasonable utilization of the bandwidth. At this time, we can use the topology to determine the number of devices to be used, and to achieve the expected control effect.

3. Judgment of the Properties of Control Commands
   By using wireless serial port command code, it can achieve the distinction of the actual property control. The serial communication mode is: N-6-1, where N is the initial digit, 6 data bits, 1stop bit, and no parity. The communication baud rate has always been in the range of 1300–114500 bps. When the data packet is sent, it starts with the command "OT02" and ends with the command "OT0d" to reduce the bit error rate as much as possible. The structure of data report command is the initial command, the remote command code, the control signal data, and the control command tail. Table 1 is a set of instructions of remote control device.

Table 1. The specific encoding of control commands.

Instruction code	Content	Meaning
CC	0~0 × 11	Open
DC	0~0 × 11	Close
DD	0~0 × 11	Range of finishing
EC	0~0 × 11	Functional disconnection
ED	0~0 × 11	Open the alarm
CC	0~0 × 11	Shut down the alarm
CD	0~0 × 11	Disconnect all functions

3 DESIGN OF REMOTE CONTROL SOFTWARE

New smart home control system software can not only indicate whether the signal is safe, but also appropriately process the signal. In order to be more user-friendly and to ensure that the design of the operating interface is more intuitive, the data will be timely and effectively improved and updated. In the software system, the main modules include data storage, data analysis and processing, data display and man-machine interface.

3.1 Design of remote signal acquisition

If the remote control signal is obtained from the sensor network, the control information can be stored and processed. The attributes of the control information include acquisition time, command characteristics, number of the device, and whether it is specific or not. Generally, the more complex control information will have the lower probability of false operation.

3.2 Design of the display of control signal's attribute

By using this module, it can be more accurate to describe the properties of the control signal, as it uses the voice to play the relevant information. In addition, this software can also put the related data directly into the log, which can be applied to the different time periods of the accurate data records of valve pressure.

3.3 Design of accurate judgment of control signal

This module is the core of the overall software design. The basis of the attribute operation of the control signal is generally related to the wavelet function, and if it is continuous, the moment when "m" order vanishes is of the integer value. When "t" is in the interval [a, b], if "f(t)" wavelet can be transformed as:

$$\left|W_{2^j} f(t)\right| \leq K\left(2^j\right)^a$$

The corresponding logarithm shall be taken as:

$$\log_2 \left|W_{2^j} f(t)\right| \leq \log_2 K + j\alpha$$

In this formula, if K is a constant, then the Lipschitz index between the [a, b] is uniformly a.
If a = 2 j, then the above formula is:

$$\left|W_{2^j} f(t)\right| \leq K\left(2^j\right)^a$$

Table 2. The range of value of the control signal's attribute.

Value range of a	Properties of control signal
$m < a < m+1$	Control signal properties are of a small change, so the m+1 times function cannot micro.
$a = 1$	In the critical range of control command, the (t) is slope or the mountain function in a certain width.
$0 < a < 1$	Under the wrong operation, the function is discontinuous.
$a = 0$	The control signal does not change significantly; it is not likely to have control command transmission.
$a = -1$	Under repair, do not send any control request

or:

$$\log_2 \left|W_{2^j} f(t)\right| \leq \log_2 K + j\alpha$$

According to the analysis of the above theory and the control signal corresponding to the attribute value range in Table 2, the control command signal changes when the maximum value of the module changes. It can distinguish the noise of the signal, the maximum value caused by misuse, and the control signal that is not affected. Then the partial wavelet coefficients of the noise can be eliminated completely, and the corresponding coefficients of the control signals can be reconstructed effectively.

4 PRACTICAL APPLICATION TEST

To test whether the design method proposed by this research is effective, it uses the calculation method to verify the system. The control signal is transmitted and collected according to the different distances from the control signal. In the validation process, completely different attributes and command signals are divided into different groups, in which 3 of them are used to complete the signal control and cooperative transmission. The control signals are sent and collected according to different types. There are two kinds of control command, which are sent and closed. Simulation of command signal is shown in Figure 1.

Based on Figure 1, the results show that this system can satisfy the requirements of a certain job on the basis of the transmission and acquisition of multiple command signals. The performance of

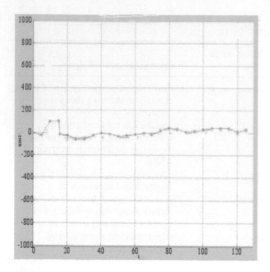

Figure 1. Waveform of the command signal.

the system can be satisfied with the general performance requirements, and can be used for receiving and sending multiple models. The performance of the test system is given, which can basically meet the requirements of setting distance, and achieve a more satisfactory effect.

Using the remote smart home control system based on the Internet of things, the simulation of the operator bears the starter in the instruction to identify the value of the received instruction information, and then controls and manages various types of accessed electrical equipment according to related instruction. In essence, the role of the simulation of the operator is similar as computer motherboard. It makes specific operation according to the instructions corresponded to the content, so as to achieve the effective control of equipment. The most obvious advantage of remote control is that it is easy to carry and operate. Users only need to integrate the control chip into the phone so that they can use the radio to control the home appliances. And in this process, the remote controller is ensuring that users can transfer the command to the signal receiver promptly and accurately, so as to facilitate the control system to make a more timely and effective response.

5 CONCLUSION

With the quickening pace of human life, people need to live and work under increasing pressure. Through the remote smart home control system based on the Internet of things, it can not only alleviates people's work pressure, but also improve their quality of life. However, in view of that the cost of remote smart home control system is relatively high and the operation is also very complex, it is still difficult to achieve a wide range of popularity. Therefore, it is necessary to put forward new design methods to promote the improvement of remote smart home control system, and ensure that the system can play a good role in our daily life.

REFERENCES

Deng Yun, Cheng Xiaohui. Design of intelligent home system for the Internet of things [J]. Journal of Guilin University of Technology, 2012, (2).

Dong Aimin, Xu Jing. Remote home sensing and control system based on the public cloud platform of Internet of things [J]. Modern Electronic Technology, 2016, (11).

Hu Xiangdong, Han Kaimin, Xu Hongru et al. Safety design and verification of the smart home Internet of things [J]. Journal of Chongqing University of Posts and Telecommunications (NATURAL SCIENCE EDITION), 2014, (2).

Meng Xiaoli. The design of intelligent home system based on cloud computing in the Internet of things platform [J]. Science Bulletin, 2016, (6).

Tu Liang, Duan Hongguang. Design of Internet of things smart home system based on 433 wireless transceiver modules [J]. TV Technology, 2012, (6).

Xin Hailiang, Zhong Peisi, Zhu Shaoqi et al. Intelligent home control system based on ZigBee [J]. Electronic Technology Applications, 2013, (12).

Xu Jinqiang. Application of BENQ-M23 GSM/GPRS module in remote intelligent home control [J]. Modern Electronic Technology, 2014, (18): 41–43.

Zhang Guiqing, Lu man, Wang Ming et al. The «spring» of smart home is coming [J]. Computer Science, 2013, (z1).

Electromechanical Control Technology and Transportation – Jia & Wu (Eds)
© *2017 Taylor & Francis Group, London, ISBN 978-1-138-06752-3*

Risk evaluation and method of complex information system

Zhong Guo
Unit No. 95899, PLAAF, Beijing, China

Hui Wang
CNCERT, Beijing, China

ABSTRACT: Based on the risk identification, according to the property, deterrent and fragility, the paper analyses the frequentness and influence of risk event, computes the risk level of the information system by means of both qualitative and quantitative methods, and provides the proof for formulating risk control policy.

1 INTRODUCTION

In the field of logistics, information system has become the main form and means of information collection and efficiency improvement over the full life cycle. Due to the critical role of information system, the occurrence of risk events will impose grave impacts on the process of commodity flow. How to effectively prevent and control risk events as well as guarantee the reliable operation of information system in a secure and efficient manner is a problem that must be addressed in the organization and application of information system.

Based on risk identification, this paper analyses the probability and consequences of the occurrence of risk events, assesses the risks by means of both qualitative and quantitative methods, computes risk levels and provides reference for formulating risk control strategies in logistics information system.

2 CONCEPT OF RISKS

2.1 Basic concept

Based on such functions of logistics information system as information collection, product tracking, transportation monitoring and storage monitoring, the information system risk is defined as: uncertainties that have impacts on the information system's effective completion of commodity flows. The existence and occurrence of such deterrent will enormously influence the validity and efficiency of commodity flows.

The risk assessment in logistics information system refers to analyzing man-made and natural deterrent facing the information system and the degree of possible harm in case of risk events from the perspective of risk management and through the methodology of qualitative and quantitative studies, based on which targeted prevention measures are proposed to minimize losses and negative impacts.

2.2 Features of risks

The fundamental feature of risk is uncertainty and potential losses. Information system risks show the following features:

a. Objectivity. Risks in the information system are objective reality. As long as the system is in operation, risks are always existing in a unique way within certain scope and occurring at a certain frequency and in different ways, causing losses to the system.
b. Universality. Risks universally exist in information system. Examples include instability of hardware in electromagnetic environment, slowing down of software after long-time operation and flaws resulting from imperfect technologies.
c. Contingency. Losses incurred by information system risks are contingent, meaning that time, place, nature and extent of the loss is nothing short of random combination and unpredictable results.
d. Harmfulness. Information system functions as the nerve system of logistics. Risks deter the successful completion of logistics flow or even lead to failure.
e. Controllability. By employing proper risk assessment methods and analyzing deterrent facing the information system through qualitative and quantitative studies, the occurrence of risk events can thus be controlled to the maximum extent.

3 ANALYSIS OF RISK

3.1 *Analysis process of risk assessment*

Risk events occur in information systems as a result of both external and internal causes, with the former being deterrent and the latter fragility. The probability of an incident can thus be calculated by assessing the deterrent to and fragility of the system. Meanwhile, since the impacts of an incident are related to property, the probability can thus be calculated by assessing the property. Therefore, R, the information system risk, can be taken as a function of property, deterrent and fragility, which is $R = g(c, t, f)$, with c referring to the impacts on property, t the frequentness of deterrent to the system, and f the severity of fragility.

The risk calculation model is shown in Figure 1:

Main procedures of risk assessment analysis include:

a. Identifying property and assigning the importance of property;
b. Identifying deterrent, describing their nature and assigning the occurrence frequentness;
c. Identifying fragility of property and assigning the specific severity of fragility;
d. Analyzing the probability of risk events based on the results of deterrent and fragility identification;
e. Calculating the losses incurred by risk events based on both the severity of fragility and importance of property;

Calculating the impacts on organization once a risk event takes place, which is the value at risk, based on its probability and losses.

3.2 *Risk identification*

Risk identification, the prerequisite of risk assessment, aims at understanding the risks objectively existing in the information system, diagnosing the root causes and analyzing the consequences and impacts of risk events. Information system generally consists of software equipment and hardware equipment as well as human participation. Taking into consideration the three elements in risk calculation model, risks can be classified as software risks, hardware risks and operational risks.

a. Software subsystem risk identification
Software subsystem risks refer to the risks that occur during the running of information system due to unpredictable software incidents or attacks to the software system and restrict the information system from functioning effectively. Software subsystem risks can be divided into system software risks, service software risks and logistics application software risks, as shown in Table 1.

b. Hardware subsystem risk identification
Hardware subsystem risks refer to the risks that occur during the running of information system due to unpredictable hardware failure or impacts of complex electromagnetic environment and restrict the information system from functioning effectively. Hardware subsystem risks can be divided into sensor equipment risks, computer

Table 1. Breakdown of software.

No.	Type of risk	Typical risks
1	System software risks	Bugs in Windows, Linux and other computer operating system software
2	Service software risks	Software exception in Database, Office and other service software that supports the operation of application software.
3	Logistics application software	Flaws in user interface, positioning and navigation, warehouse management and other application software.

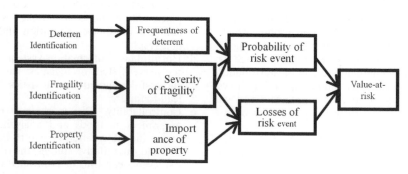

Figure 1. Risk calculation model.

Table 2.	Breakdown of hardware.	
No.	Types of risks	Typical risks
1	Sensor equipment risks	Failure in cargo scanners, conveyors and delivery units
2	Computer equipment risks	Failure in computers and servers
3	Communication equipment risks	Failure in communication network, data terminals and positioning equipment

Table 3: Levels of the frequentness.

Level	Description
1	Indicating this risk factor is extremely unlikely to occur
2	Indicating this risk factor is unlikely to occur (low possibility)
3	Indicating this risk factor is likely to occur
4	Indicating this risk factor is extremely likely to occur (high possibility)
5	Indicating this risk factor will definitely occur (highest possibility)

equipment risks and communication equipment risks, as shown in Table 2.

c. System operational risk identification

System operational risks refer to the risks occurring during the running of information system due to improper organization or other external factors that restrict the effective operation of the information system, with major factors listed as follows:

1. Design of the system anti-misoperation;
2. Psychological quality, professional skills and proficiency of the operators;
3. Other external factors, such as blackout.

3.3 Risk assessment

Risk assessment is the process based on risk identification that includes comprehensively analyzing risk factors and estimating the probability of risk events and possible losses with the help of risk assessment model, so as to determine the overall level of information system risks and provide scientific basis for risk management. The choice of assessment methods directly influences every aspect of the assessment process or even changes the final results. Therefore, specifics of the system should be taken into consideration when selecting the appropriate methods.

3.3.1 Qualitative assessment of information system

Qualitative assessment mainly depends on the knowledge, experience and lessons gained from the past of the person who assesses the risks. Without much burden of calculation, this method is relatively easier but strongly subjective, requiring some degree of experience and capability on the part of the assessor. Since some unpredictable factors are difficult to be analyzed quantitatively, the qualitative method is applicable and meaningful in some cases. Levels of the frequentness of risk events are shown in Table 3; levels of impacts on property are described in Table 4; and severity levels of fragility are described in Table 5 (Fu Yu, 2010)

Table 4. Levels of impacts.

Level	Description
1	The property is unimportant, with little or no impact.
2	The property is relatively unimportant, with little impact. (some degree of impact)
3	The property is moderately important, with moderate impact.
4	The property is relatively important, with relatively large impact.
5	The property is very important, with huge impact.

Table 5. Severity levels of fragility.

Level	Description
1	Extremely low fragility that is extremely unlikely to be exploited
2	Relatively low fragility that is unlikely to be exploited
3	Moderate fragility that is likely to be exploited
4	Relatively high fragility that is very likely to be exploited
5	Extremely high fragility that is extremely likely to be exploited

The probability of a risk event and the losses incurred to property should be analyzed when conducting qualitative analysis. Therefore, the knowledge information in Table 3, 4 and 5 should be comprehensively taken into consideration. If a risk event is extremely likely to occur (Level 4), with fragility that is very likely to be exploited (Level 4) and relatively large impact in case of occurrence (Level 4), we assume that the value-at-risk of this event is high, requiring attention and focus. In the case of the practical R & D and application of a logistics information system in an enterprise, according to experts' experience, a qualitative analysis matrix can be drawn as shown in Table 6.

Table 6. Qualitative analysis matrix.

No.	Risk Event	(Index) Frequentness	(Index) Impacts	(Index) Fragility	Management
1	Bugs in Windows, Linux and other computer operating system software	(2) unlikely	(4) large impact	(2) Low fragility	Medium risk requiring right action plan and fix within a certain time period
2	Software exception in Database, Office and other service software that supports the operating of application software.	(3) likely	(4) large impact	(2) Low fragility	Medium risk requiring right action plan and fix within a certain time period
3	Flaws in user interface, positioning and navigation, warehouse management and other application software.	(4) extremely likely	(4) large impact	(3) Moderate fragility	High risk requiring immediate implementation of right action plan
4	Unpredictable hardware failure due to complex electromagnetic environment	(2) unlikely	(5) huge impact	(2) Low fragility	Low risk requiring the person in charge of information system to fix or accept the risk
5	Failure due to the operators' psychological quality, professional skills and proficiency	(2) unlikely	(4) large impact	-	Low risk requiring the person in charge of information system to fix or accept the risk

3.3.2 Quantitative assessment of information system

Quantitative assessment method is used to study the quantitative relationship and changes of risks and present the assessment results with data. This method boasts more presentable, scientific and profound results, but suffers from complicated calculation.

a. Definition of entropy

In information science, entropy is a state parameter which is used to reflect uncertainties (ZHAO Dongmei, 2004). Each state of an information system corresponds to a kind of entropy and a risk as well. Once the entropy of a system is determined, the risk is accordingly set down. By employing entropy in risk assessment, we can master the rules of changes in system risks and thus conduct related monitoring.

b. Calculation of system risk and entropy

The functional relationship between system risk R and system entropy H is R = f(h), an increasing function, which indicates a linear rule between the two. In theory, we can identify a series of intervals of system risk R and thus adopt different level-based strategies to deal with risks in different intervals (Tang Yong-li, 2008).

Assuming that a system may be in the following n kinds of states and S_1, S_2..., S_n, P_1 refer to the probability of the system in the state of S_i, $0 \leq P_i \leq 1$, $\sum_{i=1}^{n} P_i = 1$, the entropy is:

$$H = (P_1, P_2..., P_n) = -\lambda \sum_{i=1}^{n} P_i \ln P_i \qquad (1)$$

In the function, $\lambda = \frac{1}{\ln n}$ and P_i is obtained from risk estimates with prior known data. It is defined that when, $P = 0$, $P_i \ln P_i = 0$.

With the above algorithm for entropy and the risk functional relationship R = f(h), we can get R_c, R_t and R_f, respectively impacts on property, frequentness of deterrent and severity of fragility. By taking into consideration the risks of all factors, we come to the following relationship of system risks:

$$R = g(c, t, f) = k_1 R_c + k_2 R_t + k_3 R_f \qquad (2)$$

In this function, k_1, k_2, k_3 refer to the relative importance of the three factors, and $k_1 + k_2 + k_3 = 1$.

c. Case study of quantitative assessment

According to the logistics information system organization in an enterprises, its risks are divided into software system risks R_1, hardware system risks R_2, and system operational risks R_3. Taken the software system risk R_1 as an example, we conduct quantitative assessment in the following procedures:

1. Building set of risk factors and set of assessment

A set of risk factors is built as A = $\{a_1, a_2..., a_n\}$, referring to n kinds of software risks as listed in Table 1; Sets of assessment is built as $B_c = \{b_{c1}, b_{c2}, b_{c3}, b_{c4}\}$, $B_t = \{b_{t1}, b_{t2}, b_{t3}, b_{t4}\}$ and $B_f = \{b_{f1}, b_{f2}, b_{f3}, b_{f4}\}$, referring respectively to frequentness of deterrent, impacts on property and severity of fragility of a risk event, with specific meanings laid out in Table 3, Table 4 and Table 5.

2. Evaluating membership matrix

With prior professional opinions on risk factors of the information system, P_i, the membership probability of all risk factors is evaluated to form the membership matrix P, as shown in Table 7.

3. Calculation of value-at-risk

With the definition of function (1) and risk functional relationship R = f(h), we can calculate R_c, R_t and R_f, respectively the value-at-risk of impacts on property, frequentness of deterrent and severity of fragility.

With the definition of function (2) and specifics of the information system, we can assign k_1, k_2 and k_3 to calculate R_1, the software system risks of this information system.

4. Integration of value-at-risk

Considering the relative importance of all components in the information system and comprehensive assessment, weighted average method is employed to calculate the value-at-risk

Table 8. Demonstration of grades.

R	0~0.2	0.2~0.4	0.4~0.6	0.6~0.8	0.8~1
Nature of risk	Low	Relatively low	Medium	Relatively high	High

in the information system. We respectively assign software equipment risk R_1, hardware equipment risk R_2 and system operational risk R_3: d_1, d_2, and d_3; and assume that $d_1 + d_2 + d_3 = 1$. By adopting the weighted method that is $R = d_1 R_1 + d_2 R_2 + d_3 R_3$, we can calculate the total value-at-risk of the information system.

5. Risk management

Based on the value-at-risk of the information system and risk grades shown in Table 8, we can evaluate the risk levels of the logistics information system and guide the risk management.

4 CONCLUSION

Information system risks are objective reality. How to effectively assess risks and take proper and efficient measure to manage risks is a problem that must be solved in the organization and application of information system. Based on risk identification, this paper discusses about risk assessment process, analyzes the frequentness and influence of risk events, computes the risk level through qualitative and quantitative methods, and lastly introduces a case study of a logistics information system to provide reference for formulating risk control policy.

REFERENCES

BSI.IT Baseline Protection Manual. Standard Security Measure. Version: October 2000.

Christopher Alberts, Audrey Dorofee. Managing Information Security Risk: The OCTAVE Approach.

Fu Yu. An Approach for Information Systems Security Risk Assessment on Fuzzy Set and Entropy-Weight [J]. Acta Electronica Sinica, 2010, 7(7):1489~1494.

GB/T 20984-2007. [S]. 2007.

Tang Yong-li. Information Security Risk Analysis Model Using Information Entropy [J]. Journal of Beijing University of Posts and Telecommunications, 2008 31(2):50–53.

Zhao Dongmei. Fuzzy Risk Assessment of Entropy-weight Coefficient Method Applied in Network Security [J]. Computer Engineering, 2004, 30(18):21–23.

Table 7. Membership matrix.

	b_{c1}	b_{c2}	b_{c3}	b_{c4}	b_{t1}	b_{t2}	b_{t3}	b_{t4}	b_{f1}	b_{f2}	b_{f3}	b_{f4}
a_1	0	0.1	0.2	0.5	0.1	0.4	0.3	0.1	0.1	0.5	0.2	0.2
	0	0.1	0.3	0.5	0	0.25	0.4	0.25	0.1	0.5	0.2	0.2
	0	0.1	0.2	0.4	0	0.1	0.3	0.5	0.1	0.2	0.5	0.1
...	...			...			...			...		
a_n	...			...			...			...		

Electromechanical Control Technology and Transportation – Jia & Wu (Eds)
© *2017 Taylor & Francis Group, London, ISBN 978-1-138-06752-3*

Development and application of the load simulation software for shearer drum

Rui Zeng
Department of Airfield Engineering, Air Force Logistics College, Xuzhou, China
Field Engineering College of PLA University of Science and Technology, Nanjing, China

Yong Zhang, Jinkun Sun & Jie Li
Department of Airfield Engineering, Air Force Logistics College, Xuzhou, China

ABSTRACT: On the basis of the establishment and analysis of the model of the cylinder, the generation method of a random number of different distribution laws was studied. MATLAB was used to simulate the distribution of different coal seams, and the size and position parameters of the distribution of the inclusions were output, which provided the data for the simulation of the drum cutting load. According to the factors determined by the load of the drum, the relevant program was compiled, and a window method was put forward to judge the cutting teeth of the cutting area in the process of simulation. The starting interface, working interface, roller structure parameter interface, working parameter setting interface, and calculating input coefficient interface of the simulation system were designed by using the GUI human–machine interface development tool. Three-dimensional modeling software Pro/E was used to establish a three-dimensional model of the cylinder, and the output of the simulation system was loaded into the three-dimensional model. Strain analysis, stress analysis, and modal analysis were carried out using ANSYS software. The results show that when cutting the inclusions, the cutting force can be cut off, and the fluctuation of random cutting load is larger than that of the average cutting load. In the case of the nozzle, the load of the drum cutting was obviously decreased, but the load fluctuation of the drum cutting was not greatly improved. In the case of blade cutting nozzle configuration, the cutting resistance is reduced greatly.

1 INTRODUCTION

Using computers to simplify the real system, modeling, and simulation is a common method and effective means in the field of modern scientific research. To carry out the simulation research, the real system should be simplified, and the mathematical language of the load calculation will be converted into computer language. At the same time, the programmer must fully grasp the software users in the process of software design mentality and try to provide a simple and practical software tool for users to learn. MATLAB is one of the popular mathematics application software packages in the world. It uses the original core of the numerical calculation model, which has the function of calculating, managing, and visualizing the data. Since 1984, its core used C language and the numerical calculation based on the function of the graphics processing function. Now, MATLAB is the most popular software system with its intelligent visualization and human–computer interaction system, modular calculation method, rich matrix computation, and data processing and graphics, rendering functions like a simulation and control system design.

2 SIMULATION METHOD

2.1 *Load simulation of random cutting drum*

When the spiral drum break coal, the load on the cutting gear was formed under the influence of various complex and changing factors in space. With a large number of random factors, these random factors were difficult to predict. The related sample statistics showed that the random load of the roller was the superposition of the average load component and the random component (O.Z. Hekimoglu, 2004 & E. Mustafa Eyyuboglu, 2005). It was assumed that the mean of the random load on the pick was equal to the mean of the average load. In this case, the random load of the single tooth could be determined by a certain distribution (Du Chang-long, 2008). Cutting resistance of single pick on the cut obeys the gamma distribution or Weibull distribution. The correlation between cutting resistance and traction resistance is relatively large, and its correlation coefficient (R_{zy}) is 0.67–0.85 (Liu Song-yong, 2008). The lateral resistance is not correlated with other cutting resistance, which is $R_{zx} = R_{yx} = 0.078$–0.271 (T. Muro, 2004).

Because the gamma distribution is difficult to be used in engineering practice, it is replaced by the Rayleigh distribution of two stationary random processes. Rayleigh random process can be used for two independent normal random process of the same distribution, the cutting resistance center value of the press type calculation:

$$Z_{cp}(i) = \sqrt{\xi_1(i)^2 + \xi_2(i)^2} - 1.25 \tag{1}$$

At this time, a random process of cutting resistance can be calculated by pressing:

$$Z_i = Z_{cp}\sigma_{\bar{z}} + \bar{Z}_i \tag{2}$$

According to the correlation coefficient, the center value of the traction resistance can be determined by the type:

$$Y_{cp} = \sigma_{\bar{Y}}\left(R_{zy}Z_{cp} + \sqrt{1 - R_{zy}^2}\,\eta\right) \tag{3}$$

2.2 Simulation of cutting load of coal seam with inclusion

When the load of the coal seam in the package was cut by the simulation cylinder, the coal seam of the package body was simulated first, and the size of the package body was determined. Then, the pure coal seam load simulation program was used in the process to determine not only the cutting teeth in the cutting state but also the pick and the inclusion of the contact.

When the cutting tooth tip coordinates satisfy the above inequality, we could determine the cutting inclusions. The research showed that when the cutting tooth tip was cut off from the center of the inclusion body, the maximum cutting force could be generated, and the cutting force of the cutting form could be 0.3–0.9 times the maximum load (Bemardino Chiaia. 2001).

The average cutting resistance of the iron sulfide inclusions was three to seven times of the coal, and the cutting force of the cutting force could be multiplied by a certain amplification factor on the basis of cutting the cutting of the inclusions. The amplification factor was determined by the cutting impedance of the wrapped body and the form of the cut of the cut.

2.3 Simulation of cutting force of nozzle for cutting

Water jet assisted cutting could reduce the cutting force on the cutting teeth, and the cutting force could be reduced by other cutting conditions. Under the experimental conditions, the cutting force of the cutting force was reduced and the distance is two times. However, the relevant experimental study found that in the case of other conditions, at the same time, the cutting force was changed to reduce the proportion of great change. Therefore, the reduction ratio of the cutting force in the cutting process of the water jet-assisted cutting was a factor that was affected by many factors, and the relationship between the cutting force and the cutting condition could not be determined at present.

The specific simulation process could be divided into three steps, which could be divided into the position of the nozzle, the calculation program of the drum load, and the cutting force correction. First, the system was determined according to the input nozzle coordinates, and then the load calculation program was carried out. After the calculation was completed, according to the judgment result and the cutting force, the cutting force was reduced, the corresponding cutting force was modified, and the final water jet assisted cutting drum load curve was output. In the simulation process, the cutting force was reduced by two types of input methods. The first was the direct random input, the second was the computational input, and the second input methods were generally based on the test.

3 SOFTWARE INTERFACE AND FUNCTION DESIGN

3.1 Start-up interface design

Start-up interface generally does not have the actual function; however, as the first window of human–computer interaction, the quality of its design will directly affect the users' impression of the software, so the design work are equally important.

As professional application software, the start interface of the software to the actual work screen has a background and music. When starting the software, the interface in the start gives users a refreshing feeling at the same time more prominent software features and shows its application areas.

3.2 Work interface design

Working interface is the most important one in the whole software interface, which plays an important role in the operation of the software. A reasonable layout clears the work interface to allow users to the function of software composition and the composition of the formation of rapid memory.

1. "Inclusion" radio button group
A radio button group for inclusion of coal seam condition control load simulation. Different options of "inclusion" radio button group by the

user can be simulated in the coal seam containing inclusions and drum load under the condition of not containing inclusions.

2. "Pick distribution form" radio button group

Pick the distribution form of radio button group containing "order type" and "checkerboard" radio button. Click the different arrangements of the cutting teeth, the software will call the corresponding lateral force calculation program in operation.

3. "Pick load type" radio button group

Pick load type radio button group contains the "average load" and "random" radio button. When the average load is selected, the software will call the average load simulation program during the running process.

4. "Water jet" radio button group

Water jet radio button group and used to control the water jet-assisted cutting drum load simulation.

5. "Coal" radio button group

Set the radio button group including coal "brittle", "brittle", and "toughness" radio button. At different coal qualities, the program will call different parameters to calculate the coal seam.

4 SOFTWARE APPLICATIONS

The water jet-assisted cutting software can be output in the form of the three direction force at the time of the roller load simulation. The users can use the software to obtain the data at any time by force data of drum in the process of load analysis.

Table 1. Force data of drum picks.

Stub	Angle/°	a_i/N	b_i/N	c_i/N
0	0	−907.64	−8120.71	264.95
0	60	6586.55	−4846.66	271.14
0	120	7271.43	2929.54	2.14
20	20	−5897.08	−5551.94	195.58
20	140	8035.16	−2240.36	440.88
44	80	2344.71	−8319.25	709.51
72	40	−3591.60	−7595.04	345.25
72	160	8156.07	517.17	179.79
104	100	5244.87	−7094.73	591.33
150	65	258.27	−8868.12	417.17
230	48	−2048.12	−9591.77	413.62
230	168	8648.75	2314.99	166.69
310	31	−4756.50	−8187.90	307.91
310	151	9469.18	−25.30	307.91
390	14	−6758.62	−5872.07	166.69
390	134	9425.53	−2712.22	413.62
470	118	8508.82	−5218.77	477.66
550	104	7069.48	−7146.84	507.70
630	89	4996.40	−8744.424	516.22

The use of Pro/E models and the simulation of the roller were randomly selected from a group of shearer drum cutting coal in cutting tooth stress data for meshing. The roller cutting force data and stress and strain analysis process are shown in Table 1 and Figure 1.

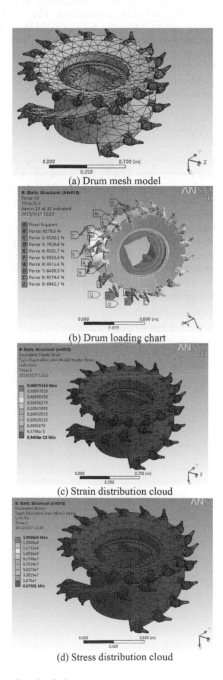

(a) Drum mesh model

(b) Drum loading chart

(c) Strain distribution cloud

(d) Stress distribution cloud

Figure 1. Analysis process.

The applied load (B-J) and constraints (A) built to solve the static finite element model of spiral drum was established by spiral roller of equivalent strain and stress distribution, as shown in Figure 1 (c) and (d). It is evident from the figure that the stress distribution was mainly in the tooth tip, blade, and hub connection and cutting tooth of the end plate connection. The tooth tip should be larger than the tooth seat, blade, disk, and drumer at the end. The maximum stress and maximum strain were approximately 150.84 and 0.00075 MPa, respectively, which were far less than the yield limit. However, the maximum stress was within the allowable range.

5 CONCLUSIONS

1. According to the establishment of the calculation model, the MATLAB software was used for the preparation of pure coal seam average cutting load simulation program and the drum cutting load simulation program.
2. According to the load simulation system, the GUI interface development tool was used to design the software startup interface, inclusion simulation interface, water jet nozzle configuration interface, the interface of drum structure parameters, and so on.
3. The selected drumer was established by using Pro/E 3D model and imported into ANSYS Workbench, which completed the analysis stress and modal strain, and indicated the practicability of load simulation system of water-jet-assisted cutting of the cutting drum.

REFERENCES

Bemardino Chiaia. 2001. Fracture Mechanisms induced in a brittle material by a hard cutting indenter[J]. *International Journal of Solids and Structures*, 38(5): 7747–7768.
Du Chang-long, Liu Song-yong, Cui Xin-xia. 2008. Study on pick arrangement of shearer drum based on load fluctuation. *Journal of university of mining & technology*, 32(4): 305–310.
Hekimoglu, O.Z., L. Ozdemir. 2004. Effect of angle of wrap on cutting performance of drum shearers and continuous miners. *Mining Technology* 113(7): 118–122.
Liu Song-yong, Du Chang-long, Cui Xin-xia. 2008. Research on the cutting force of a pick. *Mining science and technology*, 27(9): 514–517.
Muro, T., D.T. Tran. 2004. Regression analysis of the characteristics of vibro-cutting blade for tuffaceous rock. *Journal of Terramechanics*, 40(1): 191–219.
Mustafa Eyyuboglu, E., Naci Bolukbasi. 2005. Effects of circumferential pick spacing on boom type roadheader cutting head performance. *Tunnelling and Underground space Technology*. 20(3): 418–425.

Electromechanical Control Technology and Transportation – Jia & Wu (Eds)
© 2017 Taylor & Francis Group, London, ISBN 978-1-138-06752-3

Advantages and disadvantages of FBG sensors and strain gauges in the pullout test of GFRP soil nails

Zhen Song & Hongwei Zhu
Jiangxi University of Science and Technology, Gan Zhou, Jiang Xi, China

ABSTRACT: This research report focuses on the pros and cons of FBG sensors and strain gauges in the pullout test, which is used to monitor the behavior of GFRP soil nails. A lab pullout test of GFRP soil nails was carried out according to the review of FBG and soil nailing technique. The axial strain distributions data and pullout force data were extracted by using FBG sensors, strain gauges, a load cell, and LVDTs during the test. A comparison of data measured by FBG sensors and strain gauges has been carried out to find the advantages and disadvantages of FBG sensors and strain gauges in the pullout test of GFRP soil nails. The test clearly indicated that the FBG sensors performed better than strain gauges in terms of stability, lightweight, reliability, and accuracy.

1 INTRODUCTION

The soil nailing technique is a technique that uses a great deal of structures (such as steel bar and fiber glass bar), which are closely spaced into the soil in situ to increase the strength of soil slopes and excavations. Obviously, the poor corrosion resistance of the traditional materials has seriously limited its further application. On the contrary, the GFRP soil nail has many advantages over the traditional materials in aspects such as lighter weight, higher strength, and higher electromagnetic resistance, which is significantly lower in relaxation and is more economical. Strain gauges and Fiber Brag Grating (FBG) can be used to monitor the mechanical mechanism of the GFRP soil nail. However, their pros and cons are not defined clearly in previous studies. This report would obtain the advantages and disadvantages of FBG sensors and strain gauges in the pullout test of GFRP soil nails.

2 REVIEW

2.1 *Brief introduction of FBG*

A Fiber Bragg Grating (FBG) is a type of distributed Bragg reflector that can create a periodic variation in the refractive index of the fiber core. In 1978, Hill et al. firstly demonstrated the formation of permanent gratings in an optical fiber at Canada. As time goes by, the FBG has played an increasingly important role and has become the basis for a technology in the field of sensor systems and optical communications. Moreover, FBG was found to have a broad range of applications and it is used with higher efficiency.

2.2 *Soil nailing and soil nailing technique*

Soil nailing is a technique of reinforcing a slope in a passive way. It has been popular for over 30 years in retaining excavations or stabilizing slopes. Watkins and Powell carried out some research aiming to achieve slope improvement in 1992. The research showed that soil nailing is economic and flexible. And then, soil nails began to be used in France, Germany, and North America. Powell and Watkins (1990) pointed out that the soil nails could only be used as a measure to deal with limited areas and deeply weathered material or good rock before 1987. However, it is a fast and flexible construction method, thereby making it more attractive. Soil nails are commonly steel rods surrounded by the grout, thereby enhancing the bond between nails and the soil and providing corrosion protection. Soil nailed construction consists of a system of soil nails and facing. Figure 1 shows a typical arrangement of grouted soil nails.

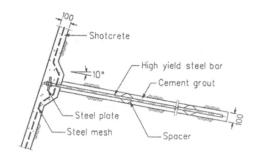

Figure 1. Typical detail of soil nail (Shiu et al., 1997).

In the early of 1960s, the soil nailing technique was developed (Clouterre, 1991; FHWA, 1998). With the widespread use of the soil nailing technique, in the early 1970s, semi-empirical designs for soil nailing began to develop. What followed is the first systematic research on the soil nailing technique (model tests and full-scale field tests) in Germany in the mid-1970s. In the 1990s in the era of the 20th century, some meaningful development work was carried out in France and the United States. In the subsequent decades, the research and the development work provided the basis and useful reference for the soil nailing technique.

3 EXPERIMENTAL INVESTIGATION

This part presents the experimental investigation on performance of the GFRP soil nail. A comprehensive experimental investigation was carried out to figure out the advantages and disadvantages of FBG sensors, and strain gauges were used to observe the behaviour of the GFRP soil nail in the pullout test. In order to obtain the characteristics of GFRP, full details of soil nailing of eight tests were conducted (eight stages of the pullout test). For the purpose of comparison, eight FGBs and additional three conventional electric strain gauges were attached to the soil nails.

The laboratory consists of the following:

- A test pipe
- Pullout device
- Drilling and grouting device
- Soil sample

3.1 Test pipe

The dimensions of a test PVC pipe are as follows: 2.5 m length, 0.3 m diameter, and 10 mm thickness. The diameter of the GFRP bars is 25 mm.

3.2 Pullout device

The maximum pullout capacity is 300 KN.

The maximum pullout displacement is 50 mm.

Moreover, the device permits the pullout loads to be applied manually.

3.3 Drilling and grouting device

Holes were drilled manually for the installation of the soil nails by using an embedded PVC pipe which is 110 mm in diameter. The pipe is divided into six parts along the longitudinal direction.

At one end of the pipe there is a plastic plate, and at the other end, there are two wooden plates (97 mm × 25 mm × 20 mm). A grouting pipe and funnel are used for pouring the grout into the hole by gravity flow.

3.4 Instruments

All of the instruments are listed in Figure 2.

3.5 Soil sample and soil properties

In order to obtain the soil sample of 13% water content, all soil samples would be mixed. In this study, CDG was used, and the sample contained 16% of fine particles, 50% sand, and 34% gravel. According to the soil classification system, this sample can be classified as yellowish brown and silty gravelly SAND, as shown in Table 1.

3.6 Nail property

In this study, a 110 mm diameter grouted soil nail was used. The 25-mm diameter GFRP bar was a ribbed bar having an elastic modulus of 45 GPa. A stiff mortar cannot fill all the voids, because the water content of the grout in contact with sand was reduced.

3.7 Test procedure

The complete experimental procedure is given as follows:

1. Filling the test pipe with soil sample including installation of FBG sensors and strain gauges.
2. Composing portal frames and placing of the top plate.
3. Installation of soil nails and curing (at least 28 days curing time before carrying out the pulling out test).
4. Checking the FBG sensors and strain gauges.

Capacity	Transducers	Nos.	Measurement
Load Cell	300kN	1	Pullout force
CDP 50 type LVDT	50 mm	1	Displacement of nail
TML Strain Gauges FLA-5-11 type		3	Strain distributions along the nail during pullout
FBG sensor		8	Strain distributions along the nail during pullout

Figure 2. List of laboratory instruments.

Table 1. Physical properties of the soil used.

Properties	Value
Gravel	34%
Sand	50%
Fine particles	16%
D_{10}	0.05 mm
D_{30}	0.56 mm
D_{60}	2.1 mm
Liquid limit	43%
Plastic limit	32%
Plasticity index	16%
Maximum dry density	1950 kg/m³
Optimum moisture content	12%

5. The first pullout test on nails should be taken conducted the self-weight of the soil.
6. Repeating Step 5 as required.
7. Checking the in situ water content and removing the soil from the test pipe.

4 TEST RESULTS

4.1 *Axial tensile force distribution along the nail measured by using FBG sensors*

Each instrumented point was kept at a distance of 25 cm, which is composed with FBG sensors. The nail axial force during the pullout test was calculated from the strain measured at different positions. The axial force at a particular position of the nail is equals to,

$$F = E \times A \times \varepsilon$$

where, F is the axial force; E is the elastic modulus of the nail; A is the cross-sectional area of the nail; and ε is the average strain.

The calculation results are in agreement with the experiment data by analysis of the calculation results and experiment data. The elastic modulus of the grout and GFRP bar are 28 GPa and 45 GPa, respectively. The change of the axial force during the pullout test at different positions of the nail has been presented. The force at the nail head is much bigger than that measured by using the load cell. Data measured with FBG sensors have a tendency of decreasing gradually (from one to eight).

The data presented in this work are only related with this research. Figure 3–Figure 7 show the test results of the axial tensile force distribution along the nail measured by using FBG sensors.

4.2 *Force distribution along the nail measured by using strain gauges*

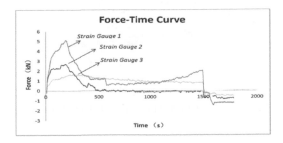

Figure 3. The force–time curve of stage 1 obtained by using strain gauges.

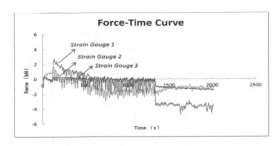

Figure 4. The force–time curve of stage 2 obtained by using strain gauges.

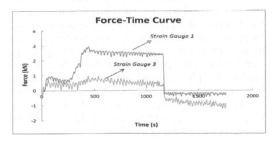

Figure 5. The force–time curve of the stage 3 obtained by using strain gauges.

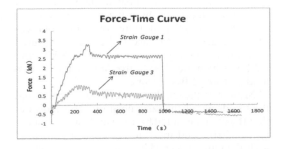

Figure 6. The force–time curve of stage 4 obtained by using strain gauges.

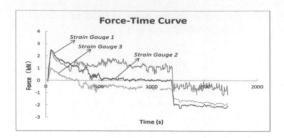

Figure 7. The force–time curve of stage 5 by using strain gauges.

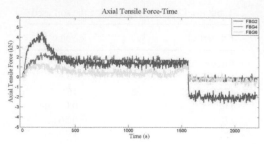

Figure 8. ATF–time curves at stage 1 are measured by using FBG 2, 4, and 6.

5 DATA ANALYSIS

5.1 *FBG sensors*

Some data were obtained through the experiments by using FBG sensors, but some of them do not have accuracy of these data. Each data of the force measured by using the first FBG sensor is much bigger than the maximum force measured by using the load cell. Therefore, the right data of No. 2 to No. 8 FBG sensors should be found out.

By observing the data, it can be understood that, at stage 1, after a dramatically increasing trend (from 0 to 200 s), there is a long stable period till 1500s. The peak number is about 4.5 kN. And then, the data dropped to a negative number sharply, because of the pre-stressing force on the installation of experimental instruments. In every stage, there is a strong rising trend, which dropped to a long stable period, and finally sharply dropped to a negative number. With respect to the different slopes, these reflect the different speeds of loading. Moreover, the strain from the end to the top of the soil nailing has been increasing gradually.

5.2 *Strain gauges*

Besides, the axial tensile force at the No. 2, 4, and 6 positions measured by using a strain gauge is roughly equal to that measured by using FBG sensors at stages 1–5. At stage 1, the forces measured by using a strain gauge are 4.8 kN, 2.5 kN, and 1.2 kN, which are close to the data that are measured by using FBG sensors.

Moreover, the numbers of forces are similar, and the trends are at the same position. This indicates that the data measured by using an FBG sensor are more reliable and more accurate when compared with the data measured by using a strain gauge. It is shown in Figures 8–12.

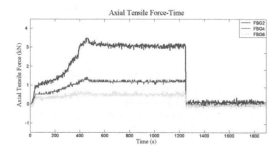

Figure 9. ATF–time curves at stage 2 are measured by using FBG 2, 4, and 6.

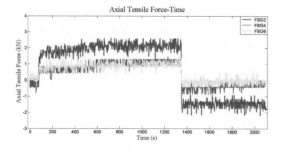

Figure 10. ATF–time curves at stage 3 are measured by using FBG 2, 4, and 6.

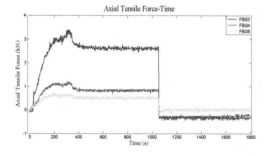

Figure 11. ATF-time curves at stage 4 are measured by using FBG 2, 4, and 6.

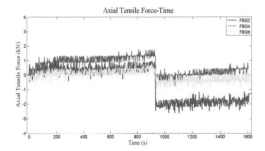

Figure 12. ATF–time curves of stage 5 are measured by using FBG 2, 4, and 6.

6 CONCLUSIONS

This research project is focused on the advantages and disadvantages of FBG sensors and strain gauges in the pullout test, which is carried out to monitor the behaviour of GFRP soil nails. The performance of FBG sensors and strain gauges are evaluated to know whether it is good at exhibiting variations of strain. All test results provide some valuable information about the advantages and disadvantages.

On the basis of the data and analysis, several conclusions can be drawn, which are as follows:

1. The FBG monitoring system and the strain gauges monitoring system have been used successfully in monitoring strain distributions during pullout tests.
2. The FBG sensors are reliable for strain monitoring of GFRP soil nails and have the advantages of light-weight and stability. The data measured by using FBG sensors are more accurate when compared with the data measured by using the strain gauge.
3. The strain from the end to the top of the GFRP soil nailing process is increasing gradually along the longitudinal direction.
4. The cracks of the cement grout might have appeared in the part of the non-pressurised grouted section while carrying out the pullout test of the GFRP soil nails.

REFERENCES

Barley, A.D., Davies, M.C.R. and Jones, A.M. "Review of current field testing methods for soil nailing". Proc. 3rd Int. Conf. On Ground Improvement Geosystems, London, England, pp. 477–483 (1997a).

Chen, P.Y.M. "Methods of test for soils in Hong Kong for Civil Engineering Purposes (Phase 1 Tests)". Geotechnical Engineering office, Civil Engineering Department, The Government of Hksar, GEO Report No. 36 (1992).

Chu, L.M. and Yin, J.H. "A Laboratory Device to Test the Pull-Out Behaviour of Soil Nails". Geotechnical Testing Journal, ASTM, 28(5), September, pp. 1–15 (2005a).

Lee, C.F., Law, K.T., Tham, L.G., Yue, Z.Q. and Junaideen, S.M. "Design of a large soil box for studying soil-nail interaction in loose fill". Soft Soil Engineering, Lee et al (eds.), pp. 413–418 (2001).

Pei H F, Cui P, Yin J H, Zhu H H, Chen X Q, Pei L Z and Xu D S2011 Monitoring and warning of landslides and debris flows using an optical fiber sensor technology J. Mount. Sci. 8 728–3.

Electromechanical Control Technology and Transportation – Jia & Wu (Eds)
© 2017 Taylor & Francis Group, London, ISBN 978-1-138-06752-3

Research on the sport city model—based on data of outdoor activities

Qiaohua Qin
South China University of Technology, GuangZhou, Guangdong, China

Ying Luo
State University of New York College of Environmental Science and Forestry, New York, NY, USA

Ye Liu
South China University of Technology, GuangZhou, Guangdong, China

ABSTRACT: Recently, outdoor activities gained their popularity in China. From a global perspective, China is only one of the followers of the outdoor activities' fever, which has spread around the world. In this study, the degree of favorite outdoor activities nationwide is illustrated by digging in a visible big data set, especially in developed countries such as the U.S, Germany, England, and Japan. Through an analysis of the initiator of global outdoor activities, the deep relationship between the urban outdoor activities' participation rate and the degree of development of cities is found. Taking the U.S as an example, it suggests to us an interaction between urban planning and public health, and an influence of the construction of the city space network on public health. A well-established city health system is required to incorporate the environmental factors in frameworks of economics, epidemiology, and environmental sciences. Hence, based on the concept "urban ecological niche", the concept "urban sport niche" is introduced and effectively incorporated into an overall framework. In addition, basic data from 21 cities of Guangdong province were employed as samples to establish research models, which provide urban planning expertise suggestions in public-health-oriented sport cities, and further nurtures the "sport city model" that can be applied to Urban Planning and the management of city health.

1 INTRODUCTION

Recently, outdoor activities have become a popular fashion. Activities like hiking, camping, and jogging have become an important part in human life. According to a research report of the COA, there are more than 380 million people in China who follow regular exercise regimes, which accounts to 27.79% of the Chinese population. Among them, 130 million people engage in outdoor activities such as hiking and trailism, which comprise 9.5% of the Chinese population. Outdoor activities such as mountain climbing and rock climbing are very popular among 60 million people, which comprise 4.38% of the Chinese population. (COA Chinese Outdoor Activities Research Report, 2013)

In addition, the numbers of marathon competitions have evolved drastically in China lately. According to the data provided by Chinese Athletics Association, in 2011, there were only 22 national marathon competitions, but in 2015, the number has increased to 64. Now metropolis cities exhibit strong willingness to hold marathon competitions than ever, because it is not only a great way to promote the city's image, but also a way to increase fiscal revenue. By estimation, in 2015, the marathon competition in held in Xiamen directly contributed approximate 207 million RMB to the local fiscal revenue, and the economic influence was over 255 million RMB (GF Securities Research Center 2015).

2 GENERAL SUMMARY OF THE DEGREE OF PARTICIPATION OF OUTDOOR ACTIVITIES GLOBALLY

The following visualized data map sets demonstrated the degree of fever for outdoor activities globally. These data maps were derived from users' data of the popular social network app "Strava" globally, which includes 77,688,848 bikers' and 19,660,163 runners' sport route data. The more intense the colored chunk is, the greater is the number of users in that area.

Judging from the global scale, European countries are outstanding, especially for UK and Germany. In other areas, Korea, Japan, U.S., and Brazil have relative good data indications. When compared to those regional outperformers, China has underperformed.

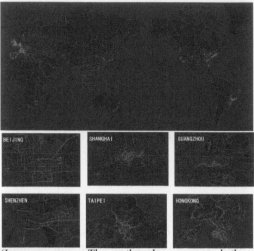

(Images source: The author has processed these images with Photoshop software based on the existing data).

Figure 1.   Heated map of Strava Chinese users' outdoor activities data.

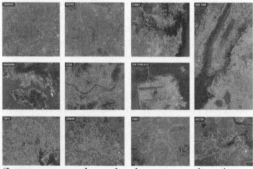

(Images source: the author has process these images with Photoshop software based on the existing data).

Figure 2.   Heated map of Strava global users' outdoor activities data.

It is easy to spot the difference from the following thermodynamic diagram (data retrieved from Nike Sport). Darker orange color represents more users in that area, as shown in Figure 1.

It is reasonable for us to conclude from the above evidences that, there is a high positive correlation between the development degree of a city and the degree of outdoor activities' participation. Cities with high participant rates are generally developed cities. In China, metropolis cities such as Beijing and Shanghai have a relatively high degree of participation, but they are still underperforming when compared with that in London, New York, and Tokyo, as shown in Figure 2.

It is also worth-mentioning that China mainland cities are also underperforming when compared with Hong Kong and Macau. Although these data are collected based on the commercial product, and users might not cover all participants in these areas, they are still representative in this demonstration.

3   THE EVOLUTION OF AMERICAN SPORT CITIES

It is logical to arrive at an argument based on our previous analysis: The degree of outdoor activities' participation of a city has a direct relationship with the city's development, economic status, environment, and even the development of its country. Therefore, what are the root causes for this phenomenon? It might be a good way to answer this question by studying the U.S, the initiator in outdoor activities' area. Our following comprehensive analysis will be based on economics, epidemiology, and environmental science as entry points.

In 1970s, in the U.S., 1 in 10 people suffered obesity, and now the number is 1 in 3. More than 2/3 of American citizens are overweight (Jan Gehl, 2010). Twenty-five percentage of young males and 40% of young females do not fit the requirements for joining the military because of overweight. (Elizabeth Kolbert, 2010) According to the data of the U.S. Disease Control Center, 1/3 of citizens born after 2000 will be diagnosed diabetes in the future. This is the first generation in the U.S that will have shorter longevity than their parents.

The government knew that dietary habit is the cause for obesity. Dietary habit is the main problem that affects one's weight. Since most Americans lack efficient exercise, their weight will increase to an extent that it becomes a direct threat to their health. When they started to solve this problem, they realized that the layout of American cities' constraint people to exercise. Without sufficient footpaths, it is nearly impossible for citizens to walk other than drive in the city.

Most Americans rely on automobiles for their daily life, but automobiles are not indispensable. In 1970s, Americans spent nearly 1/10th of their income on traffic. After that, the traffic highway numbers were doubled, and people spend approximately 1/5th of their income on traffic now. A labor family (with an annual income of 20 to 50 thousand dollars) spends more money on traffic than on housing (Patrick Condon, 2004). What will a city be if it shifts its major traffic instruments on automobiles to others? Portland might be a good example. One good example is that the

Portland government made a series of decisions in 1970s to differentiate it from other regular US cities. When other cities are expanding their borders and highways, Portland set limits on its expansion; when other cities widened their roads and cleared grasslands to increase vehicle flow, Portland narrowed automobile roads; when other cities invested in new highway roads, Portland invested in bikeways and walkways. They invested more than 60 million dollars in bicycle facilities (Heidi Garrett-Peltier, 2010). This figure seems to be substantial, but this number is an investment of 30 years in total. It is approximately 2 million dollars each year, which is only comparable to half of the cost of building a new flyover. However, this little shift changed the lifestyle of people in Portland. Their average driving mileage declined continuously from 1996, and is approximately 20% less than other areas in the U.S. A Portland resident drives 4 miles or 11 minutes less everyday than he/she used to be. Economist Joe Cortright calculated and concluded that this 4 mile or 11 minutes saved residents' income by 3.5% per year (Joe Cortright, 2007).

Gradually, well-educated young adults started to move to Portland. In the last two American Population censuses, it was observed that Portland residents with college level education have increased by 50%, which is 5 times the figure of the national average. As the city with the greatest population expansion in the west coast, Portland attracts many excellent young adults through its habitable environment and urban structure. Portland has become an "elite town" in the U.S, and possesses great allure to young people. In Portland, young men and women walk outside, and you can easily see people jogging or biking in the streets and parks. On one hand, the urban structure that is favorable for outdoor activities and people's need for outdoor activities are satisfied. On the other hand, it makes Portland a cool town that attracts more people to move in. Besides, Portland is the North American headquarters for famous sport brands such as Nike, Columbia, and Adidas, and the sports industry has become the symbolic industry of Portland. Its creativity for products are highly related with the outdoor exercise environment, people's lifestyle, and the sustainability of city development of Portland. Portland possesses these key virtues as its greatest advantages (David Brooks, 2011).

The best economic strategy for a city is not attracting multinational corporations, or developing the manufacturing industry, medical industry, and aerospace industry, but becoming a living-favourable city to attract young talents as the motivation of economic development. In the U.S., 64% of its population will choose the place to live in first, and then move on to find job positions.

This is the typical transformation paradigm of the U.S sport cities. In fact, since 1990s, the U.S government started to pay attention to public health. From 1980s to 2000s, some city governments emphasized on editing city regulations and urban planning for the sake of public health. In the late 1990s, the concept of "Healthy Neighbourhood" was formulated. Boston, California, and Indiana started the "Healthy City" plan, and boosted the U.S cities to become more favorable for outdoor activities (Jeff Speck, 2012).

4 ESTABLISHMENT OF RESEARCH STRUCTURES OF THE U.S SPORT CITIES

The example of the U.S cities illustrated the process of interaction between urban planning and public health, and the impact of the construction of the urban space network on public health in a city. A well-designed city public health system should take into account its background environment in the frameworks of epidemiology, economics, and environmental science. Thus, the concept of "urban ecological niche" was used to introduce the concept "urban sport niche" and effectively incorporate these into an overall framework, which would be used to guide the methodology for city planning to make the city more favourable to public health and outdoor activities, and further nurture the "sport city model" that is applicable in urban planning and public health management.

4.1 Introduction of "urban sport-niche"

The urban ecological niche that is essential to human beings' everyday life in any habitat represents a group of factors, which are provided to people and can can be exploited (such as water, food, energy, land, climate, construction, traffic, etc.) (Wang Ruson, 1988).

In the same essence, "urban sport-niche" means the group of factors that the habitat has provided and can be exploited in people's outdoor activities (such as land, climate, traffic, exercise facilities, street design, etc.) and interaction with social life (such as economic status, environmental capability, life quality, relationships with other cities, etc.). Urban sport-niche includes not only living conditions but also production conditions. It includes the concept of materialistic factors, energy factors, cultural factors, informational factors, and the concept of space and time. It is a representation of the current suitability degree of habitability for a city by measuring the residents' participation in outdoor activities. And it is a representation of its status, position, functions, effect on attracting new residents, and advantages and disadvantages. Parameters described above can be decisive in its attraction effect and repellent effect to different age groups people with different occupations.

Assume that each person is surrounded by a definite space, which include three elements: who, when, and where. Individualized urban sport-niche in essence is a deduction towards individuals via micro, medium, and macro-view. The micro-level incorporates the inner effect of an individual, such as the person's gender, age, and hereditary factors; the medium level refers to a person's occupation, behavior, and habits, and lifestyle also plays an important role; the macro-level involves the considerations of social epidemiology and epidemiology of the artificial natural environment, thereby focusing on the neighbourhood and city level. On the macro-level, it was concluded that social-economic status, the accessibility of service facilities, natural environment, mixed land usage, the design of the street, and walkability are important external guiding factors.

4.2 The system of the sport-city model

Based on the concept of "urban sport-niche", the sport-city model is brought up as a systematic method to integrate factors in different levels into a single model. Three different corresponding paralleled co-existing systems of "urban sport-niche" are key factors in constructing this model: The first system is the individual–group system, and it demonstrates the result of a constitutional index, and an individual is defined as an organism organized by gene and cells; the second system is the family–region–city system, and the determinants of outdoor activities are effective in this space, behaviour, and organization norms;

Table 1. A criteria table of urban sport niche.

	Target level	Determination level	Index level
Evaluation index system of urban sport-niche	Micro	Gender, age, genetic factors, etc.	A. Municipal total population, B. the number of employees at the end of the year
	Meso	Occupation, habits, lifestyle, etc.	C. Average wages of employed persons D. the population density, E. the urban population accounted for the proportion of permanent population F. per capita consumption expenditure, G. colleges, H. the number of mobile phone subscribers, I. the number of Internet broadband access users, J. private car ownership, K. buses ownership, L. college students per million people (people), M. the number of theaters and music halls, and N. the number of Public Libraries for every hundred people
	Macro	Social-economic status, services accessibility, natural environment, mixed land use, street design, and walkability	O. GDP, P. added value of the third industry Q. proportion of the third industry to the GDP, R. per capita GDP, S. built-up area, T. industrial waste gas emissions, U. industrial smoke (powder) dust emissions, V. urban green space per capita, W. urban construction land for urban area proportion, X. built-up area green coverage rate, and Y. life garbage.

the third system is management–decision system, and it involves local or even regional management of the respective lands, facilities, and urban planning.

Table 2. List of communalities.

Common factor variance

		Initial	Extraction
O (100 million yuan)	O	1.000	0.997
P (100 million yuan)	P	1.000	0.990
Q (%)	Q	1.000	0.777
R (yuan)	R	1.000	0.935
A (million)	A	1.000	0.858
E (%)	E	1.000	0.945
C (100 million yuan)	C	1.000	0.990
B (10000 people)	B	1.000	0.944
S (square kilometers)	S	1.000	0.973
D (people/sq km)	D	1.000	0.946
T (100 million cubic meters)	T	1.000	0.900
U (10000 tons)	U	1.000	0.890
V (square meters)	V	1.000	0.810
F (yuan)	F	1.000	0.953
J (unit)	J	1.000	0.986
K (units)	K	1.000	0.988
W (%)	W	1.000	0.951
H (10000)	H	1.000	0.989
I (10000 households)	I	1.000	0.951
G	G	1.000	0.953
L (people)	L	1.000	0.863
M	M	1.000	0.950
N (volume)	N	1.000	0.965
X (%)	X	1.000	0.842
Y (%)	Y	1.000	0.916

Extraction method: principal component analysis.

5 ANALYTICAL EXAMPLE USING 21 MUNICIPAL-LEVEL CITIES IN GUANGDONG PROVINCE CHINA

5.1 The grading ratios

The multiple regression model is used to study the integrated relationships between the built environment of a city and "urban sport-niche", which means the method of finding a certain discipline by using big data analysis, and constructing sport city model.

5.2 Data collection, data processing, and analysis

This study utilized 21 municipal-level cities in Guangdong province as the foundation for analysis, and the data sources are <Guangdong Statistical Yearbook 2015>, <China Cities Statistical Yearbook 2015>, and other online databases. Data are collected during the same time period and in the same format, and are transformed to ensure integrity and objectivity.

IBM SPSS Statistics 20 is used to deal with the data. The output correlation matrix table shows that many independent variables that are united of the 25 variables are highly correlated. It is not easy to make a right judgment for the urban sport-niche, and so we have to conduct an analysis on the main factors.

The output's common variable degrees are given in Table 2. The table reflects the common factors in each of the six variables. It can be explained clearly by using six factors (principal component) from X1 to X25, because these 25 variables degrees were above 0.75, and passed the KMO test; the simultaneous accompanied probability is less than 0.05, and passed the Bartlett test of sphericity.

The output's total variance explanation table is given in Table 3. It uses the eigenvalue criterion to

Table 3. Total variance explained.

Explained Total Variance

	Initial eigenvalue			Extraction square and loading	Rotating square and loading				
Component	Sum	Variance%	aggregation%	Sum	Variance%	aggregation%	Sum	Variance%	aggregation%
1	15.796	63.186	63.186	15.796	63.186	63.186	11.951	47.805	47.805
2	2.389	9.555	72.741	2.389	9.555	72.741	3.709	14.837	62.642
3	1.774	7.095	79.836	1.774	7.095	79.836	2.806	11.224	73.866
4	1.179	4.716	84.552	1.179	4.716	84.552	1.787	7.146	81.012
5	1.076	4.304	88.855	1.076	4.304	88.855	1.533	6.131	87.144
6	1.048	4.193	93.049	1.048	4.193	93.049	1.476	5.905	93.049

Extraction method: principal component analysis.

411

determine the number of main factors extracted as the initial primary factor, which means principal factors ≥1 are selected to be the initial factor and abandon main factors that are ≤1. It can be observed from Table 1 that, the first factor's characteristic is 11.951, which is approximately 47.805% of the total variance. The first six factors are extracted based on the rule of selecting eigenvalues with features ≥1 in the default process. Six factors' eigenvalues cumulatively contribute to 93.049% of total variances. Visibly, the other 19 factors explained variance taking up less than 10%, which means that the first six factors provide sufficient information about the original data. Twenty-five dependent variables may be divided into six categories to explain, and it is easy to observe the performance of each of the 21 cities in the class factor.

The output scatter plot shows that between factor 1 and factor 2, factor 2 and factor 3, factor 3 and factor 4, factor 4 and factor 5, factor 5 and factor 6's characteristic values are largely different, but other factors' differences are smaller. First, the preliminarily drawn conclusion is that six factors can summarize most of the information. Factor 1's characteristic value is obviously large.

The output factor score's coefficient matrix table is given in Table 4. The factor expression in the equation can be created with each original variable through the coefficient matrix, so as to help us to understand the meaning of factors. However, from the table it can be seen that, factor 1, 2, 3, 4, 5, and 6's load values in the original variables do not exhibit a big difference, which is not easy to explain, and further factor rotation is needed to obtain better understanding of its meaning.

The output rotation factor loading matrix table is shown in Table 5. The rotation factors coefficients show polarization. Six factors describe the differences about the urban built environment from different aspects.

Table 4.   List of component score coefficient matrices.

Component matrix[a]

		Component					
		1	2	3	4	5	6
O	O	.972	.041	−.030	.018	−.209	−.070
P	P	.963	.087	−.078	−.089	−.199	−.051
Q	Q	.821	.121	−.068	−.223	.138	.122
R	R	.869	−.008	.382	.165	−.022	.078
A	A	.687	.077	−.340	.083	−.337	.380
E	E	.763	−.244	.286	.279	.156	.346
C	C	.967	−.124	.041	−.008	−.109	−.159
B	B	.937	−.096	−.174	.030	−.034	−.157
S	S	.906	−.024	−.328	−.033	.201	−.055
D	D	.794	−.548	−.024	.085	.077	−.042
T	T	.793	.341	−.111	.169	.281	.185
U	U	−.033	.630	.071	.676	.168	−.047
V	V	.517	.449	.265	−.137	.496	−.074
F	F	.902	.029	.175	.168	.112	.260
J	J	.957	−.172	.023	.154	−.058	−.116
K	K	.980	.043	−.069	−.074	−.121	−.039
W	W	.848	−.295	−.125	−.192	.244	−.182
H	H	.982	−.047	−.085	.019	−.111	−.062
I	I	.911	.320	−.052	−.040	−.103	−.057
G	G	.702	.504	−.196	−.246	−.200	.259
L	L	.355	.749	.087	−.361	.033	−.193
M	M	.872	−.123	.260	.181	−.250	−.107
N	N	.806	−.313	.178	−.095	.273	−.320
X	X	.218	−.236	.552	−.409	.084	.510
Y	Y	−.027	.226	.825	−.019	−.354	−.242

Extraction method: principal component analysis.

a. The six components have been extracted.

412

Table 5. Urban sport niche score of 21 cities.

CITY	F1	F2	F3	F4	F5	F6	F
Guangzhou	0.73204	3.68008	1.98007	0.04169	−0.30317	0.04439	1.19
Shenzhen	3.70875	−0.63339	−0.29657	−0.33827	1.12876	−1.18551	1.74
Zhuhai	−0.29618	−0.59098	0.49179	2.72745	0.42338	0.32342	0.07
Shantou	−0.33625	0.76438	−1.11042	0.65088	−1.76566	−0.52561	−0.28
Foshan	0.67989	0.84757	−1.77399	0.31249	1.2022	2.58746	0.54
Shaoguan	−0.73715	0.02688	0.19593	0.13738	−0.26669	1.10737	−0.29
Heyuan	−0.62297	−0.28118	0.27929	−0.16883	0.33752	−1.54206	−0.42
Meizhou	−0.62931	−0.33983	0.40613	0.03972	0.02972	0.02178	−0.32
Huizhou	−0.16381	−0.34291	0.22339	0.12414	0.28616	0.77951	−0.03
Shanwei	−0.59137	−0.1496	−0.44521	0.12888	0.01441	−1.10848	−0.44
Dongguan	1.40779	−1.17638	0.58925	0.82104	−3.00716	0.37807	0.5
Zhongshan	0.0601	−0.37475	−0.24584	1.04287	0.73109	0.52901	0.1
Jiangmen	−0.36559	−0.39223	0.56101	0.49072	0.21835	−0.01543	−0.13
Yangjiang	−0.33429	−0.31159	−0.55958	−0.56191	0.67945	0.62036	−0.25
Zhanjiang	−0.40819	0.2188	0.10623	−0.4794	0.76643	−0.95051	−0.21
Maoming	−0.38637	0.43216	−0.50896	−0.83171	0.49176	−0.53492	−0.26
Zhaoqing	−0.34888	−1.16383	2.60051	−0.98344	0.76154	0.43333	−0.05
Qingyuan	−0.0743	−0.50534	−0.02521	−2.18231	−1.11266	1.3238	−0.28
Chaozhou	−0.6781	0.27736	−0.78117	0.96535	0.18812	−1.268	−0.39
Jieyang	−0.21117	0.43862	−1.45089	−1.14974	−0.91733	−0.58671	−0.4
Yunfu	−0.40464	−0.42384	−0.23573	−0.78699	0.1138	−0.43124	−0.38

5.3 Conclusion

According to the transformed data calculated by using the sum of different variance, the 21 cities' load values for the 6 factors can be obtained, which are the scores for the cities (Table 5). The comprehensive score will be the final grade on "urban sport-niche". It is clear that Guangzhou has a distinguished high score on factor 2, which represents the status of the urban size and productivity driven by higher education; Shenzhen has a relatively higher score for factor 1, which represents the level of economic development, the intensity of city development, and living styles of citizens; Zhuhai has a higher load value for factor 4, which measures the life standards of the city and citizens' need for green space; Foshan has higher load value for factors 5 and 6, which represent the urban landscape situations on green space; Zhaoqing has a higher load value for factor 3, which represents its' management of environment situations; Zhonghan has a higher value for factor 4; Shaoguan and Qingyuan achieved a higher value for factor 6 due to their superior natural tourism resources.

Heyuan, Shanwei, Shantou, Chaozhou, and Jieyang show no spectacular contribution in their cities' environment, some of which even made certain destruction to their natural environment because of the hazardous industry. Thus, they have poor performance in these above-mentioned 6 factors. The remaining cities have weak inducible properties and so these might not be discussed. The city that achieved the highest comprehensive performance in "urban sport-niche" is Shenzhen, the center of the Pearl River Delta area, which is right next to Hong Kong. Its competitiveness in economics, impetus to urban development, attraction to elite, and material space constructions have made it unparalleled and thus it has an advantage. Guangzhou stands right after Shenzhen, and this is not beyond expectation, since Guangzhou has well-developed advanced education resources, and it has multiple renowned universities in the southern China area. Its economic competitiveness is right after that of Shenzhen's, and it has abundant historic, environmental, and cultural resources, which are key attractions to young adults; Foshan ranks the third of all the cities, and it is a city that has attracted a great amount of overseas investment. Overseas investment has contributed to 26.5% of all economic scales in Foshan. Its political support, capital, resources, and services have made Foshan to be a great city with well-developed social environment, and so it is reasonable for Foshan to achieve a high score in "urban sport-niche".

It is also worth mentioning that, as a city that does not have great contributions to show in any other area, Dongguan ranked No. 4 in all 21 cities. By digging deep into the reason, we realized that there are more than 1.2 million compatriots from Hong Kong and Macau and 300 thousand overseas Chinese living in Dongguan. There is a great amount of capital in Dongguan, which has been brought by Taiwan Corporates, and over 100 thousand Taiwan business men are operating their business in Dongguan, and these people have great passion in outdoor activities than Chinese people, and have thus made a great contribution to Dongguan to achieve a high score in "urban sport-niche". Shanwei, Heyuan, Jieyang, Chaozhou, and Yunfu obtained low scores in "urban sport-niche". This is not only because they have a relative weaker economic development, but also because their policy ideology marginalized their cities. In these cities, workers on average have lower education. There are very few technical talents. So many problems hinder the development of these cities.

From our analysis of the sport-niche index of these 21 cities in Guangdong province, a preliminary idealized sport city model could be desgned:

a. Multi-mechanism industry cluster economy and labour-intensive industry cluster can continuously provide impetus to the development of the sport city.
b. The accessibility of road that has high correlation with land usage intensity and the walkability of road (local, reginal, and civic) in each level of the three coexisted systems that coordinate with "urban sport-niche".
c. Highly mixed usage of land, urban-scale in certain areas, and the accessibility of public facilities are highly related with "urban sport-niche".
d. Advanced information and communication technologies have made great contributions to the reconstitution of regional space. Diversified interactions based on the Internet platform boost the development of the outdoor activities in a city.

Although "urban sport-niche" relies heavily on the economic strength of a city, it varies across different cities. In cities like Portland, its economic power was not comparable to cities such as, New York, Washington DC, Chicago, or Boston, but it still became a habitable city that attracted more and more young American adults. With a certain degree of economic base, construction of an urban space network is more conducive to the formation of a sport city model.

What needs to be clarified is that due to the limitation of inaccessible Internet data, it is not easy to achieve a result without more valuable data to carry out research. (Some internet service providers have an anti-data retrieval mechanism in order to protect their user, which make it more difficult to get useful data.) It might be a good way to work with other media platforms, to attain a more comprehensive understanding of this issue.

6 CONCLUSION

When compared with the passive healthy urban planning in the past, urban planning now has become more and more active. With the drastic growth in the number of people enjoying outdoor activities, the city planner needs to create more space for people to carry out outdoor activities, and meet the needs of different groups with respect to work-out and exercise. The sports city is not merely a concept, but is a solution for multiple problems that have evolved during the process of city development. It is also a development strategy that focuses on sustainable growth. In this essence, it is worth introducing the idea of the sport city and to promote outdoor activities in cities.

REFERENCES

China Outdoor Association COA Chinese Outdoor Activities Research Report[M]. China Citic Publishing, 2014.
Elizabeth Kolbert, The Sixth Extinction [M]. Henry Holt Company, 2010.
GF Securities Research Center 2015 Research Report[M]. GF Securities Co., LTD, 2014.
Heidi Garrett-Peltier. Estimating the Employment Impacts of Pedestrian, Bicycle, and Road Infrastructure[M]. Political Economy Research Institute.University of Massachusetts, Amherst December 2010.
Jan Gehl, Cities for people [M]. Island Press, 2010.
Jeff Speck. Walkable City: How Downtown Can Save America, One Step at a Time[M]. Farrar, Straus and Giroux Publishing, 2012.
Joe Cortright, Portland's Green Dividend [J]. CEOs for Cities, 2007.
Patrick M.Condon. Canadaian Cites American Cities Our Differences Are the Same[J]. UBC James Taylor Chair in Landscape and Liveable Environments,2004.
Wang Rusong. Explore Urban Ecological Niche[J], Urban Environment & Urban Ecology, 1988, 1(1):20–24.

Electromechanical Control Technology and Transportation – Jia & Wu (Eds)
© 2017 Taylor & Francis Group, London, ISBN 978-1-138-06752-3

An improved algorithm of trend surface filtering based on the natural neighboring points range

Zhiheng Zhang, Rencan Peng, Wenqian Huang & Jian Dong
Department of Hydrography and Cartography, Dalian Naval Academy, Dalian, China
PLA Key Laboratory, Dalian, China

ABSTRACT: An improved algorithm of trend surface filtering based on the natural neighboring points range is put forward in this paper, to solve the problems in the aspect of gross error detection of multi-beam bathymetric data. Because uncertainty may exist in constructing a surface fitting function and the depth point may be eliminated unreasonably while using the tradition algorithm, the filtering effect of traditional algorithm is not as good as expected. The concept of the natural neighbor points range and the smallest local scope of any point in scattered depth points are introduced in the paper. By using the improved method, one unified surface model function is determined after analyzing the local approximate surface of the natural neighbor points range. To obtain the non-continuous points effectively, the mechanism of reserving the non-continuous points is carried out by using an iterative trend surface filtering method to filter and construct the actual trend surface. The experimental results show that the improved algorithm can eliminate the gross error points in multi-beam bathymetric data and improve the precision of submarine topography expression.

1 INTRODUCTION

Nowadays, the multi-beam echo-sounding system has become a major instrument to detect submarine topography. The errors of the instrument itself, the factors of sea conditions, the unreasonable mistakes of equipment parameters setting, and the influence of sea creatures in the measurement process will produce some gross errors in multi-beam sounding data and will seriously influence the authenticity of submarine topography expression (Liu, 2006). In order to improve the accuracy of multi-beam sounding data, the data should be filtered. Currently, the methods of gross error detection of multi-beam sounding data mainly include interactive filtering and automatic filtering (Dong, 2007). The former method relies on manual handling. In a three-dimensional visualization environment, the manual handling method, whose efficiency is low, relies a lot on people's experience. In order to overcome the shortcomings of interactive filtering, domestic and international scholars and experts have proposed various automatic filtering methods. Among these methods, the trend surface filtering method has attracted a great deal of attention from many researchers, because it is easy to understand its filtering process, and it has less parameter settings and simpler operation

(Dong, 2007). However, there still exist three problems in this filtering method.

i. Due to the complexity of submarine topography, the surface fitting range is not easy to be determined, so that the surface fitting function cannot reflect the real submarine topography and it will affect the filtering effect or eliminate some normal value wrongly.

ii. When the value of gross error points is large in the selected area, it will affect the normal points, thereby leading to a big deviation between the fitted trend surface and the normal points. When the gross error points relatively concentrate and appear in clusters, the filtering method cannot detect gross error points effectively. The gross error points may be reserved as normal points, while, in opposite, the real normal points are filtered.

iii. This method is under the assumption that the submarine topography is continuous. When breaking points (such as a cliff) or obstructions (such as reefs or wrecks) appear in the submarine topography, the filtering method may lead to getting rid of breaking points or obstructions.

Therefore, aiming at these three problems, in this article, an improved algorithm of trend surface filtering based on the natural neighboring points range is put forward.

2 TREND SURFACE FILTERING METHOD

The trend surface filtering method adopts the polynomial surface function to fit the trend of submarine topography by using the depth and plane position of the seabed beam footprints. And then, the deviation between the points of the study area and the fitted trend surface can be obtained. When combined with the double-fold (or triple-fold) error criterion, the points whose deviation from the trend surface exceeds a certain threshold value will be judged (Dong, 2007). Let us assume that the polynomial surface function is constructed as follows:

$$z = f(x, y) = ax^2 + by^2 + cxy + dx + ey + f \quad (1)$$

In the equation, a, b, c, d, e, and f represent the coefficients of the polynomial curved surface function, which need to be solved. In geology, equation (1) is usually adopted to fit the local topography, but in practice, if the submarine topography is more complex, and a higher order function can be chosen to fit the submarine topography (Dong, 2007).

Equation (1) can be written in the form of $AX = L$ which is given as follows:

$$A = \begin{bmatrix} x_1^2 & y_1^2 & ... & 1 \\ ... & ... & ... & ... \\ x_n^2 & y_n^2 & ... & 1 \end{bmatrix}, X = \begin{bmatrix} a \\ ... \\ f \end{bmatrix}, L = \begin{bmatrix} z_0 \\ ... \\ z_n \end{bmatrix} \quad (2)$$

According to the conditions of least squares ($\sum_{i=1}^{n}[f(x_i, y_i) - z_i]^2 = \min_{f(x,y)\in H} \sum_{i=1}^{n} \Delta_i^2$) to construct the least squares trend surface, the coefficient matrix X can be obtained as follows:

$$X = (A^T A)^{-1} A^T L \quad (3)$$

After the trend surface is fitted according to the flat coordinates (x_i, y_i) and depth values z_i in the local area, the judgment standard of the gross error is given as follows:

$$\begin{cases} z_q - f(x_q, y_q) \leq k\sigma & \text{Normal values} \\ z_q - f(x_q, y_q) > k\sigma & \text{Abnormal values} \end{cases} \quad (4)$$

In the equation, k =2 or 3, and σ is represented as the mean square deviation of each measuring point in the neighboring points range of the detection point q.

3 THE IMPROVED ALGORITHM OF TREND SURFACE FILTERING

3.1 Determining the surface fitting range and function

3.1.1 Determining the surface fitting range
The Voronoi diagram is a commonly unstructured grid. Each grid cell is called Voronoi cell. In the unstructured grids of the Voronoi diagram, the adjacent discrete points of each discrete point corresponding to its Voronoi cell can be called as the natural neighbor point (Chen, 2003). The local range composed of the natural neighboring points can be called as the natural neighboring points range. According to relevant knowledge about the submarine tracking control technology of the multi-beam sounding system, we can generally believe that there must be 3–5 continuous recording sounding points in any direction. It can determine whether there is a concave (convex) terrain at the bottom of the sea (Wu, 2001), and so the natural neighboring points range is considered to be the minimum local range of submarine topography expression. If the natural neighboring points range is used as the trend surface fitting range and when there are tiny landscapes or small obstructions (such as reefs or wrecks), which the multi-beam sounding system can detect at the bottom of the sea, it is not easy to be regarded as a gross error out. To quickly find the natural neighboring points of any discrete point, the method described in the literature (Cai, 2004) can be adopted. According to the basic principles of Delaunay triangulation constructed, the maximum side and minimum angle criteria can be used to quickly find the natural neighboring points of any point within the local area.

3.1.2 Determining the surface fitting function
Based on the literature (Wang, 2011), the local surface canonical form of the p point, which is any point on any surface S, is given as follows:

$$z^* = \frac{1}{2}(k_1(p)(x^*)^2 + k_2(p)(y^*)^2) + o((x^*)^2 + (y^*)^2) \quad (5)$$

In the equation, (x^*, y^*, z^*) are 3D coordinates of the local surface's coordinate system, $k_1(p)$ and $k_2(p)$ are two main curvatures of the p point, the origin of this local coordinate system is the p point, and the direction of the x axis and y axis are the directions of the two main curvatures; the direction of the z axis is the normal vector of the local surface, and $o((x^*)^2 + (y^*)^2)$ is the higher order dimensionless vector.

An analysis of equation (5), when $k_1(p)$ and $k_2(p)$ have same signs and are not equal to zero, the local surface is approximately an elliptic paraboloid; when $k_1(p)$ and $k_2(p)$ have different signs and are not equal to zero, the local surface is approximately a hyperbolic paraboloid; when only one is zero between $k_1(p)$ and $k_2(p)$, the local surface is approximately a parabolic cylinder. Therefore, the local surface of the p point which is any point on any surface S is the quadratic parabolic surface. If a local coordinate system is set up at the p point, the origin of the local coordinate system is the p point, the direction of the z axis is the normal vector of the local surface, and the directions of the x axis and y axis are set up optionally. The local surface of any discrete point's nature neighborhood points range is shown by using the complete quadratic parabolic in the local coordinate system, which is given as follows:

$$z^* = a(x^*)^2 + bx^* y^* + c(y^*)^2 \tag{6}$$

In the above-mentioned equation, a, b, and c represent the coefficients of the quadratic parabolic surface function, which need to be solved.

Because of the difference between the local coordinate system of equation (6) and the global coordinate system of the multi-beam sounding data, the multi-beam sounding data need to be converted into the data in the local coordinate system; equation (6) can be used to fit the surface.

A transformational matrix can be used to describe the transformation of the space coordinate system. Assume that the global coordinate system is $A(x,y,z)$ and the local coordinate system is $B(x^*, y^*, z^*)$; the p point is (x_0, y_0, z_0), the transformation process is performed by placing the origin of the global coordinate system A to the p point, thereby leading to the overlapping of two origin points and rotating the axis of the global coordinate system A around the origin of the local coordinate system and making its z axis overlap with the z axis of the local coordinate system B. The mathematical description of the process is given as follows:

$$B=TA+\Delta \tag{7}$$

In the equation, Δ is the coordinate translation matrix and T is the coordinate rotation matrix, $T = T_x \times T_y \times T_z$. The calculation equations of the matrices are as follows:

$$\Delta = \begin{bmatrix} x_0 \\ y_0 \\ z_0 \end{bmatrix} \tag{8}$$

$$T_x = \begin{bmatrix} 1 & 0 & 0 \\ 0 & \cos\alpha & \sin\alpha \\ 0 & -\sin\alpha & \cos\alpha \end{bmatrix} \tag{9}$$

In the equation, α is the shaft rotation angle of the x axis.

$$T_y = \begin{bmatrix} \cos\beta & 0 & -\sin\beta \\ 0 & 1 & 0 \\ \sin\beta & 0 & \cos\beta \end{bmatrix} \tag{10}$$

In the equation, β is the shaft rotation angle of the y axis.

$$T_z = \begin{bmatrix} \cos\gamma & \sin\gamma & 0 \\ -\sin\gamma & \cos\gamma & 0 \\ 0 & 0 & 1 \end{bmatrix} \tag{11}$$

In the equation, γ is the shaft rotation angle of the z axis.

In the process of rotation, the z axis is not rotated, and the angle α and angle β can be obtained by using the normal vector of the local surface. Firstly, the data of the natural neighboring points are fitted to the plane by using least square method. By using the normal vector $N(n_x, n_y, n_z)$ of the fitted plane instead of the normal vector of the p point approximately, the angle α and β are obtained as follows:

$$\begin{cases} \alpha = \arctan\left(\dfrac{n_y}{n_z}\right) \\ \beta = \arctan\left(\dfrac{n_x}{n_z}\right) \end{cases} \tag{12}$$

Finally, equation (6) is used to fit the surface. Because of three unknown coefficients in the process of the surface fitted, it needs at least four points to fit the surface.

For the complex submarine topography surface, if the natural neighboring points range is used as the range of the trend surface fitted, the submarine topography surface can be divided into several small local ranges; the uniform surface fitted function can be used on every small local surface to reflect the actual submarine topography, the fitted points are less (at least four points), and the first problem can be solved effectively.

3.2 The method of iterative trend surface filtering proposed

The method of trend surface filtering is carried out by using the polynomial surface function to fit the terrain surface. The main disadvantage of this method is that the points may contain gross errors, which are used to fit the submarine topography. If the gross error is large, it will affect the surrounding normal points, so that there will be a larger deviation between the fitted trend surface and the normal points. These normal points will be removed. When the gross error points are relatively concentrated and appear in clusters, the filtering method will not effectively detect gross error points. On the contrary, the gross error points may be considered as the normal points to be held and the normal points may be considered as the gross error points to be eliminated (Li, 2008). The reason for these problems is that the points which are used to fit the submarine topography contain gross error and the trend surface fitted is not the actual terrain trend surface.

Aiming to solve this problem, the thesis proposes a method of iterative trend surface filtering. As shown in Figure 1, it is two adjacent natural neighboring points ranges whose center points are P_1 and P_2, while P_3 is the center point of the next natural neighboring points range. Firstly, the gross error points in the natural neighboring points range of P_2 can be rectified by using the trend surface which is fitted by the points in the natural neighboring points range P_1. And then, all points are used in the natural neighboring points range of P_2 to fit the terrain surface, thereby finding out the points, whose deviation from the trend surface are more than twice or three times in the error, and continuing to fit the terrain surface with the rest of points. Until the points in the natural neighboring points range of the P_2 point does not contain the suspects and the trend surface fitted reaches the steady state, the final trend surface is considered as the actual trend surface. Finally, the deviation between the suspects and the trend surface is calculated and the gross error points with double (or triple) error are determined. Due to the fact that the gross error points in the natural neighboring points range have been rectified by the trend surface which is fitted by the points in the front natural neighboring points range, as shown in Figure 1, it can guarantee that there are at least four points excluding gross errors in the natural neighboring points range. Therefore,

the situation that the fitting process is terminated because of insufficient points will not appear in the iteration process.

3.3 *The discontinuous points' detection and retention*

Because of the trend surface filtering method or the above-modified method, assuming that the submarine topography is under the continuous premise, when breaking points (such as bluff) or obstacles (such as reef or sinking) appear in the submarine topography and their deviation is greater than the tolerance, the trend surface filtering method or the above-modified method may lead to the discontinuous points being deleted unreasonably, and the submarine topography cannot be accurately expressed.

In order to solve this problem, in this article, the mechanism of the discontinuous points' detection and retention based on the above trend surface filtering method is added, in order to avoid that the unreasonable deletion of discontinuous points. As shown in Figure 1, in the above-mentioned filtering process of the iteration trend surface filtering method, assuming that the P_2 point is a discontinuous point and the natural neighboring points range of the P_3 point is on the discontinuous submarine topography. When the points on the natural neighboring points range of the P_1 point are filtered, it will set the P_2 point as a gross error point and correct the P_2 point; when the points on the natural neighboring points range of the P_2 point are filtered, it will form a trend surface with a larger curvature to ensure the continuity of submarine topography; when the points on the natural neighboring points range of the P_3 point are filtered, because the natural neighboring points range of the P_3 point is on the discontinuous submarine topography, it will set the P_2 point as a gross error point again. According to this feature, the contradictory points can be set to the discontinuous points and retained in the process of filtering.

4 THE EXPERIMENTAL RESULTS AND ANALYSIS

In order to verify the correctness and feasibility of the above-mentioned improved method, in this article, the trend surface filtering method and the above-mentioned improved method with the vc++ programming are achieved; at last, the experimental results are displayed and analyzed with Surfer8.0 software. The parameters of the computer are given as follows: the processor is Intel (R) core (TM) i3, frequency is 3.4 GHz, and memory

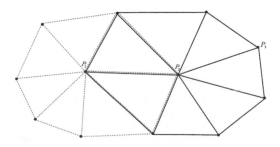

Figure 1. Schematic of the process of iterative trend surface filtering.

is 2G. The experimental data are the multi-beam data of China East Sea, which contains 716516 discrete points. In order to verify the validity of the discontinuous points detected and retained, raised cuboids are set in the plain area. In the aspect of the algorithm parameter, for the trend surface filter method, the complete quadratic polynomial is used, and the radius is confirmed by using the density of multi-beam data, which ensures that there are about 30 soundings in the selected range, and the value of k is 2 in the gross error criteria; for the improved algorithm, k is 2 in the gross error criteria, and the time of these two methods are 2414 s and 2643 s, respectively. Experimental results are shown in Figures 2–4, which are the submarine surface figures and 5 m contour maps of the original data (I), the data after trend surface filtering (II) and the improved method filtering (III), respectively. In addition, the maximum value, the minimum value, standard deviation, and point number of three data are calculated, and the results are shown in Table 1.

The analysis of Figures 3 and 4 and Table 1 gives the following results: (1) in the gross error detection, two methods both have good ability to recognize the gross error points for the single gross error point in the flat submarine topography; but in the complex areas or when the gross error points appear in clusters, the multi-beam sounding data after trend surface method filtering still contain a small number gross error points, and the improved algorithm has good recognition ability in these areas. (2) In the normal points reten-

Figure 4. The submarine surface figure and contour map of the data after carrying out the improved method filtering.

Table 1. Analysis of the experimental results.

Statistics of the depth value	(I)	(II)	(III)
Minimum	−103.33	−92.55	−98.838
Maximum	−1.55	−11.24	−11.136
Standard variance	18.54	15.27	16.92
Number of points	716516	579220	682896

tion, for the data after trend surface filtering, its submarine surface is more smooth than the that of the improved method, the tiny landforms of its contour figure have been mostly cancelled out, and the accuracy of the submarine topography is lower than that of the improved method; these indicate that a large number of normal points are cancelled out in the process of the trend surface filtering and the improved method keeps the normal points. (3) For the discontinuous part in the submarine topography, the trend surface filtering method removes the edge of the discontinuous part and makes the other part tend to be continuous; the improved method retains most edges of the discontinuous part and retains the actual shape of the discontinuous part basically.

5 CONCLUSION

Through theoretical analysis and experimental comparison, the conclusions are as follows:

i. The improved method is better than the trend surface filtering method in terms of the recognition ability of gross error points, especially in the complex areas or when the gross error points appear in clusters, while the improved method can effectively identify and eliminate the gross error points.

ii. A large number of normal points are removed in the process of trend surface filtering, while the improved method keeps the normal points and improves the accuracy of the submarine topography expression.

iii. For the discontinuous points' detection and retention, the trend surface filtering method

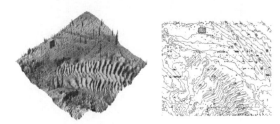

Figure 2. The submarine surface figure and contour map of the original data.

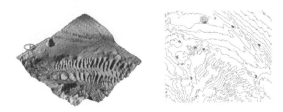

Figure 3. The submarine surface figure and contour map of the data after carrying out trend surface filtering.

will make the continuous parts be removed unreasonably, while the improved method basically retains the actual shape of the discontinuous parts.

The improvement of the trend surface filtering method is mainly concentrated on the actual trend surface of submarine topography, but in this article the gross error criteria are not discussed. It still uses the double-fold (or triple-fold) topography in the error criterion as the improved method's error criteria. This may lead to the unreasonable removal of some normal points and the retention of some smaller gross error points. The gross error criteria will be researched and improved in the next step.

REFERENCES

Cai Yongchang, Zhu Hehua. Natural Neighbor method Based on The Algorithm of Local Search [J]. Acta MechAnica Sinica. 2004(5):623–628.

Chen Jun, Zhao Renliang, Qiao Chaofei. GIS Spatial Analysis Based on Voronoi Diagram [J]. Geomatics and Information Science of Wuhan University, 2003, 28(5):32–37.

Dong Jiang, Ren Lisheng. Filter of MBS Sounding Data Based on Trend Surface [J]. Hydrographic Surveying and Charting, 2007, 27(6):25–28.

Li Zhilin, Zhu Qing. Digital Elevation Model [M], Second Edition. Wu Han: Wu Han University Press, 2008.

Liu Yanchun, Xiao Fuming, Bao Jingyang. Overview of Hydrography [M]. Beijing: Surveying and mapping press, 2006.

Wang Youning, Liu Jizhi. Lectures on Differential Geometry [M]. Beijing: Beijing Normal University Press, 2011:116–117.

Wu Yingzi. Research on Terrain Tracking and Data Processing Technology of Multi-Beam Sounding System [D]. Harbin Engineering University, 2001.

Electromechanical Control Technology and Transportation – Jia & Wu (Eds)
© 2017 Taylor & Francis Group, London, ISBN 978-1-138-06752-3

Research on dynamic energy consumption-aware task scheduling of computing resources in the cloud platform

Hui-Kui Zhou
Nanchang Institute of Science and Technology, Nanchang City, Jiangxi Province, China

Mu-Dan Gu
Jiangxi Modern Polytechnic College, Nanchang City, Jiangxi Province, China

ABSTRACT: In order to realize the low cost, high efficiency and safety of the cloud computing system, high energy consumption has become a problem that cannot be ignored. In an environment of computing resource voltage and dynamic adjustment, two energy-efficient scheduling algorithms are proposed, being the energy priority scheduling algorithm and energy genetic scheduling algorithm, which satisfy the deadline of parallel tasks and reduce the energy consumption of parallel tasks. Simulation results show that this algorithm can effectively reduce the energy consumption of parallel tasks and reduce the energy consumption of the cloud computing system.

1   INTRODUCTION

Cloud computing is defined as distributed computing, grid computing, pervasive computing, and virtualization technology; it is defined as a service concept of hybrid evolution and emerging concepts in computer science. The cloud platform is based on a large number of computers, storage devices, communication, and network to achieve virtual resources. Cloud computing is a new computing model, which has become a hot research topic in the academic and industrial fields. However, to achieve low-cost, efficient, safe, easy to use cloud computing is still a challenge. High energy consumption is one of the most serious problems of cloud computing systems. Current information and communication technologies in the field of carbon emissions account for 4% of technologies in the world. According to the current growth trend up to 2025, the average energy consumption of the information industry will reach five times, and network energy consumption will reach 13 times. This means that up to 2025, the total energy consumption of the network information industry will grow up to 43% (Yun, 2010). Power consumption has become a major barrier to the sustainable development of the network and information system. If you do not consider the energy consumption factor and the schedule does not match, it will lead to low energy consumption instead of high energy consumption. At the same time, most of the digital circuits in the processor use CMOS integrated circuits. Dynamic power consumption is the main part of the study of energy conservation, which is in the CMOS circuit power consumption. Its dynamic power is proportional to the square of the voltage (Chen, 2013).

From the literature, it can be observed that reducing the energy consumption of the scheduling problem is considered. So far, not much research has been undertaken on reducing the energy consumption scheduling problem. Therefore, in this paper, to complete the task of the parallel task of the reasonable configuration and deadline requirements are considered. In the case of the completion of the case as far as possible to reduce the power consumption of sub-tasks, the execution of parallel tasks is reduced.

2   SYSTEM MODELING AND ASSUMPTIONS

2.1   *Modeling of parallel task cloud platform*

Cloud computing nodes connecting each other are complex and diverse; these may be bus, ring, star, etc. which are regular connections, and may be without any rules of the complex network. In this paper, the general "map" describes the cloud resources system. The same type of computing resource is studied and the speed and power consumption at different voltages are calculated. Considering the precedence constraints between tasks, the Directed Acyclic Graph (DAG), a group of interdependent relations and data exchange tasks are defined as follows:

Definition 1 (parallel tasks): a parallel task can be an abstract representation, which is a Directed Acyclic Graph (DAG), DAG = (T, E, W, D), T = $\{t_1, t_2, ..., t_n\}$, and tasks are represented by using the number; the set of edges is between the task dependencies; $W = \{\omega_1, \omega_2, ..., \omega_n\}$ computation subtask sets, $\omega_i \in W$ sub-task serial computation; D is the set, which denotes the amount of communication between sub-tasks (Ding, 2015).

Figure 1 consists of eight sub-tasks in parallel computing task graphs. The number is in the circle of a node, the circle is next to the calculation amount of task nodes, and the number on the side is the amount of communication between nodes.

Definition 2 (cloud platform): a realistic cloud system can be described as the abstract graph structure, namely four tuple, $Cloud = (R, C, B, VSE)$, $B = \{b_{ij} \mid r_i, r_j \in R, c_{ij} \in C\}$ B is a collection of communication bandwidths to C. $c_{ij} = (r_i, r_j)$, $VSE = \{vse_1, vse_2, ..., v_{sek}\}$ is the collection of three elements, which is composed of the voltage, the computing speed, and the power consumption of the resource node. Figure 2 is a cloud platform, which contains seven resource nodes. The number of resources in the circle represents the communication bandwidth of the communication edge.

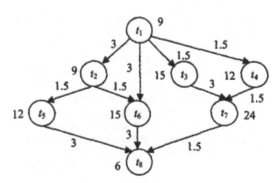

Figure 1. Schematic of the parallel task AGD graph.

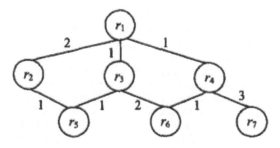

Figure 2. Schematic of the structure of the cloud platform.

2.2 The basic assumption of parallel task scheduling

For dynamic power consumption in the CMOS circuit, the power consumption model is given as follows: $P_d = ACV^2 f$, where A is the switching frequency of the circuit, C is the load capacitance, V is the supply voltage, and F is the clock frequency. For a given processor, it is generally considered that A*C is a constant, and therefore (Zhang, 2015), different supply voltages correspond to different levels of power consumption. Reducing the supply voltage of computing resources will reduce the computing speed and power consumption of the computing resources.

Parallel task scheduling in the cloud platform is based on the full consideration of the dependencies among tasks, and the process of parallel and cooperative computation of each sub-task in the graph are distributed to the resource nodes on the DAG. The false stator task is atomic and cannot be sub-divided, and the task execution is not preemptive. According to some strategy, appropriate computing resources are assigned to each task, and the scheduler and the computing resources of nodes run independently. The total energy consumption of parallel tasks is the sum of the energy consumption for each sub-task. Communication operations can be executed concurrently. If two tasks with dependencies are assigned to the same computing resource node, the communication time between these is ignored.

Different voltage calculation times and power consumption levels of each sub-task can be different, because of the difference of each computing node under different supply voltage calculation speeds and powers; the communication bandwidth between different nodes is different. The communication time between the two sub-tasks in different computing resources is not the same. In order to carry out the parallel computing task in each sub-task at the earliest start time and within the earliest completion time, the maximum power supply voltage of the execution time should be calculated, the execution time of each task should be determined, and the average bandwidth resources and the data transfer time between tasks should be calculated.

3 PARALLEL TASK AND SUB-TASK DEADLINE ASSIGNMENT METHOD

The task graph in the initial task t_1, the earliest start time $Est(t_1)$, latest finish time t_n as input, and $Lft(t_n)$ assigned a sub-deadline for each of the sub-tasks in the task graph, as long as each sub-task in the sub-deadline is complete. It will not affect the whole parallel task by using the deadline. In this paper, the computation of the deadline assignment and task when determining the fair strategy for each sub-task are proportional. Namely, for the task graph at the same level with the earliest start time for the two tasks, the calculation of the sub-task is larger, with the distribution of a longer period of time.

Define 3 (sub-deadline): the sub-task t_i and the sub-deadline $Sdl(t_i)$ are calculated as follows:

$$Sdl(t_i) = \frac{Eft(t_i) - Est(t_1)}{Eft(t_n) - Est(t_1)} \times Elt$$

Each sub-task deadline is greater than the earliest completion time and less than the latest completion time. Each sub-task in computational resources is low in power in the deadline to complete the task. Parallel tasks are completed in the final period to reduce the energy consumption of parallel task execution.

The equation description of the energy aware parallel task scheduling problem in a cloud environment is given as follows: Ct_{ij} represents the communication time of the task and t_i transmits data via the cloud platform to the task t_j. $St(t_i, r_j, v_k)$ is the task of computing resources, $T_{comm}(t_i) = \max\{Ct_{ki} | t_k \in Pred(t_i)\}$ is the time of all the parent tasks in the south of the task. In the cloud platform, $Et(t_i, r_j, v_k)$ is used to minimize the energy consumption of parallel tasks, and to meet the deadline of the parallel task scheduling problem and this can be formulated as follows:

$$\begin{cases} \min \sum_{i=1}^{n} (w_i Power(v_k) / s_k) \\ \max\{Et(t_1), Et(t_2), \cdots, Et(t_n)\} \leq Deadline \\ St(t_i, r_j, v_k) \geq 0 \\ Et(t_i, r_j, v_k) = St(t_i, r_j, v_k) + T_{Comm}(t_i) + w_i / sk \\ St(t_j, r_m, v_k) - St(t_i, r_m, v_k) \geq w_i / s_k + Ct_{ij} \\ St(t_j, r_m, v_k) - St(t_i, r_m, v_k) \geq w_i / s_k \end{cases} \quad (1)$$

The above-mentioned equation can be popularly interpreted as follows: (1) the initial task's start execution time is 0, and the other task's start execution time is greater than 0; (2) for the next two tasks, the successor task begins the execution time, which is the sum of the execution time, the task execution time, and the data transmission time; (3) the execution time of the tasks on the same computing resources cannot overlap; (4) parallel tasks over time is less than a specified cut-off time; (5) minimize the parallel task to calculate the total energy consumption.

Equation (1): scheduling problem is an NP complete problem. It is difficult to obtain the optimal solution. Generally, only the heuristic method is used to solve the near optimal solution. In this paper, an energy consumption priority based on a genetic algorithm is used.

4 EXAMPLE ANALYSIS

In order to evaluate the performance of the algorithm, the existing methods are compared with this method. The minimum load optimization method is commonly defined as a balanced scheduling method. A virtual machine is assigned and checked for a period of time, which has a physical server load. The current time load minimum physical server assigns a virtual machine. The scheduling method is mentioned in this paper, which is an improved method for minimum load optimization. Through example analysis, in this paper, the virtual machine resource is compared and the scheduling method and minimum load optimization method are balanced.

The calculation of the virtual machine is $S_{i,j}$, and resource waiting time is $wt_{i,j}$. The calculation method is as given in equation (2):

$$wt_{ij} = \left\{ int\left(\frac{queue_{i,j}}{pN_{i,j}}\right) + (1 - st_{i,j})\right\} \times t_{i,j} \quad (2)$$

The calculated index is divided into two kinds: physical servers for the average utilization rate and the physical server utilization ratio deviation. The former calculation method is described in equation (3), which reflects the physical resource utilization; the latter uses mean variance calculation methods, such as equation (4) description. Calculation results reflect the discrete degree of a data set. The balance of the degree of utilization of each physical server resource is described.

$$\mu = \frac{1}{I - I_0} \sum_{i=1}^{I} PMRate_i \quad (3)$$

$$\sigma = \sqrt{\frac{1}{I - I_0} \sum_{i=1}^{I} Y_i (PMRate_i - \mu)^2} \quad (4)$$

The minimum load optimization method and the scheduling method are used in this paper. According to the requirements of four different tasks, the calculation result is calculated respectively, and the result is shown in Figure 3 and 4, respectively.

Experiments show that the values of the different current tasks of the request are either better or worse, and the value is better or worse than that. The task scheduling method proposed in this paper is better than the general minimum load optimization method, with an increase in the

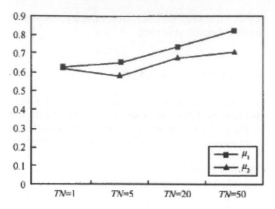

Figure 3. Average utilization curve of two scheduling strategies.

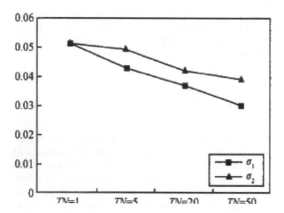

Figure 4. The calculation result curve of deviation under two scheduling strategies.

number of requests. The advantage is more obvious. The scheduling method proposed in this paper is essentially an improved minimum load optimization method.

5 CONCLUSION

In this paper, two methods are proposed to reduce the energy consumption of parallel tasks that can meet the deadline in the case of calculating the dynamic voltage. The deadline assignment problem of parallel tasks with deadline constraints is discussed, and a method of assigning the task's sub-deadline is proposed. The deadline of each sub-task is divided, so that the computing resources of each task can be achieved as low as possible, thereby reducing the total energy consumption of parallel tasks by reducing the execution power of each sub-task. A balanced scheduling algorithm for the virtual machine resource is designed, and an example is presented to show the effectiveness of the proposed method. This algorithm plays an important role in improving the balanced utilization of virtual machine resources.

ACKNOWLEDGMENT

The authors gratefully acknowledge The Educational Reform Subject (JXJG-13-27-8) project for financial support.

REFERENCES

Ding Yong, Lv Haifeng, Yu Xiaolong, Gui Feng, Xinguo remote attestation scheme based on intelligent terminals [J] password Sinica, 2015, 02: 101–112.

Hongwei Chen, Shuping Wang, Hui Xu, Zhiwei Ye, Chunzhi Wang. Automated Trust Negotiation Model based on Dynamic Game of Incomplete Information [J]. Journal of Software, 2013, 810.

Khiabani, Hamed, Idris, Norbik Bashah, Manan, Jamalul-lail Ab. Unified trust establishment by leveraging remote attestation—modeling and analysis [J]. Information Management & Computer Security, 2013, 215.

Lin Chuang, Tian Yuan, Yao Mim Green network and green evaluation:Mechanism, modeling and evaluation [J]. Chinese Journal of Computers. 2011. 34(4): 593–612.

Mezmaz M, Melab N, Kessaci Y, et al. A parallel bi—objective hybrid metaheuristic for energy-aware scheduling for cloud compuring systems [J]. Parallel Distrib Comput. 2011(7).

Shelly Salim, Sangman Moh. On-demand routing protocols for cognitive radio ad hoc networks [J]. EURASIP Journal on Wireless Communications and Networking, 2013.

Shudong Li, Lixiang Li, Yixian Yang. A local-world heterogeneous model of wireless sensor networks with node and link diversity [J]. Physica A: Statistical Mechanics and its Applications, 2011, 390(6).

Yun D, Lee J. Research in green network for future Interact [J]. Journal of KIISE, 2010, 28(1): 41–51.

Zhang Xiaowei, Wang Zheng, Chen Yongle prove a long-range program user properties [J]. Taiyuan University of Technology, 2015(02).

Electromechanical Control Technology and Transportation – Jia & Wu (Eds)
© 2017 Taylor & Francis Group, London, ISBN 978-1-138-06752-3

A classification method for artistic images on feature computation

Raoshan Xu
Nanjing College of Information Technology, Institute of Digital Arts, Nanjing, China

Zhengxing Sun & Chen Ma
State Key Laboratory for Novel Software Technology, Nanjing University, Nanjing, China

ABSTRACT: Due to the developing technique of image acquisition, generation, and processing, the number of images has been increased continuously. In this paper, authors propose an automatic classification method of artistic images based on the features computation, which can classify and visualize artistic images effectively. By using this method, the similarity of images at different properties can be calculated according to the selected features of the image, along with taking the spatial layout matching model into consideration. And then, the clustering mechanism based on MLAP is used to classify images hierarchically. As a result, experiments show the effectiveness of this proposed method for image classification.

1 INTRODUCTION

With the development of image acquisition hardware devices and computer graphics technique, many of the artistic paintings have been digitized into images for appreciation, communication, dissemination, etc. It is more and more necessary for us to classify and organize the images effectively, depending on the fast growing number of images. Artists show their subjective feelings by using different colors and drawing techniques in paintings.

We usually classify images by using the exact label information, such as author, genre, and category, when the number of images is not very large. It shows no obvious deficiencies. But when the image library is increased to a certain size, due to fact that the uncertainty of single image annotation information is not specific, the boundary of each category among images is blurred. For organizing the images effectively, an accurate classification on the basis of image annotation is required and this becomes a difficult problem. Since artistic painting classification is obviously subjective, the use of precise semantic tags describing these artistic images is not evident. In the practice scenario, without sufficient understanding of the painter, it may lead to a bias of understanding or annotation for the image, and trying to find a certain target image becomes almost insurmountable. Therefore, organizing massive artistic painting images, according to the visual properties for proper understanding have gained more and more attention. It can be achieved by using the system in accordance with the correspondence between these images, to help users classify and organize these images with high efficiency.

In this paper, according to the visual properties of the artistic image, we propose a method to classify and present images, which can solve the problem of automatic image classification. In this method, the similarity of images with different properties is selected and calculated, by making use of a novel spatial layout matching model. The visual similarity of images is incorporated into our proposed approach to generate hierarchical clustering organization, which can provide users with a hierarchical browsing structure of massive artistic images with the semantics of public aesthetics.

2 CLASSIFICATION FEATURE EXTRACTION

Different features of the image reflect the characteristics of the image from different sides. In addition to the traditional image features, because of the artistic image category concept, it can be selected by using suitable features for artistic images according to genre characteristics. Meanwhile, there are some differences between artistic and natural images, and so there are some differences in the applicability of image features. The theme, content, colors, genre, and drawing skills of the artistic image are important features for the visual distinguishing of images. Therefore, when we select the image features for computation, as many types of coverage features are taken into consideration, these features are compared from each category to decide the feature to be chosen for further processing base on their visual characteristics.

2.1 Color features

Since color features are the statistical characteristics of the entire image based on pixel features extraction, it is not sensitive to image transformations on direction, size, and rotation. Though the feature extraction is not complicated, it plays an important role in the field of image recognition and image classification.

Hue is the overall color tendency of the image, and warm color or cold color of images can be used to reveal emotions of the artist. Usually, red and yellow are defined as warm colors, and blue and green are defined as the cold ones. And so, we can use the ratio of warm and cold colors *(warm:cold)* to reflect the spirit of images. In this paper, we quantize the warm and cold colors of the artistic image, and calculate the difference for each pixel and the colors red, yellow, blue and green among the images. In other words, it is considered that the color of the pixel is warm if the difference (Euclidean distance) between red and yellow is minimal; otherwise, it is regarded as cold. Due to this quantization, the feature value of warm and cold colors can be defined as a ratio: $f\_warmcold = n\_warmpixel: n\_coldpixel$, where $n\_warmpixel$ is the number of pixels of the image that belongs to warm hue, and $n\_coldpixel$ is that of the cold hue. In the calculation, in order to reduce the ratio span, when the number of cold pixels is too small, we regard the ratios that are greater than *100* as *100* in accordance with the unified processing, and *0.01* as less than *0.01*. For the next step, to calculate the value of similarity, for which we need to use natural logarithm, the value *0* and *Inf* will be discarded. During image similarity computation, the distance between the images is calculated according to the following lemma: $L_{warmcold} = 1-1/(1 + f\_wardcold\ (I_i)-f\_warmcold\ (I_j))^2$. And then, the computation is experimented on an image library: Painting-91 (Khan, 2014), which includes 13 genres such as abstract expressionism. By calculating the mean value of various genres in the *warmcold* feature, it can be noticed that there is some distinction between each genre. Although the proportion of warm and cold colors is simple, it plays a vital role in distinguishing the artistic image genre.

Color Histogram (Stricker, 1995) describes the share of different colors in the entire image in proportion, without regard to the location of each color, which is a common low-level image feature and is widely used in a variety of image retrieval systems. Digital images are generally stored in the RGB color space, which is not coherent with human visual perception. With HSV color space,

therefore, it is possible to simulate the human visual perception better. Letters H, S, and V represent hue, saturation, and brightness, respectively. Furthermore, information in the H-channel of the image is more abundant than those in S and V relatively. In this paper, we quantify H, S, and V as 16, 4, and 4 different bins; that is to say, we quantify the whole color space as 256 aliquots to calculate the color histogram; therefore, the color histogram obtained is a 256-dimensional vector. To overcome the shortcomings for the color histogram, which can only express the color distribution without the position information of each color, the Color Coherence Vector (CCV) (Pass, 1999) was proposed as an improvement. In CCV, the color histogram of each dimension is divided into two dimensions: related and unrelated pixels, to indicate the relative position between the pixels. Each color will be divided into two color histogram statistics. If the continuous area occupied by certain color pixels is bigger than a given threshold, the pixels in this area will be treated as coherence pixels; otherwise, these are treated as non-coherence pixels. Suppose α_i and β_i represent the number of coherence and non-coherence pixels of i dimension in a color histogram, then $<\alpha_1 + \beta_1, \alpha_2 + \beta_2,..., \alpha_N + \beta_N>$ is the color histogram of the image, and $<(\alpha_1, \beta_1), (\alpha_2, \beta_2), ..., (\alpha_N, \beta_N)>$ is the color coherence vector of the image. In this paper, we calculate this feature, according to the color histogram computation, and divide each color histogram into coherence and non-coherence pixels. Instead of the 256-color sub-space, we obtained a 512-dimensional feature vector.

2.2 Texture features

Similar to color features, texture features are intuitive and visually expressive, thereby reflecting the image features of the homogeneous visual pattern, which mainly presents information about the structure and the relationship with surrounding environment. Apart from features based on pixels, texture features are the results of statistical calculations for an image overall or an area.

The Local Binary Pattern (LBP) is an image feature operator that is used to describe the local texture, with an advantage of grayscale invariance, rotational invariance, etc. (Zhao, 2012). By continuously rotating the circular area, the minimum value of the initial LBP values is obtained, which is regarded as the LBP value of this area. In statistics, these values obtained by rotation are expressed as a unified value, so as to solve the problem of rotational invariance. The supposed radius is R, and the number of sampling points is P. It would produce 2^P binary modes, which would be increased dramatically with the neighbourhood sampling points. Excessive pattern types can cause the histogram to be too sparse, which is unfavourable for texture expression. Hence, the original LBP mode dimensionality reduction is required. Ojala (Ojala, 2002) proposed the LBP equivalent model, which proposed that if an LBP corresponding cyclic binary number changed from 0 to 1 or from 1 to 0, at most twice, this LBP binary was called as the equivalent model; otherwise, it is called as the mixed mode. With this improvement, the number of patterns is from the original 2^p to $p(p-1)+2$, without losing information. In this paper, we make use of a combination of rotation invariant and equivalent modes to calculate LBP; let the radius be 2 and the sampling points be 20, and it can be found that each image is represented as a 283-dimensional feature vector.

2.3 Local features

Since the local feature is extracted based on the point or area of interest, the feature point detection is the essential step for local features. Ke (2004) proposed an efficient Scale-Invariant Feature Transform algorithm (SIFT) to describe the image's local features. It researched an extreme point in the range of spatial scales and extracted its position, scale, and rotation invariant as described features. By following the steps of constructing a scale space, extreme point detection, the key point location, direction assignment, and characteristics description, it formed the final 128-dimensional vector. For the sampling of the image information, dense sampling and interest point sampling are usually applied. Dense sampling detects all the content of the image for feature information, as shown in Figure 1(b). The interest point sampling detects the distribution numbers of interest points for feature information, as shown in Figure1(c).

Scholars have proposed the Bag-Of-Words (BOW) (Wei, 2013) model to enhance the combination with classification or clustering algorithms. Apart from feature point matching, BOW is a common method of image matching using local features, which uses a training data set to extract a large number of local features for clustering.

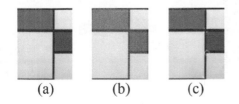

(a)   (b)   (c)

Figure 1. Schematic of the different sampling strategies for image: (a) original image, (b) dense sampling, and (c) interest point sampling.

And then, all the local features of the image can be quantized to the equal dimension histogram according to the dictionary, which will be utilized to match or retrieve images. For the application, in this paper, SIFT features (dense_sift and harris_sift) of two sampling modes were extracted and the respective effects were evaluated. For hierarchical organization of artistic images, we hope that the images with similar content or style will be clustered together, and it is not necessary for precisely identifying image objects. And so, the Bag-Of-Words model was adapted to denote the SIFT feature of images. In this paper, a training set of images from the Painting-91 Image Library was built, in which 50 artists and four images were selected for each as a total of 200. Average sampling of dense SIFT sampling in each image is 2313, thereby forming 1000 visual words. In interest point detection, the average sampling points of each image is 626, thereby forming 300 visual words; while in dense sampling on the color histogram, each image is equally divided into 686 patches, thereby forming 300 visual words.

2.4 Spatial features

Spatial features were used to describe the relative spatial position of objects in an image or area, such as the vertical and horizontal relations between areas. Spatial features are used on the basis of segmentation, which are generally developed for detecting the contents of the image by area identification. The Histogram of Oriented Gradients (HOG) is used to calculate the gradient direction histogram to form local area features, which is mainly used to detect objects in an image. In implementation, it divides the image into small areas, namely unit cells, and then calculates the histogram for each pixel of the unit cell, at the direction gradient or edge. The results obtained from these steps are then passed on to the final stage of combination, thereby putting these cell histograms together to form HOG features of the whole image. The Space Pyramids Histogram of Oriented Gradients (PHOG) (Bosch, 2007) extended the principle of HOG by adding the space pyramid matching method for achieving better spatial characteristics. In this paper, we divided 360 degrees space into 8 bins for features extraction, according to the gradient direction, and let $L=3$ for spatial pyramid matching. In this way, each image can be extracted as a 680-dimensional feature vector.

3 SIMILARITY COMPUTATIONS

In spatial layouts, artistic images are often symmetrical or divided into a three-layer structure, i.e.,

| (a) Integrated | (b) 2*2 divided | (c) 3*1 divided |

Figure 2. Pictures showing the matching model feature extraction space.

upper, middle, and lower, in accordance with the natural landscape for visual comfort. So as the feature matching is used in our method, we take into account the spatial layout of artistic images, divide the given image from top to bottom and from left to right in the middle, and divide the given image into three same-scale segments.

As shown in Figure 2, features fh_1 of the entire image can be extracted, and also the corresponding fh_{2-1}, fh_{2-2}, fh_{2-3}, fh_{2-4}, respectively of four area features, in accordance with the order from top to bottom and left to right. When feature matching is performed, the assigned weight of each feature value is $1/4$. As shown in Figure 2(c), we can extract features fh_{3-1}, fh_{3-2}, and fh_{3-3} from top to bottom, and the weight $1/3$ is assigned to each feature. Therefore, the combination of all these features of image is indicated below:

$$fh_I = [fh_{I_1}, (1/4)fh_{I_{2,1}}, (1/4)fh_{I_{2,2}}, (1/4)fh_{I_{2,3}}, (1/4)$$
$$fh_{I_{2,4}}, (1/3)fh_{I_{3,1}}, (1/3)fh_{I_{3,2}}, (1/3)fh_{I_{3,3}}] \quad (1)$$

For the histogram feature, the chi-square distance to measure the similarity between two images, I_i and I_j:

$$L_{chi}(fh_{I_i}, fh_{I_j}) = \sum_{k=1}^{n} (fh_{I_i}(k) - fh_{I_j}(k))^2 / (fh_{I_i}(k) + fh_{I_j}(k)) \quad (2)$$

For feature-matching, features from the entire image were calculated, that is 2*2 and 3*1 segments, respectively, and the final denotation features were formed in accordance with the lemma (1). And then, the chi-square distance for each pair of images on each feature is calculated by using lemma (2). The result indicates the similarity of images: the lesser the distance, the more similarity is observed between two images.

4 CLUSTERING IMAGES AND EXPERIMENTS

The key to effectiveness of artistic images clustering lies in the chosen correct feature, which can distinguish the different types of images as accurately

as possible. There are 2326 artistic images of 50 painters in Painting-91. In this paper, we attempt to classify images in accordance with the visual information, since an image of the same genre is essentially similar in visual aspects, and the image features are not able to distinguish different artists. And so, we do not classify images for different artists.

The K-means algorithm is used to cluster the image library and Painting-91 into 10, 100, and 200 classes, respectively. And then, we measure the distinguishing performance of features by calculating the ratio of intra-class and inner-class distance, for each clustering result. Because the following steps are part of a hierarchical organization on the image library, we set the clustering number to 10, 100, and 200.

We set the target of artistic image clustering to separate dissimilar images apart, according to their visual similarity. And so, within the clustering results, the smaller inner-class distance and the bigger intra-class distance exhibit better clustering quality indication. Each feature of the image separately clustering in the cluster number 10, 100, and 200 are experimentally tested, with the results shown in Figure 3.

The effectiveness of features on the experimental image library is obviously different: PHOG, dense_sift, and harris_sift are better than LBP, hsv, dense_hsv, CCV, and warmcold. Although color for genre and content distinction of the image is inferior to PHOG and SIFT features, by taking into account the human perception of an artistic image, the color is an essential tool for clustering. Hence, when clustering the experimental image library, we choose the color features for consideration. Furthermore, because harris_sift and dense_sift are closely related, we choose only the harris_sift feature. Thus, in the following classification step, PHOG, harris_sift, and dense_hsv features are chosen.

With the features chosen for classification, we adopt an appropriate hierarchical clustering

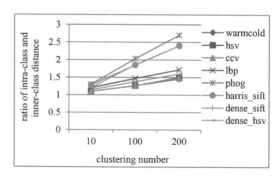

Figure 3. Performance of features in the clustering algorithm.

Table 1. Operation times comparisons of three methods.

Tasks	MLAP	groundTruth	MLKmeans
Task 1: find a single Van Gogh "Star"	3	3	5
Task 2: find 10 bright colors landscape	12	15	17
Task 3: find five CARAVAGGIO works	10	6	12

algorithm MLAP to classify images into hierarchy, thereby forming a hierarchical browsing tree, which shows a representative image in each node. MLAP is the hierarchical clustering version of AP clustering (Frey, 2007), which is a clustering algorithm based on samples, and the cluster center is a certain sample. After classification, the proposed approach is compared to a typical clustering algorithm, MLKmeans, and referenced hierarchy clustering by groundTruth, which classifies images as a browse tree according to the author and semantic of images. We designed three tasks and compared the number of operations required to complete each task, as shown in Table 1. MLAP is overall better than MLKmeans. While looking for a single image, the required operations with MLAP are equal to groundTruth. As for finding the target image, which belongs to multiple genres, with visual feature of similarity, the MLAP algorithm organizes images autonomously by their own visual similarity, while groundTruth divides different genres into different clusters; therefore, the result of MLAP algorithm is better than groundTruth. From experimental results, we can draw a conclusion that the performance of our proposed hierarchical clustering method based on the MLAP algorithm is superior to the MLKmeans method. Furthermore, in some scenarios, it is better than groundTruth.

5 CONCLUSION

A novel hierarchy-based approach is proposed to tackle the problem of organizing massive artistic images on their visual information. On the basis of typical features extraction, similarity of images can be calculated. Furthermore, by evaluating the clustering effect of each feature, we compare the applicability of these features. Finally, the method MLAP clustering, which satisfied the hierarchical classification for given images, is applied, and experiments were carried out in the Painting-91 image library. Results show that the proposed method is effective in organizing images with visual information.

REFERENCES

Bosch A, Zisserman A, Munoz X. Representing shape with a spatial pyramid kernel. Civr Proceedings of Acm International Conference on Image & Video Retrieval (2007).

Chen JY, Bouman CA, Dalton JC. Hierarchical browsing and search of large image databases. IEEE Transactions on Image Processing (2000).

Frey BJ, Dueck D. Clustering by passing messages between data points. Science (2007).

Ke Y, Sukthankar R. PCA-SIFT: A more distinctive representation for local image descriptors. Proceedings of the IEEE International Conference on Computer Vision and Pattern Recognition, 2(2004).

Khan FS, Beigpour S, Weijer VJ, et al. Painting-91: a large scale database for computational painting categorization. Machine vision and applications (2014).

Ojala T, Mäenpää T, Pietikäinen M, Viertola J, Kyllönen J, Huovinen S. New framework for empirical evaluation of texture analysis algorithms. Proc. 16th Int. Conf. Pattern Recog (2002).

Pass GR, Comapring Images using Joint Histograms. Multimedia Systems (1999).

Stricker MM. Similarity of Color Images. Proceedings of Storage and Retrieval for Image and Video Database, (1995).

Weijer VJ, Khan FS. Fusing Color and Shape for Bag-of-Words Based Object Recognition. CCIW (2013).

Zhao GY, Ahonen T, Matas J, Pietikainen M. Rotation-Invariant Image and Video Description With Local Binary Pattern Features. IEEE Transactions on Image Processing (2012).

Electromechanical Control Technology and Transportation – Jia & Wu (Eds)
© 2017 Taylor & Francis Group, London, ISBN 978-1-138-06752-3

Research and design of the paperless electronic business card

Guili Peng
Fundamental Science on Nuclear Wastes and Environmental Safety Laboratory, Southwest University of Science and Technology, Mianyang, China
School of Control and Mechanical Engineering, Tianjin ChengJian University, Tianjin, China

Tong Shen
State Key Laboratory of Geohazard Prevention and Geoenvironment Protection,
Chengdu University of Technology, Chengdu, China

Shan Jiang
Fundamental Science on Nuclear Wastes and Environmental Safety Laboratory, Southwest University of Science and Technology, Mianyang, China

Shoubin Wang
School of Control and Mechanical Engineering, Tianjin ChengJian University, Tianjin, China

Wei Li
School of Electronic Engineering, Tianjin University of Technology and Education, Tianjin, China

ABSTRACT: Environmental protection is a permanent problem in the contemporary social development situation. A business card is the most important factor in business communication. The major component of the business card is paper. A lot of paper is required to manufacture business cards, thereby resulting in a lot of waste. In order to reduce the paper waste and protect the environment, the paperless electronic business card is proposed. The main controller chip adopts STM32C8T6 MCU in this work, with ultra-low power and ultra-small size. The wireless transmission chip adopts TI company's CC2540 radio transceiver chip in this work. This chip is a low-cost chip, with a 2.4 GHz transceiver. In this article, the paperless electronic business card's design and feasibility analysis are the main areas of focus, including hardware and software design. An electronic card prototype was successfully developed, and the basic function was realized.

1 INTRODUCTION

Nowadays, the society is the place for frequent exchanges with the rapid development of market economy, business, and market competition. The business card, as a kind of principal medium of communication between people, plays an extremely significant role. A small card can show a person's status, identity, and personality traits, which can act as a second ID card. At present, the traditional business card is mainly based on paper, with the characteristics of cheaper prices, writing concisely, easy to carry, and so on. But there are a huge number of cards, and we often can't find the one we want, which would cause some economic losses. To a certain extent, it also caused paper and wood wastage. The resources are not conducive to environmental protection. The business card also lacks authenticity discrimination, and in it, there often exists the phenomenon of exaggeration.

Paper business card storage is not convenient and we need more space to store these cards.

In this work, a new kind of electronic business card is proposed, which is easy to carry, easy to use, easy to store, and easy to search. The proposed electronic business card is of great significance. It plays an important role in business contacts. The electronic business card can timely and conveniently deliver the personal information, as well as accept the relevant information you need. The information can be stored in your electronic card classification. The data one require can be found quickly. One can communicate through the business card information on the computer, and verify the authenticity of the business card through an electronic database.

The electronic business card is mainly based on SCM (Single Chip Micyoco) technology. It can completely replace the paper card. Mass storage information can be displayed through the OLED

(Organic Light-Emitting Diode). It is convenient to search and store information through the device. The exchange information mode of the electronic business card adopts Bluetooth wireless transmission, which enhances accuracy and security. The information transmission mode can be achieved with USB (Universal Serial Bus) and Bluetooth technology. The information is stored in the PC (Personal Computer) or PDA (Personal Digital Assistant).

2 THE HARDWARE DESIGN OF THE PAPERLESS CARD

The electronic paperless-card is designed, and it comprises of a power control section, STM32C8T6 MCU (MicroController Unit) part of data processing, data display and storage components, and a CC2540 (Leens, 2009) wireless transceiver. The core theories of the electronic paperless-card are embedded technology and Bluetooth wireless transmission technology. The overall structure diagram is shown in Figure 1:

2.1 The power control circuit

The electronic business card is a portable electronic instrument. Portable products need to be small-size, lightweight, easy to carry, exhibit low power consumption, and so on. For this type of product, the power supply design is particularly important, which is the premise to achieve low power consumption. The power supply design mainly comprises power management, configuration of power parameters, and low voltage alarm function. Low voltage alarm refers to the technology that, when the power supply voltage is too low, the buzzer alarm mode is opened in time to remind the user, and the system operation is set to low power consumption mode.

This power supply voltage of the device is +3.3 V, for which the +5 V battery can be used for power supply, or even can be directly utilized by using the USB power supply. Taking into account the demands of the hardware system for the power supply system, it has the characteristics of a regulator function, small ripple, and low power consumption. The power part of the hardware device TPS76033 voltage chip of the TI company is adopted to be implemented. The chip has a very tiny package, which can save space.

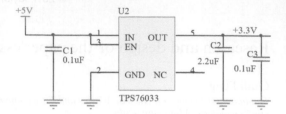

Figure 2. Power supply circuit diagram design.

Its main features are as follows: 50 mA, low pressure drop voltage regulator, mode static current less than 1 uA, and an output voltage of 3.3 V. The power supply circuit is shown in Figure 2:

This device also designs a low voltage detection function, which can provide an alarm when the power supply voltage is lower than 3.3 V, and then re-run the microprocessor at the same time. When the device is not in use, the system enters the standby mode and the STM32C8T6 single chip microcomputer enters the sleep mode. This mode can greatly reduce power consumption.

2.2 The microprocessor core modules

In this work, the electronic paperless-card needs a low power supply voltage, small size, low power consumption and high-speed microprocessor as the main controller chip. TI Company's 16 series of ultra-low power STM32C8T6MCU was chosen as the core module and STM32F103 was selected as the main control chip microprocessor.

The peripheral circuit of the device mainly comprises a keyboard control button, a switch button, a microcontroller reset circuit, and an OLED display module. This device adopts the 32768 crystal oscillator for providing the clock signal, which is used to record the time of the business card and to provide a microprocessor that runs the clock. Using the FLASH of the extensive data storage extends the storage space. An OLED display module is used to output data. The device can transmit data information through USB3.0, which can be uploaded to the PC or other device. The STM32F103C8T6 peripheral interface diagram is shown in Figure 3 (Wang, 2012; Martikainend, 2001; Satoshi, 1999; Chen, 2005):

2.3 Wireless transceiver circuit

In this work, the electronic business card executes the function of wireless data transmission, by using TI company's CC2540 Bluetooth chip. The CC2540 chip combined an excellent RF transceiver (Hager, 2003) with an industry-standard enhanced 8051 MCU, in-system programmable flash memory, 8-KB RAM, and many other powerful supporting features and peripherals. The RF transceiver data transmission

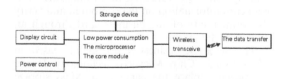

Figure 1. Overall structure of the paperless business card.

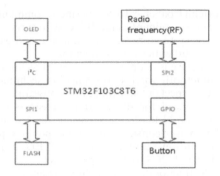

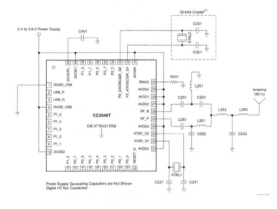

Figure 3. Schematic of the STM32F103C8T6 peripheral interface.

Figure 4. CC2540 chip circuit diagrams.

rates up to 500 kbps. The CC2540 peripheral circuit provides the functions such as data packet handling, data cache, burst data transmission, clear channel assessment, and link quality indication and electromagnetic detection. The main operating parameters could be controlled by using the SPI interface, and a microcontroller and a number of components can be used together (Khaleghi, 2007), as shown in Figure 4.

USART 0 and USART 1 are both configurable as either an SPI master/slave or a UART. These provide double buffering on both RX and TX, and hardware flow control is thus well-adapted to high-throughput full-duplex applications. Each USART has its own high-precision baud-rate generator, thus leaving the ordinary timers free for other uses. When configured as SPI slaves, the USARTs sample the input signal by using SCK directly instead of using some oversampling scheme, and are thus well-suited for high data rates.

The CC2540 chip has two additional crystals. One external 32-MHz crystal, XTAL1, with two loading capacitors (C221 and C231) is used for the 32-MHz crystal oscillator, thereby connecting between XOSC_Q1 and XOSC_Q2 feet. Another 32.768-Hz crystal, XTAL2, with two loading capacitors (C321 and C331) is used for the 32.768-kHz crystal oscillator. The 32.768-kHz crystal oscillator is used in applications where both very low sleep-current consumption and accurate wake-up times are needed. The crystal vibration amplitude can be adjusted. In order to ensure reliable launch, a series resistor may be used to comply with the ESR requirement (Liu, 2012).

3 THE SOFTWARE DESIGN OF THE PAPERLESS-CARD

The electronic business card completes the data input, data storage, and search functions mainly by using STM32C8T6 module. Through the SPI connection CC2540 Bluetooth, the microprocessor completes the data wireless transmission function, including sending and receiving signals.

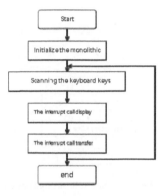

Figure 5. System overall flow chart.

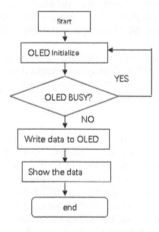

Figure 6. OLED display subroutine flow chart.

The paperless-card software system mainly includes the following features:

STM32C8T6 control module: as the main control module of the system, through the programming controls other program modules of coordination.

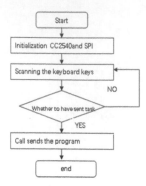

Figure 7.   Send to data subrouting flow chart.

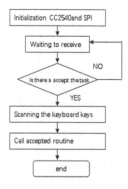

Figure 8.   Receiving data subrouting flow chart.

Function keyboard program module: it is used to implement the project requirements, and makes the SCM complete corresponding control functions.

OLED display module: it uses the character type OLED to display the required information.

Wireless transmission module: the corresponding data are transmitted through the wireless protocol.

The program flow chart is shown in Figure 5 and 6:

The paperless-card software programming language uses C language programming. The Keil MDK programming tool is used to design. Keil MDK is the most comprehensive software development solution for SCM-based microcontrollers, including all components that one needs to create, build, and debug embedded applications, as shown in Figure 7 and Figure 8.

4   CONCLUSION

In this work, the business paperless-card is based on a small-size and low-power design principle. We put forward a kind of electronic product to replace the original paper card for business communication that is dominant today. It is easy to carry, easy to use, easy to store, easy to transmit, and easily tests the authenticity discrimination.

The device is mainly based on SCM technology and Bluetooth wireless transmission technology. The two aspects of hardware and software design are introduced for the electronic business card. The key problems to be solved in the hardware are low-power consumption of the power supply circuit, button scanning circuit, OLED display circuit, STM32C8T6 MCU peripheral circuit, and CC2540 Bluetooth wireless chip transmission circuit. The software mainly solves the fundamental problems that are low-power consumption MCU power supply control program, OLED display program, and wireless transceiver program.

With the development of the socialist market economy in our country, communication is more and more closely related with the international community. The coverage of the card will be larger and larger. The replacement of paper with the paperless electronic business card is the trend. The development of the paperless business card to protect the environment against waste has a very important significance.

REFERENCES

Bluetooth SIG. Specification of the Bluetooth system version 4.0[EB/OL]. www.bluetooth.com, 2009.

Chen Xinfang, Si Yujuan, Zhao Yi, Liu Shiyong. A New AC Driving Method for Active Matrix OLED Displays. Materials Science Forum. 2005. 478–478(3): 1901–1904.

Hager C T, Midkiff S F. An analysis of Bluetooth security vulnerabilities[A]. Wireless Communications and Networking, WCNC 2003[C]. New Orleans, 2003. 1825–1831.

Khaleghi A. Dual Band Meander Line Antenna for Wireless LAN Communication[J]. IEEE Transaction on Antenna and Propagation, 2007, 55(3): 1004–1009.

Leens F. An introduction to I²C and SPI protocols[J]. Instrumentation & Measurement Magazine, IEEE, 2009, 12(1): 8–13.

Liu Hui, Lin Keye, Sun Famdian, et al. Design of inverted Fantenna for Bluetooth system based on Cadence and CST[C]. ICECC 2012, zhoushan, china: IEEE Conference Publication, 2012: 222–223.

Martikainend Li. Emerging personal wireless communications[Z]. Kluwer Academic Publishers, 2001.

Satoshi Miyaguchi, ect. Organic LED full-color passive-matrix display[J]. Journal of the Society for Information Display. 1999. 7(3): 221–226.

STM32F103xE Data Sheet[J].2010(4).

Texas Instruments Corporation. CC2540 Datasheet [EB/O]. www.ti.com/lit/ds/symlink/cc2540.pdf

Wang HX, Wu Y. Design and implementation of LCD module based on SOPC [J]. Chinese Journal of Liquid Crystals and Displays, 2012, 27: 508–514.

Research on the optimization method of the purchasing plan for fresh product e-commerce

Xin Wang, Jie Sun & Kai Chen
School of Business Administration, Shandong Institute of Commerce and Technology, Jinan, China

ABSTRACT: In order to solve the problems in the process of making a procurement plan, an indicator system based on multiple factors and multiple levels is put forward in this paper. Based on the function of the distribution center, the main factors which influence the procurement plan are discussed. According to the uncertainty and fuzziness between the factors, a procurement plan model is built, thereby making use of fuzzy evaluation. By using this model, the purchasing of fresh products in the distribution center can be forecasted and controlled.

1 INTRODUCTION

Fresh product e-commerce has attracted much attention as the last piece of Blue Ocean in the field of e-commerce. E-marketplace of the fresh product develops rapidly benefited by the market participants from the electric business platform and the industry chain of the fresh product. It has formed a relatively complete industrial structure for the past few years, but its dividend has not been fully reflected. When compared with traditional businesses, it is higher to enter the realm of the threshold, not only because of the supply of fresh category, but also warehousing and logistics. The service level determines the survival of fresh product e-commerce (Zhang, 2014). At present, the problem of declining enterprise profitability has to be solved. Its major manifestations are as follows: the non-standard of the purchasing process, an inaccurate material purchase plan, the inconformity between the planning and actual demands, the mass loss of fresh products, and out-of-stock issues. According to the analysis and construction of an index system about fresh products purchasing, a purchasing plans model is proposed based on fuzzy comprehensive evaluation.

2 CONSTRUCTION OF AN INDEX SYSTEM OF A FRESH PURCHASING PLAN

2.1 Summary

Fresh products are defined as the primary fresh chilled products for sale, which have not underwent the cooking process, including bread, cooked food, and other field processing category of goods. It is mainly divided into five categories, including aquaculture seafood, fruits, vegetables, meat and poultry and egg, milk, and dairy products (Tang, 2015). The diversity of fresh products determines the sales type of multi-species and small batches. They have different characteristics such as varieties, preservation conditions, freshness, and added value of products. In order to solve the contradiction among large quantities, few varieties in the process of purchasing and small batches, and diversification in the process of sales, the distribution center in the service of fresh electric e-commerce enterprises should have formulated processing functions, in addition to the function of collecting goods, warehousing, and distribution.

The procurement staff generally made purchasing plans according to the distribution center of radiation within the scope of population, consumption level, season, and historical order sales data plan in the traditional procurement process. The plan cannot fully reflect the profit of the distribution center according to the historical order modifiers.

The evaluation index and output index are also indispensable, such as refreshing time of the fresh products, working ability, and market orientation for distribution centers. In many cases, they play a more important role, as shown in Figure 1.

The Analytic Hierarchy Process (AHP) is a decision analysis method, which is combination of qualitative and quantitative analyses. The thought of the index system is implemented to classify indicators step by step. The structure of the system includes the target hierarchy, the rule hierarchy, and the measure hierarchy (Wang, 2015).

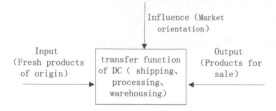

Figure 1. Schematic of influencing factors' analysis of purchasing plans.

2.2 Procurement model building based on AHP–fuzzy comprehensive evaluation

The fresh purchase index evaluation system, as given in Table 1, was designed according to the above-mentioned ideas.

1. Analysis of fresh quality and reasonable control the "input"

   The quality of the fresh product is the most intrinsic factor in the plan. A reasonable purchasing plan must fully consider the characteristics and quality of fresh products. It would optimize the sales category structure and improve the quality, according to the characteristics of fresh products.

   The capacity and the order cycle time of the fresh products reflect the state of origin and the supply position from the perspective of suppliers. The procurement plan can be increased properly if the capacity is higher or the ordering cycle is shorter.

   Refreshing time, as a reference of the plan, provides some guidance about fresh primary products and the storage process. It is conducive to retain the fresh primary stability of product quality and avoid high attrition rates.

2. Measure the resources of the distribution center to ensure the "transformation" effect

   The purchasing plan has been designed according to the market demands and storage volume, processing equipment, number of employees, and software and hardware facilities in the distribution center.

   The biggest capacity should be fully considered to ensure the preservation and fast delivery service goals.

   With an improvement in the acceptance of consumers about fresh product e-commerce, technology upgradation, and capital involved in the industry, the competition is increasingly fierce among fresh product e-commerce enterprises. One of the effective countermeasures for these enterprises is to choose the right direction, and introduce characteristic processing and fast delivery. These should promote the new products and build a trusted brand, based on the

Table 1. Indicator system of purchasing plans.

Destination layer	First grade indexes	Weight	Secondary index	Weight	Combined weight
Purchasing plan of fresh product U	Input factors U_1	a_1	Capacity U11	a_{11}	w_1
			Order cycle time U12	a_{12}	w_2
			Refreshing time U13	a_{13}	w_3
	Transforming factors U_2	a_2	Working ability U21	a_{21}	w_4
			Freezing capacity U22	a_{22}	w_5
			Storage capacity U23	a_{23}	w_6
	Outcome factors U_3	a_3	Finished goods turnover U31	a_{31}	w_7
			Product profitability U32	a_{32}	w_8
	Influence factors U_4	a_4	Market orientation U41	a_{41}	w_9

clear development orientation of the distribution center. It is also important to increase the procurement plan of high value-added fresh products and avoid high attrition rate problems. In the second place, the distribution center should measure its storage and processing capacity scientifically, and avoid increasing the product category and purchasing plan blindly based on the actual situation.

3. Analyze the sales data of the raw commodity and make plans by using the "output" feedback Aimed at the "output", sales data of the fresh products are predicted to meet the needs of the market. We should analyze and predict consumer demands for fresh primary products and recognition correctly, obtain historical data such as frequency of product sales, marketing, category correlation, customer satisfaction, forecast the demand of all kinds of fresh primary products in the area around the distribution center, determine the category and arrange the purchasing plan reasonably. In order to meet the needs about the product categories of

the consumers, the designed procurement plan is not confined to economies of the scale.

4. Timely grasping of the market orientation and focusing on the "impact"

The ultimate goal of the purchasing plan is the improvement of product quality and delivery speed based on the optimal combination of the product category, quantity, and the capacity of the distribution center. And then, the effective utilization and benefit maximization of the distribution center can be realized. Therefore, the market demand inevitably influences the purchasing plan. It is required that the plan should be predictable and forward-looking.

The distribution center should strengthen the market research and scientific prediction, and adjust the category, quantity, and cost structure of the purchasing plan, according to seasonal characteristics, distribution areas, and the market demand (Sun, 2015).

3 PROCUREMENT MODEL BUILDING BASED ON AHP–FUZZY COMPREHENSIVE EVALUATION

3.1 Procurement planning process based on AHP–fuzzy comprehensive evaluation

1. Constitute the factor set by evaluating the purchasing plan

The factor set is the data collection composed of indicators in the index system of the fresh purchasing plan. In order to obtain reasonable value in the process of the fuzzy comprehensive evaluation, the influence factors are divided into four parts: input factors U1, transforming factors U2, outcome factors U3, and influence factors U4. And then, we divide each factor mentioned above to obtain nine factors, including capacity U11, which is expressed as $U = \{u_1, u_2, ..., u_9\}$ (Wang, 2015).

2. Make the evaluation set by determining the purchase plan

The evaluation set is a collection of all evaluation results garnered from 12 experts. The purpose of the fuzzy comprehensive evaluation is the best evaluation result in the evaluation set based on the comprehensive consideration of nine factors affecting the purchasing plan. The comments can be marked in a range by using the 1–5 sign method as follows: $V = \{v_1, v_2, v_3, v_4, v_5\}$ {very satisfied, satisfied, general, dissatisfied, and very dissatisfied}. The result of the instance will be obtained by using the expert scoring method.

3. Make the factor weight set

The factor weight set represents the important degree of collection about nine factors in the indicator system. It can be represented as follows: $W = \{w_1, w_2, ..., w_n\}$, where W is the weight of each factor allocation, w_i is the weight of the i factor, and $\sum_{i=1}^{n} w_i = 1$.

The Analytic Hierarchy Process (AHP) is used to determine the weights in this paper. The importance of the evaluation index is compared to establish the judgment matrix. It has satisfactory consistency. And then, the matrix eigenvalues and corresponding eigenvectors were calculated, consistency was checked, and the weight of the indicators was finally determined.

4. Single factor fuzzy integrated assessment

The fresh purchasing plan is evaluated based on u_k, which is the k factor in the factor set U. C_{kj} is the membership degree of v_j which is the j element ($j = 1, 2, ... ,4$) in the evaluation set. And then, the result according to the evaluation can be expressed with fuzzy set, which is as follows:

$$C_k = \{c_{k1}, c_{k2}, ..., c_{k4}\}$$

C_{kj} is a kind of subordinate relation between u_k and v_j. There is also the possibility of a fresh purchase plan, which can be judged as grades, from the perspective of factor u_k.

The matrix is calculated after we evaluate all the factors and this is given as follows:

$$C = \begin{bmatrix} C_1 \\ C_2 \\ ... \\ C_9 \end{bmatrix} = \begin{bmatrix} c_{11} & c_{12} & \cdots & c_{14} \\ c_{21} & c_{22} & \cdots & c_{24} \\ ... & ... & ... & ... \\ c_{91} & c_{92} & \cdots & c_{94} \end{bmatrix}$$

The single factor fuzzy integrated assessment can be considered as the fuzzy relation between the factor set U and the evaluation set V.

5. Multi-factor fuzzy evaluation

W and C will be synthesized by using the appropriate fuzzy comprehensive evaluation model of B = WC. The evaluation set is obtained by using the fuzzy transforming method, which represents a kind of method used to obtain synthetic data about index w and c, namely, the combination of fuzzy operators. There are many fuzzy operator combinations. These constitute different evaluation models.

According to the previous theory, common matrix multiplication (the weighted average method) is used in the process of the fuzzy synthesis operation. Each factor plays a role in the model, which reflects the evaluation object. If $\Sigma b_j \neq 1$, the result should be normalized. b_j represents the membership degree of evaluation objects to v_j, thereby considering all the factors (Gao, 2007).

All weighted values can be obtained about the fresh products in the index system of the purchasing plan according to the information collection and data calculation. At the same time, the fuzzy evaluation data can also be obtained about each fresh product.

3.2 Data preparation

According to the fresh purchasing plan index system, the index matrix S_{kj} can be calculated by matching the index factors with the fresh products (assuming that there are l kinds of fresh products) in the distribution center. This matrix should be modified by using the following methods before compiling.

1. The quantifiable indicators can be used to calculate the percentage between each index and total value about fresh products. Equation 3 was used to deal with the matrix.

$$C_{kj} = \frac{S_{kj}}{\sum\limits_{j=1}^{l} S_{kj}} (k = 1, 2, ..., 8, j = 1, 2, ..., l) \qquad (1)$$

2. The expert scoring method was used to process the data from fuzzy indicators. The assessment marks of experts are modified to obtain the numerical value. The following equation was used to obtain the matrix.

$$C_{kj} = \frac{S_{kj}}{\sum\limits_{j=1}^{l} S_{kj}} (k = 9, j = 1, 2, ..., l) \qquad (2)$$

3.3 Construction of the comprehensive model of evaluation

The fresh purchasing plan model based on AHP–fuzzy comprehensive evaluation is shown in the following equation:

$$M_j = T\sum\limits_{k=1}^{n} \left(w_k \times C_{kj}\right)(j = 1, 2, ..., l) \qquad (3)$$

Among these, M_j is the cost of credit allocated to each indicator j; T is the total purchasing cost in a certain distribution center; C_{kj} is the reference index matrix, in which the k reference index matches with j kind of fresh products in the process of the allocation plan.

Constraints: $\sum\limits_{j=1}^{n} C_{kj} = 1 \qquad (4)$

W_k is the weight of index k. Constraints:

$$\sum\limits_{k=1}^{n} w_k = 1 \qquad (5)$$

4 THE INSTANCE ANALYSIS OF THE FRESH PURCHASING PLAN

4.1 AHP–fuzzy comprehensive evaluation

1. Factor sets U = {capacity, order cycle time, refreshing time, working ability, freezing capacity, storage capacity, finished goods turnover, product profitability, market orientation}.
2. Standby selected aggregate $V = \{v_1, v_2, v_3, v_4, v_5\}$ = {very satisfied, satisfied, general, dissatisfied, very dissatisfied}.
3. Determination of the weight coefficient

The index system of the fresh purchasing plan mentioned above is analyzed based on the Analytic Hierarchy Process (AHP). The weights of indicators are determined according to the results of the questionnaire. Twelve experienced staffs from warehousing and distribution department are involved in this research.

① The weight of the evaluation unit
The judgment matrix of the secondary evaluation unit is given in Table 2, by carrying out value assignment of the evaluation unit.

$$\lambda_{max} = \frac{1}{n}\sum\limits_{i=1}^{n} \frac{\left(A\bar{W}\right)_i}{W_i} = 4.108$$

$$CI = \frac{\lambda_{max} - n}{n-1} = 0.036$$

The Random Index (RI) of the fourth order matrix is 0.90; $CR = CI/RI = 0.040 < 0.1$; and the matrix has the satisfactory consistency.

② The weight of the secondary evaluation unit
The judgment matrix of the secondary evaluation unit can be obtained by carrying out value assignment of the evaluation unit, as shown in Table 3.

Table 2. $U_i - U$ matrix.

U	U_1	U_2	U_3	U_4	Eigen vector $\overline{W}$	Weight vectors $\overline{W}_i$
U_1	1.0	3.2	2.4	5.6	2.154	0.509
U_2	0.3	1.0	1.8	4.0	1.026	0.250
U_3	0.4	0.6	1.0	1.6	0.642	0.158
U_4	0.2	0.3	0.6	1.0	0.335	0.083

$$\lambda_{max} = \frac{1}{n}\sum_{i=1}^{n}\frac{(A\bar{W})_i}{W_i} = 3$$

$$CI = \frac{\lambda_{max} - n}{n-1} = 0.0002$$

The Random Index (RI) of the three degree matrix is 0.58; $CR = CI/RI = 0.0003 < 0.1$; and the matrix has satisfactory consistency, as shown in Table 4.

$$\lambda_{max} = \frac{1}{n}\sum_{i=1}^{n}\frac{(A\bar{W})_i}{W_i} = 3.013$$

$$CI = \frac{\lambda_{max} - n}{n-1} = 0.006$$

The Random Index (RI) of the three degree matrix is 0.58; $CR = CI/RI = 0.011 < 0.1$. The matrix has satisfactory consistency, as shown in Table 5.

The matrix has the satisfactory consistency when degree is less than or equal to two.

The consistency of the judgment matrix shows that the weight of the distribution is reasonable. The results are given in Table 6.

As a result, the weight of each factor can be determined as follows:

W = {0.281, 0.077, 0.151, 0.035, 0.138, 0.078, 0.063, 0.095, 0.083}

Table 3. $U_{1j} - U_1$ matrix.

U_1	U_{11}	U_{12}	U_{13}	Eigen vector $\bar{W}$	Weight vectors $\bar{W}_i$
U_{11}	1.0	3.6	1.9	1.659	0.553
U_{12}	0.3	1.0	0.5	0.453	0.151
U_{13}	0.5	2.0	1.0	0.889	0.296

Table 4. $U_{2j} - U_2$ matrix.

U_2	U_{21}	U_{22}	U_{23}	Eigen vector $\bar{W}$	Weight vectors $\bar{W}_i$
U_{21}	1.0	0.2	0.5	0.416	0.138
U_{22}	4.5	1.0	1.6	1.670	0.552
U_{23}	2.0	0.6	1.0	0.931	0.309

Table 5. $U_{3j} - U_3$ matrix.

U_3	U_{31}	U_{32}	Eigen vector $\bar{W}$	Weight vectors $\bar{W}_i$
U_{31}	1.0	0.7	0.800	0.400
U_{32}	1.5	1.0	1.200	0.600

Table 6. Weights at all levels.

First grade indexes	Weight	Secondary index	Weight	Combined weight
Input factors U_1	0.509	Capacity U11	0.553	0.281
		Order cycle time U12	0.151	0.077
		Refreshing time U13	0.296	0.151
Transforming factors U_2	0.250	Working ability U21	0.138	0.035
		Freezing capacity U22	0.552	0.138
		Storage capacity U23	0.309	0.078
Outcome factors U_3	0.158	Finished goods turnover U31	0.400	0.063
		Product profitability U32	0.600	0.095
Influence factors U_4	0.083	Market orientation U41	1.000	0.083

4.2 Data analysis

All fresh products obtained from this distribution center are divided into five categories in the paper ($l = 5$), as given in Table 7. According to the research and questionnaire data processing, matrix S_{kj} can be obtained.

1. For quantifiable indicators, matrix C_{kj} can be obtained based on equation (1) and equation (2). Take U_{11} as an example,

$$C_{1j} = \frac{C'_{1j}}{\sum_{j=1}^{l}C'_{1j}} = \{0.259, 0.167, 0.184, 0.287, 0.103\}$$

Similarly, the other matrix is obtained.

2. For fuzzy indicators, the data results, which have been obtained by quantitative treatment on the basis of investigation among 12 experts, are obtained according to equation $C_{kj} = \frac{S_{kj}}{\sum_{j=1}^{l}S_{kj}}$. Take U_{41} as an example, $C_{41} = \{0.336, 0.261, 0.218, 0.101, 0.084\}$. Similarly, the other matrix can be obtained. From this equation, processed data can be obtained in Table 8.

Table 7. Indicator data matrix.

Basic information	Quantifiable indicators					
	U_{11}	U_{12}	U_{13}	U_{21}	U_{22}	U_{23}
Fruits	4.5	5	7	3.6	0.80	4.8
Vegetables	2.9	3	3	2.4	0.60	3.2
Beef	3.2	10	10	1.9	0.79	4.0
Milk	5.0	12	15	5.6	0.85	3.7
Eggs	1.8	15	18	3	0.73	2.6

Basic information	Quantifiable indicators	Fuzzy indicators
	U32	U41
Fruits	1.50	4.0
Vegetables	1.20	3.1
Beef	0.80	2.6
Milk	0.54	1.2
Eggs	0.33	1.0

Table 8. Indicator data matrix after operation.

Basic information	Quantifiable indicators					
	U_{11}	U_{12}	U_{13}	U_{21}	U_{22}	U_{23}
Fruits	0.259	0.255	0.132	0.218	0.212	0.262
Vegetables	0.167	0.426	0.057	0.145	0.159	0.175
Beef	0.184	0.128	0.189	0.115	0.210	0.219
Milk	0.287	0.106	0.283	0.339	0.225	0.202
Eggs	0.103	0.085	0.340	0.182	0.194	0.142

Basic information	Quantifiable indicators		Fuzzy indicators
	U_{31}	U_{32}	U_{41}
Fruits	0.238	0.343	0.336
Vegetables	0.159	0.275	0.261
Beef	0.152	0.183	0.218
Milk	0.254	0.124	0.101
Eggs	0.197	0.076	0.084

4.3 Results obtained by using the model

The ordering cost is known in the distribution center, $T = \sum M_j = 1000$, $(j = 1, 2, ..., 10)$ (IU, hundred). Matrix C_{kj} and Matrix w_k can be obtained from Table 8 and Table 6. According to equation (3), $M_j = T \sum_{k=1}^{n} (C_{kj} \times w_k)$ $(j = 1, 2, ..., l)$. Take fruits as an example.

$$M_j = 1000 \sum_{k=1}^{n} (C_{k1} \times w_k)$$
$$= 1000[0.259, 0.255, 0.132, 0.218, 0.212, 0.262, 0.238, 0.343, 0.336] \cdot [0.151, 0.035, 0.138, 0.078, 0.063, 0.095, 0.083]^{T} = 245.$$

Table 9. The results of the procurement plan.

Basic information	Procurement plan	
	Rate	Cost
Fruits	0.245	245
Vegetables	0.186	186
Beef	0.185	185
Milk	0.226	226
Eggs	0.157	157
Total	1.000	1000

Similarly, the final purchasing plan can be obtained based on the model, as shown in Table 9.

According to the result of the model, it is not difficult to find the equilibrium distribution of the costs. All kinds of product allocations are reasonable, which fully embody the characteristics of storage and processing functions in this distribution center. These also reflected the influence of seasonal factors to a certain degree. The model can be modified and extended to the real decision process, thereby making the decision process more flexible.

5 CONCLUSIONS

A more detailed analytical study should be conducted not merely because there are many parameters that have no rules to follow, but also because of the characteristics of fuzziness. The accuracy of the planning process can be improved by strengthening the procurement plan evaluation index and sales data research in the future.

Warehouse logistics service for fresh product e-commerce is a systematic project. In order to realize efficient management, procurement indicators should be quantified in the process of enhancements of functions of the distribution center based on the optimization method.

REFERENCES

Gao Jianfang. A study on social responsibility evaluation system of tourism corporate [D]. Beijing Forestry University, 2007.

Sun Bo, Xiao Rucheng. Bridge Fire Risk Assessment System Based on Analytic Hierarchy Process-Fuzzy Comprehensive Evaluation Method [J]. Journal of Tongji University (Natural Science), 2015, 43(11):1619–1625.

Tang Mingxue, Guan Huiwen. Research on the supply chain of Fresh Product E-commerce [J]. Brand, 2015, (7):71–71.

Wang Jing, Xiong Ying. Study on E-business Dominated Fresh Farm Produce Supply Chain Alliance [J]. Logistics Technology, 2015, 34(11):176–178.

Zhang Xiaheng. Current Situation, Problems and Development Trend about the Logistics of Fresh Product E-commerce [J]. Guizhou Agricultural Sciences, 2014, (11):275–278.

Electromechanical Control Technology and Transportation – Jia & Wu (Eds)
© *2017 Taylor & Francis Group, London, ISBN 978-1-138-06752-3*

The CS-OMP estimation algorithm for OFDM-based LV-PLC channels

Ruiming Yuan
State Grid Jibei Electric Power Co. Ltd., North China Electric Power Research Institute Co. Ltd., Beijing, China

Junlong Guo
College of Information Science and Technology, Beijing University of Chemical Technology, Beijing, China

Zhenyu Jiang
State Grid Jibei Electric Power Co. Ltd., North China Electric Power Research Institute Co. Ltd., Beijing, China

Xuewei Wang & Shanshan Ma
College of Information Science and Technology Beijing, University of Chemical Technology, Beijing, China

ABSTRACT: In this paper, the multipath characteristics of Low-Voltage Power Line Carrier communication (LV-PLC) channels are analyzed, and the OFDM baseband transmission system and the multipath channel transmission characteristic model of LV-PLC are established. Meanwhile, the impulse response h(n) of LV-PLC channels is obtained according to the model simulation. On this basis, the OFDM-based LV-PLC multipath transmission mathematical model that meets the compressed sensing conditions is deduced, and the CS-OMP estimation algorithm is proposed for OFDM-based LV-PLC channels, which overcomes the poor precision of LS and MMSE estimation algorithms. The results show that in the presence of the same quantity of the pilot and SNR = 14 dB, when compared with a better one of the LS and MMSE estimation algorithms, the MSE and BER of the CS-OMP estimation algorithm are reduced by 16 dB and 3 times respectively.

1   INTRODUCTION

In recent years, there has been an increasing interest in studying the LV-PLC, which uses existing power lines as a communication medium that has the advantages of low cost and wide distribution, but power line channels have large attenuation, strong noise, frequency selective fading, multipath effect, and other characteristics, which have grand challenges for developing the LV-PLC technology (Zimmermann, 2002). As OFDM technology is applied to the LV-PLC, the transmission rate of LV-PLC has been greatly improved. Channel estimation is one of the key technologies of OFDM, the performance of which determines the reliability and effectiveness of the communication system.

The channel estimation problem is a widely studied research topic in the areas of communication. The main objective of the channel estimation problem is to detect the time domain and frequency domain transmission characteristics of the channel from the received distorted signal. During the past decades, channel estimation problem has been extensively investigated, and several important channel estimation algorithms have been proposed in literature including the Least Square (LS) estimation algorithm (Edfors, 1998; Barhumi, 2003), Minimum Mean Square Error (MMSE) estimation algorithm (Park, 2006; Kim, 2014; You, 2005), Discrete Fourier Transform (DFT) estimation algorithm (Kang, 2007; Pandana, 2005; Gao, 2007). However, these algorithms are not using the sparse property of multipath channels to improve the estimation performance.

Based on the signal decomposition and approximation theory, the concept of CS is proposed by Donho, Candes, and Tao (2006). With the development of CS theory, the CS reconstruction algorithm has been applied to areas of channel estimation and CS estimation algorithms for different OFDM-based channels are proposed, e.g., Mobile Wireless Channel (Ren, 2014; Taub02ck, 2008; Hou, 2014; Qi, 2011; Bajwa, 2010; Kim, 2014), Underwater Multipath Channel (Berger, 2009), and Bernoulli-Gaussian channel (Pejoski, 2015).

However, the LS and MMSE estimation algorithms for LV-PLC channel which were proposed before having large MSE and high BER; the existing CS channel estimation algorithms mainly point at the wireless communication system; different from wireless communication, LV-PLC has attenuation, fading, multipath effect, impedance change, and noise interference at the same time. The transmission characteristic model is difficult to be modeled and the parameters are complex, so the CS channel estimation algorithm of wireless communication system model does not apply to LV-PLC; the CS estimation algorithm for OFDM-based LV-PLC channels does not have related reports. In this paper, we establish the OFDM baseband transmission system and the multipath channel transmission characteristic model of LV-PLC, deduce the OFDM-based LV-PLC multipath transmission mathematical model which meets the condition of CS, do the modeling of the multipath channel estimation problem as the CS problem, and propose the CS-OMP estimation algorithm for OFDM-based LV-PLC channels.

2 OFDM BASEBAND TRANSMISSION SYSTEM

In this section, the OFDM baseband transmission system is established as shown in Figure 1. $X(k) = [X_0, X_1, ..., X_{N-1}]^T$ is the frequency domain signal that is representative of the pilot signal, where $[*]^T$ is the symbol of transpose. The time domain signal $x(n)$ is obtained after N-point IDFT of $X(k)$. In order to avoid the inter-symbol and inter-subcarrier interference, we insert the Cyclic Prefix (CP) into the transmitted signal and obtain $y(n)$ after removing the CP for the received signal. Then, $y(n)$ is transformed into $Y(k)$ by DFT. If the length of the channel $h(n)$ is L, then $Length(CP) \geq L$. The above transmission process of the OFDM baseband transmission system can be described as

$$Y = DFT(IDFT(X) \otimes h + n) \tag{1}$$

In order to facilitate computer processing, (1) can be rewritten into matrix expression form. First, (1) can be rewritten as

$$Y = DFT(IDFT(X) \otimes h + n) = DFT(x \otimes h) + n_k \tag{2}$$

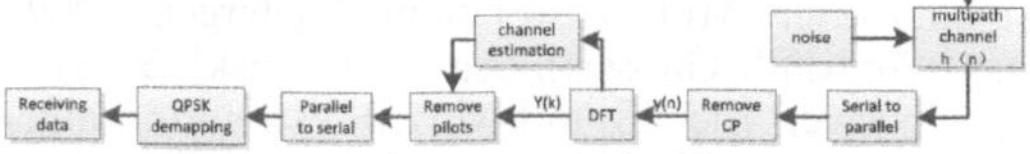

$x(n)$ is required to include the CP and parallel-serial transform before transmission, and then CP is removed after the serial-parallel transform at the receiving end. The convolution operation can be replaced by a circular convolution operation. We define F as an $N \times N$ DFT matrix. According to the properties of the DFT, we can get

$$Y = diag(X) \cdot H + n_k = diag(X) \cdot DFT(h_0) + n_k$$
$$= diag(X) \cdot Fh_0 + n_k$$

$$h_0(n) = \begin{cases} h(n), & 0 \leq n \leq L-1 \\ 0, & L \leq n \leq N-1 \end{cases}$$

$$F = \frac{1}{\sqrt{N}} \begin{bmatrix} W_N^{00} & W_N^{01} & \cdots & W_N^{0(N-1)} \\ W_N^{10} & W_N^{11} & \cdots & W_N^{1(N-1)} \\ \vdots & \vdots & \ddots & \vdots \\ W_N^{(N-1)0} & W_N^{(N-1)1} & \cdots & W_N^{(N-1)(N-1)} \end{bmatrix} \tag{3}$$

where $W_N^{nk} = e^{-j\frac{2\pi nk}{N}}$.

The widely used pilot structures are block-type pilot, comb pilot, and grid pilot. In this paper, we choose comb pilot and define M as the number of pilots. The received signal at pilots can be expressed as:

$$Y_M = S_M Y = S_M(diag(X) \cdot Fh_0 + n_k)$$
$$\simeq S_M diag(X) F \cdot h_0 = \Phi_M \cdot h_0 \tag{4}$$

$$\Phi_M = S_M diag(X) F \tag{5}$$

where S_M is the 0–1 matrix of $M \times N$ dimension, which is used to extract channel values at pilots.

3 MULTIPATH CHANNEL MODEL OF LV-PLC AND H(N)

3.1 The multipath channel transmission characteristic model of LV-PLC

Multipath effect of carrier signal transmission is shown in Figure 2. When carrier signal transmission is in complex situations (reflection, refraction, a standing wave, etc.), it will reach the node L4 by different paths (the phenomenon of multipath effect).

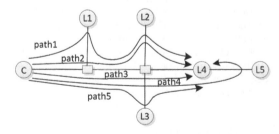

Figure 1. OFDM baseband transmission system.

Figure 2. Multipath effect of carrier signal transmission.

The N-paths channel transmission characteristic model of LV-PLC can be expressed as

$$H(f) = \sum_{i=1}^{N} g_i \cdot A(f, d_i) \cdot e^{-j2\pi f \tau_i} \qquad (6)$$

where i represents the i-path; d_i is the length of the power line; g_i is the weighting factor of i-path ($g_i <= 1$) and determined by reflection and transmission coefficient; $A(f, d_i)$ is an attenuation coefficient and increasing with length and frequency; τ_i is the delay of i-path ($\tau_i = d_i\sqrt{\varepsilon_r}/c_0 = d_i/v_p$, where c_0 is the speed of light, ε_r is the dielectric constant of the power line).

Based on transmission line theory, the expression of $A(f, d_i)$ is deduced in [1].

$$A(f, d_i) = e^{-(\alpha_0 + \alpha_1 f^k)d_i} \qquad (7)$$

where α_0, α_1 and k (value is between 0.5 and 1) are attenuation parameters.

As a result, (6) can be reformulated as

$$
\begin{aligned}
H(f) &= \sum_{i=1}^{N} g_i A(f, d_i) e^{-j2\pi f \tau_i} \\
&= \sum_{i=1}^{N} g_i e^{-(\alpha_0 + \alpha_1 f^k)d_i} e^{-j2\pi f \tau_i} \\
&= \sum_{i=1}^{N} g_i e^{-(\alpha_0 + \alpha_1 f^k)d_i - j2\pi f d_i \sqrt{\varepsilon_r}/c_0} \\
&= \sum_{i=1}^{N} g_i e^{-(\alpha_0 - \alpha_1 f^k - j2\pi f d_i \sqrt{\varepsilon_r}/c_0)d_i}
\end{aligned}
\qquad (8)
$$

3.2 The simulation of h(n)

Referring to the interpretative structural model of the LV-PLC system, which is established by authors of this paper (Wang, 2015), we analyze a carrier communication channel of the LV-PLC test system and obtain the carrier signal transmission multipath effect diagram of this channel according to test results as shown in Figure 2. It can be seen that the carrier signal passes through five different paths from node C to node L4. According to the established multipath channel transmission characteristic model of LV-PLC, we simulate the channel and obtain h(n).

Multipath channel parameters in (8) are selected to be $\alpha_0 = 0$, $\alpha_1 = 7.8 \times 10^{-10}$ s/m, k = 1, $\varepsilon_r = 3.8$, $c_0 = 3.0 \times 10^8$ m/s, and the others are listed in Table 1.

Within the simulation frequency range from 10 kHz to 10 MHz, 64 points are extracted and transformed into the time domain by IFFT. In order to facilitate research, the impulse response is optimized by set zero for small numbers, as shown in Figure 3.

Table 1. Parameters of the LV-PLC multipath channel transmission characteristic model.

Path i	Weighting coefficient g_i	Path length/m d_i
1	0.87	65
2	0.78	120.4
3	0.65	171
4	0.57	245.5
5	0.49	365.5

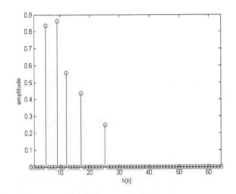

Figure 3. Impulse response h(n) of the multipath channel.

4 COMPRESSIVE SENSING AND CS-OMP ESTIMATION ALGORITHM

Based on signal decomposition and approximation theory, the concept of compressed sensing is proposed by Donho, Candes, and Tao in 2006[9]. Different from the traditional uniform sampling, the core of CS is using a specific matrix to project a sparse (or compressible) high dimensional signal onto a low dimensional space, and then based on the prior condition that the signal is sparse, we use the linear or nonlinear reconstruction model to reconstruct the original signal.

The process of CS is that a sampled signal y (Length(y) = M and M << N) is obtained after a K-sparse signal x (Length(x) = N) through the linear projection as shown in (9).

$$y = \Phi \cdot x \qquad (9)$$

where Φ is an $M \times N$ measurement matrix.

The construction of Φ is one of the three core issues of CS, which determines the accuracy and stability of the reconstructed signal. Candes proposed that the measurement matrix Φ obeys a "Restricted Isometry Hypothesis (RIP)", which is satisfied with:

Table 2. The CS-OMP estimation algorithm for OFDM-based LV-PLC channels.

Input:	Φ_M, Y_M, K
Initialize: $h_0^{(0)} \leftarrow 0$, $r^{(0)} \leftarrow Y_M$, $\Gamma^{(0)} \leftarrow \varnothing$, $\Phi_{M\Gamma^{(0)}} \leftarrow \varnothing$, $k \leftarrow 1$	
While $k < K$ **do** $g^{(k)} = \Phi_M{}^T r^{(k-1)}$ $\Phi_{\lambda^{(k)}} \leftarrow \Phi_M(:, \lambda^{(k)})$ where $\lambda^{(k)} = \arg \max_{j=1,2,\dots,N} \left\| g^{(k)}[j] \right\|$ $\Gamma^{(k)} = \Gamma^{(k-1)} \cup \{\lambda^{(k)}\}$, $\Phi_{M\Gamma^{(k)}} = \Phi_{M\Gamma^{(k-1)}} \cup \{\Phi_{\lambda^{(k)}}\}$ $h_0^{(k)} = \arg\min_h \| Y_M - \Phi_{M\Gamma^{(k)}} h_0 \|_2$ $r^{(k)} = Y_M - \Phi_{M\Gamma^{(k)}} h_0^{(k)}$, $k = k+1$	
end while **return** $\tilde{h}_0 = h_0^{(k)}$	

For an $M \times N$ measurement matrix Φ, the K-restricted isometry constant δ_k of Φ is the smallest quantity such that

$$(1 - \delta_k)\|x\|_2^2 \leq \|\Phi x\|_2^2 \leq (1 + \delta_k)\|x\|_2^2 \quad \forall x, \|x\|_0 = K \tag{10}$$

holds for all K-sparse signals x.

Comparing with the form of (4) and (9), we can see that if h_0 is a K-sparse vector ($K <= M$) and Φ_M is a measurement matrix, then (4) can be resolved using OMP reconstruction algorithm. Since h_0 is an impulse response of the multipath channel, we have $\| h_0 \|_0 \leq M$ (i.e., non-zero number less than M), therefore h_0 is a K-sparse vector. (5) represents the process that selects the rows (the same place as the pilot) from the DFT matrix in uniform distribution randomly and multiplies the corresponding pilot value in each row. If pilot values are set to 1, then Φ_M is a partly random Fourier matrix. The literature (Rudelson, 2006) proved that the partly random Fourier matrix satisfies the RIP condition of order $O(M/(\log N)^4)$. According to the analyses above, the mathematical model of (4) satisfies the condition of CS, thus, h_0 can be recovered using Orthogonal Matching Pursuit (OMP) reconstruction algorithm which is the core of the CS-OMP estimate algorithm. The proposed CS-OMP estimate algorithm is described by the pseudo code as shown in Table 2.

5 SIMULATION AND ANALYSIS OF RESULTS

In this section, we simulate the CS-OMP algorithm for an OFDM-based LV-PLC channel. In the OFDM system, the modulation is QPSK, the sub-carrier number N is 256, the length of the pilot is 128, the length of CP is 64, and the pilot is inserted in a comb way. The OFDM-based LV-PLC channel is $h(n)$ that is obtained in section 3 and the length is 64.

The following Mean Squared Error (MSE) is employed to evaluate the performance,

$$MSE = \frac{1}{N} \sum_{n=1}^{N} (h(n) - \hat{h}(n))^2 \tag{11}$$

When SNR = 14 dB, the original and estimation $h(n)$ are shown in Figure 4. It shows that the proposed CS-OMP algorithm successfully detects an OFDM-based LV-PLC channel with 5 multipath and SNR = 14 dB. Next, we will discuss and compare the performance of channel estimation among the LS, MMSE, and CS-OMP estimation algorithms.

Matlab simulation using the LS, MMSE, and CS-OMP algorithms have been conducted for an OFDM-based LV-PLC channel. In addition, the MSE and BER curves of different SNR are given in Figures 5 and 6. It can be seen that the MSE and BER performance using the CS-OMP estimation algorithm is much improved compared with the other estimation algorithms. Especially, in the presence of the same quantity of the pilot and SNR = 14 dB, when compared with a better one of the LS and MMSE estimation algorithms, the MSE and BER of the CS-OMP estimation algorithm are reduced by 16 dB and 3 times respectively, and its performance will have significant enhancement with the increase in SNR.

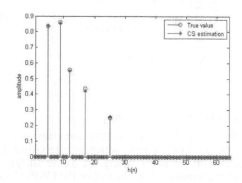

Figure 4. The reconstruction of $h(n)$ (SNR = 14 dB).

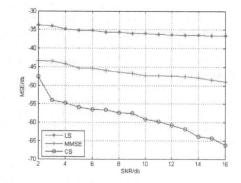

Figure 5. The MSE of the LS, MMSE, and CS-OMP estimation algorithms.

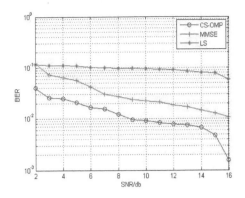

Figure 6. The BER of the LS, MMSE, and CS-OMP estimation algorithms.

6 CONCLUSIONS

In this paper, the CS-OMP estimation algorithm for OFDM-based LV-PLC channels is proposed to overcome the defect of large MSE and high BER in LS and MMSE estimation algorithms. The OFDM-based LV-PLC multipath transmission mathematical model that meets the condition of CS is deduced, which models the multipath channel estimation problem as the CS problem. Meanwhile, the multipath channel impulse response of LV-PLC is reconstructed with the CS-OMP estimation algorithm and the MSE and BER of the LS, MMSE, and CS-OMP estimate algorithms are compared and analyzed. The results show that in the presence of the same quantity of the pilot and SNR = 14 dB, when compared with a better one of the LS and MMSE estimate algorithms, the MSE and BER of the CS-OMP estimate algorithm are reduced by 16 dB and 3 times respectively, and its performance will have significant enhancement with the increase in SNR.

ACKNOWLEDGMENTS

This paper was supported by the State Grid Jibei Electric Power Co. Ltd Research Program (52018 K14001 W).

REFERENCES

Bajwa W U, Haupt J, Sayeed A M, et al. Compressed Channel Sensing: A New Approach to Estimating Sparse Multipath Channels [J]. *Proceedings of the IEEE*, 2010, 98(6):1058–1076.

Barhumi I, Leus G, Moonen M. Optimal training design for MIMO OFDM systems in mobile wireless channels [J]. *IEEE Transactions on Signal Processing*, 2003, 51(6):1615–1624.

Berger C R, Zhou S, Preisig J C, et al. Sparse Channel Estimation for Multicarrier Underwater Acoustic Communication: From Subspace Methods to Compressed Sensing [J]. *Signal Processing IEEE Transactions on*, 2009, 58(3):1708–1721.

Candes, E., J. Romberg, T. Tao. Robust uncertainty principles: Exact signal reconstruction from highly incomplete frequency information [J]. *IEEE Transactions on Information Theory*, 2006, 52(3):489–509.

Edfors O, Sandell M, Jan-Jaap V D B, et al. OFDM channel estimation by singular value decomposition [J]. *Communications IEEE Transactions on*, 1998, 46:931–939.

Gao X, Jiang B, You X, et al. Efficient Channel Estimation for Mimo Single-Carrier Block Transmission with Dual Cyclic Timeslot Structure [J]. *Communications IEEE Transactions on*, 2007, 55(11):2210–2223.

Hou W, Lim C W. Structured Compressive Channel Estimation for Large-Scale MISO-OFDM Systems [J]. *IEEE Communications Letters*, 2014, 18(5):765–768.

Kang Y, Kim K, Park H. Efficient DFT-based channel estimation for OFDM systems on multipath channels [J]. *IET Communications*, 2007, 1(2):197–202.

Kim H M, Kim D, Kim T K, et al. Frequency domain channel estimation for MIMO SC-FDMA systems with CDM pilots [J]. *Communications & Networks Journal of*, 2014, 16(4):447–457.

Pandana C, Sun Y, Liu K J R. Channel-aware priority transmission scheme using joint channel estimation and data loading for OFDM systems [J]. *Signal Processing IEEE Transactions on*, 2005, 53(8):3297–3310.

Park H, Lee Y, Noh M. Low complexity LMMSE channel estimation for OFDM [J]. *IEEE Proceedings—Communications*, 2006, 153:645–650.

Pejoski S, Kafedziski V. Asymptotic Capacity Lower Bound for an OFDM System with Lasso Compressed Sensing Channel Estimation for Bernoulli-Gaussian Channel [J]. *IEEE Communications Letters*, 2015, 19(3):379–382.

Qi C, Wu L. Optimized Pilot Placement for Sparse Channel Estimation in OFDM Systems [J]. *IEEE Signal Processing Letters*, 2011, 18(12):749–752.

Ren X, Chen W, Tao M. Position-Based Compressed Channel Estimation and Pilot Design for High-Mobility OFDM Systems [J]. *IEEE Transactions on Vehicular Technology*, 2014, 64:1918–1929.

Rudelson M, Vershynin R. Sparse reconstruction by convex relaxation: Fourier and Gaussian measurements [C]. *CISS 2006 40th Annual Conference on Information Sciences and Systems*. 2006:207–212.

Taub02ck G, Hlawatsch F. A compressed sensing technique for OFDM channel estimation in mobile environments: Exploiting channel sparsity for reducing pilots [J]. *Proc.ieee Int.conf.acoust.speech & Signal Processing Las Vegas Nv*, 2008:2885–2888.

XueWei Wang, XiuJu Xun, JunLong Guo. The model based lv-plc effect testing on household appliances [J]. *ICIC Express Letters*, 2015, 9(7):2045–2050.

You R, Li H, Bar-Ness Y. Diversity combining with imperfect channel estimation [J]. *IEEE Transactions on Communications*, 2005, 53(10):1655–1662.

Zimmermann M, Dostert K. A multipath model for the power line channel [J]. *IEEE Transactions on Communications*, 2002, 50(4):553–559.

Electromechanical Control Technology and Transportation – Jia & Wu (Eds)
© *2017 Taylor & Francis Group, London, ISBN 978-1-138-06752-3*

Application of information technology in the digitalization construction of an animation library

Yiling Tang
Jinling Institute of Technology, Nanjing, Jiangsu Province, China

ABSTRACT: This paper, with its research focusing on animation libraries, has thoroughly analysed the current status of the development of digital animation libraries in China. It also puts forward a project of how to build up a digital animation library and to solve the problems that may arise during the process based on the application of new information technology to this industry. Accordingly, a generic engineering platform is proposed as well in the paper, in hopes of providing novel thoughts and perspectives for developing animation libraries in China more efficiently.

1 INTRODUCTION

New media and new technologies are emerging, which brings opportunities and challenges for the digital construction of animation libraries. As a new type of library, it is imperative and significant to use new information technology and a means to create a unique cartoon library.

Since the beginning of the new century, the animation industry has become one of the important global cultural industries, domestic animation industry has developed rapidly, and the base is also exponential growth. In some cities, the animation industry has become a new economic growth point. From a national perspective, according to the data issued by the Ministry of Culture, the total output value of the animation industry in 2011 was 621.84 billion yuan. By the end of 2014, China's total output value of the animation industry had more than 100 billion yuan. China's animation industry output value had exceeded 100 billion yuan, compared with 87.085 billion yuan in 2013, with more than 15% growth rate, which was higher than the national cultural industry value-added growth. The development of the animation industry not only promotes the development of related industry chain, but also directly contributes to the birth of the animation library. As a new type of library, the development of the animation library will inevitably bring about the impact and challenge to the development of the traditional library. In the information age, the traditional and new type libraries are faced with the difficult task of digital construction. A few comments and suggestions are put forward on the construction of the digital animation library.

2 CURRENT SITUATION OF THE DIGITALIZATION CONSTRUCTION OF A DOMESTIC ANIMATION LIBRARY

The first in the list is the Nanjing Library, which is one of the most collections and the most library space in terms of animation resources nationally (Tang, 2013). It was established recently in September 2012 with its website, thousands of ebooks and 300 more animation CDs. The next is the animation library of Dongguan Library, established in July 2004, which was the first library in the subject of animation, electronic magazines and resources with the library space of 1200 square meters. It has more than 10,000 animation books, 70 magazines, 2000 videos and 8000 electronic animation books from the mainland, Hong Kong and Taiwan. The others are subject resource branch libraries of public libraries. To name a few, the National Library, which has the nationally and internationally excellent cartoons and Chinese digital series cartoons with collections of about 4000 for each category, provides national public animation database; the Hefei Children's Library has about 7000 of digital animation books and many interactive software of animation (Li, 2011); ChangSha City Library with collections of ebooks and animation videos; Guang Zhou Library with collections of 15,000 books of animation subjects and more than 1,600,000 ebooks; HuaiBei City Library with collections of cartoons, illustrated and audio items of about 6000; the SSTEC animation library with collections of illustrated books in animation specialty and the excellent items by eco publishers. From the development status of the animation library described above, China's animation library business is an ascendant area with great development potential.

3 THE SIGNIFICANCE OF THE DIGITIZATION CONSTRUCTION OF AN ANIMATION LIBRARY

Although a certain amount of animation libraries have been established in some parts of the country, whether from the construction area or the numbers, it cannot be mentioned in the same breath with 3000 public libraries of other kinds. Meanwhile, according to the regional animation library development plan, the rapid development of the animation and animation industry, as well as the development and prosperity of China's socialist culture, to improve the cultural soft power is in urgent need. In view of this, on the one hand, the government should attach importance to the construction of animation library entities, and the establishment of a number of professional animation library representatives; on the other hand, we should base on the existing resources, accelerate the construction of the animation virtual library, and promote the animation library as well as the harmonious development of the animation industry.

Animation library digital is the use of modern technology for the animation library resources for digital processing, the establishment of a personalized database and a variety of platforms for readers to use. With the advent of the era of large data, the reader's reading habits have changed, and the convenience of mobile reading is more popular with readers. "Cartoon" industry characteristics determine its combination with the figures which will bring a qualitative leap. It is more agile and realistic so as to meet the diverse needs of different people.

Due to the existing unique scarcity of animation library collection resource features, it is difficult to meet the needs of many readers. At the same time, it greatly weakened the effectiveness of the animation library resource service, that is, paper resources are easy to wear and tear, which is not conducive to long-term repeated use. Digitalization of the collection of cartoon resources is an effective way to solve the above-mentioned perplexities. The advantage of digital animation library includes: quickly and easily finding the animation information; enjoying the mobile reading anytime and anywhere; possibly utilizing the latest high-tech AR to allow pictures "move" in order to create new reading experiences. It can provide improvements such as protection, rescue and application of valuable resources or a model of 3D printing technology to produce popular works. Digitalization of animation library makes animation culture inheritable and animation resources shareable, which provides strong protection for the development and research of China's animation.

4 SOME IDEAS OF THE DIGITALIZATION OF AN ANIMATION LIBRARY

With the advent of the 21st century and the updating of mobile technologies, it has appeared that the solution and method for various mobile libraries will come true. The mobile libraries erected in the way of SMS, WAP, APP, etc. have sprung up, enriching the application field of mobile library constantly. At present, most libraries in domestic universities have realized the mobile library service by WAP, APP and WeChat. Given the expense of the current mobile Internet, the application system of mobile libraries ought to include more content with less data.

4.1 The special website for the animation library as a platform

In the time of overloaded information, all kinds of information have been emerging without being distinguished. The establishment of a special website as the platform of animation resources has facilitated readers to search and make use of modern science and technology. It can meet the animation enthusiasts for the need of animation works' digital reading, bringing them colourful animation and audio-visual experience. Besides, it can also stimulate the reader's interest to raise people's awareness of the significance of the development of the animation industry, and thus promote the development of the animation industry.

4.2 The animation e-book and video resource as the basic

The paper book in the library is its greatest feature. For example, most of the books in the Nanjing animation library are donated by Mr Zhihui Pan. With a strong scarcity, most of them are the Taiwan versions and they are extremely rare in the mainland. The electrification of animation paper books is able to solve the problem of resource sharing and to make full use of resources.

The organization of the animation audio and video resources platform cannot be neglected in digital construction. An animation video library can be built with the collection from the holding resource and the usage of VOD (Video On Demand) technology, which is relatively mature in the library industry, so that readers have the access to choose the online animation resources they like. Satisfying the deep requirements of information consumer in the current information age society is the mainstream method in multimedia audio.

4.3 Animation practice teaching resources as the characteristic

The rapid development of China's animation industry led to the animation education taking a hit. The lack of teaching resources is a general problem in animation teaching.

Animation museum in the digital construction should have high-quality and practical teaching resources to build the "animation teaching resource platform" through the "school-enterprise cooperation" mode or to make videos via MOOC and the like, realizing the complementation and sharing of resources. There are some lessons online, such as animation moulding design, original painting design, animation movement rules, two (three)-dimensional animation, three-dimensional character animation, online game design, virtual reality, network video creation, technology of director and so on, which help animation enthusiasts see the professional design technology inside the classic works while they are appreciating them. It is able to make full use of the leading role of animation library in reading and foster the talents for the animation industry, making these resources the highlight in the digital construction of animation library.

4.4 Establishing the platform of the digital genetic engineering for the animation industry

This platform adopts the digital resource management platform to manage and store the data through the advanced database management system in the form of server/client. The content form is based on the unique culture of animation industry and the inherent characteristics of animation resources to develop the sharing platform of digital network resources.

The animation industry digital genetic engineering, a service platform to build and share the multi-level document resources, is formed by the use of metadata technology to integrate the digital resources, the establishment of cloud services platform, and the active cooperation with CALIS, CASHL, NSTL and other national document security service system.

The digital genetic engineering of the cartoon industry digitally processes the classic paper manuscripts, books, toys and so on through the digital graphics collection and repair technology.

To keep the original state, it constructs the ergonomic distance visual display mode, architecture and digital data-viewing navigation system through the virtual display technology and interactive technology. It forms a comprehensive system for the animation industry through the smooth index channel to radiate the database and network library to the society, industry, and college. It also provides a large-scale resource platform of comprehensive services for animation industry, teaching and research. The specific program is to create the eight major gene pools.

1. Digital gene pool of animation books
   We will digitize the classic paper manuscripts and books through the digital graphics collection and repair technology. The new media technology is used to identify, classify, and establish professional VIP readers and login system of ordinary readers, which can provide services for high-level personnel and ordinary readers for research and reading.

2. Digital gene pool of animation video
   The animation in library needs to be collected, arranged, and repaired by the filming repaired technologies. The contents of the classic works such as modelling, scene, and wonderful lens are unitized to form the genre database of the animation film and television works. On the basis of this, the video of the collection of cartoon works is collected and edited.

3. Animation three-dimensional modelling gene pool
   The three-dimensional structure map, three-dimensional network map, three-dimensional chartlet map, three-dimensional units map, three-dimensional combination map and three-dimensional dynamic map are collected by the three-dimensional animation structure technology and animation three-dimensional canning system.

   According to the characters, animals, duties, architecture, the natural environment, transportation, and other animations are usually modelled to be classified and researched as service objects through the establishment of the navigation system.

4. Animation two-dimensional model gene pool
   The two-dimensional structure map, two-dimensional network diagram, two-dimensional units diagram, two-dimensional combination diagram, and two-dimensional dynamic map are collected by the paperless animation technology and the animation model scanning system.

   According to the characters, animals, duties, architecture, the natural environment, transportation, and other animation are usually modelled to be classified and researched as service objects through the establishment of the navigation system.

5. Animation derivative product gene pool
   This pool consists of a finished product library and three-dimensional database.

   The finished product gallery uses the ordinary set doubt photography technology and the stereoscopic photography technology to collect the plane or the stereoscopic video and graph and to preserve the basic condition of food, providing the accurate data and establishing the work exhibition system. The three-dimensional data gene library uses three-dimensional precision scanning technology or three-dimensional imaging technology, collecting the three-dimensional structure of the classic animation derivative data, flat map, open film plans, process maps and materials, and material data.

6. Animation master and talent gene pool
This pool is composed of the master module, the entrepreneur module, the designer module, the animation new module, inducing and researching by text, video, image, animation and other means. It also selected the representative figures to show the master's style and work status through the virtual reality technology to establish the reduction type master studio.

7. Excellent animation entrepreneur gene pool
This pool sees the animation enterprises in all provinces as the main body and radiated among the whole nation. It establishes the original works of the sub-lens sets of the gene pool, dynamic—sub-lens sets of the gene pool, character model design draft gene pool, plane scene gene pool, three-dimensional scene gene pool, animation work gene pool, and the winning works gene pool based on the two- and three-dimensional animation techniques. It establishes the special gene pool, display window, ranking list of outstanding entrepreneurs, and brilliant works.

8. Animation academic research gene pool
This pool can be built by creating databases for students' animation works, professors' academic researches, animation text books by publishers, and monographs. Platforms for exhibitions, academic researches, and national competitions may be established. In the meantime, typical animation courses may be made available to animation enthusiasts by MOOCs.

5 CONCLUSION

With the rapid development of the animation industry, the animation library will play an increasingly important role. Through the effective integration of animation resources, the genetic engineering platform of animation digitalization may be built to effectively share digital animation resources so that the development of China's animation industry is accelerated. At the same time, it helps to build effective connections to join the national digital library group by emphasizing the characteristics of digital collections of an individual library in order to maximize the advantage of the library community. It also enriches the variety and functionality of the library.

ACKNOWLEDGEMENTS

This project was supported by the Ministry of Education, "211 Project" CALIS III, 4401-JS-401 and the cultural foundation of Jiang Su 201 301 105.

REFERENCES

Changsha New Library, "ATM machine of books".
Dongguan Library Comics Museum.
Guangzhou Library, cartoon Toy Library, http://news.sohu.com/20061024/n245962764.shtml
Hefei City Children's Library Electronic Comics Museum
http://ah.anhuinews.com/system/2014/05/29/006445971.shtml.
http://e.hfslib.com:8085/bulletindetail.aspx?id = 5
http://news.xinhuanet.com/newmedia/2014-02/27/c_126198443.htm
http://reader.gmw.cn/2014- 03/12/content_10655589.htm
http://www.chinadaily.com.cn/hqcj/xfly/2014-03-18/content_11424216.html
http://www.comiclib.cn/index.html
Huaibei City Library Children's cartoon center.
Li Yang, Creative animation materials as the collection of National Library.
Tang Yiling, *On development of animation library,* New Century Library, **6,** 77 (2013).
The SSTEC National Park Animation Library.

Electromechanical Control Technology and Transportation – Jia & Wu (Eds)
© 2017 Taylor & Francis Group, London, ISBN 978-1-138-06752-3

Teaching android development class for noncomputer majors

Anbao Wang
School of Computer and Information Engineering, Shanghai Polytechnic University, Shanghai, China

ABSTRACT: Some students in our school, although not majoring in computer science or coming from any programming-related specialty, are very interested in programming for android. Because of the complexity of programming in Android, the students should first learn some fundamental programming aspects, which are not only indispensable for Android programming, but also imperative for other software developments. The fundamental programming knowledge at least includes the basis of object-oriented programming, which is generally the Java programming language. In this paper, we introduce a method for teaching non-computer-majoring students to code for Android. Some teaching methods will be depicted, and the designs for this course will be explained as well. Some examples are provided in this paper.

1 INTRODUCTION

With mobile computing becoming very common nowadays, a lot of developers are using Android to code and build apps. Hence, developing apps for mobile devices has become more and more popular, and is of great significance in programming.

As described in Mobile Development Trends (2015 Update), the top mobile platform targeted by developers in 2015 continues to be the Android smartphone, followed closely by the Android tablet. It claims that a large portion (43%) of the developers in their audience are engaged in developing mobile applications and many are doing so on multiple platforms targeted at different audience segments (Developer Media, 2015).

Java is a programming language and a computing platform; it is fast, secure, and reliable. The Android is reshaping client-side Java (Mednieks, 2012), and it is already the most widely used way of creating interactive clients using the Java language. In our school, students are from non-computer majors, so it needs some other methods or designs to organize the process of this course.

In Android, the UI is designed using a combination of Java code and XML (Extensible Markup Language), which is composed of a set of Java source and XML configuration files (ZHU, 2014). XML is used to describe data, the standard of which is a flexible way to create information formats and electronically share structured data via the public Internet as well as via corporate networks. In android development, the XML can not only be used to create a Graphical User Interface (GUI), such as a button or a scroll bar, but also used as configuration files, even for the whole Android project. It is necessary to give students some supplement knowledge of XML, which will be described in this paper.

The students from non-computer specialty show great interest in learning the Android App programming. However, along with the process of the class, they will feel confused and depressed. It is difficult for non-computer-specialty students to learn much information in Android classes, as they lack knowledge on Java, XML, the Database, the concept of MVC (Model-View-Controller), and so on, not to speak of the software engineering either. Some advice, idea, and examples will be given in this paper in order to help students learn the Android development.

2 PREPARING FOR ANDROID

Our students from non-computer specialty may have learned C/C++ as their primary programming. However, in our curriculum, a Java course is required to be obtained by the students. If we attempt to develop Android apps without having a good understanding of Java, it will be frustrating (Gerber, 2015). Hence, basic knowledge of Java is mandatory, which is acquired as follows.

A. There are some students who have never learned any of the programming languages (such as C/C++). If they want to learn the Android development, they should spend a lot of time on grasping some knowledge (Java primitive types, reference types, expression and statement, control structure (If, Switch, While, For, etc.), array) in Java. The form of group counseling was adopted for students, so no class hours will be occupied.

B. For those students who have learned C/C++, contents including Java primitive types, expression and statement, control structure (If, Switch, While, For, etc.), array, and so on, are similar to the C/C++. It takes only less time to introduce the difference between these two programming languages. These contents will only occupy 2 class hours.

C. Some class hours have to be occupied for those contents that are object-oriented programming concepts and so on. All those concepts are brand new for the students from non-computer specialty, so more time will be spent for inducing basic knowledge in object-oriented programming. A total of 8 class hours will be needed.

D. Because of the limitation of total class hours, no more time can be taken out from the class for other contents in learning the Java now. The strategy of "teaching and learning it when necessary" is taken, which helps students acquire the knowledge of dealing with the exception, the methods for event-driven programming, XML, and so on.

Android programming is based on Java programming language. If we have basic understanding on Java programming, it will be a fun to learn Android application development. It is described in this paper (Kevin, 2016) that three things are necessary: familiarity with the language, using Java in an object-oriented way, and learning the core libraries.

The development environment, which is easy to be used by the students, should be selected. The two familiar development tools that are predominantly used are android studio and eclipse.

Android Studio is the official (ZHU, 2014) integrated development environment (IDE) for Android platform development, which was announced on 16 May 2013 at the Google I/O conference. New features are expected to be rolled out with each release of Android Studio. The following features are provided in the current stable version: Gradle-based build support; Android-specific refactoring and quick fixes; Lint tools to collect data on performance, usability, version compatibility, and other problems; ProGuard integration and app-signing capabilities; Template-based wizards to create common Android designs and components; a rich layout editor that allows users to drag and drop UI components; option to preview layouts on multiple screen configurations; and so on.

Eclipse is an Integrated Development Environment (IDE) used in computer programming, and is the most widely used Java IDE. It contains a base workspace and an extensible plug-in system

for customizing the environment. Developing in Eclipse with ADT (which is a plugin for the Eclipse IDE that is designed to give a powerful, integrated environment in which to build Android application.) is highly recommended and is the fastest way to get started. With the guided project setup, as well as tools integration, custom XML editors, and debug output pane, ADT produces an incredible boost in developing Android applications.

Because the students from non-computer specialty are the beginners for programming in Android, Eclipse with ADT is the best choice for those students. First, it is very easy to install and configure. After installing and configuring the JDK, it only needs to unzip the adt-bundle-windows-x86_64_*.rar file to a fold, and everything is ready now. Second, several students have learned how to use the Eclipse IDE when they studied the C/C++ as their primary programming language. The other consideration is that changing from Eclipse to Android Studio is difficult for the students who have used the eclipse IDE as the tool to develop software application before. Finally, Eclipse is more mature and very stable.

3 COURSE STRUCTURE

In this section, first, the topics and issues to be covered in the course on Android development will be designed. Then, the structure of our course is described, including the list of topics covered in the course. Thereafter, the exercises for the experiments in this course are introduced.

The first thing to be addressed in this course is the set of topics to be covered. Because the students are not majoring computer science or some related specialty, we have to redesign the course. The difficult parts were omitted. We try to attract the attention and interest of the students about Android programming for the further studies in software development. The lists of topics in the course are as follows:

The difference between C/C++ and Java: learning how to create, compile, and run a Java program. Introduce methods, Boolean type, characters, and string objects and use them to develop programs.

The concept of OOP (Object-Oriented Programming): It includes Object and Class, Inheritance, Polymorphism, the overload for methods, and so on. It explores the differences between procedural programming and object-oriented programming (Daniel, 2015).

Other contents in Java: dealing with the exception, the methods for event-driven programming, XML, and so on. They are also introduced using the strategy of "teaching and learning it when necessary".

Activity and its life cycle: An activity represents a single screen with a user interface just like window or frame of Java. It represents the presentation layer of an Android application. The content of what are Activity and its life cycle are very important. The contents of the life cycle of activities are explained; the responsibilities that activities have during each of these states are introduced during the course as well.

User Interface design (UI): Understanding layouts is important for students to design a good Android application. In this course, the students will learn about the layouts, controls, fragments as well as dialogs and paging. In this part, the XML will be introduced as supplementary for Android development.

Event handled in Android: Events are useful ways to collect data about a user's interaction with interactive components of applications. We will teach the students what are event listeners, event listener's registration, and event handlers in Android after introducing the event-driven programming in Java.

Intents and Intent Filters: An Intent is a messaging object we can use to request an action from another App component, and it is an abstract description of an operation to be performed. An intent filter declares what an activity or a service can do and what types of broadcasts a receiver can handle.

Resource in Android: The resource files are stored separately as XML files, and graphics and raw data can be stored as resources. The value of the resource can be simple as strings and colors, which can also be complex like images.

Multimedia framework: It includes capturing and playing audio, video, and images in a variety of common media types.

There are two parts in this course: (1) teaching at the classroom for the theory in android and (2) carrying out experiment exercises in the computer room. It requires the students to finish some simple Android Apps.

4 TOPICS COVERED IN THE COURSE

The course contents have been described in the upper section, according to contents in this course. The list of topics covered in the course is presented in Table 1.

According to the topics of the course, the school hours is distributed for every topic. In our school, every school hour has 50 min. The school hours are distributed for every topic listed in Table 2.

Table 1. Topics covered in the course.

1. The difference between C/C++ and Java
   - How to create, compile, and run a Java program
   - Introduce methods, Boolean type, characters, and string object.
   - develop simple Java programs
2. The concept of OOP (Object-Oriented Programming)
   - Object and Class
   - Inheritance, Polymorphism
   - The overload for methods
3. Activity and its life cycle
   - What is Activity
   - Activity life cycle
4. User Interface design (UI)
   - Layouts, controls, fragments
   - Dialogs and paging
   - Using XML for Android development
5. Event Handling in Android
   - Event-driven programming in Java
   - Event listeners in Android
   - Event listener's registration
   - Handling UI events
   - The Model-View-Controller Framework
6. Intents and Intent Filters
   - Intent
   - Intent Filter
7. Resource in Android
   - Resource files
   - The value of the resource
8. Multimedia framework
   - Capturing and playing audio, video, and images in a variety of common media types.

Table 2. School hours distributed for every topic.

Topic	School hours
The difference between C/C++ and Java	2
The concept of OOP (Object-Oriented Programming)	8
Activity and its life cycle	6
User Interface design (UI)	4
Event handling in Android	4
Intents and Intent Filters	4
Resource in Android	2
Multimedia framework	2

5 LABORATORY EXERCISES

The exercises were organized into five simple applications: (1) the use of Android programming tools, (2) UI programming in android, (3) event handling in Android, (4) a video player application to illustrate multimedia features, and (5) the use of intent in Android programming.

The use of Android programming tools: Understanding the debugging tools (ADB and DDMS), creating the Android project using Eclipse, creating the emulator for testing the project, and understanding the files R.java and Manifest.xml in the project.

UI programming in Android: the students perform two exercises: one is using the LinearLayout, and the other is using the Relative Layout, which are shown in Figures 1 and 2, respectively.

Event handled in Android: the students need to write different event handlers for different event types. It can help students understand the exact difference in different event types and their handling methods.

A video player: the students code a simple video player, which can start, pause, and stop when playing the video.

The use of intent in Android programming: it requires the students to use three different methods to invoke other activities at the same project. The three different methods are:

- intent.setClass();
- intent.setClassName();
- intent.setComponent().

Figure 1.   Exercise using LinearLayout.

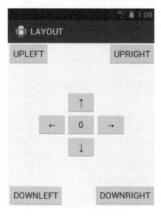

Figure 2.   Exercise using RelativeLayout.

Figure 3.   Exercise for final exam.

The students should learn how to invoke the activity in other App and transfer some data to or get some result from the other App as well.

6   TEST DESIGN

At the end of the class, the students should be rated. A time of 2 weeks is given to them for the final exercises before the end of the semester, and the score for each student is related with the quality for the exercises.

Details of the final exercise are given below.

- The UI should be implement, as shown in Figure 3, and the number key should be pressed, which can be displayed in the EditText at the top of the UI.
- Implementation of addition and multiplication
- The "map" button is pressed, and the destination will be displayed on the map.
- Implementation of calling a phone and sending message.
- Implementation of importing a big image using the button "bigpic".
- The music and video are played.
- Implementation of analysis for XML.
- The EditText is cleared.
- Other functions can be decided using buttons X1, X2, and X3.

7   CONCLUSION

The views on how to teach Android development and its concepts to non-computer students were presented. Some teaching methods were depicted, and the designs for this course were explained as well. Furthermore, some examples were provided in this paper.

ACKNOWLEDGMENT

This study was supported by "Key Disciplines of Computer Science and Technology of Shanghai Polytechnic University" (No. XXKZD1604).

REFERENCES

Android Development Tools for Eclipse, https://marketplace.eclipse.org/content/android-development-tools-eclipse, (2016).

Android Tutorial, http://www.tutorialspoint.com/android/index.htm, (2016).

Android—Event Handling, http://www.tutorialspoint.com/android/android_event_handling.htm, (2016).

Daniel Liang, K. Intro.to.Java.Programming.Comprehensive.Version.10th.Edition, (2015).

DeveloperMedia, Mobile development trends (2015 update), developermedia.com/mobile-development-trends-2015-update, retrieved (2016).

FengShan ZHU, Android mobile application development tutorial, Tsinghua university press, (2014).

Gerber; Adam Gerber; Clifton Craig, Learn Android Studio: Build Android Apps Quickly and Effectively, Apress (2015).

Kevin McCullen, an Android Application Development Class. Journal of Computing Sciences in Colleges, 31(6):11–17, (2016).

Mednieks, Z., Dornin, L., Meike, G.B., Nakamura, M., *Programming Android*, Sebastopol, CA: O'Reilly, (2012).

What is Java technology and why do I need it? https://java.com/en/download/faq/whatis_java.xml, retrieved (2016).

Wikipedia, Android Studio, https://en.wikipedia.org/wiki/Android_Studio, (2016).

XML (Extensible Markup Language), http://searchsoa.techtarget.com/definition/XML, retrieved (2016).

Electromechanical Control Technology and Transportation – Jia & Wu (Eds)
© *2017 Taylor & Francis Group, London, ISBN 978-1-138-06752-3*

Assessing the impact of design intent 3D annotations on model reusability

Qinyi Ma, Lihua Song, Dapeng Xie, Maojun Zhou & Haihua Jin
Department of Mechanical Engineering, Dalian Polytechnic University, Dalian, China

ABSTRACT: In the parametric modeling systems, design intent information is usually expressed implicitly within the model. However, there is evidence suggesting that an explicit representation of the design intent information can greatly facilitate the reusability of the model and effectively improve the efficiency of product development. On the basis of secondary development of UG, this paper proposes a 3D annotation management system, which can better manage and view the 3D annotations marked by PMI and then assesses the impact of 3D annotation search function on the product development, which is a useful tool that can express a design intent explicitly to help designers find the modeling scheme quickly and accurately. The results show statistically significant benefits of 3D annotations models, suggesting that an annotation manager is a valuable tool to help people find the right solution quickly and accurately.

1 INTRODUCTION

With the development of computer technology, mechanical manufacturing industry has gradually entered the digital development stage. As the latest phase of digital product definition, Model-Based Definition (MBD) technology is a digital definition method, which attaches all the relevant information such as design definition, process description, attributes, and management to the product three-dimensional model (Fan, 2012). The Model-Based Enterprise (MBE) concept is founded on the basis of Model-Based Definition (MBD) (Whittenburg, 2012). The 3D CAD model is the core in the MBE through defining 3D model digitally, allowing designers, analysts, manufacturing personnel, and other users to view various types of information more intuitively, understand the product, and improve the effect of product design and analysis and product reusability.

Product design mostly evolved from existing products. The difference between the so-called initial design, variables design, and adaptive design is the degree of utilization of the original product. To effectively shorten the product development cycle, designers are always more willing to transform design on proven mature product (WANG, 2013). A key factor in the new product development activities is reusing and utilizing previous knowledge and gaining the design ability from the previous process (Ullman, 2010). In order to ensure effective CAD reusability, the users need to know how and why the model is built in such a way and so on. That is, they need to understand its design intent reflected in the entire product life cycle (Xu, 2010). The reusability degree of a CAD model strongly depends on the modeling methodology, the proper definition, and communication of the geometric design intent rather than the technology (Bodein, 2014).

3D annotation is the most direct and effective way to express design intent. At present, some major 3D software have Product Manufacturing Information (PMI) modules and users can more intuitively and clearly understand the modeling information, analysis information, and manufacturing information in the product development process though annotating the product design intent on a 3D model by PMI module. Some pivotal issues, including how this type of information can be captured, represented, managed, processed, and stored, are active parts of the research (Bracewell, 2009). However, there are significant limitations on current annotation mechanism. Our previous paper proposed an annotation management mechanism based on UG secondary development, allowing users to manage 3D annotations on the basis of the design intent in UG's PMI module more conveniently. On the basis of the previous annotation management mechanism, this paper assesses the impact of design intent 3D annotations on model reusability.

2 PROTOTYPE SYSTEM

The scenarios assessed in this paper are shown in Figure 1.

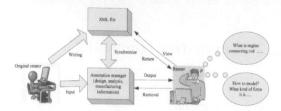

Figure 1.　User scenarios.

First, the original creator inputs product design information in the PMI module of UG, including modeling information, analysis information, and manufacturing information. Users can better manage and view the 3D annotations annotated in PMI modular through the annotation manager. Meanwhile, the original creator writes XML file, and the information is stored in an external XML file to facilitate downstream users to view. To reuse the existing product design knowledge, the users can retrieve the product models annotated by original designer though annotation manager or related annotated information with specific features. The users can also view the XML file to understand the original knowledge of product design.

The annotation manager primarily facilitates users to view the 3D model with annotated information and helps users better understand the product design intent, including design information, analysis information, and manufacturing information, throughout the product life cycle. The users can view annotations in different ways. They enter annotation keyword or model name to find all related geometries and geometric features with the annotations. On the contrary, they can also enter the name of geometric feature or geometric entity to query to the relevant 3D annotations. That is the search function. Filtering is designed to select to view the same name, feature, annotation content, type, time, and other relevant annotations. Annotation manager's grouping function is mainly designed to divide the product design information into design intent, analysis intent, and manufacturing intent for modelers, analysts, and manufacturing personnel, respectively. These three main functions can help designers to find the needed information quickly and accurately, improve design efficiency, and shorten the product development cycle.

3　EXPERIMENTAL DESIGN

The framework of this study is the application of 3D annotations to parametric modeling processes. The goal is the assessment on reference value of model with 3D annotations in reusing the existing product design knowledge. Therefore, this paper designs two research questions:

Question 1: When users need to reuse existing product design knowledge, is it a useful tool of the model search function to help users find the needed annotated model?

Question 2: When the major design goals have been identified, is it a valid expression tool with a design intent of 3D annotation?

3.1　*Experiment 1*

1. Experiment preparations

The main purpose of the first experiment is to determine whether the model retrieval is useful for developers in the model design development of new product. The study designed an experiment aiming at model retrieval. The experiments were conducted in a computer laboratory environment, where participants were equipped with a workstation and the UG software. There are 30 participants who are familiar with UG and were randomly divided into two groups. One group served as the control group (participants use the software without the annotation manager module) and the other served as the experimental group (participants use the UG software with the annotation manager).

2. Experimental procedure

First, the participants of the two groups equipped the same 3D model of an engine-connecting rod. As shown in Figure 2, the participants were asked to invert two chamfers. Obviously, there is little effect of the two chamfers sizes on the stress distribution model, so they are not clearly defined. However, if the chamfer is too large, it will affect the other modeling features of the model. The control group participants only do several attempts on the model until finding a suitable chamfer size, which is beautiful, without affecting modeling of the other features. However, the experimental group participants can retrieve a similar model with annotations through the annotations manager, and then do modeling operation. As shown in Figures 3 and 4, the participants input keywords "link" and "chamfer" in annotation manager and retrieved related annotations. It helped the participants build the chamfer quickly and accurately.

Figure 2.　Experimental Model 1.

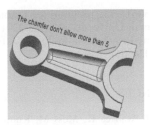

Figure 3.   Annotation chamfer 1.

Figure 4.   Annotation chamfer 2.

3. Experimental results

Assessing the results through the modeling speed and accuracy of the participants of the two groups, 20% of the chamfers in control group are too large to impact modeling of the other features. However, all the chamfers in the experimental groups meet the requirements, and most participants in the experimental group used less time than the control group participants, indicating that the model search function is a useful tool to help users find the needed annotated model.

3.2   *Experiment 2*

1. Experiment preparations

The main purpose of the second experiment is to determine whether it is useful for modeler when modeling a key step. The experiment preparations are the same as those of the first experiment, but the second experiment aims at model alteration.

2. Experimental procedure

Participants of the two groups equipped the same 3D model. The model of control group is shown in Figure 5, the experimental group model plus 5 annotations, and only one annotation in the experimental is useful. There is no assumption symmetrical because the original geometry may need to change. The experimental model is shown in Figure 6. Unrelated annotations with design issues are ignored for a clear view.

The first task is adding a rib to the model, and the shape is the same as that of the first rib. There are two ways: one is using the mirror tool and mirroring all features of existing rib; and the other is

Figure 5.   Control group model.

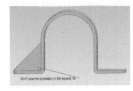

Figure 6.   Experimental group model.

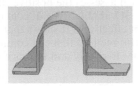

Figure 7.   Control group model. (The first alteration used mirroring).

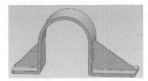

Figure 8.   Control group model. (The first alteration used sketching).

re-stretching rib feature by a sketch. Intuitively, the first approach seems easier and faster than the second one, as it requires fewer steps and no features need to be modeled from scratch. It is an efficient model as long as all features in both ribs remain symmetrical in new variations of the model. Performing the alteration to the model by creating the second rib from scratch may not seem like an efficient strategy, as it does require a number of additional steps. However, each rib can be controlled and edited separately without affecting the other.

The second task is adding the length of the bottom, but the bottom side still remains tangent with the side ribs. It shows the drawbacks of the mirror operation in the first modification and the rib side is not remaining tangent with the bottom side after the second modification, which is shown in Figure 7. However, the bottom side still remains tangent with the side ribs, which is shown in Figure 8.

3. Experimental results

Assessing the results through modeling speed and accuracy of the participants of the two groups, all the participants in the experimental and control groups modeled correctly. About 90% of the students in the control group used mirroring features to create a second rib firstly and then deleted the rib in the second alteration. It shows that the 3D annotation can effectively express design information and help designers to model quickly and accurately.

4 CONCLUSION

On the basis of the UG PMI module, this paper proposes an annotation management mechanism, which can manage 3D annotations effectively and then assesses the positive effects of this mechanism. 3D annotation can communicate the design intent effectively. It shows that 3D annotation is a useful method that can communicate the design intent effectively and improve the product reusability.

ACKNOWLEDGMENTS

This work was financially supported by the National Natural Science Foundation of China (No. 51305051), the program for Liaoning Excellent Talents in University (No. LJQ2015007), and the Liaoning Province Natural Science Foundation (No. 2014026006).

REFERENCES

Bracewell R, Wallace K, Moss M, Knott D. *Capturing design rationale.* ComputAided Des. **41**, 173–86 (2009).

Bodein, Y., B. Rose, E. Caillaud, *Explicit reference modeling methodology in parametric CAD system,* Computers in Industry. **65**, 136–147 (2014).

Fan Yuqing. *Model based definition technology and its practices.* J. Aeronautical Manufacturing Technology, **6**, 42–47 (2012).

Ullman, D.G. *The Mechanical Design Process,* fourth ed., Mc Graw Hill, Boston, MA, (2010).

Wang Junfeng, Lu Mingshang. *MBD based product assembly dataset and its applications.* J. Manufacturing Automation. **35**, 78–82 (2013).

Whittenburg M. *Model-based enterprise: an innovative technology-enabled contract management approach.* J. Contract Manage **10**, 103–12 (2012).

Xu Wang. *Study on Conceptual Modeling Technology Based on Design Intent Capture.* D. Beijing: Tsinghua University, (2010).

Electromechanical Control Technology and Transportation – Jia & Wu (Eds)
© 2017 Taylor & Francis Group, London, ISBN 978-1-138-06752-3

A CAE intelligent assisted system

Qinyi Ma, Dapeng Xie, Lihua Song, Maojun Zhou & Junli Shi
Department of Mechanical Engineering, Dalian Polytechnic University, Dalian, China

ABSTRACT: The CAE Intelligent Assisted System (IAS) is an auxiliary software system, which can help a user complete the project analysis quickly and effectively. At present, the use of CAE software is rapidly expanding in the manufacturing industry. However, the problems of CAE are more and more outstanding, such as learning difficulties, complicated operation, and lack of professional talents. Aiming at these problems, this paper presents a CAE Intelligent Assisted System (IAS). The system was usability-tested by two groups of student testers of the same major. Results show that the system can provide effective help for beginners.

1 INTRODUCTION

Nowadays, the use of CAE software is rapidly expanding in the manufacturing industry. As the number of people who use the CAE system increases, the user group has started to include not only analysts who have expertise in the Finite-Element Method (FEM) domain, but also design engineers (Akio, 2001). For design engineers, this means a burden of learning to use the CAE software. Even for CAE beginners, the learning process of the software is also very time-consuming. In the past, several similar systems had been studied. The collaborative CAE system has been proposed: Ni (2006) proposed the collaborative CAE system and implementation method. At present, some similar systems have been explored. Akio (2001) proposed a CAE interface agent system and introduced the relevant concepts, basic functions, and system prototype of the system. An implementation method of a kind of finite-element consulting system based on aerospace case was proposed by Laszlo (2015). Samira (2015) introduced a knowledge-based magnetic component design system, which has the function of designing scheme push, automatic finite-element analysis, and command interpretation. A prototype of an intelligent rule-based consultative system was introduced by Marina (2008). In this paper, we describe an IAS, many functions of which are similar to those in previous introduction. The interface is more concise, and functional modules are clearer. This paper mainly describes the basic function of the system, the realization of the prototype, and the usability test.

2 FUNCTIONS OF THE SYSTEM

Through the analysis of use experience of CAE software and interviews with FE analysis, the authors have thought that the reasons for the difficulty of CAE software are more functions and complex operation. However, CAE IAS, as an auxiliary system, can provide users with guidance analysis and knowledge interpretation to reduce the difficulties of CAE. As shown in Fig. 1, the analysis was completed by CAE software and IAS cooperation. Users put the existing CAD model into the system. The model will be reasoned and evaluated through the corresponding operation. The solution and the corresponding knowledge explanation are given according to the user's need.

This IAS is composed of many functional modules. As shown in Fig. 2, all the functions of the system are realized through mutual cooperation between various modules.

2.1 Overview of the system

The function of the system is realized by using C programming language. The whole structure is composed of man–machine interface, knowledge base, reasoning machine, database, and explanation mechanism. The basic function of the system

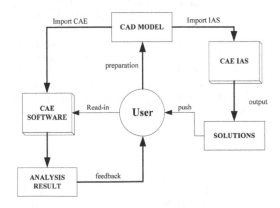

Figure 1. CAE software and IAS cooperation.

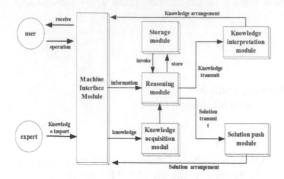

Figure 2. Functional modules of IAS.

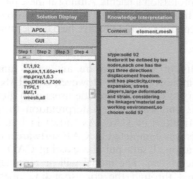

Figure 3. Basic function interface.

is realized through the mutual cooperation between the modules. The realization of the basic function interface of the system is shown in Fig. 3.

2.2 Model of the system

At present, many people learning and operating the CAE software are nonprofessional in finite element. For the crowd, learning the operation of the software and project analysis is very difficult. There is such a system that can teach them how to operate, and tell why they operate this way, and get good results, so the task can be well completed. In general, the IAS has the following functions: 1) identifying model information and make the corresponding operation; 2) calling operation results: users can call the solution as needed and set the results display; 3) knowledge interpretation; and 4) knowledge acquisition: improving their knowledge base and achieving knowledge reuse. Fig. 4 shows the basic process of the realization function.

3 USABILITY TEST

Usability test is the reliability of the test system in the practical application. The defects of the system were found to contribute to the improvement of the system. Here, tests A and B are reported.

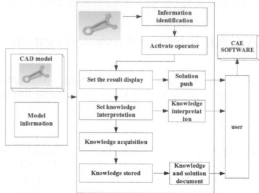

Figure 4. Realization of the function.

3.1 Objective of the test

Usability test is a form of evaluation of the system. The objective was to find the system's defects and verify its practicality.

3.2 Procedure of the test

Test A was conducted with 22 testers in August 2016; 20 of them were either senior students or Master's course students majoring in mechanical engineering. The other two are teachers. All of them knew the word FEM; 30% of them had experience with FEA; 50% had taken a FEM course; and 15% had made a FEM program.

The 22 testers were divided into 10 groups with an average of two members, with two teachers as observers participating in the test. Tests were performed with a 1 h time limit. IAS was used in each group of test A. During the test, the testers' behaviors were video-recorded. The observers observed the testers' behaviors and made appropriate related records. They were needed to answer a questionnaire in free time.

Test B was roughly similar to test A in personnel composition, while the main difference is that IAS was not used. The comparison is shown in Table 1.

3.3 Task of the test

The system was tested through a simple analysis of task in this paper. Taking the engine-connecting rod as the test object, the static analysis of the connecting rod was realized by the finite-element analysis method.

3.4 Evaluation method

The way of evaluation is based on a questionnaire prepared before the test. The first part of the questionnaire consists of questions about background knowledge and experience about the FEA and

Table 1. Comparison of testers in two tests.

Item	Test A	Test B
Tester's major	Mechanical Eng.	Mechanical Eng.
Is volunteer	Yes	Yes
Number of teams (testers)	10 (22)	10 (22)
Allowed time (h)	1	1
Has conducted FE analysis	30%	32%
Has taken FEM course	50%	45%
Has written FEM programs	10%	15%

Table 2. Results of the questionnaire.

Function	Questions	A	Function	Questions	B
Overview of IAS	Good interactive interface, and easy to operate	7.5	Overview of CAE	Function, the operation is complex, the error rate is high	6.0
Solution support	High reliability	7.5			
Knowledge interpretation	General knowledge detail	6.0			
Interface	Friendly interface, easy to operate	8.0	Interface	General interface	5.0

the tester's general skill level in computer. The latter part of the questionnaire is about the evaluation of each function of IAS with the form of scoring. Testers were requested to answer the questionnaire when they wait for the system's response so that they would not forget the details of the operation process. The total score of all problems is not less than 100. The average score of all the testers was calculated for each function to evaluate its practicability.

The advantages and disadvantages of the whole system and the practical value are judged according to the efficiency, accuracy, and the questionnaire survey of the two groups.

4 TEST RESULTS

In test A, 80% of the 10 teams arrived at the analysis result within the allowed time; 10% have obtained the solution, but the final result was not obtained because some errors occurred during the operation of CAE software; 10% have some errors during operating the IAS, which did not obtain the solution.

In test B, 30% have obtained the analysis result; 20% have half of analysis at the end of time; 50% of them did not go well, and the operation failed.

4.1 Questionnaire results

Table 2 is a summary of the tests and questionnaires. The results show that test A is better than test B in comprehensive evaluation. The practical value of IAS is confirmed.

The main difference between the two groups is the IAS. IAS can indirectly improve the knowledge reserve of testers, so it is easier to carry out the project analysis.

4.2 Observations and discussion

Through the observation, group A students obtained the solution by assisting IAS and used solutions to operate the CAE software. At the beginning of the test, group A was required to learn the operation of the IAS, so group A was slightly slower than group B. By observation, in group A, some testers did not have too much confidence in the unfamiliar system, and the CAE software was directly used without help according to their knowledge, so they can rarely complete the task within the specified time and frequently make errors. Through the questionnaire survey, the system still has some defects. For example, some solutions given by IAS are not detailed enough, such as the setting of real constant and functions explanation, so that the tester will become more passive in using the system. Some explanations were fuzzy about setting real constants and the formula interpretation so that the testers can only apply the solutions in a passive way.

In test B, the testers use the CAE software directly. At the beginning, the test was performed smoothly. The error rate was gradually increased, especially in the meshing, mesh refinement, establishment of the coordinate system, and setting of the boundary conditions. Therefore, some testers were anxious, causing many futile operations. The testers of group B reacted that the operation of CAE software was too cumbersome, and they lacked standardization of operation, leading to frequent errors.

According to the test of the two groups, IAS has played a role in general. It can be used as a guide to give the appropriate operation guidance for tester, standardize their operation, and reduce the error rate.

5 CONCLUSION

The authors proposed to verify the practicability of IAS that has a capability to support design engineers and some nonprofessional beginners in performing FEA. The experiment was carried out under the condition whether or not they used IAS, and a simple analysis was performed to the engine-connecting rod in the specified time. The IAS has been evaluated to get a good score by the questionnaire of group A. The IAS has played its role in the analysis process through the comparison of tests A and B. At the same time, some defects in the system were reflected in the test, which provided a basis for the later improvement.

ACKNOWLEDGMENTS

This work was financially supported by the National Natural Science Foundation of China (No. 51305051), the program for Liaoning Excellent Talents in University (No. LJQ2015007), and the Liaoning Province Natural Science Foundation (No. 2014026006).

REFERENCES

Akio Miyoshi, Genki Yagawa, Ryota Shimizu. An Interface Agent System for CAE, J. JSME International Journal, Series A, **44**(4), 623–630, (2001).

Laszlo Hetey, Fames Campbell, Rade Vignievic. Advisory system development for reliable FEM modeling in aerospace, J. Aircraft Engineering and Aerospace Technology: An International Journal, **87**(1), 11–18, (2015).

Marina Novak, Bojan Dolsak. Intelligent FEA-based design improvement, J. Enginnering Applications of Artificial Intelligence, (21), 1239–1254, (2008).

Samira Janghorban, Qingmai Wang. A Knowledge-based Magnetic Component Design System with Finite Element Analysis Integration, C. MARCH 2015. ResearchGate, (2015).

Xiaoyu Ni. Collaborative CAE System and Implementation Method, J. Chinese Journal of Mechanical Engineering, **42**(8), 71–77, (2006).

Electromechanical Control Technology and Transportation – Jia & Wu (Eds)
© 2017 Taylor & Francis Group, London, ISBN 978-1-138-06752-3

Research and design of a random encryption scheme based on ECC and RSA

Shuai Zhang
Computer Science and Engineering Department, Guangzhou College of Technology and Business, Foshan, China

ABSTRACT: In response to the problem that more and more account passwords need to be stored and managed safely, this paper has explored and designed a random encryption scheme based on ECC and RSA. First, a random number is generated as the select code, which is used to decide to encrypt with ECC or RSA according to its parity. Then, the confuse code is designed by the following information: the select code, the length of ciphertext, and the private key number. Finally, according to the designed confuse rule, we mix the ciphertext and the confuse code bit by bit to form a new ciphertext, which is finally saved in the database of the system. The decryption is the inverse process of the encryption. To a certain extent, the scheme in this paper can reduce the pressure of the super password and the degree of security dependence among these account passwords.

1 INTRODUCTION

The rapid development of computer and network has accelerated the building process of information management systems from all walks of life. It brings not only great convenience but also new problems to people's life. One of these problems is that more and more account passwords need to be stored and managed safely.

However, it is difficult for people to remember so much passwords. The traditional password management tools, such as memos, are neither safe nor convenient. The modern password management tools, such as all kinds of simple password management systems (Wang, 2004; Wang, 2005), also make people worried. Therefore, in this paper, we designed a random encryption scheme on the basis of ECC and RSA.

2 BASIC IDEA OF THE RANDOM ENCRYPTION SCHEME

The encryption algorithm of this random encryption scheme is shown in Figure 1, and the decryption algorithm is shown in Figure 2.

3 DETAILED DESIGN OF THE RANDOM ENCRYPTION SCHEME

3.1 *Generation of random code*

We generate a pseudo-random number as the select code, which is also called the random code.

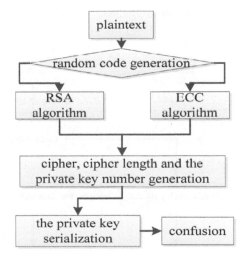

Figure 1. Encryption algorithm.

The random code is either 0 or 1. A random code of 0 means encryption with the RSA algorithm, whereas 1 means the encryption with the ECC algorithm.

3.2 *Design of the encryption and decryption system*

The plaintext will be converted to an array of bytes, and divided into m groups. The length of each group is k. The encryption process is shown in Figure 3, and the decryption process is shown in Figure 4.

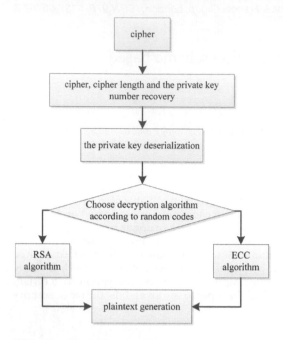

Figure 2. Decryption algorithm.

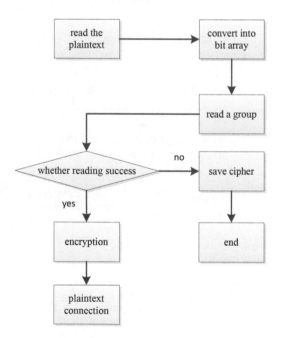

Figure 3. Detailed encryption process.

3.3 Storage and reading of private key

Each plaintext is numbered, then reused as the serial number of the corresponding private key, and stored in both a dynamic array and the confusing code. Each element in this dynamic array contains

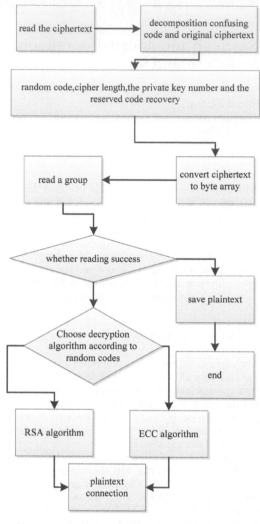

Figure 4. Detailed decryption process.

two fields: the serial number and the corresponding private key. The dynamic array will be serialized to a binary file and stored in the user's computer.

After each random encryption, users deserialize their binary file, read the dynamic array, and get the private key. If it is the first time to encrypt the plaintext, 1 is added to the biggest serial number that already exists, then the new generated serial number and the corresponding private key are inserted into the dynamic array, and the confusion code is updated. If not, the old corresponding private key is replaced by its new private key in the dynamic array. Then, the updated dynamic array is serialized into binary files, and the file is covered before.

The binary file that saves all the private keys is encrypted by the super password and the RSA algorithm and stored in the local computer.

When the user needs the private key, the binary file is decrypted first, deserialized, and read by application. Finally, the dynamic array is read. According to the serial number of the private key in confusion code, the user can search the private key in the dynamic array. In the end, the user can decrypt the cipher according to the private key and the decryption algorithm, which is determined by the random code.

3.4 Generation of the confuse code

The confuse code consists of five fields: the random code, the length of ciphertext, the serial number of the private key, the reserved code, and the complement code (optional, a random number).

3.5 Synthesis of ciphertext and confuse code

If the length of the ciphertext is greater than that of the confuse code, the complement code is added behind the confuse code. On the contrary, the complement code is added behind the ciphertext. This makes the length of the ciphertext and the confuse code be consistent.

Set m bits ciphertext is: $P = (P1, P2, P3, \ldots Pm)$, $Pi \in (0,1)$. Confuse code is: $Q = (Q1, Q2, Q3, \ldots, Qn)$, $Qi \in (0,1)$. Confuse code and ciphertext are synthesized bit by bit. It is called odd bits ciphertext filling when Qi is before Pi, as $M_{odd} = (Q1, P1, Q2, P2, Q3, P3, \ldots QN, PN)$. In the same way, it is called even bits ciphertext filling when Pi is before Qi.

4 SAFETY ANALYSIS

Compared with encryption by the ECC or RSA algorithm (Shor, 1997; Kleinjung, 2010; Cormen, 2009; Pollard, 1978; Knuth, 1999; Kocher, 1996), this scheme improves security. The following is taking RSA algorithm, for example, to analyze the reliability of the scheme with the probability thought method.

Set: a collection of encryption algorithm $E = \{E1, E2, E3, \ldots\ldots, En\}$.

The probability that the encryption algorithm is cracked $PA = \{PA1, PA2, PA3\ldots, PAn\}$.

The probability that each encryption algorithm is selected $PB = \{PB1, PB2, PB3\ldots, PBn\}$.

The original ciphertext $C = c1c2c3c4c5c6c7c8$, confuse code $X = x1x2x3x4x5x6x7x8$. The number of 0's is w in each byte of the original ciphertext and s after confusion (w, s <= 8).

The probability that every bit of ciphertext is the same before and after confusion is:

$$P = \frac{s}{8} \times \frac{w}{8} + \frac{8-s}{8} \times \frac{8-w}{8}$$

The probability that every bit of ciphertext is the same before and after confusion is:

$$P_b = P^8 = \left(\frac{s}{8} \times \frac{w}{8} + \frac{8-s}{8} \times \frac{8-w}{8} \right)^8$$

When the ciphertext length is less than the confuse code length, the length of ciphertext after confusion is:

$$l = (4 \times 4) \times 2 = 32$$

According to the provisions of the before, the probability that this scheme is cracked is:

$$P_d = PA_i \times PB_i \times P_b^l (0 < i < n)$$

Because the plaintext is random, we assume that $w = 4$. Then, we can calculate:

$$P = \frac{s}{8} \times \frac{w}{8} + \frac{8-s}{8} \times \frac{8-w}{8} = \frac{s}{8} \times \frac{4}{8} + \frac{8-s}{8} \times \frac{8-4}{8} = \frac{1}{2}$$

$$P_b = P^8 = \left(\frac{s}{8} \times \frac{w}{8} + \frac{8-s}{8} \times \frac{8-w}{8} \right)^8 = \frac{1}{2^8}$$

$$P_d = PA_i \times PB_i \times P_b^l = PA_i \times PB_i \times \frac{1}{2^{8 \times 32}}$$

When the length of the ciphertext is greater than the confuse code length, the length of the original cipher is set k. The probability that each bit matches successfully is:

$$P_i = P^8 = \left(\frac{s}{8} \times \frac{w}{8} + \frac{8-s}{8} \times \frac{8-w}{8} \right)^8 (0 < i < 2k)$$

Setting w = 4, the length of the original ciphertext is i. The probability is calculated that the ciphertext after confusion completely matches original cipher:

$$P(A \mid B_i) = \prod_{k=1}^{i} P_k = \frac{1}{2^{8 \times i}} (i \geq 16)$$

The probability is set P (Bi) when the length of the original cipher is i. Then,

$$P(A) = P(A \mid B_1) \times P(B_1) + P(A \mid B_2) \times P(B_2) + \ldots\ldots + P(A \mid B_n) \times P(B_n)$$

We can calculate the probability that the ciphertext matches successfully before and after confusion as:

$$P_d = \sum_{i=16}^{\infty} P(B_i) P(A \mid B_i) = \sum_{i=16}^{\infty} P(B_i) \frac{1}{2^{8 \times i}}$$

467

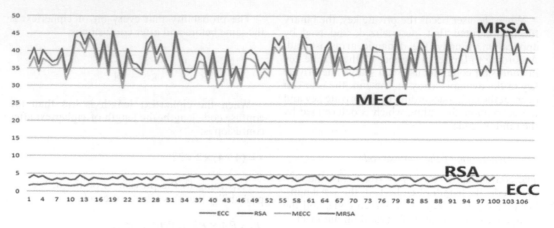

Figure 5. Comparison of time among different algorithms (unit: ms).

5 PERFORMANCE ANALYSIS

The random encryption scheme has tested 1000 times, and the length of plaintext group is 8 byte. Program running time distributed between 30 and 45 ms in general. If the code is further optimized, the time can be shortened. At the same time, we compared ECC, RSA, and MECC with MRSA algorithms, as shown in Figure 5.

It can be seen clearly in Figure 5 that the time gap between RSA and ECC is negligible because of the use of confuse code.

6 SUMMARY

Because of filling the confuse code, the length of the ciphertext of this encryption scheme has more than doubled the length of the original ciphertext. It will increase the time cost when needing to transmit on the network. However, if the scheme is applied on a local machine, or in the case of the length of the plaintext being generally short, it is feasible.

REFERENCES

Cormen, T.H. *Introduction to algorithms.* (MIT press 2009).

Kleinjung, T., K. Aoki, J. Franke etc. Factorization of a 768-bit RSA modulus [A]. In: ACC[C], (2010).

Knuth. D.E., The art of computer programming. **2** (1999).

Kocher. P.C., Timing Attacks on Implementations of Diffie-Hellman, RSA, DSS, and Other Systems[A]. In: AICC[C]. (1996).

Pollard. J.M. Monte Carlo methods for index computation(mod p). MOC [J], (1978), **32**.

Shor, P., Polynomial-time algorithms for prime factorization and discrete logarithms on a quantum computer, SIAM J. Comput, **26** (5) (1997).

Wang, X., A. Yao, F. Yao. New Cryptanalytic Results Against SHA-1 In:Weblog:Schneier on Security. https://www.schneier.com/blog/archives/2005/08/new_cryptanalyt.html

Wang, X., D. Feng, X. Lai etc. Collisions for Hash Functions MD4, MD5, HAVAL-128 and RIPEMD [A]. In: IACR Cryptology ePrint Archive [C]. **199** (2004).

Electromechanical Control Technology and Transportation – Jia & Wu (Eds)
© 2017 Taylor & Francis Group, London, ISBN 978-1-138-06752-3

Discussion of an e-commerce platform for shipping logistics enterprises

You-Lin Li, Hong-Peng Wang, Peng-Xiang Tong & Xiao-Qing Sun
Navigation College, Jimei University, Xiamen, China

ABSTRACT: The popularization of the Internet has enabled the shipping enterprises' marketing system to be extended to any demand from the world. The Internet has changed the world as well as the shipping industry. E-commerce based on Internet technology is an opportunity and a challenge for shipping enterprises. In an information-driven economic society, the integration of information elements will ultimately form a platform economy. In the future, the trading platform will become increasingly important. The combination of shipping business and e-commerce is inevitable; however, the development of shipping enterprises' e-commerce platforms still has a long way to go. In this paper, we introduce the development of the e-commerce platform and its development history worldwide, as well as analyze the status quo and existing problems of China's shipping logistics enterprises' e-commerce platform. SWOT analysis is also employed to analyze the shipping e-commerce platform, on the basis of which it proposes to improve the development strategy of the platform.

1 OVERVIEW OF THE SHIPPING E-COMMERCE PLATFORM

E-commerce is often considered a business activity with the adoption of the microcomputer technology and the network communication technology. The most authorized overview of electronic commerce, that is, to realize electronic process for the whole trading activities was concluded in an International E-Commerce Conference held by the International Chamber of Commerce in Paris in the end of the 20th century. From the aspect of business model, it can be defined as: each trading party conducts trade business through electronic ways instead of face-to-face exchanges or direct negotiation. From the aspect of technical set, it can be defined as an aggregation integrating with various technologies (Xue, 2001).

1.1 Development process of the foreign shipping e-commerce platform

At present, there are three major public information portal platforms in the global shipping and logistics industry: INTTRA, GT Nexus, and Cargo Smart. GT Nexus successfully attracted to APL, Hyundai and Korea Hanjin's sponsorship, Maersk, iron slag line of China and France's CMA CGM, have begun to introduce INTTRA. OOCL with the IT department is a subsidiary of the company and the establishment of Cargo Smart. The three major platforms contain nearly all the large liner companies. The route coverage rate is high, providing a large number of cabins and attracting a large number of paid users, as shown in Table 1.

Table 1. Typical foreign shipping e-commerce platform table.

Number	Platform name	Operation time	Main business
1	INTTRA	October 2000	Ocean sailing schedules, online booking, shipping instructions, Bill of lading tracking.
2	GT Nexus	1998	Order management, delivery management, booking, system management.
3	Cargo Smart	2000	Online booking, business process outsourcing.

1.2 Development course of the domestic shipping e-commerce platform

The electronic commerce of China's shipping companies started late but developed rapidly. In the late 20th century, COSCO, China shipping, Sinotrans, led by the three major domestic shipping companies, have established their respective portals, taking the lead to carry out e-commerce services. In the 21st century, hundreds of domestic shipping companies are competing to gain access to the Internet via the Internet to provide customers schedule inquiry, booking service, propaganda enterprise image, and so on. The research discovered that the establishment in the platform market effect is both sides (Eisenm, 2006), as shown in Table 2.

Table 2. Typical domestic shipping e-commerce platform table.

Number	Platform name	Operation time	Main business
1	EPANASIA	2012	Shipping and transportation route query, query rates, online orders, full service, etc.
2	SHIPPING BOOKING NETWORK	1 August 2013	Set of industry information, online booking and dynamic query function, etc.
3	ESHIPPING	October 2014	Support online booking, check the bill of lading, online payment, etc.
4	MATOUWANG	June 2011	Integration materials, spare parts, shipping service, excellence in the shipping, etc.

2 EXISTING PROBLEMS IN THE DEVELOPMENT OF THE SHIPPING ENTERPRISE'S E-COMMERCE PLATFORM

2.1 Ambiguous platform positioning and low utilization efficiency

The e-commerce platforms were just introduced successively without any planning, which have different standards and technologies. As a result, effective integration cannot be achieved among systems. Timeliness and accuracy of data sharing among systems were relative in poor condition.

2.2 Auxiliary service insufficient consummation

In the settlement service aspect, the special national condition causes the import–export trade to be unable to carry out the settlement through the ships proxy and the overseas ship owner. However, such situation has also caused the exportation marine transportation operation work of freight transportation proxy companies cumbersome.

2.3 Difficulty in transparent data sharing

Advantages of shipping e-commerce mainly lie in resource sharing and highly transparent data,

especially in prices. Specific promotion and group buying discount owned by e-commerce can bring terminal clients with real and tangible benefits. However, the prices issued by the shipping enterprise through different channels are often not unified. Even the marketing personnel may offer different prices for different clients (Zhou, 2015). In addition, part of information regarded as business secret by the shipping enterprise will be difficult to share. Information asymmetry will doubtlessly affect the credibility of e-commerce platform.

3 SWOT ANALYSIS OF SHIPPING E-COMMERCE PLATFORM

SWOT analysis is a basic method for analyzing and positioning resources and environment of an organization in four regions: strengths, weaknesses, opportunities, and threats (Samejima, 2006). SWOT analysis is a comprehensive method to determine the external and internal factors of the enterprise. By finding out the strengths and weaknesses of the internal environment and evaluating the opportunities and threats of the external environment, the enterprise can put forward better tactics for higher progress (Xu, 2013).

A Opportunities

The Prime Minister has proposed to support a combination of the Internet with finance, business, logistic, and others in the government work report, which means that the state will carry out slack policy on the field of e-commerce, benefiting port, shipping, and logistics. More opportunities can be provided to the development of shipping e-commerce platform. Various types of government papers have provided preferential policies on e-commerce, especially the development of cross-border e-commerce and shipping industry, strengthening integration with financial service business, expanding shipping transaction functions, improving shipping transaction information services, innovating shipping transaction service products, lowering transaction costs, and lifting service efficiency since 2015. It plays an irreplaceable role in promoting the development of business modes of shipping e-commerce platform.

B Threats

Electronic commerce uses a variety of Internet and modern information technology to carry out business activities, negotiation information, order information, payment information, and other large business information through the network in the computer system, while hacker attacks and computer viruses will be threats to the system.

C Strengths

Shipping business platform, in fact, is the "logistics" as a standard of service products. The combination of shipping and e-commerce enables many clients to check the freight of product, complete on-time order and payment, as well as check the transport condition of goods and document circulation at any time.

D Weaknesses

Using the Internet and a variety of modern information technology to carry out business activities, negotiation information, order placing, payment information, and a large number of business information are stored and handled in the computer systems through the Internet; therefore, there will be a huge risk.

4 DEVELOPMENT STRATEGIES FOR THE SHIPPING E-COMMERCE PLATFORM

4.1 Broaden terminal services

A common feature of the shipping industry and international freight forwarding industry is credit transaction. Time of credit depends on the degree of money assistance. The entry threshold of freight forwarding enterprise is relatively low. Thus, the speed of capital turnover is very important for business operation of small and medium-sized enterprises in their start-up period. Therefore, banks can access the platform to provide enterprise credit basis, combined with capital and business valuation of shipping companies.

4.2 Clear service object

In the planning of e-commerce platform, we must emphasize the "customer demand as the center", rather than "system functions as the center". To achieve the customer demand as the center, it is necessary to make clear the electronic commerce service object to carry on the subdivision of the customer. That is, for some large customers, services are provided through VIPEDI; for small and medium-sized customers, services are provided through e-commerce (Ding, 2011). Small and medium-sized customers should be further subdivided according to the target object to develop Web site functions. For new users and less-experienced users in e-commerce applications, it is necessary to introduce a simple and practical version of the client.

4.3 Strengthen the effort in personnel training

Development of shipping booking website shall be guaranteed by specialized talents. Therefore, subsequent talent team construction is of great importance, especially upon the initial layout completion of the shipping booking service network. Shipping logistics enterprises must attach great importance to the construction of corresponding human resources and determine scientifically and rationally talents problems that might be involved in the future through institutional mechanisms in a proactive manner. Great development in the e-commerce platform should be achieved through giving full play to the talent.

5 CONCLUSIONS

The development of the shipping e-commerce will be a transformation, an upgrade in operating modes of the shipping industry with a purpose of building a more rational, efficient, and win–win ecological environment for the shipping industry, which enables all shipping enterprises to search for their own space for development. If an enterprise wants to own a place in shipping e-commerce, it must find an accurate positioning for adapting to the terrain we land in. If a port city would like to better develop shipping e-commerce, then it must create its superior environment positively for the shipping electronic commerce development.

REFERENCES

Eisenm ann T, A lstyne P J. Strategies for two-sided markets. Harvard Business Review, 2006.

Samejima, M., Y. Shimizu, M. Akiyoshi, and N. Komoda, "Swot analysis support tool for verification of business strategy," in *International Conference on Computational Cybernetics*. IEEE, Aug. 2006, pp. 1–4.

Xi Ding, Hou Chun Ding, "An analysis of the present situation of the application of e-commerce in small and medium sized enterprises in China" [F], China Business & Trade, 2011, No#29, pp. 99–110.

Xue Zheng Zhang, Electronic Commerce [M], The first edition, Shanghai People's Publishing House, 2001.

Yingxin Xu, "SWOT Analysis of E-commerce Websites— Based on Dangdang.com," in 2013 *International Conference on Computational and Information Sciences*, IEEE 2013, No#01, pp. 1–3.

Zhou Fang, Zhi Qiang Cao, "Present situation and trend of the development of China's shipping e-commerce platform" [J], Containerization, 2015, No#01, pp. 22–25.

Electromechanical Control Technology and Transportation – Jia & Wu (Eds)
© *2017 Taylor & Francis Group, London, ISBN 978-1-138-06752-3*

A genetic fuzzy classification system based on computing with words

Hong Ji & Ming Ma
College of Information Technology and Media, Beihua University, Jilin, China

ABSTRACT: Genetic fuzzy classification system is a hot spot of fuzzy systems, but the contradiction of its accuracy and interpretability is still a problem. Adding computing with words during the design of the fuzzy system is an effective method to improve its accuracy and keep its interpretability. In this paper, we propose a new method to design a fuzzy classification system on the basis of computing with words. Computing with words is used to adjust the membership function of fuzzy partition during the mutate process. During the initialization process, expert selection and random selection are used to make the system convergent fast, meanwhile ensuring the diversity of the whole population.

1 INTRODUCTION

Genetic fuzzy classification system is a hybridization of fuzzy system and genetic algorithm, which has become a hot spot of fuzzy system (Dong, 2012; Huang, 2013; Huang, 2014; Huo, 2011; Sande, 2012; Le, 2012; Li, 2008). It is well known that a good genetic fuzzy classification system should be accurate and interpretable, but it is an impossible mission. Professor Zadeh (Zadeh, 1996) proposed to use computing with words to adjust the shape of the membership function, which makes the system more accurate. It does not reduce the interpretability of system because of computing with words. In this paper, we propose a new method to design a fuzzy classification system. During the population initialization, all the attributes of test samples are divided into the same average fuzzy partitions. Half of the initial fuzzy rules are generated by random selection of the fuzzy partition of all attributes, and the other half are generated by expert selection of the best fuzzy partition of randomly selected samples. During the process of mutation, parameters based on computing with words are used to adjust the shape of the membership function. Simulation experiments of Iris data set show that the new method has a higher classification accuracy.

2 RESEARCH OF GENERATE FUZZY PARTITION BY COMPUTING WITH WORDS

Professor Zadeh (1996) considered that adding language qualifiers can make a connection between

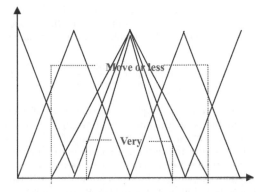

Figure 1. Computing with words application on five fuzzy partitions.

data information and computing with words. That makes the adjustment of fuzzy membership function equate to the modified language words, and the fuzzy system will be easy to interpret. Let us take Iris data set for example. If all the attributes are divided into five average fuzzy partitions, two language qualifiers ("very" and "more or less") are added to adjust its fuzzy membership function for each attribute (see Figure 1). Besides these, "don't care" condition is also added to each attribute. For each attribute of a sample, language qualifiers are added to expand the search space from $(5 + 1)^4$ to $(15 + 1)^4$. Language modifiers are added to improve the system accuracy, which reduces its convergence rate. In order to solve this problem, expert selection is used in initializing process to improve the convergence speed.

3 NEW DESIGN OF GENETIC FUZZY CLASSIFICATION SYSTEM

According to the analysis above, in order to expand the algorithm's search space while maintaining its convergence speed, in this paper, we propose a method that in the population initialization, half of the initial population is generated by random selection and the other half is produced by expert selection. During the mutate process, four mutate parameters are selected to adjust the shape of the membership function of fuzzy partition.

3.1 Population initialization

In this paper, we use pizibao code to design a genetic fuzzy classification system. Each population has L individuals $\{R_{set1}, \ldots, R_{setL}\}$. Each individual contains M rules $R_{set} = (R_1, \ldots, R_M, fit)$, where fit is the individual's fitness value.

Triangular membership function is used to represent fuzzy set, and (a, b, c) represents the vertices of the triangle. The triangular function formula is:

$$f(x; a, b, c) = \max\left(0, \min\left(\frac{x-a}{b-a}, \frac{c-x}{c-b}\right)\right) \quad (1)$$

For all training samples x_p, search space of each dimension is divided into five triangular intervals with the same size, and all the triangular function intervals form the candidate membership function set $Aset_i = \{a_{k,i}, b_{k,i}, c_{k,i}\}, (k \in [1, 5], i \in [1, n])$.

Then, we generate the whole population with two different methods, whose details are as follows.

First, M rules are randomly selected from the candidate membership function to form an individual of the population, and it is repeated for $L/2$ times to complete the random selection.

Second, M training samples are randomly selected from the training set. To each dimension x_{pi} $(i \in [1, n])$ of a sample, its compatibility U_i with candidate membership function is defined as:

$$U_I = \left| x_{pi} - b_{k,i} \right| \quad (2)$$

where $b_{k,i}$ is a vertex of $Aset_i$. Then, according to the minimum U_i, we obtained the best triangular membership function of all the dimension of a sample, which makes the sample get the maximum compatibility, which is regarded as the expert experience. It is repeated for $L/2$ times to complete the expert selection.

Finally, the $L/2$ individuals of random selection and the $L/2$ individuals of expert selection are put together to become a population. The code of R_j is $(a_{j1}, b_{j1}, c_{j1}, \ldots a_{jn}, b_{jn}, c_{jn}, C_j)$, and C_j is the consequent class generated by 3.2.

3.2 Consequent class

To each rule of an individual, its compatibility $u_j(x_p)$ with all training samples x_p is defined as:

$$u_j(x_p) = \prod_{i=1}^{n} u_{ji}(x_p) \quad (3)$$

To each consequent class, we calculate the sum of all samples that belong to it, and the formula is:

$$\beta_{classh}(R_j) = \sum_{x_p \in classh} (u_j(x_p)) \quad (4)$$

Then, the class of maximum $\beta_{classh}(R_j)$ is chosen as the consequent class of the rule R_j. If more than two classes have the same $\beta_{classh}(R_j)$, the one with less amounts of sample will be selected.

3.3 Fitness value

In this paper, we use five average fuzzy partitions to remain interpretability of fuzzy classification system; thus, the only thing we must consider during the design of fitness function is the classification accuracy. The fitness value, $fitness(R_{seti})$, is defined as:

$$fitness = \begin{cases} CP(R_{seti}) - \omega MP(R_{seti}), & MP(R_{seti}) \leq m \\ 0, & MP(R_{seti}) > m \end{cases} \quad (5)$$

where CP is the number of right classifications of the system, MP is the number of wrong classifications of the system, ω is a punishment parameter of wrong classification, which lies in [0.2, 0.5], and m is the maximum number of the wrong classifications permitted.

3.4 Fuzzy genetic operations

The selection strategy adopted in this paper is the combination of roulette wheel selection and elite selection. One-point crossover is used to select individuals (see Figure 2). The elite individual of the old generation is retained when the new one is generated.

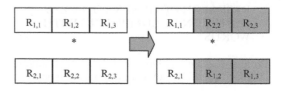

Figure 2. Crossover operation.

In this paper, one-point mutation used to each rule is set as R_{set}. When the random number is larger than the mutate rate, we randomly selected a rule R_j to mutate (see Figure 3). According to computing with words, four language qualifiers were added during the mutate process to modify the shape of the membership function: "extreme", "very", "more or less", and "little".

To each dimension of a rule R_j, there are three parameters: a_{jn_1}, b_{jn_1}, and c_{jn_1}. We keep b_{jn_1} constant and calculate the new a_{jn_1}' and c_{jn_1}' by using mutate parameters as follows:

$$a_{jn_1}' = \max\{b_{jn_1} - \delta(b_{jn_1} - a_{jn_1}), x\min_{n_1}\}$$
$$c_{jn_1}' = \min\{b_{jn_1} + \delta(c_{jn_1} - b_{jn_1}), x\max_{n_1}\} \qquad (6)$$

where δ is randomly selected from all the mutate parameters.

Figure 3.   Mutate operation.

Table 1.   Experiments on training set (M = 3).

Fuzzy partition	Training set		
	Best classification rate	Average classification rate	Worst classification rate
Five average fuzzy partitions	87.3%	87.3%	87.3%
Fuzzy partition of computing with words	96.1%	95.2%	94.1%
Our approach	96.1%	95.9%	95.1%

Table 2.   Experiments on testing set (M = 3).

Fuzzy partition	Testing set		
	Best classification rate	Average classification rate	Worst classification rate
Five average fuzzy partitions	87.5%	86.9%	85.4%
Fuzzy partition of computing with words	97.9%	95.0%	91.7%
Our approach	97.9%	94.2%	87.5%

Table 3.   Experiments on training set (M = 10).

Fuzzy partition	Training set		
	Best classification rate	Average classification rate	Worst classification rate
Five average fuzzy partitions	97.1%	96.4%	95.1%
Fuzzy partition of computing with words	100.0%	97.9%	96.1%
Our approach	99.0%	98.4%	98.0%

Table 4.   Experiments on testing set (M = 10).

Fuzzy partition	Training set		
	Best classification rate	Average classification rate	Worst classification rate
Five average fuzzy partitions	100.0%	97.1%	91.7%
Fuzzy partition of computing with words	100.0%	96.5%	93.8%
Our approach	100.0%	96.8%	89.5%

4   SIMULATION EXPERIMENTS

Simulation experiments use Iris data set: 2/3 as training set and 1/3 as testing set. Using the pizibao code, the number of initial population is 50, crossover rate is 0.9, and mutate rate is 0.1. We choose four mutate parameters, 0.4, 0.7, 1.3, and 1.6, representing "extreme", "very", "more or less", and "little", respectively. Simulation results on the training set and testing set as shown in Tables 1–4.

5   CONCLUSIONS

In this paper, we proposed a new method to design the genetic fuzzy classification system. Our approach generates half of the initial population by random selection of the samples and another half by expert selection of the randomly selected samples. Then, it adds language qualifiers during the mutation process to adjust the shape of the fuzzy membership function. The new method adapts the advantage of expertise and machine learning, which is demonstrated to be more efficient.

ACKNOWLEDGMENT

This work was supported by the Science and Technology Department of Jilin Province Natural Science Foundation (20140101185 JC).

REFERENCES

Dong Jie, Shen Guojie, "Remote Sensing Image Classification Based on Fuzzy Associative Classification [J]", Journal of Computer Research and Development, vol. 49, no. 7, (2012), pp. 1500–1506.

Huang Kaiqi, Ren Weijiang, Tan Tieniu, "A Review on Image Object Classification and Detection [J]", Chinese Journal of Computers, vol. 37, no. 6, (2014), pp. 1225–1240.

Huang Kaiqi, Tan Tieniu, "Review on Computational model for vision", Pattern Recognition and Artificial Intelligence, vol. 26, no. 10, (2013), pp. 951–958.

Huo Weigang, Shao Xiuli, "A Fuzzy Association Classification Method Based on Multi-objective Evolutionary Algorithm [J]", Journal of Computer Research and Development, vol. 48, no. 4, (2011), pp. 567–575.

Le Q.V, Ranzato M.A, Monga R., "Building high-level features using large scale unsupervised learning [C]", Proceedings of the international Conference on Machine Learning (ICML), Edinburgh, UK, (2012), pp. 07–114.

Li Jie, Deng Yi Ming, Shen Shituan, "Classification Rule Extraction Based on Fuzzy Area Distribution and Classification Reasoning Algorithm [J]", Chinese Journal of Computers, vol. 31, no. 6, (2008), pp. 934–941.

Van de Sande K., Uijlings J., Snoek C., Smeulders A., "Hybrid coding for selective search [C]", Proceedings of the workshop on PASCAL VOC, Florence, Italy, (2012), pp. 1–8.

Zadeh, L.A. "Fuzzy logic = computing with words [J]", IEEE Transactions on Fuzzy Systems, vol. 2, (1996), pp. 103–111.

Error analysis and accuracy of a 6-DOF parallel manipulator

Xiaoyong Sun
Chongqing Institute of Green and Intelligent Technology, Chinese Academy of Science, Chongqing, China
Chongqing De ling Technology Co. Ltd., Chongqing, China

Yongting Zhao & Danlu Zhang
Chongqing Institute of Green and Intelligent Technology, Chinese Academy of Science, Chongqing, China

ABSTRACT: In this paper, a 6-DOF parallel manipulator is introduced. On the basis of the analysis of error sources, the error model of the parallel mechanism is established by vector method. The accuracy of the analytical formulas is pushed by the error model. The sensitivity of various error sources is obtained by MATLAB simulation, and the maximum singular value of the matrix can be considered as impact index of the error. The results of error and accuracy analysis can be of guiding significance for the control system.

1 INTRODUCTION

The error analysis is the basis of kinematic calibration, which pointed out a key link in parallel mechanism. It can provide a clear focus and direction to improve the pose accuracy and calibration of the parallel mechanism. Many error sources exist in parallel mechanism, such as every kinematic parameter is error source. Moreover, there are many error modeling and analysis methods. Error model of parallel mechanism was established by using the motion constraints generated by a mechanical device (RAUF, 2001). Liu (2005) proposed the kinematics error model using the perturbation method, and calibration for the error model analysis on the mechanism was discussed in the MATLAB environment. The error model of a parallel mechanism was established on the basis of the relationship between input and output differential methods; the sensitivity of the error sources was analyzed; and the positioning error extremes expression of the mechanism was deduced (Li, 2009). The absolute coordinates of an arbitrary measuring point were absolute coordinates; a system of kinematics equations was built and solved optimally to identify the configuration parameters (Yan, 2012). The inverse kinematics error model of 6UPS parallel mechanism was built on the basis of vector error method. Data optimization was obtained by measurement test and error analysis (Pei, 2006). The kinematics accuracy and an improved error model were proposed (Liu, 2010), and the 3-DOF parallel mechanism was calibrated.

The conclusion of the error analysis is the accuracy of various error sources. The accuracy analysis theory and verification method of a 3-DOF

parallel mechanism were proposed, and conclusions were validated by experimental means (Briot, 2010 and Merlet, 2006). The accuracy analysis mathematical model of a parallel mechanism was built by Guo (Guo, 2006). Chang (Chang, 2007) not only proposed an error analysis method, but also solved the problem of the error analysis by using the accuracy analysis of vector set and error space.

In this paper, errors and accuracy analysis of a 6-DOF parallel mechanism will be discussed, and the impact of the error will be obtained too. This conclusion can provide a theoretical basis for the calibration and control of the parallel mechanism.

2 ERROR ANALYSIS

2.1 *Parallel mechanism*

Structure drawing of this parallel mechanism is shown in Figure 1. This mechanism is mainly composed of the moving base, the slider assembly, strut components, ball joints, the moving platform, and rod component. The slider drive rod drives the platform to achieve the movement of the solid connection on the moving platform with struts and model, which is in accordance with the given law of motion. This mechanism has a total of six-group kinematic chains, each containing the same moving parts and movement pairs that are composed of the mobile pair and the ball joints pair, S.

The purpose of error analysis is to find the error source of the parallel mechanism, to find out the causes of error and improve the accuracy. Error of parallel mechanism can be divided into static errors and dynamic errors. Static errors include conversion

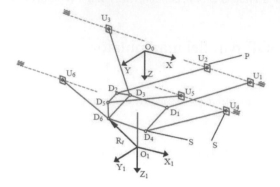

Figure 1. Structure of 6-PSS parallel mechanism.

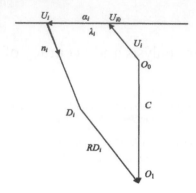

Figure 2. Branched vector schematic of one chain.

errors and quality errors; dynamic errors include elastic deformation, vibration errors, and driver error.

Conversion error is a parallel mechanism-specific error, which is caused by the difference between the inverse kinematics model in control and the actual size. Quality error is the result of elastic deformation; the stiffness coefficient is not a constant, and gravity will cause deformation of the end of actuator. The stiffness of parallel institutions depends on its natural frequency; the first natural frequency will greatly determine the dynamic performance of the parallel mechanism.

A change in the drive force, such as the frictional force, results in a change in the movement speed. It will generate the control error, which is not generated by the control system itself but by changes to the control system in external conditions.

These types of errors are the main errors affecting the accuracy of the parallel mechanism; each error has different characteristics, which is the same as the extent and scope of the accuracy effects. Conversion errors and quality errors are static, whereas driver errors and vibration errors are dynamic.

2.2 Error model

The branched vector schematic of one chain of this parallel mechanism is shown in Figure 2.

Point O_1 is the origin of the coordinate system; point O_0 is the origin of the fixed coordinate system; vector a_i is the unit vector of the slider movement; vector n_i is the unit vector of the rod axis; λ_i is the slider displacement vector; U_i is the position vector of the sphere center for the upper ball joints; U_{i0} is the initial position vector of the sphere center for the upper ball joints; RD_i is the position vector between the lower ball joints center Di and point O_1 in fixed coordinate system; R is the transformation matrix of the coordinate system to the fixed coordinate system; and l_i is the vector of the rod axis.

Figure 2 shows a single closed-loop vector chain of parallel mechanism. In a fixed coordinate system, each vector is given as below:

$$\lambda_i a_i = C - R \cdot D_i - U_i - l_i n_i \qquad (1)$$

Equation (1) can be changed into another error model by partial differential calculation:

$$\delta\lambda_i a_i + \lambda_i \delta a_i = \delta C - \delta R \cdot D_i - R \cdot \delta D_i - \delta U_i - l_i \delta n_i - \delta l_i n_i \qquad (2)$$

where R is the differential form cosine matrix from the dynamic coordinate system to the fixed coordinate system. Constraint equations (1, 2):

$$a_i \cdot \delta\lambda_i + \lambda_i \cdot \delta a_i = \delta C - \delta\theta \times D_{i0} - R \cdot \delta D_i - \delta U_i - l_i \cdot \delta n_i - n_i \cdot \delta l_i \qquad (3)$$

where vector D_{i0} is the position vector of the sphere center for the lower ball joints in the fixed coordinate system. After multiplying n_i^T on both sides, equation (3) can be transformed into:

$$n_i^T \delta C + (n_i \times D_{i0})^T \delta\theta = n_i^T a_i \delta\lambda_i + n_i^T R\delta D_i + n_i^T \delta U_i + \delta l_i + n_i^T (\lambda_i \delta n_i) \qquad (4)$$

The parallel mechanism has six branched chains, according to the equations $n_i^T n_i = 1$ and $n_i^T \delta n_i = 0$, equation (4) can be expressed as:

$$J^{-1}\delta X = \delta\Lambda + M\delta L + N_1\delta D' + N_2\delta A_0 + N_2\delta A \qquad (5)$$

where δX is the attitude error vector of the parallel mechanism, $\delta\Lambda$ is the position error vector of the slider, δL is the rod length error, $\delta B'$ and δA_0 are mounting position errors of the ball joints, and δA is the straightness error of the guide.

Figure 3 shows a schematic view for each error vector and the influence of final accuracy from all types of errors. The circle represents error range of

478

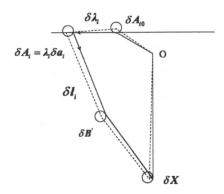

$\delta\lambda_i$ δA_{i0}

$\delta A_i = \lambda_i \delta a_i$

O

δl_i

$\delta B'$

δX

Figure 3.   Error diagram.

various sources of error; the dotted line represents a theoretical value of each point of the joint, and the solid line indicates the actual error value for each point of the joint. The closed-loop chain composed of the dotted arrow lines is equivalent to the theory movement transmission chain of the parallel mechanism; the closed-loop chain composed of the solid arrow lines is equivalent to the movement transmission chain in the case of various errors. It can be seen that the end of actuator position of parallel mechanism deviated from the original theoretical position.

If matrix $\mathbf{J}^{-1}$ is reversible, the moving platform error can be simplified to:

$$\delta X = \mathbf{E} \delta\varepsilon \tag{6}$$

Here, matrix $\mathbf{E}$ is defined as the total error of the transformation matrix $\mathbf{I} \in R^{6\times6}$.

$$\mathbf{E} = \mathbf{J}[\mathbf{I}\ \mathbf{M}\ \mathbf{N}_1\ \mathbf{N}_2\ \mathbf{N}_2] \in R^{6\times48} \tag{7}$$

Matrix $\delta\varepsilon$ is defined as the error source vector matrix:

$$\delta\varepsilon = \begin{bmatrix} \delta\mathbf{A}^T & \delta\mathbf{L}^T & \delta\mathbf{B}^T & \delta\mathbf{A}_0^T & \delta\mathbf{A}^T \end{bmatrix}^T \in R^{48\times1} \tag{8}$$

Equation (7), pointing out various errors, can not only be passed through by the transmission Jacobin matrix, but also by other means. Matrix M, N_1, N_2 which are on behalf of institutions driver error in geometry, such as rod error, rail mounting error, and position error, are determined by the shape and structure of the parallel mechanism. The reason is that it is a diagonal matrix whose mapping for each movement branched is mutually independent. When $a_i^T n_i$ is zero, the rod direction vector and the slider vector are circular, and the mechanism is in a singular location. When parallel mechanism is at or near the singular configuration, the nondrive error will be magnified, and the impact of the nondrive error is much larger than that of the end-effector position and orientation error.

3   ACCURACY ANALYSIS

The accuracy analysis takes into account various sources of error. In order to study the impact of errors on the accuracy, the input source of the error is usually defined as a unit amount. However, it is easy to overlook the error range of error sources and maximum error value in this analysis method, so the concept of maximum unit amount of various error sources is introduced:

$$\begin{aligned} \delta A &= \|\delta A\|_{max}\ \widehat{\delta A} \\ \delta L &= \|\delta L\|_{max}\ \widehat{\delta L} \\ \delta D' &= \|\delta D'\|_{max}\ \widehat{\delta D'} \\ \delta A_0 &= \|\delta A_0\|_{max}\ \widehat{\delta A_0} \\ \delta A &= \|\delta A\|_{max}\ \widehat{\delta A} \end{aligned} \tag{9}$$

Here, $\widehat{A}$ denotes unit vector A, $\|\delta A\|_{max}$ denotes the maximum value of the matrix A, $\widehat{\delta A}$ is the largest unit vector of the matrix, and the remaining matrices are also in accordance with this definition.

Rail straightness error $\|\delta A\|_{max}$ can be defined as:

$$\|\delta A\|_{max} = \lambda_{max} \|\delta A\|_{max} \tag{10}$$

Here, λ_{max} is the maximum error of the straightness of the linear guide. For the Jacobin matrix, $\widehat{\mathbf{J}}$ is the largest unit of the Jacobi matrix.

Equation (10) can be changed as:

$$\delta X = \widehat{E} \delta\varepsilon \tag{11}$$

$$\widehat{E} = \begin{bmatrix} \widehat{J}\|\delta A\|_{max} & \widehat{J}M\|\delta L\|_{max} & \widehat{J}N_1\|\delta D'\|_{max} \\ \widehat{J}N_2\|\delta A_0\|_{max} & \widehat{J}N_2\|\delta A\|_{max} \end{bmatrix} \tag{12}$$

$$\delta\widehat{\varepsilon} = \begin{bmatrix} \delta\widehat{A}^T & \delta\widehat{L}^T & \delta\widehat{D}'^T & \delta\widehat{A}_0^T & \delta\widehat{A}^T \end{bmatrix}^T \tag{}$$

Equation (11) can also be written as the general equation:

$$y = Ax \tag{13}$$

where y and x are output vectors of $m \times 1$ and $n \times 1$, respectively.

In order to analyze the important performance of transfer matrix A, the range of error sources is assumed as follows:

$$\|x\|^2 \le 1 \tag{14}$$

Equation (14) can be simplified to:

$$\|A^+ y\|^2 \le 1 \tag{15}$$

Here, matrix A^+ is the false inverse matrix of A, matrix A can be decomposed into three matrices according to the matrix singular value decomposition formula:

$$A = U\Sigma V^T \qquad (16)$$

where matrix U and V^T are both Hermit matrix, and matrix Σ is a positive definite diagonal matrix composed of a singular value of the matrix A.

Matrix A^+ can also be expressed as:

$$A^+ = V\Sigma^+ U^T \qquad (17)$$

Here, matrix Σ^+ is the false inverse matrix of Σ. According to equations (17) and (15):

$$\begin{aligned} \|V\Sigma^+ U^T y\|^2 &= (y^T U\Sigma^{+T} V^T)(V\Sigma^+ U^T y) \\ &= (U^T y)^T diag(\sigma_1^{-2}, \sigma_2^{-2}, \ldots, \sigma_m^{-2}) \\ &\quad (U^T y) \le 1 \end{aligned} \qquad (18)$$

Equation (18) shows the error relations between the error of mechanism and end of the actuator. The propagation of error transforms is based on a singular value of the error amplification. Three types of error amplification factors are supposed to be the major considerations in the design of parallel mechanism. The matrix boundary volume is proportional to Jacobi matrix in equation (18); the maximum singular value of the matrix can be considered as an influence coefficient of the largest source error.

Matrix $\hat{E}$ is the transfer matrix of total error, which can be decomposed into three corresponding error impact factors:

$$\begin{aligned} E_1 &= \sqrt{\det(\hat{E}\hat{E}^T)} = \prod_{i=1}^{6}(\sigma_{\hat{E}})_i \\ E_2 &= (\sigma_{\hat{E}})_{max} \\ E_3 &= (\sigma_{\hat{E}})_{max}/(\sigma_{\hat{E}})_{min} \end{aligned} \qquad (19)$$

where $\sigma_{\hat{E}}$ is the singular value of error propagation matrix $\hat{E}$. Three kinds of coefficients in equation (19) present the mechanism precision values. The smaller the value of the coefficients, the higher the accuracy. The coefficients calculated from equation (19) can be used as the evaluation accuracy standards.

Various error values of the parallel mechanism are given, and the accuracy can be obtained through the simulation. The upper portion of the figure is displacement error of the three axis directions, while the following sections are three angle errors. The maximum displacement errors are 0.054, 0.01, and 0.032 mm. The angle of pitch, roll angle, and yaw angle errors are -1.5×10^{-3}, -4.6×10^{-3}, and -1.5×10^{-3}, respectively.

4 CONCLUSIONS

This paper introduces an error and accuracy analysis of a 6-DOF parallel mechanism. The error model equation is obtained by means of vector representation. The sensitivity of various error sources is obtained by simulation. The results of the error and accuracy analysis have important guiding significance for the control system. The accuracy of analytical methods described in this paper can provide a common analytical method for similar institutions.

ACKNOWLEDGMENTS

The authors acknowledge partial supports provided by the Youth Science and Technology Personnel Training Program of Chongqing (cstc2014kjrc-qnrc70001).

REFERENCES

Briot S, Ilian A. B. Accuracy analysis of 3T1R fully-parallel robots. Mechanism and Machine Theory, 2010, 45: 695–706.

Chang Peng, Li Tie min, Liu Xin jun. Accuracy analysis of parallel mechanisms based on set theory. Journal of Tsinghua University (Science and Technology). 2007, 47(11): 1984–19.

Guo Zong he, Niu Guo dong, Sun Shu hua, et al. Accuracy Analysis of a 3-DOF Translational Parallel Robot Mechanism. Transactions of the Chinese Society of Agricultural Machinery, 2006, 37(1): 145–148.

Li Juan, Sun Li ning, Liu Yan jie. Errors Modeling and Analysis of a Novel Planar Parallel Positioning Mechanism. Piezoelectrics and Acoustooptics, 2009, 31(2): 284–288.

Liu Da wei; Wang Li ping; Guan Li wen. Accuracy Analysis and Calibration of a Special 3-DOF Parallel Mechanism. Chinese Journal of Mechanical Engineering, 2010, 46(9): 46–51.

Liu Hong jun, Gong Min, Zhao Ming yang. Error Analysis and Calibration of a 4-DOF Parallel Mechanism. Robot, 2005, 27(1): 6–9.Luigi T.De Luca, Propulsion physics (EDP Sciences, Les Ulis, 2009).

Merlet J.P. Jacobian, manipulability, condition number, and accuracy of parallel robots, Journal of Mechanical Design, 2006, 128 (1): 199–205.

Pei Bao qing, Luo Xue ke,Chen Wu yi, et al. Calibration of a 6UPS Parallel Machine by Vector Method[J]. China Mechanical Engineering. 2006, 17(8):854–857.

Rauf A, Ryu J. Fully autonomous calibration of parallel manipulators by imposing position const raint [C] Proceedings of International Conference on Robotics and Automation. Piscataway, NJ, USA: IEEE, 2001:238922394.

Yan Hao, Li Chang chun, Sun Hui peng. Configuration Parameter Identification and Error Analysis of Mobile 6-DOF Parallel Mechanism. Acta Armamentarii, 2012, 33(10): 1276–1280.

Traffic and transportation

Electromechanical Control Technology and Transportation – Jia & Wu (Eds)
© 2017 Taylor & Francis Group, London, ISBN 978-1-138-06752-3

A case study on the traffic impact evaluation of building constructions

Wei Yang
China Merchants Chongqing Communications Research and Design Institute Co. Ltd., Chongqing, China

ABSTRACT: With the sustained and rapid development of domestic economy and the continuously accelerating urbanization, a mass of manpower, material resources and financial resources have been invested in the construction and improvement of public facilities. The implementation of such public facilities surely brings great changes to a city. However, the negative impact on urban traffic that already results in an imbalance between supply and demand will get worse during the construction. By taking the construction of stadium-like buildings and a one-floor underground garage in a city as an example, this paper analyzes the impact of traffic conditions through the classification status survey that needs to be done prior to engineering construction, studies construction schemes and traffic characteristics at the construction area based on the construction characteristics, and elaborates with examples of the influence mechanism of building construction on traffic and related disposal measures.

1 INTRODUCTION

1.1 Project profile

The Reconstruction and Extension Project of planning exhibition gallery, science museum, local chronicles museum and library of the city is located on the central axis of the central region of the city facing a municipal government building. The southernmost end of the construction base is adjacent to a residence community, the construction site is surrounded by urban roads, the terrain of construction is flat, and the transportation is rather convenient. The project consists of two comprehensive stadium-like buildings and one-floor underground garage. The west building is a 24-meter-high multi-story building with the main function of planning exhibition gallery, among which the exhibition part includes two floors and four floors of other auxiliary function rooms. The east building is also a 24-meter-high multi-story building mainly consisting of municipal science museum, youth center, local chronicles museum and extension of library. Among these, the science museum is located at the northern end and covers three floors, and the youth center is located at the center and covers four floors with the addition of interlayer. The local chronicles museum and extension of library are located at the southern end and cover a total of five floors. The total floorage of the project is 69,173.61 m², with 50,991.45 m² above the ground and 18,182.16 m² underground. The reasonable service life of the buildings is 50 years, and the safety level is grade II.

1.2 Traffic conditions of surrounding roads

The surrounding roads related to this Reconstruction and Extension Project mainly include the third section of Huandu Avenue, the second section of Binjiang North Road, the first section of Zhenjiang West Road and Qingyuan South Road. It will have a certain negative impact on vehicles and pedestrians in the third section of Huandu Avenue, the second section of Binjiang North Road, the first section of Zhenjiang West Road and Qingyuan South Road during construction.

Huandu Avenue is a main passage connecting three districts, namely the Shunqing District, the Gaoping District and the Jialing District of the city, and it is a road with two-way six lanes plus a bicycle lane. As for daily transport, the traffic conditions at this section are rather good, and thus the average daily traffic volume is huge. During construction, the Reconstruction and Extension Project will have a certain impact on the vehicles passing through this road.

Binjiang North Road is an important longitudinal traffic trunk with two-way six lanes plus a vehicle–bicycle separator plus a bicycle lane in Shunqing District, and it intersects with Huandu Avenue at the roundabout. It is a key north-to-south fast passage. The average daily traffic volume is huge, the traffic conditions are good, and the traffic order is sound.

Located at the south of Huandu Avenue, Zhenjiang East Road and Zhenjiang West Road are responsible for diversion of vehicles for traffic trunk as secondary trunk roads. The roads have

Table 1. Traffic volume at rush hours and current service level at main sections and roads.

Road	Direction	Traffic volume at rush hour (pcu/h)	Saturation	Service level
Second section of Huandu Avenue	East to West	1860	0.44	Level I
	West to East	1521	0.36	Level I
Third section of Huandu Avenue	East to West	1723	0.41	Level I
	West to East	1415	0.34	Level I
Qingyuan North Road	South to North	783	0.20	Level I
	North to South	874	0.22	Level I
Qingyuan South Road	South to North	826	0.21	Level I
	North to South	918	0.24	Level I
First section of Zhenjiang West Road	East to West	781	0.30	Level I
	West to East	758	0.29	Level I
First section of Zhenjiang East Road	East to West	587	0.23	Level I
	West to East	571	0.22	Level I
Second section of Binjiang North Road	South to North	1264	0.32	Level I
	North to South	1338	0.34	Level I

two-way four-lanes plus a bicycle lane. The traffic conditions are rather good, the average daily traffic volume is relatively small, and the traffic order is sound.

Located at the west of theater, Qingyuan South Road and Qingyuan North Road have two-way four lanes plus a vehicle–bicycle separator plus a bicycle lane, and at the west side lies a residential community. The community is one of the concentrated residential areas of the new district under the municipal government; however, the present lodging ratio is somewhat low, so the traffic pressure from pedestrians and vehicles is relatively low.

1.3 Traffic performance

Through the field investigation and traffic survey, the statistical results of traffic volume and service level influencing such road sections in this region are described in Table 1.

The survey results of traffic volume show that the traffic volume of Huandu Avenue and Binjiang North Road at rush hours is huge, but the roads are still somewhat clear; the traffic volume of Qingyuan South Road and Zhenjiang West Road is somewhat small, and the roads are also clear. So, the service level at such sections and roads is Level I.

2 TRAFFIC IMPACT ANALYSIS DURING CONSTRUCTION

The roads subjected to the direct impact from the construction of the project are the third section of Huandu Avenue, Qingyuan South Road, Zhenjiang East (West) Road, and Binjiang North Road.

The survey and analysis of traffic conditions of the project show that Huandu Avenue and Binjiang North Road are traffic trunks with large traffic volume at morning and evening rush hours, and that the traffic volume of Qingyuan South Road, Qingyuan North Road, Zhenjiang East Road and Zhenjiang West Road at morning and evening rush hours is relatively small.

The Reconstruction and Extension Project of planning exhibition gallery, science museum, local chronicles museum and library is a building construction project. The construction area does not cover any existing road during construction, and temporary fencing is provided outside the boundary line of existing roads for separation. In general, the traffic impact on peripheral roads from construction is relatively small.

The third section of Huandu Avenue intersects with Binjiang North Road and Qingyuan South (North) Road, and it is a road with a cross-section of two-way six-lanes plus a bicycle lane. The survey data of traffic volume show that, at present, the two-way traffic volume at rush hours at the second section of Huandu Avenue is 1723 pcu/h for east to west with heavy-duty goods vehicles accounting for 31%, and at the third section of Huandu Avenue is 1415 pcu/h for west to east with heavy-duty goods vehicles accounting for 26%. The traffic service level is somewhat high; however, a relatively large proportion of vehicles that run on such sections are heavy-duty goods vehicles at present, and the increase in construction vehicles will have an impact on the traffic ability of the roads to some extent; however, rational arrangement of transportation routes of construction vehicles, enhanced traffic control, and prohibition of on-road parking will reduce the impact on traffic of the road and ensure a clear road.

The second and third sections of Binjiang North Road intersects with the second and third sections of Huandu Avenue on one plane, and it is a road with two-way six lanes plus a vehicle–bicycle separator plus a bicycle lane. At present, the two-way traffic volume at rush hours at such section is 1264 pcu/h in south to north direction, 1338 pcu/h in north to south direction, and the road is relatively clear. During construction, the increase in construction vehicles will have impact on the traffic ability of the roads to some extent; however, rational arrangement of transportation routes of construction vehicles, enhanced traffic control, and prohibition of on-road parking will reduce the impact on the traffic of the road and ensure a clear road.

Qingyuan South (North) Road intersects with the first and third sections of Huandu Avenue on one plane, and it is a road with two-way six lanes plus a vehicle–bicycle separator plus a bicycle lane. At present, the two-way traffic volume at rush hours at such section is 826 pcu/h in south to north direction, 918 pcu/h in north to south direction, and the road is relatively clear. During construction, the increase in construction vehicles will have impact on the traffic ability of the roads to some extent; however, rational arrangement of transportation routes of construction vehicles, enhanced traffic control, and prohibition of on-road parking will reduce the impact on traffic of the road and ensure a clear road.

Zhenjiang East Road intersects with Zhenjiang West Road and Qingyuan South Road, and it is a road with two-way four lanes plus a bicycle lane. At present, the two-way traffic volume at rush hours at such section is 781 pcu/h in south to north direction, 758 pcu/h in north to south direction, and the road is relatively clear. During construction, the increase in construction vehicles will have impact on the traffic ability of the roads to some extent; however, rational arrangement of transportation routes of construction vehicles, enhanced traffic control, and prohibition of on-road parking will reduce the impact on traffic of the road and ensure a clear road.

3 TRAFFIC ORGANIZATION AND MANAGEMENT SCHEMES DURING CONSTRUCTION

3.1 Traffic organization and management requirements

The traffic organization and management requirements include: (1) prohibiting all construction vehicles to park on existing roads; construction vehicles can park at other service facilities outside the boundary line of the road to reduce the impact on traffic during waiting and make the road accessible to residents along the road; (2) adjusting the bus routes and stops as appropriate, and temporary bus stops shall be arranged for passengers to get on and

get off the bus safely; and (3) making emergency traffic plans and reduce the occurrence of traffic jam due to traffic accidents or other emergencies.

3.2 Specific traffic organization and management schemes

During construction, in order to reduce the impact on road network traffic at the construction area as far as possible, the access to the construction area will be reduced. Furthermore, engineering vehicles shall be prevented from running on municipal traffic trunks such as Huandu Avenue, Binjiang North Road, etc. There are two traffic organization schemes as follows:

Scheme I: the construction area is provided with only one access, which is located at the entrance of theater parking lot at the central section of Binjiang North Road. Besides, vehicle–bicycle separator of Binjiang North Road at this specific place shall be provided with an opening so that the construction vehicles can enter and leave conveniently.

The route for construction vehicles to enter the construction area starts from Mashipu Road, through Wenyuan Street, Maoyuan North Road, Huandu Avenue, and to the entrance at Binjiang North Road. As for this route, the construction vehicles only need to turn left at the intersection of Maoyuan North Road and Huandu Avenue, and then go straight or turn right at other intersections. As for the route for construction vehicles to leave construction area, the construction vehicles turn right at the exit and go into the second section of Binjiang North Road, through Zhenjiang West Road and into Mashipu Road. The construction vehicles turn right at all intersections and will have little impact on the traffic at intersections.

Scheme II: the construction area is provided with only one access, which is located at the entrance of a theater parking lot at the central section of Qingyuan South Road. Besides, vehicle–bicycle separator of Qingyuan South Road at this place shall be provided with an opening so that construction vehicles can enter and leave conveniently.

The route for construction vehicles to the enter construction area starts from Fujing Road, through the planned Ba Road, Zhenjiang East Road, and Zhenjiang West Road, and to the entrance at Qingyuan South Road. As for this route, the construction vehicles all turn left at the intersections, and will have little impact on the traffic at intersections. As for the route for the construction vehicles to leave construction area, they will turn right at the exit and go into Qingyuan South Road, through Huandu Avenue, Wannian East Road, the planned Ba Road, and into Fujing Road. The construction vehicles only need to turn left at the intersection of the second section of Huandu Avenue and Qingfeng Road and at the intersection of Wannian Road and the planned Ba Road, and turn right at other intersections.

Both Schemes I and II have their own advantages and disadvantages. As for Scheme II, the access of construction area is arranged at the Qingyuan South Road, a road with two-way six-lanes plus a vehicle–bicycle separator plus a bicycle lane. The present traffic volume of this route is relatively smaller than that of Binjiang North Road. There are few peripheral bus routes. The lodging ratio of Ruifeng Garden Community is relatively low. There are no mature shops at the surrounding, so Scheme II will have less impact on the surrounding traffic than scheme I.

3.3 Specific traffic management measures

During project construction, it shall be ensured that the roads within the impact of the construction area are clear. Parking of construction vehicles on the roads in the vicinity of access to the construction area is prohibited. The traffic ability of related roads shall be enhanced and made use of, and some traffic diversion shall be organized. At rush hours of weekdays, that is, at morning rush hours (7:00–9:00) and evening rush hours (17:00–19:00), the construction vehicles shall be strictly monitored, and shall be controlled at any time according to the specific saturation conditions of roads so as to limit or prohibit the construction vehicles to go into urban roads; as a result, the roads will be clear for urban vehicles.

Law enforcement management shall be enhanced. In the construction area, the general intersections shall be arranged with management force that mainly relies on policemen and auxiliary police as subsidiary. Moreover, the important intersections shall be arranged with management of fixed posts held by policemen, and sections of roads shall be arranged with the management of fixed posts held by auxiliary police and mobile patrol performed by policemen, so as to enhance the response speed and traffic diversion capability. In the construction area, illegal acts that seriously interfere with traffic order such as suddenly cutting in, overtaking, traffic light violation, driving on do-not-drive roads, etc. shall be punished severely so as to ensure orderly operation of the roads that are subjected to the impact of construction. Traffic diversion roads and the video monitoring system of intersections shall be made full use of, and monitoring of road network in the region under the impact of construction shall be strengthened to ensure that traffic jam and accidents can be detected and dealt with quickly.

Improving the notification and guidance system: the important diversion intersections and busy intersections in the region under the impact of construction shall be provided with indication signs to inform in advance the traffic organization measures at the front intersection and related bypass routes of vehicles.

Publicity at the municipal and district level: the construction overview, impact, notices for going out, law observance education, etc. shall be publicized through various media at the municipal and district level, so that citizens will understand and cooperate with the construction activities.

4 CONCLUSION

This paper takes the construction of stadium-like buildings and a one-floor underground garage in a city as an example, analyzes the traffic conditions of peripheral roads of the project based on the project profile systematically, makes a detailed analysis and evaluation of the possible impact on peripheral traffic during construction, and explains the specific traffic organization and management measures for this engineering instance. The measures serve to increase the service level of road network at the impacted area, and promote the urban road network to meet related urban traffic demands simultaneously. Furthermore, they can also significantly reduce the negative impact of the construction itself and the related traffic organization measures on the traffic of the peripheral area of the construction site.

ACKNOWLEDGEMENTS

This study was supported by the open funds of the National and Local Joint Engineering Laboratory of Traffic Civil Engineering Materials of Chongqing Jiaotong University, the Chongqing Key Laboratory of Mountainous Road Structure and Material, and the Hi-tech Laboratory for Mountain Road Construction and Maintenance.

REFERENCES

Guiyan Jiang, Jiwei Li, Jiaqi Zhao. Traffic Impact Assessment Method for Large—scale. Urban traffic. 6(5): 59–63 (2008).

Sien Zhou, Chun Xu, Fengchun Han, Shaowu Qian. Traffic safety analysis and countermeasure research in construction area. J. Road Traffic & Safety. 02(2008).

Wei Zhou, Weidong Li, Traffic Impact Assessment Method for Large—scale Commercial Construction Project. J Traff Transp Eng. 4(2):93–99 (2008).

Weihua Zhu, Xueqin Niu, Y.Z., Kang. Study on Determination Method of Traffic Impact Area of Large-Scale Public Buildings. J. Railway Transport and Economy. 01(2006).

Xiaoli Ma, Yinfang Liu, Xuancai Lin, Hong Li. Discussion on Traffic Organization Design during Reconstruction. J. Highway. 07(2008).

Electromechanical Control Technology and Transportation – Jia & Wu (Eds)
© 2017 Taylor & Francis Group, London, ISBN 978-1-138-06752-3

Study on the design and protection of a subgrade side slope in the distribution area of loose accumulation bodies

Bingyang Chen
National and Local Joint Engineering Laboratory, Traffic Civil Engineering Materials, Chongqing Jiaotong University, Chongqing, China

Fa Wang
China Merchants Chongqing Communications Technology Research and Design Institute Co. Ltd., Chongqing, China

ABSTRACT: Earthquakes can cause serious geological hazards in many sections of roads, resulting in a great change in topographical, landform and geological conditions. On the basis of the investigation of a survey area, this paper briefs the design principle and prevention measures of a subgrade side slope in accordance with various types of subgrades in areas where loose accumulation bodies distribute, to ensure transportation safety to a maximum extent.

1 INTRODUCTION

Side slope is formed naturally or artificially. It is one of the most fundamental geological environments in human engineering activities, and the most common type in engineering construction. The loss of stability and collapse of the side slope, which is one of the top three geological disasters in the world, endangers the safety of people significantly. With the rapid development of infrastructure in China, a variety of fields have been involved in the issue of the side slope. So, it is a problem that must be taken into consideration by the engineering designers to understand side slope properly, thus minimizing the hazard caused by loss of stability of the side slope. There are various types of side slope. In the southwest of China, the side slope of accumulation body has higher engineering significance, which mainly includes the side slopes of landslide accumulation body, collapsed accumulation body and artificial accumulation body.

Earthquake can generate an amount of loose accumulation bodies. These bodies are mostly huge in scale. Under the influence of storm, earthquake, human engineering activities and other factors, a large loose accumulation body is apt to develop into a geological disaster, endangering the safety of humans and properties. Therefore, in the area where large loose accumulation bodies distribute, a reasonable selection of the design method and protection measures of side slope is the prerequisite to ensure the stability of loose accumulation body side slope and transportation safety.

2 OVERVIEW OF THE PROJECT

The highway project is located in the southeast of Qinghai-Tibet Plateau, on the east side of Qionglai mountain chain, with Songpan and A'ba massif on its northeast, Jintang arc structural zone on its southwest and Longmenshan geosynclinal area on its south. The geological structure belongs to the S-shaped Xuecheng-Wolong structure on the east side of the Jintang Arc fold belt in the geosynclinal area in the west of Sichuan Province. The geological structure is complicated. The structural system borders with Maowen Fault on its east and connects obliquely with the Jiudingshan Cathaysian structural system. The structural system is complicated by distortion of the mainstay, Xuelongbao Mountain, comprising a series of S-shaped and cambered linear fold and cambered compression and scissor faults. Its northeast section extends in the 60° direction, only 10–20 km in width. Its middle section, i.e. Lixian County–Xuelongbao zone, approaches the center of torsion, so it is curved into S-shape. In the section, the cambered compression and scissor fault are developed evidently, so there are especially dense folds. The southwest section extends in the 220° direction, and fans out gradually, with more than 40 km in width and more than 150 km in length. The section extends into the Xiaojin-Baoxing Range, meeting with the east side of Jintang cambered structure. Faults disperse mainly in the south and east of the S-shaped structure. The surrounding area of Xuelongbao Mountain is the one where faults concentrate, which meets obliquely with or is generally parallel to the Jiudingshan Cathaysianc tectonic line. Because of powerful crustal stress in the survey area, the brittle soft rocks break and produce fissures. In addition, the powerful weathering effect destroys the linkage of rocks, and reduces the strength of rock, causing adverse geological phenomenon in engineering, such as landslide, collapse, and falling fragments along the highway.

The outcropping stratum in the survey area includes metamorphic rocks of Devonian (D), Silurian (S) and Ordovician (O) in Paleozoic erathem. Lithology mainly includes phyllitizated schists, slates and phyllites. Because of the effect of late tectonic movement, the network structure of quartz distributes commonly in the rock masses. The accumulated layer of the Quaternary Period distributes widely in the survey area, which mainly includes alluvium and diluvium deposit, colluvial deposit, gravel soil and broken stone soil of the Holocene Series and Pleistocene Series, as well as alluvium and diluvium deposit, slopewash and colluvial deposit of the Holocene Series.

The relative elevation difference of the highway is generally 2000 m or higher. Zagu'nao River is the main water system in the survey area, and also the local erosion basis and centralized drainage zone of ground water. The project site is complex in geological structure, and vulnerable in ecological environment. There are different types of engineering geology disasters in various positions of the existing highway. The geological condition for engineering is evaluated to be moderate.

3 DESIGN AND PROTECTION MEASURES OF A SUBGRADE SIDE SLOPE

3.1 *Fill subgrade*

1. Slope ratio of fill side slope: the slope ratio of fill subgrade side slope is determined in accordance with types of filling material, height of side slope and geological condition of the subgrade base. The slope ratio of the side slope of common embankment is described below.

    Whereas the height of the side slope of embankment is ≤8.0 m, the slope ratio of 1:1.5 is adopted. Whereas the height of the side slope of embankment is 8.0 m <H≤20.0 m, the slope ratio of the upper portion of 8 m high from top is 1:1.5. A plain stage of 1.5 m or 2.0 m in width is set at the grade change point. The slope ratio of the lower portion below the plain stage is 1:1.75. Whereas the height of the side slope of embankment is H > 20.0 m, the slope ratio of the upper portion of 8 m high from top is 1:1.5, the medium portion of 20 m high is 1:1.75, and

the lower portion is 1:2. Plain stages of 1.5 m or 2.0 m in width are set respectively at the two grade change points.
2. Width of the berm: fill berm of subgrade is 0–1.0 m wide. The berm takes on the cross slope of 4% outward inclination.
3. Scope of the highway: the scope extends 1.0 m beyond the external edges of trenches at both sides of the embankment (or beyond the toes of the embankment or berm if there is no trench), or beyond the external edge of the cut-off drain on top of the through-cut slope (the place where there is no cut-off drain is the top edge).

3.2 *Through-cut subgrade*

The side slope of through-cut subgrade shall be designed by taking into consideration the natural grade of mountain mass, lithology, relationship between the attitude of tectonic fissure and route, weathering degree of rock mass, mechanical properties, excavation elevation, environmental protection and greening, earth-rock balance and other factors. Based on economical and reasonable principles, the design of the side slope shall be combined closely with protection and water drainage thereof.

The side slope of through-cut subgrade shall be designed in accordance with the topographical, hydrogeological and engineering geological conditions along the highway. In case of a favorable geological condition and the height of the side slope of through-cut subgrade less than 30 m, the following slope ratios of the side slope of through-cut subgrade can be applied:

1. Side slope of through-cut subgrade in soil and completely weathered rock zone is 1:0.75–1:1.25.
2. In case of heavily weathered to lightly weathered hard rock, and no structural face on the side slope that adversely affects the stability of the side slope of through-cut subgrade, the slope ratio of the side slope of through-cut subgrade is 1:0.5~1:1.25.
3. In case of lightly weathered and fresh hard rock, and no structural face on the side slope that adversely affects the stability of the side slop of through-cut subgrade, the slope ratio is 1:0.1~1:0.75.
4. Berm at the foot of the through-cut slope is 1.0 m wide. Some highway sections that have complete rock mass might not be equipped with berm at the foot of the through-cut slope to reduce cut, so that the slope extends to the top at the same ratio or different ratios. The through-cut side slope is staged every 8 m high, and a plain stage of 0–2.0 m wide is set for each stage. The cross slope of 4% outward inclination is adopted.

5. For the side slope that is high in cut, and "peeled off" by a thin layer of rock or earth from its surface, steep sloping is adopted in principle based on enhanced engineering protection to reduce unnecessary cut and protection.

The design of the side slope of through-cut subgrade shall be combined with the design of its protection. For the highway section with the height of the side slope higher than 30 m, the slope ratio of each stage shall be determined according to the geological condition and the change in protection type, in addition to the engineering protection to ensure the stability of the side slope.

3.3 *Protection of the side slope of through-cut subgrade*

In order to avoid new engineering geological disaster induced by excavation of the side slope, the general concept of design of the side slope of through-cut subgrade is the avoidance of excavation of soil side slope of the eluvium layer, less excavation of rocky side slope, and selective excavation of soil side slope of alluvium and diluvium layer. The through-cut side slope shall be protected in accordance with its height, overburden thickness, soil–rock interface, characteristics of rock–soil mass, and the stability of the side slope.

1. Most of the existing rock side slopes along the highway are steep, high, huge and generally stable, but broken debris due to weathering is detected locally. In accordance with the grade of the highway and investment, the general design principle is as follows:
   ① If the existing side slope is not to be excavated, and the falling rock is small and sliding along the slope face, the condition of the existing side slope shall be maintained.
   ② If the existing side slope is not to be excavated, and he rock will fall into the highway directly and endanger transportation safety, the dangerous rock shall be removed or protected locally by flexible protection nets.
   ③ If the existing side slope is not to be excavated, and the rock mass is broken and joints develop, the entire slope face shall be protected by means of shotcrete with steel bar nets, flexible protection net, facing wall, etc.
   ④ If the existing side slope is not to be excavated, and there are unfavorable structural faces (e.g. plane fault, outward inclined structural face), which endanger the safety of the slope, the side slope shall be as gentle as possible and be protected by anchor measures (e.g. anchored bar type grid beam, etc.).
   ⑤ If the existing side slope will be excavated, considering redistribution of stress in loose zones generated by blasting in construction and relaxed zones, the slope shall be graded to be 1:0.1–1:0.5 in accordance with the physical and mechanical properties of rock mass and sloping conditions (most of the slopes are not provided with precondition for grading). A plain stage of 2.0 m wide shall be set at a height of 8.0–12.0 m. Because of the vulnerable ecological environment in the project site, engineering protection shall prevail:

2. The soil side slopes of alluvium and diluvium layers, because of its dense state, good cementation and good stability, are provided with preconditions for grading. The general design principle is as follows:
   ① If the existing side slope is not to be excavated, the condition of the existing side slope shall be maintained.
   ② If the existing side slope is to be excavated, grading + greening protection is adopted in principle.

3. For the soil side slope of eluvium layer, because of the condition of its formation and engineering nature, the general design principle for the zone that is vulnerable to engineering geological disaster is to avoid excavation of such side slope as much as possible. If such zone has to be excavated because of restriction by criteria of the route, the excavation shall be reduced as much as possible. The general design principle is as follows:
   ① The section of the highway that has been subjected to engineering geological survey is defined as the adverse geological section, which is determined to be a special subgrade and designed as per special site.
   ② If the existing side slope is not to be excavated and is stable, its current conditions shall be maintained.
   ③ If the existing side slope is not to be excavated and is confirmed stable through excavation, grading plus greening protection is adopted in principle.
   ④ If the existing side slope is to be excavated and vulnerable to the hidden trouble of safety, it is determined to be a special subgrade and designed as per special site.
   ⑤ If the existing side slope is to be excavated and there is slippery face or shallow sliding on the slope, the facing wall or cut slope wall shall be set.

3.4 *Subgrade along the river*

Riverside subgrade is the key in the subgrade design in the project. The determination of the reasonable principle and measures of treatment is significant to prevent the flood damage and control the engineering cost. For these purposes, considering the characteristics of the project, the following design principles are applied:

1. The riverbed and bank are mostly made of alluvium and diluvium layer, covered with sandy pebble soil of high strength. The edge of subgrade is far away from the top edge of the slope of riverbed, and it is demonstrated by the survey that there is generally no subsidence and collapse of subgrade in the existing highway caused by flood damage. The subgrade does not locate in the wash zone of the riverbed, and the subgrade need not be treated in principle.
2. If there is underlying sandy pebble soil, there is no flood damage in the existing highway. The toe protection of mortar rubble masonry or Galfan-plated gabion cages shall be set.
3. The retaining wall along the river that is currently in good condition shall be fully utilized. If the edge of subgrade extends beyond that of the original retaining wall, the wall shall be complemented:
   ① If the exceedance is ≤30 cm, the cantilever structure is adopted: the upper portion of the original retaining wall is chiseled off, and then complemented by means of cantilever of C20-reinforced concrete.
   ② If the exceedance is >30 cm, the retaining wall is complemented externally and connected by steel bars in the middle portion.
4. If there is no flood damage in the existing highway and the cross slope of the riverbed is gentler than the 1:1 upstream face, the subgrade can be protected by Galfan-plated gabion cages.
5. If the river is wide and the direct grading of embankment does not evidently affect the flow rate, the subgrade is protected by slope protection + toe protection.
6. For the riverside subgrade adopting engineering protection, the burial depth of its foundation shall not be less than 1.0 m below the scouring line.
7. The design water level is equal to the calculated water level + backwater level + swash height of wave + 0.5 m.

3.5 Protection against collapse, falling stone, crushed debris of the side slope and dangerous rock

3.5.1 Principle of treatment

Collapse, falling stone, rock weathering and spalling along the existing highway have developed, and affected significantly the normal operation of the highway, endangering the safety of vehicles and persons. Since most altered sections of the highway are the old sections subject to widening treatment, collapse, falling stone, rock weathering and spalling still threaten the normal transportation of the highway.

The field location survey has investigated comprehensively in detail the dangerous rock, collapse, falling stones, rock weathering and spalling, debris on the slope face of the contracted section that may affect the construction and operation of the highway. In view of the dangerous rock, collapse, falling stones, rock weathering and spalling, debris on the slope face along the highway have developed, and the random damage is high. The design is specific to protect the dangerous rock, collapse, falling stones and debris on the slope face that are high in probability and serious in damage. Meanwhile, the highway maintenance authority shall clean away the collapsed rock, falling stones and debris deposit behind the gabion cage retaining the wall, cut slope wall and other protection works in time.

3.5.2 Selection of protective measures

The subgrade and side slope in sections with dangerous rock, collapse, falling stone, rock weathering and spalling shall be designed in accordance with the principle of prevention combined with protection. For the section where there is no direct hazard, prevention treatment, i.e. cleaning-up measures, shall be adopted. For the section where there is direct hazard to the safety of the highway, protection treatment, i.e. cleaning-up, SNS-active safety net system, passive safety net system, cut slope wall, stone cage retaining wall, and other reinforcing and blocking measures, shall be adopted.

Collapse and falling stones mainly develop in steep phyllite side slope and slope zone, where the slope is generally steep, the overburden is thin, the vegetation does not develop, and the dangerous rocks and collapsed rocks roll into the highway.

Prevention and protection measures, such as cleaning-up, SNS active safety net system, passive safety net system, cut slope wall, and stone cage retaining wall, can be adopted in accordance with the mode and scale of dangerous rock as well as its effect on the project.

In the zone where the highway passes through, dangerous rocks and rolling stones shall be prevented by means of stabilizing, cleaning-up and blocking. In the zone where falling and rolling stones distribute and dangerous rocks develop, the SNS safety net system shall be set against the direction of the affected highway, and rolling stones and dangerous rocks that are liable to loss of stability and falling shall be cleaned up. For the dangerous rocks and rolling stones that are not easy to be cleaned up, they shall be stabilized by means of inlaying, GPS2 SNS active protection system and anchorage.

Cleaning-up is appropriate if the dangerous rocks are not large in scale, and their mass has separated from the matrix, and its cleaning-up does not result in secondary disaster.

Cut slope wall and stone cage retaining wall are appropriate to the dangerous rock zone where the slope is less than 25–35°, and there is plain surface of some width. These measures can be used if the dangerous rocks are small in scale. The condition to build cut slope wall and stone cage retaining wall is provided under the dangerous rocks.

Active flexible protection net is appropriate to the zone where the slope is steep, the height of dangerous rocks is less than 30 m, and falling stones occur frequently. There is no buffer space under the rocks so that the stones fall into the highway.

Passive flexible protection net is appropriate to the zone where the slope ratio is gentle. There is some buffer space, and the energy level of the falling stones is lower than a criterion. The zone where the slope is steep and the falling stones shall be intercepted in restricted space in the highway.

Rock weathering and spalling mainly develop in the rock zone of the steep slope. The exposed rock mass is weathered in air into debris, which will change into sliding and slippery debris flow during rainfall to occupy the highway and encumber its normal operation. In these zones, interception of accumulation bodies is the main treatment method, i.e. cut slope wall or stone cage retaining wall is set on the internal side of the highway. The wall top is equipped with the passive protection net, combined with the flexible passive protection net set on the slope face, so that the influence of spalling rock–soil mass on the highway can be mitigated or abated.

4 CONCLUSION

Taking a highway project as the survey field and its general condition as the basis, this paper explains the design principle of the side slope of fill subgrade, through-cut subgrade and riverside subgrade and protection measures of a through-cut side slope and dangerous rocks respectively. It indicates that the side slopes of different cross profiles of the subgrade shall be designed based on multiple factors. For example, the determination of the slope ratio of the side slope of fill subgrade depends on the type of filling material of the subgrade, height of the side slope and the engineering geological conditions of the base. The side slope of through-cut subgrade shall be designed by taking into consideration the natural slope of the mountain mass, lithology, relationship between the tectonic fissure attitude and route, weathering degree, mechanical properties and excavation height of the rock mass, as well as environmental protection, greening, soil-rock balance, etc.

For the protection measures of side slopes in the zone where loose accumulation bodies distribute, this paper also elaborates the rock side slope, soil side slope of the alluvium and diluvium layers, and soil side slope of the eluvial layer, respectively. It also suggests that the subgrade and side slope in the zone where there are dangerous rocks, collapsed rocks, falling stones, or weathered and spalling rocks shall be designed in accordance with the principle of protection combined with prevention. Subgrade and side slope in the zone where the highway is not endangered directly shall be designed in accordance with the principle of prevention, while those in the zone where the highway is endangered directly shall be designed in accordance with the principle of protection.

ACKNOWLEDGEMENTS

This study was supported by the open funds of the National and Local Joint Engineering Laboratory of Traffic Civil Engineering Materials of Chongqing Jiaotong University, the Chongqing Key Laboratory of Mountainous Road Structure and Material, and the Hi-tech Laboratory for Mountain Road Construction and Maintenance.

REFERENCES

Jianjun Zhao, Nengpan Ju, Guoxiang Tu. Deformation Mechanism and Supporting Measures of Manmade Slopes in Loose Accumulation Body. J. Journal of Engineering Geology. 05 (2008).

Wenxing Jian, Kunlong Yin, Yang Wang, Lei Zheng, Linlin Yao. Mechanism Analysis and Stability Assessment of Xixipu Loose Accumulation Body in Wanzhou. J. Geol Sci Technol Inf. S1 (2005).

Xiumei Ding. A Study on the Deformation and Stability of Typical Debris & Embankment Slope with Complicated Environment in Southwestern China. D. Journal of Chengdu University of Technology. (2005).

Xuetang Yang, Fei Wang. Evaluation Method of Slope Stability and Its Developing Trend. J. Geotechnical Engineering Technique. 02 (2004).

Xuezhou Yang. Analysis of Deformation Mechanism of Sedimentary Slope after Excavation and Support Measure. D. Journal of Chengdu University of Technology. (2008).

Electromechanical Control Technology and Transportation – Jia & Wu (Eds)
© 2017 Taylor & Francis Group, London, ISBN 978-1-138-06752-3

Traffic investigation of urban intersections

Chunxia Zhu & Tiying Wang
Mechanic and Electrical Engineering College, Hainan University, Hainan Haikou, China

ABSTRACT: Traffic investigation is the basis of urban planning, urban road system planning, and urban road design work. To understand the traffic situation of a city, the most direct way is to analyze the traffic volume and traffic situation of urban road intersections. More urban intersections coupled with the unreasonable signal timing, cause vehicles to be disturbed frequently, the delay time will be long and the intersection will be a low pass. The paper analyzed the peak hour traffic of Haikou City's Renmin Road and Haidian Wuxi Road intersection. It found that Haikou City's traffic is congested and unstandardized; people do not obey traffic rules, and traffic planning is unreasonable. The paper puts forward corresponding control measures, the traffic organization optimization, signal control, intersection channelization, popularization of traffic regulations, and several aspects of congestion control. In order to improve traffic efficiency and prevent traffic accidents, this paper provides a reference for traffic accidents.

1 INTRODUCTION

The rapid development of the economy brings an imbalance of supply and demand to city traffic. The problem of urban road traffic congestion caused by rapid urban development is a problem that all the major cities in the world are facing. Urban transport being the main supporting measure of urban operation and development, its function is irreplaceable. As the first international tourism island issued by China, Hainan has resulted in the rapid development and brought the pressure of people and traffic to the capital city of Haikou. As a result, the development of urban traffic cannot keep up with the economic growth, while the continuous expansion of urban scale is brought about by the continuous increase in traffic demand, but the previous road construction of Haikou city has been unable to meet the needs of the moment, greatly affecting the impression of friends at home and abroad in Hainan. Therefore, urban traffic congestion is one of the major problems facing the development of Haikou city.

The intersection is the bottleneck in an urban road network restricting the road capacity at the throat. The intersection traffic has the characteristics of traffic flow, high traffic density, low traffic speed, conflict point and interweaving point, mutual interference of different types of traffic flow, heterogeneity of traffic volume in time, and so on. Therefore, reasonable and effective management of intersection traffic problem is the key to solving the problem of urban traffic congestion. A single-point intersection traffic signal control accounts for 90% of the urban intersection, so the study of traffic volume at the intersection is of great significance.

2 INTERSECTION TRAFFIC SURVEY ANALYSIS

2.1 *Study of the present situation of intersections*

In this study, the intersection of Renmin Road and Haidian Wuxi Road is the representative intersection of Haikou city. It is a large crossroad with large traffic volume and large turning vehicles. Renmin Road is to the south of Baishamen Park, north of Haidian Island. Haidian Wuxi Road is on the west of the Century Bridge, east of the Haixin Bridge, and circulates the Xinfu Bridge. The traffic capacity of the intersection has a greater impact on the overall traffic service level of Haidian Island.

The intersection was on the basis of the original channel through the transformation and addition of the right lane, but the actual road conditions after the transformation are not ideal. Due to the large traffic volume during peak hours, under the existing signal control, the vehicle delay time is large, and the intersection is easy to produce a congestion phenomenon.

2.2 *Geometric characteristics of intersections*

Renmin Road and Haidian Wuxi Road intersection diagram was shown as follows.

East import (Haidian Wuxi Road): four-lane (3 straight + 1 left) + right-lane;

South entrance (Haidian Wuxi Road): five-lanes (2 straight + 3 left) + right-lane;

North entrance (Renmin Road): five-lanes (3 straight + 2 left) + right-lane;

West entrance (People's Road): five-lane (3 straight + 1 left) + right-lane.

The signal level of the intersection is 4 phase control. Yellow lights are three seconds.

2.3 Intersection traffic survey

1. Suvey time
The survey was conducted during normal morning peak hours (7:30–8:30) in good weather conditions.
2. Survey locations
Renmin Road and Haidian Wuxi Road intersection of all sections.
3. Survey site
Renmin Road and Haidian Wuxi Road intersection of each section.
4. Survey methods
Using artificial counting method, the number of vehicles in each entrance road at the crossroads was investigated and a continuous investigation was conducted for one hour without interruption. The number of imported lanes, lane width, and signal timing was also investigated.
5. Investigation subgroups
There were four groups and each group was made of 10 people who were at the entrance to the division of labor count. At the same time, two students recorded traffic light cycle.

3 TRAFFIC SURVEY RESULTS

3.1 Vehicle classification

Motor vehicles: taxis, cars, public transport, buses, trucks. Non-motor vehicles: electric bicycles.
Standard models such as the conversion table in Table 1.

3.2 Analysis of the proportion of each model

In a total of 5160 vehicles that continuously passed in an hour, the data of each model was in Table 2.

Table 1. Standard model conversion coefficient.

Type of vehicle	Small car	Taxi	Truck	Bus	Electric bicycle
Conversion Coefficient	1.0	1.0	1.5	1.5	1.0

Table 2. Statistical analysis of the data of each vehicle type.

Type of vehicle	Small car	Bus	Truck	Taxi	Bus	Statistics
Data	4446	451	44	65	154	5160

Table 2 shows that cars occupy a large proportion (86%) in motor vehicles. This shows that the performance of people's living standards has improved. Taxi also occupies a certain proportion. The proportion of buses is very small which indicates that people have to change the way out to the private car. Excessive private cars can cause road congestion.

3.3 Analysis of traffic flow characteristics

Traffic conditions include traffic volume, traffic flow composition, and the proportion of various types of vehicles. The composition of traffic flow at the entrance of Renmin Road and Haidian Wuxi Road and the conversion coefficients of the number of vehicles with different types of vehicles are shown in Table 3 and Table 4, respectively.

From Table 3 and Table 4, it can be seen that the maximum traffic direction is the west, which may be due to the fact that the west entrance road received the vehicles coming down from the Century Bridge. In addition, a part of the U-turn vehicles' traffic increased. The traffic direction of the minimum traffic flow is the north intersection, the traffic flow of the south side and the east side is also maintained at a high level, the east-west traffic volume is larger, the north-south traffic volume is smaller, but the south-side is slightly higher than the north side; south-north side of the junction is one-way six lanes, the east side of the junction is one-way five lanes, the west side of the junction is one-way six lanes. Therefore, the amount of traffic on each roadside is directly related to the road width and geographical position. At the same time, it can be seen that the north-south trunk road traffic volume is not large, but the number of lanes is large.

Table 3. Equivalent number of imports after conversion.

Direction	East	South	West	North
Large car (pcu)	72	135	133.5	55.5
Small car (pcu)	1245	1102	1625	974

Table 4. Statistical data of various models.

Intersection	Turn left	Straight	Turn right	Turn around
East	153	893	271	0
South	841.5	280	115.5	0
West	171.5	501	749.5	286
North	442	413.5	174	0

4 CONCLUSIONS AND RECOMMENDATIONS

4.1 *Renmin Road and Haidian Wuxi Road intersection in traffic management and control problems*

1. The development of public transport in Haikou is relatively backward.

The existing bus routes are 92 accounting for 68% of the city; buses are 1515 accounting for 59.8% of the city, but the bus traffic is only 18%. At the crossroads of Renmin Road and Haidian Wuxi Road, the car is the main vehicle for traffic flow, which maybe one of the reasons for the traffic congestion at the intersection.

2. The existing traffic distribution is unreasonable.

In the early and late peak hours of this section, the second queuing phenomenon often occurs, and the queue length is longer, and sometimes even five minutes. Traffic congestion is serious, which has a great relation to the existing traffic lights control. Unreasonable traffic phase timing and lane transformation have an impact too.

3. Electric bicycle's interference.

A number of electric bicycles did not abide by the traffic rules and disrupted the traffic order. The phenomenon which has a greater impact on the road capacity should be regulated.

4. Sidewalk planning and design are unreasonable.

The pedestrian crossing is too long. The pedestrian crossing waiting area is lacking. Green light time is too short. Pedestrians crossing the road are at danger. Traffic light time should be adjusted to ensure pedestrians' safety.

5. Intersections lack the necessary canalization.

The intersection area is too large. Lack of the necessary channelization inevitably leads to increased vehicle conflict. Drivers and pedestrians are distracted and that is prone to traffic accidents.

4.2 *Suggestions and countermeasures*

In view of the above traffic management and control problems, the following countermeasures and solutions are proposed.

1. Strengthen road safety publicity and education.

Traffic control departments should strengthen the public's traffic knowledge and traffic laws and regulations. In peak hours, the traffic management person should be strengthened. The relevant management system should be improved. The electric bicycle should be prohibited and considered illegal driving.

2. Strengthen the public transport management.

The government should use more powerful administrative interventions to guide travelers to take active part in public transport, such as drawing up a more rational roadmap and timetable, improving the public transport capacity and service quality, and optimizing the operation mode of public transport.

3. Optimize intersection signal timing.

The signal timing of the intersection should be optimized and the traffic engineering design should be carried out. The traffic route of the vehicles should be regulated by traffic engineering facilities such as the marking line and the isolation railing. The traffic order and service level of the intersection should be improved.

4. Set up pedestrian safety facilities.

To ensure the safety of pedestrians crossing the street, security islands, safety barriers, and other safety facilities should be set up. According to relevant regulations, when the length of the crosswalk is more than 16 m, a pedestrian crossing island should be set up in the center of the pedestrian crossing. The crossing length of the pedestrian crossings are 37 m and 27 m respectively and there is no safety island, resulting in pedestrian crossing difficulties. Combined with traffic diversion island layout and signal timing control, the road zebra crossing should be set up in the appropriate location.

5. Increase the traffic police force investment.

The traffic policemen should assist the traffic system in the management of traffic flow and the order management of traffic sections. During the peak hours, the police should manually change the traffic signal according to the actual traffic conditions at the intersection.

ACKNOWLEDGMENT

This work was supported by the Natural Science Foundation of Hainan Province (20155212).

REFERENCES

Chen Yu-Feng, Mamtimin Geni, Ezatgul AKUP, etc. Mathematical Modeling and Numerical Simulation for Lntersection Traffic Flow. *J. Machinery Design & Manufacture. 2012*, (10): 256–258.

Cui Shi-kui, Gao Xue-feng, BI Xin, etc. Application of ARM and linux in traffic flow detecting. *J. Machinery Design & Manufacture. 2012*, (10): 256–258.

Liu Fei, Huang Qing-long, Mao Wei. Thinking about the planning and design of new road intersection in Sichuan city. *J. Chengdu architecture. 2015*, (3): 28–30.

Lu Ming-yu, Wang Xing. Analysis and Countermeasures of traffic congestion at intersections. *J. Journal of Changchun University of Technology. 2015*, 36 (3): 327.

Peng Fei. Optimization and Simulation Research of Signal Timing at Urban Plane Junction. *D. Beijing Jiao tong Unversity, 2012*.

People's Republic of China Ministry of Construction. GB50220–1995 Urban Road Traffic Rules Design Specification. *S. China National Standard Press, 1995*.

Wang Qiu-ping, Tan Xue-long, Zhang Shengrui. Signal Timing Optimization of Urban Single—point Intersections. *J. Journal of Traffic and Transportation Engineering,2006*, (2): 60–64.

Wu Hai-yan, Zhang Rui, Dai Ji-feng. The Overall Evaluation of Intersection. *J. Journal of Beijing Institute of Civil Engineering and Architecture. 2001*, (3):50–55.

Xu An-cheng. Improvement and optimization of urban branch function in Hangzhou. D. *Zhejiang University, 2012*. LI Yong. Study on the road traffic congestion in Guangzhou. D. *South China University of Technology, 2013*.

An experimental study on the fire extinguishing characteristics of water mist and compressed air foam in a railway tunnel rescue station

Kai Yao & Wanfu Liu
School of Mechanical Engineering, Tianjin University of Commerce, Tianjin, China

Weiping Han & Jianjun Xia
Tianjin Fire Research Institute of MPS, Tianjin, China

Ya Liu & Shengyuan Zhong
School of Mechanical Engineering, Tianjin University of Commerce, Tianjin, China

ABSTRACT: At present, research on the combination of the fixed water mist fire extinguishing system and the fixed compressed air foam system used in railway tunnel rescue stations is still at the discussion stage. In this paper, the fire extinguishing characteristics of water mist and compressed air foam were studied experimentally, and the fire extinguishing effect of water mist and compressed air foam was determined under certain fire conditions. The development of fire and the temperature field of the two kinds of fire extinguishing systems were analyzed. This study can provide a reference for the fire prevention plan in railway tunnel rescue stations.

1   INTRODUCTION

In recent years, with the rapid development of railways in China, the number of railway tunnels is increasing. A railway tunnel is a kind of structure that is built under the ground or under water and is used for laying the tracks. Because of the complex terrain and vast mountainous areas in China, a large number of tunnels are distributed across China's railway network. By the end of 2014, China had in all 11,516 tunnels, out of which 742 tunnels had a total length of over 3,000 m. Currently, the longest railway tunnel in China is the new Guanjiao tunnel of Qinghai-Tibet line, with a length of 32,690 m. Thus, the total number of railway tunnel constructions in China has been far ahead of the rest of the world.

Countries around the world set up emergency rescue station for 20 km or more of railway tunnels and tunnel group. As an important part of superlong railway tunnel, railway rescue station takes on the function of firefighting and evacuation. As a kind of safety and environmental protection agent, water mist causes no damage or pollution to the protected object and area personnel, and can be widely used in all situations.

Compressed air foam has high fire extinguishing efficiency, can minimize water consumption, and is suitable for the large-scale use of new fire extinguishing technology. The combination of these two kinds of fire extinguishing technology will be of great significance for the rescue operation.

2   WATER MIST FIRE EXTINGUISHING SYSTEM

The experiment uses the fixed high-pressure water mist fire extinguishing system developed by this project. The system is mainly composed of a centrifugal pump, a high-pressure plunger pump, a high-pressure water mist sprayer, a regulating and control system, related measuring equipment, and so on. The maximum pressure of the pump can be adjusted to 12 MPa. The maximum flow rate of the pump can be adjusted to 220 L/min. Water mist nozzle spacing is 3 m and the height is 4.5 m, according to the technical specification for the water mist fire extinguishing system. Because the top of the tunnel is provided with an electric wire, the installation method of the sprinkler head is lateral installation (see Figure 1).

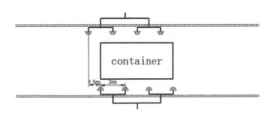

Figure 1.

3 COMPRESSED AIR FOAM FIRE EXTINGUISHING SYSTEM

Compressed air foam fire extinguishing system is mainly composed of a centrifugal pump, a foam ratio mixer, a foam generator, a filter, a control valve, and related measuring equipment. The maximum flow rate of the pump can be adjusted to 700 L/min. This model can be applied in the pre-project development of the compressed air foam pipe. The foam spray tube is arranged only on one side of the tunnel.

4 TEST SCENE AND FIRE LOCATION

4.1 *Test scene*

The test was performed in a tunnel with a length of 30 m, a width of 6 m, and a height of 6 m. It is similar to a carriage of length 6 m, width 3 m, and height 2.5 m. The carriage is a freight car, the top of which is the opening state, and is placed at the center of the tunnel (see Figure 2). The t2 ultra-fast fire refers to a category of train freight fire. The total known truck fire heat release rate is very high, so the standard combustion (built-in plastic cups) is adopted for the test.

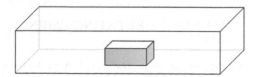

Figure 2.

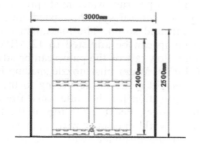

Figure 3.

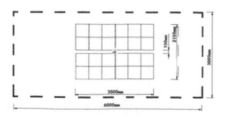

Figure 4.

4.2 *Fire location*

Ignition is located in the compartment near the ground and in the middle of four standard combustions. In addition, there are about 20 cm wooden cushions at the bottom of the standard combustion. This can indirectly simulate the elevation of the wheel in the emergency rescue station to a certain extent. The test with the ignition source was carried out according to the GB/T 26785-2011 general technical specifications for water mist fire extinguishing systems and components, in which a cotton rod was dipped in 120 mL n-heptane and then used for ignition, with the cotton rod diameter being 75 mm and the length being 75 mm. Ignition position is shown in Figure 3, and Figure 4 shows the way to ignition.

5 EXPERIMENTAL PHENOMENA

Table 1 presents the experimental phenomena at different time periods.

Table 1.

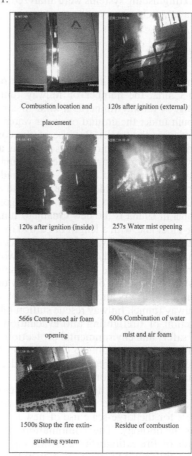

6 TEST RESULT ANALYSIS

6.1 *Analysis of heat release rate*

After the ignition, the standard combustion chamber is ignited and the measuring device begins to collect a large amount of smoke. The heat release rate starts to rise rapidly. According to the real-time display of the control side, when the heat release rate exceeds 10 MW, the water mist fire extinguishing system is turned on. The start time of water mist spray is 257 s, the start time of the air bubble is 566 s, the time of disappearance of fire is 1100 s, the stop time of water mist is 1221 s, and the close time of air foam is 1447 s. The heat release rate curve from the beginning until the end is shown in Figure 5.

As the ignition source is small, and the early stage is only the standard combustion material box, the fire heat release rate is small. Therefore, the temperature, humidity, water content, and other factors have a greater impact on the combustion process.

When the water mist began to be released, the flame was significantly suppressed, and the heat release rate began to decline rapidly. After a period of time, the fire has a tendency to rise, and when the compressed air foam is opened, the heat release rate decreases steadily. The heat release rate decreases to near zero, and the flame disappears, so the heat release rate decreases to 0 in a short time.

6.2 *Results of radiation heat flux meter*

Figure 6 shows the variation of the flame radiation. The figure shows that the box starts to burn after a short time. Then, the radiation heat increases significantly. At this time, the fire rises rapidly. When the time reaches 257 s, the radiant heat reaches a maximum value of 8.59 kw/m². At this point, as the water mist is turned on, the heat radiation begins to decline rapidly. At 566 s, when the compressed air foam is opened, the radiant heat declines rapidly and then gradually decreases to 0.

When the water mist is opened, it cools the combustion object, and the flame is produced. As a result, the radiation heat decreases. However, the flame gradually increases after a period of time, so the radiation heat gradually increases. Compared with the heat release rate measured by the calorimeter, the heat release rate measured by the heat flux meter fluctuates near 0 after 600 s. The standard upper combustion is burnt. The flame position drops to the compartment model. The heat flux meter is far away from the compartment model. The radiation is blocked by the body and thus can be changed, as shown in the figure.

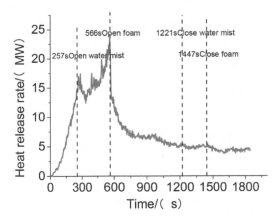

Figure 5.

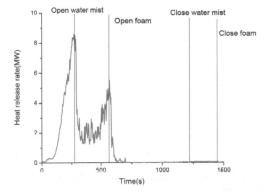

Figure 6.

6.3 *Temperature field analysis*

According to the measurement results, the temperature of the top of the tunnel and the temperature of each section change with time, as shown in Figure 7. As can be seen from the following chart, the overall trend of the temperature of the tunnel cross-section is consistent. The temperature rises rapidly in the main stage of standard combustion after being ignited during the combustion process. Then, there is a period of rapid decay. The fast decay stage is the main process of the initial control of the fire after opening the high-pressure water mist. During this stage, the fire rises rapidly. The main reason for this is that the intensity of the fire inside the carriage is larger, and the water mist in the fire process cannot reach the depths of the burning. As the internal fire continues to burn, the fire spreads again. Finally, for the rapid decay phase, the compressed air bubble is turned on and coupled with water mist. This leads to the isolation of the combustion surface and air, and successfully suppresses the fire. The ambient temperature

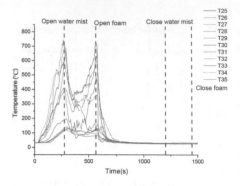

Figure 7. Temperature distribution on the central section of the tunnel.

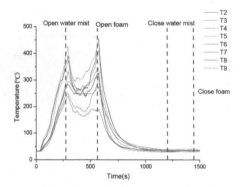

Figure 8. Temperature distribution on the north side of the tunnel.

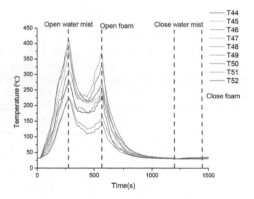

Figure 9. Temperature distribution on the south side of the tunnel.

decreases rapidly and finally decreases to ambient temperature.

In this experiment, before the application of high-pressure water mist, the temperature reaches up to 700°C. And the top of the south and north of the highest temperature reaches 400°C. When the

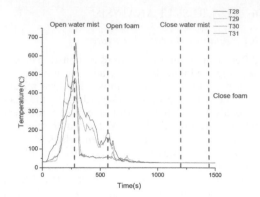

Figure 10. Temperature distribution on the south side of the tunnel.

high-pressure water mist is applied, the temperature inside the tunnel enters the fast decay stage. The temperature of the fire source decreases to 400°C, while the temperature on the south side and the north side reaches a maximum of 250°C. When the fire spreads again, the tunnel fire temperature increases. Before the application of compressed air foam, the top temperature of the fire source reaches 700°C, while the top temperature of the south side and the north side reaches a maximum of 400°C. After the application of compressed air foam, the temperature of the center of the tunnel, the central section, the top of the south side, and the southern section enters into the rapid decay phase. The temperature of the north side of the top and the north section continues to rise for a certain period of time before it begins to fall into the rapid decay phase. This is due to the application of compressed air foam from the south side to the north side of the tunnel, and the south side of the foam is relatively more accumulated. Therefore, the declining trend of the temperature is obvious. At this time, the north side of the bubble is relatively small, and only through the control of the fire, the surrounding temperature can be reduced. So, the temperature first increases and then decreases. After 500 s, the fire is extinguished and completely controlled. Then, high-pressure water mist and compressed air foam are closed, and the temperature of the tunnel is reduced to ambient temperature.

The top temperature of the carriage is shown in Figures 11 and 12. Before the application of water mist, the temperature of the top of the car reaches 500°C, and the maximum temperature reaches 700°C 1m from the top of the carriage. This is because the measuring points are arranged above the combustion object. Once again, the temperature rises before the opening of the compressed air bubble: the highest temperature reaches 700°C on the top of the car and 400°C 1m from the top of the car. Then, the temperature decreases rapidly,

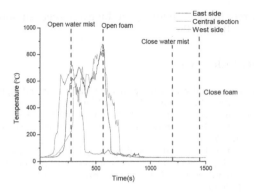

Figure 11. Temperature distribution on the top of the compartment.

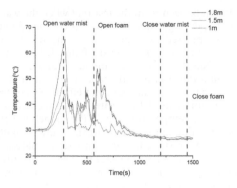

Figure 13. Temperature distribution of 0.5 m on the north side of the compartment.

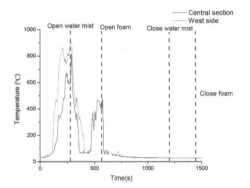

Figure 12. Temperature distribution of 1 m from the top of the carriage.

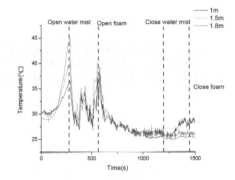

Figure 14. Temperature distribution of 2 m on the east side of the compartment.

7 CONCLUSION

followed by the temperature of the fire being in a steady state at 250 s.

Taking into account that the firefighters can be close to the fire source, the measuring point arranged in the north side of the carriage is 0.5 m and in the eastern side of the carriage is 2 m. Before opening the compressed air foam, the maximum temperature reaches 65°C and 45°C, respectively. After opening the water mist, the temperature of the north side decreases to about 40°C at 20 s, and the temperature of the east side decreases to about 20°C at 20 s. Before the opening of the compressed air foam, the temperature reaches to a maximum of 50°C on the north side and to a maximum of 40°C on the eastern side. After the compressed air foam is opened, the temperature of each measuring point is reduced to ambient temperature at 500 s. The combination of high-pressure water mist and compressed air foam can effectively reduce the temperature around the scene of the fire. The firefighters can then come around the fire location to carry out their operation.

The fire test results showed that before 237 s, the heat release rate could reach up to 10 MW, and then could exceed this rate. Before opening the fire extinguishing system, the maximum heat release rate of the fire was 18 MW. After opening the water mist fire extinguishing system, the heat release rate decreased gradually. However, after a period of time, the heat release rate gradually rose to 23 MW. At the same time, when the compressed air foam system was opened, the size of the fire was expected to set.

Through the experiment, it can be concluded that water mist can control the flame for a certain period of time. Thereafter, the intensity of the fire needs to be controlled by combining water mist with compressed air foam. At 20 minutes after the end of the air bubble, the flame disappeared. The fire was suppressed in 10 minutes without resurgence of the phenomenon. After the test, the statistics burnt out standard combustion box is about 48; the top two layers disappeared. Due to the submerged protection of the foam, it concerned only a small portion of the two layers of the standard combustion.

It can be seen from the test that when the water mist fire extinguishing system was combined with the compressed air foam system, the water mist had a defoaming effect on the foam. It had a certain influence on foam accumulation, and the overall condition was not affected. The combination of water mist and air foam can effectively control the development of fire. Water mist can effectively reduce the temperature inside the tunnel and the heat radiation in the vicinity of the source. Finally, there is no resurgence of combustible materials. Therefore, although the water mist has some influence on the foam, the combination of water mist and foam can effectively extinguish the fire under certain conditions, and also has a good effect on cooling the surrounding environment.

ACKNOWLEDGMENTS

This work was supported by the Study on application of firefighting technology in railway tunnel 2014BAK17B02 and the Study on application of firefighting technology in tunnel rescue station 2014BAK17B03.

REFERENCES

Experimental study on the effect of water mist and high temperature foam in the oil tank fire [J].

Gardiner B S, Dlugogorski B Z, Jameson G J. Rheology of fire fighting foams. Fire Safety J, 1998, 31(1): 61–75.

Lattimer B Y, Hanauska C P, Scheffey J L, et al. Behavior of Aqueous Film Forming Foams (AFFF) exposed to radiant heating. NRL Rpt Ser 6180, Naval Research Laboratory. 1999.

Mao W.Z. Fire prevention and rescue of railway tunnel [J]. tunnel construction, 2010, 30(1): 20–23.

Routley J G. Compressed air foam for structural fire fighting: A field test. TR074, Boston Fire Department, 1993.

Statistics of Transportation Bureau of China Railway Corporation.

Weaire D, Phelan R. The physics of foam. Phys Condens Matter, 1996, 8: 9519–9524.

A numerical study on heat release rate in a multichannel parallel test system for smoke spillover from the tunnel rescue station

Ya Liu & Wanfu Liu
School of Mechanical Engineering, Tianjin University of Commerce, Tianjin, China

Weiping Han & Jianjun Xia
Tianjin Fire Research Institute of MPS, Tianjin, China

Kai Yao & Shengyuan Zhong
School of Mechanical Engineering, Tianjin University of Commerce, Tianjin, China

ABSTRACT: In this study, a multichannel parallel heat release rate test system for the railway tunnel rescue station was established by using FDS (Fire Dynamics Simulator). The influence of smoke spillover on the measurement results was studied, and the feasibility of the heat release rate in the multichannel parallel testing system was proved. The results showed that the heat release rate of simulation is smaller than the actual fuel combustion measured by the oxygen consumption method because of smoke spillover. The greater the amount of leakage, the later the peak time will be reached. Compared with the actual combustion process of the fuel, the oxygen consumption method will cause the lag of the measurement results. The larger the amount of smoke leakage, the more obvious the lag of the measurement results will be.

1 INTRODUCTION

The fire dynamics simulation software FDS is a kind of fire driven by the computational fluid dynamics (CFD) model. It is calculated by using the N-S equations, which are suitable for the calculation of low-velocity heat flow, especially in the event of fire. The core algorithm aims to establish a clear predictor corrector system to ensure the accuracy in terms and space.

LES (Large Eddy Simulation) is used in FDS to deal with the problem of turbulence. FDS software, which has solved a large number of fire engineering problems, shows the results of the fire dynamics simulation by using the visualization program Smokeview.

2 ESTABLISHMENT OF A MODEL OF MULTICHANNEL PARALLEL HEAT RELEASE RATE TEST SYSTEM

2.1 Tables

The heat release rate is measured by oxygen combustible method. It is necessary to measure the flow rate of flue gas and the concentration of different substances in the flue gas. In order to measure accurately, all flue gas generated by burning should be collected through the exhaust system. According to the formula of calculating the heat release rate, the range of the heat release rate testing system is related to the flow of flue gas and the range of flue gas testing system. Therefore, the use of a multiple parallel system to collect flue gas can improve the range of the test system, which is convenient for large-scale measurement of combustion.

Based on the investigation and the relevant standards of the existing railway tunnel rescue station, a physical model of the full-size railway tunnel rescue station is established. The length of the test bench is 30 m, and the section size is 6 m × 6 m. In order to enhance the effect of flue gas collection, the smoke blocking screens are set at both ends of the bench. The bench of the railway tunnel rescue station is shown in Figure 1. Taking into account

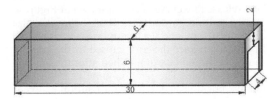

Figure 1. The bench of the railway tunnel rescue station (unit: m).

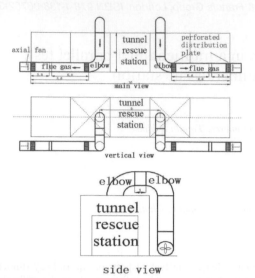

Figure 2. The schematic diagram of the smoke exhaust pipeline (unit: m).

the actual size of the bench and the ability of the smoke exhaust system, the design of the test bench range is 0–12MW and the exhaust pipe diameter is 1.4 m. The schematic diagram of the smoke exhaust pipeline is shown in Figure 2.

2.2 The establishment of the calculation model

According to the actual size of the multi-channel parallel heat release rate of the railway tunnel rescue station, a simulation model, as shown in Figure 3, is established. The wall material is fire brick (FIRE BRICK). FDS software cannot establish the circular curve module, so the exhaust pipe section is transformed into approximate square cross-sections. The length of the side is 1.24 m and the equivalent diameter is 1.4 m, which is the same as the actual exhaust pipe. The exhaust vent is arranged at the outlet of the exhaust pipe. Different smoke exhaust effects were simulated by setting different wind vents.

In the simulation process, a combustible foam is used, and it is lighted by burner. In order to simulate the range of the experimental platforms, the size of combustible is 3 m * 1.4 m * 0.2 m. The heat release rate curve of fuel is shown in Figure 4.

The ignition parameter is 1000 kw/m², the size is 0.4 m * 0.2 m, and the heat release rate is 80 kW. In order to reduce the error caused by the ignition source, it should be extinguished after the ignition of combustible. The ignition time is 100 s. The relative position of the fuel and ignition source is shown in Figure 5. The fuel is located at the center of the combustible model and the ignition source is located below the edge of combustible.

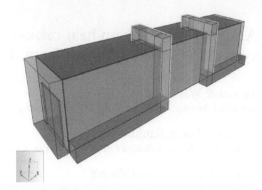

Figure 3. The simulation model of the experimental system.

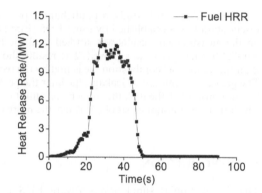

Figure 4. Heat release rate curve of fuel.

In addition, it is also an important step in the simulation process of FDS to divide the simulation computing area into the computational grid (meshes), which directly affects the accuracy of the calculation results. The small size of the grid can improve the accuracy, but it will slow down the simulation speed and increase the calculation time. The large grid size can improve the simulation speed, but it will reduce the accuracy of the simulation. Therefore, considering the factors such as model size, computer configuration and the accuracy of calculation results, the mesh size in the numerical simulation is set as 0.2 m * 0.2 m * 0.2 m.

At the same time, in order to reduce the number of grid and fully utilize the grid, the size of the mod-el is divided into seven blocks, and finally the mesh number is 155872, as shown in Figure 6.

In order to obtain the parameters of temperature distribution, gas flow and O_2 concentration in the model, the temperature slices, the velocity slices, the flow measuring points, the temperature measuring points, and the concentration measuring points are added in the model. These parameters

Figure 5.　Fuel and ignition source.

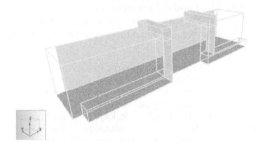

Figure 6.　The mesh generation.

are used to study the heat release rate. The results of the oxygen consumption method are compared with the actual heat release rate of combustible material. Those differences between the calculated oxygen consumption and the actual heat release rate under different conditions are analyzed.

3　ANALYSIS OF SIMULATION RESULTS

In the actual experiments, smoke overflow will occur due to the small wind speed and the big fuel heat release rate, which exceeds the experimental bench scale range. The flue gas cannot be completely collected and thus may affect the test results of the heat release rate of the oxygen consumption method. Meanwhile, it is difficult to study by physical experiments. Therefore, the influence of flue gas spillover is studied by the numerical simulation.

The flue gas leakage is related to the heat release rate of the fuel and the smoke extraction volume of the platform. The speed of the exhaust duct is changed in order to change the status of the flue gas overflow. The smaller the wind velocity in the exhaust flue, the more the flue gas will overflow into the experimental system. The analog computation scheme is shown in Table 1. The flue gas flow of the overflow platform is obtained by the flow measuring device at the two sides of the experimental platform.

The flow rate of flue gas is shown in Figure 7. The volume of flue gas is obtained by integrating the flue gas volume flow curve, and the total volume of the flue gas produced by combustible combustion is 2011.255 m³.

The experimental platform of smoke spillover is shown in Figure 8.

The overflow discharge flue gas flow rate under different wind speed conditions is shown in Figure 9. It can be seen from the Figure 9 that the smaller the exhaust air velocity, the greater the smoke leakage flow rate. The curve of the volume of the overflow flue gas is obtained by integrating the curve in Figure 9. The overflow discharge flue gas flow rate under different wind speed conditions is shown in Figure 10.

Therefore, the overflow discharge flue gas volumes under different conditions are given in Table 2.

Table 1.　Design of simulation working conditions of the flue gas spillover.

Simulation number	Fuel size	Ignition source size	Exhaust pipe wind speed
C1	3 m × 1.4 m × 0.2 m	0.4 m × 0.2 m	5 m/s + 5 m/s
C2			10 m/s + 10 m/s
C3			15 m/s + 15 m/s
C4			20 m/s + 20 m/s

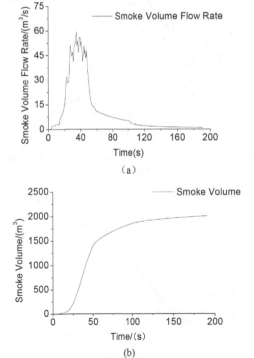

Figure 7.　Flue gas flow rate of fuel: (a) smoke volume flow rate; (b) smoke volume.

According to the results, the heat release rates under different smoke leakage quantity are shown in Figure 11. It can be seen that the greater the amount of smoke leakage, the smaller the peak value of the heat release rate when using the same combustible material. The greater the amount of leakage is, the later the peak time is reached. At the same time, the greater the smoke leakage is, the longer the burning time lasts.

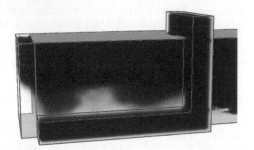

Figure 8. The experimental platform of smoke spillover.

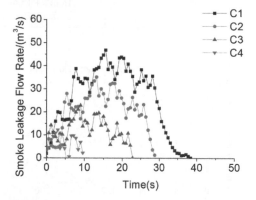

Figure 9. The overflow discharge flue gas flow rate under different wind speed conditions.

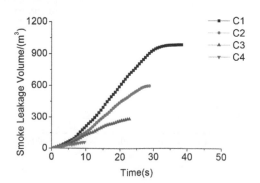

Figure 10. The overflow discharge flue gas flow rate under different wind speed conditions.

The fire loads under different smoke leakage quantities are shown in Figure 12. It can be seen from the figure that the greater the amount of smoke leakage, the slower the growth rate of fire load; however, there is no significant influence on the numeric value of the fire load.

The fire characteristic parameters under different smoke leakage quantities are given in Table 3.

The heat release rate and the growth time under different amounts of smoke leakage are shown in Figure 13(a) and (b) respectively.

Overall, the flue gas overflow system has effects on the heat release rate. First, the test results are

Table 2. The overflow discharge flue gas volume under different conditions.

Simulation number	Exhaust pipe wind speed/ $m \cdot s^{-1}$	Smoke leakage volume/m^3	Overflow flue gas volume fraction
C1	5	987.14	49.08%
C2	10	596.45	29.66%
C3	15	281.83	14.01%
C4	20	58.39	2.9%

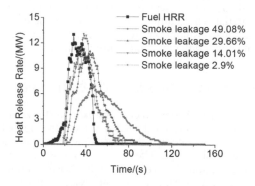

Figure 11. Different heat release rates correspond to the smoke leakage quantity.

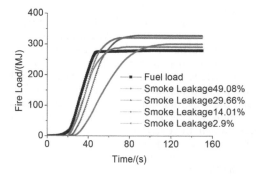

Figure 12. Different fire loads correspond to the smoke leakage quantity.

smaller than the actual combustion heat release rates of the fuel. The greater the amount of smoke leakage, the smaller the heat release rate. Second, compared with the actual combustion process of the fuel, the oxygen consumption test method will cause obvious delays in the measurement results to a certain extent. The smaller the wind speed of the flue gas, the greater the amount of smoke leakage and the greater the lag of the measurement results.

Table 3. Different fire characteristic parameters correspond to the smoke leakage quantity.

Smoke leakage/%	Peak heat release rate/MW	Growth period of time/s	Burning time/s	Rate of increase/ MW·s^{-1}
49.08	7.016	25.2	98.4	0.185
29.66	11.013	23.4	67.8	0.417
14.01	13.09	19.2	52.2	0.556
2.9	11.279	18.2	54.6	3.333

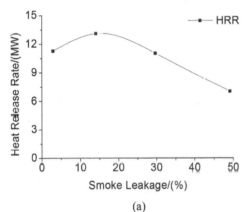

(a)

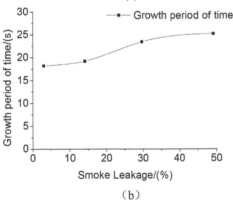

（b）

Figure 13. The relationship between leakage quantity of smoke and fire characteristic parameter: (a) relationship between the amount of smoke leakage and the peak of heat release rate; (b) relationship between the amount of smoke leakage and fire growth time.

4 CONCLUSION

In this paper, the model of a multiple parallel heat release rate test system of the railway tunnel rescue station was established by using the fire dynamics simulation software FDS. The effect of the smoke overflow is studied by changing the wind speed in the smoke exhaust. The results show that the fuel gas spillover will affect the heat release rate.

1. The greater the amount of smoke leakage, the smaller the peak value of the heat release rate when using the same combustible material. The greater the amount of leakage, the later the peak time is reached. At the same time, the greater the smoke leakage, the longer the burning time lasts.
2. Compared with the actual combustion process of the fuel, the oxygen consumption method will cause delays in the measurement results. The larger the amount of smoke leakage, the more obvious the lag of the measurement results will show. Therefore, the smoke exhaust system should be used to avoid the phenomenon of smoke leakage in the actual experimental process.

ACKNOWLEDGMENTS

This work was supported by the National Science Technology Support Program No. 2014BAK17B02 and the National Science Technology Support Program No. 2014BAK17B03.

REFERENCES

Hui Liang. Discussion on design of smoke prevention and control for subway fire [J]. Chinese Journal of Safety Science and Technology, 2010, 06: 119–122.

Kevin Mcgrattan. Fire dynamics simulator (Version 5) user's guide [M]. NIST Special Publication, 2001: 1-186.

Weicheng Fan, Qingan Wang. A brief course of fire science [M]. Hefei: University of Science & Technology China press, 1995.

Xueyi Jin, Wenying Chen, tunnel ventilation and tunnel aerodynamics [M]. Beijing: China Railway Press, 1983.

Yunlong Liu, Nathan White, et al. Water mist fire suppression of a train fire [C]. Fire Safety Sea Road Rail Conference, Melbourne, Australia, 2005.

Zhisheng Xu, Qing Zhou, Yu Xu. Operation experiment and numerical simulation of passenger train tunnel fire model [J]. Journal of the China Railway Society, 2004, 01: 124–128.

Electromechanical Control Technology and Transportation – Jia & Wu (Eds)
© 2017 Taylor & Francis Group, London, ISBN 978-1-138-06752-3

Design and application of a model carriage for a fire characteristic test

W.P. Han, X.Z. Zhang, J.J. Xia, B.J. Yang, C. Hu, Y. Chen, L.S. Jing, Z.M. Bao, L.W. Tian,
D.X. Yu & R.J. Wang
Tianjin Fire Research Institute of MPS, Tianjin, China

W.F. Liu, S.Y. Zhong, Y. Liu & K. Yao
Tianjin University of Commerce, Tianjin, China

ABSTRACT: To improve the safety level of the carriage, a model carriage was designed for a fire characteristic test. First, the parameters of the carriage were determined by spot investigation and literature research. These parameters included size, structure and material. Second, the model carriage was prepared using similar material. Finally, this model carriage with standard combustion material boxes and bags was burnt out. The collected data will be used to design a relative fire extinguishing system in the future.

1 INTRODUCTION

High-speed railway is a type of railway transport that can run significantly faster than traditional railway. The train speed could even reach 240–250 kilometers per hour (Zhai, 2013). The shape of the train was streamlined like that of a bullet train. In recent years, high-speed railway has been undergoing rapid development in China. In the 13th Five-year Plan from 2016 to 2020, the new high-speed railway length will reach 17,000 kilometers. Therefore, attention is paid to the comfort and safety level of the high-speed rail train carriage, as shown in Figure 1.

To improve the safety level of the carriage, some facilities have been installed, including video recorder, portable extinguisher and emergency alarm system. The video is at the position of the carriage's doorpost to monitor the whole carriage. Generally, there are one water base extinguisher and one dry powder extinguisher at the two ends of every carriage. In the dinning carriage and driver's cab, more extinguishers are installed. There is

Table 1. Recent accidents on a high-speed rail train.

Date	Area	General situation
2015.08.02	Lunel, French	The locomotive was burnt out Three hundred passengers were evacuated successfully
2015.08.07	Ganzhou, China	The fire on diesel tank was put out in a high-speed rail operating vehicle by firemen
2012.11.01	Jiaxing, China	The smoke and fire were detected in G54 No.11 and No.12 carriage caused by electrical problem

also an alarm system in the conjunction between two carriages. However, there are chances of a fire or other accidents occurring in this type of carriage, as described in Table 1 (Yang, 2013).

Therefore, the fire characteristics of the carriage should be studied to improve the safety level (Chen, 2014). There are many constituent parts in a high-speed rail train. These constituents together ensured the normal use of the high-speed rail train. In this paper, more attention will be paid to some constituent parameters including size, structure and material. Then, spot investigation in Tianjin Railway Station and literature research were carried out, as shown in Figure 2.

2 INVESTIGATION AND LITERATURE RESEARCH

2.1 Size parameters of the carriage

The version of the Siemens Velaro high-speed train used in China on the Beijing-Tianjin Intercity

Figure 1. The railway map of China.

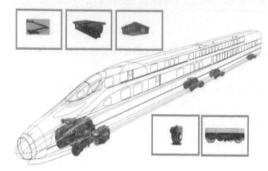

Figure 2. The constituent parts of high-speed rail train.

Railway line is the CRH3. It is capable of a service speed of 380 km/h as with Velaro E in the Spanish rail system. Compared with Velaro E, the width of CRH3 is 300 mm longer, which has an advantage of a more generous structure gauge to fit more seats. The outer and inner size parameters of the carriage are both determined in Tianjin Railway Station.

The outer size parameters of the carriage are given in Table 2. These carriages have a similar ratio of length to width. The outer size of the CRH3 carriage shell is ensured (Li, 2014).

The inner size parameters of carriage include carriage floor, carriage window and carriage luggage carrier, as shown in Figure 3. First, the carriage floor is located on the bottom, which is made of flame-retardant rubber floor and special floor leather. The aisle between two and three seats is 0.5 meter wide. Second, the carriage windows are located on both the sides, which are made of double-layer hollow toughened glass. The window is elevated above 0.9 meter from the ground floor. Third, the carriage luggage carrier is above the carriage windows, which is made of steel frame, aluminum alloy or toughened glass. The carriage luggage is elevated nearly above 2 meters from the ground floor.

The carriage seats including two seats and three seats are made of steel frame, fire-retardant cloth and sponge. Compared with common seats in the regular speed rail train, the seats in the high-speed rail train have better flame-retardant properties. The inner size parameters are given in Table 3.

In addition, there are glass partition and glass door between the neighboring carriages. This glass door can reduce the spread of fire and smoke. When there is a fire accident in a carriage, the fire cannot be put out using the portable extinguisher. The passengers will pass through the glass partition and leave this carriage or this train (Fridolf, 2014). When the evacuation is complete and nobody is left, this door will be closed to limit the fire and smoke. Therefore, the door of the model carriage should be designed considering the ventilation affected by the width of the glass door (Meng, 2014), as shown in Figure 4.

2.2 Structures and material of the carriage shell

The structure of the carriage shell was also determined through literature research, as shown in Figure 5 (Zhang, 2015). There are four parts in the carriage shell including metal plate carriage body, paint, fire proof glue and glass wool. The metal

Table 2. Size parameters of the carriage shell.

Type	Size parameters
25T	25.5 m length, 3.105 m width and 4.433 height
SRZ125Z	25.5 m length, 3.105 m width and 4.750 height
22	23.6 m length, 3.106 m width and 4.283 height
YW25B	25.5 m length, 3.104 m width and 4.433 height
CRH3	24.8 m length, 2.95 0 m width and 3.890 height

Figure 3. The total view of the CRH3 carriage with aisle, carriage window, carriage seat and carriage luggage carrier.

Table 3. Size parameters of the carriage.

Name	Size parameters
Aisle	0.5 m width
Carriage windows	1.3 m length and 0.65 m width
Two seats	0.85 m length and 0.46 m width
Three seats	1.3 m length and 0.46 m width

Figure 4. Glass partition and glass door.

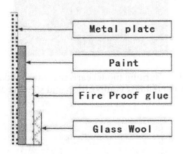

Figure 5. Four parts in a carriage shell.

Table 4. Material properties of the actual carriage.

Material	Properties
Rubber floor	Anti-skid, anti-corrosion, easy to clean, acid-alkali resistance, wear resistance
Film on the wall	Water resistance, oil resistance and acid-alkali resistance
Aluminum alloy	Low density, high strength, anti-fouling, plastic
Glass fiber reinforced plastics	Low density, high strength, plastic, impact resistance

plate can support the body frame of the whole carriage for its excellent ductility. The fire-retardant coating by paint can decrease the rate of corrosion. The fire proof glue and glass wool can be bonded together. The glass wool with high softening point has a good fire resistance. This structural fire protection is very effective. The relative materials are given in Table 4.

3 PREPARATION OF THE MODEL CARRIAGE

According to the parameters presented in Sections 2.1 and 2.2, the model carriage is finished including the outer frame part, inner part and seat part.

First, the outer frame part is finished using different types of steel. The flat surface on both sides of the model carriage is produced by a galvanized steel sheet. The top and bottom of the model carriage are produced by a checkered plate. Four columns in the four corners of the model carriage are produced by a steel plate whose thickness is greater than 3 mm. In addition, the rectangular steel and C steel are also used to improve this structural strength. According to the test requirements, the length of the model carriage is a quarter of the whole actual carriage (Li, 2014). The width of the model carriage is closer to that of CRH3. Concerning the height, the ventilation partition's height should be subtracted. Considering the width of the glass door in the conjunction between neighboring carriages, the opened door in the model carriage should also be located in the middle of the cross-section. Then, all the model carriage parameters are determined, as shown in Figure 6.

Second, the inner part is finished, as shown in Figure 7. The window is elevated above 0.9 meter from the ground floor. The width between two neighboring windows is 0.65 meter. The double-layer hollow toughened glass is used to simulate the actual carriage's windows. The carriage luggage is elevated above 1.8 meters from the ground floor. The steel rods are used to simulate the actual carriage's luggage carrier, whose diameter is greater than 8 mm. The simulated seats can be placed below the luggage carrier near the windows. The simulate bag or luggage can be placed upon the luggage carrier. In the inner carriage, the floor is covered by plywood, flame-retardant glue and PVC floor leather. Both walls are covered by flame-retardant glue and PVC floor leather instead of film. The roof is covered by glass fiber-reinforced

Figure 6. The frame of the model carriage (6*3*2.5 m).

Figure 7. The inside of the model carriage with simulated seats and bags.

Table 5. Properties of the model carriage material.

Material	Properties
7541 glass fiber—reinforced plastics	Thickness 5 mm, heat value 10 MJ/kg, gram weight 7.2 kg/m², martin heat resistance 275°C, compressive strength 1412 kg/cm², bending strength 3177 kg/cm², tensile strength 3058 kg/cm², shock strength 491 kgf cm/cm².
PVC floor leather	Thickness 0.8 mm, heat value 18.8 MJ/kg, gram weight 1.6 kg/m².
Plywood	Thickness 5 mm, heat value 18.9 MJ/kg, gram weight 2.6 kg/m².

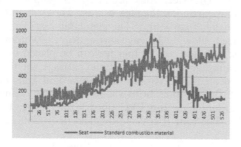

Figure 8. The heat release rate curves of actual seat and standard combustion material box composition (X axis: Time/s; Y axis: release heat rate/kW).

plastic. These parameters of material in the model carriage are given in Table 5.

Third, the standard combustion material boxes are chosen to replace the actual seat. The maximal Heat Release Rate (HRR) of composition of actual three seats in the normal speed rail train is 1000 kW, which is higher than the high-speed rail train. The maximal HRR of standard combustion material box composition is also over 800 kW as

Figure 9. The burning actual seat and burning standard combustion material box composition.

Figure 10. Burning model carriage with standard combustion material box composition and simulated bags.

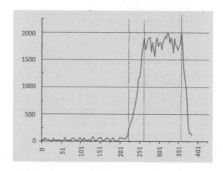

Figure 11. Burning model carriage with standard combustion material box composition and simulated bags (X-axis: Time/s; Y-axis: release heat rate/kW).

time passes. Their HRR values are quite close to one another, as shown in Figure 8, although their burning speeds are different (Babrauskas, 2016). Therefore, the standard combustion material box with higher HRR requirement can replace three seats in the high-speed rail train.

4 FIRE CHARACTERISTIC TEST

The fire characteristic test was carried out using this model carriage in a tunnel (Carvel, 2016).

The standard combustion material box was ignited by cotton. The fire spread in the vertical direction from the carriage luggage to the roof. The fire also spread in the horizontal direction from one box to the neighboring one. Then, the heat release rate was determined by the measurement system in the tunnel. The result indicated that the whole material's maximal heat release rate could reach 15 MW before extinguishing operation in 350 s, as shown in Figure 10 and Figure 11.

5 CONCLUSION

In summary, a model carriage was designed and subjected to a fire characteristic test. The standard combustion material box composition replaced the actual seat. In future, the relate data will be used to design the fire extinguishing system to improve the safety level of the railway system.

ACKNOWLEDGMENTS

This work was supported by the National Science Technology Support Program No. 2014BAK17B03, CECS Standard Program 2016.

REFERENCES

Babrauskas V, The cone calorimeter[M]//SFPE handbook of fire protection engineering. Springer New York, 2016: 952–980.

Carvel R, Ingason H. Fires in Vehicle Tunnels[M]//SFPE Handbook of Fire Protection Engineering. Springer New York, 2016: 3303–3325.

Chen J, Yao X, Yan G, et al. Comparative study on heat release rate of high-speed passenger train compartments[J]. Procedia engineering, 2014, 71: 107–113.

Fridolf K, Nilsson D, Frantzich H. The flow rate of people during train evacuation in rail tunnels: effects of different train exit configurations[J]. Safety science, 2014, 62: 515–529.

Li Y Z, Ingason H, Lonnermark A. Fire development in different scales of train carriages[J]. Fire Safety Science, 2014, 11: 302–315.

Li Y Z, Ingason H, Lönnermark A. Fire development in different scales of metro carriages[C]//11th International Symposium on Fire Safety Science, February 10–14, 2014. 2014.

Meng N, Hu L, Wu L, et al. Numerical study on the optimization of smoke ventilation mode at the conjunction area between tunnel track and platform in emergency of a train fire at subway station[J]. Tunnelling and Underground Space Technology, 2014, 40: 151–159.

Yang P, Li C, Chen D. Fire emergency evacuation simulation based on integrated fire-evacuation model with discrete design method[J]. Advances in Engineering Software, 2013, 65: 101–111.

Zhai W, Wang S, Zhang N, et al. High-speed train-track-bridge dynamic interactions-Part II: experimental validation and engineering application[J]. International Journal of Rail Transportation, 2013, 1(1–2): 25–41.

Zhang J, Zhao M, Xie S, et al. Structure Sensitivity of the CRH3 EMU Based on FEM[C]//Fifth International Conference on Transportation Engineering. 2015.

Author index